Coastal Research Library

Volume 35

Series Editor
Charles W. Finkl
Department of Geosciences
Florida Atlantic University
Boca Raton, FL, USA

The aim of this book series is to disseminate information to the coastal research community. The Series covers all aspects of coastal research including but not limited to relevant aspects of geological sciences, biology (incl. ecology and coastal marine ecosystems), geomorphology (physical geography), climate, littoral oceanography, coastal hydraulics, environmental (resource) management, engineering, and remote sensing. Policy, coastal law, and relevant issues such as conflict resolution and risk management would also be covered by the Series. The scope of the Series is broad and with a unique cross-disciplinary nature. The Series would tend to focus on topics that are of current interest and which carry some import as opposed to traditional titles that are esoteric and non-controversial.

Monographs as well as contributed volumes are welcomed.

More information about this series at http://www.springer.com/series/8795

Geoff Bailey • Nena Galanidou
Hans Peeters • Hauke Jöns
Moritz Mennenga
Editors

The Archaeology of Europe's Drowned Landscapes

Foreword by Nicholas Flemming
and Dimitris Sakellariou

Springer Open

Editors
Geoff Bailey
Department of Archaeology
University of York
York, UK

College of Humanities, Arts and Social Sciences
Flinders University
Adelaide, SA, Australia

Nena Galanidou
Department of History and Archaeology
University of Crete
Rethymno, Greece

Hauke Jöns
Lower Saxony Institute for Historical Coastal Research
Wilhelmshaven, Germany

Hans Peeters
Groningen Institute of Archaeology
University of Groningen
Groningen, The Netherlands

Moritz Mennenga
Lower Saxony Institute for Historical Coastal Research
Wilhelmshaven, Germany

ISSN 2211-0577 ISSN 2211-0585 (electronic)
Coastal Research Library
ISBN 978-3-030-37366-5 ISBN 978-3-030-37367-2 (eBook)
https://doi.org/10.1007/978-3-030-37367-2

This Springer imprint is published by the registered company Springer Nature Switzerland AG.
The registered company address is: Gewerbestrasse 11, 6330 Cham, Switzerland

COST is supported by the EU
Framework Programme Horizon 2020

COST (European Cooperation in Science and Technology) is a pan-European intergovernmental framework. Its mission is to enable breakthrough scientific and technological developments leading to new concepts and products and thereby contribute to strengthening Europe's research and innovation capacities. It allows researchers, engineers, and scholars to jointly develop their own ideas and take new initiatives across all fields of science and technology, while promoting multi- and interdisciplinary approaches. COST aims at fostering better integration of less-research-intensive countries to the knowledge hubs of the European Research Area. The COST Association, an international not-for-profit association under Belgian Law, integrates all management, governing, and administrative functions necessary for the operation of the framework. The COST Association has currently 36 Member Countries (www.cost.eu).

Foreword

This book describes the archaeology of all known submerged prehistoric remains and their regional context for the whole European continental shelf. They were inundated by the postglacial rise of global sea level. Over 2600 submerged Stone Age artefact sites, settlements, and anthropogenic indicators have been identified on the seabed of the European marginal seas, and their significance is analysed to show how the underwater remains influence our understanding of early exploitation of coastal and marine resources and early seafaring. The scope of this study is unique, since no attempt has been made previously to assemble and integrate submerged archaeological data on this geographical scale. It articulates fully the range of archaeological issues to which underwater prehistoric discoveries are relevant in Europe.

The finds extend over a time range from the earliest human presence north of the Alps in the Early Palaeolithic nearly one million years ago up to the establishment of modern sea level about 5000 years at the end of the Neolithic. Some later material is presented where coastlines have continued to subside.

The book begins with a succinct overview of the Ice Age cycles of sea-level change and the causes of submergence of prehistoric settlements and artefacts in the various geomorphologically contrasting European seas: from the Baltic, which behaves almost like a large estuary to the Atlantic margins with a huge tidal range and exposure to colossal storms, round to the Mediterranean and Black seas, where the postglacial rise of sea level is combined with local earthquakes and vertical earth movements. The reasons that prehistoric peoples lived on or crossed the exposed continental shelves differed in each case, and the regional environment, fauna, and flora influenced their culture, subsistence techniques, and their methods for exploiting the resources of the sea and coast. Submerged remains have been found on all types of coasts. The most extensive seabed archaeological deposits are on wide or undulating shelves, with relatively few on steep rocky coasts. A web-based catalogue of all recorded sites with details of the archaeology is publicly available. This provides a resource for future research.

The main thrust of the book is contained in 19 richly illustrated chapters that describe the archaeology of submerged sites in each regional sea, nation by nation. Since both archaeology and oceanography continue seamlessly across national maritime jurisdictional boundaries, there are overarching reviews of each marine basin from coast to coast. A concluding section considers the universal issues of legal and regulatory regimes and interactions with offshore industries and cultural heritage laws.

Europe has one of the longest traditions of research on the prehistoric archaeology and submerged landscapes of the continental shelf anywhere in the world, extending back to the nineteenth century in some countries, and the largest currently known concentrations of submerged prehistoric archaeological sites. This is especially the case in the western Baltic and to a lesser extent in the southern North Sea, the southwest corner of the Black Sea, and the Levant coast of the eastern Mediterranean.

Prehistoric populations depended intimately on the ecosystem of which they were a part. Thus, the understanding of how people lived, and died, is inextricably linked to the landscape, fauna, and flora that surrounded them and the seasonal variations or migrations of species. Northern and northwest Europe were occupied and abandoned several times during successive glacial cycles. Underwater finds are therefore likely to be of great significance in filling some very large gaps in the Stone Age record, and an increasing number of finds have come to light in the past two decades.

The SPLASHCOS Action described in this book involved more than 120 experts from prehistoric archaeology and the marine geosciences working together. The issues of the Quaternary landscape of the exposed continental shelf, and its variability through time and space, are analysed in a separate volume that has already been published. This book is thus a twin volume, and the two books enable researchers and students to study the details of the archaeological record in the context of the immediate local environment and the wider landscape of the previously exposed shelf.

The types of remains and their frequency of occurrence vary between regions. In northern Europe, the low sea levels exposed thousands of square kilometres of undulating plains, marshes, estuaries, and lakes. In the Mediterranean, where steep mountainous coasts are more common, and the shelf narrower, a drop of sea level does not produce so much new habitable land except in the Adriatic and Aegean seas and the western Black Sea shelf. Notwithstanding this limitation, important remains have been found in caves in the south of France and in Greece. In areas of extensive sedimentation, such as the southern Baltic parts of the Mediterranean coast of Croatia and Israel and the Bulgarian coast of the Black Sea, groups of settlements have been preserved with stratigraphic integrity, including organic remains such as leather, cloth, rope, burials, and food remains, including items which are seldom preserved on dry land. In the southern Baltic, finds include fish weirs, dugout canoes, bows, spear shafts, fibres, and woven fabric preserved in anaerobic sediments. Bows are made of elm wood and fish hooks are made from bone. Food remains include seals, dolphin, and whale. Unique examples survive of different types of wood for making different tools and hearths with burnt wood and food remains. In Britain, the submerged Mesolithic site at a depth of 11 m off Bouldnor Cliff near the Isle of Wight exhibits worked timbers, hut foundations, twine, and food remains. Where large numbers of sites are known, it is advantageous to organise regional integration and multidisciplinary studies so as to derive the spatial interaction between changes of climate, sea level, and the human response. The best example of this is the SINCOS project in the southern Baltic where researchers found 50 new sites and excavated 8 of them underwater.

Most of the documented prehistoric remains have been found in shallow water, not near the shelf edge. Sea level 12,000 years ago was 60 m below present and had risen to within a metre of present sea level by 5000 years ago. This time span embraces the final stages of the Palaeolithic, with predominantly mobile hunter-gatherer communities largely lacking evidence of constructed habitations, no pottery, no agriculture, and no metal tools, through to the final Neolithic and Early Bronze Age when there were small towns, sophisticated decorated pottery, complex utensils made of bronze, and early stages of writing. These dramatic changes in the capability of human communities took place against a background of warming European climate, melting ice caps, and rising sea level. The context of social evolution against a background of changing climate, ecosystems, and rainfall has been examined by archaeologists for over one hundred years, but the interaction with rising sea level has been much less studied. This is for the simple reason that most of the evidence is on the present sea floor, and the missing data are brought together and recorded here for the first time. In some cases, the submerged discoveries reveal that technical skills, food types, and materials were already in use thousands of years before their earliest occurrence on the present dry land.

A few sites described in this book are older than 20,000 years, and thus predate the Last Glacial Maximum, and some are even older, going back through several glacial cycles. These early sites tell us how Palaeolithic communities interacted with the sea and its resources. The ability to cross sea straits in some form of floating craft certainly existed tens of thousands of years ago. Submerged sites in the time range 20,000 to 120,000 years, that is, pre–Last Glacial Maximum, confirm that prehistoric material in context can survive multiple marine transgressions. These include a remarkable cave in the south of France in which there are wall paintings, a settlement preserving flint tools and evidence of flint knapping at a depth of 20 m off the north coast of France, and a concentration of early Palaeolithic materials beneath sediments in the North Sea.

Contrary to popular expectation, a large number of prehistoric materials have survived in the intertidal zone on coasts where they are exposed to storm waves and tidal currents. Among the most distinctive of these intertidal finds are the human and animal footprints preserved in intertidal sediments in many locations, especially at Goldcliff East. The earliest, at Happisburgh, are dated between 0.78 and one million years ago. Large numbers of intertidal records survive around the coast of Britain and northern France. The British sites are often beneath submerged forests or drowned peat beds, while the French remains are mostly megaliths and stone tombs on a rocky coast with a large tidal range.

Major discoveries have resulted from collaboration between archaeologists and commercial operations, especially during the building of the Great Belt Fixed Link in Denmark and excavation of the harbour basin Maasvlakte 2 in the North Sea at Rotterdam. While industrial projects bring powerful machinery to bear on the seabed, controlled by acoustic positioning and monitoring equipment, academic research projects are increasingly using high-technology positioning and survey equipment underwater, resulting in large data sets of both palaeoenvironmental data and archaeological finds. Three-dimensional photogrammetry has been valuable at many sites, and there is increasing use of remotely operated vehicles (ROVs).

One of the problems that occur repeatedly is the difficulty of first finding and then confirming the nature of a presumed prehistoric feature. There is no technology available at present that can detect at a distance of a few metres deposits of wood and worked flint contained within sediments of sand, mud, or gravel. Visually, anthropogenic materials can be detected on the surface by high resolution photography or directly by observations by divers, but these methods can only search small areas slowly. As a result, most submerged prehistoric sites to date have been found by chance, by trawlers, dredgers, coastal engineering, or geophysical coring carried out for other reasons. As soon as the first artefacts are identified and reported, archaeologists respond with the appropriate technology and controls.

The application of modern high technology has a vital role to play. Marine geoscientific surveys of the continental shelf that measure the properties of Pleistocene sediments or processes should be conducted with the prehistoric implications in mind. In some cases, projects may conduct such surveys specifically for archaeological purposes, but more often there are other reasons, often industrial or commercial, and in these cases the archaeological benefits should be extracted from the archived data.

When a prehistoric deposit has been identified on the sea floor, the application of trained divers supported by technology, depending on the depth, becomes essential. It is important to note that some phases of work can be conducted quite cheaply, and the apparent cost of the most sophisticated projects should not be interpreted as a deterrent. As the knowledge of existence of a site is confirmed by the first finds, so the application of technology can be ramped up to include comprehensive analysis of the surrounding landscape, topography, drowned river drainage, ecosystem indicators, ancient sedimentary DNA, 3D photogrammetry, and so on. The volumes of data obtained, or archived, describing all aspects of palaeolandscapes, palaeoclimate, and palaeoecosystems, plus the archaeology and preserved artefacts, are truly enormous, requiring the most advanced techniques of data retrieval and data management.

Seabed prehistoric research has made huge advances in the last two decades, and results are presented in this book. A combination of new intellectual insights and goals with new technology guarantees that this progress will continue in future. We hope that early-stage researchers will adopt this research, with increasing support from academic and commercial institutions.

National Oceanography Centre
Southampton, UK
Nicholas Flemming

Hellenic Centre for Marine Research
Anavyssos, Greece
Dimitris Sakellariou

Preface

The purpose of this volume is to provide, for the first time, a comprehensive review and interpretation of the archaeological material relating to human occupation of the now-submerged landscapes of Europe's continental shelf and to evaluate its significance and the prospects for future research. It was planned as one of two volumes intended to complement each other as the principal published output of SPLASHCOS—an acronym for *Submerged Prehistoric Archaeology and Landscapes of the Continental Shelf*—a 4-year research network that ran, formally, from November 2009 to November 2013 as COST Action TD (Trans-Domain) 0902. The first volume was published by Wiley Blackwell (2017) as *Submerged Landscapes of the European Continental Shelf: Quaternary Paleoenvironments*, edited by Nicholas Flemming, Jan Harff, Delminda Moura, Anthony Burgess, and Geoff Bailey, and is concerned primarily with the geological and environmental processes that have shaped the character of the continental shelf as a human habitat during periods of low sea level, and with the factors that have determined the preservation and visibility of archaeological features on these submerged landscapes.

The present volume is also intended as a complement to an online database of known underwater archaeological sites throughout Europe, known as the SPLASHCOS Viewer, which has grown to be another major output of SPLASHCOS and is publicly available at http://splashcos.maris2.nl/ and http://splashcos-viewer.eu/. The information in the database has been collected by the authors of the national overview chapters in this volume, and these chapters draw on the records of the SPLASHCOS Viewer and provide the scientific description, analysis, and evaluation of them.

We have, therefore, aimed in this volume to facilitate cross-referencing with the SPLASHCOS Viewer and with the chapters in the *Quaternary Paleoenvironments* volume, with amplification of issues relating to sea-level change, landscape setting, and geological and oceanographic conditions of archaeological site preservation as well as a full description of the archaeological finds themselves. We hope and intend that this combination of volumes and the SPLASHCOS Viewer will provide a rich mine of information to stimulate future research.

COST is a European funding mechanism that supports Actions (the COST term for a research network) in all subject areas and is open in principle to participation from all European countries and "near neighbours" in North

Africa, the Near East, and the former Soviet Union (http://www.cost.eu/about_cost). Each participating country nominates two national representatives, appointed through its own national research organisation, who form the Management Committee, with provision to recruit additional members as appropriate. The work of the Action is supported by a Memorandum of Understanding that sets out the main aims and objectives of the Action and the details of the research programme. Funding is designed to facilitate communication across national boundaries and, in the case of Trans-Domain Actions, across disciplinary boundaries as well. It covers expenses for meetings, workshops, conferences, training, and publications. It does *not* provide funds for new research, such as salaries and fieldwork or laboratory expenses, but is intended to complement, stimulate, and add value to national, bilateral, and EU sources of research funding through the development and promotion of new ideas, discussions, plans, collaborations, and grant applications.

The vision of the SPLASHCOS Action as expressed in the COST Memorandum of Understanding of 11 June 2009 (COST 245/09 Annex; https://www.cost.eu/actions/TD0902/#downloads) was to promote research on the archaeology, climate, and palaeoenvironment of the drowned landscapes of the European continental shelf; to stimulate collaboration across national and disciplinary boundaries; to bring together interested parties from the worlds of academic science, commerce, and government; to encourage participation and training of early-stage researchers; and to facilitate exchange of ideas, planning of research projects, application for research funds, and dissemination through publications and other media. Target audiences included not only the many scientific disciplines concerned with researching the continental shelf—archaeology, marine geology, geophysics, biology, climatology, palaeoenvironment, and oceanography—but also museum curators and heritage managers, government officials responsible for the marine environment and its cultural heritage, industrial operators working on the seabed, funding agencies, school children, and a wider public.

During its lifetime, SPLASHCOS grew to include over 120 archaeologists, marine scientists, and members of museums and heritage agencies from 25 countries, including Israel, the Ukraine, and the Russian Federation (see the SPLASHCOS website at https://www.splashcos.org/ and the COST website at https://www.cost.eu/actions/TD0902/#tabs|Name:parties for the full list). The Management Committee organised 8 major meetings, 3 international conferences, and 16 training missions involving 65 early-stage researchers and stimulated €20 million of successful grant applications and a wide range of outputs including publications and online technical reports. Full details of the history of SPLASHCOS and the outcomes of its activities are given on the project website at http://splashcos.org.

Not all the participating countries are represented in the chapters of national overviews in this volume, principally because of limited or absent data, though relevant information is referred to wherever possible in the editorial introductions for each of the major marine basins that form the major subdivisions of the volume. Other countries eligible for participation, notably Albania, Turkey, and the countries of North Africa bordering the

Mediterranean, are also not represented despite attempts to recruit their participation to SPLASHCOS, though some information from Turkey has subsequently become available and is referred to in this volume, along with the addition of a chapter on Cyprus.

We have also taken the opportunity to add a group of chapters, expanded from conference papers originally presented at the SPLASHCOS meeting in Esbjerg, Denmark, in March 2013, organised by Thijs Maarleveld as an international workshop with representatives of North Sea industry. These chapters expand on the legal, regulatory, and historical frameworks affecting the relationship between archaeology and industrial offshore activity, a relationship which has come to play a leading role in the exploration and discovery of new material.

Four Working Groups carried forward the work of the Action: Archaeological Data and Interpretations (WG1) led by Anders Fischer; Environmental Data and Reconstructions (WG2) led by Jan Harff; Technology, Technical Resources and Training (WG3) led by Ole Grøn and Tine Missiaen; and Commercial Collaboration and Outreach (WG4) led by Julie Satchell. The results of the work by WGs 2, 3, and 4 are published elsewhere: WG2 in the volume edited by Nicholas Flemming and co-editors referred to above; and WG3 and WG4 online on the SPLASHCOS website at https://www.splashcos.org/resources/methods and https://www.splashcos.org/sites/splashcos.org/files/downloads/SPLASHCOS_Marine_Industry_Guide.pdf, and in individual chapters in other SPLASHCOS outputs published as edited volumes.

Principal among these other volumes are *Geology and Archaeology: Submerged Landscapes of the Continental Shelf*, edited by Jan Harff, Geoff Bailey, and Friedrich Lüth and published in 2016 by the Geological Society of London, and *Under the Sea: Archaeology and Palaeolandscapes of the Continental Shelf*, edited by Geoff Bailey, Jan Harff, and Dimitris Sakellariou and published in 2017 by Springer International. Both are collections of chapters covering a wide range of themes, methods, and case studies, including examples from other parts of the world as well as Europe, and were outcomes of conference sessions, the first arising from a session at the 34th International Geological Congress in Brisbane, Australia, in June 2012 and the second from the final showcase conference of SPLASHCOS held at the University of Szczecin, Poland, in September 2013. Also to be mentioned is a volume aimed at a general audience and published in 2018 as *Oceans of Archaeology*, edited by Anders Fischer and Lisbeth Pedersen and published by the Jutland Archaeological Society, with a selection of examples from Denmark and other parts of Europe.

In preparing this volume, Geoff Bailey has led the editorial team in the main tasks of structuring the contents, selecting reviewers, and writing, editing, and coordination. Nena Galanidou and Hans Peeters have added their archaeological expertise and perspectives from different parts of Europe to the work of writing and editing, and Hauke Jöns has guided the development of the SPLASHCOS Viewer from the beginning and coordinated the work of data collection by the national contributors. Moritz Mennenga has taken on the arduous task of collating and inputting the national records of underwater archaeological sites and preparing the digital maps of site distributions that appear throughout the volume. We anticipate that this database will continue to be added to and improved as a major research tool for the benefit of future research.

Although the formal term of COST funding ceased in 2013, SPLASHCOS has continued to be active informally, through the completion of publications planned and initiated during the life of the Action, and through new research based on existing collaborations and the development of new ones. In 2013 towards the end of the SPLASHCOS Action, the European Marine Board invited the formation of a Working Group (SUBLAND) under the chairmanship of Nicholas Flemming to publish a major policy document on the new field of continental shelf prehistoric archaeology for European funding bodies, government agencies, and other stakeholders, published in 2014 as *Land Beneath the Waves: Submerged landscapes and sea level change. A joint geoscience-humanities strategy for European Continental Shelf Prehistoric Research.*

New research is now in progress in many parts of Europe, and further afield in other parts of the world, and it is likely that new techniques will be developed, new strategies devised, and new discoveries made in the coming decades to provide better answers to the questions already on the research agenda and to open up new questions that we cannot yet begin to define with any clarity or even to imagine. We hope that the information presented in this volume will provide the foundation and the inspiration for many new investigations, new collaborations, new directions of research, improved management of the underwater cultural heritage through scientific, governmental, and industrial cooperation, and new participants in this growing field of underwater archaeology.

York, UK — Geoff Bailey
Rethymno, Greece — Nena Galanidou
Groningen, The Netherlands — Hans Peeters
Wilhelmshaven, Germany — Hauke Jöns
Moritz Mennenga

Acknowledgements

In bringing this volume to completion, we thank first and foremost all the participants in the meetings of WG1 that took place during the lifetime of the COST Action for all their hard work and contribution of data and ideas, especially Anders Fischer as Chairman of WG1, who led the discussions and played an important role in laying the foundations for the resulting publications and online database. Most of our authors in this volume participated in these original discussions while others were recruited later to fill critical gaps in geographical coverage and expertise. We thank them all for their patience and perseverance in dealing with a succession of editorial interventions and requests for revisions. All the chapters have been subjected to independent peer review, and we thank the many colleagues who agreed to act as anonymous peer reviewers (some are named with their agreement in individual chapters) and for their constructive comments.

Special thanks are due to the Lower Saxony Institute for Historical Coastal Research, Wilhelmshaven, Germany, for making available resources and personnel for the development of the SPLASHCOS Viewer and for providing a long-term home for its future curation.

We also thank the institutions that hosted the SPLASHCOS plenary sessions and Working Group meetings: the University of York, UK, in March 2010; the Hellenic Centre for Marine Research, Rhodes, Greece, in October 2010; the Museum of Asian Art, Berlin, in April 2011; the University of Zadar, Croatia, in October 2011; the Cultural Heritage Agency of the Netherlands, Amersfoort, in April 2012; the University of Rome "La Sapienza" in September 2012; the University of Southern Denmark, Esbjerg, in March 2013; and the University of Szczecin, Poland, in September 2013. Smaller meetings took place in the intervals of the plenary sessions to advance particular strands of the work, and we thank in particular the German Archaeological Institute, Berlin, and the Danish Agency for Culture and Palaces, Copenhagen, for hosting these meetings.

To all our colleagues in SPLASHCOS, we extend our thanks for their participation, energy, and ideas, and especially to our colleagues who organised major meetings: Dimitris Sakellariou, Hellenic Centre for Marine Research, Athens; Friedrich Lüth, German Archaeological Institute, Berlin; Irena Radić Rossi, University of Zadar; Henk Weerts, Cultural Heritage Agency of the Netherlands; Francesco Latino Chiocci, University of Rome "La Sapienza"; Thijs Maarleveld, University of Southern Denmark; and Andrzej Witkowski, University of Szczecin.

We owe a special debt of gratitude to Nicholas Flemming, Friedrich Lüth, and Dimitris Sakellariou, who have been stalwart pillars of support from the very beginning of SPLASHCOS, and especially for their supporting role and good advice in helping to bring this volume to completion. We also thank Petra van Steenbergen and Hermine Vloemans at Springer International, and their successors Robert Doe and Carmen Spelbos, for unfailing support and guidance through the many difficulties and delays that attended the completion of this project.

We also thank the following organisations and individuals for contributions towards the cost of open access publication: the Hellenic Centre for Marine Research; Historic England; the Lower Saxony Institute for Historical Coastal Research; Lund University; the Museum of Cultural History, University of Oslo; the University of Crete; the University of Groningen; the University of Southern Denmark; the University of York; Wessex Archaeology; Jon Adams; Pablo Arias; Nuno Bicho; Helen Farr; Ehud Galili; Timmy Gambin; Irena Radić Rossi; and Kieran Westley.

In conclusion, we thank the COST Office for their financial and administrative support, and especially the following individuals: the successive Science Officers who guided our work—Julia Stamm, Geny Piotti, Andreas Obermaier, and Luule Mizera; the COST Administrative Officer Leo Guilfoyle; and the COST Rapporteurs Dr. Ipek Erzi (Scientific and Technological Research Council of Turkey, TUBITAK) and Prof Daniela Koleva (Sofia University "St. Kliment Ohridski", Bulgaria), who attended meetings and gave invaluable advice, and Cynthianne DeBono Spiteri, who acted as grant holder and administrative secretary of SPLASHCOS at the University of York and provided with infinite patience and good humour the central coordination of the complex network of communications, paperwork, and logistics required in the organization of meetings, training programmes, and progress reports.

University of York
York, UK
Geoff Bailey

University of Crete
Rethymno, Greece
Nena Galanidou

Groningen Institute of Archaeology
Groningen, The Netherlands
Hans Peeters

Lower Saxony Institute for Historical Coastal Research
Wilhelmshaven, Germany
Hauke Jöns

Lower Saxony Institute for Historical Coastal Research
Wilhelmshaven, Germany
Moritz Mennenga

Contents

Contributors

Jon Adams The Centre for Maritime Archaeology, University of Southampton, Highfield, Southampton, UK

Luc Amkreutz National Museum of Antiquities, Leiden, The Netherlands
Faculty of Archaeology, Leiden University, Leiden, The Netherlands

Albert J. Ammerman Department of the Classics, Colgate University, Hamilton, NY, USA

Søren H. Andersen Moesgaard Museum, Højbjerg, Denmark

Fabrizio Antonioli Italian National Agency for New Technologies, Energy and Sustainable Economic Development (ENEA), Rome, Italy

Pablo Arias Instituto Internacional de Investigaciones Prehistóricas de Cantabria, Universidad de Cantabria-Gobierno de Cantabria-Santander, Santander, Spain

Geoff Bailey Department of Archaeology, University of York, York, UK
College of Humanities, Arts and Social Sciences, Flinders University, Adelaide, SA, Australia

Martin Bell Department of Archaeology, School of Human and Environmental Sciences, University of Reading, Reading, UK

Jonathan Benjamin College of Humanities, Arts and Social Sciences, Flinders University, Adelaide, SA, Australia

Elena Flavia Castagnino Berlinghieri Dipartimento di Scienze Biologiche, Geologiche e Ambientali, Università di Catania, Catania, Italy
Soprintendenza per i Beni Culturali e Ambientali di Siracusa, Sicily, Italy

Nuno Bicho ICArEHB, Universidade do Algarve, Faro, Portugal

Andrew Bicket Wessex Archaeology, Edinburgh, UK
Department of Geography, Loughborough University, Loughborough, UK

Cyrille Billard DRAC-SRA Normandie, Caen, France
UMR 6566 Centre de Recherche en Archéologie, Archéosciences, Histoire, Caen, France

Yves Billaud Département des Recherches Archéologiques Subaquatiques, Marseille, France

Catherine Bizien-Jaglin Centre Régional d'Archéologie d'Alet, Rennes, France

Valerija Butorac Faculty of Sciences, Department of Geography, University of Zagreb, Zagreb, Croatia

Antoine Chancerel Musée National de Préhistoire, Les Eyzies, France

Dominique Cliquet DRAC-SRA Normandie, Caen, France
UMR 6566 Centre de Recherche en Archéologie, Archéosciences, Histoire, Caen, France

Marie-Yvane Daire CNRS, UMR 6566 Centre de Recherche en Archéologie, Archéosciences, Histoire, Rennes, France
Association Manche Atlantique pour la Recherche Archéologique dans les Îles, Rennes, France

Maikel De Clercq Renard Centre of Marine Geology, Ghent University, Ghent, Belgium

Katerina Dellaporta Hellenic Ministry of Culture and Sports, Byzantine and Christian Museum, Athens, Greece

Ine Demerre Agentschap Onroerend Erfgoed, Brussels, Belgium

Sarah Dromgoole School of Law, University of Nottingham, University Park, Nottingham, UK

Vered Eshed Israel Antiquities Authority, Jerusalem, Israel

Mina Weinstein Evron Zinman Institute of Archaeology, University of Haifa, Haifa, Israel

R. Helen Farr The Centre for Maritime Archaeology, University of Southampton, Highfield, Southampton, UK

Nicholas Flemming National Oceanography Centre, Southampton, UK
Guildford, Surrey, UK

Nathalie Fourment DRAC-SRA Grande Aquitaine, Bordeaux, France

Nena Galanidou Department of History and Archaeology, University of Crete, Rethymno, Greece

Ehud Galili Zinman Institute of Archaeology, University of Haifa, Haifa, Israel
Israel Antiquities Authority, Haifa, Israel

Timmy Gambin Department of Classics and Archaeology, Archaeology Centre, University of Malta, Msida, Malta

Henri Gandois UMR 6566 Centre de Recherche en Archéologie, Archéosciences, Histoire, Caen, France
UMR 8215, Université Paris 1, Panthéon, Sorbonne, Rennes, France

Håkon Glørstad Museum of Cultural History, University of Oslo, Oslo, Norway

Julia Goldhammer State office for Cultural Heritage Baden-Wuerttemberg, Gaienhofen-Hemmenhofen, Germany

Michael J. Grant Coastal and Offshore Archaeological Research Services (COARS), Ocean and Earth Science, National Oceanography Centre Southampton, University of Southampton, Southampton, UK

Jostein Gundersen Directorate for Cultural Heritage, Oslo, Norway

Alex Hale Historic Environment Scotland, John Sinclair House, Edinburgh, UK

Anton Hansson Department of Geology, Lund University, Lund, Sweden

Karen Hardy Department of Archaeology, University of York, York, UK
ICREA, Barcelona, Catalonia, Spain
Departament de Prehistòria, Facultat de Filosofia i Lletres, Universitat Autònoma de Barcelona, Barcelona, Catalonia, Spain

Sönke Hartz Museum for Archaeology Schloss Gottorf, Schleswig, Germany

Israel Hershkovitz Department of Anatomy and Anthropology, Faculty of Medicine, Tel-Aviv University, Tel Aviv, Israel

Liora Kolska Horwitz National Natural History Collections, Faculty of Life Science, Berman Building, E. Safra Campus, The Hebrew University, Jerusalem, Israel

Briagell Huet UMR 6566 Centre de Recherche en Archéologie, Archéosciences, Histoire, Caen, France

Leandro Infantini ICArEHB, Universidade do Algarve, Faro, Portugal

Hauke Jöns Lower Saxony Institute for Historical Coastal Research, Wilhelmshaven, Germany

Sergey Kadurin Department of Engineering Geology and Hydrogeology, Odessa I.I. Mechnikov National University, Odessa, Ukraine

Ivor Karavanić Faculty of Humanities and Social Sciences, Department of Archaeology, University of Zagreb, Zagreb, Croatia
Department of Anthropology, University of Wyoming, Laramie, WY, USA

Hans-Joachim Kühn State Archaeology Department of Schleswig-Holstein, Schleswig, Germany

Frode Kvalø Norwegian Maritime Museum, Oslo, Norway

Marine Laforge UMR 6566 Centre de Recherche en Archéologie, Archéosciences, Histoire, Caen, France
ÉVEHA, Vezin-Le-Coquet, Rennes, France

Loïc Langouët Association Manche Atlantique pour la Recherche Archéologique dans les Îles, Rennes, France
Centre Régional d'Archéologie d'Alet, Rennes, France

Luc Laporte CNRS, UMR 6566 Centre de Recherche en Archéologie, Archéosciences, Histoire, Rennes, France

Jean-Marc Large Groupe vendéen d'études préhistoriques, membre associé UMR 6566 Centre de Recherche en Archéologie, Archéosciences, Histoire, Rennes, France

Frédéric Leroy Département des Recherches Archéologiques Subaquatiques, Marseille, France
Sous-Marines, Marseille, France

Elías López-Romero UMR 6566 Centre de Recherche en Archéologie, Archéosciences, Histoire, Junior Chair (Neolithic societies), Université de Bordeaux, Bordeaux, France

Thijs J. Maarleveld Maritime Archaeology Programme, University of Southern Denmark, Esbjerg, Denmark

Svea Mahlstedt Lower Saxony Institute for Historical Coastal Research, Wilhelmshaven, Germany

João Marreiros ICArEHB, Universidade do Algarve, Faro, Portugal

Chloé Martin CNRS, UMR 6566 Centre de Recherche en Archéologie, Archéosciences, Histoire, Rennes, France
Association Manche Atlantique pour la Recherche Archéologique dans les Îles, Rennes, France
Centre Régional d'Archéologie d'Alet, Rennes, France

Léopold Maurel DRAC-SRA Grande Aquitaine, Bordeaux, France

Moritz Mennenga Lower Saxony Institute for Historical Coastal Research, Wilhelmshaven, Germany

Tine Missiaen Flanders Marine Institute (VLIZ), Ostend, Belgium

Garry Momber Maritime Archaeology Trust, National Oceanography Centre, Southampton, UK
Department of Archaeology, University of York, York, UK

Jean-Laurent Monnier CNRS, UMR 6566 Centre de Recherche en Archéologie, Archéosciences, Histoire, Rennes, France

Björn Nilsson Department of Archaeology and ancient History, Lund University, Lund, Sweden

Pål Nymoen Norwegian Maritime Museum, Oslo, Norway

Christopher Pater Historic England, London, UK

Hans Peeters Groningen Institute of Archaeology, University of Groningen, Groningen, The Netherlands

Preslav Peev Institute of Oceanology, Bulgarian Academy of Sciences, Varna, Bulgaria

Marnix Pieters Agentschap Onroerend Erfgoed, Brussels, Belgium

Pierre Régaldo DRAC-SRA Grande Aquitaine, Bordeaux, France

Anne Ropars DRAC-SRA Normandie, Caen, France

Baruch Rosen Independent researcher, Atlit, Israel

Irena Radić Rossi Department of Archaeology, University of Zadar, Zadar, Croatia

Dimitris Sakellariou Hellenic Centre for Marine Research, Institute of Oceanography, Anavyssos, Greece

David Simpson University Museum of Bergen, University of Bergen, Bergen, Norway

Arne Sjöström Department of Archaeology and ancient History, Lund University, Lund, Sweden

Birgitte Skar NTNU University Museum, Norwegian University of Science and Technology, Trondheim, Norway

Vladimir Slavchev Museum of Archaeology, Varna, Bulgaria

Olena Smyntyna Department of Archaeology and Ethnology of Ukraine, Odessa I.I. Mechnikov National University, Odessa, Ukraine

Pierre Stéphan UMR 6554 LETG-Brest Géomer CNRS – Institut Universitaire Européen de la Mer, Brest, France

Fraser Sturt Centre for Maritime Archaeology, University of Southampton, Southampton, UK

Lauren Tidbury Museum of London Archaeology, Mortimer Wheeler House, London, UK

Louise Tizzard Wessex Archaeology, Portway House, Salisbury, Wiltshire, UK

Luc Vallin DRAC-SRA Hauts-de-France, Lille, France

Sven Van Haelst Flanders Marine Institute (VLIZ), Ostend, Belgium

Kieran Westley School of Geography and Environmental Sciences, Ulster University, Coleraine, Northern Ireland, UK

Peter Woodman (Deceased) Department of Archaeology, University College Cork, Cork City, Republic of Ireland

Valentina Yanko-Hombach Department of Physical and Marine Geology, Odessa I.I. Mechnikov National University, Odessa, Ukraine
Avalon Institute of Applied Science, Winnipeg, MB, Canada

Abbreviations and Acronyms

ADS	Archaeological Data Service
AEZ	Archaeological Exclusion Zone
APSL	Above Present Sea Level
ASL	Above Sea Level
BMAPA	British Marine Aggregate Producers Association
BPSL	Below Present Sea Level
BSL	Below Sea Level
BSMAP	Black Sea Maritime Archaeology Project
CADW	Historic Environment Service of the Welsh Government
CAL BP	Calibrated years before present (1950); usually used when converting radiocarbon years to calendar years
CAL BC	Calibrated years BC (before Christ); usually used when converting radiocarbon years to calendar years
COST	European Cooperation in Science and Technology
COWRIE	Collaborative Offshore Wind Research into the Environment (UK)
CUA	Centre for Underwater Archaeology (Sozopol, Bulgaria)
DAC	Data Archive Centre
DCGH	Department of Culture, Heritage and the Gaeltacht (Republic of Ireland)
DEM	Digital Elevation Model
DRAC	La Direction Régionale des Affaires Culturelles
DGPC	Direção Geral do Património Cultural
DRASSM	Département des Recherches Archéologiques Subaquatiques et Sous-Marines
EEZ	Exclusive Economic Zone
EIA	Environmental Impact Assessment
ELC	European Landscape Convention
EMODNET	European Marine Observation and Data Network
ESL	Ice-Equivalent Sea Level
EUA	Ephorate of Underwater Antiquities
GIA	Glacio-isostatic Adjustment
HCMR	Hellenic Centre for Marine Research
HE	Historic England
HER	Historic Environment Record
HES	Historic Environment Scotland

INFOMAR	Integrated Mapping for the Sustainable Development of Ireland's Marine Resource
INRAP	Institut National de Recherches Archéologiques Préventives
IPA	Instituto Português de Arqueologia
ka	Thousand years (date)
kyr	Thousand years (duration)
LGM	Last Glacial Maximum (dated approximately to the period 29–16 ka cal BP)
LOSC	United Nations Law of the Sea Convention
MAP-SDU	Maritime Archaeology Programme, Southern Denmark University
MEDIN	Marine Environmental Data and Information Network
MIBACT	Ministero per i Beni e le Attività Culturali e del Turismo
MIS	Marine Isotope Stage
MSP	Marine Spatial Planning
NMR	National Monuments Record
NMRW	National Monuments Record of Wales
NRHE	National Record of Historic England
NSPRMF	North Sea Prehistory Research and Management Framework
ODN	Ordnance Datum Newlyn (mean sea-level height in Great Britain)
OHCCMAPP	Outer Hebrides Coastal Community Marine Archaeology Pilot Project
OSL	Optically Stimulated Luminescence
P5CANT	El Paleolítico Sumergido: Paisajes y Poblamiento Prehistóricos en la Plataforma Continental Cantábrica
PPN	Pre-Pottery Neolithic (A, B, C)
RCE	Rijksdienst voor het Cultureel Erfgoed (Agency for Cultural Heritage, the Netherlands)
REC	Regional Environmental Characterisation
RCAHMW	Royal Commission on the Ancient and Historical Monuments of Wales
ROV	Remotely Operated Vehicle
SCUBA	Self-Contained Underwater Breathing Apparatus
SEA	Strategic Environmental Assessment
SGU	Sveriges Geologiska Undersöknung (Geological Survey of Sweden)
SINCOS	Sinking Coasts
SLIP	Sea-Level Index Point
SPLASHCOS	Submerged Prehistoric Archaeology and Landscapes of the Continental Shelf
SRA	Service Régional de l'Archéologie
UCH	Underwater Cultural Heritage
UNESCO	United Nations Educational, Scientific and Cultural Organization
UCHC	United Nations Convention on the Protection of the Underwater Cultural Heritage
VLIZ	Vlaams Instituut voor de Zee (Flanders Marine Institute)
WSI	Written Scheme of Investigation

1 The Archaeology of Europe's Drowned Landscapes: Introduction and Overview

Geoff Bailey, Nena Galanidou, Hans Peeters, Hauke Jöns, and Moritz Mennenga

Abstract
This is the final volume of the SPLASHCOS research network and marks the completion of an international and collaborative survey to compile and evaluate the known records of underwater material around Europe's coastlines, deposited during periods of low sea level and subsequently drowned by sea-level rise at the end of the Last Glacial Period. Some of these underwater finds, especially those exposed in the intertidal zone, have been known about since the last century but have languished in obscure publications, museum archives, or national sites and monument records. Systematic underwater investigation and excavation did not begin until the 1970s and only then in a few research centres, notably in Denmark, Bulgaria, and Israel. Only since the turn of the new millennium has this field of underwater research come into focus as a significant missing part of the archaeological record, and many new investigations have begun or gained new momentum during the past decade, not least because of improved technologies of marine exploration and remote sensing, intensified industrial exploitation of the seabed, and collaboration between offshore industries, government agencies, and commercial archaeologists. The material reviewed here covers 20 countries bordering all the major marine basins from the Atlantic coasts of Ireland and Norway to the Black Sea and from the western Baltic to the eastern Mediterranean. Nearly 3000 underwater find spots are known, ranging from isolated artefacts to in situ settlements and spanning a period from c. 1 million years ago to the establishment of modern sea level between 7000 and 2000 years ago. In this introductory chapter, we outline the history of investigations, highlight the importance of this underwater record for a better understanding of European (and world) prehistory, explain the rationale for the research and the way we have structured the volume, and summarise some key issues. The chapters are organised in four parts, and each part is prefaced by its own

G. Bailey (✉)
Department of Archaeology, University of York, York, UK

College of Humanities, Arts and Social Sciences, Flinders University, Adelaide, SA, Australia
e-mail: geoff.bailey@york.ac.uk

N. Galanidou
Department of History and Archaeology, University of Crete, Rethymno, Greece
e-mail: galanidou@uoc.gr

H. Peeters
Groningen Institute of Archaeology, University of Groningen, Groningen, The Netherlands
e-mail: j.h.m.peeters@rug.nl

H. Jöns · M. Mennenga
Lower Saxony Institute for Historical Coastal Research, Wilhelmshaven, Germany
e-mail: joens@nihk.de; mennenga@nihk.de

G. Bailey et al. (eds.), *The Archaeology of Europe's Drowned Landscapes*, Coastal Research Library 35, https://doi.org/10.1007/978-3-030-37367-2_1

editorial introduction summarising key themes for that group of chapters.

Keywords
Sea-level change · Submerged landscapes · Underwater archaeology · Cultural heritage · Stone Age

1.1 Background

It has been well understood for many decades that sea level has been much lower than the present for most of human history over the past one million years and more, repeatedly reaching to a depth of more than 100 m below present sea level and making extensive areas of land available for human settlement and dispersal over many millennia. In Europe, the amount of new land exposed at lowest sea level during the Last Glacial Maximum 20,000 years ago has been estimated at c. 2.5 million km^2, representing some 40% of the present-day land mass (Fig. 1.1).[1] Much of this submerged territory was well-watered and well-stocked with plants and animals, often providing more attractive environments for human exploitation than the contemporaneous hinterlands and arguably, in some cases, environments with no known modern analogue (Bailey and Flemming 2008). Nearly all this land was progressively lost to sea-level rise over a period of about ten millennia as the continental ice sheets melted, creating an ever-changing mosaic of palaeogeographic and palaeoenvironmental conditions until sea-level rise reached the modern position.

Archaeologists, however, have for a long time been reluctant to focus on this missing record or to do anything about it. In part this stems from a belief that nothing of much significance could have survived the destructive impact of inundation and that whatever has survived would in any case make little difference to our understanding of world prehistory. Other factors are the perceived difficulties and costs of mounting research expeditions to search for whatever minimal remains might still be found on or beneath the seabed and the high risk of failure. A further disincentive is the deep-rooted belief that the seashore and the sea beyond were, at best, of marginal significance to the main trajectories of human evolution and social development until a very late period in recent millennia and that all the major social, economic, and technological transformations that created the foundations of early civilisation were largely driven by the exploitation of terrestrial resources on dry land—scavenging, hunting, plant gathering, domestication of plants and animals, and village farming (Erlandson 2001; Bailey and Milner 2002). Typically, this belief provides a rationalisation for not engaging with underwater research, and that lack of engagement in its turn reinforces existing preconceptions, creating a perpetual and self-reinforcing circularity of argument that can stifle new research initiatives.

Four factors have gradually brought about a change in the climate of opinion (for overviews and recent discussion, see Flemming 2004; Benjamin et al. 2011, 2017; Evans et al. 2014; Harff et al. 2016, 2017a; Bailey et al. 2017; Flemming et al. 2014, 2017a; Fischer and Pedersen 2018; Sturt et al. 2018; Wickham-Jones 2018).

1. The steady accumulation of underwater finds in many parts of the world demonstrating that a great deal of prehistoric material, including palaeontological remains of Ice Age mammals as well as archaeological finds, has survived on the seabed, some of it with preservation of organic materials, food remains, and items of material culture sealed beneath marine sediments in anaerobic sediments, items that rarely survive in archaeological deposits on the present-day land surface.
2. Better information about sea-level change, bathymetry and sea-bed morphology that has

[1]These are very approximate figures and are based on a land area that largely excludes European Russia. The total area given here is lower than the 4 million km^2 cited in Flemming et al. (2017a, p.1), because the latter figure was arrived at by taking 40% of the current European land mass *including* European Russia.

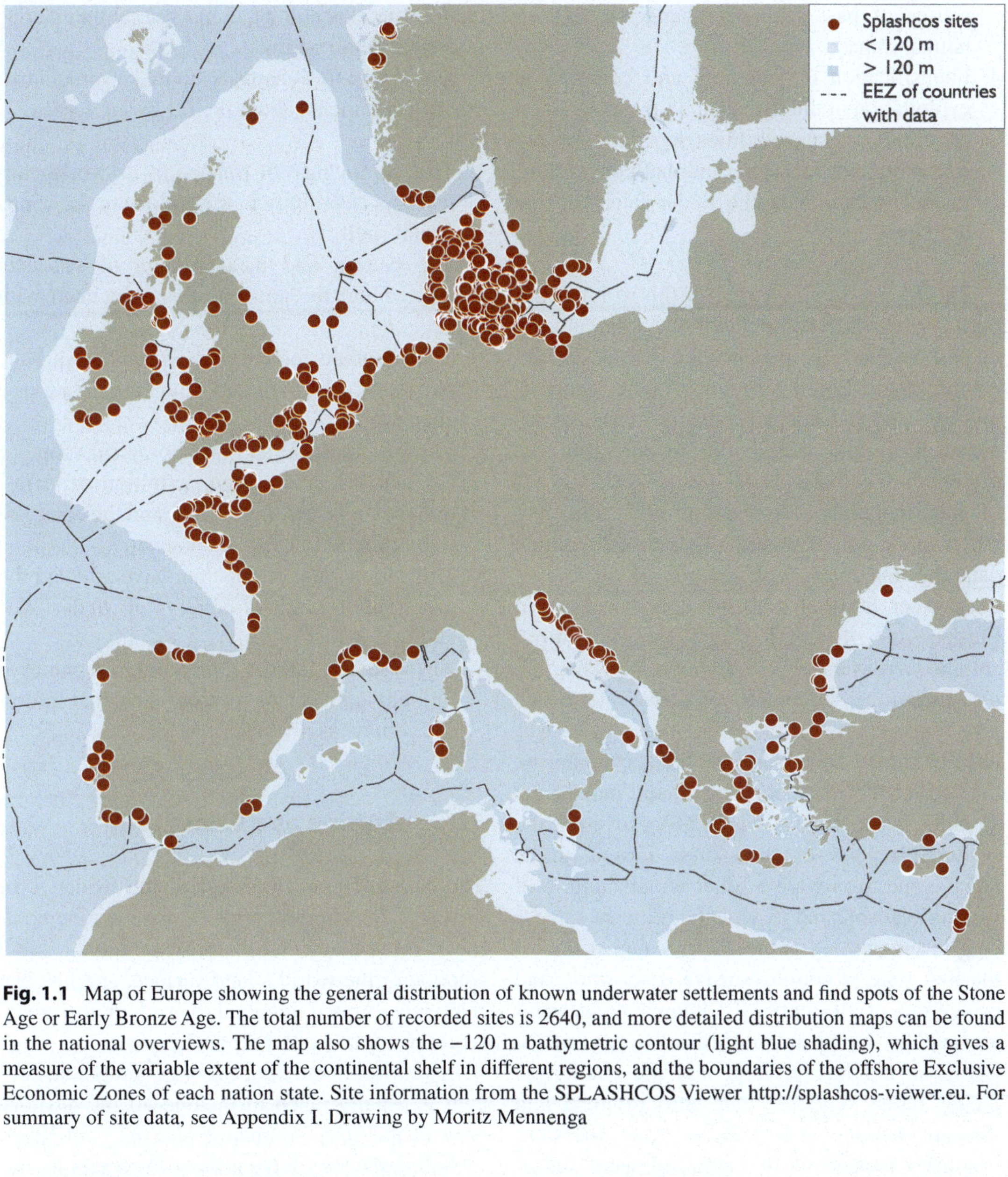

Fig. 1.1 Map of Europe showing the general distribution of known underwater settlements and find spots of the Stone Age or Early Bronze Age. The total number of recorded sites is 2640, and more detailed distribution maps can be found in the national overviews. The map also shows the −120 m bathymetric contour (light blue shading), which gives a measure of the variable extent of the continental shelf in different regions, and the boundaries of the offshore Exclusive Economic Zones of each nation state. Site information from the SPLASHCOS Viewer http://splashcos-viewer.eu. For summary of site data, see Appendix I. Drawing by Moritz Mennenga

demonstrated more clearly the pattern of sea-level change, the length of time that low sea levels have persisted, and how much land has been lost since the Last Glacial Maximum. This is coupled with a growing realisation that this extensive terrestrial territory harboured concentrations of plant and animal life, water supplies, and human settlement, with major consequences for social, economic, and demographic change as land gradually sunk beneath the waves.

3. A growing imperative to study sea-level change, stimulated by the threat of global warming and sea-level rise in the coming centuries, and an interest in harnessing the extensive archives of palaeoenvironmental and archaeological information on the continental shelf to better understand the pattern and

causes of past sea-level changes and their human impact.
4. Intensification of industrial and commercial exploitation of the seabed and the extension of government responsibilities, legislation, regulatory regimes, and international conventions to the protection of the underwater cultural heritage.

The research culminating in this volume has been carried out in parallel with two other closely related projects as part of the SPLASHCOS Action: a pan-European review of the geomorphology and palaeoenvironments of the continental shelf and the conditions affecting the preservation and discovery of prehistoric archaeological materials (Flemming et al. 2017a) and the compilation of a publicly available online database and maps with records of all known underwater finds of prehistoric archaeology in Europe—the SPLASHCOS Viewer (http://splashcos-viewer.eu/; http://splashcos.maris2.nl/). In Europe, the number of underwater Stone Age finds reported from different countries was already known to be many hundreds even before the SPLASHCOS Action began, and part of the stimulus to the project was to bring together the scattered reports of such finds into a single archive and to provide a fuller understanding of what is available and its wider significance.

The aim of this volume, then, is to provide, for the first time, a comprehensive description and evaluation of the currently known underwater archaeology on the submerged landscapes of the European continental shelf and the challenges facing future research. Our chronological and thematic focus is principally on sites that were originally located on dry land but were subsequently inundated by sea-level rise. In practice, this refers to Stone Age material before about 6000–5000 BP on most coastlines, but it includes Bronze Age or even Iron Age sites in those regions where metal technologies appeared early or relative sea level has continued to rise because of vertical land movements or continued melting of continental ice into the world oceans. We discuss issues of sea-level change and define and qualify our chronological and spatial boundaries in greater detail below.

Our aim in this chapter is to make some introductory comments about the following topics:

- The significance of Europe in its world context as a case study in the history of the continental shelf.
- The pattern and measurement of sea-level change and the uncertainties associated with it.
- The rationale for the way in which we have structured the volume and organised the chapters.
- The geographical and chronological scope of the work, including the definition of terms, their benefits and limitations, and the compilation of the SPLASHCOS Viewer.
- The variable conditions that have affected the preservation and discovery of underwater sites.
- Key research themes concerning human evolutionary and social developments that inform the volume as a whole.

1.2 Europe as a Case Study

The case for concentrating on the evidence of Europe's coastlines deserves some opening comment, given that Europe represents quite a small fraction of the world's land surface, and just one relatively limited regional record in the much larger panorama of world prehistory.

By convention, Europe as a geographical concept is usually taken to include all the territory west of the Ural Mountains and the Ural River, including the very large area of European Russia with coastlines on the Arctic Ocean and the Caspian Sea. In this volume we confine our attention to those coastal regions that border the Baltic Sea in the north, the Atlantic Ocean to the west, and the Black Sea and Mediterranean Sea to the south. In practice this excludes most of the western zone of the Russian Federaton (European Russia), apart from its short sections of coastline at the eastern end of the Baltic Sea and on the northern coastline of the Black Sea. We also include Turkey within these boundaries, though it has very few underwater finds, Cyprus, and Israel, which has a long history of underwater investigations and one of the most important and well-studied groups of underwater sites in the world. We do not include other countries bordering the eastern Mediterranean (Syria, Lebanon, Israel, and the Palestinian National Authority) nor all European countries with a coastal border (notable absences are Albania, Bosnia-Herzegovina, and Montenegro), because of lack of underwater investigation and absence of information.

From an archaeological point of view, one reason for focussing on Europe is the simple fact that it has one of the longest traditions of research on the prehistoric archaeology and submerged landscapes of the continental shelf anywhere in the world, extending back to the nineteenth century in some countries, and some of the largest currently known concentrations of prehistoric archaeological sites. This is especially the case in the western Baltic and to a lesser extent in the southern North Sea, the south-west corner of the Black Sea, and the eastern Mediterranean. Whether this concentration is the product of unusual conditions of site preservation and visibility peculiar to European and Mediterranean conditions, or rather to high levels of expert and public interest in these regions and sustained monitoring and investigation over many decades, is a question to be explored and is addressed in the relevant chapters of this volume.

Also, the long-term history of human occupation in Europe witnessed a number of developments that are of relevance to the some of the biggest themes in world prehistory. These include the earliest dispersal of our human ancestors into Europe at least one million years ago from a presumed centre of origin in Africa, most likely via the Levantine corridor but also very likely taking advantage of favourable territories for settlement and dispersal on now submerged landscapes especially in the Aegean region and possibly also involving sea crossings of the Mediterranean; early developments in the exploitation of marine foods, the resources of offshore islands, and seafaring; the colonisation or recolonisation of extensive new territories exposed by the retreating ice; and the early development and dispersal of agricultural and metal-using societies including some of Europe's earliest complex stratified societies such as the Minoan civilisation on Crete and the Varna culture of Bulgaria. All of these developments were taking place at a time when sea levels were lower than present; many most likely had a coastal or maritime aspect; and it is likely that the investigation of the now-submerged coastal regions and palaoeshorelines of Europe and the Mediterranean will throw entirely new light on these developments and new information of interest and relevance at the world scale.

Moreover, the environments available for human occupation on Europe's submerged landscapes ranged from periglacial tundra in the north to Mediterranean woodland in the south and included dry lands, wetlands, lakes, rivers, estuaries, marshes, and coastlines with intertidal zones and offshore conditions ranging from shallow marshy shorelines and extensive mudflats to rocky shorelines in tideless basins, many of them subject to quite rapid and dramatic changes resulting from changes in sea level and the dynamic processes of geomorphological change at the coast margin. These would have offered great variety of opportunities to prehistoric societies as well as challenges to be overcome or succumbed to.

This is not to overlook the fact that traditions of underwater research, the discovery of underwater sites, and the investigation of their relevance to themes in the early development of human society have been going on in other parts of the world. Perhaps the largest concentration of research after Europe is in North America, especially on the coastlines of the Northwest Pacific and the Gulf of Florida, with more sporadic finds or new work only recently getting under way elsewhere and in other parts of the world (see in particular chapters in Masters and Flemming 1983; Benjamin et al. 2011; Evans et al. 2014; Harff et al. 2016; Bailey et al. 2017; Veth et al. 2019). Nevertheless, the concentration of finds, expertise, scientific research, and public interest in Europe is exceptional.

There are other reasons why Europe offers unusual interest and opportunities for underwater research. Although it has a continental hinterland, Europe has a strongly maritime character with its many peninsulas and large offshore islands, its large semi-enclosed marine basins, its convoluted coastlines and archipelagos, and a very high ratio of coastline to land area, especially if one takes into account the many thousands of smaller offshore islands. It also has a great variety of coastal and offshore geomorphologies and oceanographic conditions within a relatively small area of the Earth's surface, especially as these affect the likely preservation and accessibility of submerged landscapes and archaeological materials. These conditions range from the virtually tideless basins of the Baltic, the Mediterranean, and the Black Sea, many of their coastlines with limited wind fetch or exposure to storms, to exposed coastlines facing the Atlantic Ocean and the North Sea in the west with very high tidal ranges.

Offshore, conditions range from shallow and extensive areas of continental shelf with high rates of sedimentation and coastal peat formations, as, for example, in the North Sea and the NW Black Sea, to narrow and rocky coastlines with steep drop offs and narrow shelves as in much of the Mediterranean. The Mediterranean, because of its unusual geological history and its extensive areas of limestone, has a unique karstic geomorphology of erosion and cave formation along many of its coastlines. The low sea-level stands of the Pleistocene accentuated cliff-bound shorelines and creation of coastal caves resulting from intensified groundwater movements and carbonate dissolution. These features are not unique to the Mediterranean but amplified earlier processes of erosion that took place during the Messinian salinity crisis between c. 5.96 and 5.33 million years ago, when the Mediterranean was cut off from the Atlantic and high rates of evaporation turned it into a deep and dry basin (Flemming and Antonioli 2017). This has resulted in thousands of coastal caves, many of which are now submerged below present sea level, and some of which clearly have potential for preservation of terrestrial deposits and archaeological material.

All of this makes for a very wide range of different environmental conditions that would have faced prehistoric societies and to which they would have had to adapt, different geomorphological conditions affecting the survival of landscape features and archaeological sites in the face of inundation by sea-level rise, and different opportunities for comparative analysis, with results that are likely to impact on some of the major themes in world prehistory.

1.3 Sea-Level Change

Understanding of sea-level change is fundamental to everything discussed in this volume. Generalised diagrams of sea-level change over the Last Glacial cycle, from the Last Interglacial period of high sea level 130,000 years ago through to the establishment of modern sea level in the mid-Holocene, have been used to illustrate archaeological syntheses of European coastal prehistory for over 35 years (Bailey 1983; Shackleton et al. 1984), following the demonstration that $\delta^{18}O$ measurements of foraminifera in deep-sea sediment cores provide a continuous proxy record of changes in ocean volume (Shackleton and Opdyke 1973; Shackleton et al. 1983). Typically, these early archaeological studies produced maps plotting sea-level variations against modern bathymetry to show the extent of submerged territory at the Last Glacial Maximum or other selected periods.

Since then, measurements of past sea-level change and understandings of the geological and climatic processes involved have, of course, greatly improved but so too has understanding of the scientific complexities underpinning such measurements and the margins of error inherent in them. Margins of error are broadly of two kinds, errors of dating and errors in the vertical positioning of sea level at any given date relative to the modern position. For some archaeological purposes, quite general and imprecise patterns are sufficient to provide useful insights. However, as new archaeological questions have led to the demand for greater precision, so too the need to appreciate the uncertainties of the underlying science has become more imperative. This is not always apparent in archaeological texts, which are liable to give a misleading impression of precision by citing specific figures without clarifying the uncertainties involved or to extrapolate global sea-level curves to the region of archaeological interest. Conversely, specialists frequently engage in technicalities that seem of little relevance to archaeological concerns and that many archaeologists are ill-equipped to comprehend.

This situation is further confused by the fact that sea-level change is of interest to several different disciplines, each with different objectives that require different types of data and different degrees of precision. Archaeologists generally want to know to a considerable degree of precision the date and geographical position of palaeoshorelines in their region of interest; palaeoclimatologists are more interested in the global pattern of sea-level variation through time, usually to the highest possible level of measurement, while geologists and tectonic geomorphologists are primarily interested in vertical movements of sea level relative to land in particular regions but not necessarily with the chronological precision required by palaeoclimatologists or the spatial precision required by archaeologists (see Inglis et al. 2019).

From an archaeological point of view, it is important to appreciate two basic problems. The first is the degree of uncertainty involved in measuring changes in sea level that result from changes in ocean volume, that is, changes in the amount of water in the world oceans resulting from the expansion and contraction of the continental ice sheets. This is a global measure, commonly referred to as the ice-equivalent sea level, or esl for short, and roughly corresponds to what is commonly referred to as eustatic sea level. The difference between esl and eustasy is that the latter incorporates changes in the geometry and capacity of the world's oceans to hold seawater as well as the volume of ice-equivalent seawater. The difference between the two measures in the time ranges of interest in this volume, particularly for the late Pleistocene, is small and can be ignored, and the two terms are often used interchangeably.

The difficulty in arriving at reliable measures of global esl is that dated palaeoshorelines provide only fragmented and discontinuous records, compounded by vertical movements of the Earth's crust. More useful are deep-sea sediment records, which provide a continuous sequence of marine sediments and proxy measurements of sea-level variation in the form of $\delta^{18}O$ variations in the carbonate skeletons of planktonic and benthic foraminifera. They are dateable, and the longest sediment cores extend back over many hundreds of thousands of years. However, they are difficult to relate to the heights of palaeoshorelines without calibration against dated palaeoshoreline features such as submerged or elevated corals terraces, beachrock and peat, or measurements of ice volume. They are also subject to other distortions such as temperature effects on isotopic composition, differences of circulation in different marine basins, bioturbation of sediments, dating errors, and so on. One of the most detailed sea-level curves of this type is from the Red Sea and is based on a variety of proxy measures cross-checked against other palaeoclimate and geological records to construct a sea-level curve over the past 500,000 years (Fig. 1.2; Siddall et al. 2003; Grant et al. 2012, 2014).

As Grant et al. (2014, p. 4) point out, their sea-level curve measures relative sea-level (rsl)—relative to a particular geographical region. They produce evidence to suggest that there is an offset between rsl and esl of as much as 10 m in the Red Sea, especially at glacial maxima, corroborated by detailed studies of vertical land movements (Lambeck et al. 2011).

This raises the second problem for archaeologists and others, and that is that the Earth's crust is not fixed but has been subjected to sustained vertical movements over many geographical scales, some of them in recent geological and archaeological time. These take the form either of isostatic effects associated with loading and

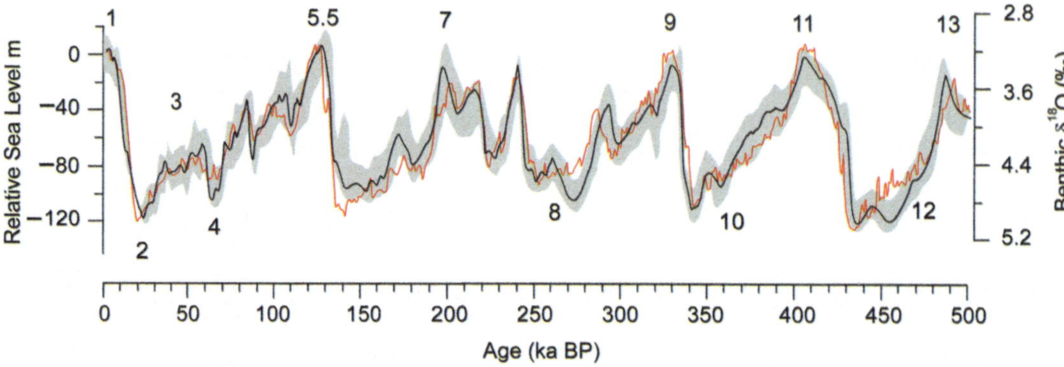

Fig. 1.2 Relative sea-level curve for the Red Sea over the past 500,000 years, based on a variety of proxies including the $\delta^{18}O$ signal derived from planktonic foraminifera in deep sea sediment cores (black line). The light grey shading indicates the probable margin of uncertainty, including margins of error associated with radiometric dating and the estimation of sea-level position from proxy data. The curve for the past 150,000 years is further constrained by dated palaeoshoreline features and is considered to be the most accurate part of the sequence. The red line is the continuous curve of variation in the $\delta^{18}O$ signal of benthic foraminifera in deep sea cores from a variety of sources. Also shown are some of the principal marine isotope stages. Data from Siddall et al. (2003) and Grant et al. (2012, 2014). Drawing by Geoff Bailey

unloading of ice sheets and to a lesser extent seawater acting on the continental shelf—GIA (glacio-isostatic adjustment) for short—or tectonic effects, notably at plate boundaries and rifts, where they are associated with uplift or subsidence. Both effects are present in coastal regions of Europe, GIA especially in the northwest in regions of former glaciation, and tectonics in parts of the central and eastern Mediterranean. Harff et al. (2017b) give an extended discussion of these processes with reference to Europe, and examples are also discussed in the relevant chapters of this volume. GIA can extend very widely, well beyond regions of former glaciation albeit with attenuated effect, but is especially prevalent in northern Britain and Scandinavia, where sea level went down initially after the LGM (Last Glacial Maximum) because the land was rebounding faster than sea level was rising. Because of time lags in the Earth response, the land is still rebounding today in areas of Scandinavia formerly located beneath the thickest mass of ice, and the amount of vertical uplift can be hundreds of metres.

Separating out the different contributions of tectonics, GIA, and changes in sea-water volume is problematic because of the difficulty of obtaining fully independent measures of each variable and the combined margins of error associated with each type of measurement. One approach is to look at palaeoshoreline indicators in 'far-field' regions which are tectonically stable and far from GIA effects. A variety of modelling procedures incorporating far-field data have been developed to isolate and refine measures of global esl and to develop models of sea-level change in other regions incorporating whatever dated palaeoshoreline information is available in the region of interest (see Lambeck et al. 2014, for a succinct introduction to the problems of sea-level measurement). Even these models are subject to remaining uncertainties such as the viscosity of the underlying mantle in different regions. Using these modelling procedures and a more detailed data set than Grant et al., Lambeck et al. (2014) have produced an esl curve for the past 35,000 years (Fig. 1.3). Of special note is the evidence for a maximum lowering of sea level of 134 m at 21,000, the persistence of sea levels at or lower than c. 120 m below present for a period of 13,000 years between 29,000 and 16,000, and an offset of 10–20 m between their esl curve and the Red Sea rsl curve during this period.

Also of note is the evidence that sea level did not stabilise at about the present position until 2000 years ago. The commonly cited date of

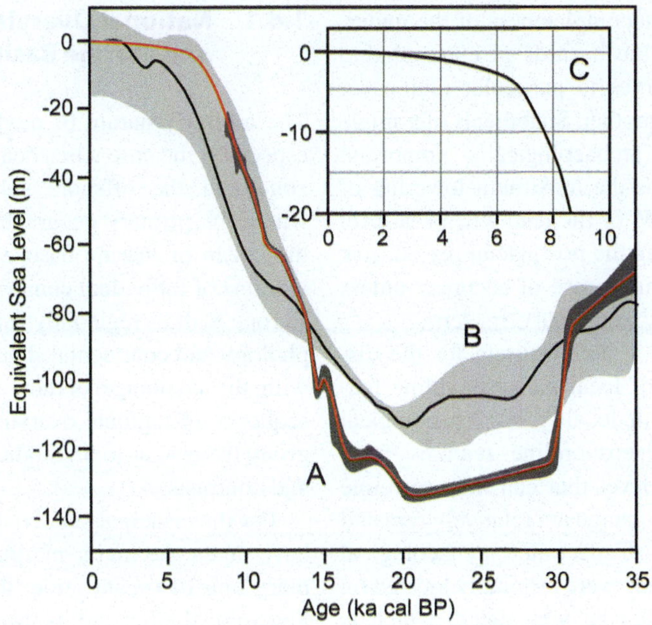

Fig. 1.3 Comparison of the global ice-volume equivalent (eustatic) sea level (esl) with the relative sea-level curve from the Red Sea for the last 35,000 years. A (red line): the esl curve with the statistical margin of measurement error at the 95% probability level (dark grey). B (black line): the relative sea-level curve from the Red Sea with its margin of uncertainty (light grey). C (Inset): the esl curve for the past 9000 years (the legend for the figures on the axes is the same as for the main image). Note that the maximum depth of sea level in A is −134 m at 21 ka and that a period of low sea level at ≥120 m lasted from 29 to 16 ka. Note also that the two curves correspond quite well within the margins of uncertainty except for the period of the LGM, when there is an offset of 10–20 m. Note also that the rate of rise in the esl drops sharply between 8.2 and 6.7 ka, corresponding to the final melting of the North American ice sheet, but that the esl continues to rise more slowly through another 3 m before fully stabilising. See text for further discussion. Data in A and C are taken from Lambeck et al. (2014) and in B from Fig. 1.2. Drawn by Geoff Bailey

6800 cal BP for the final melting of the northern hemisphere continental ice sheets is sometimes used as a proxy measure for the date when sea level stopped rising. However, the esl data in Fig. 1.3, while they show a marked drop in the rate of sea-level rise at that date, also demonstrate a continued slow increase of sea level through about 3 m until 2000, most probably because of continued melting of ice from the Antarctic and Greenland ice sheets.

In addition to the issues discussed above, there are yet additional variables that need to be taken into account when assessing relative sea-level position and the configuration of palaeoshorelines at a regional or local scale. In the regional, semi-enclosed seas of Europe, for example, the effect of variations in wind strength and direction and barometric pressure can affect sea level (Harff et al. 2017b). Shorelines of softer materials such as Pleistocene sediments are also vulnerable to lateral shifts in shoreline position because of erosion or progradation, especially during periods when sea level remains at about the same level for long periods, processes that have been well studied in the Holocene (Harff et al. 2017b). Changes in the depth of the seafloor because of accumulation or erosion of marine sediments since the time of archaeological interest can also affect the reconstruction of palaeogeographic contours and shorelines from present-day bathymetry.

The upshot of the above discussion is that archaeologists interested in plotting the impact of sea-level change on their archaeological data and their region of interest should be aware of the margins of uncertainty involved in the primary measurements, the incompleteness of the underlying science, and the hazards of extrapolating global sea-level curves to particular regions or from one region to another. Statements of a single measurement for archaeologically important benchmarks, such as the maximum lowering of sea level at the LGM or the date when sea level stopped rising during the postglacial period, risk imputing a misleading sense of accuracy unless they are qualified by a suitable error term.

Ultimately there is no substitute for the discovery and dating of local palaeoshoreline features in the region or locality of archaeological interest. Dated palaeoshoreline features from below modern sea level that can help to refine models of sea-level change are relatively rare and much sought after. Underwater archaeological sites, especially if they were originally located on the shoreline and can be well dated, which is often the case, are therefore of great interest to sea-level modellers, and archaeologists and sea-level specialists can make common cause in the search for such data to mutual benefit.

1.4 Organisation and Scope

The initial compilation of data for the SPLASHCOS Viewer and the subdivision of tasks undertaken by the members of WG1 was organised by each modern nation state, reflecting the practical reality that scientific and academic research institutions, government agencies responsible for cultural heritage, and the conduct of field research are largely organised at national level. This has carried through to the chapters that form the basic building blocks of this volume, which present a series of national overviews. However, modern state boundaries are a poor fit with the environmental features that are of primary relevance in the interpretation of the underwater archaeological record. They risk placing data into arbitrary categories that inhibit supra-national integration and obscure the understanding of variables that go across modern political boundaries.

1.4.1 National Overviews and Marine Basins

The arbitrary nature of modern state borders is especially the case when dealing with the coastal zone and the offshore submerged landscape, where the primary geographical unit of study is the ocean or marine basin to which the coastal margins of individual countries relate. Moreover, Europe with its regionally varied coastal geomorphology and continental shelves has presented us with the challenge of how best to arrange the sequence of national overviews and how best to group them so as to highlight relevant similarities and contrasts.

For the principal subdivisions of the book, we have taken the major marine basins as our primary unit of organisation: the Baltic, the North Sea and Atlantic, and the Mediterranean and the Black Sea. Each has its own distinctive characteristics that provide a coherent whole and a unifying focus for transnational comparisons, and these categories also facilitate comparison with the more detailed information on Quaternary palaeoenvironments presented in the Flemming et al. (2017a) companion volume, which is structured along similar lines. A case could be made for further geographical subdivision, for example, between the North Sea and the Atlantic coasts of Western Europe, between the different sectors of the Mediterranean, or between the Mediterranean and the Black Sea, but this has to be tempered by the quantity and detail of the available archaeological data. This is heavily concentrated around the coastlines of a small number of countries in Northwest Europe, especially those bordering the North Sea and the western Baltic, which have the longest histories and the best-developed traditions of underwater investigation and collaboration with offshore industries, whereas the archaeological data elsewhere is much patchier (Fig. 1.1).

A complicating factor is the fact that some countries have two coastlines. France, for example, has an Atlantic and a Mediterranean coastline and Denmark a North Sea and a Baltic one.

Conversely, some marine basins are only partially represented in our coverage: the Mediterranean, in particular, is represented only by its northern and eastern shorelines, and the North African coastline is absent, despite its importance in prehistory. This reflects both the absence of known underwater prehistoric remains and the lack of participants from the North African countries in the SPLASHCOS Action.[2] Similarly, the Black Sea is represented by only two national contributions (Bulgaria and Ukraine). Also, the Baltic is only represented by the countries at the western end (Germany, Denmark, and Sweden). Again, this reflects, in part, the near absence of underwater archaeological data to the east and the north, but more importantly the impact of glacio-isostatic rebound, which means that the archaeological evidence for the earliest settlement of coastlines in these areas after the retreat of the Scandinavian ice sheet is now mostly uplifted above modern sea level.

With all these factors in mind, we have used three primary groupings to organise the national overviews: the Baltic and Scandinavia (including Norway) (Part I), the North Sea and the Atlantic (Part II), and the Mediterranean and the Black Sea (Part III). These are broad groupings that reflect a compromise between the ideal subdivisions derived from environmental considerations and the relative proportions of archaeological data and investigation distributed across the continent. We recognise that this grouping introduces arbitrary boundaries of its own, and we have dealt with this by introducing each of our main groupings with an editorial introduction, which highlights the features of the national overviews within that grouping, addresses relevant aspects of within-basin variability and inter-basin boundary issues, and incorporates such information as is available from isolated underwater archaeological finds elsewhere and not represented in the national overviews.

[2]COST rules certainly cover the funding of representatives from these countries, but we were unsuccessful in our attempts at recruitment.

Two chapters are from countries (Malta and the Ukraine) that currently have no known underwater finds (apart from a worked flint in a core recovered from the north-west shelf of the Black Sea). The reason for their inclusion is that they are all countries where underwater survey has been carried out with archaeological questions in mind and with prospects for future discoveries.

We have chosen to present the groupings and their constituent chapters in geographical order, beginning with the Baltic, and working in an anticlockwise direction through the North Sea and Atlantic coastlines, and thence through the Mediterranean Sea and the Black Sea working from west to east.

The choice of the Baltic as our starting point reflects the fact that this region has the best developed tradition of underwater prehistoric archaeology in Europe—and indeed the World—the largest number of practitioners, and by far the largest number of currently known prehistoric underwater sites. These include settlement sites and burial grounds in Denmark and Germany such as Tybrind Vig, Møllegabet II, and Timmendorf-Nordmole that demonstrate the quality and detail of preservation in sediments rapidly submerged under the advancing sea, with remains of wooden artefacts, fish traps, dugout canoes, paddles, plant fibres, wooden structures, and other organic materials. It is also a region that highlights the dramatic interplay between sea-level rise at the end of the Last Glacial Period and the glacio-isostatic adjustments associated with ice melting—uplift of the territory formerly depressed under ice cover in Scandinavia and sinking of territory and coastlines around the former ice margin—resulting in alternations between periods when damming of the Baltic turned it into a freshwater lake basin isolated from the North Sea and periods of reconnection involving dramatic episodes of flooding and complex patterns of palaeogeographical and ecological reconfiguration; these must have had a correspondingly dramatic impact on the pre-existing human populations of the region. Because of its glacial history, evidence for human activity earlier than about 15,000 years ago is absent and at this early date only in Denmark. Elsewhere, earliest dates are closer to 10,000 years ago or soon after, sites of this age being present both on uplifted coastlines (as in Norway) and on submerged coastlines (in southern Denmark and Sweden).

Our second regional grouping (Part II) is dominated by finds from around the North Sea, especially in British and Dutch territorial waters and to a lesser degree along both coastlines of the English Channel/La Manche and the coast of Belgium. The finds from this region extend the underwater prehistoric record into deeper water and further back in time, at least as far back as 300,000 years ago, and in the case of human footprints in intertidal sediments in Britain back to the earliest known human presence in the region nearly one million years ago.

These finds include a large and important group of sites exposed in the intertidal zones of Britain and along the Atlantic coastlines of northern France including partially submerged human footprints and megalithic structures, and some systematic excavations of fully submerged sites, notably the Mesolithic site of Bouldnor Cliff on the Isle of Wight and, directly opposite across the Channel, the Middle Palaeolithic site of La Mondrée, Fermanville, on the Normandy coast of France.

Further offshore, large quantities of Ice Age fauna, numerous stone artefacts, and lesser quantities of human bones including part of a Neanderthal skull have been dredged up from the seafloor by fishing and gravel extraction. Although little is known about the provenance of many of these finds, efforts are now being devoted to pinpointing their location and obtaining new information. In addition, new and important material has been recovered through large-scale operations made possible by collaboration with industrial companies working offshore.

One good example is the systematic investigation during the past decade of the submerged Mesolithic site of Maasvlakte-Yangtze harbour on the Dutch coast, associated with the construction of new harbour facilities (Moree and Sier 2015). This marked an important turning point in demonstrating the possibilities for discovery of prehistoric remains buried beneath marine sediments. Here a stepwise process of acoustic and geological survey identified likely target areas, followed up by coring of sediments for traces of human activity and large-scale excavation in the most promising localities. Another good example is the A240 site off the east coast of England (Tizzard et al. 2014; 2015), where discovery of artefacts in the gravel spoil from aggregate extraction led to sampling of the sediments in the vicinity of their original discovery and the recovery of larger quantities of artefacts and palaeoenvironmental data from dateable sediments (see chapters in Part II and Part IV for further details).

In addition, analytical techniques such as radiocarbon dating and stable isotope analysis of palaeodiets can be applied to chance finds dredged up from the seabed, regardless of their provenance, revealing new information about the environmental features of the submerged landscape and the nature of human adaptations to them. The recovery of ancient DNA from buried sediments (sedaDNA) offers a new technique for gaining insight into the plants and animals present in the pre-inundation landscape.

The Mediterranean-Black Sea region is our most disparate and widely scattered collection of material (Part III). It includes an important group of finds on the Mediterranean coastline of France, including the famous and partially submerged rock-art site of Cosquer Cave and other submerged karstic caves with evidence of preserved terrestrial sediments containing artefacts and fauna. Elsewhere, there is a wider scatter of submerged sites in Italy, Croatia, Greece, Cyprus, and Bulgaria, mostly submerged or partially submerged in shallow water, and including the important submerged sites of Pakostane and Zambratija in Croatia, Agios Petros, and Pavlopetri in Greece, submerged and well-preserved Chalcolithic and Early Bronze Age village settlements in Bulgaria, and another concentration of expertise and discoveries on the Carmel coast of northern Israel including the underwater Pre-Pottery Neolithic Village of Atlit Yam, one of the most intensively studied underwater sites in the world. Israel has a tradition of researching underwater prehistoric sites, which, as in Denmark, goes back to the 1970s. The number of sites, though relatively small, and extending in time only back to about 9000 years ago, demonstrates the extraordinary range and quality of evidence that can be recovered from systematic underwater survey and excavation extending over a period of many years. It is fitting that the finds from these opposite ends of our geographical range should provide the opening and closing chapters, respectively, of the national overviews.

1.4.2 The Industrial, Legal, and Commercial Context

Most of the national overviews with underwater finds include a section at the end of the chapter on the management of the underwater cultural heritage and the responsible governmental agencies or other institutions. This is an issue of major importance, given the growing threats from a variety of causes, especially intensifying commercial and industrial exploitation of the seabed, but also including natural processes of erosion, some of which are being accelerated by recent climate change. From its very beginnings early in the twentieth century through to the present day, the history of underwater investigation of the submerged landscape has been intimately connected with the progress of offshore commercial and industrial activity (Sturt et al. 2017). Sometimes this leads to tensions and conflicts between the needs of scientific and archaeological research and the regulations and resources available to protect and manage the underwater cultural heritage. Accordingly, Part IV brings together a group of chapters that examine legal, historical, and managerial issues associated with offshore and underwater work and its relationship to commercial and industrial exploitation of the seabed, issues that are of paramount importance to the present practice of underwater archaeology and the prospect of future discoveries.

Much of the underwater research in NW Europe is being carried out by commercial archaeological companies in response to the requirements imposed by legislation on offshore industries to undertake initial investigation and impact assessments in advance of offshore developments such as wind farms, construction of port facilities, and building of bridges and tunnels. In some cases, these have resulted in high-profile projects involving collaboration with industrial companies and major discoveries that would not otherwise have been possible. Notable examples, as mentioned earlier, are the A240 Project, in collaboration with gravel extraction companies in the United Kingdom, and the expansion of the Yangtze harbour in collaboration with the Port of Rotterdam authority in the Netherlands. Some of these projects provide funding and equipment on a scale that far exceeds the normal research budget, and the results are of benefit to all concerned, producing important archaeological discoveries as well as good publicity and wider public interest. Increasingly it is archaeologists in the commercial sector of the community who are developing the skills and experience and who are able to command the resources to conduct such investigations. Typically, this type of work involves a three-way relationship between the industrial companies, archaeologists, and government agencies responsible for managing the underwater cultural heritage. The issues that arise from this relationship are considered from a variety of perspectives in this final group of chapters.

1.5 Boundaries and Definitions

An important issue is the question of the uppermost chronological boundary for the finds that we include in our survey. At an early stage, we agreed that work should concentrate on underwater archaeological finds no later in date than the time when the ocean-volume equivalent (or eustatic) sea level stabilised after the melting of the continental ice sheets, that is to say at about 6000–5500 cal BP, a date that marks the approximate boundary on many European coastlines between the end of the Stone Age and the beginning of the Bronze Age. Our aim, after all, is to highlight the significance of now-submerged territory and to promote its study. However, this is an arbitrary boundary and raises two important issues.

1.5.1 Chronological Issues

The first issue is that, as discussed earlier, relative sea-level rise continued later than this limit of 5500 years ago on some coastlines, whether from continued addition of glacial meltwater to the oceans or continuing land subsidence resulting from isostatic or tectonic processes. Tectonic effects associated with relative plate motions in the eastern Mediterranean, especially subduction in the Aegean, are causing stretching of the Earth's crust and long-term subsidence. Some important later prehistoric settlement sites have been partially or totally submerged in consequence and provide relevant and interesting examples of what survives the process of inundation in shallow water.

There is also an interesting category of finds that represents the deliberate and ritual deposition of votive offerings under water, rather than materials accidentally lost overboard during fishing expeditions or sea crossings. These are well known from Bronze Age times onwards, or at least more easily identifiable because of the types of metal artefacts found, but this practice may have existed in the Stone Age, when it is more difficult to distinguish from the stone tools whether they represent the byproduct of day-to-day subsistence activities or ritually deposited material.

We have encouraged the authors of the national overview chapters to take a flexible attitude to this chronological boundary, while emphasising that the primary focus should remain on pre-inundation landscapes of the Stone Age and archaeological evidence that informs on the occupation and use of those landscapes when sea level was lower than the present.

Many coastlines of Northwest Europe harbour later evidence of maritime activities and material culture in partially submerged or waterlogged coastal wetlands, or on shores that have become covered by marine sediments because of progressive landward erosion, sometimes with excellent preservation of features such as wooden fish traps, trackways, landing stages, and remains of wooden boats, extending in date right through to recent historical periods. Similarly, in the tectonically active regions of the Eastern Mediterranean, notably in Italy and Greece, there is a plethora of stone-built harbours, landing stages, fish tanks, and other shoreline buildings and facilities of Classical and later date that have undergone partial or total submergence because of land subsidence. We consider these to be outside the scope of this volume, although they may receive a brief mention in some of the national overviews.

1.5.1.1 Radiocarbon Chronologies

A subsidiary issue that should be mentioned here is the conventions for citing dates, which are the subject of ongoing confusion. Within the time range of radiocarbon dating, many dates are radiocarbon ages. However, it is well known from dendrochronological calibration that radiocarbon ages deviate significantly from calendar ages, especially during periods critical to our enquiry, such as the late glacial and early postglacial periods, when the deviation may be as much as 2000 years, with occasional plateaux or even reversals. Radiocarbon ages therefore need to be calibrated if they are to be compared with calendrical or other dating schemes. We have therefore encouraged our authors to stipulate whether ages derived from radiocarbon dating are radiocarbon ages, usually expressed as a figure with a ± statistical margin of error indicating one standard deviation (a 68% probability that the true value lies within the stated margin), or calibrated dates, usually indicated by the suffix cal (cal BP or cal BC as the case may be) and sometimes expressed as a range of dates at two standard deviations (95% probability).

The situation is further confused by the preference amongst archaeological specialists of Neolithic and later periods for the BC (or BCE)/AD convention, and amongst many Palaeolithic and Mesolithic specialists for the BP convention, while palaeoenvironmental specialists prefer to use BP (without the cal prefix) when referring to calibrated dates. In older literature before calibration was widely applied, dates are given as BC or BP without qualification but usually meaning that they are uncalibrated radiocarbon years. Other conventions are 'years ago' or 'ka' (thousands of years ago), which usually refer in a general way to years before the present, where closer accuracy is not required. There is at present no agreed solution to these different conventions, and wherever possible dates in this volume are expressed in a way that hopefully makes clear their derivation.

1.5.2 The Ambiguity of the Land-Sea Boundary and Issues of Definition

The question of chronology raises a second important boundary issue and that is the issue of how we define 'under water', where we draw the line between dry land and submerged land and hence between terrestrial archaeology and underwater archaeology. Strictly speaking, all investigation of past human landscapes is terrestrial archaeology, whether these landscapes are now below modern sea level or above it, and ideally both should be treated together as a seamless whole, with the present-day shoreline as an arbitrary and largely irrelevant boundary until sea level reaches the modern level. However, there are practical issues that complicate this simple vision.

The first complication arises from how we define an underwater site. A minimum definition for our purposes is any archaeological material originally deposited on a dry land surface that has subsequently been inundated by relative sea-level rise. This includes archaeological material in the intertidal zone or shallow water up to and including the high-water mark at the time when these pre-inundation land surfaces were in use, for example, abandoned dugout boats, fish weirs, or domestic refuse discarded into shallow water by people living on settlements located on the shoreline.

This in its turn raises a second complication, and that is how we are to deal with evidence found in the modern intertidal zone. Here there is a major contrast between different regions. The Atlantic and North Sea shorelines of Western Europe are exposed to large tidal ranges—as much as 14 m in some parts of Britain—with intertidal zones that can extend at low tide for many hundreds of metres or even kilometres out from the present-day shoreline. In the Baltic, Mediterranean, and Black Sea Basins, the tidal range is almost everywhere tiny by comparison, measured in tens of centimetres, though the amplitude may be increased a little by other effects such as variations in winds and barometric pressure.

In basins with minimal tidal ranges, the boundary between land and sea is quite sharp and invariant, and it is usually obvious what constitutes underwater archaeology deposited on a land surface that was subsequently inundated and what constitutes terrestrial archaeology above modern sea level. Underwater sites may occur at varying depths; some are in water shallow enough that one can stand on the seabed without putting one's head under water, but it is clear that the material has been permanently submerged since the time of inundation (with rare exceptions) and that closer investigation and excavation requires divers, other standard methods of underwater investigation, and assessment of how deposits and materials have been affected by inundation.

In the case of Atlantic coastlines, the boundary between land and sea is much less sharply defined, and the question of what constitutes terrestrial as opposed to underwater archaeology is highly ambiguous. Many artefacts have been recovered from these intertidal zones and may represent (a) in situ material on an old land surface that has been covered by marine sediments and then re-exposed by stripping away of the overlying sediments by tidal action or storms, (b) material that has been eroded out from a more deeply submerged land surface and washed ashore, or (c) material that has been eroded out from terrestrial deposits on the landward side of the modern shoreline such as cliffs or raised beaches and redeposited on the modern beach. Clearly category (c) material cannot count as evidence for the use of a submerged land surface and should therefore be excluded from our statistics for underwater sites. However, it is often not possible to distinguish between these possibilities without careful and detailed investigation. With many isolated specimens, it may prove impossible to identify which of the above categories they belong to, in which case the finds should be excluded from the underwater category.

Moreover, archaeological investigation of these extensive intertidal zones in Northern Europe, notably in France and Britain, is usually carried out as an extension of land-based archaeology, with survey on foot and excavation of material during periods of exposure at low tide using conventional land-based techniques. Yet, some of the ancient land surfaces and archaeology exposed at extreme low tides in Britain that have been investigated in this way are earlier in date, at greater distance from the modern shoreline, and deeper below mean sea level, than some of the sites in Denmark, Germany, and Sweden that are permanently under water and have been investigated by divers. In both cases the archaeology refers to the use of a landscape that was subsequently inundated by sea-level rise. In the former case, however, the material is treated as an extension of archaeological investigation on land, and the low-water mark is the limit beyond which investigation does not go; in the latter case, the material is treated as part of the underwater record, and the high-water mark defines an upper limit below which investigation may proceed to any depth and distance from the shore that is within reach of SCUBA technology or acoustic survey. The British and Scandinavian traditions of investigating archaeology at the land-sea boundary are, thus, highly divergent, reflecting the very different tidal regimes and research traditions in each region. Both have resulted in major projects leading to information about the use of submerged landscapes, but the British projects are centred around the investigation of coastal wetlands and the Danish projects around the investigation of underwater settlements.

A further difficulty is the fact that some coastal sites on dry land that have never been inundated by sea-level rise nevertheless do not make sense as places of settlement or activity except in relation to a landscape that is now submerged. They may contain important archaeological evidence for the exploitation of resources, particularly marine resources, on that now-submerged land surface. Coastal caves above modern sea level are especially relevant here because they sometimes contain stratified deposits with deeper layers that extend back in time to a period of lowered sea level. These are present on many European coastlines especially in Southern Europe and are referred to in the relevant national overviews even though they do not constitute underwater sites in the strict definition of that term.

Until recently, specialists of the Palaeolithic and early Mesolithic periods have been at risk of building false interpretations around a body of data that represents the truncated hinterland fragment of a wider landscape, most of which is now submerged and is therefore ignored. It is equally important that, in the interests of promoting underwater research, we do not fall into the opposite trap of excluding information relevant to understanding the human use of now-submerged landscapes by defining too strictly the limits of study and confining our attention only to archaeological material that is fully submerged below present sea level.

Another category of archaeological interest that blurs the boundary between land and sea is 'wetlands' archaeology, which has a long tradition of study especially in Europe (e.g. Coles 1984, 1992; Menotti and O'Sullivan 2012). Here the emphasis is on waterlogged material, with its unusual opportunities for preservation of organic materials, especially wood, whether in underwater contexts on submerged coastlines, in shallow marine inlets and fjords as in Denmark that have been drained in more recent times but retain waterlogged sediments and preserved organics remains, or in inland lake basins. It is no coincidence that Grahame Clark, the founder of Mesolithic studies in Britain and the economic approach to prehistory, immediately recognised the significance of the Colinda harpoon dredged up from the North Sea in 1932, the same year that he founded the Fenland Research Committee, or that he was simultaneously engaged in researching the submerged landscapes along the Essex coastline in the 1930s alongside the wetlands of the East Anglian hinterland (Smith 1997; Sturt et al. 2017; Bailey et al., Chap. 10, this volume).

There is no simple solution to these definitional dilemmas other than to recognise that the boundary between underwater and abovewater archaeology is a blurred one, and to refer to the wider context of the terrestrial archaeological record where it is relevant to the primary focus of this volume, which is an understanding of the submerged landscapes and the underwater archaeology belonging to earlier periods of prehistory when sea level was lower than present.

1.6 The SPLASHCOS Viewer

The SPLASHCOS Viewer is a simple digital inventory of underwater archaeological sites around Europe's coastlines, linked to a digital map of Europe. At present, it contains 2640 individual records, ranging from find spots that consist of a single artefact to underwater settlements that have been systematically excavated and generated thousands of finds (Appendix I; Fig. 1.1). It is publicly available as an online GIS and includes summary data about each site and its location so that users can filter information in accordance with their own research interests. Online access is intended to support scientific research and to promote the ideal of open science. Data are organised by nation state and have been compiled by the authors of the national overviews presented in this volume. The Viewer is the source of information for the distribution maps of archaeological sites presented throughout this volume and for the tabulation of site information presented in some of the chapters. Full details about the compilation of the data in the Viewer and the accompanying maps are presented in Appendix I, together with summary statistics for Europe as a whole, and for the major marine basins which form the structure for the chapters in this volume.

The SPLASHCOS Viewer has a permanent home at the Lower Saxony Institute for Historical Coastal Research, Germany, where the data are collated and managed, and the online information is on servers maintained under EU auspices and available at http://splashcos.maris2.nl/ and at http://splashcos-viewer.eu/. Arrangements are in place to include new data as it becomes available. By providing a permanent and publicly available source of scientific data, and one that can be linked to other online platforms, the intention is to promote the agenda of scientific research on submerged landscape archaeology, raise public awareness and interest, and contribute to the management and protection of the underwater cultural heritage.

1.7 Conditions of Preservation and Discovery

One of the striking features of the map shown in Fig. 1.1 is the very marked variation in the number and distribution of archaeological sites in different regions, with the largest concentrations in the western Baltic and around the coastlines of the North Sea and the English Channel/La Manche. This naturally provokes the question of why this is so, and the extent to which it is due simply to differences in the intensity of investigation and the availability of skills and training, or rather to variable conditions of preservation and exposure, or to a combination of factors.

There is no doubt that the concentration of finds in Northwest Europe is in part the result of a long history of interest and development of skills and methods of underwater investigation over the past 40 years. Other related factors are the importance of postglacial sea-level rise and palaeogeographical changes in understanding the colonisation of the region after glacial retreat at the end of the Last Ice Age; the importance of marine resources, coastal travel, and shoreline settlements in the early development of the Stone Age cultures that moved into this new territory; high levels of awareness and interest amongst a wider public, with reporting of finds by beach combers, recreational divers, and fishermen; and in the past decade industrial collaborations that have made possible large-scale investigations in deeper water with some impressive results.

Equally, one could argue that this high level of early interest is itself in part due to the abundance of underwater archaeological material and its exposure by erosion in shallow water and in the intertidal zone where it is easily visible to the casual observer. However, similar levels of public interest, professional training in underwater archaeology and recreational diving are present in other regions, notably in the Mediterranean. Here, however, the focus of interest has been on submerged harbours and other shoreline features of Greek, Roman, or later date, on shipwrecks, and on refining measurements of the interplay between sea-level change, isostasy, and tectonic instability, rather than the submerged landscapes of the Stone Age.

Differences in preservation conditions resulting from variations in offshore topography, sedimentary regimes, and oceanographic conditions are also implicated. A detailed examination of these matters can be found in Flemming et al. (2017b), and more specific data on the conditions of preservation and discovery associated with archaeological discoveries are presented in the national overview chapters in this volume. This issue is of obvious relevance under water, given the expectation that so much material has likely been disturbed, destroyed, or submerged by sea-level rise. But it applies equally to archaeology on land and has given rise to a growing interest in 'landscape taphonomy'—analysis of the variable conditions associated with the initial deposition of archaeological material and the subsequent processes of preservation, burial, disturbance, exposure, and destruction. Lest the terrestrial archaeologist be tempted to consider the search for submerged archaeology as a lost cause, it should be emphasised that these problems of differential preservation and visibility apply with equal force on land, and perhaps more so, because of extensive disturbance and destruction of the land surface by agricultural and industrial activities, to say nothing of erosion by wind and rain. These processes, of course, simultaneously destroy or obscure but can also expose to archaeological discovery the buried remains of past human activity, and this point applies with equal force above and below modern sea level.

One of the paradoxes of discovering submerged remains of Stone Age archaeology is that the best chances of preservation are where material is covered fairly rapidly by a protective layer of marine sediment. However, that process in itself removes the archaeological material from view and continues to bury it under ever-thickening layers of sediment unless the material is subsequently disturbed by storm activity, submarine erosion, or other intrusive methods such as dredging, trawling, coring and drilling, or industrial-scale excavation. In basins of heavy sediment accumulation, such as those that rim many of the southern shorelines of the North Sea, the accumulated thickness of recent marine sediments is so thick that the prospect of discovering the buried land surface may be very low indeed. Accumulation of protective sediment cover is thus a key factor in long-term preservation, but subsequent exposure by erosion or other means is also a necessary accompaniment to discovery. The presence of so many finds in shallow water near the modern shore or in the intertidal zone probably has as much to do with the erosional effects of wave action and shallow-water currents as with their relative ease of accessibility.

Two fallacies about the underwater survival of archaeological remains should be mentioned here: one is the belief that the best chances of survival are during periods of rapid sea-level rise or—which may amount to the same effect—slower sea-level rise on very gently shelving shorelines; the second is the belief that the worst chances of survival are on open coastlines exposed to the full force of waves, winds, and turbulent currents. Neither is true as a generalisation, and there are many examples in this volume that give the lie to such statements. No rate of sea-level rise can be so rapid that it spares the archaeological finds from a prolonged period of exposure to the potentially destructive effects of wave action in the surf zone or to the turbulence of shallow-water currents. Even the steady lapping of gentle currents against shorelines in protected bays can very quickly undermine and displace unconsolidated archaeological deposits. The one positive consequence of rapid inundation that should be mentioned is that it can cover perishable organic materials such as wood or other delicate materials and features before they are destroyed by bacterial attack and subaerial erosion and thus helps to ensure their long-term preservation in anaerobic conditions.

As for storm-exposed beaches, the very storminess that at first sight seems so destructive actually contributes to the accumulation of sediments that protect the older land surface and then strips away parts of this protective cover to re-expose the earlier surface. The number of submerged forests, peat beds, and archaeological features discovered in the exposed intertidal zones and shallow waters of Northwest Europe, and the stone structures in the eastern Mediterranean, are sufficient testimony to this process.

Although generalisations should be treated with caution, there are nevertheless some general variations at the European scale that influence the likelihood of preservation and discovery and the types of underwater deposits and archaeological materials available for investigation. There are major differences in the width and gradient of the continental shelf, and the accumulated thickness of sediments, especially between Northwest Europe and the Mediterranean (Fig. 1.1). The prevalence of submerged forests and peat beds is a characteristic feature of the shallower shelves of Northern and Western Europe, and many sites in the shallow waters and extensive intertidal zones of this region have been protected and preserved by these organic deposits. The concentration of archaeological finds in the inner archipelagos of Denmark and the southern Baltic most likely relates in part to a relatively shallow sea floor, relatively calm and protected sea conditions with limited exposure and tidal amplitude, and extensive deposits of submerged peat and relatively easy visibility of eroding materials in shallow water at the shore edge. Many of the coastlines around the North Sea and along the coasts of the English Channel/La Manche are also basins of sediment accumulation because of long-term subsidence and the volume of sediments introduced from the catchments of the major European river systems that drain into these basins.

Conversely, much of the Mediterranean coastline, particularly in the more tectonically active regions of the central and eastern Mediterranean, comprises steep and rocky coastlines, with hilly or mountainous onshore topography, coastal cliffs, steeply sloping offshore shelves, and more limited accumulations of offshore sediment. In these circumstances, coastal caves, both onshore and below sea level, offer prospects for obtaining information about the use of the submerged landscape. It is also of interest and reflects to some extent the same sort of coastal geology and geomorphology that some of the best-known underwater sites in the region, such as those of Greece and Israel, have become visible because of remains of stone structures easily spotted on the seabed, rather than from exposure of submerged peats, stone artefacts, or faunal remains. There are, of course, exceptions to these broad generalisations, notably in the Golfe du Lion off the Mediterranean coasts of Northeast Spain and Southern France, the Northern Adriatic, and the Western Black Sea, where shallow basins of sediment accumulation and peat formation are present, associated with the estuaries of large rivers such as the Rhone, the Po, and the Danube. These areas have good conditions of preservation, have already yielded evidence of submerged settlements, and have good potential for future discoveries.

However, these are very broad differences and should be compared with more detailed information in the individual chapters of this volume and in Flemming et al. (2017a, b). The key to the prospects for archaeological preservation and discovery lies as much in localised features as in the more general characteristics of coastal zones, and that can only come from detailed investigations at a regional and local level.

Another factor that may be relevant to the differential distribution of underwater finds across Europe is the availability of funding and equipment, some of which is expensive and highly specialised, especially where large ships are required. This is not the place to elaborate on the techniques of underwater investigation. Details of methods, equipment, and strategies can be found elsewhere (Galili et al. 2017; Holmlund et al. 2017; Missiaen et al. 2017; Uldum et al. 2017) and some examples of their application in the following chapters. Nevertheless, it is worth emphasising that many of the finds reported in this volume were discovered or more intensively explored and excavated by relatively simple means, on foot at low tides on the more extensive intertidal zones of the north-west, or in shallow water with small boats, simple equipment and small teams of SCUBA divers as pioneered in Denmark (Fischer 1995, 2004). Research in deeper water in collaboration with industrial partners is also more advanced in Northwest Europe, a partnership that can supply big ships, complex equipment, support teams, and engineering solutions, often at minimal cost. Arguments against conducting underwater research on grounds of cost or technical complexity have been largely superseded by the growing number of projects based on one or other of the above strategies.

1.8 Research Themes

Deployment of resources and personnel, especially where significant funding is involved, if it is to be justified, does of course require intellectual justification in terms of worthwhile problems that require underwater investigation and that cannot be solved in any other way. A variety of ideas and themes are presented throughout this volume, and we offer here only the briefest summary.

Major themes include the earliest colonisation of Europe and the major routes and pathways by which our hominin ancestors entered Europe from Anatolia and Africa. Large areas of submerged landscape are likely to have offered extensive lowlands dotted with lake basins, springs, and stream channels, especially in the eastern Mediterranean, offering attractive resources for hunting and gathering around lake basins, and along palaeocoastlines and now-submerged river channels. These would have provided broader and more easily accessible and traversable land connections around many coastal regions of the Mediterranean and the Black Sea than is the case in the present-day coastal topography. Similar comments apply to the extensive shelf regions of Northwest Europe and their role in the earliest colonisation of the region or its recolonisation after periods of glaciation.

Early sea travel, whether involving simple methods of rafting or more organised seafaring with early canoes or boats, is also a current topic of high interest, with evidence that sea crossings and the use of islands in the eastern Mediterranean were being accomplished in the late Palaeolithic period well before the spread of pioneer farmers, and perhaps as early as the Middle Palaeolithic (Ammerman and Davis 2013, 2014). The key lies in archaeological evidence for human presence on islands, and some of the claims of earliest dates remain controversial. Regardless of these controversies, it is clear that evaluation must depend on reconstruction of palaeoshorelines and sea-crossing distances at periods of low sea level when the palaeogeographic configuration of coastlines and islands would have been quite different from the present-day pattern. Moreover, the earliest locations of landfall and departure and the archaeological evidence for them must, by definition for periods when sea level was lower than present, lie on now-submerged shorelines.

As with early population dispersals, so with the more recent spread of farming in the Neolithic from centres of origin in Anatolia and the Near East, coast-wise patterns of dispersal figure prominently in interpretations, and the period of earliest agricultural dispersal was at a time when sea levels and coastlines were still somewhat lower than the present, again reinforcing the need for underwater research and the likelihood that present understandings are limited by the rarity of such investigations.

Another major theme is the time depth of human interest in the use of marine resources and the establishment of permanent settlements on the coast sustained by the abundance of marine foods. By the time that Mesolithic coastlines become visible on the postglacial shorelines of Northwest Europe, it is clear from food remains in archaeological deposits and stable isotope signatures in human skeletons that marine foods and permanent settlement on the coast were a major feature of the period. Moreover, it is clear from the uplifted coastlines of Norway and Western

Sweden that seafaring and hunting of sea mammals were a necessary accompaniment to earliest human dispersal in these higher latitudes and the provision of storable marine foods with their high fat content, without which survival during the long northern winters would have been impossible. Comparable evidence from earlier periods and regions further to the south is absent, apart from hints of connections with the submerged sea coast from occasional sea shells or other marine indicators present in hinterland sites.

It was long assumed that this florescence of maritime cultures in Mesolithic Europe, and comparable developments in other parts of the world at about the same time, was the result of population growth and new social, economic, and technological developments at that time. However, given the coincidence of this explosion of evidence with the period when eustatic sea levels stopped rising, an equally plausible hypothesis is that the pattern of evidence is just as likely to result from differential visibility and preservation. Comparable patterns of coastal settlement may well extend much further back in time to earlier periods of low sea level, but the evidence is now hidden from view. How far back in time we can trace these maritime patterns, and how much they differed from those visible in more recent millennia, is, of course, the key question, and it is one that can only be answered by underwater investigations, identification of submerged coastlines, and the search for underwater shoreline settlements.

A further topic of growing interest is the human impact of sea-level change. As more information has become available on the nature and extent of the submerged landscapes of the Last Glacial Period and the evidence of their exploitation, so attention has begun to focus on the social, economic, and demographic consequences of sea-level change, particularly sea-level rise and drowning of previously productive hunting grounds and wetland environments. At present, only limited investigations aimed at charting the socio-economic impact of sea-level rise are available and mostly only for the final stages of shoreline change in the Holocene (e.g., Harff and Lüth 2007, 2011). For the inundation of land exposed at earlier periods and lower sea levels, and especially for extensive areas such as the southern North Sea basin, there is, at present, insufficient evidence to go very much beyond speculation about the social consequences of long-term sea-level rise and the resulting and often-dramatic changes in ecological and geographical reconfiguration of coastal environments. However, simply to think about these consequences is to raise questions that should inform future research.

Perhaps the biggest challenge for the future is the purposeful and successful search for underwater prehistoric archaeological sites, especially further offshore and in deeper water. Mapping of the submerged landscape by remote sensing for features worth targeting with more focussed coring or diving, taking known finds dredged up by chance from the seabed as a starting point for more detailed exploration, use of predictive modelling such as Fischer's well-known fishing site model to identify site locations (Fischer 1993), closer engagement with offshore commercial and industrial operations, and development of acoustic techniques for identifying small finds and features typical of Stone Age cultures are just some of the strategies that are now being employed to further advance underwater investigations.

References

Ammerman AJ, Davis T (eds) (2013) Island archaeology and the origins of seafaring in the eastern Mediterranean. Proceedings of the Wenner-Gren Workshop held at Reggio Calabria on October 19–21, 2012. Part One. Eurasian Prehistory 10

Ammerman AJ, Davis T (eds) (2014) Island archaeology and the origins of seafaring in the eastern Mediterranean. Proceedings of the Wenner-Gren Workshop held at Reggio Calabria on October 19–21, 2012. Part Two. Eurasian Prehistory 11

Bailey GN (1983) Economic change in Late Pleistocene Cantabria. In: Bailey G (ed) Hunter-gatherer economy in prehistory. Cambridge University Press, Cambridge, pp 149–165

Bailey GN, Flemming N (2008) Archaeology of the continental shelf: marine resources, submerged landscapes and underwater archaeology. Quat Sci Rev 27:2153–2165

Bailey GN, Milner N (2002) Coastal hunters and gatherers and social evolution: marginal or central? Before

Farming: The Archaeology of Old World Hunter-Gatherers 3–4(1):1–15
Bailey GN, Harff J, Sakellariou D (eds) (2017) Under the sea: archaeology and palaeolandscapes of the continental shelf. Springer, Cham
Benjamin J, Bonsall C, Pickard C, Fischer A (eds) (2011) Submerged prehistory. Oxbow, Oxford
Benjamin J, Rovere A, Fontana A, Furlani S, Vacchi M, Inglis RH, Galili E, Antonioli F, Sivan D, Miko S, Mourtzas N, Felia I, Meredith-Williams M, Goodman-Tchernov B, Kolaiti E, Anzidei M, Gehrels R (2017) Late quaternary sea-level changes and early human societies in the central and eastern Mediterranean Basin: an interdisciplinary review. Quat Int 449:29–57
Coles J (1984) The archaeology of wetlands. Edinburgh University Press, Edinburgh
Coles B (ed) (1992) The wetland revolution in prehistory, WARP occasional paper, vol 6. The Prehistoric Society and WARP (Wetland Archaeology Research Project), Exeter
Erlandson J (2001) The archaeology of aquatic adaptations: paradigms for a new millennium. J Archaeol Res 9:287–350
Evans AM, Flatman JC, Flemming NC (eds) (2014) Prehistoric archaeology on the continental shelf: a global review. Springer, New York
Fischer A (1993) Stone Age settlements in the Småland Bight: A theory tested by diving. Miljøministeriet, Skov- og Naturstyrelsen, København (In Danish and English)
Fischer A (ed) (1995) Man and sea in the mesolithic: Coastal settlement above and below present sea level. In: Proceedings of the International Symposium, Kalundborg, Denmark 1993, Oxbow monograph 53. Oxbow, Oxford
Fischer A (2004) Submerged stone age – Danish examples and North Sea potential. In: Flemming NC (ed) Submarine prehistoric archaeology of the North Sea. Research priorities and collaboration with industry, CBA Research Report, vol 141. Council for British Archaeology, York, pp 21–36
Fischer A, Pedersen L (eds) (2018) Oceans of archaeology, Jutland Archaeological Society Publications Vol. 101. Jutland Archaeological Society, Højbjerg
Flemming NC (ed) (2004) Submarine prehistoric archaeology of the North Sea. Research priorities and collaboration with industry. CBA Research Report, vol 141. Council for British Archaeology, York
Flemming N, Antonioli F (2017) Prehistoric archaeology, palaeontology, and climate change indicators from caves submerged by change of sea level. In: Campbell PB (ed) The archaeology of underwater caves. Highfield Press, Southampton, pp 23–38
Flemming NC, Cagatay MN, Chiocci FL, Galanidou N, Jöns H, Lericolais G, Missiaen T, Moore F, Rosentau A, Sakellariou D, Skar B, Stevenson A, Weerts H (2014) Land beneath the waves: submerged landscapes and sea level change. A joint geoscience-humanities strategy for European continental shelf prehistoric research. Chu NC, McDonough N (eds) position paper 21 of the European Marine Board, Ostend, Belgium
Flemming NC, Harff J, Moura D, Burgess A, Bailey GN (eds) (2017a) Submerged landscapes of the European continental shelf: quaternary paleoenvironments. Wiley, Chichester
Flemming NC, Harff J, Moura D (2017b) Non-cultural processes of site formation, preservation and destruction. In: Flemming NC, Harff J, Moura D, Burgess A, Bailey GN (eds) Submerged landscapes of the European continental shelf: quaternary paleoenvironments. Wiley, Chichester, pp 51–82
Galili E, Benjamin J, Herskovitz I, Weinstein-Evron M, Zohar I, Eshed V, Cvikel D, Melamed J, Kahanov Y, Bergeron J, Ruggles C, Ronen A, Kolska Horwitz L (2017) Atlit-yam: a unique 9000 year old prehistoric village submerged off the Carmel Coast, Israel – the SPLASHCOS Field School (2011). In: Bailey GN, Harff J, Sakellariou D (eds) Under the sea: archaeology and palaeolandscapes of the continental shelf. Springer, Cham, pp 85–102
Grant KM, Rohling EJ, Bar-Matthews M, Ayalon A, Medina-Elizalde M, Bronk Ramsey C, Satow C, Roberts AP (2012) Rapid coupling between ice volume and polar temperature over the past 150 000 years. Nature 491:744–747
Grant KM, Rohling EJ, Ramsey CB, Cheng H, Edwards RL, Florindo F, Heslop D, Marra F, Roberts AP, Tamisiea ME, Williams F (2014) Sea-level variability over five glacial cycles. Nat Commun 5:1–9. https://doi.org/10.1038/ncomms6076
Harff J, Lüth F (eds) (2007) SINCOS – sinking coasts: geosphere, ecosphere and anthroposphere of the Holocene southern Baltic Sea. Bericht der Römisch-Germanischen Kommission 88
Harff J, Lüth F (eds) (2011) SINCOS II – sinking coasts: geosphere, ecosphere and anthroposphere of the Holocene southern Baltic Sea. Bericht der Römisch-Germanischen Kommission 92
Harff J, Bailey GN, Lüth F (eds) (2016) Geology and archaeology: submerged landscapes of the continental shelf, Special publication 411. Geological Society, London
Harff J, Furmanczyk K, Von Storch H (eds) (2017a) Coastline changes of the Baltic Sea from south to east. Springer, Cham
Harff J, Flemming NC, Groh A, Hünicke B, Lericolais G, Meschede M, Rosentau A, Sakellariou D, Uścinowicz S, Zhang W, Zorita E (2017b) Sea level and climate. In: Flemming NC, Harff J, Moura D, Burgess A, Bailey GN (eds) Submerged landscapes of the European continental shelf: quaternary paleoenvironments. Wiley, Chichester, pp 11–49
Holmlund J, Nilsson B, Rönnby J (2017) Joint explorations of the sunken past: examples of maritime collaboration between industry and academia in the Baltic. In: Bailey GN, Harff J, Sakellariou D (eds) Under the sea: archaeology and palaeolandscapes of the continental shelf. Springer, Cham, pp 53–63

Inglis RH, Bosworth W, Rasul NMA, Al-Saeedi AO, Bailey GN (2019) Investigating the palaeoshorelines and coastal archaeology of the southern Red Sea. In: Rasul NMA, Stewart ICF (eds) Geological setting, palaeoenvironment and archaeology of the Red Sea. Springer, Cham, pp 533–581

Lambeck K, Purcell A, Flemming N, Vita-Finzi C, Alsharekh A, Bailey GN (2011) Sea level and shoreline reconstructions for the Red Sea: isostatic and tectonic considerations and implications for hominin migration out of Africa. Quat Sci Rev 30(25–26):3542–3574

Lambeck K, Rouby H, Purcell A, Sun Y, Sambridge M (2014) Sea level and global ice volumes from the last glacial maximum to the Holocene. Proc Natl Acad Sci U S A 111(43):15296–15303. https://doi.org/10.1073/pnas.1411762111

Masters PM, Flemming NC (eds) (1983) Quaternary coastlines and marine archaeology. Academic, London

Menotti F, O'Sullivan A (eds) (2012) The Oxford handbook of wetland archaeology. Oxford University Press, Oxford

Missiaen T, Sakellariou D, Flemming NC (2017) Survey strategies and techniques in underwater geoarchaeological research: an overview with emphasis on prehistoric sites. In: Bailey GN, Harff J, Sakellariou D (eds) Under the sea: archaeology and palaeolandscapes of the continental shelf. Springer, Cham, pp 21–37

Moree JM, Sier MM (eds) (2015) Interdisciplinary research programme Maasvlakte 2, Rotterdam. BOOR, Rotterdam, pp 7–350

Shackleton NJ, Opdyke ND (1973) Oxygen isotope and paleomagnetic stratigraphy of Pacific core V28-238: oxygen isotope temperatures and ice volumes on a 10^5 and 10^6 year scale. Quat Res 3:39–55

Shackleton NJ, Imbrie J, Hall MA (1983) Oxygen and carbon isotope record of East Pacific core V19-30: implications for the formation of deep water in the late Pleistocene North Atlantic. Earth Planet Sci Lett 65:233–244

Shackleton JC, van Andel TH, Runnels CN (1984) Coastal paleogeography of the central and western Mediterranean during the last 125,000 years and its archaeological implications. J Field Archaeol 11:307–314

Siddall M, Rohling EJ, Almogi-Labin A, Hemleben C, Meischner D, Schmelzer I, Smeed DA (2003) Sea-level fluctuations during the last glacial cycle. Nature 423:853–858

Smith PJ (1997) Grahame Clark's new archaeology: the fenland research committee and Cambridge prehistory in the 1930s. Antiquity 71(271):11–30

Sturt F, Dix J, Grant MJ (2017) The history of industry-linked research in English waters: lessons for the future. In: Bailey GN, Harff J, Sakellariou D (eds) Under the sea: archaeology and palaeolandscapes of the continental shelf. Springer, Cham, pp 425–436

Sturt F, Flemming NC, Carabias D, Jöns H, Adams J (2018) The next frontiers in research on submerged prehistoric sites and landscapes on the continental shelf. Proceedings of the Geological Association. https://doi.org/10.1016/j.pgeola.2018.04.008

Tizzard L, Bicket AR, Benjamin J, de Loecker D (2014) A middle Palaeolithic site in the southern North Sea: investigating the archaeology and palaeogeography of area 240. J Quat Sci 29:698–710 https://doi.org/10.1002/jqs.2743

Tizzard L, Bicket AR, de Loecker D (2015) Seabed prehistory: investigating the palaeogeography and Early Middle Palaeolithic archaeology in the southern North Sea. Wessex archaeology report 35. Wessex Archaeology, Salisbury

Uldum O, Benjamin J, McCarthy J, Feulner F, Lübke H (2017) The late mesolithic site of Falden, Denmark: results from underwater archaeological fieldwork and a strategy for capacity-building based on the SPLASHCOS mission. In: Bailey GN, Harff J, Sakellariou D (eds) Under the sea: archaeology and palaeolandscapes of the continental shelf. Springer, Cham, pp 65–84

Veth P, McDonald J, Ward I, O'Leary M, Beckett E, Benjamin J, Ulm S, Hacker J, Ross PR, Bailey G (2019) A strategy for assessing continuity in terrestrial and maritime landscapes from Murujuga (Dampier archipelago), north west shelf, Australia. Journal of Island and Coastal Archaeology. https://doi.org/10.1080/15564894.2019.1572677

Wickham-Jones C (2018) Landscape beneath the waves: the archaeological investigation of underwater landscapes. Oxbow, Oxford

Part I

The Baltic and Scandinavia

The Baltic and Scandinavia: Introduction

2

Geoff Bailey and Hauke Jöns

Abstract

This section deals with some of the largest concentrations of underwater sites in Europe. Because of the complex history of isostatic rebound and periodic damming back of the Baltic Sea associated with the retreat of the Scandinavian ice sheet, these chapters also present a wide range of preservation conditions and palaeoenvironmental changes, ranging from shorelines that have been lifted clear of sea-level rise in Norway and parts of Sweden but without organic preservation, to sinking shorelines on the Baltic coastlines of Denmark and Germany with little overburden of marine sediments and spectacular organic preservation, to open-coast conditions along the North Sea shorelines of Denmark and Germany where sites are mostly buried under thick marine deposits. Many sites in Denmark and Germany have benefited from systematic excavation, demonstrating that the majority of finds and features have been recovered from marine refuse areas in shallow water along the shoreline and that the adjacent settlement areas originally located on dry land have been largely eroded away or badly disturbed by subsequent sea-level rise. The abundance of finds and excellent preservation in many cases reflect the high marine productivity of the Littorina Sea and the concentration of settlements on the shoreline with a heavy emphasis on marine resources, the presence of coastal topography conducive to trapping of fish and sea mammals, the deposition of cultural material directly into shallow water with fine-grained sediments where items were quickly buried and protected from bacterial attack or marine erosion and the occurrence of settlements in relatively shallow water easily accessible to diver investigation. There is considerable scope for new investigations and new discoveries, not only in Denmark and Germany but also in Norway, where study of the numerous Stone Age sites on uplifted shorelines has overshadowed investigation of the many regions with inundated coastlines, in Sweden, where the Early Mesolithic period, almost non-existent on the submerged shorelines of Denmark and Germany, is represented by an extensive and relatively shallow submerged landscape with excellent preservation conditions, and further east in the Baltic on the coastlines of Poland, Lithuania and Latvia, with offshore conditions similar to those of

G. Bailey (✉)
Department of Archaeology, University of York, York, UK

College of Humanities, Arts and Social Sciences, Flinders University, Adelaide, SA, Australia
e-mail: geoff.bailey@york.ac.uk

H. Jöns
Lower Saxony Institute for Historical Coastal Research, Wilhelmshaven, Germany
e-mail: joens@nihk.de

G. Bailey et al. (eds.), *The Archaeology of Europe's Drowned Landscapes*,
Coastal Research Library 35, https://doi.org/10.1007/978-3-030-37367-2_2

Germany and Denmark and the likelihood that numerous submerged sites of high scientific value await discovery.

Keywords
Early Mesolithic · Early Neolithic · Ertebølle · Marine refuse area · Isostatic rebound · Excavation

2.1 Introduction

The four regions covered by the chapters in this section together have the largest number of underwater sites in Europe, accounting for fully 1968 (75%) of the finds recorded in the SPLASHCOS Viewer and a correspondingly large number of in situ culture layers. The majority of these sites are in Denmark (1699), but there are significant numbers in the other regions in this section: 142 sites in Germany, 83 in Norway and 44 in Sweden. Many sites, of course, are single finds or unstratified material that may have been subjected to post-depositional disturbance, but the number of sites with in situ cultural layers, and especially ones that have been tested or more fully excavated, is unusually high. Three chapters—Denmark, Sweden and Germany—deal with a closely related group of regions in the south-western Baltic, which share many similarities in terms of geological and hydrological history, archaeological sequences and preservation conditions. Norway, properly speaking, falls within the North Sea and Atlantic group of chapters and is briefly referred to later in the introduction to Part II (Peeters et al., Chap. 7, this volume), but its geological history of isostatic rebound and its archaeological sequence have more in common with the Baltic group, and we discuss the results from Norway here. Denmark and Germany also have coastlines bordering the North Sea, with coastal conditions and submerged landscapes that have much in common with other coastlines around the southern rim of the North Sea. They have received rather less attention and yielded far fewer underwater finds than their Baltic counterparts, largely it seems because of the greater difficulties of accessing the buried Stone Age land surface.

All the regions discussed in this section have been dominated by geological processes associated with the expansion and contraction of the Scandinavian ice sheet. These processes include major reshaping of the physical landscape by ice action during the Last Glacial Period and vertical movements of the Earth's crust associated with glacio-hydro-isostatic changes in mass loading. These processes in their turn have interacted with eustatic changes in ocean volume to create a complex history of changing coastlines, with very significant implications for the differential preservation and visibility of coastal settlements and underwater finds.

All four regions have other features in common in addition to the impact of isostatic movements. These include the following:

- A relatively short archaeological sequence compared to the other regions of Europe, beginning with earliest human entry into new territory made available by the retreat of the Scandinavian ice sheet, and therefore only very few finds, whether on land or under water, earlier than the Late Glacial (from, at earliest, c. 16,000 years onwards).
- A kaleidoscope of rapidly and quite dramatically changing Late Pleistocene and Early Holocene climates, coastlines, vegetation conditions and available terrestrial and marine food resources.
- Excellent conditions of accessibility and preservation of underwater material with some of the best examples available of organic artefacts made from wood and plant fibres and unique insights into the nature of the conditions in which underwater archaeological deposits are preserved.
- Heavy emphasis on coastal settlement and exploitation of marine resources from the earliest period of human colonisation, resulting from a combination of productive marine environments and hinterlands of relatively limited extent or productivity.

In this introduction, we summarise the Late Glacial and Early Postglacial geological history of the region and highlight the principal archaeological findings from each country with reference to the geological processes affecting the visibility of palaeoshorelines, exploitation of marine resources, conditions of underwater preservation, the reasons for the large number of finds and the impact of industrial collaboration.

2.2 Geological, Palaeogeographical and Climatic History

The geological process that dominates all the areas examined here is glacio-isostatic adjustment (GIA)—the isostatic rebound of the land following the melting of the Scandinavian ice sheet. The rate of uplift or subsidence varied in different regions and interacted with eustatic sea-level rise to create complex patterns of palaeogeographic change including submergence or elevation of shorelines and periodic isolation of the Baltic Basin to form a freshwater lake isolated from the world oceans (see Jöns et al. 2011; Jöns and Harff 2014; Rosentau et al. 2017, for general overviews, and the chapters in this group for regional variations specific to the underwater archaeological record).

At the Last Glacial Maximum (LGM), the ice sheet covered all of Norway, Sweden and Finland, extending eastwards into northern Russia and westwards across the North Sea. Its southern margin covered parts of Denmark, northern Germany and Poland, most of the Baltic States (Lithuania, Latvia and Estonia) and Russian territory abutting the Baltic coastline. The maximum thickness of the ice sheet was 2700 m, centred over NE Sweden and the northernmost shores of the Baltic Basin (The Gulf of Bothnia), thinning out towards the edges. As the ice sheet melted, the mass loading on the Earth's crust changed, previously glaciated areas began to rebound and areas at and beyond the southern and western ice margins subsided. These adjustments happened slowly and continue in many areas to the present day.

The greatest rates of isostatic uplift occurred in the areas where the ice was thickest, especially northern Norway and north-east Sweden, and uplift of shorelines has reached a maximum elevation of 270 m above present sea level in these regions. Here, uplift consistently outpaced the eustatic rise of sea level, with the result that relative sea level has dropped continuously from the onset of deglaciation to the present day, and the earliest shorelines have been lifted above the present sea level (see Glorstad et al., Chap. 6, Fig. 6.2, this volume). Along parts of the southern and western margins, land has subsided, and the sea has steadily encroached on the land, amplifying the eustatic rise of sea level and submerging shorelines, even those of relatively late date (see Jöns et al., Chap. 5, Fig. 5.3 and 5.4). In intermediate regions, relative sea-level rise shows a more complex pattern of rise and fall, especially on shorelines affected by the periodic damming back of the water in the Baltic Basin and its reconnection with the world oceans, notably in southern Sweden (see Nilsson et al., Chap. 4, Fig. 4.3).

2.2.1 The Baltic Sea

The Baltic Sea is one of the largest inland seas in the world and today is a brackish-water basin connected to the Atlantic Ocean via the North Sea through the narrow straits and islands that separate the Jutland Peninsula of Denmark from southern Sweden (from west to east the Little Belt, the Great Belt and the Öresund). It was originally created by the gouging out of deep basins by ice action, a process which has largely erased evidence of surface deposits earlier than the LGM. As the ice sheets melted after the LGM, the Baltic went through four stages between c. 16,000 and 8000 years ago (16 ka and 8 ka cal BP): two periods as an isolated freshwater lake and two periods of connection with the world ocean.

In the first stage, known as the Baltic Ice Lake, from c. 16 ka to 11.7 ka cal BP, meltwaters began to fill the southern Baltic basin but were dammed back by the persistence of an ice barrier at the south-western end and later by isostatic land uplift, and this lake remained covered with ice for most of the year (see Bailey et al., Chap. 3, Figs. 3.2b and 3.3c, this volume). During this period, isostatic uplift kept pace with eustatic sea-level rise, maintaining the isolation of the Baltic as a freshwater lake except for a narrow outlet through the Öresund. As the land in the Öresund area uplifted, so the outlet of water in the Öresund cut down through glacial deposits, maintaining the level of the Baltic lake at about the same height as sea level. However, at some point, the downcutting reached the more resistant limestone, which blocked the outlet. Continued uplift and continued melting of the ice sheet resulted in a rise in the Baltic water level above that of the contemporaneous sea level (Nilsson et al., Chap. 4, this volume).

Between c. 13 ka and 10.7 ka, there were two short-lived periods of reconnection with the sea through the southern Swedish lowlands (Rosentau et al. 2017, Fig. 8.12), separated by a brief period of renewed damming by ice readvance during the Younger Dryas. The second of these is known as the Yoldia Sea stage, dating between c.11.7 and 10.7 ka cal BP (see Bailey et al., Chap. 3, Fig. 3.3d, this volume). These episodes of reconnection with the sea resulted in a rapid drop of water level of at least 25 m as Baltic waters drained into the North Sea, exposing new areas of land in the south-west Baltic (Nilsson et al., Chap. 4, this volume). Renewed blockage of the Baltic outlet resulted in a new and short-lived episode of lake conditions, the Ancylus Lake period (c.10.7–9.8 ka cal BP), and a renewed rise in lake level (Bailey et al., Chap. 3, Fig. 3.3e, this volume), with inundation of coastal lowlands in the south-west of the Baltic basin. Limited exchange of water with the North Sea through a river system in Denmark maintained brackish-water conditions in the Ancylus Lake. Meanwhile, isostatic uplift had lifted coastlines clear of rising water levels in western Sweden, south-east Norway and the eastern Baltic. After 9.8 ka cal BP, the continued eustatic rise of sea level finally saw the inundation of the Danish Straits and the entry of seawater into the south-western Baltic— the Littorina Transgression beginning at c. 8.5 ka cal BP (Bailey et al., Chap. 3, Fig. 3.3f, this volume). Sea level continued to rise in this period until about 6 kc cal BP, causing further inundation of land in the south-west Baltic, amplified by continued land subsidence.

Alongside these complicated changes in land-sea relationships, climate was changing from periglacial tundra traversed by reindeer migrations to mixed oak forests and temperate fauna. The fertility of lake water and sea water, and their capacity to support aquatic and marine resources available for human exploitation, would also have varied substantially, ranging from the largely infertile waters of the Baltic ice-dammed lake to high levels of productivity in the warm, saline waters of the Littorina period in the Danish Straits and the south-west Baltic and on Late Glacial Norwegian coastlines where the incoming Gulf Stream mixed with meltwater runoff from the land.

2.3 Palaeoshorelines and Coastal Settlement

Because of the large regional variations in uplift and subsidence, palaeoshorelines occur at many different elevations. Some have been lifted clear of eustatic sea level rise, opening up a window into the very early use of shoreline settlements and potential use of marine resources on dry-land archaeological sites, notably in Norway and Sweden; other palaeoshorelines have been lifted clear, only to be partly submerged again as sea-level rise overtook uplift of the land, which is the case for parts of the Norwegian coast, yet other palaeoshorelines have remained fully submerged since the time of inundation, notably in southern Sweden, Denmark and the Baltic coast of Germany. These regional differences provide 'windows' into the study of coastlines of different periods. The earliest such periods are on uplifted coastlines in Norway and western Sweden (Bohuslän) and are of Late Palaeolithic and Early Mesolithic date (see Bailey et al., Chap. 3, Table 3.1, for terminology and chronology of archaeological periods). The earliest fully submerged shorelines, of Early Mesolithic date, are in southern Sweden (Scania and Blekinge), while the most easily accessible and best-studied submerged sites in Denmark are mostly Middle and Late Mesolithic and in Germany mainly Late Mesolithic and Neolithic. Denmark is of particular interest because the tilting of the land by isostatic adjustment has resulted in concentrations of on-land sites of the Late Mesolithic Ertebølle period on slightly uplifted shorelines in the north and sites of similar age on now-submerged shorelines in the south.

The uplifted shorelines are present along parts of the Norwegian coastline; west, central and northern Sweden; and in Estonia. In most cases Stone Age sites are present on these shorelines from the earliest date that they became visible above present sea level (apsl) or became free of ice and available for occupation. In areas of substantial uplift, the sites are now marooned well inland of the present-day coastline, such as the Early Mesolithic sites of the Södertörn Peninsula near Stockholm in Sweden, which are now 40–80 m apsl or the Norwegian site of Stunner near Oslo at c. 165 m apsl (Jöns 2011). The earliest such sites are on the uplifted shorelines of Bohuslän in western Sweden where large concentrations of sites of the Late Palaeolithic Hensbacka culture dating from c. 12 ka cal BP are associated with very high levels of marine productivity (Schmitt et al. 2006, 2009). Further north in Norway, coastlines became ice-free as early as 13.8 ka cal BP, but coastal sites are not present until 11.2 ka cal BP because access from the south was blocked by an ice barrier until that time (Glorstad et al., Chap. 6, this volume). Once the Norwegian coastline became accessible, colonisation was rapid—almost instantaneous within the margins of error of radiocarbon dating—with an economy based on seal hunting, some hunting of elk and reindeer in the hinterland and settlements with dwellings comprising stone-built foundations at sites such as Nyhamna and Vega (Bjerck 2008).

The presence of archaeological sites on uplifted shorelines has dominated the study of the earlier Stone Age in Norway and Sweden. However, as Glorstad et al. (Chap. 6, this volume) point out, there are significant gaps in this on-land archaeological record that can only be filled by underwater investigations. The acid soils associated with so much of the regional geology mean that sites on the present-day land surface have very few if any organic remains; details of subsistence economy can be reconstructed only by indirect means, mainly from inferences about the functions of stone tools, the location of sites and ecological reconstructions of potential resource availability. Bone remains, let alone wooden artefacts, are rare or absent. Moreover, on the outer coastlines of southern and northern Norway, coastal sites are missing from the Early Mesolithic period because the coastlines of this period are now largely submerged. This is of particular significance in Norway because settlement was always concentrated on or close to the coastline because of the harsh climate and the mountainous nature of the immediate hinterland. Underwater investigations therefore have an important role to play, both in filling otherwise unexplained gaps in coastal site distributions and as sources of evidence for subsistence economy and the organic component of the material culture.

In Norway underwater investigation has been overshadowed by the emphasis on terrestrial archaeology and by the much more abundant data from underwater sites in neighbouring Denmark and Germany. Nevertheless, the number of finds is surprisingly large—83 formally recorded sites. Most are isolated finds of stone tools, but animal bones have been recovered in some locations. At the Early Mesolithic Hummervikholmen underwater site, human skeletal remains are well preserved. Stable isotope analysis of the bones has revealed a heavy reliance on marine resources, and a DNA analysis has identified a genetic origin in populations from both Western Europe and Eastern Europe. These results hint at the potentially very rich additional information that can be obtained from underwater finds with good organic preservation, and it is clear that there are many stretches of the Norwegian coastline with submerged Stone Age shorelines in relatively shallow and easily accessible water awaiting investigation.

In Sweden, the situation is quite similar in that emphasis has been devoted to sites on land rather than to underwater investigations, apart from pioneer discoveries of underwater Stone Age finds in the Öresund Strait in the 1970s (Nilsson et al.,

Chap. 4, this volume). In the past decade, renewed geological and archaeological investigations have revealed favourable conditions for the preservation of underwater landscapes and archaeological sites, especially in the period between c. 11.5 and 8.5 ka (coinciding with the Early Mesolithic period) when some 3500 km^2 of the Baltic seafloor was exposed as dry land before being inundated by the Ancylus and Littorina transgressions. This landscape is now at depths ranging from c. 22 m to 10 m bpsl with abundant evidence of submerged forests, river estuaries, lakes, bogs and palaeoshorelines. At least 44 underwater sites have been recorded. and many have been test-excavated, though larger-scale excavations comparable to those conducted in Denmark and Germany have yet to be undertaken.

Notable finds in Sweden are the materials found at the site of Haväng. Wooden artefacts include remains of a stationary fish weir at 9–8.8 ka cal BP, making this the earliest fish weir in the region, earlier than the earliest examples in Denmark or Germany discussed below and the earliest so far known anywhere in the world. Cut-marked animal bones and wooden artefacts are also found in earlier deposits at this site, dating back to at least 10.7 ka cal BP. In situ deposits of a similar age and with similar potential for preservation of organic materials are present on other coastlines in southern Sweden, including remains of human burials. Also, shipworm (*Teredo* spp.), a wood-boring marine mollusc capable of damaging and ultimately destroying submerged wood, is, so far, absent because of the low salinity of the Baltic seawater in this region.

The Swedish material is of exceptional interest because it is the only region that provides easy access to deposits associated with a submerged Early Mesolithic coastal landscape, sites located on palaeoshorelines adjacent to a productive marine environment, and excellent conditions of preservation. Coastlines of this period are either absent elsewhere, too deeply submerged to be easily accessible, associated with unproductive marine environments, or on uplifted shorelines without organic preservation. As Nilsson et al. (Chap. 4, this volume) point out, the apparent evidence for a transition to greater emphasis on permanent coastal settlement and marine resources from the Middle Mesolithic period onwards in Sweden may be illusory, because it has been based on a comparison between Maglemosean (Early Mesolithic) sites originally located some distance inland from their contemporaneous coastlines and Ertebølle (Middle and Late Mesolithic) sites located on the shoreline.

Similar comments apply to Denmark and Germany. In Denmark the great majority of underwater finds are from the inner straits and archipelagos at the entrance to the Baltic. Very few of these finds are of Early Mesolithic (Maglemosean) date, and those that have been recovered lie at the very end of this period with radiocarbon dates of 9.4–8.3 ka cal BP (Bailey et al., Chap. 3, this volume). This appears to reflect the relatively late establishment of marine conditions in this region associated with the Littorina transgression, dating back at earliest to 9.8 ka cal BP and to the fact that shorelines of this period are quite deeply submerged or have been eroded away. Similarly in Germany, submerged coastal sites on the Baltic coastline are associated with the Littorina transgression, and the earliest radiocarbon-dated sites date from c. 8.4 ka onwards (Jöns et al., Chap. 5, this volume). Systematic excavation and recovery of materials from Early Mesolithic sites in Sweden have yet to be undertaken but will undoubtedly help to fill what remains a very large gap in our current understanding of long-term trends in coastal settlement and exploitation of marine resources in Scandinavia and the Baltic.

2.4 Conditions of Discovery and Preservation

The best insights into these issues come from Denmark and Germany, where investigations over a long period have resulted in a large archive of information including systematic excavations—four principal excavations in Denmark and eight in Germany—that have produced abundant finds and details of preservation conditions, stratigraphic associations and depositional environments. An important point that emerges from these excavations is that the culture layers that have preserved the great majority of organic finds are refuse dumps of materials that were originally deposited in shallow water at the shore edge, whereas the adjacent settlement areas on dry land have been disturbed or largely destroyed by marine inundation and erosion.

In Denmark, the site of Tybrind Vig offers the classic example of preservation conditions on underwater sites. This is a site that was systematically excavated over a 10-year period and is one of the best-known and best-reported underwater excavations in Europe. The main settlement area extended along the shore for c. 130 m and extended over an area of 2600 m^2, judging by the concentration of stone tools recovered from the original land surface (Bailey et al., Chap. 3, this volume). However, apart from two human burials, which probably survived the destructive action of waves and water currents because they were dug below the original land surface and covered over, other settlement features such as evidence of dwellings and pits, or preservation of organic materials, are absent from this area. All the organic finds of bone, antler and wood, and a number of stone tools, come from a refuse area in front of the site. This refuse area extended over a similar area (c. 2000 m^2), of which 192 m^2 was excavated. The finds recovered in this area mostly represent artefacts discarded there after use in the adjacent settlement area or redeposited from their original position and materials such as logboats, fish traps and fishing equipment abandoned in situ in the shallow-water zone after use.

Similar features are present at the other excavated sites in Denmark, with a land area that has been disturbed or eroded and a refuse area in the adjacent marine zone where most of the organic material has been preserved and recovered. However, not all is destruction on the original land surface, and pockets of well-preserved material are occasionally present, notably stone-lined hearths, one at Ronæs Skov with small branches of wood used for the fire and a tinder fungus and another at Argus Bank with charred food remains (Bailey et al., Chap. 3, this volume).

In addition to implements of stone, bone and antler, this refuse zone at Tybrind Vig yielded a large collection of wooden artefacts including leister prongs, dugout logboats, paddles, handles for hafted axes, shafts for spears and arrows, bows, the remains of a wicker fish trap and fish weirs comprising wooden stakes and softer withies woven together to make a stationary structure. Other organic materials include textiles, cordage and uncharred remains of food plants. These finds revealed a hitherto little-known aspect of Ertebølle material culture, much of it associated with fishing or other maritime activities. This is consistent with the faunal remains at Tybrind Vig, which include bones of cod, seal and porpoise, and with the location of the site as one of a group of sites clustered around the narrow mouth of a shallow bay, ideally situated to trap quantities of fish and other marine animals. Bones of land mammals such as red deer, roe deer and wild boar are also present, and sites of the same period are present on rivers and lake edges up to 10 km inland, indicating a terrestrial and hinterland component to the coastal economy.

One of the outstanding features of the Danish evidence is the remains of stationary fish weirs. The largest and best preserved of these are from the early Neolithic period. At Oleslyst, a complete panel was recovered, 5.5 m long and 1.7 m high, comprised of stout vertical stakes at half-metre intervals with long, flexible hazel withies woven horizontally through the stakes to make a tightly woven mesh, believed to have been designed to trap large quantities of eels during their seasonal migrations. The evidence at this site indicates that the original fence comprised a number of these panels extending for at least 30 m out from the shore. At Nekselø, a line of stakes extended out to at least 250 m. Similar remains, though more fragmentary, are present from the Ertebølle period, and the earliest is from a pre-Ertebølle context at Kalø Vig dated at c. 8.4 ka cal BP. As noted above, an earlier example has been dated at 9–8.8 ka cal BP at Haväng in Sweden. Hundreds of vertical stakes, mostly of hazel, would have been needed in the construction of these fences, and thousands of withies. The quantities of hazel rods required for fence construction and their uniform shape are clear evidence of coppicing.

The working of wood for a wide variety of purposes shows a sophisticated understanding of wood technology, including an appreciation of the properties of different types of wood and their suitability for different purposes—hazel for long uprights, ash for spear shafts, guelder rose for arrow shafts, elm for bows, applewood for axe handles and lime for logboats. The fish weirs are testament both to the communal labour required in their construction and maintenance and to the large quantities of food made available and their important role in sustaining relatively large communities at permanent settlements on the shoreline. The logboats are also an important feature of the technology, the largest known being nearly 10 m long. These boats would have played an important role in offshore fishing, in transportation and in social communication.

In Germany, similar evidence has been recovered from the excavated sites on the Baltic coastline, which span a date range from the Kongemosen culture (Middle Mesolithic), through the various phases of the Ertebølle to the Early Neolithic (FNB)—c. 8.4–4.2 ka cal BP (Jöns et al., Chap. 4, this volume). At Neustadt in Lübeck Bay, for example, with one of the more extensive areas excavated, at 110 m^2, a very large assemblage of artefacts and organic remains was recovered from a marine refuse area, while the on-land area of the settlement had been destroyed by marine erosion. At Timmendorf-Nordmole I in the Wismar Bay, a similar refuse area provided most of the finds, while the on-land area had been eroded except for a pit with a sediment fill containing some organic remains and the timbers of a collapsed structure—the structure itself was considered by the excavators to be too small for a dwelling. Preservation on land here, as at the Danish sites, appears to be the result of features that had protected the deposit in question from the worst ravages of marine erosion, in this case the presence of a sediment-filled pit. A hearth area was also recorded at Breetzer Ort in Rugen, where it had been protected by two fallen tree trunks. Hearth areas sealed beneath peat and marine sediments were also recovered at Jäckelberg-Huk and Timmendorf-Normole II in Wismar Bay. In general, though, preservation of features on the original land is the exception, and the vast majority of finds come from marine refuse areas in front of the sites.

The range of material culture in the German Baltic sites, particularly items made of wood, is very similar to the Danish sites, with remains of fish weirs, logboats, leister prongs, paddles, wicker work and cordage.

In addition, there are two distinctive features worth noting. The first is the way in which the sites of the Wismar Bay track the retreat of the shoreline with progressive inundation by the Littorina Transgression and land submergence, demonstrating just how much land was lost and how people adapted to the changing conditions. The earliest site in the sequence, and the deepest at 8.5 m bpsl, dated at 8.4–8.0 ka cal BP is Jäckelberg-Huk. The remains of aquatic food are dominated by freshwater fish such as pike and perch and the migratory eel, which spends part of its life in freshwater, in keeping with an early stage in the Littorina transgression when saltwater had not yet penetrated far into the Baltic. One thousand years later, at the sites of Jäckelberg-Nord and Timmendorf-Nordmole II, marine fish are dominant in the food remains along with continuing exploitation of eel.

The second feature is the presence of underwater Neolithic sites. In Wismar Bay, the youngest site in the sequence and the shallowest at 2 m bpsl is the Early Neolithic Funnel Beaker settlement of Timmendorf-Tonnenhaken dated at 5.2–4.7 ka cal BP. All the bone remains at this site are of domestic cattle and pig, demonstrating that coastal locations remained important for Neolithic settlement even where marine resources were not apparently exploited. At Neustadt, there is continuity of occupation from the late Ertebølle to the Early Neolithic Funnel Beaker culture. Here too, domestic animals are the main species represented in the faunal remains of the Neolithic deposits, but analysis of residues on the inner surface of the Neolithic potsherds shows the presence of aquatic foods and acorns, and both resources are present on the residues of the Ertebølle potsherds from this site, demonstrating some continuity of economic practices across the Mesolithic-Neolithic boundary alongside the introduction of domestic animals.

2.5 Reasons for Site Preservation

The reasons for the survival of so many submerged sites and the preservation of rich assemblages of organic materials are similar throughout the regions discussed here. They are especially clear in Denmark and Germany and are due to three principal factors. The first is the high productivity of the marine environment and a coastal topography of marine channels and inlets that facilitated the easy capture in large quantities of fish and sea mammals during their seasonal migrations, a factor further enhanced by the construction of fish weirs, and one that resulted in concentrations of settlement along the immediate shoreline.

A second factor is the relatively sheltered marine conditions of coastlines in the south-western Baltic and the Danish Straits, with limited tidal movement and wind fetch, conditions conducive to the preservation of archaeological material. However, even in this protected environment, archaeological deposits are not invulnerable, as is clearly demonstrated by the erosion and destruction by marine action of most of the on-land sectors of the submerged settlements that have been excavated. Rather, the key factor is the fine sediments associated with these marine conditions and their slow but steady accumulation with progressive rise in relative sea level, producing deposits of soft, fine-grained, clay-like gyttja. Materials discarded into these deposits or engulfed by them quickly sink below the surface and are rapidly protected from subaerial degradation and the disturbing or destructive impact of waves and marine currents. Moreover, burial in these sediments followed by permanent submergence under water provides anaerobic conditions that protect organic material from bacterial attack. It is these depositional conditions, combined with the large quantities of cultural material abandoned in the shallow-water zone next to settlements located on the shoreline, that are the key to understanding the number of underwater sites discovered in Denmark and Germany and the quality of organic preservation. Similar conditions are associated with many of the Swedish sites and are also present in Norway.

As always, when considering issues of underwater preservation, there is a delicate balance between sediment accumulation that is sufficiently rapid to provide protection of land surfaces and archaeological materials, but not so rapid as to bury them deeply beyond the reach of easy observation. This is the major contrast between the Baltic and the North Sea coastlines of Denmark and Germany. In the North Sea, the huge volumes of sediment poured into the basin by the large rivers draining the continental land mass have resulted in thick accumulations of recent sediment. In these circumstances, Stone Age material is likely to be deeply buried, and discovery subject to the vagaries of exposure by storms that periodically remove some of the sediment cover or intrusive industrial activities, a point well illustrated by the sites on the German coastline (Jones et al., Chap. 5, this volume).

The third factor is the resources available for investigation. In Denmark the many reported finds are in large part the product of a long history of public and professional interest in which sports divers have made a significant contribution including a major role as volunteers in the underwater excavation of Tybrind Vig (Andersen 2013, pp. 7–8). Also, protective legislation for seabed archaeology has been in place since 1984, overseen by the national heritage agency, with the protection of underwater Stone Age sites as an explicit theme. The predictive 'fishing site' model developed by Anders Fischer and sponsored by the national heritage agency as a tool for management and protection of the underwater cultural heritage has also contributed to the discovery of many new sites (Fischer 1993, 2007). This model is testament to what can be achieved at relatively limited cost using easily accessible data on seabed bathymetry and geology, simple acoustic-survey methods and diver inspection. However, research funding has been limited, especially for larger-scale work, and most investigation has taken place as rescue work in response to industrial activity or the threat of erosion.

In Germany, in contrast, results have been achieved mainly through research-led funding from regional or central government agencies. The SINCOS project, funded through a major grant from the central government funding source (Deutsche Forschungsgemeinschaft) has played a major role in new discoveries. Although not explicitly informed by a predictive model for locating sites in the same way as Fischer's Danish work, this project combined acoustic survey and diver inspection systematically to identify and explore palaeoshorelines and conduct excavations at selected sites, resulting in the discovery of 50 new underwater Stone Age sites, eight excavations, and an unusual level of integration across disciplinary boundaries. It is an outstanding example of what can be achieved through large-scale research-led collaboration between marine geophysicists, geologists, climatologists and archaeologists.

2.6 Industrial Collaboration

A significant contribution to the development of research on submerged prehistory and the discovery of new finds throughout this region is the degree of collaboration between government heritage agencies, industrial companies working offshore and archaeological and scientific researchers. All the countries in this group are members of the European Union and subject to EU legislation that requires Environmental Impact Assessments (EIAs) to be conducted by industrial companies before embarking on offshore development (see also Parts II and IV of this volume). The relationship between these different stakeholders has evolved somewhat differently in different countries, but in general, underwater cultural heritage is usually included in the impact assessment and has resulted in important discoveries.

An important early example of collaboration is the Great Belt project of Denmark, which resulted in numerous offshore investigations and discoveries of submerged landscape features and archaeology during the construction of the road bridge and rail tunnel between the islands of Funen and Zealand in Denmark (Pedersen et al. 1997; Bailey et al., Chap. 3, this volume). Other examples are the collaboration with a commercial survey company in the 'Landscapes Lost' project in Sweden, which played a key role in the discovery of the finds at Haväng (Holmlund et al. 2017; Nilsson et al., Chap. 4, this volume), and the coring of the seabed during windfarm construction offshore of Norway, which provided key palaeoenvironmental data about palaeoshoreline development and the potential for human occupation on the eastern edge of Doggerland during the Late Glacial and early Postglacial period (Glørstad et al. 2017, and Chap. 6, this volume).

Other recent projects are the construction of the Fehmarn tunnel between the Danish island of Lolland and northern Germany and the Nord Stream pipeline project. In the Fehrmanbelt project, submerged landscape research and heritage management formed a completely integrated part of the planning and implementation of the construction project from its inception in 2008. As part of the EIA, extensive geological, geophysical, biological and archaeological investigations were carried out including seismic and side-scan survey, coring of sediments and geological, geochemical and palaeobotanical analysis. All the costs of survey and analysis were covered by the construction company. The results demonstrated the existence of a freshwater lake before the Littorina transgression and submerged archaeological remains on the Danish side (Dirk Enters et al. 2015).

The Nord Stream pipeline is a gas supply route from Russia through the Baltic Sea to Western Europe, extending for 1200 km from Portovaya Bay near Vyborg in north-western Russia and ending in Lubmin near Greifswald in north-eastern Germany. The national archaeological authorities in each of the countries involved, principally Russia, Denmark and Germany, were fully integrated in the planning and construction process. Geophysical survey was conducted in a 2 × 2 m grid and a 2 km wide corridor along the route, with ROV and diver inspection of anomalous features, with the project bearing the full costs of these investigations.

Industrial collaborations of this sort are obviously limited to target areas where offshore projects are to be carried out. However, with sufficient planning and organisation, they can provide technical and analytical resources on a scale that is rarely possible in research-led projects and are likely to play an increasingly important role in the future.

2.7 Conclusion

In reviewing these chapters, two things stand out. The first is the large number of sites and the quality of preservation compared to other areas of Europe, particularly along the coastlines of the Baltic and the Danish Straits. The second is that, despite this wealth of existing information and perhaps because of it, a range of new challenges and opportunities are coming more clearly into focus. Most of the existing information has come from sites in shallow water, mostly <5 m, easily accessible to SCUBA divers, and hence from sites of relatively late date (mostly from the Middle Mesolithic after c. 8.5 ka cal BP onwards). Even in this period, relatively few sites have received more than a test excavation. Most records refer to individual finds, and it remains unclear in many cases whether these are genuinely isolated specimens or surface indications of more extensive deposits beneath the seabed and if the latter whether they refer to in situ culture layers or redeposited material. For earlier periods, the submerged palaeoshorelines of southern Sweden clearly offer opportunities for investigations of the Early Mesolithic period (c. 11.7 to 8.4 ka cal BP) at depths within relatively easy reach of divers, and the same appears to be likely for some of the submerged coastlines of Norway.

Looking further afield, the gyttja sediments associated with the best conditions of preservation at the known excavated sites are more widely distributed in the Baltic and in the North Sea, according to evidence from geological cores and more extensive evidence of sediments and submerged forests. This is certainly the case further east in the Baltic (Damusyte et al. 2004; Uścinowicz et al. 2011). Judging from isolated finds of artefacts and preliminary palaeoenvironmental investigations, there is considerable potential for the discovery of new submerged sites in Poland, Lithuania and Latvia extending back at least to the Late Palaeolithic period (Kaube 1985; Kalniņa and Bērziņš 2010; Miotk-Szpiganowicz et al. 2010; Žulkus and Piličiauskas 2010).

On the North Sea coastlines of Germany and Denmark, finds are far fewer and are either relatively late in date or subject to chance exposure because of the much greater thickness of marine sediments compared to the Baltic. Nevertheless, the Maglemosean artefact dredged up from the Danish sector of the Dogger Bank (Bailey et al., Chap. 3, this volume, Fig. 3.4) and the use of a distinctive red flint in the Late Upper Palaeolithic industries of northern Germany with its source on the offshore island of Heligoland (Jöns et al., Chap. 5, this volume, Fig. 5.7) are clear indications that the extensive lowland landscape of Doggerland was an important zone of human activity at periods of lower sea level. It is here that settlements of significantly earlier date in the Late Palaeolithic period or earlier will have to be sought, and there is no reason why submerged palaeoshorelines with topography and depositional conditions comparable to the Mesolithic sites of the Baltic should not be present further out into the North Sea Basin. The technical challenges and costs of finding such features in deep water and most likely buried beneath marine sediments are, of course, much greater, but these challenges are already being addressed by offshore expeditions using research ships with palaeolandscape reconstructions and archaeological objectives in view (e.g., Gaffney et al. 2017; Hepp et al. 2017) and through palaeoenvironmental examination of the very large archive of existing sediment cores from inshore waters (Jöns et al., Chap. 5, this volume).

References

Andersen SH (2013) Tybrind Vig. Submerged Mesolithic settlements in Denmark. Jutland Archaeological Society Publications, vol 77. Jutland Archaeological Society, Højbjerg

Bjerck H (2008) Norwegian Mesolithic trends: a review. In: Bailey G, Spikins P (eds) Mesolithic Europe. Cambridge University Press, Cambridge, pp 60–106

Damusyte A, Bitinas A, Kiseliene D, Mapeika J, Petrodius R, Pulkus V (2004) The tree stumps in the south eastern Baltic as indicators of Holocene water level fluctuations. In: IGC-IUGS-UNESCO (ed) 32nd

International Geological Congress (Abstracts II), 20th–27th August 2004, Florence, p 1167
Enters D, Wolters S, Blume K, Lücke A, Segschneider M, Theuerkauf M (2015) Multiproxy-Analysen zur spät- und postglazialen Entwicklungsgeschichte des Fehmarnbelts. Siedlungs- und Küstenforschung im Nordseegebiet 38:43–58
Fischer A (1993) Stone Age settlements in the Småland Bight: A theory tested by diving. Miljøministeriet, Skov- og Naturstyrelsen, København (In Danish and English)
Fischer A (2007) Coastal fishing in stone age Denmark – evidence from below and above the present sea level and from human bones. In: Milner N, Craig OE, Bailey GN (eds) Shell middens in Atlantic Europe. Oxbow, Oxford, pp 54–69
Gaffney V, Allaby R, Bates R, Bates M, Ch'ng E, Fitch S, Garwood P, Momber G, Murgatroyd P, Pallen M, Ramsey E, Smith D, Smith O (2017) Doggerland and the lost Frontiers project (2015–2020). In: Bailey GN, Harff J, Sakellariou D (eds) Under the sea: archaeology and palaeolandscapes of the continental shelf. Springer, Cham, pp 305–319
Glørstad H, Gundersen J, Kvalø F (2017) The northern coasts of Doggerland and the colonisation of Norway at the end of the Ice Age. In: Bailey G, Harff J, Sakellariou D (eds) Under the sea: archaeology and palaeolandscapes of the continental shelf. Springer, Cham, pp 285–303
Hepp DA, Warnke U, Hebbeln D, Mörz T (2017) Tributaries of the Elbe palaeovalley: features of a hidden palaeolandscape in the German Bight, North Sea. In: Bailey GN, Harff J, Sakellariou D (eds) Under the sea: archaeology and palaeolandscapes of the continental shelf. Springer, Cham, pp 211–222
Holmlund J, Nilsson B, Rönnby J (2017) Joint explorations of the sunken past. Examples of maritime archaeological collaboration between industry and academia in the Baltic. In: Bailey GN, Harff J, Sakellariou D (eds) Under the sea: archaeology and palaeolandscapes of the continental shelf. Springer, Cham, pp 53–63
Jöns H (2011) Settlement development in the shadow of coastal changes – case studies from the Baltic rim. In: Jöns H, Björck S, Hoth P (eds) The Baltic Sea basin. Springer, Berlin/Heidelberg, pp 301–336
Jöns H, Harff J (2014) Geoarchaeological research strategies in the Baltic Sea area: environmental changes, shoreline-displacement and settlement strategies. In: Evans AM, Flatman JC, Flemming NC (eds) Prehistoric archaeology on the continental shelf: a global review. Springer, New York, pp 173–192
Jöns H, Björck S, Hoth P (eds) (2011) The Baltic Sea basin. Springer, Berlin/Heidelberg
Kalniņa L, Bērziņš V (2010) Latvia's former coastlines: the state of research on palaeolandscapes and Stone Age archaeology. SPLASHCOS meeting, 10–11 March 2010, York. https://www.splashcos.org/events
Kaube A (1985) Harpun z poroża wyłowiony z Bałtyku. Materiały Zachodniopomorskie 31:409–412
Miotk-Szpiganowicz G, Pomian I, Uścinowicz Sz, Was M, Witkowski A (2010) Underwater research in Poland. SPLASHCOS meeting, 10–11 March 2010, York. https://www.splashcos.org/events
Pedersen L, Fischer A, Aaby B (eds) (1997) The Danish Storebælt since the ice age: man, sea and forest. A/S Storebæltsforbindelsen, Copenhagen
Rosentau A, Bennike O, Uscinowicz S, Miotk-Szpiganowcz G (2017) The Baltic Sea basin. In: Flemming NC, Harff J, Moura D, Burgess A, Bailey GN (eds) Submerged landscapes of the European continental shelf: quaternary paleoenvironments. Wiley, Chichester, pp 103–133
Schmitt L, Larsson S, Schrum C, Alekseeva I, Tomczak M, Svedhage K (2006) 'Why they came': the colonization of the coast of western Sweden and its environmental context at the end of the last glaciation. Oxf J Archaeol 25(1):1–28
Schmitt L, Larsson S, Burdukiewicz ZJ, Svedhage K, Zamon J, Steffen H (2009) Chronological insights, cultural change, and resource exploitation on the west coast of Sweden during the late Palaeolithic/early Mesolithic transition. Oxf J Archaeol 28(1):1–27
Uścinowicz S, Miotk-Szpiganowicz G, Krąpiec M et al (2011) Drowned forests in the Gulf of Gdańsk (southern Baltic) as an indicator of the Holocene shoreline changes. In: Harff J, Björck S, Hoth P (eds) The Baltic Sea basin. Springer, Berlin/Heidelberg, pp 219–231
Žulkus V, Piličiauskas G (2010) Stone Age archaeology below sea level in Lithuania: on the shelf and at the coast. SPLASHCOS meeting, 10–11 March 2010, York. https://www.splashcos.org/events

Denmark: Mesolithic Coastal Landscapes Submerged

3

Geoff Bailey, Søren H. Andersen, and Thijs J. Maarleveld

Abstract

Denmark has the richest concentration of underwater Stone Age finds in the world, thanks to a combination of factors that include favourable conditions of preservation and a long tradition of professional and amateur interest in underwater prehistory. The majority of finds are from the central areas of the Danish Straits and date back to about 8500 years ago, when rising sea level finally created a marine connection between the North Sea and the Baltic Sea. Most are in shallow water, c. 2–5 m in depth, but sites at greater depth have also been identified. Most fall within the Ertebølle period, between 7400 and 5900 years ago, and include excavations at shoreline settlements such as Tybrind Vig, Ronæs Skov and Møllegabet II. These sites have yielded large assemblages of material including wooden artefacts such as remains of fish weirs, dugout canoes, bows, spear shafts and leister prongs, as well as remains of fibres and woven fabric. These materials owe their excellent state of preservation to the deposition in fine-grained marine sediments alongside settlement areas located on the shoreline, and to subsequent sea-level rise because of marine inundation, which has maintained the material in permanently waterlogged and anaerobic conditions. This chapter examines the conditions that have given rise to this unusual concentration of underwater sites, provides an overview and illustration of some of the most distinctive finds, discusses their wider significance and addresses future challenges.

Keywords

Boats · Fish weirs · Submerged settlements · Shell mounds · Wooden artefacts · Mesolithic · Kongemose · Ertebølle

G. Bailey (✉)
Department of Archaeology, University of York, York, UK

College of Humanities, Arts and Social Sciences, Flinders University, Adelaide, SA, Australia
e-mail: geoff.bailey@york.ac.uk

S. H. Andersen
Moesgaard Museum, Højbjerg, Denmark
e-mail: sha@moesgaardmuseum.dk

T. J. Maarleveld
Maritime Archaeology Programme, University of Southern Denmark, Esbjerg, Denmark
e-mail: t.maarleveld@sdu.dk

3.1 Introduction

Denmark has one of the largest concentrations of Stone Age underwater sites in Europe or indeed anywhere else in the world. Estimates of the total number vary, ranging from 2300 (Fischer 2004, p. 23) to as many as 4000 (Andersen 2013, p. 14),

G. Bailey et al. (eds.), *The Archaeology of Europe's Drowned Landscapes*,
Coastal Research Library 35, https://doi.org/10.1007/978-3-030-37367-2_3

with a formal record of 1686 in the national records maintained by the responsible Danish government agency (http://www.kulturarv.dk/fundogfortidsminder).[1] The majority are represented by finds of single artefacts, but this may result from a lack of further investigation or limited exposure of the underlying land surface. Sites subject to more intensive survey and collection have generated hundreds to thousands of artefacts. Most are in shallow water (c. 2–5 m) and younger than 10,000 cal BP (10 ka),[2] mostly Mesolithic or Early Neolithic. Many are the result of reports by sports divers and amateur archaeologists from the 1950s onwards or chance finds by offshore industrial and commercial activity going back to the nineteenth century. Some of the larger assemblages of material represent cultural layers that are still in situ and have remained largely intact following submergence, while others represent material that has been redeposited by marine action during and after submergence. Distinguishing between primary and secondary deposits is a major issue in the interpretation of underwater sites in Denmark (Andersen 2009, p. 13, 2013, p. 29).

Systematic survey began in the early 1970s, with bursts of activity in the 1980s and 1990s. Much of the most detailed survey work concentrated around the shorelines of the island of Fyn and its immediate neighbours (Skaarup 1983, 2001, 2004; Fischer 1993). Systematic excavation took place at a small number of sites, notably at Tybring Vig (Andersen 1980, 1985, 2013), Møllegabet II (Grøn and Skaarup 1993; Skaarup and Grøn 2004), Argus Grund (Fischer 1987) and Ronæs Skov (Andersen 2009), resulting in the development of novel methods of underwater excavation and equipment (see Dal 2013 for details).

Another important group of remains on both sides of the Great Belt (Storebælt) was recovered during construction of the road and rail link between Fyn and Zealand (Pedersen et al. 1997).

These sites, with their extraordinary record of organic remains preserved in anaerobic sediments, have dominated the evaluation of the Danish record and remain key sources of information and points of reference (see also Fischer and Vang Petersen 2018 for a summary of key finds).

In addition to archaeological finds, there are also numerous remains of submerged forests, including tree stumps in situ and fallen logs, identified at all depths on the Danish seafloor including the North Sea sector (Fischer 2004). These provide valuable information on changes in shoreline position and rates of sea-level change, targets for closer survey for archaeological sites and the potential for high resolution dating based on dendrochronology (Christensen 1997, 2013; Fischer 1997).

This unusual concentration of finds is the product of many factors:

- A geomorphological and palaeogeographical history that has created extensive shallow and sheltered areas with accumulations of gyttja, a peaty clay conducive to organic preservation
- Ecological and topographic conditions favourable to palaeoeconomies heavily dependent on marine resources and to concentrations of settlement on the shoreline, especially in areas conducive to the trapping of fish and sea mammals during their seasonal migrations
- An abundance of flint sources for making artefacts that are easily visible eroding out of submerged layers on the shallow seabed or along the modern shoreline
- Informed interest amongst sports divers, amateur archaeologists and other members of the public in reporting finds, which intensified in the 1950s with the wider availability of SCUBA equipment
- A relatively long history of systematic investigation by professional archaeologists with

[1]The differences are due to the fact that the higher figures include estimates of the number of sites that should be present along the submerged shorelines of southern Denmark by extrapolation from the density of finds of the same period known from the uplifted shorelines of northern Denmark.

[2]In this chapter, ka (thousands of years ago) is used to mean thousands of calibrated radiocarbon years before present. For conversion to cal BC, subtract 2000 years. For example, 6 ka is equivalent to 6000 cal BP or 4000 cal BC and these conventions are used interchangeably throughout the text.

dive training and experience and the development of new skills and knowledge in underwater exploration and excavation
- A national heritage agency (currently known as the Danish Agency for Culture and Palaces) committed to the recording and management of underwater sites, enshrined in national legislation since 1984
- Recovery of material by industrial activities that go back to the nineteenth century, such as harbour construction, dredging of waterways, commercial extraction of shell, sand and gravel and, in the past two decades, major bridge and tunnel projects across the Danish Straits accompanied by systematic underwater investigation

By common consent, Denmark has acquired a reputation as the 'capital' of underwater Stone Age archaeology, a phenomenon that archaeologists elsewhere look to for standards of best practice and examples of what might be found on other submerged landscapes. A key question is whether this concentration of sites is unique to conditions in Denmark, and more specifically to the region of the Danish Archipelago and the Danish Straits, or can be expected elsewhere, either as visible remains on the seafloor or embedded in deeper sediments. In fact, comparable finds have been found in the adjacent regions of Sweden (Nilsson et al., Chapter 4) and northern Germany (Jöns et al. Chapter 5, this volume) and in other parts of Europe as described elsewhere in this volume; nevertheless, the concentration of material in Denmark is exceptional.

The aim of this chapter is to provide the palaeogeographical and archaeological context for the Danish material, to examine conditions of underwater preservation and visibility, to provide an analysis of finds with illustrations of some of the best-preserved examples, to evaluate the significance of the Danish material in relation to its wider context and the prospects for underwater research elsewhere and to outline some directions for future research.

3.2 Geographical and Archaeological Setting

Denmark today has one of the highest ratios of shoreline to land area of any country considered in this volume, with c. 7400 km of coastline including over 400 named islands, and a land area of c. 44,000 km^2 of low relief comprising glacial moraines and lake basins with elevations no higher than 170 m (Nokkentved et al. 2018). No place on land is more than 52 km from its nearest coastline. The land comprises the Jutland Peninsula, divided in the north by a stretch of water known as the Limfjord, and an archipelago of closely adjacent islands between Jutland and southern Sweden, the largest of which are Zealand, Fyn, Lolland and Falster, and the more distant eastern outlier of Bornholm (Fig. 3.1).

Together, the inshore waters of the east Jutland Peninsula and the Danish Archipelago define a transitional marine zone between the fully oceanic conditions of the North Sea and the brackish water of the Baltic, connected through a series of sea channels that separate Denmark from Norway and Sweden. These channels become progressively shallower from north-west to south-east, and progressively less saline, and the sea bottom consists of soft sediments of clay, sand and gyttja. To the north, the Skagerrak is a c. 100 km-wide channel, >500 m deep, an eastward extension of the Norwegian Trench, which remained an open sea channel throughout the Last Glacial period and an insurmountable barrier to human movement. To the south-east, the Kattegat is another wide channel, but much shallower, mostly less than 50 m bpsl (below present sea level), with an extensive shallow shelf of <20 m extending out from the east coast of Jutland. At its southern end, the Kattegat separates into a series of narrow and shallow channels (the Danish Straits), the Little Belt between Jutland and Fyn, the Great Belt between Fyn and Zealand and the Øresund between Zealand and southern Sweden. Here, the seabed is generally less than 20 m bpsl (Rosentau et al. 2017, Fig. 5.1). These channels, in their turn, open out eastwards into the wider expanses of the western Baltic Sea.

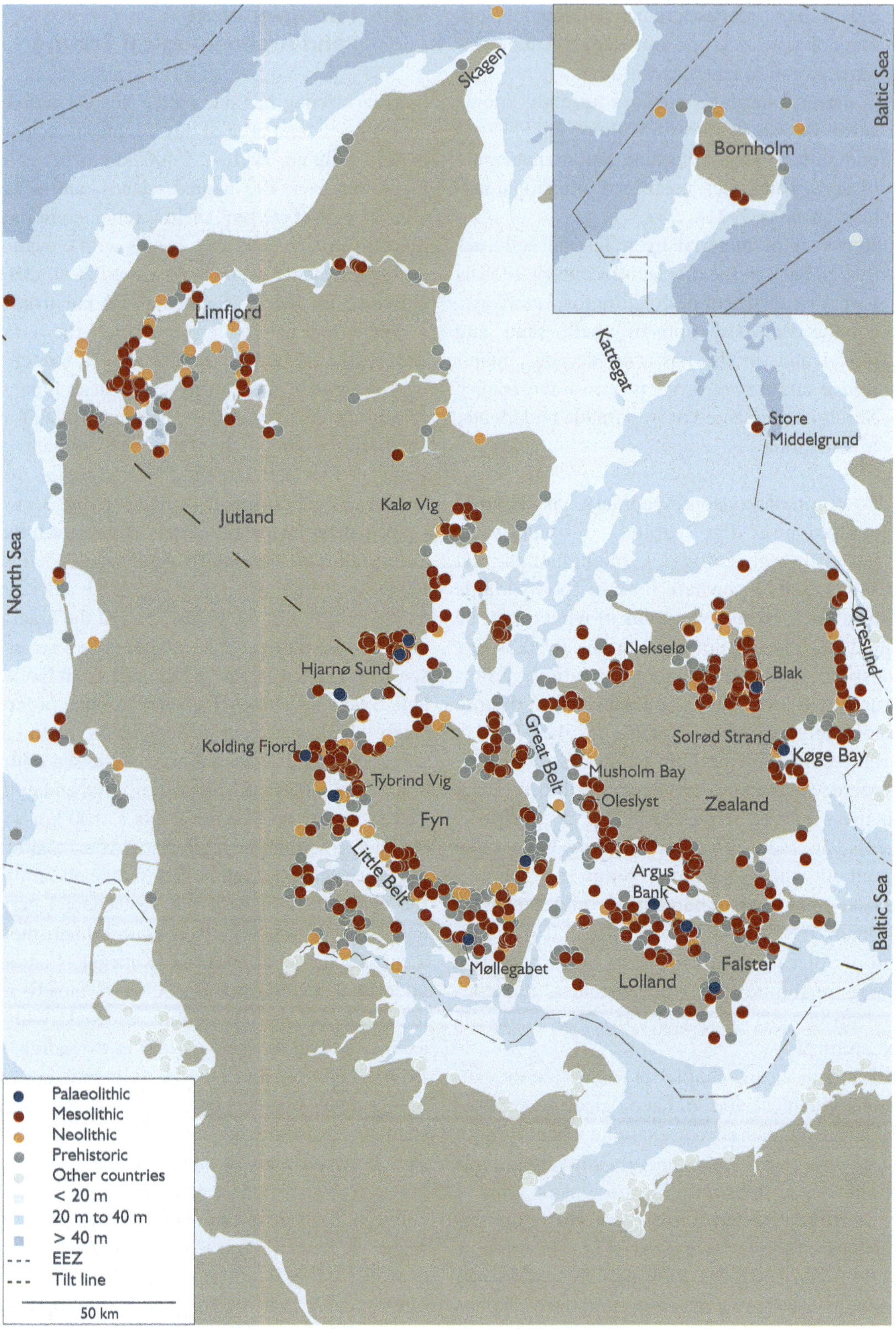

Fig. 3.1 Map of Denmark showing distribution of underwater sites, key names and other places mentioned in the text. Inset shows the island of Bornholm. Site information from the SPLASHCOS Viewer http://splashcos-viewer.eu. Drawing by Moritz Mennenga

3.2.1 Archaeological Sequence

At the LGM at c. 20 ka, Denmark was uninhabitable. Much of it was covered by the southern edge of the Scandinavian ice sheet, apart from a strip of land in western Jutland—a barren plain of outwash sands and gravels. The earliest traces of human occupation date from c. 15 ka in southern Jutland and Zealand, after the retreat of the ice front. The sites represent reindeer hunters belonging to the tanged point cultures of northern Europe, successively Hamburgian, Federmesser, Bromme and Ahrensburgian (Table 3.1), and are linked to hunting territories extending across an open tundra landscape with lakes and valleys carved out by glacial action and subsequent channel outflow from the Baltic, which constrained seasonal reindeer migration routes on a north–south axis from northern Germany to southern Sweden, and perhaps also westwards onto the now-submerged plain of the North Sea (Holm 1991; Vang Petersen and Johansen 1991; Fischer 2004). These tanged point assemblages show similarities over large areas of lowland northern Europe extending westwards across the North Sea Plain to northern Britain, indicating wide-ranging seasonal movements and long-distance contacts (Momber and Peeters 2017; see also Jöns et al., Chapter 5, this volume, Bailey et al., Chapter 10, this volume).

These tanged point cultures were succeeded by the Maglemose culture (c. 11–8.4 ka) with new equipment including axes and microlithic arrowheads adapted to the spread of more forested conditions and associated mammalian fauna—principally elk (*Alces alces*), aurochs (*Bos primigenius*), red deer (*Cervus elaphus*), roe deer (*Capreolus capreolus*) and boar (*Sus scrofa ferus*). Settlements were located around the edge of lake basins, with evidence of fishing in the form of barbed bone points. Similarities on either side of the North Sea are still apparent in this period. There are hints of contact with the contemporaneous coastline in the form of occasional seal bones in inland settlements (Degerbøl 1943), but no known Maglemose coastal settlements on the modern shoreline, which would still have been some distance inland from the contemporaneous shoreline.

From c. 8.4–7.4 ka, a new type of microlithic arrowhead appeared that defines the Kongemose culture, showing much continuity with the preceding period, but with the appearance for the first time of shoreline settlements and evidence of marine exploitation. This in turn is succeeded by the Ertebølle culture (c. 7.4–6 ka) with its well-

Table 3.1 Chronological chart showing the main archaeological periods and changes in palaeogeography. Chronological boundaries are approximate

Cal BP	Archaeological period	Archaeological subdivision	Palaeogeographical stage	Cal BC
4400–3700	Late Neolithic		Littorina Sea	2400–1700
4800–4400	Middle Neolithic	Single Grave Pitted Ware		2800–2400
5300–4800		Funnel Beaker		3300–2800
5900–5300	Early Neolithic			3950–3300
6300–5900	Late Mesolithic	Late Ertebølle		4300–3950
6800–6300		Middle Ertebølle		4800–4300
7400–6800		Early Ertebølle		5400–4800
7700–7400	Middle Mesolithic	Late Kongemose		5700–5400
8000–7700		Middle Kongemose		6000–5700
8400–8000		Early Kongemose		6400–6000
9800–8400	Early Mesolithic	Maglemosian	Initial Littorina Sea	7800–6400
10,700–9800			Ancylus Lake	8700–7800
11,000–10,700			Yoldia Sea	9000–8700
11,700–11,000	Late Palaeolithic	Ahrensburgian		9700–9000
12,500–11,700			Baltic Ice Lake	10,500–9700
14,000–12,500		Bromme/Federmesser		12,000–10,500
14,700–14,000		Hamburgian		12,700–12,000

Data from Pedersen et al. (1997) and Rosentau et al. (2017)

known evidence of coastal settlement, large shell mounds (kitchen middens), marine exploitation and new items of material culture including transverse microlithic arrowheads and ceramic vessels alongside evidence of ongoing hunting and gathering on land. From 6 ka, Neolithic settlements appeared with new types of pottery and evidence of farming, most likely involving some degree of population replacement, but with evidence for the continued exploitation of marine resources, especially shell middens, at least in the first 4–500 years.

Taken at face value, this sequence suggests a progressive shift from emphasis on big-game hunting and large territories to a broader spectrum economy including some aquatic resources, to a highly diversified economy with major emphasis on marine resources, smaller territories and permanent settlements and ultimately to replacement by a mixed farming economy. It is tempting to suppose that this sequence indicates intensification associated with population growth and related sociocultural developments. However, this would be to take the evidence at face value without consideration of other variables. Changes in the earlier part of the sequence are clearly related to climatic and vegetational changes associated with de-glaciation, later in the sequence to regionalisation associated with geographical fragmentation by rising sea level, and at the end of the sequence to the introduction of domestic crops and livestock from outside the region. The apparently progressive growth in the number of coastal settlements and importance of marine resources during the Mesolithic period is the main focus of interest in this chapter and needs to be calibrated against the changing configuration and ecology of coastlines and the conditions of site preservation and discovery associated with relative sea-level change.

3.3 General Conditions of Preservation and Visibility

Cohen et al. (2017) and Rosentau et al. (2017) set out the general conditions that affect the survival of submerged land forms and archaeological material for the North Sea and the Baltic Sea, respectively. Denmark, because of its location on the boundary between these two basins, has some distinctive patterns of variability in space and time that we examine here. As with all discussions of underwater preservation, we emphasise that local features may also be of great importance and over-ride general conditions in the wider area. These local conditions will be examined more closely later when discussing individual archaeological sites.

3.3.1 Oceanographic Variables

Oceanographic variables are of significance in highlighting the distinctiveness of the inner waters of the Danish Archipelago in contrast to the more typical conditions that prevail on North Sea and Atlantic coastlines.

On the exposed western and northern coasts of Jutland facing the North Sea and the Skagerrak, the tidal range is relatively large by Danish standards, being c. 1 m, although this is at the lower end of the spectrum compared to other coastlines of the North Sea and the Atlantic margins (Westley 2017). A coastline of sandy beaches and embayments is protected in places by offshore sand bars, but wave fetch is large, extending without interruption for hundreds of kilometres from the west and north-west, and the coastline is vulnerable to storm erosion (Sistermans and Nieuwenhuis 2018). In the Danish Archipelago and the Limfjord, in contrast, tidal range is c. 10 cm (though this may be increased by local conditions of wind and currents), and wind fetch may be as little as 1–5 km, resulting in low rates of erosion of the seabed even in shallow water (Rosentau et al. 2017).

During the earlier part of the Littorina Sea period, coinciding with the Late Mesolithic (Table 3.1), there is evidence for greater tidal inflow into the Danish Straits than today and a greater tidal range (Petersen and Rasmussen 1995; Petersen 2013). At about the Mesolithic–Neolithic transition, conditions appear to have changed, with a reduction in tidal amplitude and salinity and increased rates of sedimentation in shallow fjords and estuaries. These changes are relevant to issues of underwater site preservation and visibility, though their timing, causes and consequences are complex (e.g., Lewis et al. 2016) and may have varied according to local circumstances.

3.3.2 Palaeogeographical Change

Between the Last Glacial Maximum at 20 ka and the cessation of sea-level rise generated by ice melting at about 5 ka, changes in regional geomorphology and palaeogeographical configuration show a complex interplay between ice-retreat, isostatic land adjustment, eustatic sea-level rise and intermittent damming of the Baltic as a freshwater lake (Björck 1995; Rosentau et al. 2017; Astrup 2018) (Figs. 3.2 and 3.3). This has very significant implications for understanding where palaeoshorelines were located at different periods of the archaeological sequence and the conditions that have variously favoured or inhibited the preservation and exposure of underwater material.

Initially, rivers drained into the North Sea and later into the Skagerrak as the ice began to retreat (Fig. 3.2a, b). With further ice retreat, the present-day territory of Denmark and southern Sweden opened up as a continuous area of lowland territory that acted as a barrier between a Baltic Basin that filled with glacial meltwater and a marine inlet that occupied the deeper, eastern part of the Kattegat Basin > 40 m bpsl.

The Baltic remained largely isolated as a freshwater basin, its waters dammed back initially by ice and later by slowly uplifting land rebounding after the removal of ice cover, but with sufficient outflow of water to erode the valleys that now form the Danish Straits (Fig 3.3c). A sea connection was briefly established to the north between the Baltic and the Skagerrak at c. 11.7–10.7 ka, the Yoldia Sea period (Fig. 3.3d). However, the shallow shelf along the east coast of

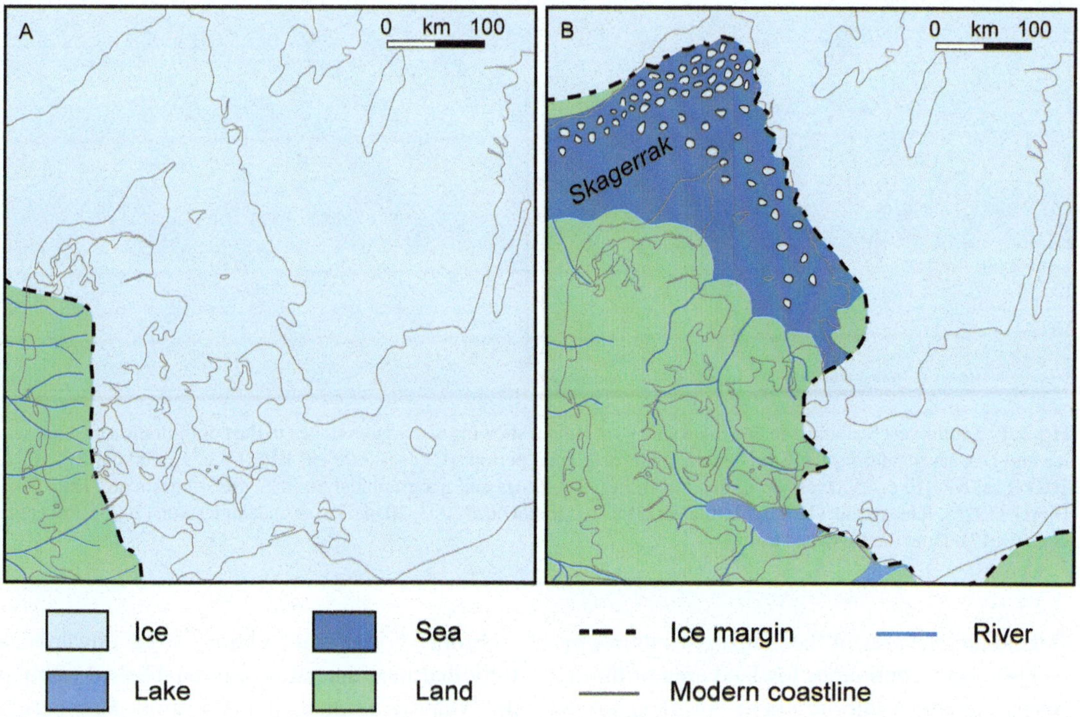

Fig. 3.2 South-west Scandinavia and the south-west Baltic, showing changes in the configuration of land, sea and ice during and immediately after the Last Glacial Maximum: (A) c. 22,000 cal BP; (B) c. 16,000 cal BP. Data from Houmark-Nielsen and Kjær 2003; Rosentau et al. 2017. Drawing by Geoff Bailey

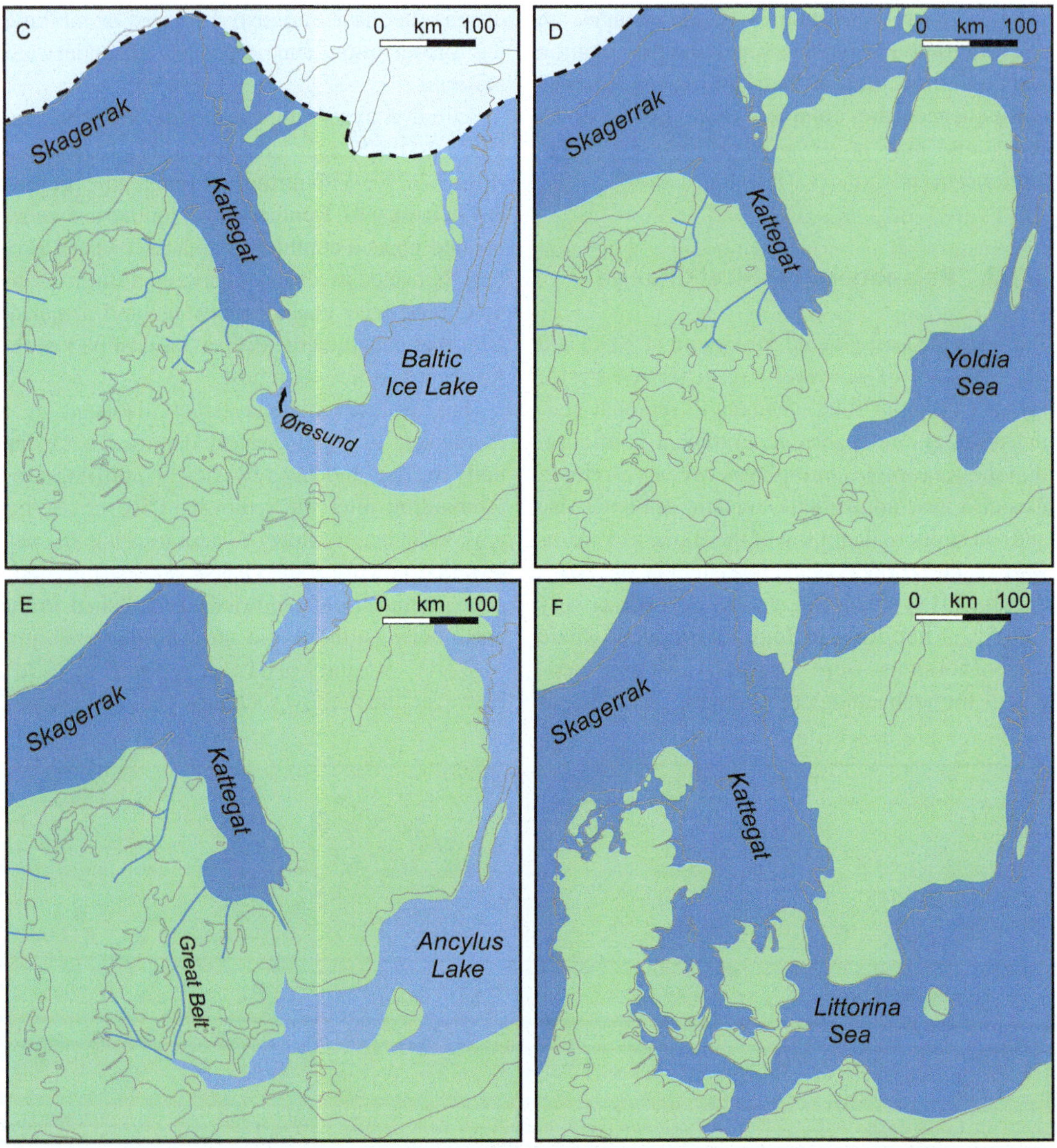

Fig. 3.3 South-west Scandinavia and the south-west Baltic, showing successive stages in the configuration of land, sea, ice and lake basins during the Late Glacial and Postglacial period: (C) c. 14,000 cal BP; (D) c. 11,000 cal BP; (E) c. 10,000 cal BP; (F) c. 8500 cal BP. Legend as in Fig. 3 2. Dates and geographical outlines are approximate. Data from Björck (1995), Jensen et al. (1997), Mathiasen (1997), Uścinowicz (2003, 2014), Påsse and Andersson (2005), Rosentau et al. (2017). Drawing by Geoff Bailey

Jutland and the area of the Danish Straits continued to form a continuous lowland area of limited relief extending into southern Sweden, dotted with small lake basins and traversed by river valleys formed by the outflow from the Baltic (Fig. 3.3e).

From shortly after about 10 ka onwards, a fully marine connection was established through the Danish Straits—the Littorina Sea period (Table 3.1). Sea level rose from about −30 m to near the present level in a matter of c. 1000 years between c. 9 and 8 ka with a slower rise towards the present level thereafter (Christensen 1995). Between about 7.5 and 5.5 ka, the final rise of sea level took place, with minor fluctuations that created a series of four high-sea-level stands, separated by periods of minor sea-level regression. These are the so-called Littorina shorelines. They most likely represent episodes of isostatic uplift interacting with a final, slow and steady rise of eustatic sea level from −2 m to reach the present sea level at about 5 ka, rather than eustatic sea-level fluctuations (Gehrels et al. 2006). At the time of the highest Littorina sea level, north-eastern Denmark was fragmented into an archipelago with broader openings between the North Sea, the Skagerrak and the Kattegat and a stronger inflow of water from the North Sea (Fig. 3.3f).

After 5 ka, isostatic effects continued, with uplift in the north-east and submergence in the south-west. The boundary between uplift and submergence runs from south-east to north-west (Fig. 3.1), and the Littorina shorelines show progressively higher elevations and earlier dates as one moves north, with the highest shorelines in northern Jutland at 12 m apsl (above present sea level) with dates of c. 7.2 ka and progressive submergence in the south to depths of −8 m (Christensen 1995; Christensen and Nielsen 2008; Astrup 2018).

For the earlier part of the sequence, most of the area between Denmark and Sweden was above the contemporaneous sea level, and now-submerged sites are likely to be hunting stations for monitoring or trapping animals, or settlements along rivers or lake edges. Once the Littorina Sea entered the Danish Straits, marine shorelines would have been present from the later Maglemose or early Kongemose period onwards at depths dependent on the local interaction between sea-level rise and isostatic adjustment. Shorelines of the Maglemose period are everywhere submerged below the present-day sea level or eroded away. Shorelines of the Kongemose period are mostly submerged or covered by later sediments, being present above sea level only in north-eastern Denmark (Fischer 1993, p. 60), while shorelines of the Ertebølle period are at or above modern sea level in north-eastern Denmark and submerged in areas to the south-west.

The main effect of isostasy is that shorelines of the same age are now at a relatively higher elevation in the north than the south, an effect best illustrated in the Ertebølle period. Many hundreds of Ertebølle coastal sites are located on raised shorelines in the Limfjord region of northern Jutland, the east coast of Jutland and the northern coasts of Fyn and Zealand. Andersen (2000, p. 362) refers to over 500 Ertebølle shell mounds in this region, with perhaps a comparable number of contemporaneous coastal sites that are not shell middens. To the south, in the coastal regions that have undergone submergence, there is a comparable number of sites that are now under water. Underwater shell midden deposits, however, are very rare. It is the effect of this submergence that is key to understanding the abundance, character, state of preservation and visibility of underwater sites in the central and southern region of the Danish Archipelago.

3.4 Archaeological Analysis

Currently there are 1699 archaeological finds recorded in the SPLASHCOS Viewer (http://splashcos-viewer.eu/), of which 1686 are registered in the records of the Danish Agency for Culture and Palaces (http://www.kulturarv.dk/fundogfortidsminder/).[3]

[3]This is less than the total number referred to in the published literature. In 2004, Fischer referred to 2003 prehistoric sites and noted that this number of finds was probably 'just a few percent of what is actually existing in the national sea territory' (Fischer 2004, p. 23). It is clear that we are dealing here with a minimum number of finds, but equally that a great many find spots are single artefacts or material that has been disturbed or re-deposited by marine erosion. The analysis in this section assumes that the available records are broadly representative. The online records of the Danish Agency are in Danish and give information on location, brief details of finds, their history of discovery and references to the published literature where available. It is also important to note that many prehistoric sites were located around shallow bays that subsequently silted up and were drained to create modern farm land. These sites are not, strictly-speaking, 'underwater' but they include material that was deposited under water or later became inundated by sea-level change, and we include reference to some of them, notably from the Halsskov Fjord discussed later in this chapter.

The great majority, 1307 (79%), are recorded only as single finds, while sites indicating evidence of settlement comprise 293 records (17%) (Table 3.1). A 'settlement' in this context means clear evidence of human occupation and artefact production, usually involving a considerable density of cultural material (Fischer 1993, p. 72). This may take the form of features such as a hearth, the presence of debitage from tool-working, burnt flint and abandoned tools, remains of stationary fish weirs and graves as indicated by the presence of human bones. Artefacts by themselves, even in considerable quantity, are not necessarily a sufficient indication of in situ activity or a settlement with dwelling structures, since they may have been redeposited by marine action or represent more ephemeral activities or specialised functions. Categorisation in many cases is a matter of subjective judgement and the experience of the archaeological observer. In cases of doubt, material is placed in the category of unstratified material (12 records, 0.7%).

Conversely, the recording of a single find does not rule out the presence of a settlement and indicates only what was observable on or recovered from the surface of the seabed or along the modern coastline. The quantity of material recovered is also in part a function of the time spent examining a given locality and the conditions of visibility, especially when diver inspection is involved. It is also clear from the general literature (e.g. Fischer 2004, p. 28; Fischer and Vang Petersen 2018) that material recorded as single finds may include items accidentally lost at sea such as amber pendants, fish hooks and harpoon heads as well as items discarded on a now-submerged land surface or archaeological settlement.

Two other categories to note are votive deposits, some 12 records (0.7%), mostly of Neolithic or Bronze Age date, but including at least one Ertebølle example, identified by the presence of artefacts presumed to be of great value because of their careful workmanship such as bifacially worked flint daggers and battle axes, and miscellaneous finds (26 records, 1.5%). The latter include isolated remains of fish traps, large stone objects of uncertain function or indeterminate artefacts.

Many sites with settlement remains were discovered by Fischer's 'fishing-site model' (Fischer and Sørensen 1983; Fischer 1993, 1995, 2007). This is a predictive model based on the observation that a number of already-discovered sites were located in topographic conditions ideal for trapping fish during their seasonal migrations: at the mouths of stream inlets, at the entrance to semi-enclosed shallow bays or on the end of small peninsulas jutting out into a marine channel, especially where a small offshore island funnelled fish movements into a narrow strait between the island and the facing shoreline. Use of bathymetric contours supplemented by simple acoustic surveys to identify similar topographic conditions on submerged palaeoshorelines resulted in a high degree of success in the location of new sites.

It is clear from the map (Fig. 3.1) and the data in the SPLASHCOS Viewer that the great majority of finds (1437) are from the protected shorelines of the inner Danish Archipelago, including the sheltered coastlines of east Jutland from Aarhus Bay southwards and the coastlines of the major islands, especially southern Fyn and southern Zealand. The next largest category (155) is in the Limfjord region of northern Jutland, which would have offered similar protected conditions during the Littorina Sea period. Just 50 sites are recorded from the more exposed coastlines of the North Sea, the Skagerrak and the Kattegat, with a small number at some distance offshore and at depth. The latter include a pressure flaker made from red deer antler of typical Maglemosian type from the Dogger Bank some 250 km west of the Danish coastline (Fig. 3.4).

In terms of chronology and disregarding the 805 records that can only be identified as 'prehistoric', by far the largest number of finds are Mesolithic in date, 676 records (39.8%), followed by 122 Neolithic finds, a smaller number (39) that may be Mesolithic or Neolithic, 45 Bronze Age and 15 Late Palaeolithic records (Table 3.2).

The earliest underwater archaeological finds currently known from Danish waters are worked reindeer bones recovered from Køge Bay in east Zealand, one of which, from Solrød Strand, is

Fig. 3.4 A pressure flaker made from red deer antler, dredged up from a depth of 30–40 m on the Dogger Bank some 250 km west of the Danish coastline. This specimen is typical of the younger Maglemose period, a date confirmed by radiocarbon dating of the artefact itself to c. 7040–6700 cal BC (Andersen 2005). Scale in cm. Photo courtesy of Moesgaard Museum Photography Department

Table 3.2 Distribution of underwater finds by type of site and period. Miscellaneous other includes remains of fish weirs, isolated ground stone objects and other indeterminate artefacts or remains of structures

Types of sites	Late Palaeolithic	Mesolithic	Mesolithic-Neolithic	Neolithic	Bronze Age	Indeterminate Prehistoric	Total N	Total %
In situ cultural features settlement data	–	160	19	6	2	113	300	*17.6*
Unstratified material multiple finds	2	2	1	–	–	7	12	*0.7*
Graves (human bone)	–	14	–	2	4	29	49	*2.9*
Single unstratified find	12	496	19	107	27	645	1306	*76.6*
Votive deposits	–	2	–	4	3	3	12	*0.7*
Miscellaneous other	1	9	–	3	9	5	27	*1.5*
Total N	**15**	**683**	**39**	**122**	**45**	**802**	**1706**	***100***
%	*0.9*	*40.0*	*2.3*	*7.2*	*2.6*	*47.0*		
Presence of organics (wood and/or bone)	3	72	4	10	3	54	**146**	***8.6***

Data from the SPLASHCOS Viewer at http://splashcos-viewer.eu, and from Danish government records at http://www.kulturarv.dk/fundogfortidsminder

dated at 14 ka (Vang Petersen and Johansen 1991, p. 36, endnote 28). These were found in sand and gravel deposits at 6–10 m bpsl near the outflow at the western end of the ice-dammed Baltic Lake and are most plausibly explained as evidence for the use of the local topography to ambush animals during their seasonal migrations.

The next earliest sites are broadly labelled as Early Mesolithic, generally associated with the Maglemose period. However, there are very few that fall certainly within this period, and these only towards the end, namely, Svalerumpen on the east coast of Zealand facing the Øresund Strait, with a single flint tool, a small number of animal bones and a radiocarbon date of 8225 ± 95 (Fischer 1993; http://www.kulturarv.dk/fundog-fortidsminder/Lokalitet/164010/), equivalent to c. 9.5–9.1 ka (7400–7100 cal BC; see Table 3.1), and sites in Aarhus Bay, Jutland, notably Fløjstrup Skov, with typical microliths of the Maglemose tradition and radiocarbon dates ranging from 7200 to 6300 cal BC, most probably indicating a mixture of artefacts from different periods (Dencker and Jensen 2000; Astrup 2018). Material of similar date has been found on the Swedish side of the Øresund indicating settlements on or close to the shoreline at a time when seawater had entered the northern Øresund but before a marine connection had been established to the Baltic Basin (Larsson 2017; Nilsson et al., Chapter 4, this volume).

The other Early Mesolithic sites, notably the settlement of Musholm Bay in the Great Belt (Fischer and Malm 1997), Kalø Vig on the east coast of Jutland (Fischer 2007), and Blak in Roskilde Fjord (Sørensen 2017), all of which are radiocarbon-dated, lie on the boundary of the Maglemose and early Kongemose periods associated with the onset of the Littorina period, the entry of seawater into the Kattegat and the Danish Straits, and the establishment of a marine connection with the Baltic. From this time onwards, sea level rose rapidly. Sites of successively later periods occur at shallower depths and provide a precise measure of the rate of sea-level rise and the changing configuration of the coastline. At Musholm Bay, for example, an early Kongemose settlement originally located on the contemporaneous shoreline is now at a depth of −8 m and 2.5 km distant from the present shoreline. In little more than 1000 years, by the Middle Ertebølle period, sea level was within half a metre of the present level, and the shoreline had retreated to about its present position. The relationship between the age of coastal archaeological sites and sea-level depth can, of course, be reversed, and inspection of bathymetric charts used to good effect both to reconstruct coastal topography and shoreline configuration at different periods and to predict the location of new sites taking into account the most favourable locations for trapping fish (Fischer 1993, 1997).

Table 3.3 Distribution of underwater finds by archaeological period and time duration in thousands of years (kyr)

Archaeological Period	Number of Recorded Finds	Duration kyr	Recorded finds per kyr
Neolithic	159	2.2	72
Late Mesolithic	488	1.5	325
Middle Mesolithic	95	1.0	95
Early Mesolithic	22	2.6	8
Late Palaeolithic	15	3.7	0.2

Data from http://splashcos-viewer.eu

The number of dateable underwater finds shows an exponential increase from the Late Palaeolithic period through the Early and Middle Mesolithic period, reaching a maximum in the Late Mesolithic (Ertebølle) period, and reducing again in the Neolithic period (Table 3.3; Fig. 3.3). The peak in the Ertebølle period most likely represents, in part, the fact that underwater sites in this period are in shallow water where they are easily discoverable from material eroding out in shallow water or along the modern shoreline, compared to the more deeply submerged sites of earlier periods. The reduction in the Neolithic period may reflect the difficulty of distinguishing on typological grounds between Ertebølle and Neolithic artefacts from isolated finds and also the fact that more shorelines and coastal settlements of this period are above present sea level and therefore not represented in the statistics for underwater finds in Table 3.3.

3.5 Underwater Settlements

According to Table 3.2, there are some 160 Mesolithic site records with evidence of settlement activity, namely, evidence of a cultural layer with indications of in situ activity. However, this number should be treated with caution; very few of these have been excavated and the nature of the activities represented and the question of whether they are genuinely in primary position as opposed to materials redeposited by marine action remains uncertain in many cases. The small number of sites that have been extensively excavated provide important information on the quality of the evidence that has survived and insights into the variable conditions of preservation at different locations, and they are examined below.

3.5.1 Tybrind Vig

This is the most intensively investigated underwater site in Denmark, excavated over a 10-year period from 1978 to 1987 (Andersen 1980, 1985, 2013; Malm 1995) It is one of five submerged settlements within a 10 km radius on both sides of the Little Belt, the others being Kolding Fjord, Gudsø Vig, Helligkilde and Ronæs Skov (Fig. 3.5). At Tybrind Vig itself, there are, in fact, four closely co-located settlements with in situ or partially in situ material, found at depths ranging from −5 to −2.5 m below present sea level (Fig. 3.6). Three settlements are on a sheltered south-west facing shoreline at the mouth of what would originally have been an almost totally enclosed and shallow bay. They are successively shallower and younger deposits representing the landward movement of the shoreline with sea-level rise. The fourth is on an exposed beach ridge on the seaward-facing spit of land on the opposite side of the bay mouth, and represents a different type of site, most likely one with a specialised function. In addition, there are five other locations nearby with redeposited artefacts.

None of these sites represents a completely intact settlement. In all cases the original dry-land area of settlement where domestic activities and burials were concentrated has been largely eroded away, leaving only traces of the original land surface and its material remains, or redeposited materials that have been washed out from their original place of deposition. What remains in situ is the dump area just offshore of the main settlement area, where many of the by-products of daily life were originally thrown away, or where material remains were abandoned in situ, representing activities originally located offshore such as fish weirs extending out from the shoreline, or logboats.

Together, these sites represent a cumulative palimpsest of activities representing repeated use of the area and most probably year-round settle-

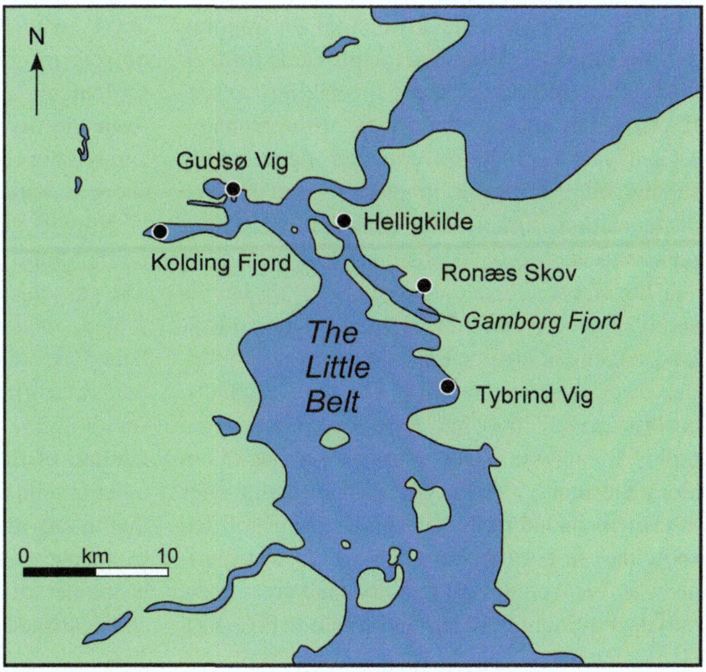

Fig. 3.5 Map of the Little Belt, showing the position of Tybrind Vig and Ronæs Skov. The three other sites shown are also submerged and contemporaneous with the other two. Like Ronæs Skov, they yielded assemblages with harpoons, sea mammal bones and ceramic blubber lamps. This and their locations suggest that they were optimally positioned to hunt porpoises during their seasonal migrations through the Little Belt by diverting them into narrow side fjords. After Andersen 2013, Figs 7.1 and 7.2, redrawn by Geoff Bailey

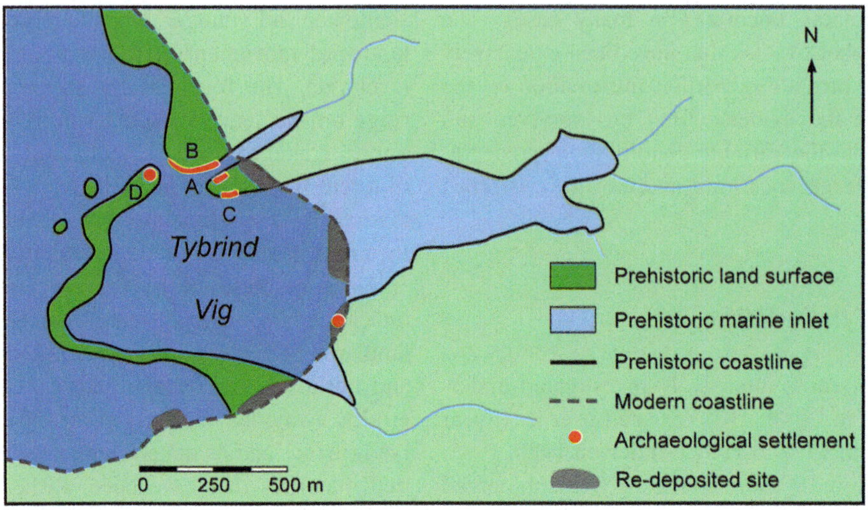

Fig. 3.6 The Tybrind Vig Cove showing the location of archaeological sites and the contemporaneous coastline. After Andersen 2013, Fig. 1.6, redrawn by Geoff Bailey

ment throughout a 1400-year-period between 7.4 and 6 ka (5400–4000 cal BC), equivalent to almost the full span of the Ertebølle period After 6 ka, activities ceased, and sea level continued to rise by a further 2–2.5 m because of isostatic submergence; indeed, this additional sea-level rise may have been a significant factor causing the abandonment of the sites by inundating the spit of land that had previously created the enclosed bay with its topographic advantages for trapping fish and creating shelter. On the plus side from an archaeological point of view, this additional sea-level rise has ensured that many of the archaeological materials have remained permanently submerged in anaerobic conditions since the time of deposition, significantly contributing to their preservation.

The largest site, and also the latest in date, is site B (Figs. 3.6 and 3.7). Here, the original dry-land settlement area on the shoreward side, judging from the concentration of stone artefacts, covered some 2600 m^2, forming an elongated outline 20 m wide and extending for about 130 m along the shore, a form typical of Ertebølle sites elsewhere including those known from uplifted shorelines in northern Denmark. Only two burials were recovered from this dry-land area, one a burial of a young woman and a child (Fig. 3.8), the other a grave with two adults. Both burials are radiocarbon dated to the Early Ertebølle and therefore belong to an earlier phase of settlement than the rest of the material discussed below. Other material had been eroded or washed away. Excavation concentrated on the marine deposits, and trenches extending over a total of 192 m^2 were excavated, representing about 10% of the total marine refuse area (estimated at 1500–2600 m^2). The material recovered includes objects originally dumped here at the time of settlement, material subsequently eroded out from the dry-land area and redeposited, and in situ remains of fish weirs, logboats and other offshore activities.

Notwithstanding the emphasis of excavation on the dump area, the full range of material culture associated with the Late Ertebølle is represented, including a flint technology of blades, flaked axes and transverse microliths (some 431 items), worked bone and antler (302) and potsherds (825) all from pointed-base vessels. The number of flints is relatively small compared to other Ertebølle sites, but that may reflect the fact that many more were discarded on dry land. Conversely the range of bone and antler artefacts is greater than usual, including axe heads and shafts shaped from red deer antler, a small num-

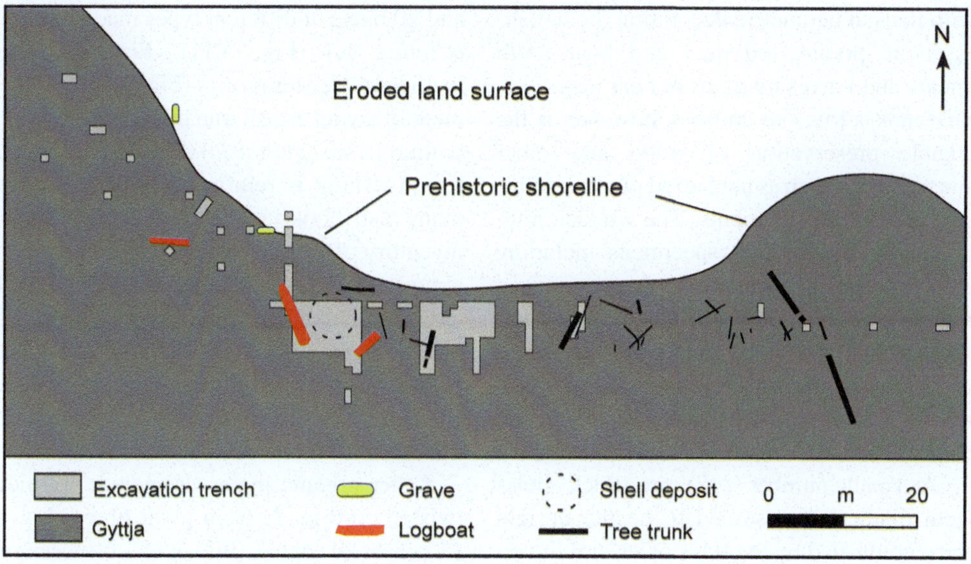

Fig. 3.7 The plan of Tybrind Vig showing the layout of the excavation trenches and other features. After Andersen 2013, Fig 1.10, redrawn by Geoff Bailey

Fig. 3.8 Tybrind Vig burial of a young woman, 14–16 years of age, in situ on the seabed and partially exposed by marine erosion before excavation, showing the skull and bones of the upper arms. The skeleton was fully extended in a shallow grave. Also present were the partial remains of an infant 0–3 months of age. Photo by Hans Dal, courtesy of Moesgaard Museum Photography Department

ber with incised geometric decoration, bone fish-hooks, bone points, red deer and boar tooth ornaments and knives made from boar tusks.

What makes this site unusual, however, is the remarkable preservation of wood and other organic materials such as uncharred plant remains, cordage and textile fragments. The wooden artefacts amount to over 500 specimens including pieces of split timber and wood chips produced during implement manufacture. The largest number are some 250 rods of hazel (*Corylus avellana*) at least 2 m in length with a sharpened tip, representing the remains of fish weirs. Their uniform dimensions indicate systematic coppicing of hazel. A small number (48) are thick poles (4–8 cm diameter) presumed to be the upright supports, confirmed by the finds of several in vertical position, and the remainder are thinner branches or withies presumed to be the horizontal members. Other artefacts include 67 ash shafts (*Fraxinus excelsior*) up to 2 m long used as spear shafts, 6 wooden axe handles (Figs. 3.9 and 3.10) and 20 bows of different types made of elm wood (*Ulmus* sp.) (Fig. 3.11). Fishing equipment includes 67 leister prongs (Fig. 3.12) and the fragment of a wicker fish trap made from woven alder withies (*Alnus glutinosa*) (Fig. 3.13). This emphasis on fishing is reinforced by the discovery of many fish hooks made from animal bone. The inventory of wooden artefacts is completed by an almost intact dugout canoe 9.5 m in length and large parts of at least three other canoes, all made of lime wood (*Tilia* sp.), and some 14 paddles or paddle fragments of different shapes made of ash wood, including examples with geometric decoration and coloured inlays.

Other organic materials include 33 pieces of twisted cord made from plant fibres (Fig. 3.14), 14 pieces of textile (Fig. 3.15), 6 pieces of rope and 11 pieces of fibrous bark material mostly of

Fig. 3.9 Axe handle from Tybrind Vig made of Pomoideae wood. The base is shaped to fit in the hand. At the other end, the wood is tapered and shaped to mount a flint chisel on the under surface. See also Andersen (2013, p. 131). Scale in cm. Photo courtesy of Moesgaard Museum Photography Department

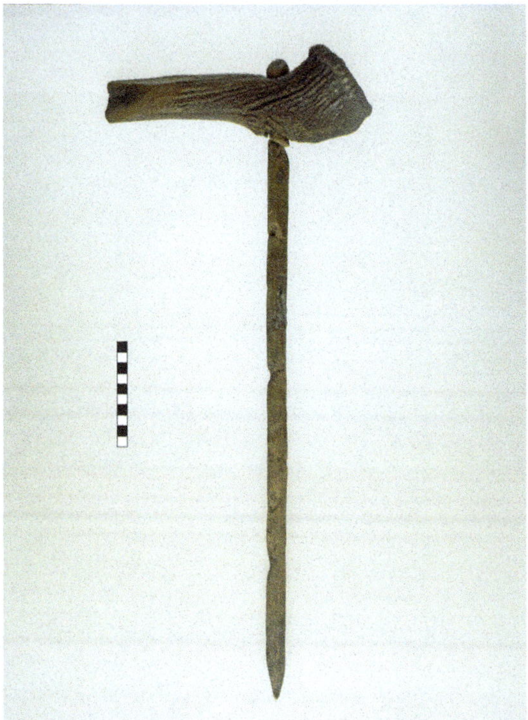

Fig. 3.10 Shaft-hole axe made of red deer antler at Tybrind Vig with a fully preserved wooden handle made from a straight hazel rod just under 2 cm in diameter. Observe that the shaft terminates in a tongue-shaped point, which might indicate a function as a digging stick. Scale in cm. Photo courtesy of Moesgaard Museum Photography Department

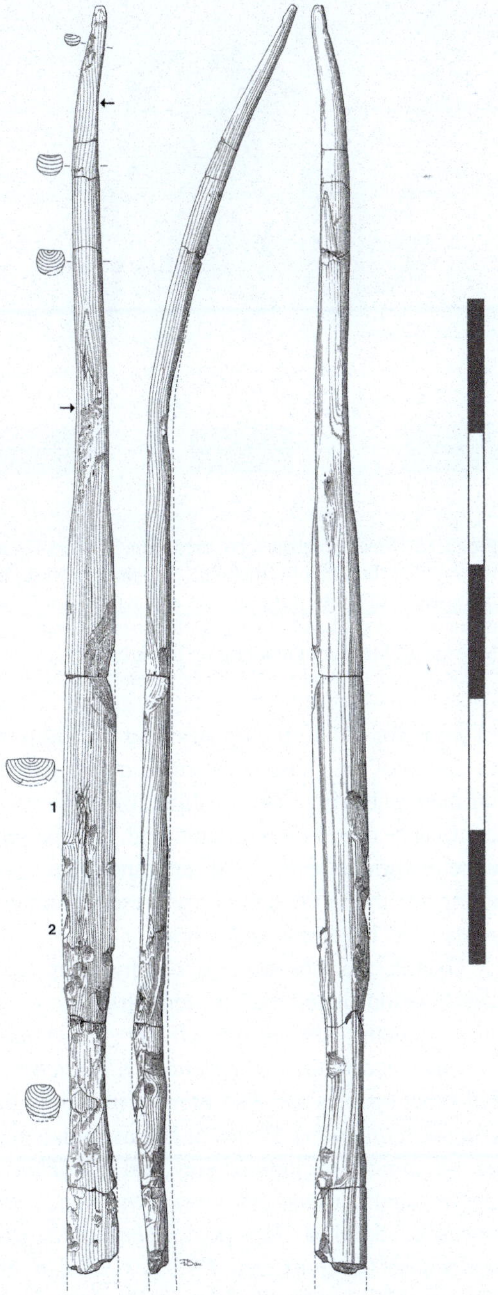

Fig. 3.11 Bow made of elm wood. It is narrowed at the waist to form a grip and has been broken just below. This is the most common type of bow with ten examples at Tybrind Vig with more than half of the bow intact. The original length would have been 1.5–1.6 m. Scale in 10-cm units. Drawing by F. Bau, courtesy of Moesgaard Museum Photography Department

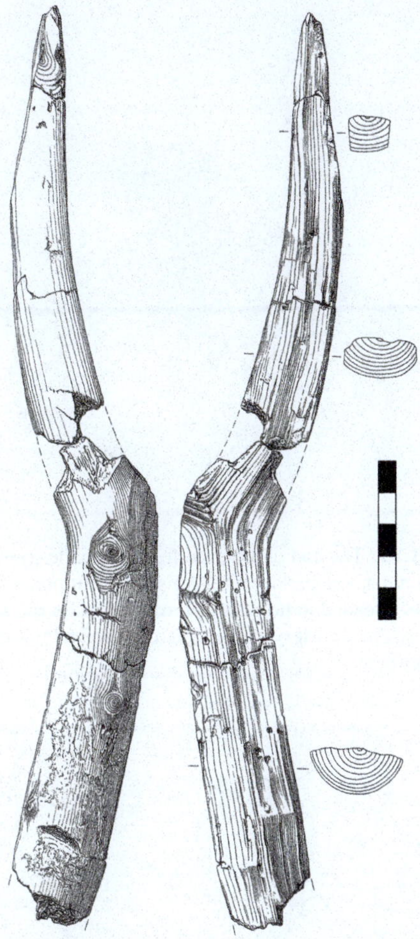

Fig. 3.12 Roughout for a leister prong at Tybrind Vig. It was made from a piece of bent hazel, split through the centre to create two pieces that are mirror images. They would have been bound together at the base and attached to a shaft to create a two-pronged leister. Scale in cm. Drawing by E. Morville, courtesy of Moesgaard Museum Photography Department

willow, pieces of tinder fungus and uncharred remains of sea-beet roots and seeds of manna grass likely used as food. There are also remains of charred food crusts attached to the inner surface of some of the potsherds incorporating evidence of fish bones and plant remains.

These organic remains are remarkable in a number of ways: for the quantity and range of materials and artefacts preserved, which include

Fig. 3.13 The end of a woven fish trap made from withies of alder and guelder rose. A fist-sized stone was found inside, intended to hold the fish trap down on the seabed. Scale in cm. Courtesy of Moesgaard Museum Photography Department

Fig. 3.15 Piece of fabric recovered from Tybrind Vig and woven with 'buttonhole stitching'. The threads were spun from fibres of grass (Gramineae) and willow (*Salix* sp.) using a Z (right-hand) twist. Scale in cm. Courtesy of Moesgaard Museum Photography Department

Fig. 3.14 The preservation conditions for plant raw materials at Tybrind Vig were extremely favourable, and many different types of cord, rope and textiles were found. The image shows a stake of hazel wood c. 22 cm long found in situ in the Tybrind Vig deposit before excavation with a piece of cord tied round it. The cord was made from plant fibres of lime or willow. Photo by Hans Dal, courtesy of Moesgaard Museum Photography Department

examples not previously known about; for details about manufacturing methods and selection of materials, particularly selection of different types of wood for different purposes; and for additional evidence of the subsistence economy, in particular the great emphasis on fishing, the evidence of equipment used in hunting on land, and the presence of various seeds, roots and other plant foods alongside the more commonly preserved charred remains of hazelnuts and acorns.

Overall, the subsistence economy comprised a combination of marine and terrestrial resources. Fish remains are dominated by cod (*Gadus morhua*) and flatfish (*Pleuronectes*, *Platichthys*), but other species are also present including eels (*Anguilla anguilla*). Bones of sea mammals indicate hunting especially of grey seal (*Halichoerus grypus*) and porpoise (*Phocoena phocoena*), with bones of dolphin (*Lagenorhynchus albirostris*) and whales (*Orcinus orca*, *Physeter catodon*) also present. Gathering of marine molluscs included the ubiquitous oyster (*Ostrea edulis*). On-land hunting focussed principally on red deer, boar and roe deer and fur-bearing animals, especially pine marten (*Martes martes*); gathering included a variety of plant foods including roots, berries and nuts. These are represented both by charred remains, notably sea beet (*Beta vulgaris* spp. *maritima*), hazelnuts and acorns, and by uncharred seeds, notably of manna grass (*Glyceria fluitans*), raspberry (*Rubus idaeus*) and strawberry (*Fragaria vesca*). On the marine side, the combination of food remains and organic artefacts confirms fishing as the major activity and clearly a key determinant of the site location.

3.5.2 Ronæs Skov, Argus Bank and Møllegabet II

About 5 km to the north of Tybrind Vig is the Late Ertebølle site of Ronæs Skov, excavated in eight seasons between 1992 and 2004 using the same excavation and recovery techniques developed at Tybrind Vig. It is a smaller settlement than Tybrind Vig, and was used over a shorter period, between 6.4 and 6 ka during the Late Ertebølle period at a time of slowly rising sea level, but in other respects shows a comparable range of artefacts, preservation conditions and subsistence practices, including fishing, hunting on land and shellgathering. It is the largest of 14 settlements located around the shores of the 3-km long Gamborg Fjord, and like its southerly neighbour was on a south-west facing shoreline opposite a narrow channel, well placed to trap fish or sea mammals entering the fjord.

The dry-land area of the settlement had largely been eroded away, and little was left apart from several stone-lined hearths, three thick vertical wooden poles and a single charred human bone suggesting cremation. One of the hearths contains branches of hazel and hawthorn used as fuel, with a tinder fungus lying alongside (Fig. 3.16), demonstrating both the quality of fragile, organic remains that are preserved in these deposits and the fact that intact features can survive despite the impact of marine erosion (Andersen 2009, p. 46).

The remaining finds, and especially the organic materials, were recovered from a shallow offshore dump area as at Tybrind Vig. Remains of fish weirs in the form of pointed hazel rods, some still in vertical position, and wooden leister prongs, some with impressions left by lashing, indicate the importance of fishing. Wooden bows, spear shafts, axe handles and remains of paddles and canoes, antler axes and a rich ceramic inventory are all present. As at Tybrind Vig, charred food crusts on potsherds give additional insights into the plant materials used in subsistence (Fig. 3.17). Distinctive wooden finds are a single wooden arrow shaft (made from guelder rose, *Viburnum opulus*) with a transverse microlithic flint held in place by pitch (Fig. 3.18) and some

Fig. 3.16 Photo of Ronæs Skov hearth (left) in situ on the seafloor with a ring of soot-blackened stones, two pieces of partially burnt wood, and to the left a tinder fungus, also shown in side view on the right. Tinder fungus is ideal material for lighting fires. Photo on left by Hans Dal. Both photos courtesy of Moesgaard Museum Photography Department

Fig. 3.17 Potsherd from Ronæs Skov with charred food crusts embedded on the inner surface including fragments of mistletoe (*Viscum* sp.). Left: general view. Right: close up. The sherd is c. 10 cm long. Photos courtesy of Moesgaard Museum Photography Department

notched sticks interpreted as cross-braces in the construction of fish weirs.

The faunal inventory is comparable to Tybrind Vig, with a slightly greater preponderance of sea mammals (seal, dolphin and whale). This together with the presence of harpoons including refuse from their manufacture and a preponderance of blubber lamps in the ceramic assemblage suggests that the occupants of the site may have taken advantage of the coastal topography not only to trap fish but also to channel sea mammals during their seasonal migrations into the fjord where they could easily be despatched. This interpretation is reinforced by other sites with similar finds on the Jutland shoreline of the Little Belt. Andersen (2009, pp. 213–214, 2013, p. 303) suggests from the dates of these finds an intensification of sea mammal hunting in this region in the Late Ertebølle period, taking advantage of the narrowest and most constricted part of the Little Belt and the 'blind alley' fjords on either side of it. A similar tradition is recorded in historical times in the same region, when up to 36 men in 12 boats engaged in winter hunting of porpoises and dolphins, producing catches of up to 1500 animals (Andersen 2009, p. 214). Even at a conservative estimate, that is enough calories to supply the needs of 500 people for 6 months of the year. The Ertebølle hunt may not have been so intensive and the animal fat was important for heating and lighting as well as for storable food, but these figures nevertheless underline the incalculable importance of sea mammals and the presence of coastal topography that facilitated their capture.

Two other sites to mention here are Argus Bank and Møllegabet II. Both sites are of interest in demonstrating the preservation of intact features on the dry-land surface of the original settlement as well as the presence of offshore dump areas in marine sediments with flint, wood and antler artefacts.

Argus Bank had long been known as a concentration of finds extending over an area of about 50 × 100 m at a depth of 4–6 m bpsl as a result of gravel extraction, and a more systematic rescue survey and a small trial excavation were carried out between 1984 and 1986 (Fischer 1987; Fischer et al. 2007a). The majority of the finds (some 3000 artefacts and faunal remains) were recovered during gravel extraction rather than by systematic techniques of archaeological recovery but nevertheless appear to represent a homogeneous assemblage of the Kongemose period. Like many other coastal sites, the Argus site was located on a promontory overlooking a narrow fjord ideal for trapping fish. The main interest of the site is the recovery in the excavated

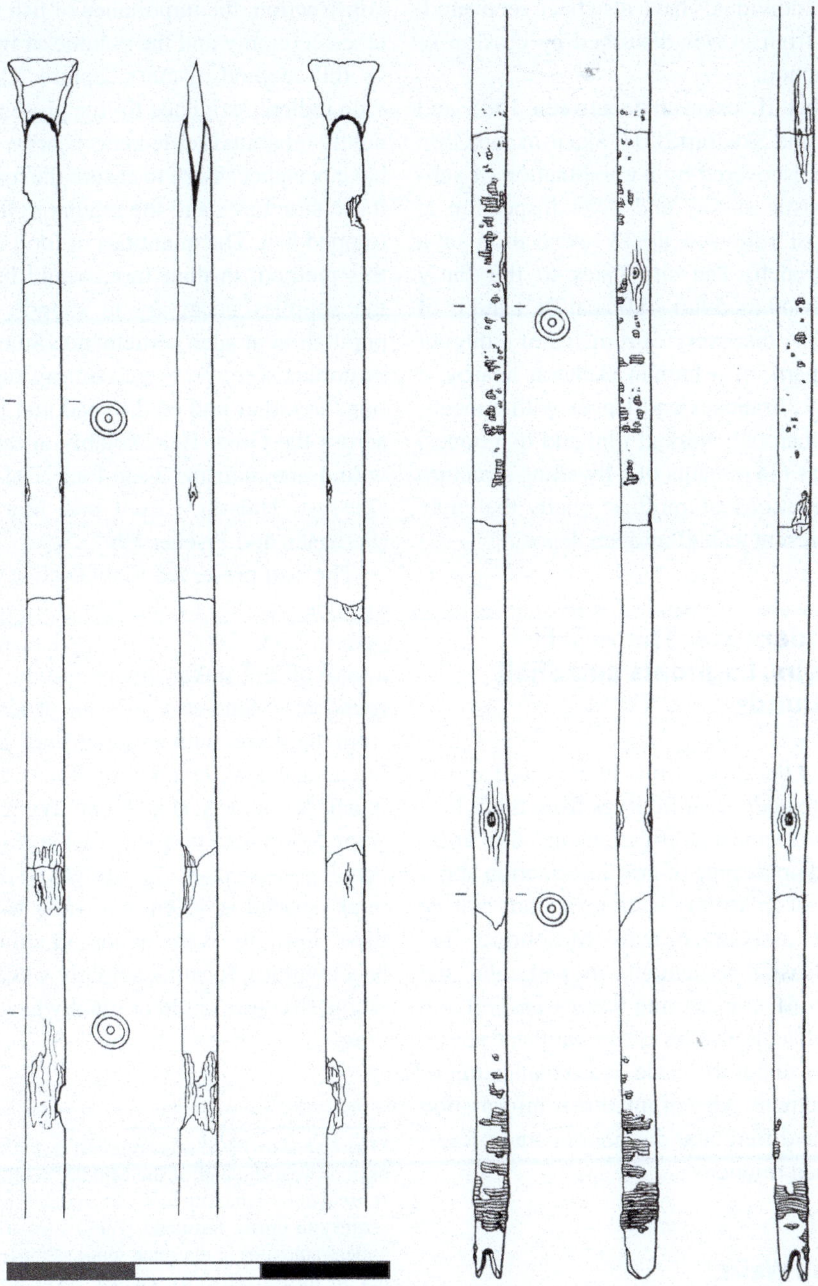

Fig. 3.18 Drawing of a complete arrow shaft from Ronæs Skov, shown in two parts and from three angles. The shaft was found in situ with a transverse arrowhead of flint fixed at the tip with a black pitch-like substance. Traces of a similar material are present at the nock end, presumably for the attachment of feathers. Scale in 10-cm units. Drawing by O. Svendsen, courtesy of Moesgaard Museum Photography Department

area below a sand layer of an intact stone-lined hearth, about 0.5 m in diameter, with charred branches, ash, burnt flint and charred food remains including hazelnuts, pips of raspberry and blackberry and fish bones, with a date of 5970–5570 cal BC. Redeposited human bones of at least four individuals have also been recovered, originating from graves disturbed by erosion or gravel extraction.

Møllegabet II, excavated between 1988 and 1993 (Grøn and Skaarup 2004; Skaarup and Grøn 2004), was discovered by a combination of sub-bottom acoustic survey and diver inspection at 4.5–5 m bpsl followed up by excavation of a 6 × 4 m trench. The site dates to the Early Ertebolle period (c. 5400–4800 cal BC) and is of interest for the discovery of remains of a dugout canoe and parts of a human skeleton nearby, a layer of bark, branches and leaves with concentrations of charcoal, worked flint and fish bones, interpreted as the remains of a dwelling structure and concentrations of mollusc shells that may perhaps represent a shell midden deposit.

3.6 Underwater Themes: Fish Weirs, Logboats and Shell Mounds

The results of the excavations summarised above, together with additional findings from more limited or more recent investigations at other sites, have opened up a range of new information about the nature of maritime activities and marine exploitation associated with Mesolithic and Neolithic coastal settlement, in particular the importance of fish weirs and water transport and what the archaeological evidence for these activities indicates about the scale and organisation of coastal settlement. Also of interest is information regarding the differential distribution and preservation of shell middens.

3.6.1 Fish Weirs

The wooden stakes and wattles used in the construction of fish weirs are some of the most numerous of the wooden artefacts preserved at these underwater coastal settlements. They are of special interest in demonstrating the considerable investment of communal labour in their construction and maintenance, skills in forest management required to produce the wood for their construction, the importance of fish in the subsistence economy and the volume of food produced by this method of mass capture. They also provide indirect evidence for the presence of permanently inhabited settlements nearby, which would have been necessary to ensure the maintenance of these structures and the regular collection of the trapped fish. The quantities of food obtainable by this method, in their turn, would have provided the supplies necessary to support the resident population at such settlements. Some of the best examples were recovered during rescue excavations at either end of the road and rail link built across the Great Belt, notably at three locations within the narrow 2-km-long Halsskov Fjord: Oleslyst, Halsskov itself and Margrethes Næs (Pedersen and Fischer 1997).[4]

The best preserved is at Oleslyst with two sets of finds dated to the Early and Middle Neolithic period, 3500–2600 cal BC (Pedersen 1997). Here a total of 259 stakes driven into the ground were found to extend in a line for at least 30 m out from the shore, with a shorter line of stakes forming an acute angle at the seaward end. An almost intact panel was also found lying on its side, some 5.5 m long and 1.7 m wide (Fig. 3.19). The panel consisted of 12 upright stakes, 2–4 cm thick, spaced at 45 cm intervals. Woven through these uprights were about 50 thinner, flexible hazel withies, forming a tightly woven mesh. The whole structure could originally have been 200 m long.

[4]The Halsskov Fjord was originally a shallow marine inlet until at least the end of the Middle Neolithic period, but subsequently silted up and was drained in the nineteenth century to create farmland. These sites are not therefore underwater sites in the same sense as those discussed earlier in this chapter. We include them here because the remains were originally deposited on the seabed in shallow water at the shore edge and have remained encased in marine sediments ever since. They are also clearly coastal in nature and are integral to the wider pattern of sites investigated as part of the Great Belt project, which include fully underwater remains. They are a good example of the arbitrary nature of the division between terrestrial and underwater sites and the need sometimes to integrate material on both sides of that the boundary according to the context, as discussed in Bailey et al. (Chapter 1).

Fig. 3.19 The Oleslyst fishing weir panel, dated to the Early Neolithic. This panel was discovered lying flat on marine sediments and largely intact, except for damage at the left-hand side caused by a machine during excavation. There are 12 uprights consisting of poles of lime or hazel wood 2–4 cm in diameter with pointed bases for driving into the seabed. More flexible hazel withies, 1–3 cm in diameter, were used as the horizontal members. The longest is 4.5 m long. These were woven through the uprights to create a tightly meshed panel. Two of the uprights are forked poles, providing additional support for the horizontal withies, and holding the ends of the withies at the right-hand side in a tight grip to facilitate attachment to the neighbouring panel. Isolated wooden stakes representing the uprights of other panels were also found further out from the shore and their distribution shows that these structures were built out for many tens of metres into the sea, with a shorter section constructed at an acute angle at the seaward end to create, in plan, a V-shaped trap, culminating in a pot-shaped fish trap of woven wickerwork. No such trap was found in situ at Oleslyst, but fragments have been found elsewhere (see Fig. 3.13 for an example from Tybrind Vig). After Pedersen (1997, Fig. 6, p. 127), original by L Johansen and L Pedersen, redrawn by G Bailey

Less complete remains of structures interpreted as fish weirs and dating to the Ertebølle period were found at Halsskov, with two examples dated, respectively, at c. 5400 and 4800 cal BC, and at Margrethes Næs, dated at c. 5700 cal BC.

Another well-preserved example is on the island of Nekselø, offshore from the north-west coast of Zealand, with evidence of several different stages of weir construction. The full range of radiocarbon dates suggest a time span from the Late Ertebølle through to the Middle Neolithic with radiocarbon dates of c. 4490–2900 cal BC (Pedersen et al. 2017). However, only two out of 18 radiocarbon dates are Late Ertebølle, and the others cluster in the Early Neolithic and Early Middle Neolithic period, c. 3500–2900 cal BC. Here a line of stakes can be traced out from the shoreline over a distance of 250 m to a depth of 4–5 m bpsl, and there are indications that it may have extended further out again (Fischer 2007). A 40-m length of wattle panels was found lying in horizontal position, showing a construction of vertical uprights some 5 cm in diameter with narrower and perfectly straight horizontal hazel withies (Fig 3.20). Estimates for the construction of the 250-m long structure suggest that it would have needed hundreds of vertical poles up to 6 m long for the main supports and some 6000 straight hazel withies up to 4 m long for the rest of the structure. This testifies both to the labour-intensive requirements of the construction and to the extent of the coppiced hazel woods that would have to have been maintained to provide the raw materials.

This mesh of hazel woodwork was so tightly constructed as to be almost impenetrable, and Fischer (2007, p. 63) suggests that the design was intended to facilitate the mass capture of European eels, which pass through the Danish Straits in huge numbers during their autumn spawning runs from the Baltic to the Atlantic. Similar traps made from hazel wood were recorded in use in the nineteenth century. The resulting large quantities of eel could have been easily stored for long periods to provide important food reserves for the winter months.

In total, remains of fish weirs are recorded at 17 sites extending through the full chronological span of Mesolithic and Neolithic coastal settle-

Fig. 3.20 Photo of part of a panel from the Nekselø fish weir, showing closely spaced horizontal hazel withies 2–3 cm thick. Photo by Anne Marie Eriksen, reproduced with permission of The National Museum of Denmark,

ment, from Kalø Vig on the late Maglemose–early Kongemose boundary at 6450 cal BC through to Ølby Lyng at c. 2900–2350 cal BC. This combination of evidence indicates a 4000-year tradition of construction. However, the Mesolithic examples appear to be simpler and cruder structures made only from hazel wood, whereas the Neolithic fish weirs appear to be larger and stronger with more tightly woven panels and the use of several different types of wood, mainly lime but also including hazel, ash, maple (*Acer* sp.), elm and oak (*Quercus robur*). This suggests a development of new and more elaborate construction techniques in the Neolithic period compared to the Ertebølle.

The absence of remains in subsequent millennia may be due to the reduction in accumulation of fine sediments that provide a preserving medium for wooden artefacts, rather than to the abandonment of such fishing practices (Fischer 2007, p. 60).

3.6.2 Water Transport

Remains of dugout logboats (canoes) and paddle blades have been recovered from 19 Mesolithic (Ertebølle) sites in Denmark (Andersen 2013, p. 200). Those from Tybrind Vig represent one of the largest assemblages including an almost complete example of a logboat and provide new details about size, construction methods and probable functions (Andersen 2012, 2013).

The complete canoe is 9.5 m long and c. 0.5 m wide, with sides that are 2–3 cm thick (Fig. 3.21). When first discovered it was completely flat, but in its original state the sides would have been 30 cm high. It was made from a single trunk of a lime tree that must originally have weighed over three tonnes. It was hollowed out by chopping and wedging, reducing the finished weight to about 200–300 kg. The stern was open, presumably with a removable bulkhead, and a hearth next to the stern, indicated by an oval deposit of sand and clay, suggests the possibility of using flares at night to attract fish. There was a smaller hearth indicated by charred material near the front, perhaps indicating use for on-the-spot cooking, and a large 30 kg stone in the body of the boat which is thought to have acted as ballast or a stabiliser (Fig. 3.21). Experimental work suggests that it would have taken 1–4 weeks of labour for two skilled people to shape such a logboat. The parts of several other canoes show evidence of splitting and the presence of regularly spaced holes around the cracks that indicate the attachment of patches to repair the damage (Fig. 3.22). Similar examples of extensive repair work are well illustrated by the more fragmentary remains of a canoe at the Margrethes Næs site (Myrhøj and Willemoes 1997, p. 160). All of this indicates the investment of time and effort involved in construction and maintenance.

Some 14 paddles or fragments of paddles are present at Tybrind Vig and are made from split timbers of ash to shape the handle and the blade in a single piece. The paddles are about 1 m long

Fig. 3.21 The largest logboat at Tybrind Vig, with a length of 9.5 m. Upper photo: as it looked after removal from its underwater location to dry land. In its original position, it had become flattened, and the sides had collapsed inward. The view is from the stern, with a grey hearth area in the foreground near the stern, and a large ballast stone towards the middle of the boat. Lower image: an artist's reconstruction of the boat shown in the photo and a smaller logboat recovered from the partially submerged site of Lystrup Enge (north of Aarhus), with two hearths. Drawing by O. Svendsen, courtesy of Moesgaard Museum Photography Department

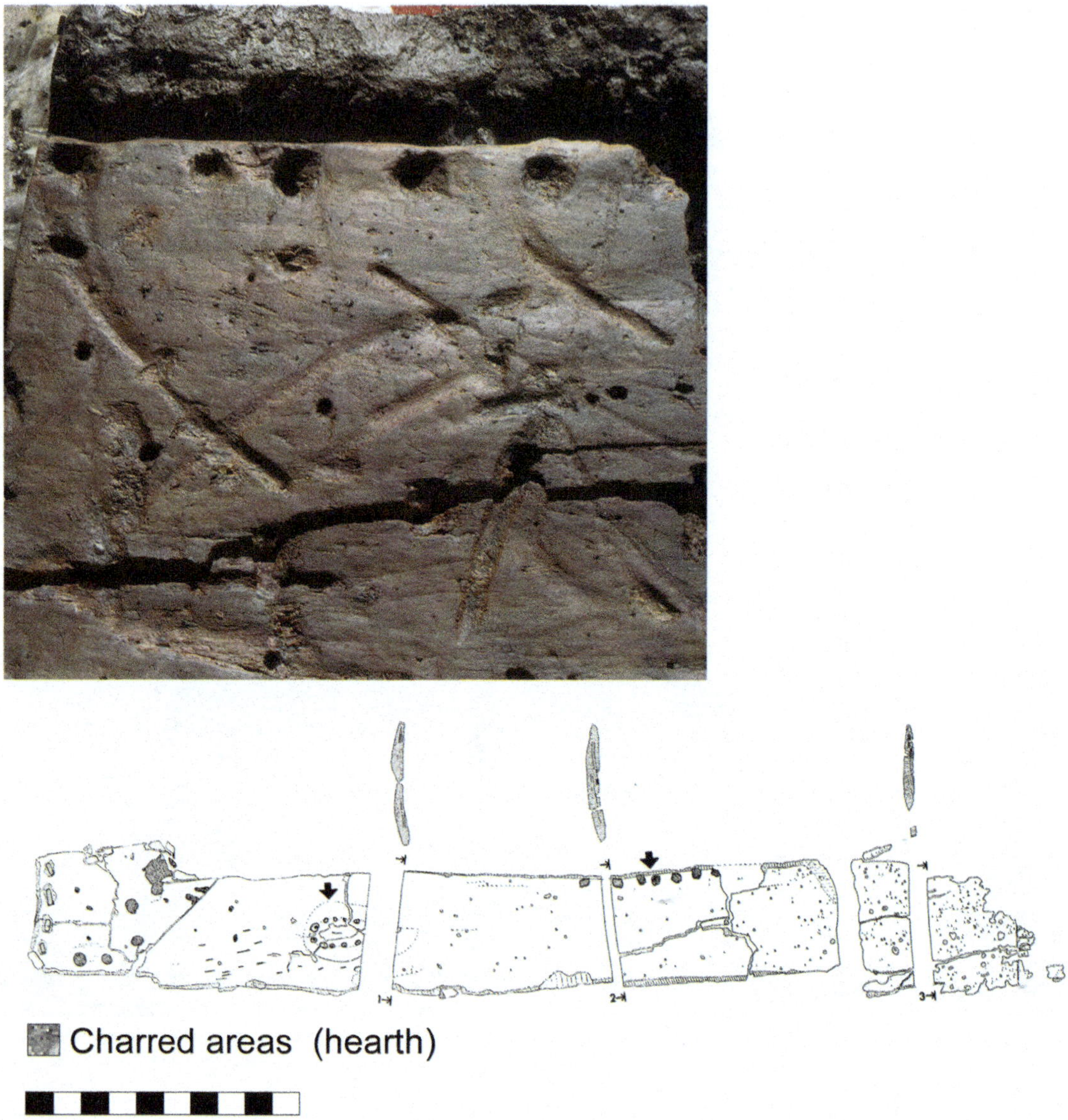

Fig. 3.22 A smaller logboat from Tybrind Vig showing evidence of repair. Lower: the plan of the boat, with the stern on the left and evidence of hearths near the stern. Holes made to assist the repair of cracks are shown by arrows and include the repair of an oval hole originating from a knot in the wood near the stern and a split in the middle section of the boat. Fibres in the holes near the stern suggest that a patch was sewn onto the damaged section. Upper: photo of the holes alongside the split seen on the interior of the boat. Scale in 10-cm units. Plan drawn by J. Dencker and O. Svendsen, plan and photo courtesy of Moesgaard Museum Photography Department

with a stout handle and an oval- or spade-shaped blade not larger than about 20 × 25 cm (Fig. 3.23). Some of the paddle blades are richly decorated with carved geometric patterns of lines and dots in different designs, some of which resemble a human face (Fig. 3.24). Some show colouring matter pressed into the inlaid surface. Paddles from other sites show different designs. As visible items, especially when viewed from a distance, these variations in design may have acted as important symbols of ownership and communal identity. The more richly decorated paddles may have served for ceremonial occasions as much as for daily use.

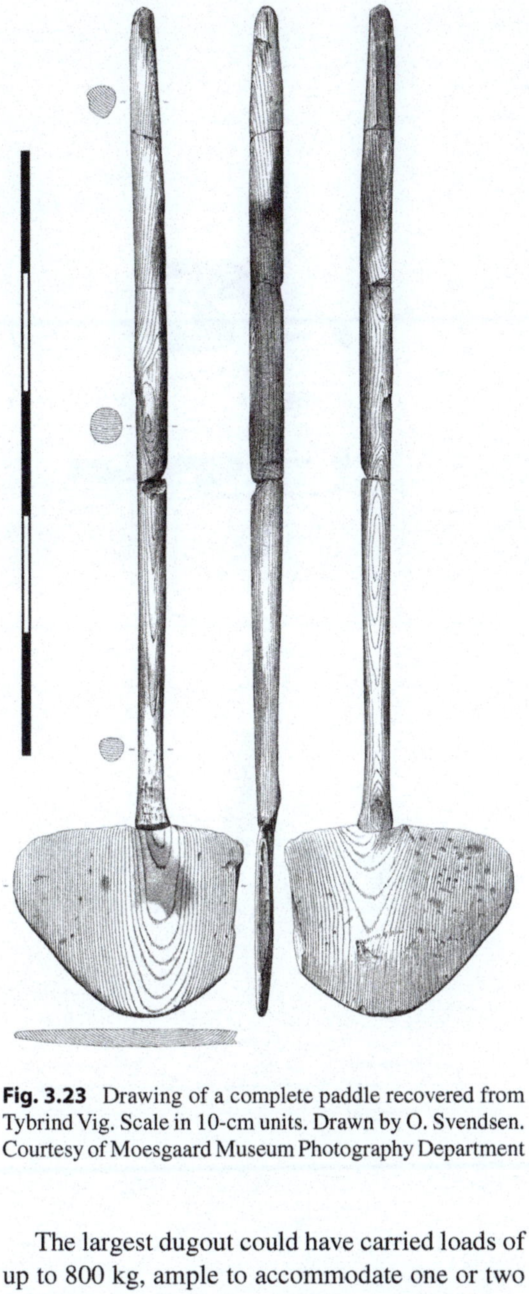

Fig. 3.23 Drawing of a complete paddle recovered from Tybrind Vig. Scale in 10-cm units. Drawn by O. Svendsen. Courtesy of Moesgaard Museum Photography Department

The largest dugout could have carried loads of up to 800 kg, ample to accommodate one or two families of up to eight people and their belongings, or to carry large loads of fish or other resources. They are well suited to short-distance movements in the sheltered waters of the Danish archipelago where they must have played an important role in fishing, sea mammal hunting, the maintenance of fish weirs and the transportation of the catch to the shore, as well as in social communication. They could also be used in more open sea conditions as shown by an experimental crossing of the Øresund in a dugout canoe, which took 5 h to cover the 15-km distance (Christensen et al. 1979).

The 'boat burial' at Møllegabet II, claimed as evidence of a body in a boat with a bark covering, also suggests the possibility of a ritual use. Whether the Møllegabet example is an authentic boat grave, or the fortuitous co-location on the seabed of human bones and the remains of a boat remains unclear (Skaarup 2004, p. 35; Andersen 2013, pp. 202–203), but better examples are available from the inland bogs of the Neolithic period. Also, stains in the soil around a few graves at Ertebølle coastal sites on dry land have been interpreted as evidence of boats used as containers or coverings for the corpse. The idea of sinking a canoe containing the corpse of an important person covered in a bark shroud remains a plausible if unproven Mesolithic ritual.

3.6.3 Shell Mounds

One of the defining characteristics of the Late Mesolithic Ertebølle period is the shell mounds or køkkenmøddinger (kitchen middens)—also known as 'shell middens' or 'shell-matrix sites' (Claassen 1998)—which form such a visually impressive feature of many coastal settlements on the uplifted Littorina shorelines of north-east Denmark (Andersen 2000). These are deposits dominated by shells of marine molluscs discarded as by-products of food consumption, principally oysters (*Ostrea edulis*) but also including other species such as cockles (*Cerastoderma edule*) and mussels (*Mytilus edulis*). The largest middens are several hundred metres long, 30–40 m wide, and 2 m thick, with a volume of up to 8000 m^3. Many are smaller. Mounds of varying size often form clusters around fjords and inlets. In total, there are over 500 Mesolithic shell middens in north-east Denmark, with the majority dating to the Ertebølle period between c. 4900 and 3900 cal BC. The oldest date back to c. 5600–5400 cal BC. Others are known from Neolithic and later periods albeit with cockles or

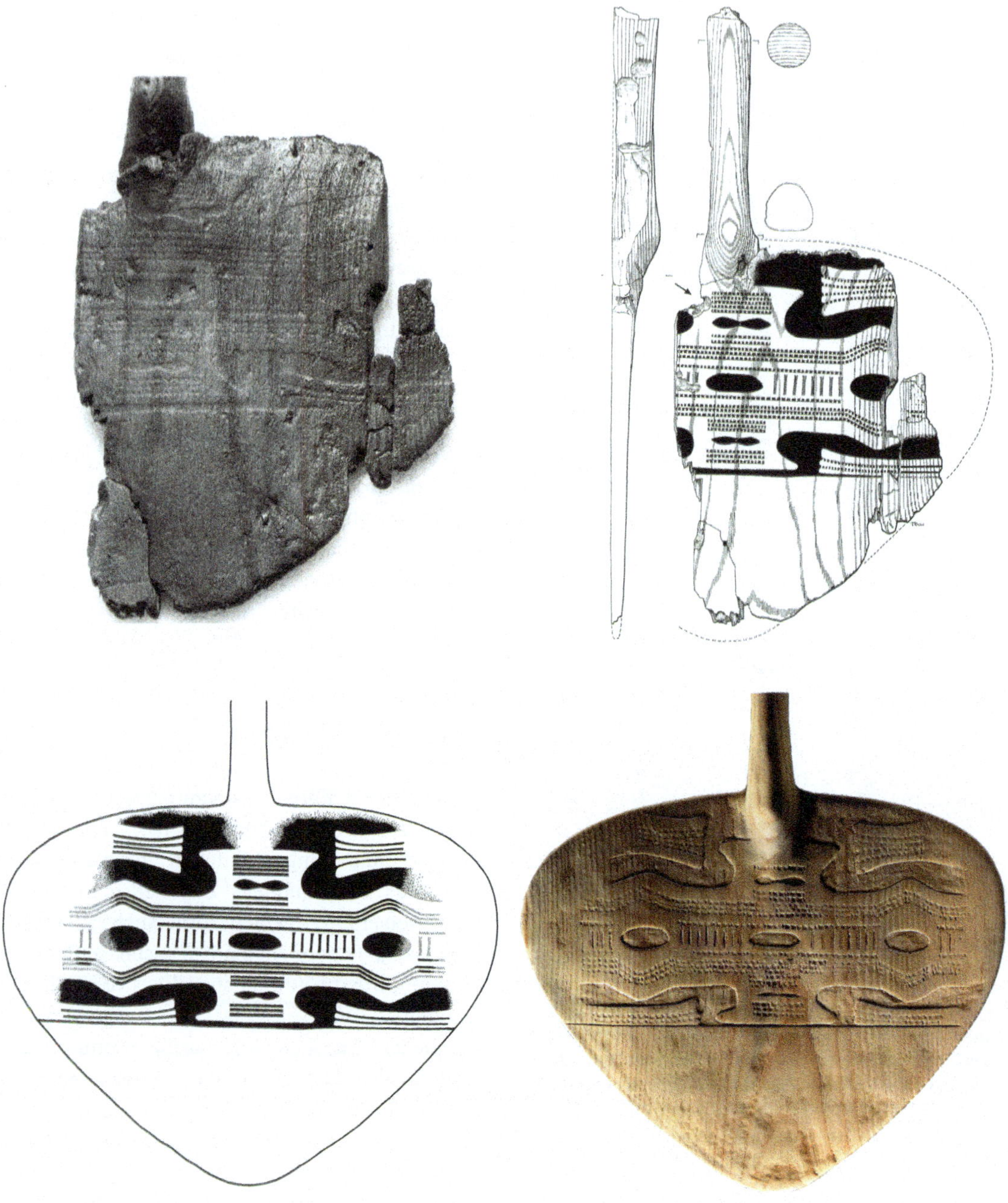

Fig. 3.24 Photograph and drawings of a damaged paddle blade from Tybrind Vig. Top left: the original. Top right: drawing of the decorative patterns. Bottom: a modern reconstruction. Scale in cm. Drawing by F. Bau, courtesy of Moesgaard Museum Photography Department

mussels as the dominant mollusc species rather than oysters (Andersen 2007).

It is now well understood that these Ertebølle shell mounds are coastal settlements with a wide range of artefacts and other food remains including marine and terrestrial vertebrates and that the oysters contributed a relatively small component of the subsistence economy, though an important one, especially in the late winter and early spring when other food supplies were in short supply

(Bailey 1978; Milner 2001, 2002). Moreover, contemporaneous coastal settlements lacking shells are just as common in northern Denmark, and in some local fjord regions they are more common than the shell mounds, the presence or absence of shells being determined by the patchy distribution and local availability of the natural shell beds (Andersen 2000, p. 365). Conversely, on the submerged shorelines of southern Denmark, despite the many hundreds of underwater finds that have been recovered, most of them dating to the same time range as the Ertebølle shell mounds of the north, shell middens are almost unknown or subject to uncertainties over their cultural status.

These large, mounded shell deposits are a highly visible marker of coastal settlement and intensive use of marine resources, and similar deposits appear in their hundreds of thousands across the world at about the same time as the Ertebølle sites or slightly earlier in the mid-Holocene. Comparable deposits are almost unknown from earlier periods, either because earlier populations exploited marine resources less intensively, or because the sites have been drowned and destroyed. Whether similar shell mounds accumulated at earlier periods of low sea level, and, if so, whether they could have survived the erosive power of waves and water currents during inundation by sea-level rise, is thus a question of wide interest. Despite some searches for underwater shell mounds in other parts of the world, and much speculation about whether they could survive the effects of inundation (Bailey 2014; Faught 2014; Nutley 2014), very few have been identified, a notable exception being the recently discovered Econfina site in the Gulf of Mexico (Cook Hale et al. 2018). The Danish underwater evidence therefore provides an important opportunity to investigate this problem.

In fact, edible mollusc shells are present in at least three of the excavated underwater sites discussed above, forming in some cases what looks like a shell midden deposit. However, a major difficulty is determining whether these are genuine in situ shell midden deposits, as opposed to material that has been eroded from the landward side of the shoreline and been redeposited in the dump area where other materials were disposed of, or natural death assemblages of shells—oysters in particular attach themselves to the shells of dead molluscs, resulting in time in quite thick shell banks. There is the additional possibility that shells that have died from natural causes have become mixed with food debris.

At Tybrind Vig, for example, two concentrations of shells are present in and amongst the other remains (see Fig. 3.7; Andersen 2013). The larger deposit covers an area of 5 × 10 m and is 50 cm thick, dominated by oyster shells, along with shells of other edible species. The deposits include some worked flints and animal bones, pieces of worked wood, small twigs, charcoal, bark and hazelnuts, all mixed in with sand and gravel. However, the investigators concluded from the species and age composition of the molluscs, and from the fact that the bivalve shells were often present as paired and closed valves aligned horizontally, that this was a natural shell deposit (Andersen 2013, p. 44). At Ronæs Skov, one of the deposits is a thick layer of oysters and other shells, which Andersen (2009, pp 33–34, p. 210) interprets as an amalgamation of a redeposited shell midden mixed with shells that died naturally. Similar ambiguities apply to the interpretation of the shell deposits described as middens at Møllegabet.

The matter is further complicated by the possibility that oysters may have been less abundant on the now-submerged shorelines of southern Denmark because of progressively reduced salinity and tidal range with increasing distance from the North Sea. Oysters are sensitive to lower salinity, and some indication of this is the evidence for larger and thicker oyster shells in the most northerly shell mounds or at the mouths of fjords, compared to those further south or in inner fjords. However, a wide range of biological indicators demonstrates that the tidal range in the south was greater than today (Petersen 2013) and that the widespread replacement of oysters by cockles as the dominant food species in the upper levels of many Ertebølle shell mounds in the north probably reflects regional or local infilling of bays and estuaries by soft sediments rather than reduction in salinities (Lewis et al. 2016).

In short, absence of sufficient quantities of available molluscs seems the least likely hypothesis to account for the rarity of underwater shell middens in the south. The causes must lie elsewhere. Regional variation in food preferences is a possibility but we know that at least some molluscs were certainly collected for food in the south, as demonstrated by the evidence of heating on some of the Tybrind Vig shells. Another possibility noted by Fischer (1995, p. 382) is that during a period of rapidly rising sea level, the shoreline would not remain in one place for long enough to allow the stabilisation of shell beds and accumulation of sufficient shells in one place to create a thick deposit. But this cannot apply to the Tybrind Vig site, where the shoreline scarcely moved position during the 1400-year period of occupation. The most likely remaining hypothesis is that shell midden deposits have become degraded and dispersed by marine erosion or are otherwise hidden beneath overlying marine sediments.

Perhaps the best case for a submerged shell midden comes from the recently excavated site of Hjarnø Sund (Fig. 3.25; Skriver et al. 2017; Astrup et al. 2019). Here shell deposits up to 1 m thick have been discovered beneath gyttja and eroding out at the surface of the seabed (Fig. 3.26). The accumulation of oyster shells looks like a typical midden deposit as known from terrestrial examples on the uplifted shorelines of north-east Denmark rather than a natural shell bank, and the deposits contain worked flints, abundant fish bones, mammal bones and concentrations of charcoal indicating a settlement area with a hearth. Investigations are ongoing at this site and it should be possible to develop suitably sensitive techniques for discriminating between cultural and natural shell deposits found in underwater settings.

Fig. 3.25 Aerial view of the Hjarnø underwater site. Upper: general view with the location of the site marked by a red arrow. Lower: close up of site during excavation, showing pontoon with equipment and two divers excavating under water. Photos courtesy of Peter Moe Astrup and Jonathan Benjamin

3.7 Conclusion

The number of underwater finds of Mesolithic age from Danish waters is by any standards extraordinary. Where detailed work has been undertaken, it is clear that almost the whole shoreline represents a palimpsest of archaeological sites and materials of different ages, different types and in varying states of preservation distributed over many kilometres. Notable examples of underwater site density are the Gamborg Fjord (14 sites within a 2-km radius) and the Tybrind Vig cove (seven sites within a 1-km radius). If we include the Halsskov Fjord referred to earlier, the site density is nine within a 1 km radius. This compares with 33 sites within a 5-km radius on the uplifted shoreline around the somewhat larger Bjørnsholm Fjord in northern Jutland (Andersen 2000, p. 365). Some of the underwater finds are single artefacts, dropped or lost during daily subsistence activities, others are permanent settlements with a wide range of resources and year-round settlement associated with fish weirs, others again are smaller sites used for particular purposes such as flint-knapping or mooring of dugout canoes, and some are collections of redeposited material disturbed

Fig 3.26 Section through an underwater shell midden deposit at Hjarnø. Photo by Peter Moe Astrup

by marine erosion. Similar conditions are found in fjords and inlets around many coastlines of the Danish archipelago. It is little surprise that the number of underwater prehistoric sites is estimated to be many thousands, and that no comprehensive or accurate inventory has yet been completed. Moreover, recently undertaken work driven by research questions rather than rescue operations underlines the fact that much of what is already known about or recorded represents only what is most obviously visible and most easily reached from the shoreline and in shallow water, or most visibly at risk from destruction, and that much more may remain hidden beneath marine sediments (Uldum et al. 2017).

3.7.1 Preservation Conditions Revisited

This brings us back to the question raised at the beginning of this chapter about the factors that have contributed to this extraordinary richness of underwater material, how it has altered archaeological interpretation and whether similar conditions and concentrations of finds might be found on coastlines in other areas of Europe or beyond. In addition to the general factors noted in the introduction, two other features stand out from the detailed examination of individual sites presented above. The first is the dense concentration of population and settlement along the shoreline, a fact that reflects the very high biological productivity of an archipelago environment with straits and fjords, mixing of currents and high levels of marine fertility, sheltered conditions for sea travel and shallow channels and bays with narrow entrances highly favourable to the trapping of migratory fish and sea mammals. In fact, the Danish Straits as a whole represent a large-scale funnel or trap between the Kattegat to the north-west and the Baltic to the south-east, providing unusual conditions of ecological diversity and productivity of marine resources and ease of access to them, especially to the whales, fish and eels that would have migrated in large numbers between the Baltic and the North Sea in search of seasonal feeding and breeding grounds. Similar conditions are present on other coastlines in Europe and elsewhere or may have existed at earlier periods when sea level was lower than present, though perhaps nowhere with quite the same configuration and concentration of opportunities for human exploitation. There is, therefore, every reason to expect comparable evidence in other areas with similar topographic and ecological conditions, subject only to the vagaries of post-depositional destruction and visibility.

A second factor is the shallow offshore topography and patterns of sedimentation associated with Danish sites and patterns of refuse disposal or in situ activity that resulted in the primary deposition of large quantities of material in shallow water. Excavation sections typically show an upward fining of sediments from coarser sand and gravel to fine clay or gyttja, reflecting a slight but slowing rise in sea level. Gyttja provides an ideal sediment for ensuring the rapid burial of discarded materials and their encasement in anaerobic conditions, which are maintained indefinitely as long as the material remains submerged and permanently waterlogged.

These soft sediments and anaerobic conditions have preserved large numbers of wooden artefacts. These include fish weirs and other equipment used in fishing, a factor previously underestimated, large dugout canoes with a considerable load-bearing capacity of high importance in facilitating subsistence activities at the coast and social interaction and communication in an otherwise heavily forested landscape, and new aspects of art.

However, preservation is far from perfect. Even material rapidly buried in gyttja remains vulnerable to piddocks, molluscs that bore holes into submerged wood and stone, and wooden materials can be compressed and damaged by the pressure of overlying sediments. Also, large areas of these underwater settlements have been damaged by marine erosion during the process of inundation particularly the dry-land areas on the landward side of the shoreline, with dispersal, redeposition and destruction of material. This emphasises the point that, even in the relatively gentle marine conditions of these Danish shorelines, sea-level rise is fundamentally a destructive process. Yet, not everything is destroyed. The hearths at Ronæs Skov and Argus Bank have been noted above, and there are similar examples from other sites, such as Halsskovholmen and Horsekær in the Great Belt (Pedersen et al. 1997). It is likely that some of the small items and more fragile remains found in gyttja deposits such as pieces of nutshell or charcoal and traces of hearths represent in situ activities originally carried out on the gyttja surface at low tide, which were then rapidly covered by sediment (Andersen 2013, p. 69). Traces of a possible dwelling structure have also been identified, which may represent a short-lived structure located on the foreshore. Graves on the original land surface with their human skeletons sometimes survive intact, perhaps because they were already partly protected by burial in the subsoil. The chances of survival on these land surfaces must depend on how rapidly material was covered by protective marine sediments, but the conditions that favour the preservation and discovery of these localised pockets of survival admit of no easy generalisations. The taphonomic processes of archaeological site formation and deformation in underwater settings are nothing if not complex.

An additional insight into preservation conditions is provided by the effect of eel grass (*Zostera marina*), which can form an extensive cover of vegetation on the seabed in shallow water and consolidate the underlying sediments. Over recent decades, the eel grass has been affected by disease and begun to disappear, making the underlying sediment more vulnerable to erosion and in the process exposing much of the underlying archaeology and bringing it to archaeological attention (Fischer 2004; Andersen 2013, p. 7). As this has consequences beyond archaeology, these changes are subject to intense research with reference to future management of the archaeological resource as well (Höffle et al. 2011). This process highlights the double-edged effect of sediment cover, which can both protect the original land surface but also hide it from view, and the equally double-edged effect of sediment removal, which can reveal the presence of submerged archaeology but also accelerate its destruction.

Overall, the preservation and investigation of the submerged Danish coastal landscape is exceptional in terms of the concentration and geographical extent of the archaeological material present, but there is no reason to suppose that similar conditions cannot be found elsewhere or on more deeply submerged shorelines, though they may be more isolated and patchier in their distribution.

3.7.2 Future Challenges

One of the positive outcomes of underwater research in Denmark over the past four decades is the demonstration that underwater archaeological deposits can be excavated with the same thoroughness and accuracy as archaeological sites on dry land, with techniques suitably adapted to working with divers under water (Dal 2013). Even so, despite the many records of underwater sites in Denmark, very few have been subjected to test pitting let alone to full-scale excavation as at Tybrind Vig. Moreover, work has mainly concentrated on shallow-water sites and therefore by definition those in the most recent (Late Ertebølle) period of the Mesolithic sequence. This risks creating an impression of increased population growth and intensification of maritime activities through time that may be quite false.

Meanwhile, national funding of necessity is focussed on development-led rescue work associated with offshore constructions or sites obviously at risk of erosion, but without the resources to sustain post-excavation analysis and publication to the same standards as research-led investigations, except in rare cases such as the Great Belt project (Pedersen et al. 1997), further compounding the sampling bias and the backlog of incompletely recorded material (see Uldum et al. 2017, pp. 69–70).

Traces of fish weirs, offshore refuse areas and wooden materials have been identified from some of the earliest sites that have been investigated, back to c. 6500 cal BC. Moreover, it is clear from stable isotope reconstructions of human diet from inland skeletons and other indicators that marine foods were on the menu from at least as early as the Maglemose period (Fischer et al. 2007b). The presumption must be that people began to colonise the coastlines of the Danish archipelago as soon as the region was inundated by sea-level rise, perhaps as early as c. 7500 cal BC (9500 cal BP), and that shorelines with their access to marine resources favoured major concentrations of population and settlement as in later periods. But that is a proposition that can only be tested by systematic excavation of more deeply submerged sites, and this is an obvious target for future research.

Similar comments apply to the very extensive submerged landscape offshore of Denmark's Atlantic coastline. This region has been rather overshadowed by the focus on the Danish Straits, but we know from occasional finds dredged up by chance including the Maglemose artefact from the Danish sector of the Dogger Bank noted above (see Fig. 3.4), that this was an important prehistoric landscape. And it is here that evidence for use of coastlines in significantly earlier periods—Late Palaeolithic or earlier—will have to be sought.

Another realisable target is the development of a digital database of existing records. These represent a rich source of material for analysis and for planning of future research. The SPLASHCOS Viewer represents a first step in that direction, but the records are still incomplete and organised in the form of a spreadsheet rather than a searchable database. Further development will require resources and skilled personnel and most likely research-led funding.

These targets have been on the agenda for many years (e.g. Fischer 2007) but with little significant progress. The costs and logistical complexities of underwater work, especially in deeper water, should not be overlooked. But a new generation of acoustic technologies and trained archaeological divers are now available to tackle such challenges. Multibeam bathymetry, subbottom profiling and side-scan sonar are routine techniques applied in offshore and deep-water surveys and could be deployed to apply predictive models like the fishing-site model—and others—in more deeply submerged terrain. International collaboration is now firmly on the agenda and European or international funding increasingly available to facilitate research-led investigations, with some recent examples in Denmark (Uldum et al. 2017; Benjamin et al. 2018; Astrup et al. 2019). The existing underwater work in Denmark provides an excellent platform for reaching out to these more challenging objectives. Institutional infrastructure with the potential to combine scientific and archaeological facilities is available in Copenhagen and Aarhus through their respective universities and associated museums (Roskilde and Moesgaard). This in its turn offers the potential to create a world centre of expertise and training. The increasing threat of destruction to underwater material by natural processes and commercial activities adds urgency to the development of a new era of Danish-led research into submerged landscape archaeology.

3.8 Management of the Underwater Cultural Heritage

Since the establishment of the interdisciplinary 'kitchen midden committee' in 1848, the development of Danish prehistoric archaeology has integrated issues of climate change and coastal protection for which coastal shell-mounds form such a distinctive marker (Sørensen 1898; Brinch Petersen 2015, p. 209). In early legal protection, no clear distinction was made between finds or sites on land and under water. It was only in 1969 that underwater prehistoric remains were explicitly brought under protective legislation, and 1984 saw the extension of protective regulations to the seabed up to the 24-mile limit (Kristiansen 1985; Lund 1987). In doing so, Denmark effectively established an unprecedented cultural heritage protection zone in accordance with the Contiguous Zone regulations of the UN Law of the Sea Convention (Lund 1987; Dromgoole, Chapter 25, this volume). In accordance with these responsibilities, the central government agency for nature and cultural heritage conservation created a special unit for the inventory, registration and inspection of underwater finds (Thomsen 1987). This work constituted the basis for a central database, which integrated information collected through the intricate network of regional and local museums (Smed 1987). The contributions of sports divers, the general public and amateur archaeologists to this work have traditionally been substantial (Fischer and Sørensen 1983; Skaarup 1983; Dal 2013).

At present, the Agency for Culture and Palaces under the Ministry of Culture remains the central authority responsible for government decisions relating to specific underwater sites or to underwater cultural heritage in general. It is also the central clearing house for the administration of relevant data and management of the central database. The acquisition of data, other than through agreement with other offices, is largely decentralized to museums that are recognized by the State under the 'Museum Law'. In total, Denmark has 97 such museums, of which 5 have been assigned a seabed area for which they are archaeologically responsible. The Viking Ship Museum in Roskilde is responsible for the waters around Zealand and Bornholm; Langelands Museum is responsible for the archipelagic waters around Fyn, whereas Moesgaard Museum and the two museum organizations that manage the coastal museums in northern Jutland and on the west-coast share responsibility for the North Sea, the Skagerrak and the Kattegat.

The National Museum, with a remit to advise on conservation matters, houses a small unit addressing issues related to degradation and in situ preservation of underwater sites, as well as conservation and restoration of waterlogged finds (Gregory 2017; Strætkværn et al. 2013). Activities by this unit in the field are always undertaken in close cooperation with the responsible museum. The same holds true for research undertaken by universities. Of these, Copenhagen University is relevant through its prehistoric-archaeological, geological and seismological research, while Aarhus University cooperates closely with Moesgaard Museum for research in their area of responsibility. The Maritime Archaeology Programme at the University of Southern Denmark (MAP-SDU) focuses specifically on the training of archaeological staff for development-led archaeology at sea and maintains close ties with the other institutions mentioned (Maarleveld 2009). It is likely that Aarhus University will take over this role when MAP-SDU will be closed in the near future.

As in many other countries, the professionalisation of archaeological interventions that are development-led and paid for by the developer has led to a lesser involvement of sports divers, amateur archaeologists and others lacking the necessary professional qualifications for offshore work and a reduced impetus for research-led investigations. Moreover, not all research archaeologists have embraced development-led practice as a practicable strategy for their research, and not all mitigation contracts stipulate adequate research strategies. Although the consequences of this shift are hard to gauge, it can nevertheless be mentioned that new finds not related to development-led surveys continue to be reported, although not necessarily through the five earmarked museums (Lauridsen 2013). Furthermore, it remains a challenge that the distribution of responsibilities makes for very small units, including one or two maritime archaeologists only, and hence gives little leeway for proactive deployment offshore. A further issue with decentralisation— not unique to Denmark— is that regional or local government generally has little interest in what happens far out at sea, whether inside or outside the cultural heritage protection zone (Maarleveld 2007). Moreover, Denmark has not ratified the 2001 UNESCO Convention. Strict registration and a focus on preservation limit the possibilities for proactive research, even in the context of development-led interventions. Nevertheless, research that leads to better understanding of the past has been and still is the pre-eminent reason for the protection and management of the non-renewable scientific resource, especially in or on the seabed, where other possibilities to enjoy heritage are limited (Lund 1995, p. 416).

Acknowledgements We are grateful to Peter Moe Astrup and Harry Robson for a critical reading of the text, to Peter Moe Astrup and David Gregory for assistance with images and to Anders Fischer for assisting with the compilation of site data for the SPLASHCOS Viewer.

References

Andersen SH (1980) Tybrind Vig. Foreløbig meddelelse om en undersøisk stenalderboplads ved Lillebælt. Antikvariske Studier 4:7–22

Andersen SH (1985) Tybrind Vig. A preliminary report on a submerged Ertebølle settlement on the west coast of Fyn. J Dan Archaeol 4:52–69

Andersen SH (2000) Køkkenmøddinger (Shell middens) in Denmark: a survey. Proc Prehist Soc 66:361–384

Andersen SH (2005) En glittestok fra Dogger Banke i Nordsøen. Kuml:9–15

Andersen SH (2007) Shell middens ("*Køkkenmøddinger*") in Danish prehistory as a reflection of the marine environment. In: Milner N, Craig OE, Bailey GN (eds) Shell middens in Atlantic Europe. Oxbow, Oxford, pp 31–53

Andersen SH (2009) Ronæs Skov. Marinarkæologiske undersøgelser af en kystboplads fra Ertebølletid, Jysk Arkæologisk Selskabs Skrifter 64. Jysk Arkæologisk Selskab, Højbjerg

Andersen SH (2012) Ertebølle canoes and paddles from the submerged habitation site of Tybrind Vig, Denmark. In: Benjamin J, Bonsall C, Pickard C, Fischer A (eds) Submerged prehistory. Oxbow, Oxford, pp 1–20

Andersen SH (2013) Tybrind Vig. submerged mesolithic settlements in Denmark, Jutland Archaeological Society Publications Vol. 77. Jutland Archaeological Society, Højbjerg

Astrup PM (2018) Sea-level change in Mesolithic southern Scandinavia: Long- and short-term effects on society and the environment, Jutland Archaeological Society Publications Vol. 106. Jutland Archaeological Society, Højbjerg

Astrup PM, Skriver C, Benjamin J, Ward I, Stankewicz F, Ross P, McCarthy J, Baggaley P, Ulm S, Bailey G (2019) Underwater shell middens: excavation and remote sensing of a submerged Mesolithic site at Hjarnø, Denmark. J I Coast Archaeol 1–20. https://doi.org/10.1080/15564894.2019.1584135

Bailey GN (1978) Shell middens as indicators of postglacial economies: a territorial perspective. In: Mellars PA (ed) The early postglacial settlement of northern Europe. Duckworth, London, pp 37–63

Bailey GN (2014) New developments in submerged prehistoric archaeology: an overview. In: Evans AM, Flatman JC, Flemming NC (eds) Prehistoric archaeology on the continental shelf. A global review. Springer, New York, pp 291–300

Benjamin J, O'Leary M, Ward I, Hacker J, Ulm S, Veth P, Holst M, McDonald J, Ross PJ, Bailey GN (2018) Underwater archaeology and submerged landscapes in Western Australia. Antiquity Project Gallery. Available at https://doi.org/10.15184/aqy.2018.103

Björck S (1995) Late Weichselian to early Holocene developments of the Baltic Sea – with implications for coastal settlements in the southern Baltic region. In: Fischer A (ed) Man and sea in the Mesolithic: coastal settlement above and below present sea

level. Proceedings of the International symposium, Kalundborg, Denmark 1993. Oxbow Monograph 53. Oxbow, Oxford, pp 23–34
Brinch Petersen E (2015) Between the 1820s and 1900: Discovering the Køkkenmøddings and the Older Stone Age (Mesolithic) of Denmark. In: Bicho N, Detry C, Price TD, Cunha E (eds) Muge 150th: the 150th anniversary of the discovery of Mesolithic shellmiddens – Volume 2. Cambridge Scholars, Cambridge, pp 207–218
Christensen C (1995) The Littorina transgressions in Denmark. In: Fischer A (ed) Man and sea in the Mesolithic: coastal settlement above and below present sea level. Proceedings of the International symposium, Kalundborg, Denmark 1993. Oxbow Monograph 53. Oxbow, Oxford, pp 15–22
Christensen K (1997) Oak trunks from the Halsskov Fjord. In: Pedersen L, Fischer A, Aaby B (eds) The Danish Storebælt since the Ice Age: Man, sea and forest. A/S Storebæltsforbindelsen, Copenhagen, pp 37–44
Christensen K (2013) Dendrochronological analysis of oak trunks from Tybrind Vig. In: Andersen SH, Tybrind Vig. Submerged Mesolithic settlements in Denmark. Jutland Archaeological Society Publications Vol 77. Jutland Archaeological Society, Højbjerg, pp 365–376
Christensen C, Nielsen AB (2008) Dating Littorina sea shore levels in Denmark on the basis of data from a Mesolithic coastal settlement on Skagens Odde, northern Jutland. Polish Geological Institute Special Papers 23:27–38
Christensen C, Grønnow B, Hansen CV, Jøhnsson JH, Malmros C, Petersen PV (1979) Stammebåden. Et eksperiment med udhugning og sejlads. Nationalmuseets Arbedjsmark 1979:89–94
Claassen C (1998) Shells. Cambridge University Press, Cambridge
Cohen KM, Westley K, Erkens G, Hijma MP, Weerts HJT (2017) The North Sea. In: Flemming NC, Harff J, Moura D, Burgess A, Bailey GN (eds) Submerged landscapes of the European continental shelf: quaternary paleoenvironments. Wiley, Chichester, pp 147–186
Cook Hale JW, Hale NL, Garrison EG (2018) What is past is prologue: excavations at the Econfina Channel site, Apalachee Bay, Florida, USA. Southeast Archaeol. https://doi.org/10.1080/0734578X.2018.1428787
Dal H (2013) The marine-archaeological investigations 1978–1987. Organisation, equipment and excavation techniques. In: Andersen SH, Tybrind Vig. Submerged Mesolithic settlements in Denmark. Jutland Archaeological Society Publications Vol 77. Jutland Archaeological Society, Højbjerg, pp 505–527
Degerbøl M (1943) Om dyrelivet: Aamosen i Stenalderen. In Matthiassen T et al (eds) Stenalderbopladsen i Aamosen. Nordiske Fortidsminder III Bind, 3 Hefte. Kongelige Nordiske Oldskriftselskab, pp 165–206
Dencker J, Jensen SO (2000) Nye stenalderfund fra havets dyb. Fløjstrup Skov, Århus Bay. Marinarkæologisk Nyhedsbrev fra Roskilde 13:18–19
Faught MK (2014) Remote sensing, target identification and testing for submerged prehistoric sites in Florida: process and protocol in underwater CRM projects. In: Evans AM, Flatman JC, Flemming NC (eds) Prehistoric archaeology on the continental shelf. A global review. Springer, New York, pp 37–52
Fischer A (1987) Stenalderbopladsen på Argusgrunden – En 7000-årig bebyggelse på bunden af Smålandsfarvandet. In: Kristiansen K, Fischer A (eds) Fortidsminder og Kulturhistorie, Antikvariske Studier 8. Skov- og Naturstyrelsen, København, pp 11–58
Fischer A (1993) Stone Age settlements in the Småland Bight: a theory tested by diving (In Danish and English). Miljøministeriet, Skov- og Naturstyrelsen, København
Fischer A (1995) An entrance to the Mesolithic world below the ocean. Status of ten years' work on the Danish sea floor. In: Fischer A (ed) Man and sea in the Mesolithic: coastal settlement above and below present sea level. Proceedings of the International symposium, Kalundborg, Denmark 1993. Oxbow Monograph 53. Oxbow, Oxford, pp 371–384
Fischer A (1997) Drowned forests from the Stone Age. In: Pedersen L, Fischer A, Aaby B (eds) The Danish Storebælt since the Ice Age: Man, sea and forest. A/S Storebæltsforbindelsen, Copenhagen, pp 29–36
Fischer A (2004) Submerged Stone Age – Danish examples and North Sea potential. In: Flemming NC (ed) Submarine prehistoric archaeology of the North Sea. Research priorities and collaboration with industry, CBA Research Report 141. Council for British Archaeology, York, pp 23–36
Fischer A (2007) Coastal fishing in Stone Age Denmark – evidence from below and above the present sea level and from human bones. In: Milner N, Craig OE, Bailey GN (eds) Shell middens in Atlantic Europe. Oxbow, Oxford, pp 54–69
Fischer A, Malm T (1997) The settlement in the submerged forest in Musholm Bay. In: Pedersen L, Fischer A, Aaby B (eds) The Danish Storebælt since the Ice Age: Man, sea and forest. A/S Storebæltsforbindelsen, Copenhagen, pp 78–86
Fischer A, Sørensen SA (1983) Stenalder på danske havbund. Antikvariske Studier 6:104–126
Fischer A, Vang Petersen P (2018) Denmark – a sea of archaeological plenty. In: Fischer A, Pedersen L (eds) Oceans of archaeology. Jutland Archaeological Society. Aarhus University Press, Aarhus
Fischer A, Richards M, Olsen J, Robinson DE, Bennike P, Kubiak-Martens L, Heinemeier J (2007a) The composition of Mesolithic food. Evidence from the submerged settlement on the Argus bank, Denmark. Acta Archaeol 78(2):163–178
Fischer A, Olsen J, Richards M, Heinemeier J, Sveinbjörnsdóttir AE, Bennike P (2007b) Coast–

inland mobility and diet in the Danish Mesolithic and Neolithic: evidence from stable isotope values of humans and dogs. J Archaeol Sci 34:2125–2150

Gehrels WR, Szkornika K, Bartholdy JR, Kirby R, Bradley SL, Marshall WA, Heinemeiere J, Pedersen JBT (2006) Late Holocene sea-level changes and isostasy in western Denmark. Quat Res 66(2):288–302. https://doi.org/10.1016/j.yqres.2006.05.004

Gregory D (2017) Out of sight but not out of mind: research into the *in-situ* preservation of marine archaeological sites. SDU, Esbjerg

Grøn O, Skaarup J (1993) Møllegabet II. A submerged Mesolithic site and a "boat burial" from Ærø. J Dan Archaeol 10:38–50

Höffle H, Thomsen MS, Holmer M (2011) High mortality of *Zostera marina* under high temperature regimes but minor effects of the invasive macroalgae *Gracilaria vermiculophylla*. Estuar Coastal Shelf Sci 92(1):35–46

Holm J (1991) Settlements of the Hamburgian and Federmesser cultures at Slotseng, South Jutland. J Dan Archaeol 10:7–19 Odense

Houmark-Nielsen M, Kjær HK (2003) Southwest Scandinavia, 40–15 kyr BP: palaeogeography and environmental change. J Quat Sci 18(8):769–786

Jensen JJ, Bennike O, Witkoski A, Lemke W, Kuijpers A (1997) The Baltic Ice Lake in the southwestern Baltic: sequence-, chrono- and biostratigraphy. Boreas 26(3):217–236

Kristiansen K (1985) Fortidsmindebevaring i Danmark. Status og fremtidsperspektiver. Antikvariske Studier 7:142–164

Larsson L (2017) Submerged settlements in the Öresund, western Scania, southernmost Sweden. In: Bailey GN, Harff J, Sakellariou D (eds) Under the sea: archaeology and palaeolandscapes of the continental shelf. Springer, Cham, pp 165–175

Lauridsen JG (2013) Bearbejdet oldtidsrav fra kysten omkring Blåvandshuk. Arkæologi i Slesvig. Archäologie in Schleswig 14:9–17

Lewis JP, Ryves DB, Rasmussen P, Olsen J, Knudsen K-L, Andersen SH, Weckström K, Clarke AL, Andrén E, Juggins S (2016) The shellfish enigma across the Mesolithic-Neolithic transition in southern Scandinavia. Quat Sci Rev 151:315–320

Lund C (1987) Beskyttelse af historiske skibsvrag og fortidsminder på den danske havbund. Antikvariske Studier 8:135–150

Lund C (1995) Legislation, protection and management of underwater cultural heritage – the Danish model. In: Fischer A (ed) Man and sea in the Mesolithic: coastal settlement above and below present sea level. Proceedings of the International symposium, Kalundborg, Denmark 1993. Oxbow Monograph 53. Oxbow, Oxford, pp 415–418

Maarleveld TJ (2007) Maritime management matters. In: Satchell J, Palma P (eds) Managing the marine cultural heritage: Defining, accessing and managing the resource, CBA Research Report 153. Council for British Archaeology, York, pp 49–58

Maarleveld ThJ (2009) Maritime archaeology – status and identity? In Müller U, Kleingärtner S, Huber F (eds) Zwischen Nord- und Ostsee 1997-2007 Zehn Jahre Arbeitsgruppe für maritime und limnische Archäologie (AMLA) in Schleswig-Holstein. Rudolf Habelt, Bonn pp 51–62

Malm T (1995) Excavating submerged Stone Age sites in Denmark – the Tybrind Vig example. In: Fischer A (ed) Man and sea in the Mesolithic: coastal settlement above and below present sea level. Proceedings of the International symposium, Kalundborg, Denmark 1993. Oxbow Monograph 53. Oxbow, Oxford, pp 385–396

Momber G, Peeters H (2017) Postglacial human dispersal and submerged landscapes in North-West Europe. In: Bailey G, Harff J, Sakellariou D (eds) Under the sea: archaeology and palaeolandscapes of the continental shelf. Springer, Cham, pp 321–334

Mathiasen DR (1997) The changing landscapes of the Storebælt, from the retreat of the ice to the sea flood. In: Pedersen L, Fischer A, Aaby B (eds) The Danish Storebælt since the Ice Age: Man, sea and forest. A/S Storebæltsforbindelsen, Copenhagen, pp 22–28

Milner N (2001) At the cutting edge: using thin sectioning to determine season of death of the European Oyster, Ostrea edulis. J Archaeol Sci 28:861–873

Milner N (2002) Incremental growth of the European oyster Ostrea edulis: seasonality information from Danish kitchenmiddens, Oxford: British Archaeological Reports International Series 1057. Archaeopress, Oxford

Myrhoj M, Willemoes A (1997) Wreckage from the Early Stone Age. In: Pedersen L, Fischer A, Aaby B (eds) The Danish Storebælt since the Ice Age: Man, sea and forest. A/S Storebæltsforbindelsen, Copenhagen, pp 157–166

Nokkentved C, Anderson RT, Folke H, Linton MIA, Anderson SV (2018) Denmark. https://www.britannica.com/place/denmark. Updated 27 February 2018

Nutley D (2014) Inundated site studies in Australia. In: Evans AM, Flatman JC, Flemming NC (eds) Prehistoric archaeology on the continental shelf. A global review. Springer, New York, pp 255–273

Påsse T, Andersson L (2005) Shore-level displacement in Fennoscandia calculated from empirical data. GFF 127(4):253–268

Pedersen L (1997) They put fences in the sea. In: Pedersen L, Fischer A, Aaby B (eds) The Danish Storebælt since the Ice Age: Man, sea and forest. A/S Storebæltsforbindelsen, Copenhagen, pp 124–143

Pedersen L, Fischer A (1997) Stone Age fishers at Halsskov Fjord. In: Pedersen L, Fischer A, Aaby B (eds) The Danish Storebælt since the Ice Age: Man, sea and forest. A/S Storebæltsforbindelsen, Copenhagen, pp 108–205

Pedersen L, Fischer A, Aaby B (eds) (1997) The Danish Storebælt since the Ice Age: Man, sea and forest. A/S Storebæltsforbindelsen, Copenhagen

Pedersen L, Fischer A, Gregory DJ (2017) Fletværket ved Nekselø – skovdrift og storstilet fiskeri i bondestenalderen. Nationalmuseets Arbejdsmark 2017:134–145

Petersen KS (2013) The geological conditions in and around Tybrind Vig at the time of the Ertebølle settlement, as revealed by the marine molluscan fauna. In: Andersen SH, Tybrind Vig. Submerged Mesolithic settlements in Denmark. Jutland Archaeological Society Publications Vol. 77. Jutland Archaeological Society, Højbjerg, pp 355–361

Petersen KJ, Rasmussen KL (1995) Late Weichselian and Holocene changes in the marine environment – with examples from north West Denmark. In: Fischer A (ed) Man and sea in the Mesolithic: coastal settlement above and below present sea level. Proceedings of the international symposium, Kalundborg, Denmark 1993, Oxbow monograph, vol 53. Oxbow, Oxford, pp 35–38

Rosentau A, Bennike O, Uscinowicz S, Miotk-Szpiganowcz G (2017) The Baltic Sea Basin. In: Flemming NC, Harff J, Moura D, Burgess A, Bailey GN (eds) Submerged landscapes of the European continental shelf: quaternary paleoenvironments. Wiley, Chichester, pp 103–133

Sistermans P, Nieuwenhuis O (2018) Erosion Case Study: Western Coast of Jutland. Published online at: http://copranet.projects.eucc-d.de/files/000111_EUROSION_West_coast_of_Jutland.pdf. Accessed 2018

Skaarup J (1983) Submarine stenalderbopladser i Det sydfynske øhav. Antikvariske Studier 6:137–161 København

Skaarup J (2001) Stenalderbopladser på havbunden – status og fremtid. Det Arkæologiske Nævn, Roskilde

Skaarup J (2004) Introduction and background. In: Skaarup J, Grøn O (eds) Møllegabet II. A submerged Mesolithic settlement in southern Denmark, BAR International Series 1328. Archaeopress, Oxford, pp 1–3

Skaarup J, Grøn O (2004) Møllegabet II. A submerged Mesolithic settlement in southern Denmark, BAR International Series 1328. Archaeopress, Oxford

Skriver C, Borup P, Astrup PM (2017) Hjarnø Sund: an eroding Mesolithic site and the tale of two paddles. In: Bailey GN, Harff J, Sakellariou D (eds) Under the Sea: archaeology and palaeolandscapes of the continental shelf. Springer, Cham, pp 131–143

Smed P (1987) Undersøiske fortidsminder – registreing af fund fra den danske havbund. Antikvariske Studier 8:161–187

Sørensen W (1898) Hvem er opdageren af steenalders affaldsdynger ("Kjøkkenmødingerne")? En historisk redegjørelse. Thaning & Appel, Kjøbenhavn

Sørensen SA (2017) The Kongemose Culture. Nordiske Fortidsminder, Serie B, Bind 27. The University Press of Southern Denmark, Odense

Strætkværn K, Gøtche M, Pentz P, Hjelm Petersen A, Solow M, Nørlem Sørensen J (2013) Undervejs med verdens længste vikingeskib. Nationalmuseets Arbejdsmark 2013:24–37

Thomsen B (1987) Registrering af fortidsminder og historiske vrag på havbunden – en foreløbig status. Antikvariske Studier 8:151–160

Uldum O, Benjamin J, McCarthy J, Feulner F, Lübke H (2017) The Late Mesolithic site of Falden, Denmark: results from underwater archaeological fieldwork and a strategy for capacity-building based on the SPLASHCOS mission. In: Bailey GN, Harff J, Sakellariou D (eds) Under the Sea: archaeology and palaeolandscapes of the continental shelf. Springer, Cham, pp 65–84

Uścinowicz S (2003) Relative sea level changes, glacio-isostatic rebound and shoreline displacement in the southern Baltic, Polish Geological Institute Special Papers 10. Polish Geological Institute, Warsaw

Uścinowicz S (2014) The Baltic Sea continental shelf. In: Chiocci FL, Chivas AR (eds) Continental shelves of the world. Geological Society, Memoirs 411, pp 111–144

Vang Petersen P, Johansen L (1991) Sølbjerg I – an Ahrensburgian site on a reindeer migration route through eastern Denmark. J Dan Archaeol 10(1):20–37. Odense. https://doi.org/10.1080/0108464X.1991.10590051

Westley K (2017) The northern shelf. In: Flemming NC, Harff J, Moura D, Burgess A, Bailey GN (eds) Submerged landscapes of the European continental shelf: quaternary paleoenvironments. Wiley, Chichester, pp 135–145

Sweden: Submerged Landscapes of the Early Mesolithic

4

Björn Nilsson, Anton Hansson, and Arne Sjöström

Abstract

In the Swedish waters of the south-western Baltic Sea, large areas (3500 km^2) of inundated land are present. Due to the complex patterns of past sea-level oscillations, combined with cold and brackish water, conditions for preserved Early Mesolithic sites (11,500–8000 cal BP) are good. So far, only 44 archaeological sites, or clusters of sites, have been found. Recent geoarchaeological and diving surveys have resulted in new information about now-submerged terrestrial remains of forests, river valleys, lakes and bogs and suggest that Swedish waters hide many unrevealed sites with a high scientific potential. From both an environmental and a cultural heritage perspective, these remains are important, but they have limited protection, since many are not defined as archaeological sites. Novel interdisciplinary research efforts combining ecological, environmental and archaeological perspectives are now being devoted to changing the way submerged landscapes are treated and protected.

Keywords

Submerged landscapes · Baltic Sea · Mesolithic · Yoldia Sea · Ancylus Lake · Littorina Sea · Hanö Bay · Haväng

4.1 Introduction

Marine archaeology is a thriving field of research in Sweden, and Swedish waters, especially in the Baltic Sea, are well known for their abundance of shipwrecks, not least because of the good preservation of organic materials. Within the Swedish territory, 16,000 heritage underwater sites are registered by the National Board of Antiquity including shipwrecks and material of all periods. From the Stone Age, however, only 44 archaeological sites—or clusters of sites—are known (Fig. 4.1; Table 4.1; http://splashcos-viewer.eu/). This relatively small number most likely results from lack of research. During a span of 3500 years (11,500–8000 cal BP) at least 3500 km^2 of the Baltic Sea was dry land. Despite its high scientific potential and the presence of some archaeologically unique sites (Nilsson et al. 2018), only a small part of this now-inundated landscape has been thoroughly surveyed (Holmlund et al. 2017).

B. Nilsson (✉) · A. Sjöström
Department of Archaeology and ancient History, Lund University, Lund, Sweden
e-mail: bjorn.nilsson@ark.lu.se; arne.sjostrom@ark.lu.se

A. Hansson
Department of Geology, Lund University, Lund, Sweden
e-mail: anton.hansson@geol.lu.se

G. Bailey et al. (eds.), *The Archaeology of Europe's Drowned Landscapes*, Coastal Research Library 35, https://doi.org/10.1007/978-3-030-37367-2_4

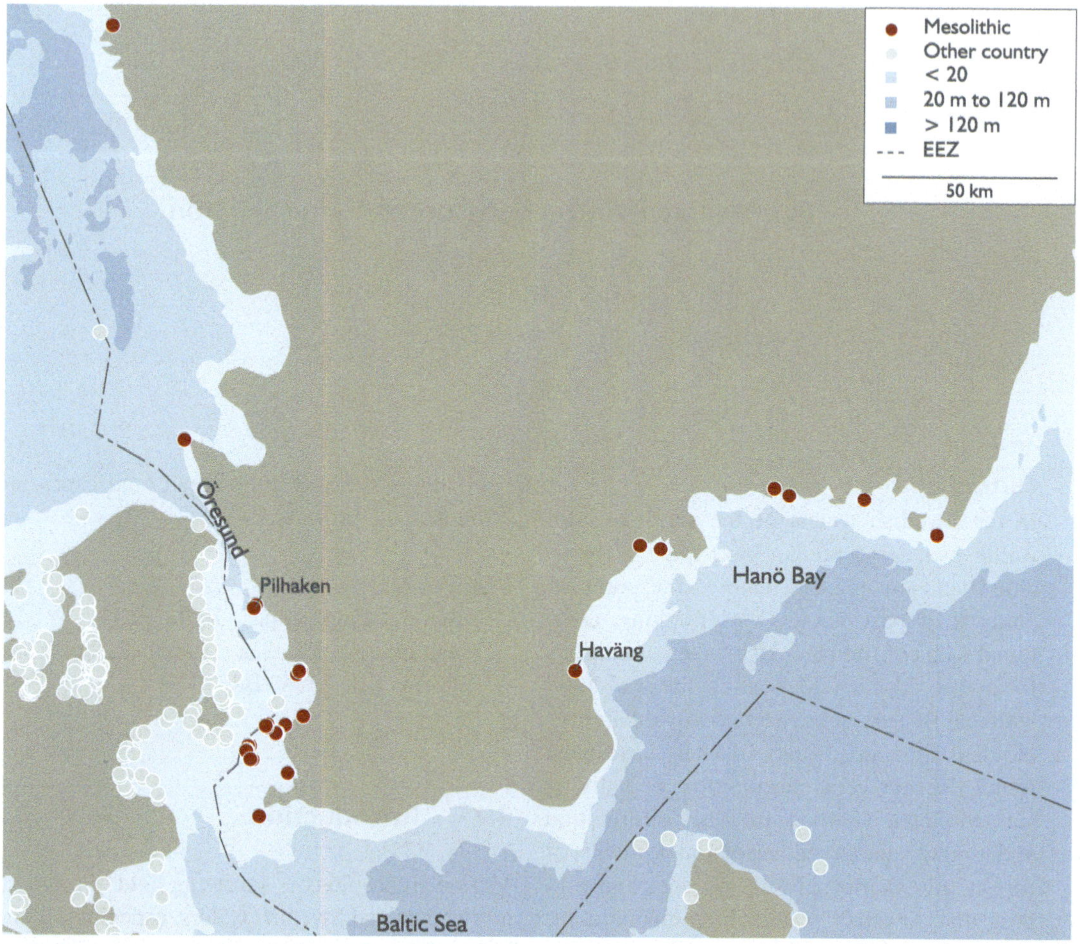

Fig. 4.1 Map of Sweden showing submerged archaeological sites and other places mentioned in the text. Site information from the SPLASHCOS Viewer http://splashcos-viewer.eu. Drawing by Moritz Mennenga

It is a long-established scientific fact that large areas of the Swedish sea bed in the southern Baltic Sea—the areas of the Öresund (Øresu nd in Danish), Ystad and Hanö Bay—were dry land for long periods of time (e.g. Wilcke 1850; Bruzelius 1870; Isberg 1927). The scientific investigation of these inundated landscapes by Quaternary geologists and archaeologists has become more important in recent decades. This is due, amongst other reasons, to the expansion of industrial activity on the continental shelf, including aggregate mining, building of offshore constructions such as wind farms and the growing need for communication and energy transfer in marine areas. Consequently, researchers, policy-makers and other stakeholders have become aware of the need for improved understanding of the ecology and environment of the sea bed in order to better manage both its natural and its cultural resources.

In the southern Baltic, due to the complex history of relative sea-level change and the lack of the common shipworm, there are unique possibilities for the preservation of archaeological and environmental data, with potentially valuable information on climatic and culture-historical change, especially from the early Holocene (11,600–8200 cal BP) and the Maglemose period (see Bailey et al., Chap. 3, Table 3.1, this volume, for chronological chart). Studies of Quaternary geology and environmental history have been devoted for some decades to the study of Holocene shore-level displacement and the wider

Table 4.1 Distribution of underwater finds by type and period. Data from the SPLASHCOS Viewer at http://splashcos-viewer.eu

Types of sites	Late Palaeolithic	Mesolithic	Mesolithic–Neolithic	Neolithic	Bronze age	Indeterminate Prehistoric	Total N	Total %
In situ cultural features-Settlement data	–	11	–	–	–	–	11	*25*
Unstratified material-Multiple finds[a]	–	22	7	–	–	1	30	*69*
Graves (human bone)	–	1	–	–	–	–	1	2
Single unstratified find	–	1		–	–	–	1	2
Votive deposits	–	–	–		–	–	–	2
Miscellaneous other	–	1	–	–	–	–	1	
Total N	**0**	**36**	**7**	**0**	**0**	**1**	**44**	***100***
%		*82*	*16*			*2*		
Presence of organics (wood and/or bone)	–	11		–	–	–	**11**	**25**

[a]This category contains areas with more than one archaeological site

history of the Baltic (for a review, see Berglund and Björck 1994; Björck 2008; Berglund and Sandgren 2010). Some of these studies specifically targeted remains of the submerged landscape in Hanö Bay during the 1980s (Hansen 1985, 1986, 1995; Björck and Dennegård 1988; Gaillard and Lemdahl 1994).

Nevertheless, archaeological investigations of Swedish submerged landscapes and their incorporation within governmental policies for management of the cultural heritage have been quite limited in comparison with other West European countries, such as Denmark, Germany, the UK and the Netherlands. This is despite the favourable circumstances for the preservation of underwater material and the strong tradition of heritage management in Sweden. There are of course exceptions. Some Stone Age sites, mostly sites of Late Mesolithic age in the Öresund region, were excavated during the late twentieth century from the late 1970s onwards (Larsson 1983; Fischer 1993, 1996).

Since 2010, there has been renewed scholarly interest thanks to new archaeological findings and the initiation of the SPLASHCOS Action, with a cluster of new research projects at Lund University (Nilsson 2012a, b; Nilsson and Sjöström 2013; Hansson 2018). This research programme includes a wide range of researchers from the disciplines of geology, bioarchaeology, marine archaeology, digital heritage and marine survey technology (e.g. Holmlund et al. 2017; Nilsson et al. 2018).

The aim of this chapter is to give an overview of some of the studies already made, highlight important findings and discuss future opportunities and strategies.

4.2 Geological and Archaeological Context

4.2.1 Shoreline Displacement

After southernmost Sweden became ice free at about 16,000–15,000 cal BP (Bailey et al., Chap. 3, this volume), the shoreline of the Baltic Ice Lake was situated at about 50–70 m asl (above sea level). At 11,700 cal BP, this lake drained into the Atlantic with a catastrophic drop in water level of 25 m in just a few years, exposing large areas of coastal land in the southern Baltic Basin during the subsequent Yoldia Sea and Ancylus Lake Stages (Björck 1995). In the Hanö Bay, the coast was situated several km offshore from the present-day coastline, as is evident from abundant remains of submerged forests found on the sea floor (Hansson et al. 2017).

At Haväng, a key locality situated on the Scanian coast of Hanö Bay and described further below, remains of rooted tree stumps and lagoon sediments have been used to reconstruct the level of the Yoldia Sea lowstand and the timing of the subsequent Ancylus Lake transgression (Hansson 2018; Hansson et al. 2018a), revealing 3500 years of submerged landscape history (Fig. 4.2).

The deepest-rooted pine stump at Haväng was found at 20.7 m bpsl (below present sea level), indicating that the Yoldia Sea lowstand was situated at 24–25 m bpsl. This corresponds to a lowstand of approximately 20 m depth in the northernmost part of the Hanö Bay (the Blekinge Archipelago) and 28 m depth in southernmost Sweden (the Ystad area, Scania).

Based on the radiocarbon dates of rooted pine stumps obtained at 5–20 m bpsl in the Haväng area, the timing of the onset of the Ancylus Lake transgression can be dated to 10,800 cal BP (Hansson et al. 2018a). The transgression continued for 500 years with a constant rate of rise in the water level of the Ancylus Lake of c. 4 cm per year until a maximum of c. 5 m bpsl was reached at 10,300 cal BP. The Hanö Bay and the south-western part of the Baltic Basin then experienced a regression when the Ancylus Lake found a southern outlet after 10,300 cal BP, which low-

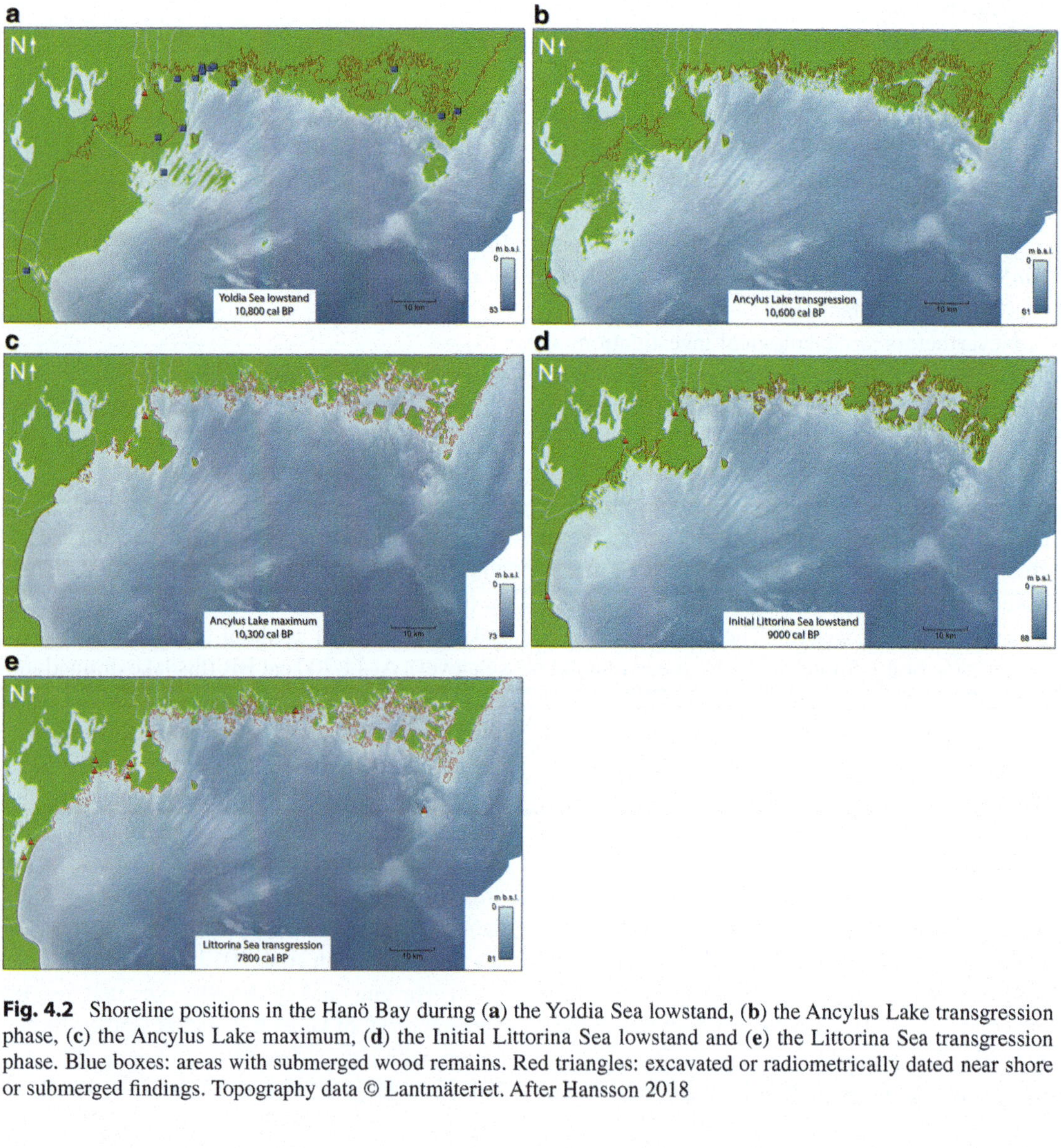

Fig. 4.2 Shoreline positions in the Hanö Bay during (**a**) the Yoldia Sea lowstand, (**b**) the Ancylus Lake transgression phase, (**c**) the Ancylus Lake maximum, (**d**) the Initial Littorina Sea lowstand and (**e**) the Littorina Sea transgression phase. Blue boxes: areas with submerged wood remains. Red triangles: excavated or radiometrically dated near shore or submerged findings. Topography data © Lantmäteriet. After Hansson 2018

ered the level of the Baltic Basin to ocean levels (Björck et al. 2008; Andrén et al. 2011). During the succeeding phase, the Initial Littorina Sea Stage (9800–8500 cal BP), the innermost parts of the submerged landscape at Haväng were characterised by thick compacted gyttja sediments formed in a highly productive freshwater lagoonal environment (Hansson et al. 2018b). This gyttja is found down to a depth of 12 m bpsl. Given that gyttja formation needs a water depth of a few metres, this suggests that the water level was approximately 10 m bpsl during the Initial Littorina Sea lowstand (9800 cal BP). Several stationary fish traps used in the lagoon are found from this period. At about 8500 cal BP, ongoing global sea level rise, combined with a decrease in the rate of isostatic uplift, reactivated the Öresund Strait, resulting in increasing levels of salinity. In southernmost Sweden, the Littorina transgression reached 5–10 m above present sea level in the period 8000–6000 cal BP (e.g., Berglund et al. 2005). From this period on, all subsequent shorelines are situated above the present shoreline (Fig. 4.3).

The waters of south-western Sweden, such as the Öresund Strait between Denmark and Sweden,

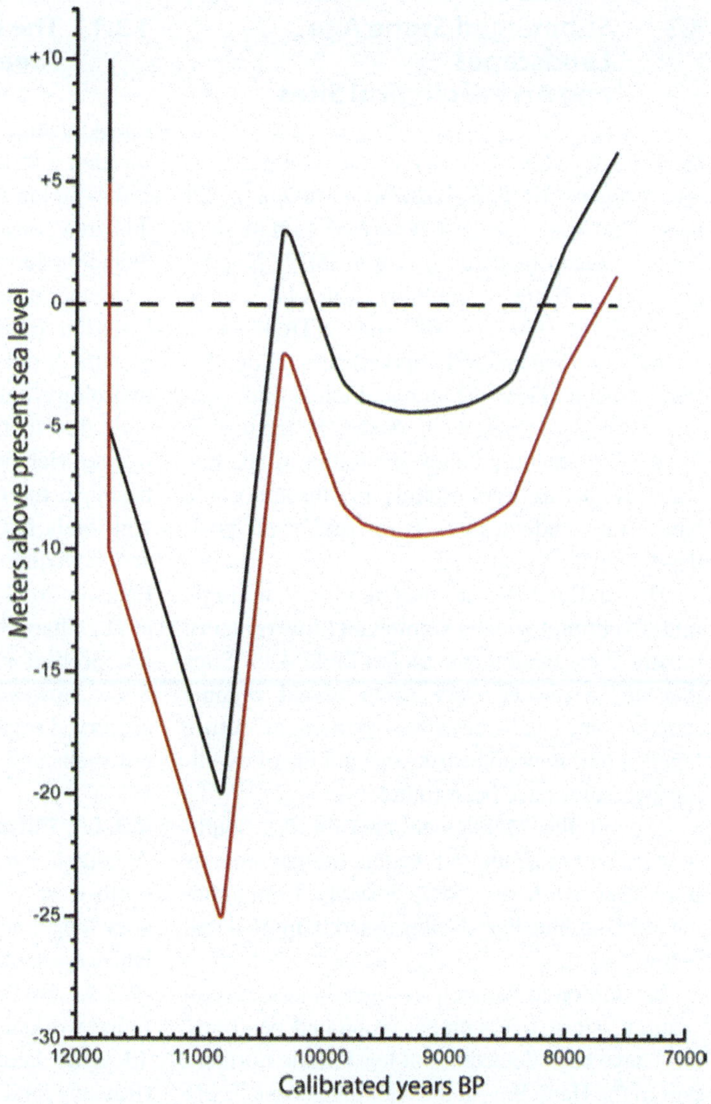

Fig. 4.3 Late Glacial and Early–Mid Holocene shoreline displacement diagram from Hanö Bay. Red: Haväng (Hansson et al. 2018b); Black: western Blekinge (Hansson et al. 2018c). The trends are similar in both areas, but the levels are different due to differences in land uplift

and the southernmost Kattegat Sea, are not part of the Baltic Sea proper, and have a different shore-line displacement history, as these areas have been connected to the Atlantic Ocean since the deglaciation. In Swedish waters, the exact level of the postglacial lowstand is based on models of glacio-isostatic adjustment rather than empirical field data, and these indicate that the lowest shore-lines in the southern part of Kattegat and the northern part of Öresund were formed around 11,000 cal BP and are found at 22–18 m bpsl. After this time, the relative sea level rise reached present levels at approximately 8000 cal BP.

4.3 Submerged Stone Age Landscapes and Archaeological Sites

In Sweden, 44 underwater archaeological sites are known (http://splashcos-viewer.eu/). Of these, 25% have been test excavated with modern documentation methods (Larsson 2017; Nilsson et al. 2018). Some of the known sites have to be regarded as clusters of sites (Tredenborg, Vendelsöfjorden and Måkläppen), and taking this into account, at least 62 archaeological find spots have been discovered in Swedish territories. Most of the recorded sites are found in shallow water. Some are not certainly in situ and could have been redeposited as a result of post-depositional erosion.

The majority of the sites represent multiple finds of chronologically significant lithic remains, mostly flint scatters and tools (Table 4.1). Sites that are excavated very often reveal organic remains and well-preserved settlement refuse layers. Until now no hearths or remains of dwelling structures have been found.

Around the Måkläppen area in the south-western part of southern Sweden, human remains have been washed ashore, possibly from inundated Mesolithic burials, discussed in more detail below.

In addition to the archaeological sites, several 'natural' sites (tree stumps, inundated peatbogs, etc.) have been detected, not least in the northern part of the Hanö Bay, at the Blekinge coast. These are not protected by culture heritage legislation but are an equally important record for cultural and natural history studies.

Due to variations in topography, geology, hydrology and shore-line displacement, submerged Early Holocene landscapes can be divided into three main geographical areas: the Kattegat and Öresund area of south-western Sweden, the riverine and lacustrine palaeo-landscapes of south-eastern Sweden and the Hanö Bay and the archipelago of Blekinge. Below is a brief presentation of these areas and the archaeological remains.

4.3.1 The Öresund and Kattegat Area

The Öresund is the strait that divides modern Sweden and Denmark and is about 10 km long and 4–30 km wide. During low sea-level periods, this area constituted the land bridge between the Scandinavian Peninsula and the European continent and became a marine seaway during the Littorina transgression 8500 years ago. Since then the Strait has played an important role in the hydrology and ecology of the Baltic Sea, not least due to the important influx of saline and oxygen-rich water into the brackish Baltic Sea basin. In the Öresund, several submerged Stone Age finds have been made on both the Danish and Swedish sides (Fischer 1993, 1996; see Larsson 2017 for a full review; see also Bailey et al., Chap. 3, this volume). Most of these consist of flint scatters eroded from nearby sites. Some, however, are found in peat deposits with preserved organic debris such as bone fragments, nut shells and charcoal.

4.3.1.1 Pilhaken

A notable example is Pilhaken, a shallow bank north-west of Landskrona (Larsson 2017), an area long known for its remains of submerged landscapes, mainly tree stumps and peat deposits (Wilcke 1850). During the late1930s, zoologists found worked flint in sub-bottom samples, which 40 years later captured archaeological interest (Rausing and Larsson 1977). After systematic archaeological sampling and subsequent diver investigation, some areas were test excavated with Danish colleagues during the 1990s (Larsson 2017, Fig. 11.3). The result showed that Pilhaken comprised a set of several sites. Radiocarbon analysis of hazelnut shells and oak dates the oldest site now located at 7–8 m depth to c. 9000 cal BP. At that time the settlement was situated on the estuary of the former Saxå River, a 'typical' location for larger late Maglemose coastal sites (Nilsson et al. 2018). The palaeochannel is clearly visible today, and several smaller sites are located along this channel or in nearby shallow areas, but these are heavily eroded.

South of Pilhaken, and along the coast down to Malmö, several sites are found, but none of these have been excavated or more thoroughly surveyed. In Malmö City, due to harbour expansion and railway construction, interesting findings have been made. Here, a transgression sequence from 8100 to 7500 cal BP is recorded. The evidence of a rapidly drowned oak and alder forest, submerged by the opening of the Öresund during the Littorina transgression around 8500 cal BP, gives a good picture of the ecological impact of the inundation and how landscapes were subject to sudden transformations. In the subsequent period, the forest was transformed into a fishing ground.

Extremely well-preserved wicker cages and weirs, wooden leister shafts and other wooden artefacts were found in the transgression sediments. During a couple of hundred years, some few generations only, the forested landscape was transformed into a shallow marine bay, suitable for fishing and hunting (Hammarstrand Dehman 2009; Nilsson 2011).

During this time, there was a socio-economic shift and increased dependence on marine resources (Fischer 2005; Larsson 2017). The cultural adaptation was swift and far-reaching. Here, and on both sides of the Strait, we have rich evidence from the Middle and Late Mesolithic (see Bailey et al., Chap. 3, Table 3.1, this volume for chronology) of burial areas, fishing grounds, large settlements and middens. However, to fully understand this environmental and cultural transition, more focus must be put on older sites that are now submerged. Not only will these tell a different story than the one we are used to, due to their relatively older age, but there is also a high probability of finding well-preserved organic material. From this point of view, the coastal archaeology of the Öresund area is of international importance, providing one of the few areas where accessible deposits of the right age and with good conditions of preservation provide a window into the nature of the earliest coastlines of the Littorina Sea and their associated archaeology.

Recent excavations in Swedish waters north of Öresund, at Vendelsöfjorden in the Kattegat outside of Varberg, Halland, bear evidence of several submerged sites. The extent and date of these remains are not thoroughly understood; the indisputable archaeological evidence contradicts existing geological shoreline displacement models (Von Arbin 2008), and this is an issue that needs to be more thoroughly examined in the future.

4.3.2 The Southern and Eastern Coasts of Scania

The shallow waters on this coastline have long been known for the presence of submerged bogs and tree stumps (Bruzelius 1870; Isberg 1927; Nilsson 1961). Trawler fishing and near-shore eel fishing have revealed several areas where Early Holocene terrestrial remains are abundant, especially offshore of today's river estuaries.

4.3.2.1 Haväng

One of the most famous sites, which has been examined on several occasions during the last 50 years, is Haväng near the Verkeån River north of the city of Simrishamn (Fig. 4.4; Nilsson 1961; Hansen 1985, 1995; Björck and Dennegård 1988; Gaillard and Lemdahl 1994; Nilsson et al. 2018; Hansson 2018).

The earliest scientific investigations at Haväng concentrated on the submerged pine forest. In the 1970s and early 1980s, some 40 logs and stumps, many of which were rooted, were dragged ashore from water depths between 13 and 80 m bpsl (Fig. 4.5). Stratigraphic investigations were also

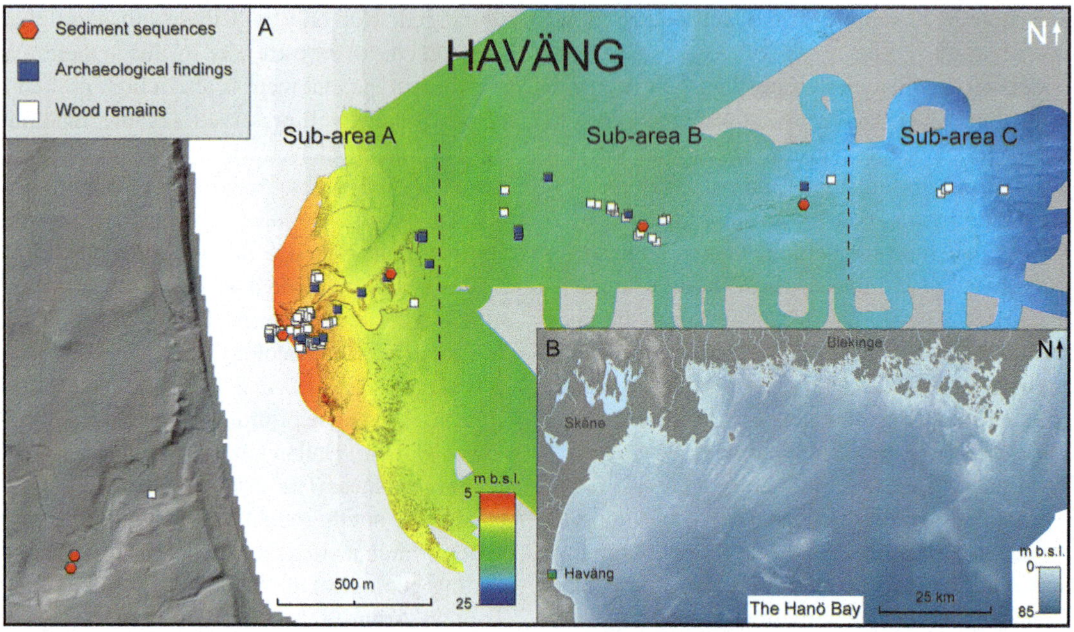

Fig. 4.4 Overview of the Haväng study area and the Hanö Bay (after Hansson 2018). (**a**) Topographic and bathymetric composite map of Haväng, showing the subdivision of the study area and the locations of the sediment sequences (red hexagons), archaeological findings (blue boxes) and tree stumps and trunks (white boxes). Not shown are the locations of the discrete organic sediment samples obtained immediately adjacent to each sampled stump or trunk. (**b**) Topographic and bathymetric composite map of the Hanö Bay region, showing the location of Haväng and the main rivers and lakes. Bathymetry data: Marmori, Sjöfartsverket, MMT, Landscapes Lost/Blue Archaeology project, Lund University. Topography data © Lantmäteriet

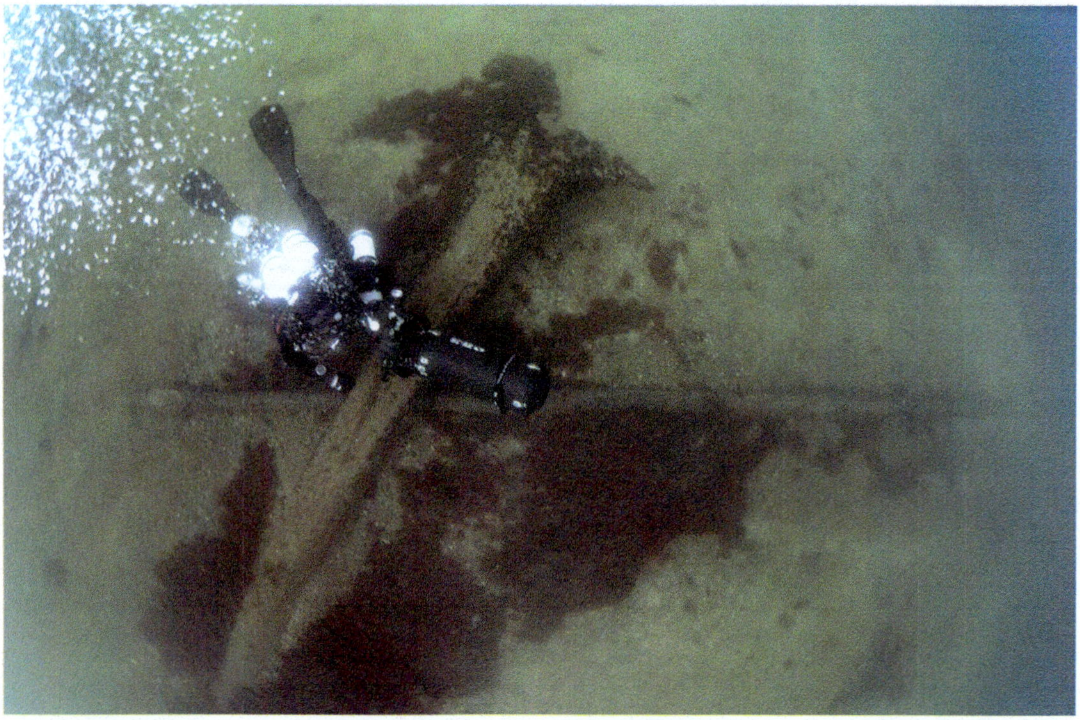

Fig. 4.5 Fallen trees and diver Beesham Soogrim at the Haväng site. Photo by Arne Sjöström

carried out on organic deposits found in shallower deposits associated with remains of pine trees (Hansen 1985, 1986; Björck and Dennegård 1988; Gaillard and Lemdahl 1994). The sediments were analysed for pollen and macroscopic plant and insect remains, and tree-ring records were measured, establishing some fragmentary, floating dendrochronologies. Twelve of the trees were dated by radiocarbon to c. 11,000–10,600 cal BP, i.e. the end of the Yoldia Sea stage and the onset of the Ancylus Lake stage. Since then, it has been assumed that the pine forest was flooded by the Ancylus Lake transgression. The Hanöbukten project, led by Hansen, was most active during the mid-1980s and the mid-1990s. Due to lack of funding, this high-profile project was discontinued. From 2010, a research program directed from Lund University (Landscapes Lost Project, 2010–2016; the Blue Archaeology Project, 2017) has continued and expanded the research. Geoarchaeological sampling and coring (Hansson 2018) have been complemented by acoustic surveys and diver investigation (Nilsson et al. 2018). This has resulted in important new results and indicated considerable potential for further archaeological discoveries over a wider area.

Fig. 4.6 Hazelwood strips from a panel of a fish weir at Haväng. The hazelwood is dated at 8800–9000 cal BP and the diver is Beesham Soogrim. Photo by Arne Sjöström

We now know that the submerged archaeological sites at Haväng span a long period —from 11,000–8500 cal BP—and extend over a large area. The sites are clustered along the now-submerged estuary and its lagoon formations, traceable today from the remains of ridges of sediment up to 4-m thick along the old river course. Several subareas (A, B, C) within the 3000 x 500 m inundated river bed and lagoon have yielded artefacts of bone, antler and wood. Fish traps, fences, wattle structures and weirs have been found, all dating to 9000–8500 cal BP (Fig. 4.6). They are some of the oldest known wooden stationary fishing traps in the world. None of these have been excavated so far. (For excavated examples of later date in Denmark, see Bailey et al., Chap. 3, this volume.) The bone material consists solely of mammals and covers a period from at least 10,700 to 8400 cal BP. All bones show marks of human use. There are also many examples of worked wood, but due to conservation costs, none of these have been sampled so far.

The archaeological findings can be divided into two chronological phases—the Early Littorina (Hansson et al. 2018a, Nilsson et al. 2018) and an older phase belonging to the period of the late Yoldia Sea and Ancylus Lake phase (Hansson et al. 2018b,c). The latter is primarily found in sub-area B, even though older findings have been recovered in the lower part of the sediments in sub-area A. No excavation has yet been conducted, however, only geological sampling, during which archaeological material has occasionally been found. The archaeological material is well preserved and largely comprises redeposited refuse material and debris displaced from a nearby settlement that has now been completely eroded away (Fig. 4.7).

One prominent finding, with evident human traces, consists of an aurochs (*Bos primigenius*)

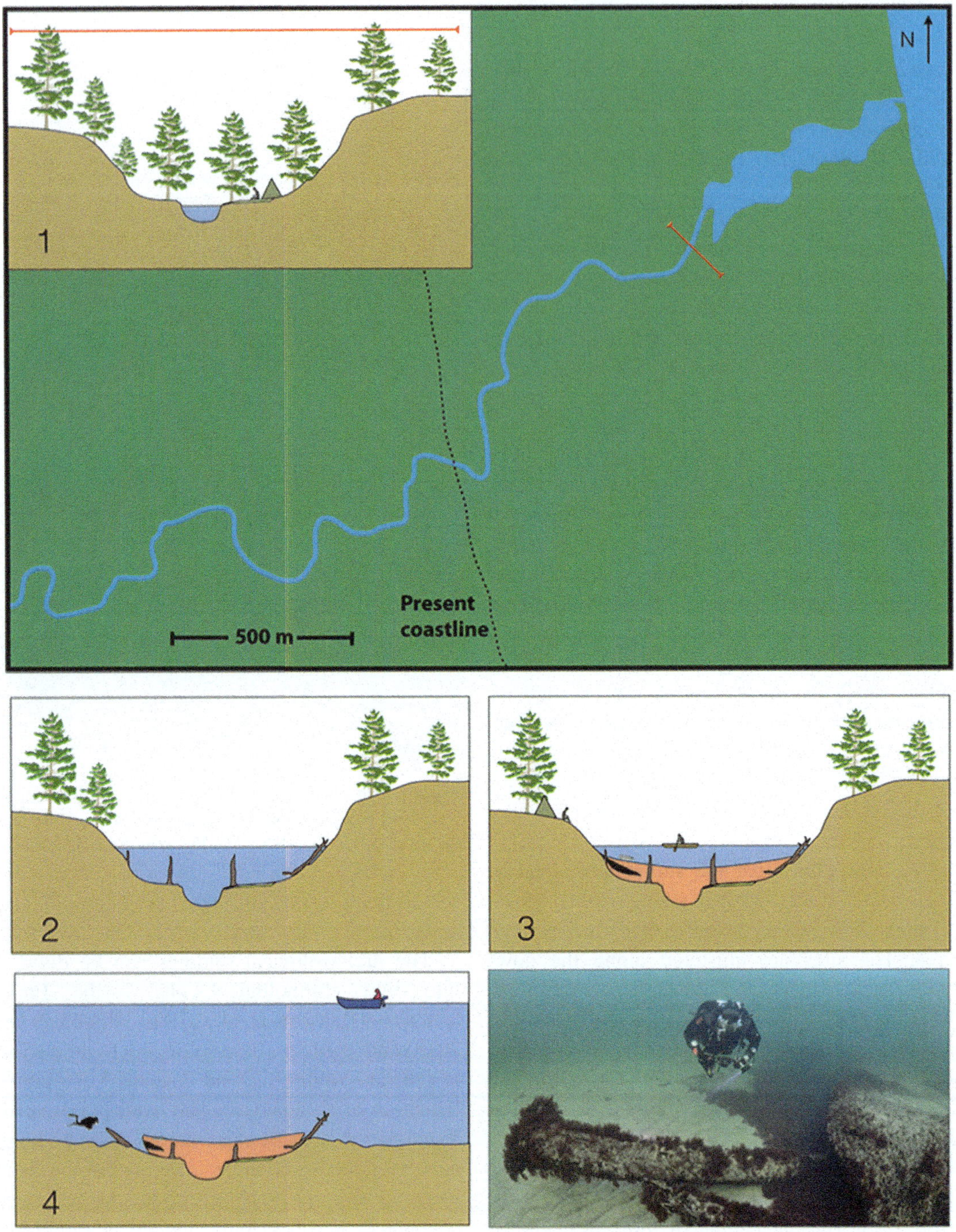

Fig. 4.7 Possible sedimentological and archaeological model of how the Haväng remains have originated. Schematic N–S section of the gyttja remains at sub-area A. Upper image shows a schematic valley and cross section. (1) at c. 10,500 cal BP, showing a river valley with pine forests, situated some kilometres from the coast, with evidence of an Early Maglemose settlement. Lower image shows successive transformations with sea level rise. (2) At 10,300 cal BP, the area is now submerged due to the Ancylus transgression. (3) At 9000 cal BP, during the Littorina lowstand, the section is situated within the estuary and with a Late Maglemosean settlement in the vicinity. Layers of gyttja (in red) fill the near-shore lagoon. Stationary fishing traps (in black) in the lagoon are important features of the settlement. (4) Present day. The sandy deposits along the river have been completely eroded away by the Littorina transgression (see photo at bottom right). What remains is the lagoon sediments with remains of fish traps and refuse layers from the Early Mesolithic settlement

Fig. 4.8 A metatarsal from an adult aurochs situated on top of the organic sediments at Haväng at 15.4 m bpsl. The bone has a distinct groove made by a burin. Possibly this is a rejected piece of raw material. Around the bone there are several pieces of worked wood. The bone is dated to c. 10,600 cal BP and the sediments are a hundred years older. Photo by Arne Sjöström

metapodial with a cut mark made by a burin (Fig. 4.8). It was found exposed in a gyttja which yielded ages of 11,800–10,600 cal BP. The bone itself is dated to 10,650–10,400 cal BP and is most likely to have been discarded from a bone-working area at a nearby settlement. Wooden debris is also found in the vicinity; some of it evidently worked. In the western part of sub-area B, an aurochs metapodial and several bones (tibia, fibula and ribs) from the European beaver (*Castor fiber*) were found a few metres apart at a depth of 15 m bpsl. The beaver is one of the oldest found in Scandinavia and dates to 10,700–10,500 cal BP. A worked antler of red deer (*Cervus elaphus*) found in the grey silty gyttja in sub-area A dates to c. 10,200–9900 cal BP. Not far from this find, but higher up in the sediments, a large pick axe made from a polished and decorated elk antler was found exposed due to heavy erosion presumably caused by a winter storm (Fig. 4.9). A hollow pine trunk located at 19 m bpsl and dated to 11,100–10,900 cal BP appears to have been used as a log fire (Fig. 4.10).

The site of Haväng is not only important thanks to its rich and unusual inventory of material. The site also spans the entire Early Mesolithic and gives a unique insight into the marine and lacustrine economy of this period. So far, very few fish and marine mammal remains have been recovered, which stands in contrast to contemporaneous on-land sites on the Swedish west coast (see Boethius 2018 for a recent review). Proper excavations at the site might of course change this. At the same time, the site is vulnerable to loss of information from destructive agencies, especially damage caused by marine traffic (Fig. 4.11).

The recent research at Haväng has created a methodological protocol on how best to use acoustic techniques and how to find new sites along the coast. So far this has been successful. In the past 2 years, at least two more lagoon

Fig. 4.9 A pick axe made from elk antler found in an organic sediment ridge at Haväng at 5.6 m bpsl. The imprint of the broken tip can be seen in the sediment. The pick axe measures 20 cm. and is dated to 8800 cal BP. Diver Anton Hansson. Photo by Arne Sjöström

sediment-ridge complexes have been found. These are situated east of Ystad on the southern coast (offshore of the Nybroån and Kabusaån estuaries). No clear evidence of archaeological remains has been found so far; however, at one of the sites (Nybroån), there are indications of fish traps. The remains look similar to Haväng, with both sediment banks and visible pine stumps. The sites are not radiocarbon-dated, but due to uneven isostatic land uplift, the lowstand is c. 3–4 m deeper than at Haväng. Consequently, the submerged landscape is potentially more extensive, and the near-shore sediments could also be slightly younger than the Haväng sediments. From an archaeological perspective, it would be very interesting if we could find sites in the Baltic Sea proper that are contemporaneous with the Pilhaken site. Not only would this offer a possibility to study the transition in slightly brackish to saline waters (9000–8000 cal BP) but also the potential effects of the 8.2 kyr climate event (for a discussion see Groß et al. 2018).

Two more finds along the Scanian Baltic coast should be mentioned. First is the interesting but enigmatic occurrence of human remains from the Måkläppen area (Fig. 4.1). The bones have given a range of radiocarbon dates in the period from c. 8000 to 7000 cal BP (Larsson 2017). The younger-dated bones must result from erosion of nearby sites on land (or represent drowned individuals). The older bones could originate from now-submerged burials, being thus some of the oldest burial grounds in Southern Sweden.

Secondly, there are recent finds from the mouth of the Skräbeå River, 50 km north of Haväng (Fig. 4.2 d, western red triangle). Here, dredging has yielded large quantities of Early and Middle Mesolithic artefacts and bones; a red deer bone (*Cervus elaphus*) is dated to 8800 cal BP. Surveys at periods of extreme low water show that lithic scatters and very large flint nodules—a type not locally present—can be found down to at least 1.5 m depth. These deposits are in situ. The bones are dredged from the river bed at depths down to c. 4 m. Judging from the findings so far, this site shows similarities with the Pilhaken site complex. The depth and the occurrence of chronologically diagnostic lithic material and good organic preservation make this a highly promising prospect for excavation.

4.3.3 The Northern Part of Hanö Bay and the Blekinge Archipelago

The province of Blekinge is well suited for shoreline displacement investigations, as the coastline runs parallel with the uplift isobases of the region, which enables data from the many basins and bays in the undulating landscape to be combined into detailed shoreline displacement records. Consequently, Blekinge has a long tradition of Baltic Basin sea-level research, starting already in the 1800s. Since then, the focus has mainly been on the Late Weichselian shoreline displace-

Fig. 4.10 A hollow and burnt pine trunk located at 19 m bpsl at Haväng and dated to 11,100–10,900 cal BP. The trunk was possibly used as a log fire. Up to now this type of fire has been recorded only in northern Sweden. Photo by Arne Sjöström

Fig. 4.11 Modern destruction of sediment banks due to fishing or anchoring at Haväng. The fish trap pictured in Fig. 4.6 is located only some metres from here. Photo by Arne Sjöström

ment, i.e. before 11,700 cal BP, and the Littorina transgressions, after 8000 cal BP, providing good evidence of fluctuations at levels above present sea level (cf. Hansson et al. 2018c).

Pine stumps dating to the Yoldia and Ancylus stages have been found sporadically in the Blekinge Archipelago. Diving and acoustic surveys throughout the last 5 years have revealed several remains of a submerged landscape in the study area (Holmlund et al. 2017), and amateur divers have located several new areas with preserved tree stumps and terrestrial palaeosurfaces in adjacent areas (Hansson et al. 2018c). Furthermore, a recently revised shoreline displacement record for Blekinge (Hansson et al. 2018c) suggests that there exists a significantly long period (c. 9500–8500 cal BP) with a relatively stable sea level during the Initial Littorina Stage at 4 m depth, well below today's shoreline (Fig. 4.3, black line).

Submerged sites from Blekinge are quite few. Recent surveys and excavations have yielded several interesting coastal sites (Fig. 4.1, Fig. 4.2). They are mainly situated some 0–2 m below the modern shoreline. To date, most of the findings are flint scatters from submerged settlements. Some of the artefacts have been polished by wave action, but there are also 'fresh' specimens indicating the presence of eroding near-shore sites located further out in the sea. The sites are difficult to date typologically, but in the few cases where this has been possible, the dates range from late Early to early Middle Mesolithic (c. 9500–8000 cal BP), i.e. during the rather long lasting Littorina low-stand period (Hansson et al. 2018c).

The Blekinge Archipelago is strikingly different to the two other areas discussed above. The potential to find well-preserved settlement sites from the Early Mesolithic is high. In marked contrast to Haväng, where refuse layers are preserved, but the settlement itself has been washed away during later transgressions, the protected environment in Blekinge could yield more complete settlement remains. The abundance of exposed palaeosurfaces from the Ancylus stage, and indication of flints washed ashore, gives a hint of what one can expect. However, this remains to be tested by underwater excavations.

4.4 Swedish Submerged Prehistories: Future Strategies

The field of underwater cultural heritage in Sweden has long focused on objects, especially shipwrecks. However, the increased interest in the prehistoric archaeology of the shallow sea floors of the Baltic Sea has promoted new thinking and emphasised the need for a broader approach that integrates the joint investigation of cultural and natural resources and their protection. Other factors promoting change are advances in marine spatial planning (MSP), the European Landscape Convention (ELS), notions of 'seascape' and the need for better scientific methods and new knowledge about submerged landscapes to underpin Environmental Impact Assessments in marine planning. In Sweden, policy discussions are currently under way at regional and national level and are still at an early stage.

In recent years, research efforts have more than doubled the number of known archaeological and geological submerged sites, including the discovery of finds that are unique in character. The Blue Archaeology project has focussed not only on discovering new underwater sites and their further implications for understanding prehistoric cultural developments but on enhancing the role and practice of underwater cultural heritage management and emphasising the need for integration of cultural and natural processes of change.

One important outcome of the Blue Archaeology project has been to highlight the value of Mesolithic seascape as a concept with which to critically examine the Mesolithic record. The new results have highlighted the likely significance of the coastline and marine resources in the Early Mesolithic period. Until recently, knowledge of this period in southern Sweden was based mainly on the study of inland settlements and areas that are on or near the modern coastline. Socio-economic and demographic interpretations have therefore been founded on settlements with land-based economies. However, it is clear that the new information from submerged coastlines is highlighting the early existence of maritime economies and that this will require a revision of views about the Early Mesolithic. Permanent fishing installations found at the Haväng site suggest large catches, which in turn calls for a new analysis of how these societies were organised (Nilsson et al. 2018). Similar trends are apparent in Denmark (Bailey et al., Chap. 3, this volume) and Germany (Jöns et al., Chap. 5, this volume) with the difference that in Sweden the submerged landscapes and archaeology of the Early Mesolithic period (equivalent to the Maglemosian in Denmark and northern Germany) are more extensive and more easily accessible and therefore fill a gap in the archaeological record that cannot easily be addressed from the underwater record elsewhere.

This concept of a Mesolithic seascape is not only 'good to think with' when addressing the interpretation of the archaeological record but also when bringing an archaeological perspective to bear on the modern situation with its settled and land-based society (Nilsson 2011). The discovery and exploration of new submerged sites of the Stone Age in Scandinavian waters are not only changing the way we conceive of the Stone Age but is also highlighting the value of the Baltic Sea as a cultural and natural resource in the modern world as well as its vulnerability.

4.5 Management of the Underwater Cultural Heritage

As in many European countries, archaeological competence and legislative power in marine matters are spread across several institutions and government agencies. National, regional, local and academic actors work side by side, and together they constitute the key organisations. Institutions with high marine archaeological competence such as academic institutes at universities, or even single scholars, have a small advisory function.

Offshore exploitation of the seabed is governed by the Act concerning the Continental Shelf under the executive direction of the Geological Survey of Sweden (Sveriges Geologiska Undersökning, SGU). If it is suspected that industrial activity could interfere with cultural heritage sites or culturally protected areas, the SGU involves the Swedish National Heritage Board (Riksantikvarieämbetet) and the Swedish National Maritime Museums (Sjöhistoriska Museet), both operating under the Swedish Ministry of Culture. The Swedish National Maritime Museums is the main institution with competence in maritime archaeology and has a professional archaeology unit that undertakes mitigation work associated with offshore developments.

Marine cultural heritage management in areas which are near-shore, or already governed by any regional or local marine planning, is treated in the same way as heritage on land, i.e. as a regional matter. This means that the County Administrative Boards (Länsstyrelse) handle the case according to the Heritage Conservation Act. The County Administrative Board assigns the proper agency if archaeological surveys or excavations are needed. This may be either the Maritime Museum (part of the Swedish National Maritime Museums), a regional museum, or private archaeological consultants. When it comes to management of submerged landscapes—a fairly new discipline—scholars from academic institutions are often appointed as advisors, either for the County Administrative Board, or the contracted archaeology institution. In 2018, only two regional museums have marine archaeology as a contracting speciality, both in southern Sweden.

Today there is a lack of professional archaeological divers, and often private companies have to be assigned as consultants to fulfil the diving regulations. Currently, no academic institute in Sweden offers training in scientific diving or an academic programme in marine archaeology.

Acknowledgements The authors wish to thank the rest of the Landscapes Lost and Blue Archaeology teams: Svante Björck, Johannes Edvardsson, Dan Hammarlund, Giacomo Landeschi, Hans Linderson, Ola Magnell, Peter Norman and Mats Rundgren. We thank divers Mikael Björk and Beesham Soogrim together with volunteers and Marinarkeologiska Sällskapet and SSDK, Karlshamn. We also thank Anders Fischer and Lars Larsson for valuable comments on earlier drafts, and the County Administrative Board of Antiquities, Blekinge, and Blekinge Museum for their support of this study in its earlier stages. For funding, we thank the Crafoord Foundation, the Royal Physiographic Society in Lund, the Swedish National Heritage Board (FoU 3.2.2-3104-2013) and the Swedish Research Council (2016-02027).

References

Andrén T, Björck S, Andrén E, Conley D, Zillén L, Anjar J (2011) The development of the Baltic Sea during the last 130 ka. In: Harff J, Björck S, Hoth P (eds) The Baltic Sea basin. Springer, Heidelberg, pp 75–98

Berglund BE, Björck S (1994) Late Weichselian and Holocene shore displacement in Blekinge, SE Sweden. Acta Universitatis Nicolai Copernici, Geografia XXVII, Nauki Matematyczno-Przyrodnicze, Zeszyt 92:75–95

Berglund BE, Sandgren P (2010) Strandförskjutningen I Blekinge – från istid till nutid. Strandhugg. Blekingeboken årgång 88 Årsbok för Blekinge Hembygdsförbund och Blekinge museum. Blekinge Hembygdsförbunds förlag, Karlskrona

Berglund BE, Sandgren P, Barnekow L, Hannon G, Jiang H, Skog G, Yu S-Y (2005) Early Holocene history of the Baltic Sea, as reflected in coastal sediments in Blekinge, southeastern Sweden. Quat Int 130:111–139

Björck S (1995) A review of the history of the Baltic Sea, 13.0-8.0 ka BP. Quat Int 27:19–40

Björck S (2008) The late Quaternary development of the Baltic Sea basin. In: The BACC Author Team (ed) Assessment of climate change for the Baltic Sea basin. Springer, Berlin/Heidelberg, pp 398–407

Björck S, Dennegård B (1988) Preliminary stratigraphic studies on the late Weichselian and Holocene development of the Hanö Bay, southeastern Sweden. Geogr Pol 55:51–62

Björck S, Andrén T, Jensen JB (2008) An attempt to resolve the partly conflicting data and ideas on the Ancylus-Littorina transgression. Polish Geol Inst Spec Pap 23:21–26

Boethius A (2018) Fishing for ways to thrive: integrating zooarchaeology to understand subsistence strategies and their implications among Early and Middle Mesolithic southern Scandinavian foragers. Acta Archaeologica Lundensia Series altera in 8°, no 70. Studies in osteology 4. Lund University, Lund

Bruzelius NG (1870). Fynden i Ystads hamn år 1868–69. Lund

Fischer A (1993) Öresundundersökningarna del 1. Kulturarvsstyrelsen, Köpenhamn

Fischer A (1996) Öresundsundersökningarna del 2. Kulturarvsstyrelsen, Köpenhamn

Fischer A (2005) Mennesket og havet i ældre stenalder. In: Bunte K (ed) Arkeologi och naturvetenskap. Gyllenstiernska Krapperupstiftelsen, Lund, pp 276–297

Gaillard MJ, Lemdahl G (1994) Early-Holocene coastal environments and climate in Southeast Sweden: A reconstruction based on macrofossils from submarine deposits. The Holocene 4:53–68

Groß D, Zander A, Boethius A, Dreibrodt S, Grøn O, Hansson A, Jessen C, Koivisto S, Larsson L, Lübke H, Nilsson B (2018) People, lakes and seashores: studies from the Baltic Sea basin and adjacent areas in the early and mid-Holocene. Quat Sci Rev 185:27–40

Hammarstrand Dehman K (2009) Våtmarksarkeologiska undersökningar i Malmö på senare år. Fornvännen 2009/1:1–16

Hansen L (1985) Resultat och redovisning av marina undersökningar mellan den 2–16 juni 1985 Hanöbuktens forntida miljö Meddelande 1. Malmö

Hansen L (1986) Hanöbuktens forntida miljö: forskningsprogram. Malmö

Hansen L (1995) Submerged Mesolithic landscapes. Preliminary results from the Hanö Bay, southern Baltic. In Fischer A (ed) Man & sea in the Mesolithic: Coastal settlement above and below present sea level. Proceedings of the International Symposium, Kalundborg, Denmark 1993. Oxbow Monograph 53. Oxbow, Oxford, pp 409–414

Hansson A (2018) Submerged landscapes in the Hanö Bay. Early Holocene shoreline displacement and human environments in the southern Baltic Basin. Lund University, Faculty of Science, Department of Geology, Quaternary Sciences, Lund

Hansson A, Björck S, Linderson H, Rundgren M, Nilsson B, Sjöström A, Hammarlund D (2017) Early Holocene landscape development and Baltic Sea history based on high-resolution bathymetry and lagoonal sediments in the Hanö bay, southern Sweden. In: Bailey GN, Harff J, Sakellariou D (eds) Under the sea: archaeology and palaeolandscapes of the continental shelf. Springer, Cham, pp 197–209

Hansson A, Nilsson B, Sjöström A, Björck S, Holmgren S, Linderson H, Magnell O, Rundgren M, Hammarlund D (2018a) A submerged Mesolithic lagoonal landscape in the Baltic Sea, southeastern Sweden – early Holocene environmental reconstruction and shore-level displacement based on a multiproxy approach. Quat Int 463:110–123

Hansson A, Björck S, Heger K, Holmgren S, Linderson H, Magnell O, Nilsson B, Rundgren M, Sjöström A, Hammarlund D (2018b) Shoreline displacement and human resource utilization in the southern Baltic Basin coastal zone during the early Holocene: new insights from a submerged Mesolithic landscape in South-Eastern Sweden. The Holocene 28:721–737

Hansson A, Hammarlund D, Landeschi G, Sjöström A, Nilsson B (2018c) A new early Holocene shoreline displacement record for Blekinge, southern Sweden, and implications for underwater archaeology. Boreas 48:57. https://doi.org/10.1111/bor.12339

Holmlund J, Nilsson B, Rönnby J (2017) Joint explorations of the sunken past. Examples of maritime archaeological collaboration between industry and academia in the Baltic. In: Bailey GN, Harff J, Sakellariou D (eds) Under the sea: archaeology and palaeolandscapes of the continental shelf. Springer, Cham, pp 53–63

Isberg O (1927) Beitrag zur Kenntnis der post-arktischen Landbrücke. Geografiska Annaler, Vol 9, Lundiana

Larsson L (1983) Mesolithic settlement on the sea floor in the strait of Øresund. In: Masters PM, Flemming NC (eds) Quaternary coastlines and marine archaeology: towards the prehistory of land bridges and continental shelves. Academic, London, pp 283–301

Larsson L (2017) Submerged settlement in the Öresund, Western Scania, southernmost Sweden. In: Bailey GN, Harff J, Sakellariou D (eds) Under the sea: archaeology and palaeolandscapes of the continental shelf. Springer, Cham, pp 165–175

Nilsson T (1961) Ravlundafältets geologi. Skånes Natur 1961:73–106

Nilsson B (2011) Fångstförsök. In: Andrén A (ed) Den äldre stenålderns passadiska och nomadiska globalitet. Förmodern Globalitet. Nordstedts, Stockholm, pp 25–44

Nilsson B (2012a) Forskningsprojekt undersöker världens äldsta fasta fiskeredskap. Marinarkeologisk tidskrift 3:9–10

Nilsson B (2012b) Flooded stone age – towards an overview of submerged settlements and landscapes on the continental shelf. Session report. European Archaeologist 38:84–86

Nilsson B, Sjöström A (2013) An på havets botten: geologisk bakgrund och arkeologisk framtid för Havängsundersökningarna. Österlent 1:19–21

Nilsson B, Sjöström A, Persson P (2018) Seascapes of stability and change: the archaeological and ecological potential of early Mesolithic seascapes, with examples from Haväng in the southern Baltic, Sweden. In: Persson P, Riede F, Skar B, Breivik HM, Jonsson L (eds) The ecology of early settlement in northern Europe – conditions for subsistence and survival. Equinox, Sheffield, pp 335–352

Rausing G, Larsson L (1977) Pilhaken, en stenåldersboplats under Öresund. Ale Historisk tidskrift för Skåneland 1977/2:1–3

Von Arbin S (2008) Sjöledning för naturgas längs den svenska västkusten. Särskild marinarkeologisk utredning, etapp 2. Bohusläns Museum Rapport 2008:40, Uddevalla

Wilcke JC (1850) Historiska och physiographiska underrättelser om Landskrona stad och hamn sammandragne a:o 1770. Stockholm

5 Germany: Submerged Sites in the South-Western Baltic Sea and the Wadden Sea

Hauke Jöns, Friedrich Lüth, Svea Mahlstedt, Julia Goldhammer, Sönke Hartz, and Hans-Joachim Kühn

Abstract

Only in the last 20 years have German research institutes and heritage agencies turned their attention to the investigation of the settlements that were inundated by rapid sea-level rise during the Holocene. Over 142 sites have been recorded so far, the majority on the Baltic coastline, and mostly of Late Mesolithic and Early Neolithic date. Underwater excavations on the Baltic coast of sites such as Timmendorf-Nordmole, Neustadt and Strande demonstrate the presence of large and well-preserved assemblages of stone, antler and wooden artefacts and other cultural features, comparable to the better-known underwater settlements of Denmark. Research within the framework of the SINCOS project has led not only to the discovery of numerous Stone Age settlements but also to a high-resolution reconstruction of the changes in coastal palaeogeography associated with the Littorina transgression. It has also raised the profile of the submerged Stone Age as a significant feature of the cultural heritage and demonstrated the value of and the need for multi-disciplinary collaboration. Fewer finds have been recovered on the North Sea coast, and this reflects the different environmental history of marine transgression, the greater thickness of marine sediments masking the prehistoric land surface and the greater technical challenges required to access it. Most finds here have been disturbed, a notable exception being the Late Neolithic votive deposit of an aurochs at Hamburger Hallig. The situation is also influenced by the legal and structural requirements imposed on research and the protection of sites by the relevant authorities in the various federal states: Mecklenburg-West Pomerania (Baltic Sea), Schleswig-Holstein (North Sea and Baltic Sea) and Lower Saxony (North Sea).

H. Jöns (✉)
Lower Saxony Institute for Historical Coastal Research, Wilhelmshaven, Germany
e-mail: joens@nihk.de

F. Lüth
German Archaeological Institute, Berlin, Germany
e-mail: Friedrich.Lueth@dainst.de

S. Mahlstedt
Lower Saxony Institute for Historical Coastal Research, Wilhelmshaven, Germany
e-mail: mahlstedt@nihk.de

J. Goldhammer
State office for Cultural Heritage Baden-Wuerttemberg, Gaienhofen-Hemmenhofen, Germany
e-mail: Julia.goldhammer@rps.bwl.de

S. Hartz
Museum for Archaeology Schloss Gottorf, Schleswig, Germany
e-mail: soenke.hartz@landesmuseen.sh

H.-J. Kühn
State Archaeology Department of Schleswig-Holstein, Schleswig, Germany

G. Bailey et al. (eds.), *The Archaeology of Europe's Drowned Landscapes*, Coastal Research Library 35, https://doi.org/10.1007/978-3-030-37367-2_5

Keywords
Littorina transgression · Mesolithic · Neolithic · SINCOS project · Submerged landscapes · Submerged sites · Underwater archaeology

5.1 Introduction

Germany, like neighbouring Denmark, has both a Baltic and a North Sea coastline, each with very different records of marine transgression and submerged archaeology. Systematic underwater archaeological investigations have focussed mainly on the Baltic coast, and though they began later than in Denmark, they have produced information of equivalent interest and importance, mostly of Late Mesolithic or Early Neolithic date. Currently there are 142 sites in the SPLASHCOS Viewer, the majority (100) from the Baltic coast (http://splashcos-viewer.eu/) (Fig. 5.1). Almost all these sites are shallow-water finds (<10 m). Some were chance discoveries by divers, many were discovered by systematic research-led investigations. Collaboration with offshore development projects has taken place in recent years through Environmental Impact Assessments conducted in advance of industrial development. Notable examples are the construction of a tunnel between Lolland in Denmark and Fehmarn in Germany, and the Nord Stream gas pipeline, which passes along the Baltic seabed from NW Russia to NE Germany. Both projects have involved geophysical and geological investigations of submerged landscapes but so far without significant discoveries of archaeological material except on the Danish side of the Fehmarnbelt tunnel (Enters et al. 2015; Sørensen 2016).

The beginnings of underwater investigations of submerged Stone Age sites in Germany go back to the nineteenth century, when occasional

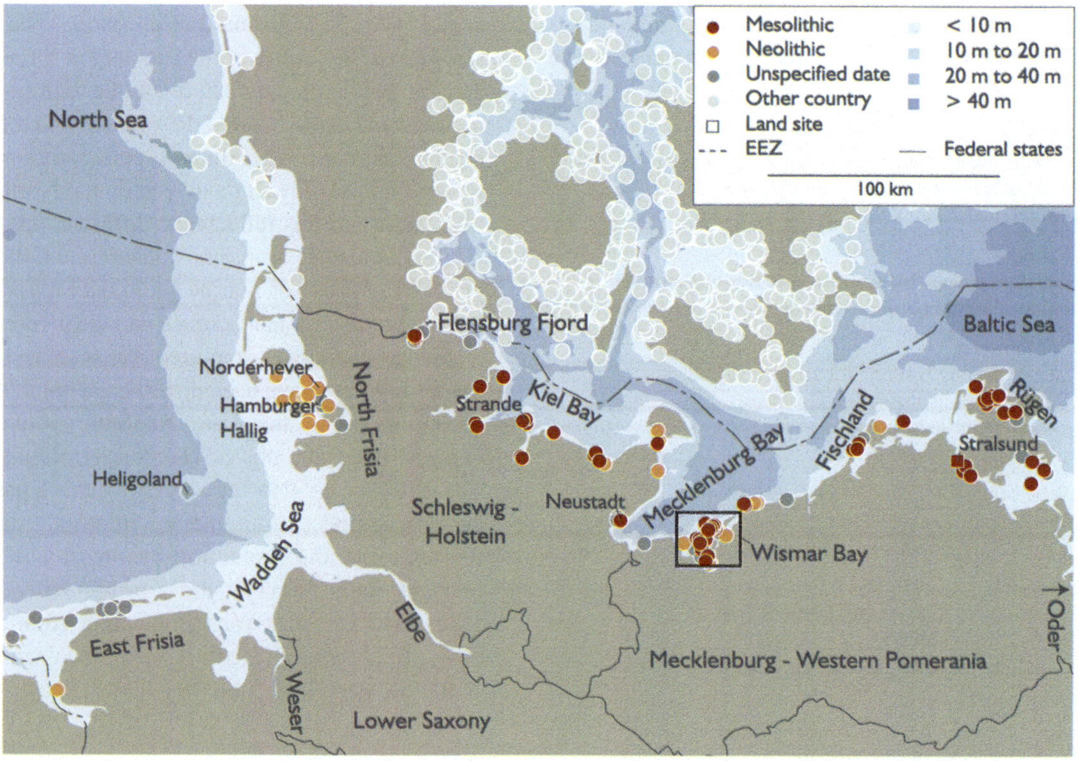

Fig. 5.1 Map of Germany showing distribution of underwater sites, key names and other places mentioned in the text. Site information from the SPLASHCOS Viewer http://splashcos-viewer.eu/. Drawing by Moritz Mennenga

Mesolithic finds brought up by offshore dredging were first recognised, and to closer collaboration of geologists and archaeologists in the mid-twentieth century, who began to unravel the history of Postglacial sea-level rise in the Baltic and its relationship to the Mesolithic-Neolithic transition (Jöns et al. 2007). However, systematic underwater research did not begin until the 1990s, inspired by the publication of the impressive results of work going back to the 1970s in neighbouring Denmark (e.g. Skaarup 1995) and by the culmination of research on the pile dwellings of Lake Constance in southern Germany (Schlichtherle 1997). The earliest investigations of submerged sites took place at the western end of the Mecklenburg Bight on the Baltic coastline of Schleswig-Holstein (Harck 1985).

An important boost to underwater research followed German reunification in 1990. Before then, the Baltic coastline of Mecklenburg-West Pomerania had been part of East Germany (the German Democratic Republic) since 1949 and civilian underwater research was prohibited. After 1990, investigations began to reveal underwater Mesolithic and Neolithic finds. These discoveries involved close cooperation between archaeological museums, heritage authorities and geological research institutes, a collaborative network that culminated in the multi-disciplinary SINCOS (Sinking Coast) project funded by the German Research Foundation (DFG) from 2002 to 2010 (Harff and Lüth 2007, 2011). This project with its collaboration between specialists in geology, archaeology, palaeobotany, archaeozoology, dendrochronology and palaeoclimatology resulted in systematic investigation of submerged shorelines and archaeology over wide areas of the Mecklenburgian and Pomeranian Bights and the discovery of some 50 new underwater archaeological sites (see Jöns 2011; Hartz et al. 2011). This must count as one of the largest concentrations of underwater prehistoric sites in Europe discovered as a result of research-led systematic exploration. These discoveries, along with excavations at some of the newly discovered sites, have produced evidence that is comparable, in terms of quantity and quality of preservation, to the famous underwater Mesolithic sites of Denmark (see Bailey et al., Chap. 3, this volume).

On the North Sea coastline, underwater sites are fewer and mainly comprise isolated finds recovered by chance following exposure of buried material by storms. This difference is largely due to the very different history of marine inundation and sediment accumulation as compared to the Baltic Sea and different conditions for the exposure and discovery of underwater material (Jöns 2010).

In this chapter we summarise the changing coastal geography and environment associated with sea-level rise since the Last Glacial Maximum (LGM), outline the archaeological context of the underwater finds, synthesise the archaeological results of underwater research with reference to key excavations, examine the variable conditions of underwater archaeological preservation and outline the prospects and challenges for future discoveries.

5.2 Palaeogeography and Sea-Level Change

The German coastline today is c. 3500 km, with c. 1500 km of coastline along the Wadden Sea facing the North Sea and c. 2000 km on the Baltic (Policy Research Corporation 2009).[1] It is bordered by three states—Lower Saxony on the North Sea coast, Schleswig-Holstein with both a North Sea and a Baltic coastline and Mecklenburg-West Pomerania bordering the Baltic—and it receives drainage from four major rivers: the Ems near the Dutch border and the Weser and the Elbe Rivers, all draining into the North Sea, and the Oder River, which forms the frontier with Poland and flows into the Baltic.

Both coastlines have a history of highly dynamic changes associated with Late Pleistocene and Early Holocene climate change and sea-level

[1] Most sources give a length of 2389 km, but these figures are notoriously vulnerable to variation because of fractal effects, and we prefer the larger figure here because it is given in a document specifically concerned with the impact of sea-level rise on the modern coastline.

rise, with marked differences between the North Sea and the Baltic at the broad scale, and differences at a regional scale along the Baltic coastline between Mecklenburg Bay in the west and Rügen Island in the east, and along the Wadden Sea between North Frisia and East Frisia (see Fig. 5.1). The nature of these palaeoenvironmental changes and their impact on prehistoric occupation and on the visibility and preservation of archaeological sites are discussed at the broad scale by Rosentau et al. (2017) for the Baltic and by Cohen et al. (2017) for the North Sea. More localised detail is presented for Denmark by Bailey et al. (Chap. 3, this volume), which has a similar palaeoenvironmental and archaeological sequence to the German coast following the LGM, but with some differences in the German case that are discussed further below.

5.2.1 The Baltic

The Baltic Sea today is a mainly brackish-water basin with a connection to the North Sea through the straits of Denmark. Periodically during the Last Glacial period, it was a fully enclosed freshwater basin cut off from the North Sea by a combination of ice barriers and isostatically rebounding land (see Jöns and bailey, Chap. 2, Bailey et al., Chap. 3, and Nilsson et al., Chap. 4, this volume). A fully marine connection to the North Sea was only finally established during the Littorina Transgression at about 8000 cal BP (6000 cal BC). Since then, sea level has continued to rise along parts of the Baltic coastline because of ongoing subsidence associated with glacio-isostatic adjustment following the retreat of the Scandinavian ice sheet (Harff et al. 2007, 2017; Groh et al. 2011; Rosentau et al. 2017; see also Glorstad et al., Chap. 6, this volume).

In central and northern Scandinavia, the Earth's crust and the more viscous material of the underlying mantle were pressed downwards by the main mass of the overlying ice sheet. Conversely, the areas of the Earth's crust around the southern periphery of the ice sheet were pushed upwards by the southward flow of material in the mantle to form a 'forebulge'. As the ice sheets retreated and the downward pressure on the crust and mantle was released, the land beneath slowly rebounded, and the forebulge subsided. Because of the time lag in the response of the crust to these changes in loading, these processes of uplift and subsidence continued long after the final retreat of the ice sheet, and in some areas they continue today. Where the land has rebounded, earlier shorelines have been uplifted, in some areas of northern Scandinavia by as much as 200 m or more, and there has been a relative drop in sea level. In areas of subsidence, the amount of vertical movement is less ($\leq$ 20 m), and here there has been a steady rise in relative sea level and progressive inundation of earlier shorelines.

The rate of relative sea-level regression associated with land rebound is highest in central and northern Scandinavia, where the ice sheet was thickest and the Earth's crust most deeply depressed (Fig. 5.2). The rate of uplift becomes progressively less towards the south, and the boundary between uplift and subsidence passes through southern Denmark and along the western section of the Baltic coast with the highest rates of subsidence and relative sea-level rise in Kiel Bay and Mecklenburg Bay. To the east, this boundary moves on shore, with some degree of uplift and relative sea-level regression further east along the Baltic coastline. This has resulted in different sea-level histories between Wismar Bay in the west and Rügen in the east (Fig. 5.3; Lampe et al. 2007).

More detailed mapping and dating of palaeoshorelines in the Wismar Bay were made possible by combined archaeological and geophysical exploration as part of the SINCOS project and show the changing configuration of the coastline at selected time slices during this period and how much land was lost by progressive encroachment of the sea between 6200 cal BC and AD 800 (Fig. 5.4).

5.2.2 The North Sea

Compared to the Baltic, the changing configuration of topography and coastlines in the North Sea basin with sea-level rise after the LGM

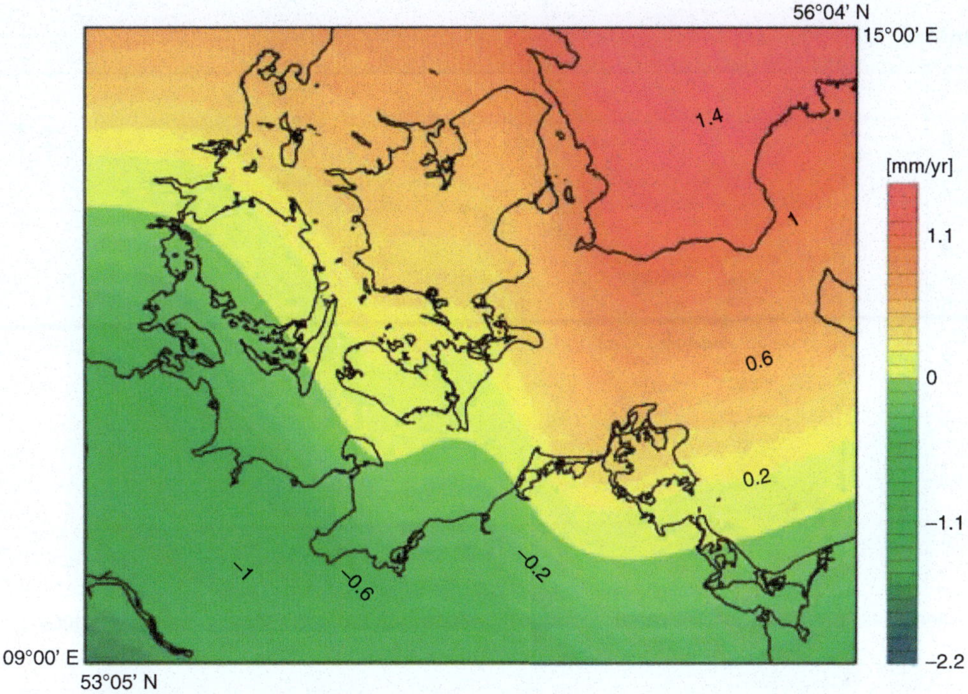

Fig. 5.2 The southwestern Baltic Sea, showing the rate of recent glacio-isostatic adjustment (GIA) based on relative sea-level change data reduced by an eustatic sea-level rise of 1.0 mm per year. After Harff and Meyer (2007, Fig. 14). Reproduced with permission

followed a different history, and the processes affecting the preservation and visibility of shoreline features and archaeological material, or their destruction, are different.

At the present day, the coastline is one of offshore barrier islands, tidal mudflats and coastal marshes, with one of the world's largest continuous mud-flat areas, up to 40 km wide—the Wadden Sea (Wattenmeer in German, Waddenzee in Dutch and Vadehavet in Danish), so named because of the ability to 'wade' through the mud at low tide. This zone is regularly inundated and then exposed again in an approximately 6-hour tidal rhythm. The tidal range, and hence the width of the intertidal zone and its impact on the potential for destruction or burial of onshore features, varies from 1.5 m in the north (North Friesland) to 4 m on the Lower Saxony coast (Fig. 5.5; Behre 2003). Storm surges can further amplify the destructive effects of marine inundation. They are no more frequent than on the Baltic coast, but the wave fetch is much larger, extending for hundreds of kilometres out into the North Sea, and generates waves more than twice as high as in the Baltic (Policy Research Corporation 2009). These destructive effects are mitigated by barrier islands formed by erosion and redeposition of earlier sediments, which can create bays and lagoons protected to some extent from the full force of waves in the open sea.

When sea level was at its lowest, land extended more than 300 km to the north and west of the modern German coastline, an area that has come to be known as Doggerland (Coles 1998), named after the area of the Dogger Bank, which would have formed a low hill within an otherwise lowland plain of limited relief, with lakes, marshes and river channels, especially those created by the extension of major rivers such as the Rhine, the Elbe and their various tributaries. A combination of seismic surveys, geophysical modelling of eustatic sea-level rise in relation to isostatic warping of the Earth's crust and coring of sediments has provided considerable detail, if patchily dis-

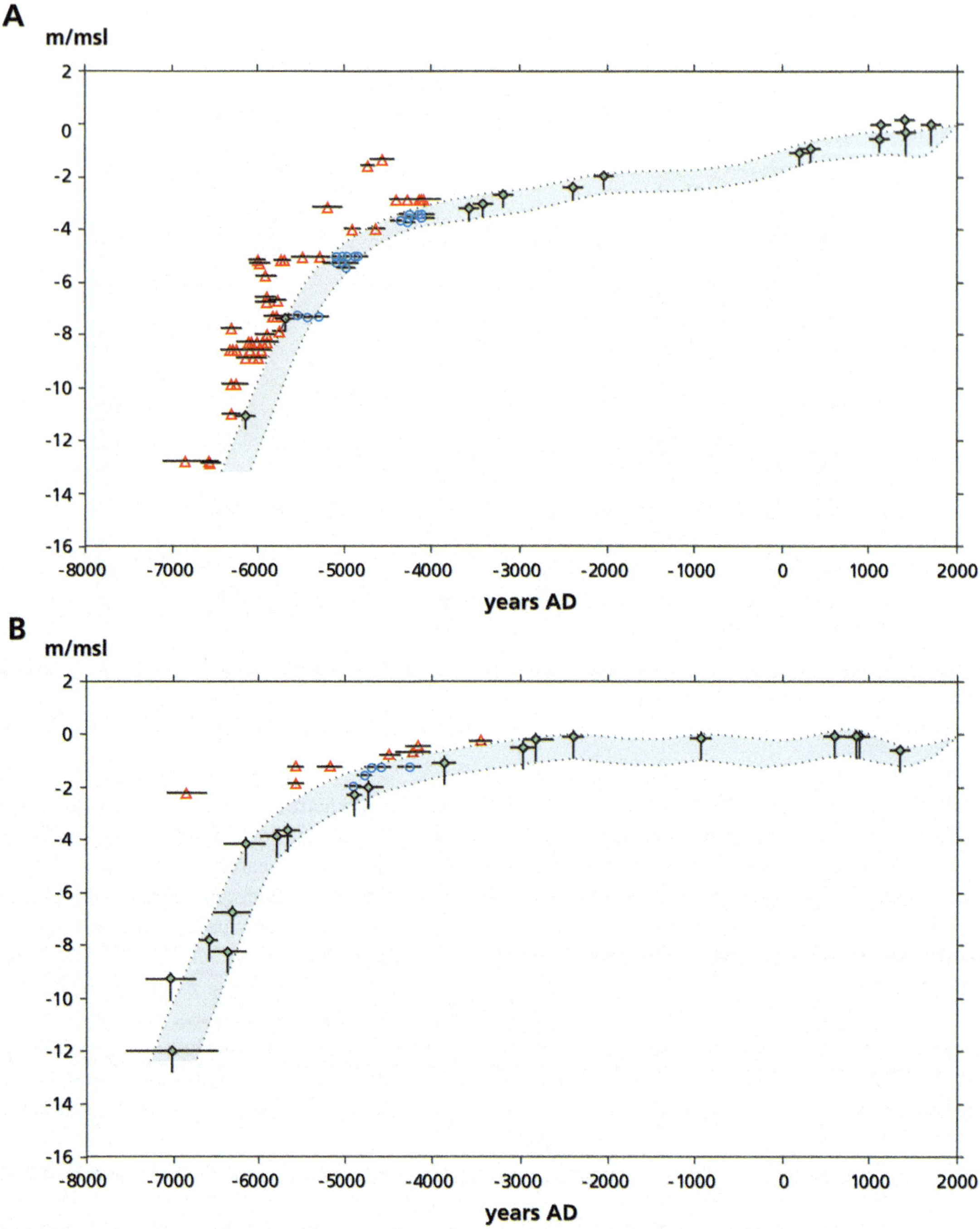

Fig. 5.3 Radiocarbon-dated relative sea-level curves for the Wismar Bight (**a**) and north Rügen/Hiddensee (**b**). Red triangles: terrestrial deposits; green diamonds, peat deposits; blue circles, archaeological finds in marine nearshore deposits. Horizontal bars represent 2σ, vertical bars indicate estimated vertical error of sea level position. After Lampe (2007, Fig. 4). Reproduced with permission

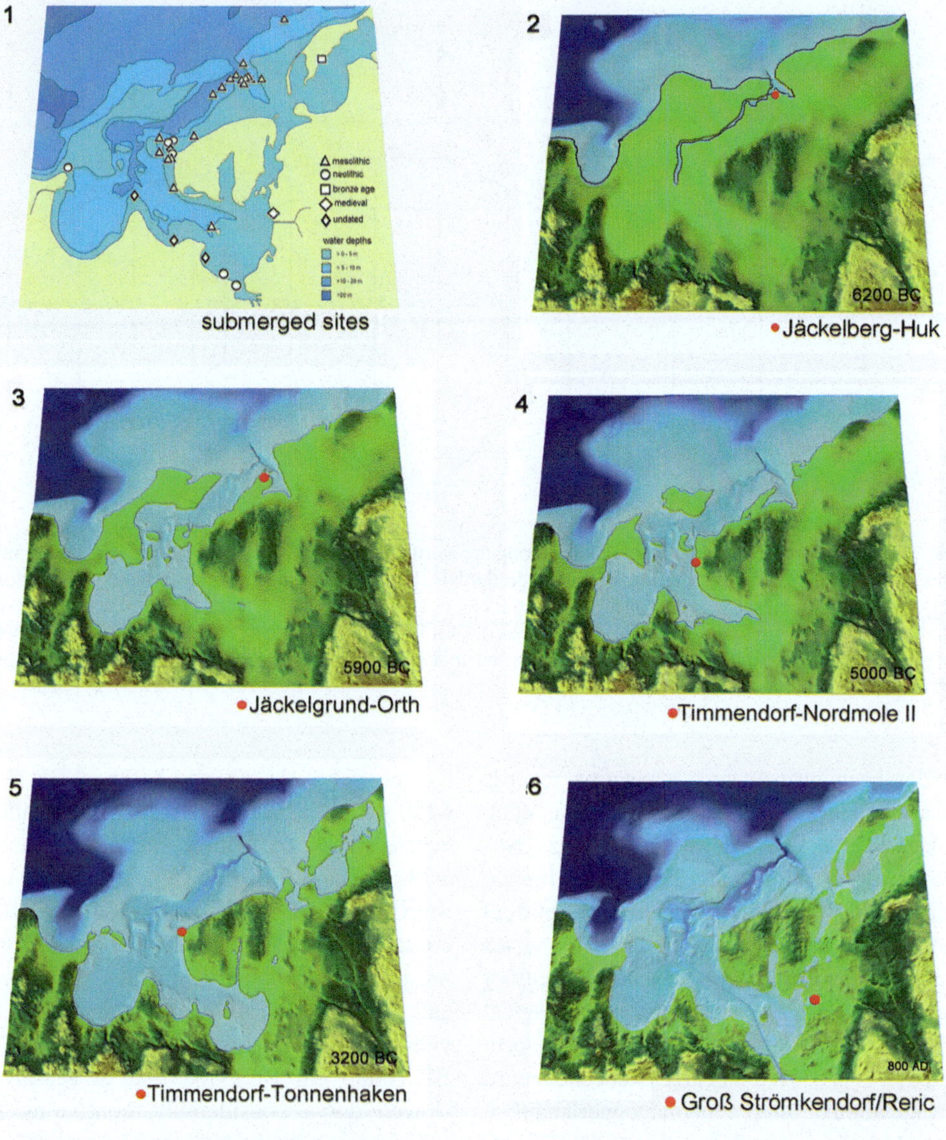

Fig. 5.4 Series of maps showing the changing configuration of shorelines and the locations of submerged archaeological sites in the Wismar Bight at different stages of marine transgression from the late seventh millennium BC to the first millennium AD. 1: location and date of sites and bathymetry; 2, 6200 cal BC; 3, 5900 cal BC; 4, 5000 cal BC; 5, 3200 cal BC; 6, AD 800. After Jöns 2011

tributed, about these topographic and environmental features and how they changed with progressive sea-level rise (Shennan et al. 2000, Streif 2004; Cohen et al. 2017 and references therein; see also Peeters and Amkreutz, Chap. 8; Bailey et al., Chap. 10, this volume). In the German sector, geophysical survey and coring have recently mapped the course of the Ems River offshore, showing that it was a meandering tributary of the Elbe River, with a width of 500–1500 m and a channel depth of 7–14 m cutting into the Pleistocene deposits below the seafloor (Hepp et al. 2017).

Information on vegetation development across Doggerland is at present still fragmentary. This is particularly true for the LGM. Deposits of this date suitable for vegetation reconstruction have

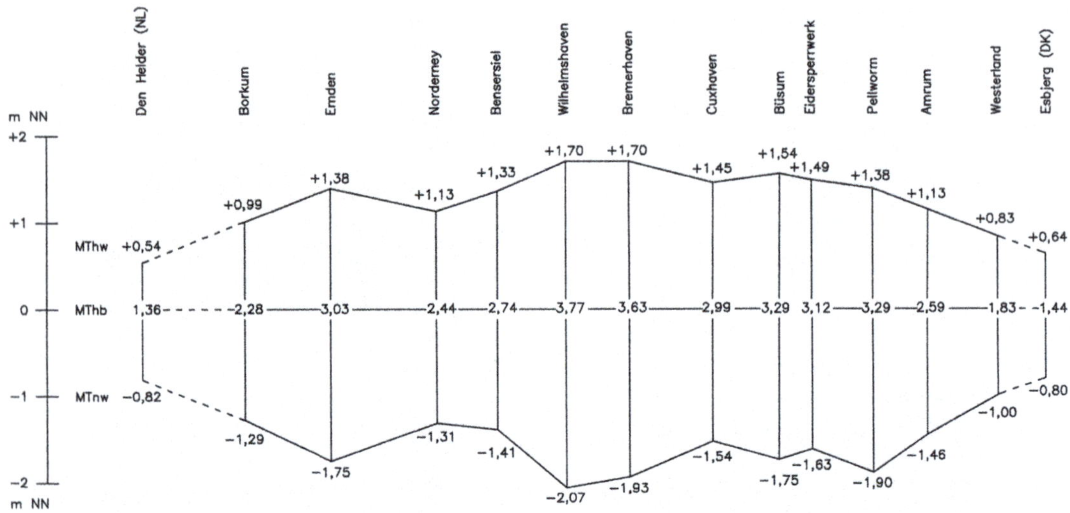

Fig. 5.5 Diagram showing variation in tidal range along the North Sea coastline from the Netherlands border in the SW (Die Helder at left of diagram) to the Danish border in the NE (Esbjerg at right of diagram). Note that the tidal range (MThb) is greatest in the central sector of the Lower Saxony coastline near the estuaries of the Weser and Elbe Rivers, and steadily reduces in both direction, in the island areas of East Frisia (Borcum to Bensersiel) and North Frisia (Büsum to Westerland). MThw: mean high water mark; MTnw, mean low water mark. After Behre (2003, Fig. 1). Reproduced with permission

not so far been investigated. Vegetation records from Late Glacial sediments are rare (Krüger et al. 2017; Brown et al. 2018). However, there is a better record for the Holocene. Thanks to climatic warming at the onset of the Holocene (c. 11,600 cal BP) and the influence of a continuously rising sea level, fen peat growth on the present-day seafloor commenced on a large scale, and palynological analysis of sediment cores from these peat sequences has produced an early Holocene record of the vegetation development on Doggerland (e.g. Behre and Menke 1969; Behre et al. 1984; Gearey et al. 2017; Wolters 2009; Wolters et al. 2010). The vegetation succession follows the general trend observed on the present-day mainland, with the development of pine-birch forests interspersed with aspen from c. 11,600 cal BP followed by expansion of hazel at around 10,700 cal BP and the successive spread of the typical deciduous elements of the mixed oak forest, such as elm (*Ulmus* sp.), oak (*Quercus* sp.), lime (*Tilia* sp.) and ash (*Fraxinus* sp.). Habitats close to the groundwater table were mostly occupied by eutrophic carr vegetation with willow (*Salix* sp.) and later mostly alder (*Alnus* sp.).

Between c. 21,000 and 8400 cal BP (19,000–6400 cal BC), sea level rose steadily and at an accelerating rate from a glacial lowstand of c. −130 m to −25 m (Fig. 5.6; Behre 2003, 2007; Streif 2004). During this period, the shoreline continuously shifted landwards with erosion of pre-existing shoreline features such as low cliffs and sand barriers, followed by deposition of fluvial and marine sediments. After about 8000 cal BP (6000 cal BC), the rate of sea-level rise slowed down considerably, continuing to rise slowly up to the present day. During this period the shoreline was subject to a series of minor regressions and transgressions, the regressions occurring during periods when the growth of peat bogs outpaced the long-term trend of sea-level rise. Thick sediments as much as 25 m thick or more also accumulated during this period, comprising complex and laterally variable sequences of peat, intertidal, brackish-lagoonal, alluvial and marine sediments.

Throughout the period of sea-level rise, the coastal landscape most likely comprised some combination of sand barriers, intertidal mudflats and coastal marshes. In the earlier part of the

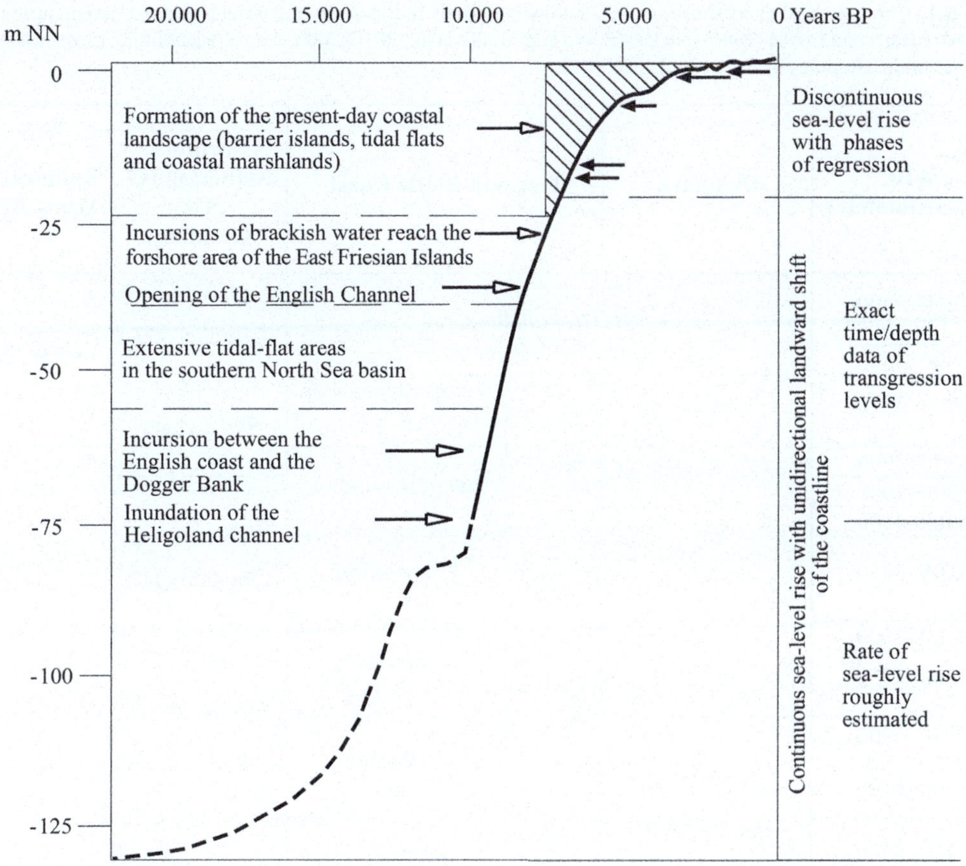

Fig. 5.6 Time/depth diagram of sea-level rise on the North Sea coastline of Germany from its lowstand during the Last Glacial Maximum to the present day. Depth is related to the German Ordnance Datum, NN, which is close to mean sea level. Note that ages are given in conventional radiocarbon years BP. Approximate conversion to cal BP is as follows: 20,000 = 23,930; 15,000 = 18,250; 10,000 = 11,550; 5000 = 5760. After Streif (2004, Fig. 3). Reproduced with permission

marine transgression when sea-level rise was relatively rapid, sediment sequences would have been disturbed by erosion and redeposition, resulting in a fragmentary record. Elevated features such as levees along river courses and barriers of sand along the coast would have provided attractive places for human settlement, but equally these features would have been vulnerable to destruction by erosion or hidden from view by burial under subsequent accumulations of sediment. During the Holocene, the reduced rate of sea-level rise and the accumulation of thick layers of sediment produced a smooth and shallow offshore gradient with extensive tidal mudflats, though not necessarily with the same character as the present-day Wadden Sea, because prior to the building of dykes, the coastline was constantly changing as a result of repeated transgressions and regressions (Jöns et al. 2013).

5.3 Archaeological Context and Underwater Sites

The northern German lowland was partly covered by ice during the LGM and uninhabited until the Late Glacial period, when ameliorating climatic conditions allowed reindeer hunters of the Hamburgian and Ahrensburgian cultures to enter the region (Table 5.1). Red flint from Heligoland

Table 5.1 Chronological chart showing archaeological periods in the German coastal areas and neighbouring countries. After Hartz and Lübke (2006), Niekus (2006) and Müller et al. (2010). Yellow: late Palaeolithic; green, Mesolithic; blue, Neolithic. Graphics by Svea Mahlstedt NIhK

cal BC (approximately)	Denmark	German Baltic Coast		Northern Netherlands / NW Germany	Radiocarbon Years BP
3250–3400	Fuchsberg/Wolkenwehe1			TRB – Brindley 1–3	4600–5000
3400–3500				Late Swifterbant	
3500–3800	Oxie	early TRB	Satrup/Siggeneben Süd		
3800–3900	Volling/Svaleklint		Wangels/Flintbek		5000–6000
3900–4000				Middle Swifterbant	
4000–4100	Late Ertebølle				
4100–4450		Ertebølle	Timmendorf		
4450–4500			Jarbock		
4500–4750				Early Swifterbant (ceramic)	
4750–4900	Early Ertebølle (aceramic)		Rosenfelde		
4900–5100				Late Mesolithic (Trapezes)	6000–7000
5100–5500			Jäckelberg		
5500–6500	Kongemose				7000–8000
6500–6600	Maglemose				
6600–9200				Early Mesolithic	8000–9000 9000–10000
9200–10600	Ahrensburgian				10000–11000

was exploited by these groups, and artefacts made of this raw material are found in the northern German hinterland some 200 km distant from the source, demonstrating movements out onto the now-submerged North Sea area (Fig. 5.7; Fries and Veil 2014).

With reforestation, Mesolithic hunter-gatherers became established in the region (Grimm et al. (2020), and the sequence can be structured into several phases with some variation between the Baltic and the Wadden Sea coastal regions. The Mesolithic cultural sequence on the Baltic coast is comparable to Denmark with Maglemose, Kongemose and Ertebølle culture periods—Early, Middle and Late Mesolithic, respectively (Brinch Petersen 1967, 1973; Holst 2014; see also Bailey et al., Chap. 3, Table 3.1, this volume, and Nilsson et al. Chap. 4, this volume). Information on the Late Mesolithic comes mostly from underwater sites, and the quality and detail of evidence have allowed a high-resolution chronology and further subdivision into cultural sub-phases (Table 5.1; Hartz and Lübke 2006).

In north-western Germany on the North Sea side, the sequence is similar to the Netherlands with Early and Late Mesolithic phases (Lanting and Van der Plicht 2000; Niekus 2006, Mahlstedt 2015), followed by material with similarities to the Swifterbant (Stapel 1991; Raemaekers 1999; Ten Anscher 2015). The Swifterbant, like the Late Ertebølle to the north, is a Mesolithic culture with ceramics. In the Netherlands it began earlier than the ceramic Ertebølle; conversely

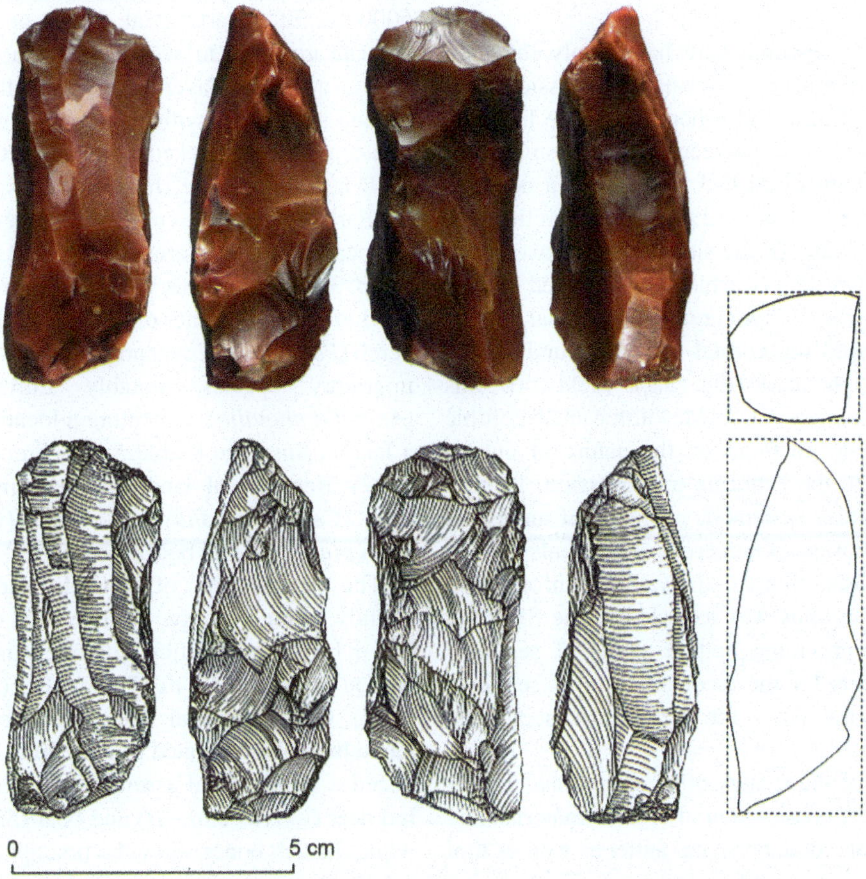

Fig. 5.7 The find of a Late Palaeolithic core made of typical red Heligoland flint found in Damme, district of Vechta, proves that the island was still connected to the mainland at the time. After Fries and Veil 2014, Fig. 3

Neolithisation took place earlier in the German Baltic region, where it is associated with the Early Funnel Beaker culture.

The currently known underwater finds are shown in Table 5.2. The majority (100) are from the Baltic and occur in water depths of 2–10 m. Mesolithic and Neolithic sites dominate (47% and 24%, respectively), and this reflects in part the fact that systematic survey in the Baltic region concentrated on these periods of the sequence, in part on the fact that sites in this depth range are the most easily accessible to underwater investigation. In situ cultural material from settlements is relatively rare (18%), the remaining sites comprising mainly collections of unstratified material (49%) or single finds (31%).

5.4 Sites on the Baltic Coast

Underwater research here has mainly focussed on the period of the Littorina transgression from 6000 to 2000 cal BC, when the people living in the maritime zone between the Oder estuary and the Flensburg Fjord had to adapt to a continuously rising sea level and a shifting shoreline (see Fig. 5.4). Many of the sites were discovered by the SINCOS project, which concentrated on systematic survey in two areas, the Wismar Bay in the west and the coastal waters around Rügen Island in the east (Fig. 5.1). Fieldwork was restricted to sites of Late Mesolithic and Neolithic date in order to focus on the nature of human response to the Littorina transgression. In both research areas, systematic geophysical survey led to the discovery of numerous settlements located on palaeoshorelines. All available information about these finds was recorded in the SINCOS database, providing both a series of sea-level index points for the reconstruction of sea-level rise and the basis for further archaeological investigation.

Some of these sites offer exceptional conditions for the preservation of organic material, and artefacts of wood, bone and antler as well as flint tools were recovered in large quantities during excavations, together with other settlement features such as remains of timber structures and hearth areas.

5.4.1 Wismar Bay

One of the most important concentrations of sites is in the Wismar Bight, where well-preserved archaeological deposits were discovered, some of which have been excavated (Jöns et al. 2007; Hartz et al. 2011; Lübke et al. 2011). A systematic programme of joint geophysical and archaeological survey was able to identify a succession of palaeoshorelines and archaeological sites associated with progressive marine transgression and transformation from freshwater to fully marine conditions (see Fig. 5.4). Here we describe the key excavations in chronological order starting with the earliest.

The oldest site is **Jäckelberg-Huk** on the north side of the present-day island of Poel, found at a depth of 8.5 m below present sea level (bpsl) and radiocarbon-dated between 6400 and 6000 cal BC (Hartz et al. 2011, pp. 94–104). Here an area of 31 m^2 was excavated and revealed part of a cultural layer sealed beneath peat and marine sediments, with remains of two hearths indicated by fire-cracked stones. Microlithic flint arrowheads comprise rhombic forms typical of the Early Kongemose culture, consistent with the radiocarbon dates (Sørensen 2017; Bailey et al. Chap. 3, Table 3.1, this volume). Faunal remains are dominated by bones of freshwater fish such as perch (*Perca fluviatilis*) and pike (*Esox lucius*) or migratory species, notably European eel (*Anguilla anguilla*), indicating a location next to a lake at this time (Table 5.3; Schmölcke et al. 2007). Remains of land mammals include red deer (*Cervus elaphus*), roe deer (*Capreolus capreolus*) and wild boar (*Sus scrofa*).

The nearby site of **Jäckelgrund-Orth** is slightly younger in age, discovered at a depth of 7–8 m bpsl in association with tree stumps radiocarbon dated at 6000 to 5700 cal BC (Jöns et al. 2007). Only terrestrial fauna was present here. **Jäckelberg-Nord**, dated at c. 5200 cal BC, produced a small faunal assemblage with bones of red deer (*Cervus elaphus*) and boar (*Sus scrofa*), while the fish bones show the first appearance of marine species such as cod (*Gadus morhua*), herring (*Clupea harengus*), flatfish (Pleuronectidae) and garfish (*Belone belone*) alongside freshwater species (Hartz et al. 2011, pp. 110–111, Table 10).

Table 5.2 Distribution of underwater finds by type of site and period. Miscellaneous other includes graves, pile settings and palaeosurfaces. Data from the SPLASHCOS Viewer at http://splashcos.maris2.nl

Type of site		Palaeolithic	Palaeolithic-Mesolithic	Mesolithic	Mesolithic – Neolithic	Neolithic	Bronze Age	Indeterminate Prehistoric	Total	
									N	%
In situ cultural features – settlement data				20	2	2		1	25	*17.6*
Unstratified material – multiple finds			1	42	2	5		19	69	*48.6*
Single find (unstratified)		1		4		24	2	13	44	*31.0*
Votive deposits						1			1	*0.7*
Miscellaneous other				1		2			3	*2.1*
Total	**N**	**1**	**1**	**67**	**4**	**34**	**2**	**33**	**142**	***100.0***
	%	*0.7*	*0.7*	*47.2*	*2.8*	*23.9*	*1.4*	*23.2*		*99.9*
Presence of organics (wood and/or bone)		1		15	2	2		2	**22**	***15.5***

Table 5.3 Faunal remains from Poel 42 and Jäckelberg-Huk (6300–6000 cal BC). Numbers of identified animal bones (NISP) and their percentage frequency. The fish species are classified as marine (M), migratory (F/M) and freshwater (F). After Hartz et al. (2011, Table 5)

Fish	NISP	%	Mammal	NISP	%
Perca fluviatilis (F)	1340	*61.0*	*Arvicola terrestris*	37	*45.1*
Esox lucius (F)	386	*17.6*	*Cervus elaphus*	22	*26.8*
Cyprinidae (F)	304	*13.8*	*Capreolus capreolus*	13	*15.9*
(*Tinca tinca*	3	*0.1)*	*Sus scrofa*	2	*2.4*
(*Rutilus rutilus*	5	*0.2)*	*Erinacaeus europaeus*	2	*2.4*
(*Scardinius* erythrophthalmus	2	*0.1)*	*Felis silvestris*	2	*2.4*
Anguilla anguilla (F/M)	122	*5.6*	Phocidae	2	*2.4*
Pleuronectidae (M)	7	*0.3*	(*Halichoerus grypus*	1)	*1.2*
Gadus morhua (M)	4	*0.2*	*Bos primigenius*	1	*1.2*
Coregonus sp. (F/M)	24	*1.1*			
Total	**2197**	***100***		**82**	***99.8***

The site of **Timmendorf-Nordmole II** is located only a few kilometres south of Jäckelberg at a depth of 5 m bpsl. Excavation over an area of 12 m^2 revealed a land surface alongside the edge of a stream bed. The stream bed had been filled with fluvial and estuarine sediments and peat, and the land surface had been partly covered and sealed by these deposits. Most of the finds were found in the marine sediments of the stream fill where preservation conditions for organic material were excellent, with recovery of wooden artefacts including several leister prongs and remains of a fish weir originally built across the mouth of the stream. Analysis of the material culture indicates that the site belongs to an aceramic phase of the Ertebølle culture, and radiocarbon dates place the main cultural layer between 5100 and 4800 cal BC (Hartz and Lübke 2006; Hartz et al. 2011, p. 111). By this period, marine species were dominant in the fish-bone assemblage (80%), with migratory eel dominating alongside cod and mackerel (*Scomber scombrus*). Beneath this main cultural layer, a second, smaller culture layer was found containing mammal bones dated to 5900–5600 cal BC. On the terrestrial surface of the adjacent shore and sealed beneath peat and marine sediment, there was a hearth area comprising a concentration of charcoal and white ash dated to c. 5400 cal BC (Jöns et al. 2007, p. 99, Table 2).

The neighbouring site of **Timmendorf-Nordmole I** is also of great scientific interest, representing an extensive scatter of cultural remains over an area some 250 m x 100 m at a depth of 2.5–3.5 m bpsl. The site was originally located on the shore of an inlet with a settlement area on dry land and an offshore 'dump' area in shallow water, similar to the locations of some of the underwater sites in Denmark (see Bailey et al., Chap. 3, this volume). Three trenches covering a total area of c. 42 m^2 were excavated, including part of the shore zone on former dry land and the offshore dump zone (Hartz et al. 2011, pp. 121–127, and references therein). Artefacts were also found scattered on the surrounding surfaces. The cultural material belongs mostly to the Late Ertebølle culture with radiocarbon dates between c. 4400 and 4100 cal BC. The settlement surface on dry land had been badly eroded by subsequent marine inundation, but in one location a pit, 3.5 m × 1.8 m at its surface and 0.9 m deep, had survived the effects of erosion. The pit contained a number of logs and poles thought to be the remains of a collapsed roof structure (Fig. 5.8), but the pit appears to be too small to represent a pit dwelling. The sediment fill contained numerous artefacts including a truncated blade hafted at right angles to a hazelwood handle with a binding of thin twisted cord and lime bast (Fig. 5.9; Lübke 2001, 2003). Elsewhere, the culture layer extended offshore with layers of peat and palaeosols sealed beneath marine sand. These contained numerous unpatinated flint artefacts, animal bones, wooden

Fig. 5.8 The pit at Timmendorf-Nordmole I in Wismar Bay with the timbers of a collapsed wooden structure. Photo courtesy of State Heritage Agency of Mecklenburg-West Pomerania

Fig. 5.9 Truncated flint blade hafted in a hazelwood handle, found at Timmendorf-Nordmole I, showing details of the cordage and lime bast used to attach the flint to the handle. A: dorsal surface; B, ventral surface; C, closeup of ventral surface. Scale in centimetres. Photo by Sabine Suhr, State Heritage Agency of Mecklenburg-West Pomerania

artefacts including leister prongs and logboat fragments and a small number of Ertebølle potsherds, some with charred food crusts. Here too, eel is a dominant species in the faunal remains with some cod and flatfish. Terrestrial fauna is also present, and the site has been interpreted as a specialist site for terrestrial-game hunting and eel fishing (Hartz et al. 2011, pp. 126–127).

Another site nearby discovered at 2 m bpsl is **Timmendorf-Tonnenhaken**, located on a former peninsula. Here a cultural layer with well-preserved artefacts made of stone, bone and antler was recovered as well as potsherds of the Early Neolithic Funnel Beaker culture. The latter are consistent with radiocarbon dates of c. 3200 and 2700 cal BC and the fact that all the bone material is from domesticated animals such as cattle and pigs (Lübke 2002).

This combination of sites is important in a number of ways. It provides a detailed reconstruction of development from the Late Mesolithic to the Early Neolithic between 6000 and 4000 cal BC for the region. It documents the progressive encroachment of the sea on the land and the pattern of shifting coastlines associated with the Littorina transgression (see Fig. 5.4). Moreover, this spatial separation of sites enforced by moving shorelines has allowed the reconstruction of a more detailed chronological subdivision of the material culture than would otherwise have been the case. The abundant faunal remains also demonstrate the importance of marine resources alongside terrestrial hunting and the progressive shift from freshwater to marine species with rising sea level (Schmölcke et al. 2007; Lübke et al. 2011). Shoreline locations were important for the Mesolithic settlements, especially on the banks of inlets, where fish could be easily trapped, but remained important during the Early Neolithic, when the economy had become more dependent on domestic animals.

5.4.1.1 Wooden Artefacts

Most of the wooden artefacts in the Wismar Bay excavations are associated with fishing, perhaps reflecting the fact that they were discovered in what were originally shallow water areas close to the shoreline where such material would most likely have been concentrated and well preserved (Klooß 2015). The most common items are sections of roundwood, sometimes with the remains of the cordage used to tie them together, leister prongs, again with remains of cordage showing how they were tied together to form leister spears most likely used for eel fishing, fragments of basketry from fish traps and paddles (Fig. 5.10). Some wooden items are thought to be the remains of fish weirs, and fragments of logboats have also been recovered. Wooden handles for antler and ground stone axes have also been found, and the distinctive hafted implement referred to above (see Fig. 5.9) and found in the pit fill at Timmendorf-Nordmole I.

The manufacture of the wooden tools implies detailed knowledge of the different types of trees appropriate to different purposes and skill in woodland management, especially the coppicing of hazel to produce straight and uniform rods.

5.4.2 Rügen

Research on the Late Mesolithic and Early Neolithic cultures of Rügen Island and the coastal areas of the adjacent mainland reaches back to the early twentieth century (Terberger 2007, p. 241; Hartz et al. 2011, p. 135). Early discoveries included numerous lithic and organic artefacts, several human skulls and artefacts of central European provenience dredged up from the seafloor, material attributed to a regional variant of the Ertebølle Culture, known as the Lietzow Culture, and indicating good prospects for the recovery of well-preserved underwater cultural deposits. In 2002, the SINCOS project began a programme of research on the existing archives of material, AMS radiocarbon dating and new underwater geophysical survey and excavation (summarised in Jöns et al. 2007, p. 153; Hartz et al. 2011).

By coincidence, large-scale rescue excavations by the state heritage authorities took place in advance of construction work in Stralsund City and at Baabe on Rügen Island over the same period as the SINCOS project. These produced abundant additional evidence of coastal sites and sea-level rise in the period 5500 to 3500 cal BC,

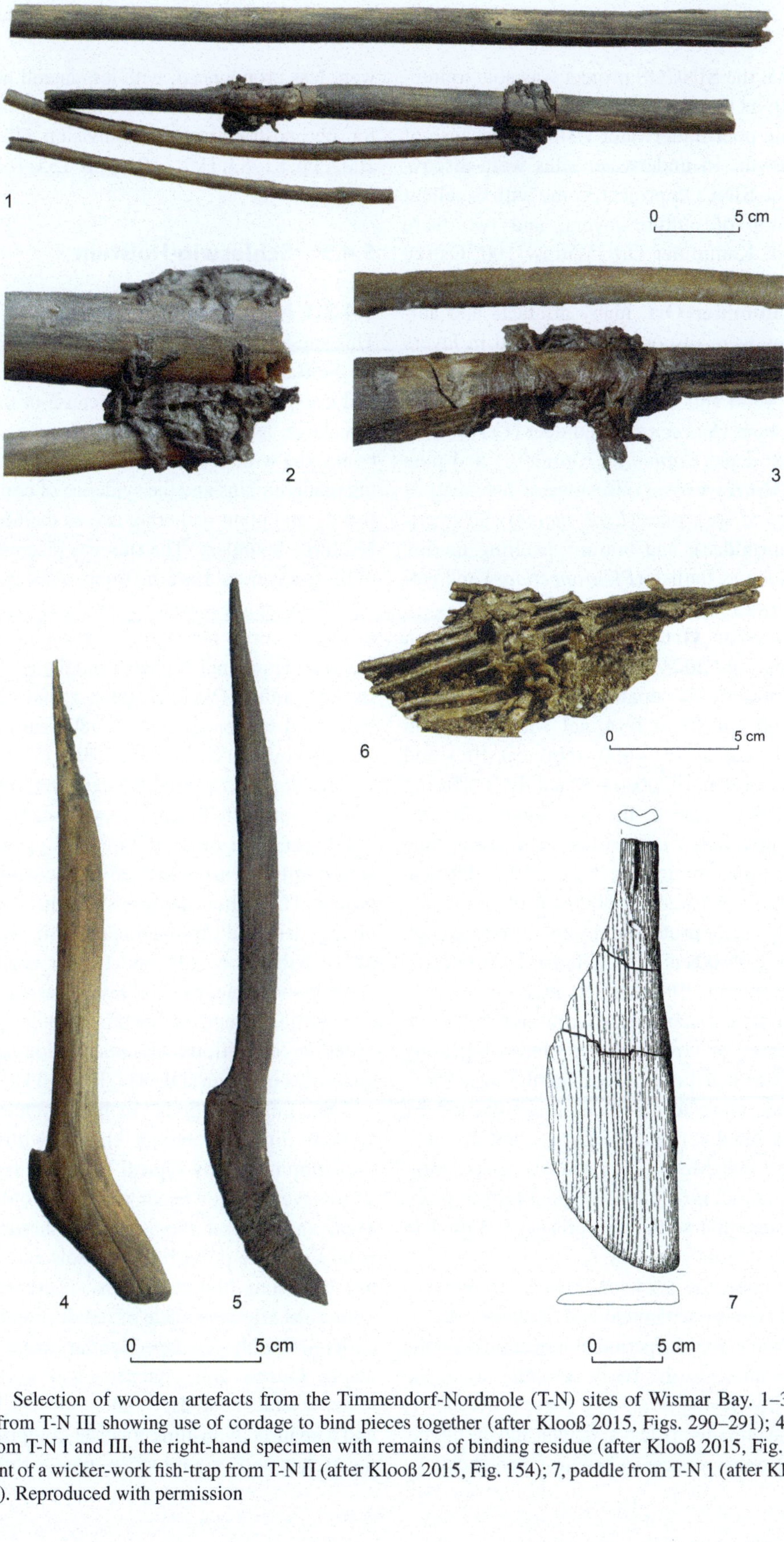

Fig. 5.10 Selection of wooden artefacts from the Timmendorf-Nordmole (T-N) sites of Wismar Bay. 1–3: wooden artefacts from T-N III showing use of cordage to bind pieces together (after Klooß 2015, Figs. 290–291); 4–5, leister prongs from T-N I and III, the right-hand specimen with remains of binding residue (after Klooß 2015, Fig. 288 a, b); 6, fragment of a wicker-work fish-trap from T-N II (after Klooß 2015, Fig. 154); 7, paddle from T-N 1 (after Klooß 2015, Taf. 36, 1). Reproduced with permission

results that the SINCOS project was able to integrate with its own research.

Despite poor underwater visibility because of algae growth, 15 underwater sites were discovered by the SINCOS project, some with excellent preservation of culture layers, and two were excavated, Kamminer Ort (Wittow 100, Ostsee VI) and Breetzer Ort (Bergen 24, Ostsee VI).

At **Kamminer Ort**, many artefacts and animal bones were discovered embedded in layers of peat and marine gyttja, dating to the period between 4900 and 4700 cal. BC. Animal bones included boar (*Sus scrofa*), red deer (*Cervus elaphus*), roe deer (*Capreolus capreolus*) and pine marten (*Martes martes*), freshwater fish such as perch (*Perca* sp.), pike (*Esox lucius*) and cyprinids (Cyprinidae), and bottom-dwelling marine species such as flatfish (*Pleuronectidae*) and eelpout (*Zoarces viviparus*).

At **Breetzer Ort**, excavation campaigns in 2004, 2007 and 2008 demonstrated the presence of three phases of occupation: Late Mesolithic (c. 5600–5500 cal BC), Terminal Mesolithic and Middle Ertebølle (c. 4800–4500 cal BC) and Final Mesolithic (4200–4100 cal BC). During this 1500-year period, the site underwent progressive inundation by sea-level rise during the Littorina transgression. A Late Mesolithic fire place submerged at about 5000 cal BC remained intact because of protection by two collapsed oak trunks. At c. 4800 cal BC, settlement refuse comprising numerous flint artefacts and fragments of wooden artefacts such as leister prongs and arrow shafts started to cover the fire place, indicating that the site had become part of the near-shore dump zone close to a settlement on the nearby shoreline. Finally, remains of a wooden structure from the Final Mesolithic,probably a fish weir, were found, but no more settlement refuse, probably because the settlement area had shifted in response to rising sea level.

In summary, the results in this region suggest a similar pattern of settlement and cultural and economic change over the period of transition from the Late Mesolithic to the Early Neolithic as in the Wismar Bay to the west. The main difference is that faunal remains are less abundant and the effects of marine transgression and environmental change were less pronounced, with less lateral movement of the shoreline and therefore reduced possibilities for chronostratigraphic subdivision (Jöns et al. 2007, pp. 81–83; Hartz 2011, pp. 135–152).

5.4.3 Schleswig-Holstein

5.4.3.1 Neustadt

This site in the Bay of Lübeck was one of the first on the Baltic coast to be systematically excavated and is especially important because of the unusually large assemblages of artefacts and animal bones, the wide range of organic materials including plant remains and the evidence of continuity in fishing and plant gathering across the Mesolithic-Neolithic boundary. The site was discovered by a military diver in 1999 in front of the harbour at 3–4 m bpsl and excavated from 2000 to 2006, producing material belonging to the Late Ertebølle Culture (Timmendorf phase) and the Early Funnel Beaker Culture (Wangels phase) with a radiocarbon-dated sequence of 4500–3800 cal BC (Hartz and Glykou 2008; Hartz et al. 2011).

The site was originally located on the shore of a lagoon protected from the open sea by a barrier beach. An area of c. 110 m^2 was excavated, of which only 10 m^2 belong to the occupation area proper. The culture layer on land has been completely destroyed by marine erosion, but in front of the habitation area, and further out from the ancient shoreline, the find layer merges into peat and gyttja formed in brackish water, providing excellent conditions of preservation and large quantities of cultural debris (Fig. 5.11).

The cultural finds include some 40,000 worked flints (excluding chips), with finished tools dominated by typical Ertebølle types such as transverse arrowheads and trapezoidal flaked axes, and several thousand pot sherds, mostly from Ertebølle point-based vessels and a smaller number from blubber lamps. There were also numerous artefacts made of red deer and roe deer antler—notably T-shaped antler axes, harpoon heads, chisels and punches—and a variety of bone artefacts—awls, points, chisels, daggers and a spatula. Non-flint material comprises hammerstones and ground stone axes including two

Fig. 5.11 Some of the finds from Neustadt showing the abundance and good preservation of bone and antler. Photo by Sönke Hartz, Museum for Archaeology Schloss Gottorf

Fig. 5.12 A shaft hole axe from Neustadt made of amphibolite. The source of the amphibolite is the uplands of central Europe, and these artefacts were widely in use by the earliest Neolithic cultures of northern and central Europe, notably the Linearbandkeramik (LBK). Isolated specimens are occasionally found in Late Ertebølle sites in the western Baltic, indicating a distribution network between the farmers of central and northern Europe and their Ertebølle hunter-fisher-gatherer neighbours to the north. Photo by Sönke Hartz, Museum for Archaeology Schloss Gottorf

Danubian shaft hole axes made of amphibolite imported from the south (Fig. 5.12), underlining the special role of East Holstein in the distribution of these imports.

Wooden tools comprise a great variety of types and raw materials. There are some 26 leister prongs made of dogwood (*Cornus sanguinea*), hazel (*Corylus avellana*), ash (*Fraxinus excelsior*), pome fruit tree (Maloideae) and oak (*Quercus* sp.), four groups of fragments of logboats made of lime (*Tilia* sp.), 13 paddles of ash wood (Fig. 5.13), 35 fragments of spear shafts made of ash, two angled handle shafts of hazelwood for axes, four wooden handles for antler axes, the handle section of a bow made from elm wood (*Ulmus* sp.), remains of six arrow shafts made from ash, hazel (Fig. 5.14 left) and guelder rose (*Viburnum opulus*) and 60 pointed sticks and poles, mostly made from ash and representing the remains of fish weirs. Also discovered were a net float made of bark, several fragments of wicker-

Fig. 5.13 Left: Neustadt arrowhead with club-shaped tip made from hazelwood (*Corylus avellana*). Right: Neustadt pieces of cord made from twisted fibres of lime bast (*Tilia* sp.). Photos by Sönke Hartz, Museum for Archaeology Schloss Gottorf

Fig. 5.14 Neustadt paddle made of ash (*Fraxinus excelsior*). Photo by Sönke Hartz, Museum for Archaeology Schloss Gottorf

work fish traps, numerous braided and twisted cords made from tree bast (Fig. 5.14 right) and the fragments of a woven piece of textile or a net (Hartz et al. 2011, pp. 90–93).

There are many thousands of animal bones of mammals, birds and fish (Glykou 2014). Mammals and birds account for 3874 identified specimens with approximately equal representation of marine and terrestrial mammals. Water fowl (Anatidae) are dominant amongst the bird bones. Land mammals include red deer (*Cervus elaphus*), roe deer (*Capreolus capreolus*), wild boar (*Sus scrofa*) and aurochs (*Bos primigenius*). Bones of marine mammals are dominated by seals (mostly harp seal, *Pagophilus groenlandicus*, together with grey seal, *Halichoerus grypus*, and ringed seal, *Pusa hispida*). Bones of porpoise (*Phocoena phocoena*) are also quite common, and there are a few bones of dolphin (*Delphinus delphis*). In a fish bone assemblage of some 13,000 specimens, marine species account for 96% of identifiable bones, mostly from cod (Gadidae) but also including migratory species such as eel (*Anguilla anguilla*) and salmon (Salmonidae).

The Neolithic material is represented by sherds of Funnel Beaker pottery and a small number of bones of domestic animals (ovicaprids and cattle).

Charred food crusts are present on the interior surfaces of many of the sherds. Bulk stable isotope analysis of these residues shows that aquatic foods had been processed in both Ertebølle and Funnel Beaker vessels, whilst evidence of dairy products was present on several Funnel Beaker sherds. These data demonstrate that aquatic resources continued to play an important role in the Neolithic economy alongside agricultural products (Craig et al. 2011). Further evidence of continuity in economic practices across the Mesolithic-Neolithic transition comes from the analysis of starch granules in the food crusts, which show an emphasis on the processing of acorns in both Ertebølle and Funnel Beaker sherds (Saul et al. 2012). Preserved phytoliths in these carbonised deposits also demonstrate the use of garlic mustard (*Alliaria petiolata*) during the Late Mesolithic and Early Neolithic (Saul et al. 2013).

5.4.3.2 Strande

Located at a depth of 6 m in Kiel Bay, this site was discovered in 2011 as a result of a chance discovery by divers investigating collapsed oak-tree trunks (Fig. 5.15). Given the presence of stone artefacts and bone fragments nearby, a 5 m^2 test excavation was carried out in 2012, which revealed the presence of a well-preserved cultural layer in organic sediments beneath a protective cover of marine sand (Goldhammer and Hartz 2015, 2017). This yielded many stone artefacts including waste products of tool manufacture and wooden implements, notably two leister prongs. Other organic materials, interpreted as food remains, include bones of marine and freshwater fish, mainly cod (*Gadus morhua*), some of which are charred, marine and terrestrial mammals including bones of red deer (*Cervus elaphus*) and roe deer (*Capreolus capreolus*) with cut marks, water birds (mostly Anatidae) and shells of acorns and hazelnuts. A tool made from the tusk of a wild boar shows use-wear probably from fur or leather

Fig. 5.15 Fallen oak trees at the Strande site in Kiel Bay, which were the first indication that an archaeological site might be preserved here. Photo by Gerald Lorenz

processing (Glykou et al. 2014). Two pieces of human jaw bone with molar teeth and two single human teeth were also recovered. The jaw bone belongs to a female in her 20s, while one of the single human teeth is from a second older individual (Peter Tarp and Prof Dr. JL Boldsen, Odense University, Denmark, pers. comm., 2018). Analysis of $\delta^{13}C$ values indicates a diet based on marine food (Goldhammer et al. 2020).

The site is important because of its date and the extent of the deposits. The lithic assemblage is characteristic of the aceramic Ertebølle, and this is consistent with radiocarbon and tree-ring dating, which indicate a time span of 5450–4750 cal BC, a period which is poorly represented in other underwater sites in the region. Test pitting at regular intervals along a series of transects radiating out from the excavation area shows that sediments with preservation of organic materials and the presence of flint artefacts extend over an area of at least 2500 m² beneath the sand layer. Further investigations are ongoing, including scientific analyses of the human remains.

5.5 Sites in the Wadden Sea

The main challenges to the discovery of material here are the result of the relatively rapid accumulation of marine sediments and the shallow offshore topography, as described earlier. Stone Age materials are quite deeply buried beneath thick layers of later marine sediment, and the only prospect of discovery is when storms erode some of the overlying sediment and expose the underlying land surface. This happens quite frequently and has resulted in the exposure of many prehistoric artefacts and sometimes more extensive cultural layers and features (Kühn 2007). However, once such materials have been exposed, there is the additional challenge that any systematic investigation has to be carried out in the short intervals of exposure at low tide, with all the attendant logistical problems of access and the risk that freshly dug excavation pits will be filled with sediment by the rising tide (Fig. 5.16).

At present, 42 Stone Age sites are known in the German part of the Wadden Sea (see Fig. 5.1). They are found in the mudflats associated with the North Frisian Islands of Schleswig-Holstein and the East Frisian Islands of Lower Saxony.

5.5.1 North Frisia

Most sites in North Frisia are of Neolithic age, demonstrating that this now-submerged landscape was part of the settlement area of the Early Neolithic Funnel Beaker Culture (Segschneider 2009). Finds include more than 20 assemblages of Neolithic flint artefacts and isolated megalith graves. Occasionally, Bronze Age finds have been recorded, as well as the remains of 'drowned' forests, which are important sources of information on the history of local settlement and topography (Kühn 2009; Kühn et al. 2013).

5.5.1.1 Hamburger Hallig: A Late Neolithic Aurochs Sacrifice

Particularly significant is the discovery of a sacrificial pit, in which the skull and bones from the extremities of a Late Neolithic aurochs had been deposited. It came to light in 2003, to the west of the island of Hamburger Hallig in North Frisia, on the edge of the deep Norderhever channel in the mudflats. When the tide was particularly low, a section of the mudflats was briefly exposed, with the tips of two partly eroded horns protruding above the surface (Fig. 5.17). An archaeological investigation revealed a pit in which the skull of a male aurochs, a vertebra and several leg bones had been buried, the latter cracked open, presumably for marrow extraction (Kühn 2009). The extremities had been placed at the bottom of the pit with the skull laid on top, suggesting a ritual deposition of specially selected parts of the carcase. Radiocarbon dating of the skull indicates deposition in the decades around 2300 cal BC. The surface of the pit is now c. 3 m below mean high water and had been dug into a land surface that has now been partly eroded away.

Analysis of plant macrofossils from the pit fill by Helmut Kroll (University of Kiel), recorded numerous stem fragments of common reed

Fig. 5.16 Archaeological survey on the East Frisian mudflats of the Wadden Sea, showing the wide expanse of the intertidal zone exposed at low tide. Photo by Sunhild Kleingärtner, courtesy of Lower Saxony Institute for Historical Coastal Research

Fig. 5.17 The Late Neolithic aurochs skull at Hamburger Hallig in North Frisia during the course of its recovery. Photo by Linda Hermannsen, Archaeological State Service of Schleswig-Holstein

(*Phragmites australis*) and seeds and fruits of nettle (*Urtica dioica*), black mustard (*Brassica nigra*), spear orache (*Atriplex patula*-type) and thistle (*Cirsium arvense*-type). This indicates a reed-dominated plant community which developed under freshwater conditions, demonstrating that despite rising sea level, the site had not yet been affected by marine conditions and was probably protected from the sea by a barrier of Pleistocene sediments (Kühn 2009, p. 67).

5.5.2 East Frisia

A total of 11 Mesolithic and Neolithic assemblages or single finds are known in this region, mostly stone tools and some antler implements. These were found along the channels between the East Frisian Islands and in river estuaries—areas of strong erosion and shifting sediments, where most of the finds, especially the single finds, are not in their original location. Of particular interest is the fragment of a Late Neolithic bell beaker, which was found in the 'Ostfriesisches Gattje' region of the Ems estuary (Niederhöfer 2016, p. 100).

An additional number of finds have been identified in the files held by various authorities and museums. These include more than 20 probable Stone Age artefacts, mostly single finds. The records mention a reindeer antler, three flint axes, a flint dagger as well as numerous unidentified flint tools, an antler pick and many aurochs horns. It is no longer possible to determine the original location of any of these objects, so they cannot be placed on any find maps. Nevertheless, they, too, are evidence that the area now covered by mudflats was inhabited in prehistoric times.

5.5.2.1 New Projects

Two recent projects have been funded to expand the investigation of offshore sediments and submerged landscapes in this region. The first is *Settlement and Cultural History of the Lower Saxony Wadden Sea*, 2012–2015 (Jöns et al. 2013; Goldhammer and Karle 2015; Karle and Goldhammer 2017). This project analysed all existing data from geological and archaeological archives, including 30,000 sediment cores, to produce palaeogeographical maps of different time intervals and to identify areas with a high potential for archaeological remains. Most of the currently known archaeological finds are Iron Age or later in date or earlier finds that have been disturbed due to reworking of older sediments.

The second project, building on the first, is the *WASA* project (*The Wadden Sea Archive of Landscape Evolution, Climate Change and Settlement History: Exploration – Analysis – Predictive Modelling*), 2016–2020. This is a multi-disciplinary and multi-institutional project combining existing archival data with new geophysical surveys, coring and sediment analysis to chart changes in sea level and landscape evolution since the Late Glacial period and to model human reactions to these changes.

5.6 Preservation Conditions and Future Challenges

As outlined above, the Baltic and Wadden Sea coastlines present different conditions of archaeological site preservation and visibility, and these conditions require different sorts of survey strategies. In both cases, it should be emphasised that archaeological finds become visible on the seabed surface mainly because of erosional processes.

In the Baltic, investigation has so far proved more successful for a variety of reasons. These include the greater accessibility and visibility of Stone Age palaeoshorelines, a longer history of research with a strong focus on reconstructing the pattern of sea-level rise associated with the Littorina transgression and the opportunity to elaborate research strategies suited to the conditions. The combination of different approaches in the SINCOS project has not only improved knowledge about changes in relative sea level and resulted in the discovery of a large number of Stone Age settlements but has also raised the public profile of research on underwater prehistory and landscapes and established survey routines for underwater archaeological investigation that can now be employed more widely in the management of the underwater cultural heritage and the conduct of investigations in advance of offshore industrial development.

Even so, most of the known submerged Stone Age sites along the German Baltic coast were originally located on the shoreline, and it is only the items of material culture that were deposited in the shallow water area in front of the sites that have been preserved. This preservation in its turn is the product of immediate submergence of the material under water and the fact that it was further sealed by a protective covering layer of marine sediment. Almost nothing is left in situ on the original dry-land surface of the settlement area, where materials have been disturbed or eroded away, except in unusual pockets of preservation such as the pit at Timmendorf-Nordmole I.

Along the Wadden Sea coastline, the shallow offshore gradients and relatively rapid accumulation of marine sediments require different methods and approaches. Most Stone Age sites in this region have been located around the North Frisian Islands, and this is because this area was drowned much later than the area of the East Frisian Islands, and its Stone Age landscapes have not been so deeply buried under marine sediments. As a general principle, the earlier the material, the more deeply buried beneath sediment it is likely to be. The currently known finds in both areas are mostly single artefacts or stray finds that have been exposed by erosional processes, and their original context is now impossible to reconstruct. In the rare case of in situ finds such as the Hamburger Hallig aurochs, these are likely to be in locations that were well inland from their contemporaneous coastline at the time of deposition, rather than in locations on or near the shoreline that could provide evidence of relative sea-level rise and exploitation of marine resources.

A systematic survey along former palaeoshorelines has not yet been carried out in the German sector of the North Sea. The major challenge in this region is the relatively thick overburden of marine sediment. If a methodology for the discovery of buried archaeological materials and a management strategy for the underwater prehistoric cultural heritage is to be developed, this will need a better knowledge of the coastal dynamics of sediment accumulation and erosion. This, in its turn, will require joint palaeoecological and archaeological investigations and the deployment of appropriately equipped research vessels over a more extensive area than has hitherto been possible. In the absence of such research investigations, industrial developments offshore in this region will be operating 'blind' without prior knowledge of areas or deposits of potential cultural significance that should be investigated in advance or avoided. Of course, offshore development may have the beneficial outcome of revealing cultural features of the buried land surface. But realising that benefit will depend on a planning process that makes adequate provision to integrate palaeoecological and archaeological investigations with planned industrial activity (see below and Pater, Chap. 27, this volume).

5.7 Management of the Underwater Cultural Heritage

In Germany unlike other European countries, a national legal framework for the management of the cultural heritage does not exist. Instead it is organised at the level of the Länder (the 16 states that make up the Federal Republic of Germany), with the result that there are 16 different heritage laws. For the management of the submerged sites in the Baltic waters of Germany, the responsible state heritage agencies are Mecklenburg-West Pomerania (https://www.kulturwerte-mv.de/Landesarchaeologie) and Schleswig-Holstein (https://www.schleswig-holstein.de/DE/Landesregierung/ALSH/alsh_node.html). The latter organisation is also in charge of research and site management of the underwater cultural heritage in the northern part of the German North Sea (North Frisia), whereas the sites in the southern part of the German North Sea (East Frisia and Lower Saxony) are the responsibility of the Ministry of Science and Culture of Lower Saxony (https://www.mwk.niedersachsen.de) advised and supported by the Lower Saxony Institute for

Historical Coastal Research (http://www.nihk.de) and the State Agency for Cultural Heritage (http://www.denkmalpflege.niedersachsen.de/).

Underwater archaeological research in Germany is restricted to scientific divers trained by one of eight certified institutions following agreed safety education standards (http://www.forschungstauchen-deutschland.de/). Most of the recent underwater archaeological research projects along the German coast have been hosted by the Heritage Agencies of Mecklenburg-West Pomerania and Schleswig-Holstein and by the Lower Saxony Institute for Historical Coastal Research.

Because Germany has no national heritage law, underwater heritage that lies outside the area of responsibility of one of the federal states depends for protection on international laws and treaties such as the 1972 UNESCO Convention Concerning the Protection of the World's Cultural and Natural Heritage and the 1982 United Nations Convention on the Law of the Sea (see Dromgoole, Chap. 25, this volume). For offshore industrial projects such as drilling for oil and gas, sand and gravel extraction or the construction of wind farms, the 2011 EU 2011/92 directive on the assessment of the effects of certain public and private projects on the environment is of special importance. According to article 3 of this directive, Environmental Impact Assessments have to include identification, description and assessment of material assets and the cultural heritage. At the moment, it is the responsibility of the national government agency that gives planning approval to ensure that the direct and indirect effects of the planned project on cultural heritage are considered. This situation will be improved and clarified when Germany accepts and ratifies the 2001 UNESCO Convention on the Protection of the Underwater Cultural Heritage.

Acknowledgements This chapter has its origins in the SINCOS project (Sinking Coasts – Geosphere, Ecosphere and Anthroposphere of the Holocene Southern Baltic Sea). Therefore we would like to thank all the colleagues who have been part of the SINCOS team. Special thanks are due to Prof. Dr. Jan Harff, who directed this research in the friendly and consistent way that is so typical of him. In addition we would like to thank the German Research Foundation for funding the SINCOS project as well as the Strande project; without their help, it is true to say that the results of submerged landscape research in German marine waters would not have been possible. In addition, we have to thank the state authorities for the management of the archaeological heritage of Mecklenburg-West Pomerania, Schleswig-Holstein and Lower Saxony who gave us access to their heritage databases, so that we could use the data stored there as a base for our research. Finally our thanks go to two external reviewers and to Geoff Bailey for many important hints, editorial advice and extensive support during the writing of this contribution and for improvement of the English.

References

Behre K-E (2003) Eine neue Meeresspiegelkurve für die südliche Nordsee: Transgressionen und Regressionen in den letzten 10000 Jahren. Probleme der Küstenforschung im südlichen Nordseegebiet 28:9–63

Behre K-E (2007) A new Holocene sea-level curve for the Southern North Sea. Boreas 36:82–102

Behre K-E, Menke B (1969) Pollenanalytische Untersuchungen an einem Bohrkern der südlichen Doggerbank. Beiträge zur Meereskunde 24(25):122–129

Behre K-E, Dörjes J, Irion G (1984) Ein datierter Sedimentkern aus dem Holozän der südlichen Nordsee. Probleme der Küstenforschung im südlichen Nordseegebiet 15:135–148

Brinch Petersen E (1967) Klosterlund – Sønder Hadsund – Bøllund. Les trois sites principaux du Maglémosien ancien en Jutland. Essai de typologie et de chronologie. Acta Archaeologica 37:77–185

Brinch Petersen E (1973) A survey of the late Palaeolithic and the Mesolithic of Denmark. In: Kozłowski SK (ed) The Mesolithic in Europe. Papers read at the international archaeological symposium on the Mesolithic in Europe, Warsaw, May, 7–12, 1973. Warsaw University Press, Warsaw, pp 77–127

Brown A, Russel J, Scaife R, Tizzard L, Whittaker J, Wyles SF (2018) Lateglacial/early Holocene palaeoenvironments in the southern North Sea Basin: new data from the Dudgeon offshore wind farm. J Quat Sci 33(6):597–610

Cohen KM, Westley K, Erkens G, Hijma MP, Weerts HJT, Westley K (2017) The North Sea. In: Flemming NC, Harff J, Moura D, Burgess A, Bailey GN (eds) Submerged landscapes of the European continental shelf: quaternary paleoenvironments. Wiley, Chichester, pp 147–186

Coles BJ (1998) Doggerland: a speculative survey. Proc Prehist Soc 64:45–81

Craig OE, Steele VJ, Fischer A, Hartz S, Andersen SH, Donohoe P, Glykou A, Saul H, Jones DM, Koch E, Heron CP (2011) Ancient lipids reveal continuity in culinary practices across the transition to farming in Northern Europe. In: Proceedings of the National Academy of Sciences, USA 108(44):17910–17915

Enters D, Wolters S, Blume K, Lücke A, Segschneider M, Theuerkauf M (2015) Multiproxy-Analysen zur spät- und postglazialen Entwicklungsgeschichte des Fehmarnbelts. Siedlungs- und Küstenforschung im Nordseegebiet 38:43–58

Fries JE, Veil S (2014) Fernkontakte späteiszeitlicher Jäger und Sammler in der norddeutschen Tiefebene: Ein Klingenkern aus rotem Helgoländer Flint von einem Oberflächenfundplatz am Dümmer bei Damme, Ldkr. Diepholz. Siedlungs- und Küstenforschung im südlichen Nordseegebiet 37:125–140

Gearey BR, Hopla E-J, Smith D, Marshall P, Fitch S, Griffiths S, Tappin DR (2017) Multi-proxi palaeoecological approaches to submerged landscapes: a case study from `Doggerland´, in the Southern North Sea. In: Williams M, Hill T, Boomer I, Wilkinson IP (eds) The archaeological and forensic applications of microfossils: a deeper understandig of Human History. Geological Society, London, pp 35–53

Glykou A (2014) Late Mesolithic-early Neolithic sealers: a case study on the exploitation of marine resources during the Mesolithic-Neolithic transition in the south-western Baltic Sea In: Fernandes R, Meadows J (eds) 'Human exploitation of aquatic landscapes' special issue. Internet Archaeology 37. https://doi.org/10.11141/ia.37.7

Glykou A, Goldhammer J, Hartz S (2014) Versunkene Steinzeit – Taucharchäologische Untersuchungen an der schleswig-holsteinischen Ostseeküste. In: Huber F, Kleingärtner S (eds) Gestrandet, versenkt, versunken. Faszination Unterwasserarchäologie. Wachholtz Verlag Neumünster, Hamburg, pp 76–97

Goldhammer J, Hartz S (2015) The Ertebølle settlement Strande LA 163, District of Rendsburg-Eckernförde, and the Littorina transgression – submarine surveys and investigations. Siedlungs- und Küstenforschung an der südlichen Nordseeküste 38, pp 29–41

Goldhammer J, Hartz S (2017) Fished up from the Baltic Sea – a new Ertebølle site near Stohl cliff (Kiel Bay, Germany). In: Bailey G, Harff J, Sakellariou D (eds) Under the sea: archaeology and palaeolandscapes of the continental shelf. Springer, Cham, pp 145–154

Goldhammer J, Karle M (2015) Geoarchaeological research in the Wadden Sea area of Lower Saxony. Siedlungs- und Küstenforschung im Nordseegebiet 38:59–70

Goldhammer J, Müller AB, Brandt L, Wolters S, Hartz S (2020) Subsistence strategies, settlement structure, and communication in the terminal Mesolithic – a new DFG-project in Kiel Bay. In: Gehlen B, Zander A (eds) From the early Preboreal to the Subboreal period – current Mesolithic research in Europe. Vom frühen Präboreal bis zum Subboreal – Aktuelle Forschungen zum Mesolithikum in Europa. Kölner Studien zur Prähistorischen Archäologie, Rahden/Westf

Grimm S, Groß D, Gerken K, Weber MJ (2020) On the onset of the early Mesolithic on the north German plain. In: Zander A, Gehlen B (eds) From the early Preboreal to the subboreal period – current Mesolithic research in Europe. Vom frühen Präboreal bis zum Subboreal – Aktuelle Forschungen zum Mesolithikum in Europa. Kölner Studien zur Prähistorischen Archäologie, Rahden/Westf

Groh A, Dietrich R, Richter A (2011) Geodetic evidence and modelling of sea-level changes and load-induced crustal deformations in the southern Baltic Sea. Bericht der Römisch-Germanischen Kommission 92:17–40

Harck O (1985) Submarine Archäologie in Schleswig-Holstein. Offa 42:431–446 (Feschrift Karl Kersten)

Harff J, Lüth F (2007) Preface to SINCOS I – sinking coasts: geosphere, ecosphere and anthroposphere of the Holocene southern Baltic Sea. Bericht der Römisch-Germanischen Kommission 88:9–14

Harff J, Lüth F (2011) Preface to SINCOS II – sinking coasts: geosphere, ecosphere and anthroposphere of the Holocene southern Baltic Sea. Bericht der Römisch-Germanischen Kommission 92:11–16

Harff J, Meyer M (2007) Changing Holocene coastal zones of the Baltic Sea – a modelling approach. Bericht der Römisch-Germanischen Kommission 88:241–266

Harff J, Lemke W, Lampe R, Lüth F, Lübke H, Meyer M, Tauber F, Schmölcke U (2007) The Baltic Sea coast—a model for interrelations among geosphere, climate, and anthroposphere, Special paper 426. Geological Society of America, New York, pp 133–142

Harff J, Flemming NC, Groh A, Hünicke B, Lericolais G, Mesched M, Rosentau A, Sakellariou D, Uścinowicz S, Zhang W, Zorita E (2017) Sea level and climate. In: Flemming NC, Harff J, Moura D, Burgess A, Bailey GN (eds) Submerged landscapes of the European continental shelf: quaternary paleoenvironments. Wiley, Chichester, pp 21–49

Hartz S, Glykou A (2008) Neues aus Neustadt. Ausgrabungen zur Ertebølle- und frühen Trichterbecher-Kultur in Schleswig-Holstein Archäologische Nachrichten aus Schleswig-Holstein 14:17–19

Hartz S, Lübke H (2006) New evidence for a chronostratigraphic division of the Ertebølle culture and the earliest Funnel Beaker culture on the Southern Mecklenburg Bay. In: Kind CJ (ed) After the Ice age: settlements, subsistence and social development in the Mesolithic of Central Europe: proceedings of the international conference 9th to 12th of September 2003, Rottenburg/Neckar, Baden-Württemberg, Germany. Materialhefte zur Archäologie in Baden-Württemberg 78:61–77

Hartz S, Jöns H, Lübke H, Schmölcke U, von Carnap-Bornheim C, Heinrich D, Klooß S, Lüth F, Wolters S (2011) Prehistoric settlements in the southwestern Baltic Sea area and development of the regional stone age economy. Bericht der Römisch-Germanischen Kommission 92:77–210

Hepp DA, Warnke U, Hebbeln D, Mörz T (2017) Tributaries of the Elbe palaeovalley: features of a hidden palaeolandscape in the German Bight, North Sea. In: Bailey GN, Harff J, Sakellariou D (eds) Under the sea: archaeology and palaeolandscapes of the continental shelf. Springer, Cham, pp 211–222

Holst D (2014) Subsistenz und Landschaftsnutzung im Frühmesolithikum. Nussröstplätze am Duvensee. Monographien des Römisch-Germanischen Zentralmuseums 120, Römisch-Germanisches Zentralmuseum, Regensburg

Jöns H (2010) The Wadden Sea as a cultural landscape and an archive of common history. In: Marencic H., Eskildsen K., Farke H, Hedtkamp S (eds) Science for nature conservation and management: the Wadden Sea ecosystem and EU directives. Proceedings of the 12th International scientific wadden Sea symposium in Wilhelmshaven, Germany, 30 March–3 April 2009. Wadden Sea Ecosystem No. 26. Common Wadden Sea Secretariat, Wilhelmshaven, Germany, pp 163–164

Jöns H (2011) Settlement development in the shadow of coastal changes. Case studies from the Baltic rim. In: Harff J, Björck S, Hoth P (eds) The Baltic Sea basin. Springer, Berlin/Heidelberg, pp 301–336

Jöns H, Lübke H, Lüth F, Terberger T (2007) Prehistoric settlements and development of the regional economic area. Archaeological investigations along the Northeast-German Baltic Seacoast. Bericht der Römisch-Germanischen Kommission 88:149–188

Jöns H, Karle M, Kleingärtner S (2013) Das Nordseebecken und der Wattenmeerraum als Forschungsgebiet. Methodische Überlegungen, Strategien und aktuelle Forschungsprojekte. Offa 69/70:71–80

Karle M, Goldhammer J (2017) The Wadden Sea of North-West Germany: an intertidal environment of high archaeological research potential. In: Bailey GN, Harff J, Sakellariou D (eds) Unter the sea: archaeology and palaeolandscapes of the continental shelf. Springer, Cham, pp 223–231

Klooß S (2015) Mit Einbaum und Paddel zum Fischfang. Holzartefakte von endmesolithischen und frühneolithischen Küstensiedlungen an der südwestlichen Ostseeküste, Untersuchungen und Materialien zur Steinzeit in Schleswig-Holstein und im Ostseeraum 6. Wachhotz/Murmann Publishers, Kiel/Hamburg

Krüger S, Dörfler W, Bennike O, Wolters S (2017) Life in Doggerland – palynological investigations of the environment of prehistoric hunter-gatherer societies in the North Sea Basin. E&G Quaternary Sci J 66:3–13

Kühn HJ (2007) Jenseits der Deiche. Archäologie im nordfriesischen Wattenmeer. In: Von Carnap-Bornheim C, Radtke C (eds) Es war einmal ein Schiff. Archäologische Expeditionen zum Meer. Mareverlag, Hamburg, pp 251–285

Kühn HJ (2009) Flintdolche im Schlick. Archäologische Nachrichten aus Schleswig-Holstein 2009:66–70

Kühn HJ, Meier D, Mertens C (2013) Dokumentation der Kulturspuren. In: Meier D, Kühn HJ, Borger GJ (eds) Der Küstenatlas. Das schleswig-hoslteinische Wattenmeer in Vergangenheit und Gegenwart. Boyens Buchverlag, Heide, pp 49–56

Lampe R, Meyer H, Ziekur R, Janke W, Endtmann E (2007) Holocene evolution of the irregularly sinking southern Baltic Sea coast and the interactions of sea-level rise, accumulation space and sediment supply. Bericht der Römisch-Germanischen Kommission 88:15–46

Lanting JN, Van der Plicht JH (2000) De ^{14}C-chronologie van de Nederlandse pre- en protohistorie: II. Mesolithicum. Palaeohistoria 39(40):99–162

Lübke H (2001) Eine hohlendretuschierte Klinge mit erhaltener Schäftung vom endmesolithischen Fundplatz Timmendorf-Nordmole, Wismarbucht, Mecklenburg-Vorpommern. Nachrichtenblatt des Arbeitskreises für Unterwasserarchäologie 8:46–51

Lübke H (2002) Steinzeit in der Wismarbucht. Ein Überblick Nachrichtenblatt des Arbeitskreises für Unterwasserarchäologie 9:75–88

Lübke H (2003) New investigations on submarine Stone Age settlements in the Wismar Bay area. In: Kindgren H, Knutsson K, Larsson L, Loefler D, Akerlund A (eds) Mesolithic on the move: papers presented at the sixth international conference on the Mesolithic in Europe, Stockholm 2000. Oxbow, Oxford, pp 633–642

Lübke H, Schmölcke U, Tauber F (2011) Mesolithic hunter-fishers in a changing world: a case study of submerged sites on the Jäckelberg, Wismar Bay, north-eastern Germany. In: Benjamin J, Bonsall C, Pickard C, Fischer A (eds) Submerged prehistory. Oxbow, Oxford, pp 21–37

Mahlstedt S (2015) Das Mesolithikum im westlichen Niedersachsen. Untersuchungen zur materiellen Kultur und zur Landschaftsnutzung, Frühe Monumentalität und soziale Differenzierung 7. Habelt, Bonn

Müller J, Brozio J-P, Demnick D, Dibbern H, Fritsch B, Furholt M, Hage F, Hinz M, Lorenz L, Mischka D, Rinne C (2010) Perdiodisierung der Trichterbecher-Gesellschaften. Ein Arbeitsentwurf. J Neolithic Archaeol 12. Available at: https://doi.org/10.12766/jna.2010.58

Niederhöfer K (2016) Archäologische Fundstellen im ostfriesischen Wattenmeer. Siedlungsgeschichte einer untergegangenen Landschaft bis 1570. Beiträge zur Archäologie in Niedersachsen 18. Archäologische Kommission für Niedersachsen, Rahden

Niekus M (2006) A geographically referenced ^{14}C database for the Mesolithic and the early phase of the Swifterbant culture in the northern Netherlands. Palaeohistoria 47(48):41–99

Policy Research Corporation (2009) The economics of climate change adaptation in EU coastal areas. Country reports: Germany. Available at: https://ec.europa.eu/maritimeaffairs/documentation/studies/climate_change_en. Accessed 18 Dec 2018

Raemaekers DCM (1999) The articulation of a "New Neolithic": the meaning of the Swifterbant culture for the process of neolithisation in the western part of the North European Plain (4900–3400 BC). Leiden University Press, Leiden

Rosentau A, Bennike O, Uścinowicz S, Miotk-Szpiganowcz G (2017) The Baltic Sea Basin. In: Flemming NC, Harff J, Moura D, Burgess A, Bailey

GN (eds) Submerged landscapes of the European continental shelf: quaternary paleoenvironments. Wiley, Chichester, pp 103–133
Saul H, Wilson J, Heron CP, Glykou A, Hartz S, Craig OE (2012) A systematic approach to the recovery of and identification of starches from carbonized deposits on ceramic vessels. J Archaeol Sci 39:3483–3492
Saul H, Madella M, Fischer A, Glykou A, Hartz S, Craig OE (2013) Phytoliths in pottery reveal the use of spice in European prehistoric cuisine. PLoS One 8(8):e70583
Schlichtherle H (ed) (1997) Pfahlbauten rund um die Alpen. Archäologie in Deutschland Sonderheft, Stuttgart
Schmölcke U, Glykou A, Heinrich D (2007) Faunal development in the southwestern Baltic area. Bericht der Römisch-Germanischen Kommission 88:205–218
Segschneider M (2009) Die Nordsee – eine jahrtausendealte Kulturlandschaft; wo gestern noch Lärm und lustiger Tisch. Archäologie in Deutschland 6:28–31
Shennan I, Lambeck K, Flather R, Horton B, McArthur J, Innes J, Lloyd J, Rutherford M, Wingfield R (2000) Modelling Western North Sea palaeogeographies and tidal changes during the Holocene. In: Shennan I, Andrews J (eds) Holocene land-ocean interaction and environmental change around the North Sea, London, pp 299–319
Skaarup J (1995) Excavation of a Mesolithic dwelling in a submerged landscape. Preliminary report. News in WARP: the newsletter of the Wetland Archaeology Research Project 17:28–31
Sørensen SA (2016) Syltholm: Denmark's largest Stone age excavation. Mesolithic Miscellany 24(2):1–10
Sørensen SA (2017) The Kongemose culture. Nordiske Fortidsminder, Serie B, Bind 27. Copenhagen
Stapel B (1991) Die geschlagenen Steingeräte der Siedlung Hüde I am Dümmer. Veröffentlichungen der Urgeschichtlichen Sammlungen des Landesmuseums zu Hannover 38. Hildesheim
Streif HJ (2004) Sedimentary record of Pleistocene and Holocene marine inundations along the North Sea coast of Lower Saxony, Germany. Quat Int 112:3–28
Ten Anscher TJ (2015) Under the radar: Swifterbant and the origins of the funnel beaker culture. In: Kabaciński J, Hartz S, Raemaekers D (eds) The Dąbki site in Pomerania and the Neolithisation of the North European Lowlands (c. 5000–3000 cal BC). Verlag Marie Leidorf, Rahden, pp 335–358
Terbeerger T (2007) Neue Forschungen am steinzeitlichen Fundplatz Ralswiek-Augustenhof, Kr. Rügen. In: Speitel E (ed) Terra Prehistorica. Festschrift für Klaus-Dieter Jäger zum 70. Geburtstag, Beiträge zur Ur- und Frühgeschichte Mitteleuropas, vol 48. Beier & Beran, Langenweißbach, pp 241–252
Wolters S (2009) Neue Daten zur Vegetationsgeschichte der südlichen Nordsee. Pollenanalytische Untersuchungen an in-situ Torflagern und Torfgeröllen. Nachrichten des Marschenrates zur Förderung der Forschung im Küstengebiet der Nordsee 45:53–57
Wolters S, Zeiler M, Bungenstock F (2010) Early Holocene environmental history of sunken landscapes: pollen, plant macrofossil and geochemical analyses from the Borkum Riffgrund, southern North Sea. Int J Earth Sci 99:1707–1719

Norway: Submerged Stone Age from a Norwegian Perspective

6

Håkon Glørstad, Jostein Gundersen, Frode Kvalø, Pål Nymoen, David Simpson, and Birgitte Skar

Abstract

The presence of submerged Stone Age sites along the Norwegian coast has been recognised for a long time. Until recently, however, they have not been treated as a topic of interest. From shallow waters, more than 80 submerged Stone Age sites are known in Norway, one of which is a probable ritual site with bones of several humans. Due to complex patterns of shoreline displacement and uplift after the last deglaciation, the present-day shallow water areas along the Norwegian coast contain sites from periods which are also represented by numerous sites on dry land. The submerged sites can provide better preservation quality for cultural remains of organic materials than settlement sites on dry land. This allows for a dynamic and informed study of Stone Age coastal adaptations, interactions and lifeways.

Keywords

Submerged Stone Age · Postglacial · Colonisation · Mesolithic · Hummervikholmen · Isostasy

6.1 Introduction

The sea has always been central to Norwegian Stone Age research, not least because the vast majority of all known Stone Age sites are situated close to present or ancient coastlines. Norway today has a productive coastline 83,281 km long, including islands (Statistics Norway 2013); also during the Stone Age period the coast in practice constituted most of the habitable land of the country because of a mountainous or ice-bound hinterland which even today offers only a very small percentage of land suitable for cultivation. The close connection between ancient habitation patterns and the sea therefore comes as no surprise. Site location, dietary evidence, and, recently, various isotope and other chemical analyses all point towards heavy dependence on the marine ecosystem

H. Glørstad (✉)
Museum of Cultural History, University of Oslo, Oslo, Norway
e-mail: hakon.glorstad@khm.uio.no

J. Gundersen
Directorate for Cultural Heritage, Oslo, Norway
e-mail: jog@ra.no

F. Kvalø · P. Nymoen
Norwegian Maritime Museum, Oslo, Norway
e-mail: frode.kvalo@marmuseum.no; pal.nymoen@marmuseum.no

D. Simpson
University Museum of Bergen, University of Bergen, Bergen, Norway
e-mail: david.simpson@uib.no

B. Skar
NTNU University Museum, Norwegian University of Science and Technology, Trondheim, Norway
e-mail: birgitte.skar@ntnu.no

G. Bailey et al. (eds.), *The Archaeology of Europe's Drowned Landscapes*, Coastal Research Library 35, https://doi.org/10.1007/978-3-030-37367-2_6

throughout the Stone Age. Consequently, variations in sea level throughout prehistory have had a large impact on human occupation and the preservation of Stone Age sites. A combination of changes in global sea level (eustasy) and isostatic rebound following the melting of the Scandinavian ice shield in the Holocene has created not only a varied and complicated history of the interplay between humans and the sea but also a complicated preservation history of Stone Age sites. Along the coastline, all variants of this dynamic interplay can be demonstrated, from coastlines that have undergone continuous postglacial rebound and uplift where the ice sheet was at its thickest, to areas dominated by transgression and inundation.

The fact that Stone Age sites can be found associated with all variants of these geological processes has long been acknowledged in Norwegian archaeology. This awareness has also grown as more precise shoreline displacement curves have been presented by geologists and pollen analysts. Nevertheless, most scientific attention has focussed on Stone Age sites preserved on dry land.

There are many reasons for this terrestrial focus. Land uplift attracted more scientific interest than inundation at the beginning of the twentieth century, and consequently settlement history was also directed towards elevated coastlines (for instance Brøgger 1905). Technical limitations connected to underwater archaeology were also an obvious limitation until approximately 1955–65, when diving equipment became more easily available. Sites on land also provided a rich foundation for establishing information on the chronology, typology, settlements and adaptations of the Norwegian Stone Age. The interest in a submerged settlement history, however, remained active—not least in a fascination with the lost plains of the North Sea Continent.

6.2 The North Sea Continent and the Postglacial Colonisation of the Norwegian Coastline

In the archaeological literature, the now submerged North Sea Plains, popularly known as Doggerland, are considered to be of great importance for early human settlement in North-Western Europe (Reid 1913; Coles 1998; Gaffney et al. 2017), and a possible bridgehead for the colonisation of Norway at the end of the last Ice Age (e.g., Nummedal 1923; see Bjerck 1994, 1995 and Glørstad et al. 2017 for a critical discussion). So far, there is no evidence for prehistoric human settlement in the Norwegian sector of the North Sea Plains lying to the west of the Norwegian Trench (Fig. 6.1). Even if human populations were present there during the Last Glacial Maximum, it appears unlikely that this was the source region for the pioneer colonisation of the Scandinavian Peninsula since this would have required a sea crossing over what would always have been a wide and inhospitable expanse of open sea. It is more probable that the first inhabitants entered the country via the Swedish west coast (Bjerck 2008; Hafeez et al. 2012; Glørstad 2012; Glørstad and Kvalø 2012; Glørstad et al. 2017).

Despite the fact that most of the Norwegian coastline had been ice-free from the Allerød Oscillation (11,800–10,600 cal BC), access was blocked either by large distances of open sea or by the Scandinavian Glacier (Glørstad 2014). Only after about 9200 cal BC did an ice-free passage open up from Bohuslän in Sweden to the Oslo fjord area. By this route humans could travel in sheltered waters, navigating with land constantly in sight. Consequently, the whole Norwegian coast became easily accessible for settlement. Bohuslän was probably one of the most densely populated areas in Europe at the

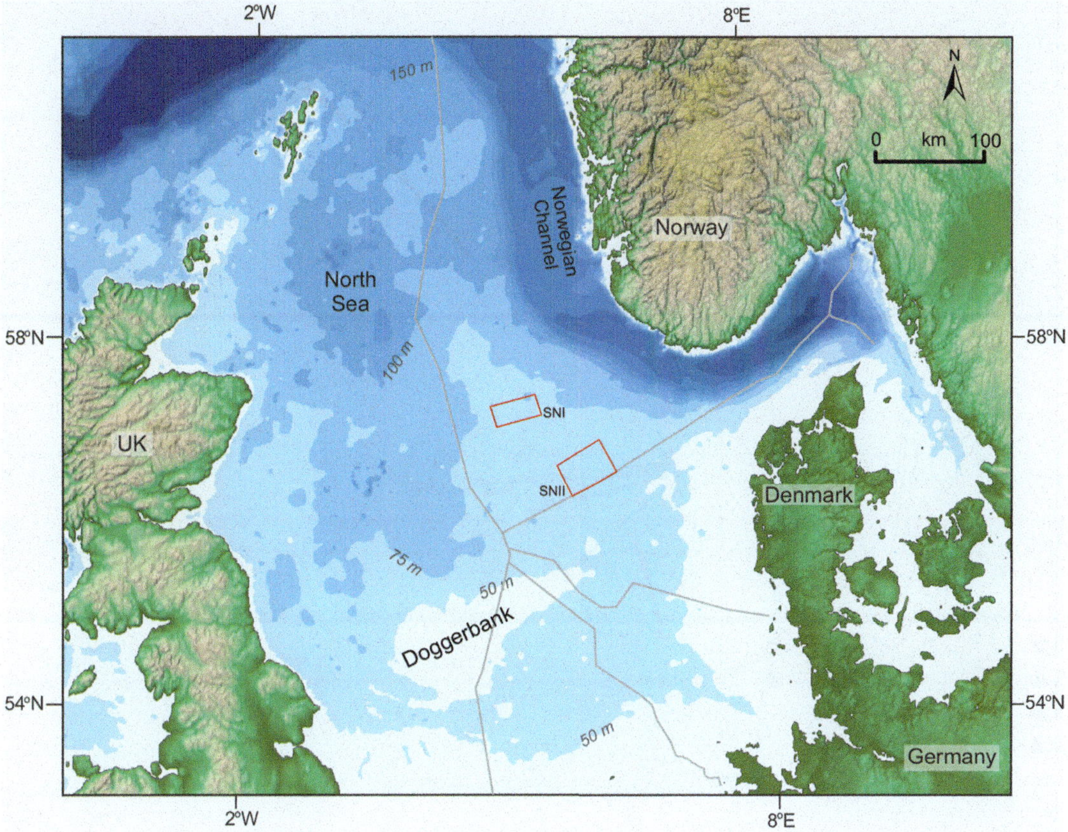

Fig. 6.1 Hypothetical drainage channels and maximum extent of Doggerland at the Late Glacial Maximum (from Glørstad et al. 2017, figure 19.2). SNI and SNII indicate locations where coring in advance of wind-farm development has demonstrated the former land surface of Doggerland buried beneath marine sediments associated with the inundation by sea-level rise before the end of the Last Ice Age. The Norwegian trench would always have been a barrier to movement into the ice-free coastal regions of Norway to the north, while the ice sheet would have blocked access from southern Norway and Sweden to the south until the final stages of ice retreat at the very end of the Last Ice Age

end of the Pleistocene because of a very rich marine ecosystem (Schmitt et al. 2006). This relatively large population would have enabled a rapid colonisation of the Norwegian coast in the Preboreal period (Berg-Hansen 2017). There is also growing evidence at the beginning of the Boreal period for the spread of new immigrants into northern Norway entering from the North East (Sørensen et al. 2013; Damlien 2014, 2016; Günther et al. 2018).

6.3 The Scandinavian Peninsula—An Archive of Early Coastal Settlements

The allure of the North Sea Continent clearly demonstrates an early if speculative interest in the submerged Stone Age in Norway. Ancient coastal settlements preserved on dry land have perhaps rightfully received much more scien-

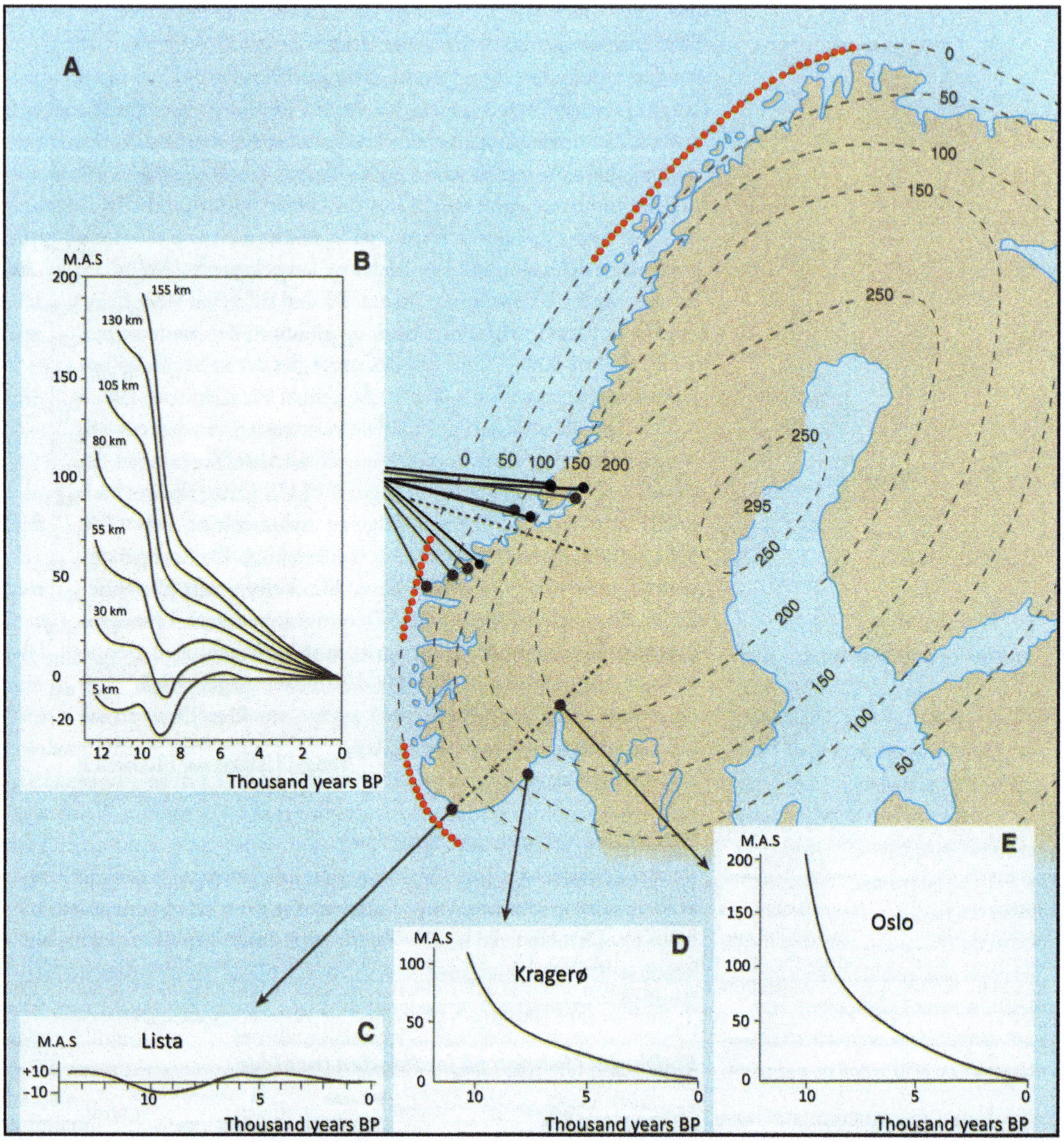

Fig. 6.2 The combined effect of sea-level rise (eustasy) and land uplift (isostasy). Broken contour lines in the main figure (**a**) show uplift, and red dotted lines show the three areas with submerged Stone Age shorelines. So far, only the two southernmost areas have yielded underwater Stone Age finds. Figures **b**, **c**, **d** and **e** present relative sea-level change in different areas. Figure **b** is a suite of seven sea-level curves derived from nine shoreline diagrams indicated by a black dot on map **a**. The curves in figure **b** were constructed for every 25 km along the projection plane indicated with a stippled line perpendicular to the uplift lines on map a. The stippled line between the Lista and Oslo curves is the projection plane in this area. Modified from Andersen 2000

tific attention, because the Scandinavian Peninsula offers rather unusual possibilities for the study of marine lifeways in a long-term perspective. In most parts of the world, Early Holocene coastal areas are now inundated. This also goes for most of Europe, meaning that potential early shore-bound settlements are now submerged. Due to the great thickness of the Pleistocene ice shield, the Scandinavian Peninsula underwent rapid isostatic uplift, parallel with and in some areas outstripping the sea-level rise as the weight from the ice diminished during the Late glacial period, resulting in elevated ancient shorelines.

The general picture is one of relative and continuous land rise, a process that is still on-going (Andersen 2000) (Fig. 6.2). Because of this geological process, whole coastal landscapes with Stone Age settlement sites, dating back to the beginning of the Holocene, are easily accessible to archaeology above water. This situation has made it possible for Norwegian and Swedish archaeologists to contribute very significantly to the understanding of the importance of early coastal adaptations in European prehistory (e.g. Brøgger 1905; Bagge 1951; Fredsjö 1953; Mikkelsen 1975; Bjerck 1986; Olsen, 1994; Fischer 1995; Åkerlund 1996). Coastal settlements in Scandinavia are among the oldest well-documented, maritime societies in the world, going back to the final part of the Ice Age (Younger Dryas) (Schmitt et al. 2009). The continuous occupation of an uplifting coastline has enabled Scandinavian archaeologists to establish a very precise chronological development of the Stone Age, contrary to the situation further to the south. The documentation of this long-lived Scandinavian maritime tradition is an important contribution to our understanding of human reliance on marine resources and coastal communication. Consequently, this is of crucial importance for the future investigation of submerged Stone Age sites in Europe.

Looking closer into the different regions of Norway, a more varied and perhaps challenging picture appears, because land rise was not equally marked all over the country. Already at the end of the nineteenth century it was demonstrated that the Scandinavian Peninsula was subjected to a rise in sea level approximately 10,000 years ago—the so-called Tapes transgression (De Geer 1888). Dependent on local isostasy, the effect of this global sea-level rise varies considerably in Norway.

In south-east Norway, a land rise of c. 50 m is recorded around the Oslo Fjord in the earliest settlement phase, c. 9500–8000 cal BC (Jaksland and Persson 2014), similarly in the south Swedish Bohuslän region, a eustatic regression rate of 2.5–3.0 m/100 year is recorded during the period c. 10,000–9000 cal BC (Schmitt et al. 2009; Schmitt and Svedhage 2016). The land continued to rise in the succeeding millennia, though the shore-displacement curves gradually flattened out (Skar and Breivik 2018). In the Oslo Fjord area, the highest marine limit is 270 m above the present sea level (Holtedahl 1953), demonstrating the very strong impact of isostasy in this area.

Steep shore-displacement curves are also found in central/western and northern Norway. In parts of central/western Norway, reconstructions show a drop in relative sea level of as much as 60 m between 9500 and 8000 cal BC (Svendsen and Mangerud 1987). The region was, however, subjected to the Tapes transgression between c. 8200–4900 cal BC (9000–6000 BP, Svendsen and Mangerud 1987). Preboreal shorelines were spared in many places, but other coastal stretches were heavily eroded or covered by the rising sea (Fig. 6.3). Some parts of the ancient coast are still inundated today and can thus be compared with the situation in other parts of Europe (Bjerck 2008; Nymoen and Skar 2011a; Skar and Breivik 2018). These areas can be found at the outer coasts of southern, western and northern Norway (Fig. 6.3).

Consequently, even if land uplift is the general dominating characteristic of the Norwegian coastline, there are important areas that have been subject to a relative sea-level rise in parts of the Holocene, especially during the Tapes transgression, c. 8200–4900 cal BC (Svendsen and Mangerud 1987).

In areas where the ice cover was thin and even partially absent in the final parts of the Ice Age, parts of the Mesolithic coastline are today submerged (Fig. 6.3). These submerged coastal landscapes have received little attention in Norwegian archaeology, although the question of the effect of the Tapes transgression has gained interest through field surveys (Nummedal 1923, 1933; Bjerck 1986). It has been demonstrated that marine deposits from this transgression have covered older sites that are now preserved on dry land (Bang-Andersen 1995) due to continuous land uplift following the transgression. In other areas, movements of land and sea have cancelled each other out in such a way that Mesolithic sites are preserved at the present shoreline or in the intertidal zone (Bjerck and Olsen 1983).

Fig. 6.3 Map showing the areas where submerged early Holocene coastlines (earlier than 9050 cal BC) are present (pink shading) and known submerged sites (purple dots). Yellow indicates areas where early Holocene shorelines lie above the current shoreline but may have been damaged by the Tapes transgression. Green indicates areas where early Holocene shorelines are above the current shoreline and the Tapes transgression level (but younger periods may have been affected by the transgression). Blue indicates areas where early Holocene sites are above the current shoreline and where there has not been a subsequent marine transgression. Further detail is shown in Fig. 6.4. Drawing by David Simpson based on information from Svendsen and Mangerud (1987), Bondevik et al. (1998), Møller and Holmeslet (1998) and Simpson (2003)

The submerged coastlines of the Early Mesolithic (c. 9300–8300 cal BC), and to some extent of the Middle Mesolithic (c. 8300–6300 cal BC), are situated in relatively shallow waters (less than 10 m depth). This can be of significance to European archaeology because coastlines of these periods elsewhere are situated in much deeper water, and consequently they are more difficult to investigate archaeologically with today's technology. Potentially, the submerged Norwegian coastlines can provide well-preserved coastal sites of the Early Mesolithic, capable of giving a unique insight into the formative processes of Mesolithic subsistence strategies, settlement patterns, population diversity and material culture during this decisive period of human history.

6.4 The Submerged Sites

The total number of known submerged Stone Age sites in Norway is uncertain. As long as archaeologists have been aware of the Tapes transgression, they have also been attentive to the possibility of sites below present-day sea level. However, possibly due to the thousands of sites

Fig. 6.4 Map of Norway showing the location of underwater Stone Age find spots currently known from the Norwegian seabed. In total, there are 86 find spots. Potential areas shaded in green are those areas of submergence where underwater sites may be expected to occur. Site information from the SPLASHCOS Viewer http://splashcos-viewer.eu. Drawing by Moritz Mennenga

Fig. 6.5 Left: The inlet at Hummervikholmen, showing the location of Mesolithic human skeletal remains. Photo by Pål Nymoen. Copyright Norwegian Maritime Museum 2010. Right: Aerial view showing the location of the site in its wider context (http://kart.kystverket.no.)

on dry land, submerged sites have been given little attention. Hence, when more than 50 partly submerged sites were recorded in the early 1980s, the fact that they were submerged during full tide was not recorded as very significant, except as a complicating factor when recording and excavating (Olsen 1981; Bjerck and Olsen 1983).

All currently known sites from the Norwegian seafloor are situated in two of the three areas where the regression minimum was below the present sea level (Figs. 6.3 and 6.4). All sites are found in shallow waters, no deeper than 10 m below present average sea level. Most sites are only partly submerged, or only submerged during full tide. Since these depths are continuously exposed to wave action and tidal cycles, the sites are often substantially eroded and difficult to discover.

6.4.1 Southern Norway

The first reported submerged site in southern Norway was found in 1958 by a summerhouse owner who found a small collection of flint and rock-crystal artefacts while bathing on his holiday at Frivold in Mandal municipality (catalogue number C29741, Museum of Cultural History, University of Oslo). Later, several other finds were reported from the same bay (id no 29084, 48661 and 89,642 in the national registry *Askeladden*).

The most famous submerged Stone Age site in Norway was also found by a summer-house owner, some 15 km east of the 1958 site. In 1994 a cabin owner at Hummervikholmen, southernmost Norway, found human skeletal remains in his small harbour while removing sediments (Fig. 6.5). This led to the discovery of a possible ritual site from the Mesolithic, which was later excavated with interesting results (Sellevold and Skar 1999; Nymoen and Skar 2011a; Eggen et al. 2014; Nymoen 2014). Human skulls and limb bones were flushed out of the seafloor when the owner was deepening his small harbour with the aid of the water stream from the propeller of his outboard motor. The County heritage administration soon concluded that the bay revealed the oldest human skeletal remains so far found in Norway (Sellevold and Skar 1999; Nymoen and Skar 2011a). The find is comprised of the very well-preserved bones of at least three, but perhaps as many as five, individuals (Fig. 6.6). One skull is from a female, c. 35–40 years of age at death. It is robust, with a high upper face and rather low eye sockets, resembling other Scandinavian Mesolithic female skulls (Bennike and Alexandersen 1997; Sellevold and Skar 1999). A second skull is from a man of approximately the same age (Sjöling 2014).

The Hummervikholmen individuals have been ^{14}C dated to c. 7732–7011 cal BC.[1] Chemical analyses ($\delta^{13}C$ and $\delta^{15}N$ measurements) indicate that more than 80% of the dietary protein of the individuals was derived from marine mammal resources

[1] 2σ range (95.4% probability), due to the heavy marine diet an offset of 380+/−30 radiocarbon years was applied, following Mangerud et al. 2006.

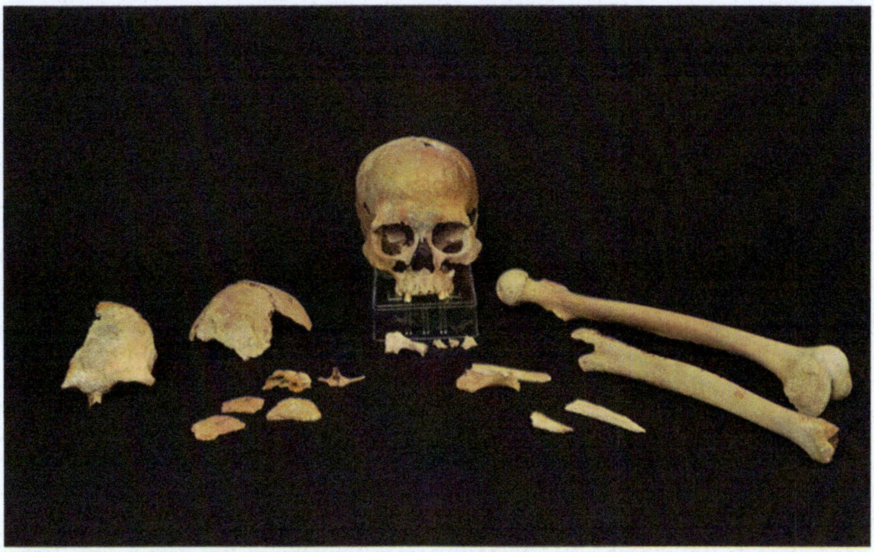

Fig. 6.6 Surviving fragments from Middle Mesolithic individuals from the seafloor of Hummervikholmen. Photo by Beate Kjørslevik. Copyright Norwegian Maritime Museum 2010

(Sellevold and Skar 1999; Skar et al. 2016). Analyses of aDNA of the female and of a male skull have confirmed a roughly 40:60% admixture of genes derived from Western hunter gatherer versus Eastern hunter gatherer populations, thus substantiating a second Early Mesolithic migration into Norway from the East (Günther et al. 2018).

The body parts seem to have been deposited on or near the beach at a time when the sea was approximately 1.2 m lower than today. The site was subsequently transgressed by the rising sea (Tapes transgression) (Nymoen and Skar 2011a; Eggen et al. 2014; Nymoen 2014) and covered by an oyster bed. The human remains from Hummervikholmen may represent ritual activity, though no grave goods or other artefacts have been recorded. Surveys have not revealed any contemporaneous settlement sites on the island.

Today a dozen submerged sites are known from this southern part of the country. A few are only single finds of artefacts, without any certain context or nearby finds (Fig. 6.4) (Nymoen and Skar 2011a). Some are adjacent to sites on land and are therefore probably part of larger sites that are only partially submerged (e.g. Nymoen and Skar 2011b). Such sites can potentially be very important for research if it turns out that the submerged parts of the settlement have good conditions for preservation of organic matter. If so, the distribution of lithic artefacts can easily be studied on land, while the submerged parts can provide the kind of objects normally lost in the acid soils of the Norwegian mainland.

So far, only Middle and Late Mesolithic (c. 8300–3900 cal BC) sites are known in shallow waters, but there are good chances that even Early Mesolithic (c. 9300–8300 cal BC) sites can be recorded in deeper waters. The archipelagos of southern Norway create several sheltered zones where sediments can be deposited with little or no erosion from wind and waves. Early Holocene transgressions happened quite fast, thus Early Mesolithic sites would have been quickly sealed by sediments.

Along the coast of southern Norway where the submerged Early Mesolithic coastlines are situated, few if any Early Mesolithic settlement sites have been found on dry land. North and south of these areas, Early Mesolithic coastal sites are found in abundance (e.g. Høgestøl 1995; Reitan and Sundström 2018). Most Early Mesolithic sites are coastal sites and the presumed reason for the absence of such sites in areas with submerged coastlines could be that the sites are situated under water. The high frequency of Early Mesolithic sites on dry land gives promise of an equivalent distributional pattern along the submerged coastlines as well.

6.4.2 Western Norway

Most of the known submerged Stone Age sites in Norway are found on the western coast (Figs. 6.3 and 6.4). The sea-level curves of the area (Fig. 6.2) show that the sea was as much as 10 m below present sea level at c. 8300 cal BC, and the area probably contains thousands of sites, even if the known distribution today clusters around two separate areas. All known sites on the western coast have been found by archaeologists, but the total number is still very uncertain.

In the early 1980s archaeologist Asle Bruen Olsen was studying the use and distribution of diabase from a Mesolithic stone axe quarry close to the city of Florø on the western coast of Norway. Olsen mapped close to 100 'beach sites' containing diabase waste and artefacts from the quarry in a limited area along the outer *Eikefjord* (Olsen 1981). Olsen's definition of a 'beach site' was that the site was found less than 2 masl. Approximately 50 of these sites are partly or fully submerged at full tide, but Olsen's concern was not if the sites were submerged or not, and the fact that he had discovered the largest known cluster of submerged sites in Norway was given little attention.

Most of Olsen's sites are probably landing sites for offloading quarried diabase from the Stakalleneset (Stakaneset) quarry from a period when the sea level was the same or just below present. The diabase waste may therefore have been lost or discarded in shallow waters while offloading boats. The only site which has been fully excavated, however, contained more than 3600 finds of diabase, flint and quartzite, and is definitely a settlement site. The site (Askeladden id 107330) was totally submerged at full tide, and the traditional 'terrestrial' excavation could only take place during a few hours every day (Bjerck and Olsen 1983).

The quarried diabase, especially the many axe roughouts, is a distinguishing feature on all of Olsens's beach sites. This is important, because the fist-sized diabase is easily recognisable among other beach cobbles without sieving through the gravel as one has to do to find the thumb-sized pieces of worked flint or quartzite. The diabase sites are therefore highly visible compared to other kinds of Stone Age sites from the same period and thus clearly may be over-represented among the known submerged Stone Age sites in Norway.

A handful of sites without diabase were later found some 80 km further north along the western coast, just south of the city of Ålesund. All these sites have been found by archaeologists doing environmental impact assessment studies for minor development projects. Three of the sites were found via surface collecting or test excavation at the edge of the water at full ebb (Simpson 2006; Nøttveit 2009; Sundsdal 2011), while two sites were detected by fully submerged test-pits at a depth of up to 7 m below present sea level.

Today's distribution of known submerged sites on the western coast, with two distinctive clusters 80 km apart, does not reflect the settlement pattern in the Stone Age. Rather it is the result of intensity in mapping and archaeological focus, and hundreds of more sites are most likely present in the area.

6.4.3 Northern Norway

The third area with a potential for traces of submerged Stone Age habitation is in the northern part of the country (Fig. 6.3). So far, no systematic investigations for submerged sites have been carried out in this huge area, even if a few areas have been subject to smaller investigations without any verified finds. In the early 1990s, diving archaeologists reported hut-like structures on a possible sunken beach ridge 4–5 m below present sea level at Sørøya in Finnmark. No actual artefacts were found, and later research on the relative sea-level changes in the area (Romundset et al. 2011) precludes the possibility for submerged sites at this depth. At present, no confirmed submerged sites or stray finds from Northern Norway have been reported to the national archive of archaeological finds and sites.

Parts of the Vesterålen and Lofoten archipelago in Northern Norway were ice-free during the Ice Age and tree species such as spruce and birch seem to have survived throughout the Pleistocene (Kullman 2002, 2006; Parducci et al. 2012). The reason for this is that the Gulf Stream reached the outer parts of the Lofoten-Vesterålen archipelago during the Ice Age because the continental shelf is quite narrow in this part of Norway (Andersen 2000, pp. 101–102). It is not unlikely that this area constituted an attractive biotope for humans in the Early Holocene, because the marine ecosystem could be supplemented by a terrestrial refuge biotope. Submarine peat, found at 2–5 m depth, dated to approximately 7400 cal BC is also known from Lofoten (Moe 1977; Møller 1984, p. 4; Vorren and Moe 1986). An important task for future research must therefore be to start looking for submerged sites in this area.

6.5 Palaeoecological Evidence

Archaeological finds of organic material are, so far, only known from two of the 86 sites from the seafloor. Both are situated in southern Norway. One of the sites is the Hummervikholmen, already described. From the island of Hidra in Vest-Agder county, a beautifully decorated pickaxe head made of bone or antler has been recovered from disturbed sediments in the Kirkhavn harbour (Fig. 6.7). The pickaxe is radiocarbon-dated to 7725–7535 cal BC (8540 ± 75 BP, TUa 1583). Similar objects made of stone are found on dry land, dated to the same period (Glørstad 2010 and references therein). From the Ostorf cemetery in Germany similar objects made of antlers have been dated to the Middle Neolithic (Lübke et al. 2007), indicating that such objects are among the tools used for a very long period of the Stone Age of Northern Europe.

Even though most of the submerged sites in Norway have not been subject to detailed archaeological investigations, the appearance of well-preserved organic artefacts indicates a clear potential for finding organic remains by more systematic excavations. This stands in sharp contrast to sites on land—where mainly lithic remains are found (see also Bjerck 2008).

In the same areas where the submerged Stone Age sites are found, several spots with rich organic sediments have been detected on the seabed. Examinations of these layers indicate that some of them have been accumulated above sea level, while others have been deposited in shallow water (Nymoen and Skar 2011a). Such sites are preserved only in very sheltered waters, protected by islands and lagoons.

There are good chances that such deposits or other accumulations of organic residue on the seabed can provide crucial information about flora and fauna at the end of the Pleistocene and at the beginning of the Holocene. In Petvik, Lofoten, peat layers contain large amounts of floral remains, among them birchwood dated to c. 7400 cal BC (Møller 1984).

Under the Middle Mesolithic deposits at Hummervikholmen, where the human remains

Fig. 6.7 Engraved bone pickaxe head with socket, 29 cm long, found during dredging in shallow water in the harbour of Kirkhavn (Scale in photo shows 1 cm units). Dated to 7725–7535 cal BC. Isotope analysis ($\delta^{13}C$) indicates that the bone is from a marine animal (probably whale). In type and decoration, it is unique—and thus exemplifies that exploration of the submerged Stone Age landscape holds the potential to retrieve items of material culture that are not available from archaeological sites on land. Photo by Pål Nymoen. Copyright Norwegian Maritime Museum 2010

were found, quite large pieces of pine were detected in the clay deposits. They are dated to the transitional Preboreal–Boreal period, 8451–7846 cal BC (Eggen et al. 2014). Gyttja dated to the Oldest Dryas is documented at even deeper levels. These kinds of deposits could therefore provide new and vital information about the earlier vegetation history of the region.

From Kaupanes, Egersund in south-western Norway, 13 antler and cranial fragments of reindeer were found in sediments in shallow water. The bones are dated to the final part of the Ice Age, 10,336–9864 cal BC (10,255 ± 80 BP, Bang-Andersen 2017). It is not clear how these bones ended up on the seabed. They may represent reindeer that drowned outside the continental plain and floated across the Norwegian trench to end up in south-western Norway. Even today it is not unusual that coastal currents transport carcases from Denmark to Norway (see also Jonsson 1995). A second possibility could be that the reindeer drowned on their way to pastures in Norway. The present reindeer population on Spitsbergen must have crossed large distances of sea ice, moving from island to island across Franz Josef Land to reach their present habitat (Gravlund et al. 1998; Yannic et al. 2014; Kvie et al. 2016). Similar migration patterns could have brought reindeer to Norway at the end of the Ice Age. A third possibility could be that the reindeer died on dry land and deposits of clay or silt preserved them when the sea rose. Either way, such finds are of extraordinary importance for understanding population dynamics because finds of organic remains on the bed of the North Sea from the final Pleistocene are extremely rare.

The Norwegian Cultural Heritage Act only protects sites with an established culture-historical content. Underwater deposits of natural origin, like turf, peat or gyttja, containing vegetation or faunal material, should however also be subject to a more conscious management policy as potential heritage sites and important archives for natural history (Nymoen and Skar 2011a). The Egersund and Hummervikholmen cases prove that such sites potentially can be of great importance for the understanding of the natural and cultural history of Europe.

6.6 Future Perspectives for Research on the Submerged Stone Age of Norway

This brief introduction to submerged Stone Age in Norway demonstrates the important potential for developing this field of research in this part of Europe. The review also demonstrates that this subject is still at an initial stage of development. Only recently has the fact that sites are submerged along the Norwegian coast gained interest as a topic in its own right. A steadily growing level of precision concerning details in the trajectory of the local isostasy and good computer simulations of this geological process enables archaeologists to predict precisely the areas with high potential for finding submerged sites. The coastline of mainland Norway is favourably situated for discovering early Holocene sites in relatively shallow waters. Elsewhere, deep waters cover such early sites, making archaeological examination more difficult. On the southern side of the Norwegian trench there is a potential for finding sites predating the Holocene (Hammer et al. 2016). Initial investigations (Hafeez et al. 2012; Glørstad and Kvalø 2012; Glørstad et al. 2017) have, however, revealed that finding such sites involves considerable challenges.

The growing interest in submerged Stone Age in Norway and the extraordinary possibilities for studying marine occupation from the Holocene on dry land in Norway certainly define the specific knowledge potential in this heritage: The submerged sites can reveal important additional information concerning marine settlements along the Norwegian coast. Secondly, and presumably of even greater importance, the submerged Stone Age sites will likely, in favourable situations, offer excellent conditions for preservation of organic remains from human occupation, thus contributing important information on the variety and complexity of human material culture and subsistence in this part of the world. The finds from Hummervikholmen, Kirkhavn and Kaupanes clearly demonstrate the potential.

6.7 Submerged Sites and Cultural Heritage Management in Norway

The Directorate for Cultural Heritage in Norway (Riksantikvaren) is responsible for the practical implementation of the Norwegian Cultural Heritage Act and the objectives laid down by the Norwegian Parliament and the Ministry of Climate and Environment. Thus, it has a general responsibility for the management of all kinds of submerged prehistoric sites.

Norway is divided into five large museum districts (Fig. 6.8). In the Tromsø and Trondheim districts, submerged Stone Age sites and underwater cultural heritage are managed by the University museums in the respective regions. In the Bergen, Stavanger and Oslo museum districts, this heritage is managed both by the Maritime museums and the University museums. The Maritime museums are public foundations, responsible for surveying and excavating archaeological heritage. The University museums are responsible for the finds and relevant documentation from the sites.

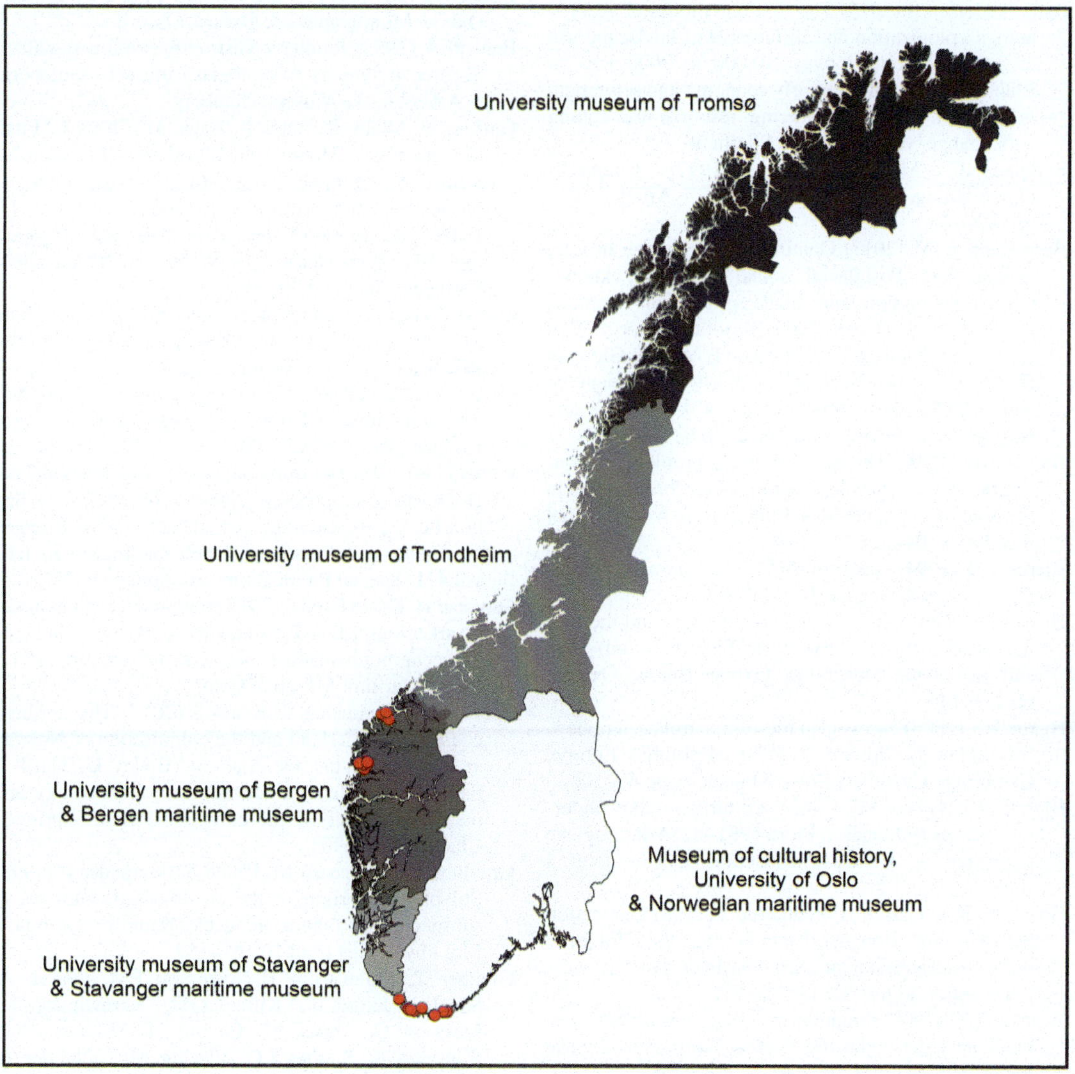

Fig. 6.8 Museum districts in Norway with institutions responsible for submerged archaeological remains. Known submerged archaeological sites are indicated by red dots

References

Åkerlund A (1996) Human responses to shore displacement. Living by the sea in eastern middle Sweden during the Stone Age. Arkeologiska Undersøkningar Skrifter nr 16. Riksantikvarieämbetet, Stockholm

Andersen B (2000) Istider i Norge: landskap formet av istidenes breer. Universitetsforlaget, Oslo

Bagge A (1951) Fagervik. Ein Rückgrat für die Periodeneinteilung der ostschwedischen Wohnplatz- und Bootaxtkulturen aus dem Mittelneolithikum. Acta Archaeol XXXII:57–118

Bang-Andersen S (1995) Mesolithic man and the rising sea spotlighted by three tapes-transgressed sites in SW Norway. In: Fischer A (ed) Man and sea in the Mesolithic. Oxbow Books, Oxford, pp 113–121

Bang-Andersen S (2017) The first one thousand years—human colonization and differentiated landscape use in South-Western Norway, 10,000 to 9000 BP. In: Blankholm HP (ed) The early economy and settlement in Northern Europe: pioneering, resource use, coping with change, vol 3. Equinox, Sheffield

Bennike P, Alexandersen V (1997) Danmarks Urbefolkning. Nationalmuseets Arbejdsmark, pp 143–156

Berg-Hansen IM (2017) Continuity and change in Late Glacial and Postglacial social networks; knowledge transmission and blade production methods in Ahrensburgian and Early Mesolithic North West Europe. In: Knutsson K, Knutsson H, Apel J, Glørstad H (eds) The Technology of Early Settlement in Northern Europe – transmission of knowledge and culture, vol 2. Equinox, Sheffield, pp 63–98

Bjerck HB (1986) The Fosna-Nøstvet problem. A consideration of archaeological units and Chronozones in the South Norwegian Mesolithic period. Nor Archaeol Rev 19(2):104–121

Bjerck HB (1994) Nordsjøfastlandet og pionerbosetningen i Norge. Viking LVII:25–58

Bjerck HB (1995) The North Sea Continent and the pioneer settlement of Norway. In: Fischer A (ed) Man and sea in the Mesolithic. Oxbow Books, Oxford, pp 131–144

Bjerck HB (2008) Norwegian mesolithic trends: a review. In: Bailey G, Spikins P (eds) Mesolithic Europe. Cambridge University Press, Singapore, pp 60–106

Bjerck HB, Olsen AB (1983) Kulturhistoriske undersøkelser på Botnaneset, Flora 1981–82. Arkeologiske rapporter 5, Historisk museum, Universitetet i Bergen

Bondevik S, Svendsen JI, Mangerud J (1998) Distinction between the Storegga tsunami and the Holocene marine transgression in coastal basin deposits of western Norway. J Quat Sci 13:529–537

Brøgger WC (1905) Strandliniens beliggenhed under stenalderen i det sydøstlige Norge. Norges geologiske undersögelser no 41. Aschehoug, Kristiania

Coles BJ (1998) Doggerland: a speculative survey. Proc Prehist Soc 64:45–81

Damlien H (2014) Eastern pioneers in westernmost territories? Current perspectives on Mesolithic hunter-gatherer large-scale interaction and migration within Northern Eurasia. Quat Int 419:5–16

Damlien H (2016) Between tradition and adaptation. Long-term trajectories of lithic tool-making in South Norway during the postglacial colonization and its aftermath (c. 9500-7500 cal BC). Unpublished PhD thesis, University of Stavanger, Stavanger

De Geer G (1888) Om Skandinaviens nivåforandringar under Quartorperioden. Geologiska Föreningens i Stockholm Forhandlingar, 10:366–379, 12:6–110

Eggen I, Nymoen P, Hammer Ø, Sjöling, E (2014) Excavation report and analyses from Hummervikholmen, Hallandvik, Søgne k, Vest-Agder. Museum of Cultural History, University of Oslo

Fischer A (ed) (1995) Man and sea in the Mesolithic, Oxbow Monographs 53. Oxbow, Oxford

Fredsjö Å (1953) Studier i Västsveriges äldre stenålder. Skrifter utgivna av Arkeologiska museet i Göteborg; 1. Arkeologiska Museet, Göteborg

Gaffney V, Allaby R, Bates R, Bates M, Ch'ng E, Fitch S, Garwood P, Momber G, Murgatroyd P, Pallen M, Ramsey E, Smith D, Smith O (2017) Doggerland and the lost frontiers project (2015–2020). In: Bailey G, Harff J, Sakellariou D (eds) Under the sea: archaeology and palaeolandscapes of the continental shelf. Springer, Cham, pp 305–321

Glørstad H (2010) The structure and history of the Late Mesolithic societies of the Oslo Fjord Area 6300-3800 BC. Bricoleur Press, Gothenburg

Glørstad H (2012) Where are the missing boats? The pioneer settlement of Norway as long-term history. Nor Archaeol Rev 46(1):57–80

Glørstad H (2014) Deglaciation, sea-level changes and the Holocene colonisation of Norway. In: Harff J, Bailey G, Lüth F (eds) Geology and archaeology: submerged landscapes of the continental shelf, Geological Society, vol 411. Special Publications, London, pp 9–25

Glørstad H, Kvalø F (eds) (2012) Havvind – paleogeografi og arkeologi. Kunnskapsinnhenting om bosetning i det holocene nordsjølandskapet. Arkeologisk rapport 12. Norsk Maritimt Museum, Oslo

Glørstad H, Gundersen J, Kvalø F (2017) The northern coasts of Doggerland and the colonisation of Norway at the end of the Ice Age. In: Bailey G, Harff J, Sakellariou D (eds) Under the sea: archaeology and palaeolandscapes of the continental shelf. Springer, Cham, pp 285–303

Gravlund P, Meldgaard M, Pääbo S, Arctander P (1998) Polyphyletic origin of the smallbodied, high-Arctic subspecies of tundra reindeer (Rangifer tarandus). Mol Phylogenet Evol 10:151–159

Günther T, Malmström H, Svensson EM, Omrak A, Sánchez-Quinto F, Kılınç GM, Krzewinska M, Eriksson G, Fraser M, Edlund H, Munteres AR, Coutinho A, Simões LG, Vicente M, Sjölander A, Sellevold BJ, Jørgensen R, Claes P, Shriver MD, Valdiosera C, Netea MG, Apel J, Lidén KB, Skar B, Storå J, Götherström A, Jakobsson M (2018)

Population genomics of Mesolithic Scandinavia: investigating early postglacial migration routes and high-latitude adaptation. PLoS Biol 16(1):e2003703. https://doi.org/10.1371/journal.pbio.2003703

Hafeez A, Planke S, Hammer Ø (2012) Post-glacial paleogeography of the Southern Norwegian North Sea, Report VBPR. University of Oslo, Oslo

Hammer Ø, Planke S, Hafeez A, Hjelstuen BOB, Faleide JI, Kvalø F (2016) Agderia – a postglacial lost land in the southern Norwegian North Sea. Nor Geol Tidsskr 96(1):43–60

Høgestøl M (ed) (1995) Arkeologiske undersøkelser i Rennesøy kommune, Rogaland, Sørvest-Norge. Arkeologisk Museum, Stavanger

Holtedahl O (1953) Norges geologi. Aschehoug, Oslo

Jaksland L, Persson P (eds) (2014) E18-brunlanesprosjektet bind I. Forutsetninger og kulturhistorisk sammenstilling. Museum of Cultural History, University of Oslo, Oslo

Jonsson L (1995) Vertebrate fauna during the Mesolithic on the Swedish west coast. In: Fischer A (ed) Man and sea in the Mesolithic. Oxbow, Oxford, pp 147–160

Kullman L (2002) Boreal tree taxa in the central Scandes during the late-glacial: implications for late-quaternary forest history. J Biogeogr 29:1117–1124

Kullman L (2006) Late-glacial trees from Arctic coast to alpine tundra: response to Birks et al. 2005 and 2006. J Biogeogr 33:376–378

Kvie KS, Heggenes J, Anderson DG, Kholodova MV, Sipko T, Mizin I et al (2016) Colonizing the high Arctic: mitochondrial DNA reveals common origin of Eurasian archipelagic reindeer (*Rangifer tarandus*). PLoS One 11(11):e0165237. https://doi.org/10.1371/journal.pone.0165237

Lübke H, Lüth F, Terberger T (2007) Fishers or farmers? The archaeology of the Ostorf cemetery and related Neolithic finds in the light of new data. In: Larsson L, Lüth F, Terberger T (eds) Innovation and continuity – non-megalithic mortuary practices in the Baltic. Verlag Phillip von Zabern, Mainz am Rhein, pp 307–338

Mangerud J, Bondevik S, Gulliksen S, Hufthammer AK, Hoisater T (2006) Marine 14C reservoir ages for 19th century whales and molluscs from the North Atlantic. Quat Sci Rev 25:3228–3245

Mikkelsen E (1975) Mesolithic in South-Eastern Norway. Nor Archaeol Rev 8(1):19–36

Moe D (1977) Nytt funn av undersjøisk torv i Norge. Naturen 1:21–23

Møller J (1984) Holocene shore displacement at Nappstraumen, Lofoten, North Norway. Norsk Geologisk Tidsskrift 64:1–5

Møller J, Holmeslet B (1998) Sealevel Change ver 3.51 (Sealev32.exe), a program for extrapolating sea level displacement curves for northern Norway, Southern Norway as well as Fennoskania and northwestern Russia. University of Tromsø, Tromsø

Nøttveit OM (2009) Marinarkeologiske registreringer. Reguleringsplan Varden, Ulstein. Report. Stiftelsen Bergens Sjøfartsmuseum, Bergen

Nummedal A (1923) Om flintpladsene. Nor Geol Tidsskr 7:90–141

Nummedal A (1933) Kan det finnes flintplasser på kyststrekningen mellom Kristiansand og Ålesund? Grieg, Bergen

Nymoen P (2014) Utgravningsrapport fra Hummervikholmen. Norsk Maritimt Museum, Oslo

Nymoen P, Skar B (2011a) The unappreciated cultural landscape – indications of submerged Mesolithic settlement along the Norwegian Southern Coast. In: Benjamin J, Bonsall C, Pickard C, Fischer A (eds) Submerged prehistory. Oxbow Books, Oxford, pp 38–54

Nymoen P, Skar B (2011b) Paradis – og andre indikasjoner på sub-marine lokaliteter fra mesolittisk tid langs den norske Sørlandskyst. Norsk Maritimt Museums Årbok 2011:91–115

Olsen AB (1981) Bruk av diabas i Vestnorsk steinalder. Unpublished magister thesis, University of Bergen

Olsen AB (1994) Bosetning og samfunn i Finnmarks forhistorie. Universitetsforlaget, Oslo

Parducci L, Jørgensen T, Tollefsrud MM, Elverland E, Alm T, Fontana SL, Edwards ME (2012) Glacial survival of boreal trees in northern Scandinavia. Science 335(6072):1083–1086

Reid C (1913) Submerged forests. Cambridge University Press, Cambridge

Reitan G, Sundström L (2018) Steinalderen i Aust-Agder. Arkeologiske undersøkelser i forbindelse med ny E18 Tvedestrand-Arendal. Cappelen-Damm, Oslo

Romundset A, Bondevik S, Bennike O (2011) Postglacial uplift and relative sea level changes in Finnmark, Northern Norway. Quat Sci Rev 30:2398–2421

Schmitt L, Svedhage K (2016) Chronological aspects of the Hensbacka – a group of hunter- gatherers/fishers on the west coast of Sweden during the Pleistocene/Holocene transition: an example of early coastal colonization. Danish J Archaeol 4(1):75–81

Schmitt L, Larsson S, Schrum C, Alekseeva I, Tomczak M, Svedhage K (2006) "Why they came" the colonization of the coast of western Sweden and its environmental context at the end of the last glaciation. Oxf J Archaeol 25(1):1–28

Schmitt L, Larsson S, Burdukiewicz J, Ziker J, Svedhage K, Zamon J, Steffen H (2009) Chronological insights, cultural change, and resource exploitation on the west coast of Sweden during the Late Palaeolithic/Early Mesolithic transition. Oxf J Archaeol 28:1–27

Sellevold BJ, Skar B (1999) The first lady of Norway. In: Gundhus G, Seip E, Ulriksen E (eds) NIKU 1994–1999. Kulturminneforskningens mangfold. NIKU, Oslo, pp 6–11

Simpson D (2003) SeaLevelCurvesSunm-STrond_v2.xls. Bergen Museum, University of Bergen, Bergen

Simpson D (2006) Arkeologiske undersøkelser på Eika, Sundet gnr. 71, Ulstein kommune, Møre og Romsdal. Report, Bergen Museum. University of Bergen, Bergen

Sjöling E (2014) Osteologisk analys, Obrända människoben/menneskeknokler från tidigmesolitikum,

Humervikholmen, Søgne kommune, Vest-Agder. SAU rapport 2014:5. Museum of Cultural History, University of Oslo, Oslo

Skar B, Breivik HM (2018) Environment and adaptation of forager pioneers on Northwestern regions of Europe. In: Persson PÅ, Riede F, Skar B, Breivik HM, Johnsson L (eds) The ecology of early settlement in northern Europe. Conditions for subsistence and survival. Equinox Publishing, Sheffield, pp 1–16

Skar B, Lidén K, Eriksson G, Sellevold B (2016) A submerged Mesolithic grave site reveals remains of the first Norwegian seal hunters. In: Bjerck HB (ed) Marine ventures. Archaeological perspectives on human–sea relations. Equinox, Sheffield, pp 225–239

Sørensen M, Rankama T, Kankaanpää J, Knutsson K, Knutsson H, Melvold SA, Eriksen BV, Glørstad H (2013) The first eastern migrations of people and knowledge into Scandinavia: evidence from studies of Mesolithic technology, 9th-8th millennium BC. Nor Archaeol Rev 46(1):19–56

Sundsdal K (2011) Rapport fra marinarkeologiske registreringer. Vannledning fra Borgarøy til Hatløy, Ulstein kommune. Stiftelsen Bergens Sjøfartsmuseum, Bergen

Svendsen JI, Mangerud J (1987) Late Weichselian and Holocene sea-level history for a cross-section of western Norway. J Quat Sci 2:113–132

Vorren KD, Moe D (1986) The early Holocene climate and sea-level changes in Lofoten and Vesterålen, North Norway. Norsk Geologisk Tidsskrift 66:135–143

Yannic G, Pellissier L, Ortego J, Lecomte N, Couturier S, Cuyler C, Dussault C, Hundertmark KJ, Irvine RJ, Kolpashikov L, Mager K, Musiani M, Parker KL, Røed KH, Sipko T, Þórisson SG, Weckworth BV, Guisan A, Bernatchez L, Côté SD (2014) Genetic diversity in caribou linked to past and future climate change. Nat Clim Chang 4(2):132–137

Part II

The Atlantic Margin and the North Sea

The Atlantic Margin and the North Sea: Introduction

7

Hans Peeters, Fraser Sturt, and Kieran Westley

Abstract

The North Sea and Atlantic margins of Europe encompass a vast area of seabed and coastline, and a correspondingly large area of potential submerged landscape. This once-seamless landscape has been divided by modern political geography, leading to different research traditions, management regimes and, consequently, necessitating division into national summary chapters. This chapter presents a synthesis of the national summaries that follow by focusing on four common themes. Within each, we explore the variability and commonalities between countries. Firstly, we assess the overall archaeological and palaeoenvironmental evidence base. It is clear that the evidence is temporally deep, incorporates material ranging from in situ human footprints and wooden structures to derived and isolated lithic finds, contains indications of both aquatic and terrestrial activity, and can be found across the study area, albeit unevenly distributed, with a bias to the North Sea. Secondly, we examine preservation and discovery conditions. In addition to spatially variable taphonomy caused by regional geological, palaeoenvironmental (e.g. sea-level change, glaciation) and hydrodynamic conditions, there are also national differences in methods of investigation, notably the role of systematic investigation versus chance finds. Based on work done in the North Sea, thoughts are suggested as to possible methodological ways forward. Thirdly, we address the research potential of the wider region. Extant research frameworks have identified numerous common themes, but sub-regional themes also exist. In both cases, these may cross-cut existing borders and require transnational collaboration. Research potential also extends to a range of evidence types, including derived as well as in situ archaeology and palaeoenvironmental records. Finally, we look at the management context, highlighting the key role played by historic environment management and offshore industry in data collection and methodological advances. Overall, this synthesis demonstrates that much progress has been made, though concentrated in certain areas (e.g. the North Sea). Still more remains to be done, in terms of extending approaches to less-studied parts of the Atlantic margin but also in improving the quality of data collected.

H. Peeters (✉)
Groningen Institute of Archaeology, University of Groningen, Groningen, The Netherlands
e-mail: j.h.m.peeters@rug.nl

F. Sturt
Centre for Maritime Archaeology, University of Southampton, Southampton, UK
e-mail: f.sturt@soton.ac.uk

K. Westley
School of Geography and Environmental Sciences, Ulster University, Coleraine, Northern Ireland, UK
e-mail: kl.westley@ulster.ac.uk

G. Bailey et al. (eds.), *The Archaeology of Europe's Drowned Landscapes*, Coastal Research Library 35, https://doi.org/10.1007/978-3-030-37367-2_7

Keywords
Submerged prehistory · Submerged landscapes · North Sea · Atlantic margin · Intertidal

7.1 Introduction

The present-day Atlantic margin and North Sea basin represent marine environments with highly diverse characteristics in terms of hydrodynamics, seafloor and coastal geomorphology as well as sub-surface lithology (Cohen et al. 2017; Dawson et al. 2017; Farr et al. 2017; Westley and Edwards 2017). These result from an equally diverse set of interrelated geogenetic processes, driven by climate fluctuation and alternating palaeoenvironmental conditions. Cycles of Quaternary marine transgression and regression and accompanying variation in the balance of sedimentation versus erosion continuously rearranged the physical characteristics of these areas, sweeping from full terrestrial to full marine conditions and back. The dynamics of such processes were far from uniform; environmental change occurred on many different spatial and temporal scales, ranging from the supra-regional to local, and from the millennial to human lifetime (or generational) scale. Hence, the human occupation history and context of human behaviour in these now-submerged parts of the continental shelf can only be understood within the framework of these variable environmental dynamics. In fact, to come to an understanding of how prehistoric people interacted with the 'ebb and flow' of palaeoenvironmental dynamics in the (near-) coastal zone also involves a deeper understanding of the representativeness of the onshore terrestrial archaeological record. These are complementary sources of information.

In this section, we will develop a synthesis of the national chapters that are concerned with the Atlantic margin and North Sea basin: Ireland (Westley and Woodman, Chap. 11, this volume), Great Britain (Bailey et al., Chap. 10, this volume), Norway (Glørstad et al. Chap. 6, this volume), the Netherlands (Peeters and Amkreutz, Chap. 8, this volume), Belgium (Pieters et al., Chap. 9, this volume), the French (Billard et al., Chap. 12, this volume) and Spanish Atlantic margins (Arias, Chap. 13, this volume), and Portugal (Bicho et al. Chap. 14, this volume). Our main objective is to explore and identify the background of the variable insights as presented in the national chapters, based on four themes. First, we will look at the actual evidence: archaeological finds, as well as palaeoenvironmental evidence, which is just as important for our understanding of prehistoric human behaviour as the artefacts left behind. Next, we address aspects of preservation and discovery in order to evaluate the significance of the known record. This in turn leads to further discussion of the area's research potential including identification of national and transnational research themes. Finally, we look at the management context, which involves the role of industry and institutions, as well as the maintenance of collections.

7.2 Archaeological and Palaeoenvironmental Evidence

The first thoughts about the potential importance of the submerged archaeological record date back to the nineteenth and early twentieth centuries, when observations of megalithic monuments below the high-water mark along the Atlantic margin indicated the presence of submerged cultural phenomena (Billard et al., Chap. 12, this volume). Similarly, observations of tree stumps in the intertidal zone along the British and Irish coasts provided evidence for drowned forests, which led to a first hypothetical reconstruction of the drowned landscape in the present-day North Sea (Reid 1913), now frequently referred to as 'Doggerland' (Coles 1998). The implications of these early observations of both submerged cultural and environmental remains are tremendous: they made, and make, clear that the archaeological record exposed on land today reflects the modern land/sea boundary and not that seen in the past. This in turn impacted (and continues to impact) on how we interpret issues of connectivity and social interaction as they played out in the past.

Now, more than a century later, the body of archaeological and palaeoenvironmental evidence from the Atlantic margin and North Sea has grown considerably. The national chapters make abundantly clear that this evidence consists of a wide array of phenomena, ranging from isolated objects deriving from unknown contexts, to undisturbed sites and palaeolandsurface remnants 'betraying' human presence in the form of footprints. However, the evidence presented is as variable between the national overviews as is the nature and dating of this evidence. Table 7.1 shows that the archaeological remains from the sub- and intertidal zones consist of artefacts made from diverse materials, as well as built structures, which functioned in settlement as well as ritual contexts. The built structures mostly date to the Neolithic or Bronze Age, whereas artefacts cover the Palaeolithic to Iron Age.

As far as can be derived from the inventories presented in the national chapters and descriptions provided in the SPLASHCOS Viewer (http://splashcos-viewer.eu/), it seems that the phenomena overwhelmingly relate to onshore activities, at least in the sense that the range of evidence is also known from onshore archaeological sites. As such, these phenomena represent submerged 'terrestrial' archaeology. For the Palaeolithic and Mesolithic, we are mostly dealing with lithic artefacts deriving from—most probably at least—the sort of settlement contexts known onshore. This is supported by research conducted on a few subtidal sites (Bouldnor Cliff; Rotterdam-Yangtze Harbour) off the British and Dutch coasts (Momber et al. 2011; Momber and Peeters 2017; Moree and Sier 2015; Peeters and Momber 2014). Examples dating to the Neolithic are various built structures, as observed in the intertidal zones along the coasts of the British Isles and France. The majority consist of megalithic monuments (tombs; menhirs), as well as wooden henges, which can be consid-

Table 7.1 Summary of archaeological evidence from the subtidal and intertidal zones of the Atlantic margin and North Sea, as presented in the national chapters

	Ireland		United Kingdom		Norway		Netherlands		Belgium		France		Spain		Portugal	
	Stray/re-worked	In situ	Stray/re-worked	In situ	Stray/re-worked	In situ	Stray/re-worked	In situ	Stray/re-worked	In situ	Stray/re-worked	In situ	Stray/re-worked	In situ	Stray/re-worked	In situ
Faunal remains	PL		PL		PL		PL/M	M	PL		PL	PL/H				
Forest remains		H		P/M			H					H				M
Animal tracks				P/M												
Human footprints				P/M/N								P/N				
Fishing structures		M		N								M/N				
Shell middens				M								N				M
Human remains	N			M/N/LP			P/M/LP									M
Burials				N		M						N				
Wooden monuments				LP												
Megalithic monuments		N										N				
Ceramic artefacts									LP			N				N/LP
Organic artefacts	M/N		P/M/N/LP	P/M/N/LP	M		P/M	M	LP			N				
Lithic artefacts	M/N		P/M/N/LP	P/M/N/LP	M		P/M	M	P/N		P/M/N	P/M/N				P/N

PL Pleistocene, *H* Holocene, *P* Palaeolithic, *M* Mesolithic, *N* Neolithic, *LP* Late Palaeolithic, *Blue* Subtidal, *Pink* Intertidal, *Grey* Subtidal or Intertidal, *White* No information ered as the 'organic' counterparts of megalithic henges.

However, this does not rule out evidence for an aquatic orientation of behaviour. Palaeolithic faunal assemblages, artistic depictions and stable isotope values from Cantabrian Spain provide evidence for the exploitation of the marine environment (Arias, Chap. 13, this volume). The numerous fish weirs (dating back at least to the Neolithic and possibly earlier) on the French Atlantic margin (Billard et al., Chap. 12, this volume) connect directly to the exploitation of aquatic resources, as do the North Wall Quay (Dublin) fish traps (Westley and Woodman, Chap. 11, this volume), possible remains of fish traps at Goldcliff, and intertidal shell middens at Westward Ho! on the British coast (Bailey et al., Chap. 10, this volume) and São Julião on the Portuguese coast (Bicho et al., Chap. 14, this volume). Finds at Bouldnor Cliff have been interpreted in connection with the possible making of canoes (Momber et al. 2011), a Neolithic logboat has been found in the intertidal zone of Strangford Lough (Westley and Woodman, Chap. 11, this volume) while boat use could be suggested by a possible wooden paddle from the Raversijde beach in Belgium (Pieters et al., Chap. 9, this volume). The remarkable finds of Middle Neolithic polished axe blades from the Brown Bank area in the Dutch sector of the North Sea also provide evidence for seafaring (as this area would have been flooded by this time), and possibly symbolic meaning of 'places' in a seascape (cf. Van de Noort 2011). The find of ritual depositions of axe blades in the intertidal zone of Brittany (Billard et al., Chap. 12, this volume) possibly connects to the symbolic meaning of the zone between land and water, and fits patterns of deposition found in onshore wetland environments all over NW Europe.

Of particular interest to the role of the shoreline are the human footprints found at various locations on the British and French coasts. Those at Happisburgh (Ashton et al. 2014; Bailey et al., Chap. 10, this volume) belong to multiple individuals (adults and children) and are among the scanty evidence for the earliest occupation of NW Europe 0.8–1 million years ago. It is likely that shorelines played an important role in the colonisation of this part of the world by early humans (Cohen et al. 2012). Coastal exploitation is also evident from the Mesolithic and Neolithic footprints at Goldcliff (Bell 2007) and Formby (Huddart et al. 1999). Detailed analysis of these prints suggests age, and maybe, gender-related task division (Bailey et al., Chap. 10, this volume). The size of human footprints encountered at the fish-weir site of Saint-Jean-le-Thomas on the French coast—dating from the Neolithic to Bronze Age transition—shows the presence of children and adults (Billard and Bernard 2016). Information on age and gender-related task division can normally not be derived from material remains directly, where interpretations strongly depend on contextual information, or the interpretation of diet-related isotope data from human bones.

The footprints and animal tracks directly link humans to the environment, as individuals who crossed the landscape. We cannot get any closer to humans and animals moving in their environment than this. The aforementioned hypothesis on the submerged landscape in the present-day North Sea (Coles 1998; Reid 1913) is, in view of these 'snapshots', somewhat revolutionary. Observations of palaeoenvironmental phenomena—at the time of unknown age—triggered an idea about the physical layout of a forested landscape. The first archaeological finds reported from this submerged landscape came from the British and Dutch sectors of the North Sea and were of Mesolithic age (Louwe Kooijmans 1971). No more than a decade ago, the first images of palaeolandscape features became available, based on the analysis of 3D seismic data (Gaffney et al. 2007). However, as yet, little is known about how this landscape looked as a living environment, hosting diverse plant and animal communities, and indeed, humans. The 'Europe's Lost Frontiers Project' (Gaffney et al. 2017) is expected to provide important new data in this respect.

Nonetheless, submerged and intertidal landscapes, in particular the North Sea, have already delivered a vast body of palaeoenvironmental evidence. Analysis of high-resolution core samples from the submerged site of Rotterdam-Yangtze Harbour provided evidence for local human interference (clearance?) in wetland vegetation, possibly to permit movement between the settlement location on a sand dune and the nearby riverbank (Moree and Sier 2015, p. 307). Considerable quantities of mammalian remains dredged from the North Sea floor are most probably of Mesolithic age, and this is partly corroborated by radiocarbon dates (Glimmerveen et al. 2006). Most human remains from the North Sea also date to Mesolithic times, and stable isotope data indicate a diet consisting of terrestrial and freshwater foodstuffs (Van der Plicht et al. 2016). However, there seems to be little evidence for marine resources as important items of diet. In the perspective of the Mesolithic drowning history of the southern North Sea this might appear somewhat peculiar at first, but it might also suggest that the relationship between humans and (potential) environmental resources is more complicated. And returning to the footprints, what about socially differentiated diets?

Similar research is less well-developed outside the North Sea, though in some cases peat layers in the intertidal zones of the Atlantic margin have permitted insights into vegetational changes, mainly during the Holocene. One thing that is clear from the national chapters is that though submerged forests and organic layers are mostly known from the British coast, such palaeolandscape vestiges have in fact been encountered nearly everywhere. Thus the full potential of this evidence base away from the North Sea remains to be explored.

The environmental evidence is not limited to the Mesolithic and later prehistoric times, but extends well into the Palaeolithic, principally represented by mammal remains. Apart from incidental finds of Pleistocene mammals from the Atlantic margin and Channel area (Billard et al., Chap. 12, this volume; Westley and Woodman, Chap. 11, this volume), the overwhelming majority of finds originates from the North Sea (Glimmerveen et al. 2004, 2006; Mol et al. 2006). Bone assemblages consist of material dating to the Lower and Late Pleistocene; material securely dated to the Middle Pleistocene is less though certainly not absent (Cohen et al. 2017, p. 168). Since these finds were brought up in fishing nets, or derive from aggregate extraction, little is known about the original context. Targeted surveying during recent sand extraction off the Dutch coast, however, permitted insight into the stratigraphic origin of palaeontological material (Kuitems et al. 2015). One notable result concerns the establishment of a North Sea highstand during the early Weichselian (MIS 5d–a), based on mollusc evidence corroborated by OSL dates (Peeters and Amkreutz, Chap. 8, this volume). The importance of this finding is that it puts the AMS-based dates of sea mammals (e.g. Beluga, in Mol et al. 2006) in the Weichselian in a new perspective: it does make a difference whether sea levels were 70 m or 20 m lower than today when we consider archaeological issues, as the shoreline was positioned differently in the North Sea basin, and species like Beluga can be expected to have been present in parts of it.

Such insight is also crucial for our understanding of the human occupation of NW Europe during the Pleistocene. Handaxes and other flint artefacts come from various locations in the North Sea (Area 240; Middeldiep; Eurogeul: Bailey et al., Chap. 10, this volume; Peeters and Amkreutz, Chap. 8, this volume), submerged Middle Palaeolithic assemblages associated with faunal remains are documented off the Normandy and Armorican coast (Billard et al. Chap. 12, this volume), and intertidal Middle Palaeolithic assemblages are known from the Portuguese Atlantic coast (Bicho, Chap. 14, this volume). These too need to be placed in an appropriate palaeogeographic and palaeoenvironmental context if we are even to begin addressing questions of how past humans/hominins used this landscape and responded as it changed over time.

7.3 Preservation and Discovery

The summary above provides some insight into the range of archaeological and palaeoenvironmental evidence. The diversity of this evidence not only relates to variability in the occupation history of this vast region, but also to variability in taphonomy and discovery contexts. Long-term natural dynamics had drastic impact on the preservation of archaeological and palaeoenvironmental materials and layers. These dynamics also influence discovery potential. As succinctly highlighted by Bailey et al. (Chap. 10, this volume) in relation to intertidal forests, there is a delicate balance between preservation by burial and exposure by erosion to enable discovery. Modern economic exploitation of resources, most notably dredging and bottom trawling, can also lead to erosion of sedimentary layers and removal of archaeological material or palaeoenvironmental evidence from its stratigraphic context. The specific technology involved largely determines which materials can potentially be discovered, whereas the involvement of individuals with variable knowledge about remains of archaeological or palaeoenvironmental interest determines which materials are actually collected. In order to make the evidence meaningful for ongoing research, to assess the research potential and define research themes, it is therefore important to take a closer look at these factors.

Table 7.2 shows that surveys and controlled research result in relatively few discoveries, and these are best represented in Great Britain and France. Taking the study area as a whole, the discovery of finds by chance in fishing nets and on waste piles at wharves outnumbers discoveries resulting from systematic work. Trends are, however, not the same for the listed nations, and many factors may underlie the differences. Erosion of the steeper parts of the British and French coast exposes archaeological sites that can readily be accessed and investigated in the intertidal and subtidal zone. In contrast, prehistoric levels along the gently sloping Dutch and Belgian coast are covered with metres of sediment. Water depth and visibility are also factors to keep in mind, as are traditional foci of interest, notably shipwrecks, among sports divers and professionals alike.

Aspects of erosion and sedimentation are, however, of prime importance with regard to preservation and discovery potential. Ward and Larcombe (2008) have proposed a conceptual framework for the assessment of geomorphological structures and lithological units in terms of archaeological potential. This is useful at a scale when more detailed information is available about the geological structure below the seafloor (Peeters and Momber 2014; Cohen et al. 2017). At a coarser scale, it remains worthwhile to bear in mind the general trends in the geological development of (parts of) sea basins. The southern and central North Sea, for instance, is to a large extent a subsiding basin with an infill of hundreds of metres of vertically stacked Quaternary sediments. This leaves room for an accumulation of archaeologically relevant strata, covering at least the entire time span for which we possess evidence of human presence in NW Europe. Conversely, areas on the 'shoulders' of the basin, where subsidence is minimal or absent exhibit greater stratigraphic complexity, with

Table 7.2 Overview of discovery contexts and their relative degree of application: incidental (+), occasional (++), frequent (+++)

	Chance finds			Surveys		Controlled research	
Nation	Exposure	Dredging	Trawling	Grab sampling	Targeted fishing	Visual assessment	Excavation
Ireland	+	+				+	+
Great Britain	+++	++	+	++		++	++
Norway		+				+	+
Netherlands		+++	+++		++		+
Belgium		+	+				
France	+++		++			++	++
Spain	+	+				+	
Portugal	+					+	++

highstand deposits of multiple ages superimposed at broadly the same level. The lack of subsidence in these areas also makes them more prone to erosion, exemplified by the exposure of Lower Palaeolithic levels (including the key early sites of Pakefield and Happisburgh) on the eastern coast of Britain (Cohen et al. 2017).

Similarly, erosive regimes on the Atlantic margin do not necessarily imply absence of archaeological remains. Without doubt, large-scale erosion in the Channel/La Manche due to catastrophic Pleistocene flooding events (Gupta et al. 2007) swept away parts of older landscape structures and their associated archaeology. In other parts, old sediment layers will have survived, whilst new sediment layers may have been deposited (and may have been eroded afterwards). The Middle Palaeolithic sites on the Armorican and Normandy coast (Billard et al., Chap. 12, this volume), as well as on the Spanish and Portuguese coasts (Arias, Chap. 13, this volume; Bicho et al., Chap. 14, this volume) emphasise the interest and importance of palaeoshorelines. At the same time, we must bear in mind that shorelines are vulnerable to erosion, particularly so when sediment supply is limited, coasts are steep and wave fetch is large. Conceptually at least, this would appear to rule out large parts of Europe's Atlantic-facing coast as likely to preserve submerged prehistoric sites and would seem to be borne out by the overall site distribution shown in the SPLASHCOS viewer. Clearly, there are many more sites clustered within the North Sea compared to the open Atlantic margin.

However, this is not to say that such zones are of no interest. Local conditions, such as protected bays and inlets, like the ones encountered along the Irish, Norwegian and French Atlantic coasts (Billard et al., Chap. 12 this volume; Glørstad et al., Chap. 6 this volume; Westley and Woodman, Chap. 11, this volume), may have favoured preservation of archaeological and palaeoenvironmental remains. Hence, the potential presence of such 'capsules' is important because of the possibilities it offers to collect at least tiny pieces of the puzzle, and to gain some idea of what is missing. These same issues apply to the erosive impact of glaciation, often seen as resulting in the removal or reworking of the archaeological record. Within the North Sea and Atlantic margin, notwithstanding potential limits to hominin environmental tolerances, this may be a contributing factor to the lack of a pre-Last Glacial Maximum (LGM) record in regions most strongly affected by glaciation: Scandinavia, northern Britain and Ireland. However, glacial removal of evidence is not universal, as exemplified by preservation of Lower Palaeolithic/Cromerian deposits under water-laid glacial sediments in eastern England or even the role of glacial depressions (e.g. tunnel valleys, kettle holes) in which post-glaciation sediment can accumulate (Cohen et al. 2017).

From this, it follows that geoarchaeological assessment of any area under consideration is a first step. We simply cannot rely on the generally coarse subsurface models available. As outlined in the 'North Sea Prehistory Research and Management Framework' (NSPRMF; Peeters et al. 2009, p. 11), we find ourselves confronted with a high level of perceived need for data, and in connection with this a high level of risk/fear of making significant mistakes. In order to lower these levels of uncertainty, there is need for more secure data points, which will improve the value of data in general and the models built on it. Hence, there is a critical need for testing geoarchaeological models of site preservation. Paradoxically, invasive economic activities provide opportunities for such model testing (Peeters et al. 2009, p. 11). However, one such activity, which provided a major source of information until recently, is now legally prohibited EU-wide: beam trawling (see also Maarleveld, Chap. 25, this volume). As a result, the fishing industry can no longer make a contribution to resolving the aforementioned problem. This is all the more unfortunate, since it has become clear that targeted fishing for palaeontological remains—and why not archaeological remains for that matter—is promising (Mol 2012).

Discovery potential remains an aspect that requires continuous attention. Which strategies and technologies are chosen for the detection of archaeological remains is (partly) one of the national traditions. In the Netherlands, onshore strategies strongly rely on core sampling in the survey stage of investigations. In a next step, extensive coring and/or test pitting is conducted to assess potential significance and subsequent decision-making. This approach has been translated to the offshore environment and executed in the context of the Rotterdam harbour extension, although here it was preceded by seismic profiling of the subsurface (Vos et al. 2015). The strategy proved to be successful, since a Mesolithic site was discovered. Subsequent stratigraphically controlled grab sampling delivered interesting results (Moree and Sier 2015; Peeters and Amkreutz, Chap. 8, this volume), but given the rather low quantities of finds, the discovery of some flint knapping debris and burnt bone in a core must be considered a lucky strike.

Despite the fact that systematic grab sampling is more destructive than core sampling, its chances of finding archaeological remains are better, if the artefact-bearing layers are located at, or near the seabed surface. Such an approach has shown its success in the case of Area 240 in British waters (Tizzard et al. 2015). Clearly, sample size (in terms of the grab sampler capacity) and sample density on the seabed are critical factors when it comes to the detection of archaeological sites (Wessex Archaeology 2008). At the same time, however, it is important to be able to put finds in a stratigraphical and palaeoenvironmental context —and this is where grab sampling is inherently limited—otherwise we are left with collections of items which have restricted information value. Hence, it appears that a combined approach to the problem bears the best chances of delivering quality data and will move us forward in steps which are successively refined in focus based on previous results (e.g. Vos et al. 2015). Geophysical techniques provide possibilities to map the seabed and subsurface, and in combination with coring information, permit geoarchaeological models, which can be tested by means of grab sampling. In a next stage, high-quality core sampling can focus on the collection of palaeoenvironmental and fine-grained stratigraphical data. In a final stage, targeted and controlled grab sampling, or perhaps if conditions are conducive, diver-led sampling, can be conducted to obtain quality data on the archaeology.

7.4 Research Potential and Themes

The range of archaeological phenomena discussed in the national chapters for the Atlantic margin and North Sea mirrors a myriad of underlying behaviours. This, in itself, demonstrates a high research potential. As discussed, preservation conditions and discovery potential are diverse at variable spatial scales. This has implications for the research potential in various parts of the Atlantic margin and North Sea Basin.

For the (southern) North Sea, the NSPRMF identifies several research themes and topics (Table 7.3). The scientific background for this document (Peeters et al. 2009), which resulted from a joint effort of researchers and heritage professionals to define a common perspective on the problem of submerged prehistoric archaeology in the sector (Salter et al. 2014), was published as a special volume of the 'Netherlands Journal of Geosciences' (Peeters and Cohen 2014). Furthermore, a national maritime archaeological research agenda exists for English waters, which outlines many similar research themes (Ransley and Sturt 2013).

All themes and topics are still relevant in the current state of affairs. In consideration of the archaeological evidence, the same (or at least very similar) themes and topics are equally applicable to the Atlantic margin. The focus on the Palaeolithic and Mesolithic can easily be expanded to later prehistoric times. Also, there is room for (sub-) regional differentiation, through inclusion of specific research topics. For instance, evidence from the Norwegian coast would lend itself to discussion of postglacial colonisation of new landscapes by sea, while evidence from around the western seaways of the British Isles would be conducive to addressing questions surrounding the nature of Mesolithic and Neolithic networks and connections between Great Britain, Ireland, mainland Europe and smaller outlying island groups (e.g. the Western Isles, Scilly Isles, Isle of Man).

Table 7.3 Research themes and topics from the NSPRMF (Peeters et al. 2009). Despite the fact that theme G primarily focusses on post-LGM hunter-gatherers, topic G.2 is broadly defined, and is of equal relevance to theme D

Theme	Topics
A. Stratigraphic and chronological frameworks	A.1: Lithostratigraphic classification and chronological anchoring
	A.2: Sea level change and glacio-isostacy
	A.3: Survival of deposits of archaeological significance
	A.4: Biostratigraphies and absolute dating
B. Palaeogeography and environment	B.1: Middle/Late Pleistocene reshaping of topography and river drainage
	B.2: Development of the Weichselian/Devensian landscape
	B.3: Palaeogeographic evolution after the Last Glacial Maximum (LGM)
	B.4: Quaternary palaeoecology
C. Global perspectives on intercontinental hominin dispersals	C.1: North Sea coastal dynamics and human uses of the coastal zone
	C.2: Pleistocene North Sea level oscillations and population of islands
D. Pleistocene hominin colonisations of northern Europe	D.1: Early human exploitation strategies in changing environments
	D.2: Natural barriers for hominin expansion
E. Reoccupation of northern Europe after the Last Glacial Maximum (LGM)	E.1: Post-LGM occupation flux
	E.2: Occupation strategies
F. Post-glacial land use dynamics in the context of a changing landscape	F.1: Changing landscape structure
	F.2: Behavioural diversity among hunter-gatherers
	F.3: Maritime archaeologies of the North Sea
G. Representation of prehistoric hunter-gatherer communities and lifeways	G.1: Spatial perspectives on North Sea palaeolandscapes
	G.2: The distributional nature of early hominin communities
	G.3: Enculturated hunter-gatherer landscapes

The brief review presented above makes clear that there exist several transnational topics, which offer potential opportunities to develop joint research programmes. One such topic concerns the role of individuals in the exploitation of the coastal zone. The discovery of human footprints from various periods along the British and French coast permits us to gain unique insights into the presence of individuals of different age and physical condition, who spent time on the intertidal mudflats for various purposes. This information source is highly vulnerable to erosion, which is expected to increase with climate change and rising sea levels. Joining forces to develop ways to safeguard this information rapidly and effectively is crucial. In addition, how can further information about these individuals, and the environments wherein they operated, be extracted? For instance, what are the possibilities for sedaDNA analysis? After all, the fact that such imprints are preserved in the first place means that they were rapidly covered with a protecting layer of sediment; and this may have been favourable for the preservation of DNA.

The potential for archaeological and palaeoenvironmental research of in situ contexts appears to be most favourable in the intertidal zone. Subtidal contexts are more difficult to identify and investigate, notably in deep water. Where potentially important layers are expected at some depth below the seafloor, the situation is even more complicated. This is not to say that high-resolution excavation underwater is not possible. The work at Bouldnor Cliff in British waters makes a good example (Momber et al. 2011; Bailey et al., Chap. 10, this volume). However, it is also clear that underwater excavation comes with restrictions on the spatial extent of investigations. High-resolution data will provide information on small sections of sites, the greatest part of which cannot be investigated. Hence, the question of representativeness is highly relevant—as it is onshore. The design of smart strategies is therefore critical, but in helping to move the discipline forward we may have to accept some initial loss of information compared to the possibilities for data collection onshore.

It is equally important to bear in mind that finds from less well documented—or even completely undocumented—contexts carry intrinsic information of potential scientific importance. A good example is the combined use of radiocarbon dates and stable isotope data to draw inferences about diet from Mesolithic human remains from the North Sea (Van der Plicht et al. 2016). Initial results from DNA analysis of bone samples from the North Sea also appear promising (Dr Luc Amkreutz, pers. comm.). Hence, scientifically valuable data can be extracted from individual finds even when the original context is unknown.

Similarly, such data can also provide a starting point for new research. Radiocarbon dating of Pleistocene mammal remains, for instance, demands a critical approach to methodology and sampling. Busschers et al. (2014) were able to show that molluscs securely dated to the Eemian (MIS 5: >75 ka) based on multi-proxy data (biostratigraphy; lithostratigraphy; OSL and U-Th dating; age-depth relations to sea level), return radiocarbon ages where they should not, given the technique's ~50 ka limit. This appears to result from the effects of the secondary introduction of carbonate, due to bacterial activity, which caused inter-crystalline secondary carbon precipitation. This raises the question to what extent radiocarbon dates obtained on Pleistocene mammal bones dredged from the North Sea floor are reliable (Busschers et al. 2014, p. 1162). There is no fundamental reason to doubt the validity of such dates for remains of animals ('cold fauna') which are known to have roamed the Weichselian landscape. However, in other cases problems arise. For instance, the claimed Weichselian age for the remains of a sabre-toothed cat (Mol et al. 2006, 2008), which was long believed to have become extinct at c. 300 ka, is surprising. Is there a possibility that the obtained date is a result of taphonomic contamination? Along the same lines, obtained dates on fauna that seem to fit a potentially correct age (e.g. woolly mammoth, c. 40 ka) are not necessarily reliable.

Obtaining insights into the age and time-depth of palaeontological bone assemblages is of direct archaeological significance. On the one hand, such information provides an important building block for our understanding of Pleistocene environments in which various hominin species lived, whilst on the other hand it paves the way for the development of new ideas and hypotheses. For instance, the vast quantity of mammoth remains dredged from the Dutch seafloor has led to ideas about the behaviour and mortality of these animals from a palaeontological perspective (Mol 2012). The same finds have made others hypothesise about the potential presence of Early Upper Palaeolithic mammoth hunting and construction of dwellings from mammoth bones, similar to those found in Eastern Europe (Grøn 2015). In both cases, insight into the chronological integrity of assemblages becomes important. The improvement of compound AMS dating opens new directions for shedding light on such matters.

From this it follows that the submerged evidence from the Atlantic margin and North Sea basin has an enormous research potential. The range of research themes and topics is wide, and addresses issues of diverse nature (archaeology; palaeontology; palaeolandscape; taphonomy; methodology). From a purely quantitative perspective, the body of evidence known and potentially present is not the problem. More important is the question as to how to make the most out of this evidence, and in which ways we can improve the possibilities to collect qualitatively sound sets of materials and data, which can be fully connected to and interpreted in the context of the onshore record that we are used to working with, and which dominates archaeological models of the human past.

7.5 Industry, Institutions and Collections

One of the key points which can be drawn both from the national summaries and the recent literature (e.g. Benjamin et al. 2011; Evans et al. 2014; Bailey et al. 2017; Flemming et al. 2017) is that submerged prehistory is now well-accepted across the region. We are now beyond Coles' (1998) 'speculative survey'. While there are still considerable gaps in our knowledge of submerged landscapes, concerted efforts have been made, or are in progress, to fill these.

Importantly, this acceptance and willingness to engage with the submerged record is not just purely research-led. Crucially, submerged prehistory is now an intrinsic part of cultural resource/historic environment management, hence is included within coastal and offshore licencing or environmental impact assessment regimes (e.g. Sturt et al. 2017, for England; see also Pater, Chap. 27, this volume), or at least steps are underway to include within national regulatory and management frameworks (e.g. Missiaen et al. 2017, for Belgium). This is important since it is largely through major offshore industry projects that the vast majority of seabed activity takes place. With the exception of a handful of research-led projects (e.g. Europe's Lost Frontiers Project: Gaffney et al. 2017), these are the only approaches with the finances capable of acquiring the extensive seabed and sub-seabed geophysical and geotechnical data required to build the detailed geoarchaeological models described above. Therefore, it should be no surprise that some of the biggest advances of the last decade have come through this engagement. Examples are the 3D-seismic-based reconstructions of the southern North Sea (Gaffney et al. 2007, 2009; Van Heteren et al. 2014), the Area 240 finds and follow-up survey (Tizzard et al. 2011, 2015), and most recently the Rotterdam Harbour Mesolithic investigations (Moree and Sier 2015). Even at a much smaller scale, the national summaries provide examples of material identified via commercial/pre-development investigation, such as the North Wall Quay fish traps (Westley and Woodman, Chap. 11, this volume) and finds from western Norway (Glørstad et al., Chap. 6, this volume).

Given the potential spatial and temporal extent of the submerged prehistoric record of the North Sea and Atlantic Margin, and the transnational nature of many of the potential sedimentary deposits of interest, it is imperative that this engagement is maintained both nationally and internationally if the discipline is to progress. It is, however, also important to ensure that engagement does not stand still, and that existing 'best practice' approaches do not become fixed. If available, innovative approaches and technologies should be introduced to improve the data collected (thus enabling more 'good' data points as described above) and to start addressing new questions as they come up (see also Peeters et al. 2009; Ransley and Sturt 2013). Despite the importance of results presented in the national chapters, it will be necessary to find ways to get a tighter grip on the archaeology that is actually resting on and under the seafloor. As noted, Coles' (1998) speculations have turned into evidence-based models of potential, underpinned with a handful of concrete archaeological observations. Now, we should engage in targeted attempts to turn these models of potential into archaeological realities.

References

Ashton N, Lewis SG, De Groote I, Duffy SM, Bates M, Bates R, Hoare P, Lewis M, Parfitt SA, Peglar S, Williams C, Stringer C (2014) Hominin footprints from Early Pleistocene deposits at Happisburgh, UK. PLoS ONE 9(2):e88329. https://doi.org/10.1371/journal.pone.0088329

Bailey GN, Harff J, Sakellariou D (2017) Under the sea: archaeology and palaeolandscapes of the continental shelf. Springer, Cham

Bell M (2007) Prehistoric coastal communities: the Mesolithic in Western Britain. CBA research report 149. Council for British Archaeology, York

Benjamin J, Bonsall C, Pickard C, Fischer A (eds) (2011) Submerged prehistory. Oxbow, Oxford

Billard C, Bernard V (2016) Pêcheries de Normandie: Archéologie et histoire des pêcheries littorales du département de la Manche. Presses Universitaires de Rennes, collection Archéologie et Culture, Rennes

Busschers FS, Wesselingh FP, Kars RH, Versluijs-Helder M, Wallinga J, Bosch JHA, Timmer J, Nierop KGJ, Meijer T, Bunnik FPM, De Wolf H (2014) Radiocarbon dating of Late Pleistocene marine shells from the southern North Sea. Radiocarbon 56(3):1151–1166

Cohen KM, MacDonald K, Joordens JCA, Roebroeks W, Gibbard PL (2012) Earliest occupation of north-west Europe: a coastal perspective. Quatern Int 271:70–83

Cohen KM, Westley K, Erkens G, Hijma MP, Weerts HTJ (2017) The North Sea. In: Flemming NC, Harff J, Moura D, Burgess A, Bailey G (eds) Submerged landscapes of the European continental shelf. Quaternary paleoenvironments. Wiley-Blackwell, Chichester, pp 147–186

Coles BJ (1998) Doggerland: a speculative survey. Proc Prehist Soc 64:45–81

Dawson S, Bates R, Wickham-Jones C, Dawson A (2017) Northern North Sea and Atlantic Northwest approaches. In: Flemming NC, Harff J, Moura D, Burgess A, Bailey GN (eds) Submerged landscapes of the European continental shelf: quaternary paleoenvironments. Wiley, Chichester, pp 187–209

Evans A, Flatman J, Flemming N (eds) (2014) Prehistoric archaeology on the continental shelf. A global review. Springer, Dordrecht

Farr RH, Momber G, Satchell J, Flemming N (2017) Paleolandscapes of the Celtic Sea and the Channel/La Manche. In: Flemming NC, Harff J, Moura D, Burgess A, Bailey G (eds) Submerged landscapes of the European continental shelf: quaternary paleoenvironments. Wiley-Blackwell, Chichester, pp 211–241

Flemming NC, Harff J, Moura D, Burgess A, Bailey GN (eds) (2017) Submerged landscapes of the European continental shelf: quaternary paleoenvironments. Wiley, Chichester

Gaffney V, Thomson K, Fitch S (eds) (2007) Mapping Doggerland: the Mesolithic landscapes of the southern North Sea. Archaeopress, Oxford

Gaffney V, Fitch S, Smith D (2009) Europe's Lost World, the rediscovery of Doggerland, CBA research report 160. Council for British Archaeology, York

Gaffney V, Allaby R, Bates R, Bates M, Ch'ng E, Fitch S, Garwood P, Momber G, Murgatroyd P, Pallen M, Ramsey E, Smith D, Smith O (2017) Doggerland and the Lost Frontiers Project (2015–2020). In: Bailey GN, Harff J, Sakellariou D (eds) Under the sea: archaeology and palaeolandscapes of the continental shelf. Springer, Cham, pp 305–319

Glimmerveen J, Mol D, Post K, Reumer JWF, Van der Plicht H, De Vos J, Van Geel B, Van Reenen G, Pals JP (2004) The North Sea project: the first palaeontological, palynological, and archaeological results. In: Flemming NC (ed) Submarine prehistoric archaeology of the North Sea: research priorities and collaboration with industry. CBA research report 141. Council for British Archaeology, York, pp 43–52

Glimmerveen J, Mol D, Van der Plicht J (2006) The Pleistocene reindeer of the North Sea — initial palaeontological data and archaeological remarks. Quatern Int 142/143:242–246

Grøn O (2015) A potential for submerged Late Weichselian mammoth-hunter sites in the Scandinavian North Sea sector? https://www.academia.edu/15411538/

Gupta S, Collier JS, Palmer-Felgate A, Potter G (2007) Catastrophic flooding origin of shelf valley systems in the English Channel. Nature 448:342–346

Huddart D, Roberts G, Gonzalez S (1999) Holocene human and animal footprints and their relationship with coastal environmental change, Formby Point, NW England. Quatern Int 55:29–41

Kuitems M, Van Kolfschoten T, Busschers F, De Loecker D (2015) Geoarchaeological and palaeontological research in the Maasvlakte 2 sand extraction zone and on the artificially created Maasvlakte 2 beach: a synthesis. In: Moree JM, Sier MM (eds) Interdisciplinary research programme Maasvlakte 2. Rotterdam. BOOR, Rotterdam, pp 351–398

Louwe Kooijmans LP (1971) Mesolithic bone and antler implements from the North Sea and from the Netherlands. Berichten Rijksdienst voor het Oudheidkundig Bodemonderzoek 20(21):27–73

Missiaen T, Pieters M, Maes F, Kruiver P, De Maeyer P, Seys J (2017) The SeArch project: towards an assessment methodology and sustainable heritage policy for the archaeological heritage of the North Sea in Belgium. In: Bailey GN, Harff J, Sakellariou D (eds) Under the sea: Archaeology and palaeolandscapes of the continental shelf. Springer, Cham, pp 415–424

Mol D (2012) Two days of trawling in the Eurogully: Did trawler OD7 find a mammoth graveyard? Dep Mag 30:13–15

Mol D, Post K, Reumer JWF, Van der Plicht J, De Vos J, Van Geel B, Van Reenen G, Pals J-P, Glimmerveen J (2006) The Eurogeul — first report of the palaeontological, palynological and archaeological investigations of this part of the North Sea. Quater Int 142/143:178–185

Mol D, De Vos J, Van Geel B, Glimmerveen J, Van der Plicht J, Post K (2008) Mammoeten, neushoorns en andere dieren van de Noordzeebodem. Kleine encyclopedie van het leven in het Pleistoceen. Veen Magazines, Diemen

Momber G, Peeters H (2017) Postglacial human dispersal and submerged landscapes in North-west Europe. In: Bailey GN, Harff J, Sakellariou D (eds) Under the sea: Archaeology and palaeolandscapes of the continental shelf. Springer, Cham, pp 321–334

Momber G, Tomalin D, Scaife R, Satchell J, Gillespie J (eds) (2011) Mesolithic occupation at Bouldnor Cliff and the submerged prehistoric landscapes of the Solent. CBA research report 164. Council for British Archaeology, York

Moree JM, Sier MM (2015) Twenty meters deep! The Mesolithic period at the Yangtze Harbour site – Rotterdam Maasvlakte, the Netherlands. Early Holocene landscape development and habitation. In: Moree JM, Sier MM (eds) Interdisciplinary research programme Maasvlakte 2, Rotterdam. BOOR, Rotterdam, pp 7–350

Peeters JHM, Cohen KM (eds) (2014) North Sea submerged landscapes and prehistory. Geology, prehistoric archaeology and research potential of the southern North Sea. Netherlands J Geosci 93(1/2):3–6

Peeters JHM, Momber G (2014) The southern North Sea and the human occupation of northwest Europe after the Last Glacial Maximum. Netherlands J Geosci 93(1/2):55–70

Peeters H, Flemming N, Murphy P (eds) (2009) North Sea Prehistory Research and Management Framework (NSPRMF). Rijksdienst voor het Cultureel Erfgoed, Amersfoort

Ransley J, Sturt F (eds) (2013) People and the sea: a maritime archaeological research agenda for England. CBA research report 171. Council for British Archaeology, York

Reid C (1913) Submerged forests. Cambridge University Press, Cambridge

Salter E, Murphy P, Peeters H (2014) Researching, conserving and managing submerged prehistory: national approaches and international collaboration. In: Evans A, Flatman J, Flemming N (eds) Prehistoric Archaeology on the Continental Shelf. A global review. Springer, Dordrecht, pp 151–172

Sturt F, Dix J, Grant MJ (2017) The history of industry-linked research in English waters: lessons for the future. In: Bailey GN, Harff J, Sakellariou D (eds) Under the sea: archaeology and palaeolandscapes of the continental shelf. Springer, Cham, pp 425–436

Tizzard L, Baggaley PA, Firth AJ (2011) Seabed prehistory: investigating palaeolandsurfaces with Palaeolithic remains from the southern North Sea. In: Benjamin J, Bonsall C, Pickard C, Fischer A (eds) Submerged prehistory. Oxbow, Oxford, pp 65–74

Tizzard L, Bicket AR, de Loecker D (2015) Seabed prehistory: investigating the palaeogeography and Early Middle Palaeolithic archaeology in the southern North Sea. Wessex Archaeology Report 35. Wessex Archaeology, Salisbury

Van de Noort R (2011) North Sea archaeologies. A maritime biography 10,000BC – AD 1500. Oxford University Press, Oxford

Van der Plicht J, Amkreutz L, Niekus MJLT, Peeters H, Smit B (2016) Surf 'n Turf in Doggerland: Dating, stable isotopes and diet of Mesolithic human remains from the southern North Sea. J Archaeol Sci Rep 10:110–118

Van Heteren S, Meekes JAC, Bakker MAJ, Gaffney V, Fitch S, Gearey BR, Paap BF (2014) Reconstructing North Sea palaeolandscapes from 3D and high-density 2D seismic data: an overview. Netherlands J Geosci 93(1/2):31–42

Vos PC, Bunnik FPM, Cohen KM, Cremer H (2015) A staged geogenetic approach to underwater archaeological prospection in the Port of Rotterdam (Yantzehaven, Maasvlakte, the Netherlands): a geological and palaeoenvironmental case study for local mapping of Mesolithic lowland landscapes. Quatern Int 367:4–31

Ward I, Larcombe P (2008) Determining the preservation rating of submerged archaeology in the post-glacial southern North Sea: a first-order geomorphological approach. Environ Archaeol 13:59–83

Wessex Archaeology (2008) Seabed in prehistory: gauging the effects of marine aggregate dredging. Final report, volume I–VIII, ref. Wessex Archaeology, Salisbury, p 57422

Westley K, Edwards R (2017) Irish Sea and Atlantic margin. In: Flemming NC, Harff J, Moura D, Burgess A, Bailey GN (eds) Submerged landscapes of the European continental shelf: Quaternary paleoenvironments. Wiley, Chichester, pp 241–280

The Netherlands: Probing into the Submerged Prehistoric Archaeology, Landscapes and Palaeontology of the Dutch Continental Shelf

8

Hans Peeters and Luc Amkreutz

Abstract

The Dutch sector of the North Sea is an important source for archaeological and palaeolandscape data from prehistoric times. A vast body of artefacts and palaeontological remains, dating from the Palaeolithic and Mesolithic, has been dredged and trawled from the seafloor. Contacts with industry, and fishermen in particular, permitted private collectors and professionals to assemble valuable collections for research. Although the overwhelming majority of finds derive from unknown contexts, these are of scientific importance. Firstly, they demonstrate the potential for the presence of well-preserved submerged archaeological sites and palaeolandscape contexts. Secondly, there is a lot of 'intrinsic' information that can be extracted from individual specimens, notably for radiocarbon dating, diet reconstruction (stable isotopes) and aDNA. Work in recent years has been increasingly concerned with the contextualisation of these data. Geoarchaeological investigations off the Dutch coast have permitted insight into the stratigraphical origin of Middle Palaeolithic flint artefacts, and the fragment of a Neanderthal skull. Targeted geoarchaeological research in the extension of the Rotterdam harbour has provided an opportunity to partially investigate a Mesolithic site at 20 m below sea level. This has led to increasing awareness among stakeholders that this submerged heritage is valuable and needs to be taken care of. Several initiatives have been taken to anticipate the potential presence of important archaeological and palaeolandscape remains in zones of economic interest.

Keywords

Submerged prehistoric · Submerged landscape · Palaeontological · Rotterdam harbour · Mesolithic · North Sea · Palaeolithic

8.1 Introduction

The publication of a paper by Louwe Kooijmans (1970/1971) describing a number of prehistoric implements caught in fishing nets from the North Sea floor first led to a general awareness that prehistoric landscapes extended into the present-day North Sea. In addition, the exten-

H. Peeters (✉)
Groningen Institute of Archaeology, University of Groningen, Groningen, The Netherlands
e-mail: j.h.m.peeters@rug.nl

L. Amkreutz
National Museum of Antiquities, Leiden, The Netherlands

Faculty of Archaeology, Leiden University, Leiden, The Netherlands
e-mail: l.amkreutz@rmo.nl

G. Bailey et al. (eds.), *The Archaeology of Europe's Drowned Landscapes*, Coastal Research Library 35, https://doi.org/10.1007/978-3-030-37367-2_8

sion of the Rotterdam harbour (Maasvlakte) in the late 1970s led to the discovery of hundreds of Mesolithic bone and antler implements (mostly points) from mudspoil used for land reclamation (Verhart 1988). Unfortunately, the exact stratigraphic, sedimentary, and broader archaeological context of these finds was largely unknown. In addition, the layers of archaeological interest were covered with about 20 m of sediment and water and were out of reach for detailed investigation. Consequently, initiatives to collect further information about the submerged prehistoric environment had to rely on an increasing number of finds without any clear context (Verhart 1995).

Cooperation between palaeontologists and fishermen since the 1980s has permitted systematic collection of palaeontological materials and artefacts, in addition to the registration of find zones (Glimmerveen et al. 2004, 2006; Mol et al. 2006; see also Maarleveld, Chap. 27, this volume). Also, targeted expeditions were organised on a yearly basis to 'fish for bones' (Glimmerveen et al. 2004). For the first time it was possible to identify the geographical location of finds to some degree of precision. The vast number of palaeontological and archaeological materials retrieved from fishing nets triggered initiatives to develop heritage management approaches (Maarleveld and Peeters 2004; Peeters et al. 2009; Peeters 2011). Over the past few years, academic and development-driven research has turned towards efforts to provide context for hitherto context-less finds.

Below, we provide an overview of finds and sites (Fig. 8.1). We underline that this overview will be of a somewhat impressionistic nature, as 'hard' data are few, while a systematic inventory is not yet available at this stage. In addition to the discussion of quantitative and qualitative aspects of finds and sites, we will discuss the research potential, and current scientific and management approaches in connection with the 'North Sea Prehistory Research and Management Framework (NSPRMF)' (Peeters et al. 2009), which sets a baseline for developments in the Netherlands.

8.2 Geological Context

The complex geology of the Dutch part of the continental shelf is affected by river dynamics, glacial processes, sea-level fluctuation and tectonics. Due to its importance for the Holocene occupation history of the Netherlands, relative sea-level rise has received particular attention in the work of geologists and archaeologists (e.g., Louwe Kooijmans 1974; Jelgersma 1979; Hijma 2009). However, factors such as palaeogeography, and sedimentation and erosion are equally important to take into consideration to gain insight into the human use of landscapes during the Pleistocene and Holocene, and the representativeness of the archaeological record (Peeters et al. 2009).

This record, however, is not a single 'container' of finds, as materials were left behind by very different people, at different times, and under variable conditions. In order to explain some of the emergent patterns, emphasis is placed on research that provides a geological context for finds (Hijma et al. 2012; Cohen et al. 2014; Peeters and Cohen 2014; Peeters and Momber 2014; Roebroeks 2014). At this stage there is still a lack of sufficiently detailed studies on the occurrence of Pleistocene and Holocene sediments to identify potential zones of archaeological interest within the Dutch part of the southern North Sea. Nonetheless, research conducted in the Middeldiep area (Middle Palaeolithic) and in the Port of Rotterdam (Mesolithic) demonstrates that there is potentially a lot to gain by an integrative approach. Also, future collaboration between geologists and archaeologists is planned with the aim of detailing and further dating the submergence of the North Sea basin at potentially informative core-drilling locations.

More recently, geological data on palaeogeographical developments in the North Sea basin have been reconsidered from a perspective that seeks to integrate prehistoric occupation dynamics and the formation of the archaeological record (Cohen et al. 2017). Similar work has been done for the Channel/La Manche and Celtic Sea (Farr et al. 2017). These are important steps to reach a

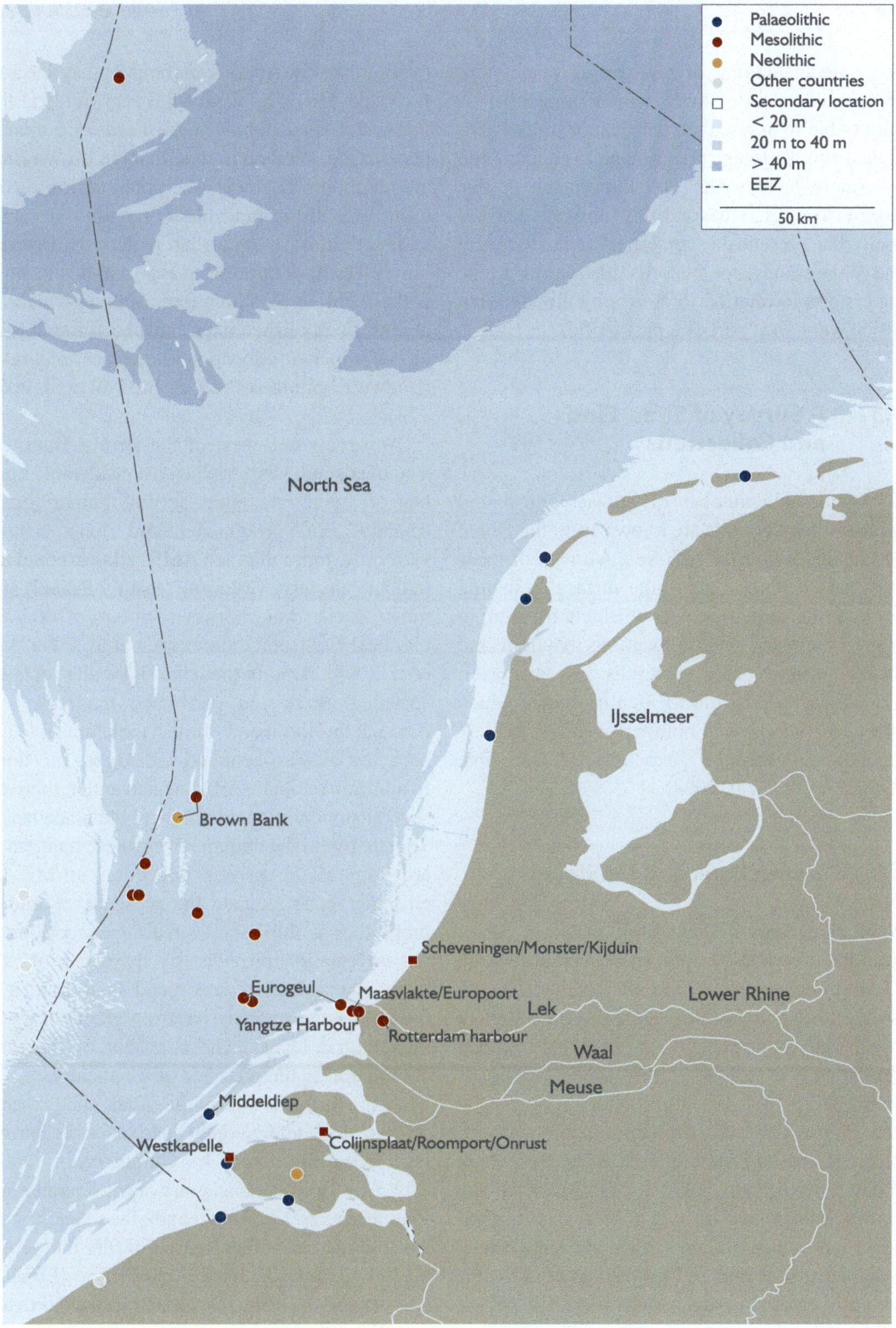

Fig. 8.1 Map showing the geographical location of the find zones mentioned in the text. A: Brown Bank and De Stekels; B: Eurogeul; C: Middeldiep; D: Westkapelle; E: Colijnsplaat, Roompot and Onrust; F: Maasvlakte-Europoort, Maasvlakte II-Yangtze harbour and Hoek van Holland; G: Scheveningen, Monster and Kijkduin. Site information from the SPLASHCOS Viewer http://splashcos-viewer.eu. Drawing by Moritz Mennenga

better understanding of taphonomy at a wider geographical scale, and without which it is impossible to assess the significance of archaeological and landscape phenomena as encountered at a more local scale. The importance of this work also reaches beyond the submerged world itself: the present-day terrestrial archaeological and palaeolandscape records also need to be understood in relation to areas now flooded, but which were once part of a past world.

8.3 A Survey of Sites, Finds and Collections

A considerable number of archaeological and palaeontological finds are known from the Dutch part of the southern North Sea. As most of these finds have been brought to the surface by fishing or dredging activities it is difficult to refer to 'sites' in terms of precise locations that delivered clear concentrations of materials. In some cases, however, it is possible to identify zones which produced finds over a number of years, making targeted 'fishing' expeditions possible (Glimmerveen et al. 2004).

8.3.1 Brown Bank and De Stekels

A small number of Mesolithic artefacts derive from the Brown Bank area, an elongated ridge of 35 km in length at *c.* 80 km off IJmuiden. Its top is only 19 m below water level. The Brown Bank is generally interpreted as an erosional remnant consisting of Late Pleistocene deposits (Busschers et al. 2007; Hijma et al. 2012). Gyttja-like deposits of Early Holocene age have been found in the surroundings of the Brown Bank (Louwe Kooijmans 1970/1971). Most of the Brown Bank finds are larger pieces of worked bone and antler, among which are large shaft-hole picks, socketed and pointed axes, adzes as well as (perforated) antler tines (Fig. 8.2). Waste products with saw and cut marks (Louwe Kooijmans 1970/1971) provide evidence for local production of tools. Some pieces show marks that may result from gnawing by dogs (Verhart 1995). Apart from bone and antler artefacts, the Brown Bank area has also yielded lithic finds such as a cobble mace head with conical perforation. While it is clear that the Brown Bank area harbours an interesting potential for uncovering Mesolithic (and older) remains of camps and activities, the current picture is severely biased. Another interesting aspect of this location is the fact that Neolithic finds, among which are at least two polished axes, have been found here, one of which at a location known as '*kolenboot*' (Louwe Kooijmans 1985, p. 14; Mol et al. 2008, 173).

An area south-west of the Brown Bank has also over some years yielded a considerable number of finds, including several human bones (mainly skull fragments and long bones). Currently, more than ten AMS radiocarbon dates indicate an Early Holocene (mainly Boreal) age; most dates fall between about 8700 and 6900 cal BC (see Glimmerveen et al. 2006; Mol et al. 2008). Among the bone and antler tools are socketed adzes and perforated mattocks; one mattock has the wood of the handle preserved. A socketed adze is decorated with zigzag incisions. In addition to objects of bone and antler, there are several stone tools including cobble mace heads. Together with the human bones, these finds (most are kept in the private collection of Mr. Jan Glimmerveen) indicate the presence of one or more sites within a restricted area according to the information provided by fishermen. As the bone material is in very good condition, it is likely that the finds derive from gradually eroding cultural layers. The character of the finds could point to the presence of a cemetery.

Apart from the Mesolithic finds, the presence of several Middle Neolithic axe blades is intriguing (Maarleveld 1984; Glimmerveen 2007) (Fig. 8.3). Both axe blades are of considerable size (one measures 19 cm in length, the other 32 cm), and skilfully manufactured out of flint that probably originates from southern Limburg (Netherlands). During the Neolithic, the North Sea coastline was roughly situated at its modern-day position, and sea levels were only a few metres lower than today. As far as we can tell, major highs such as the Dogger Bank and Brown Bank were by

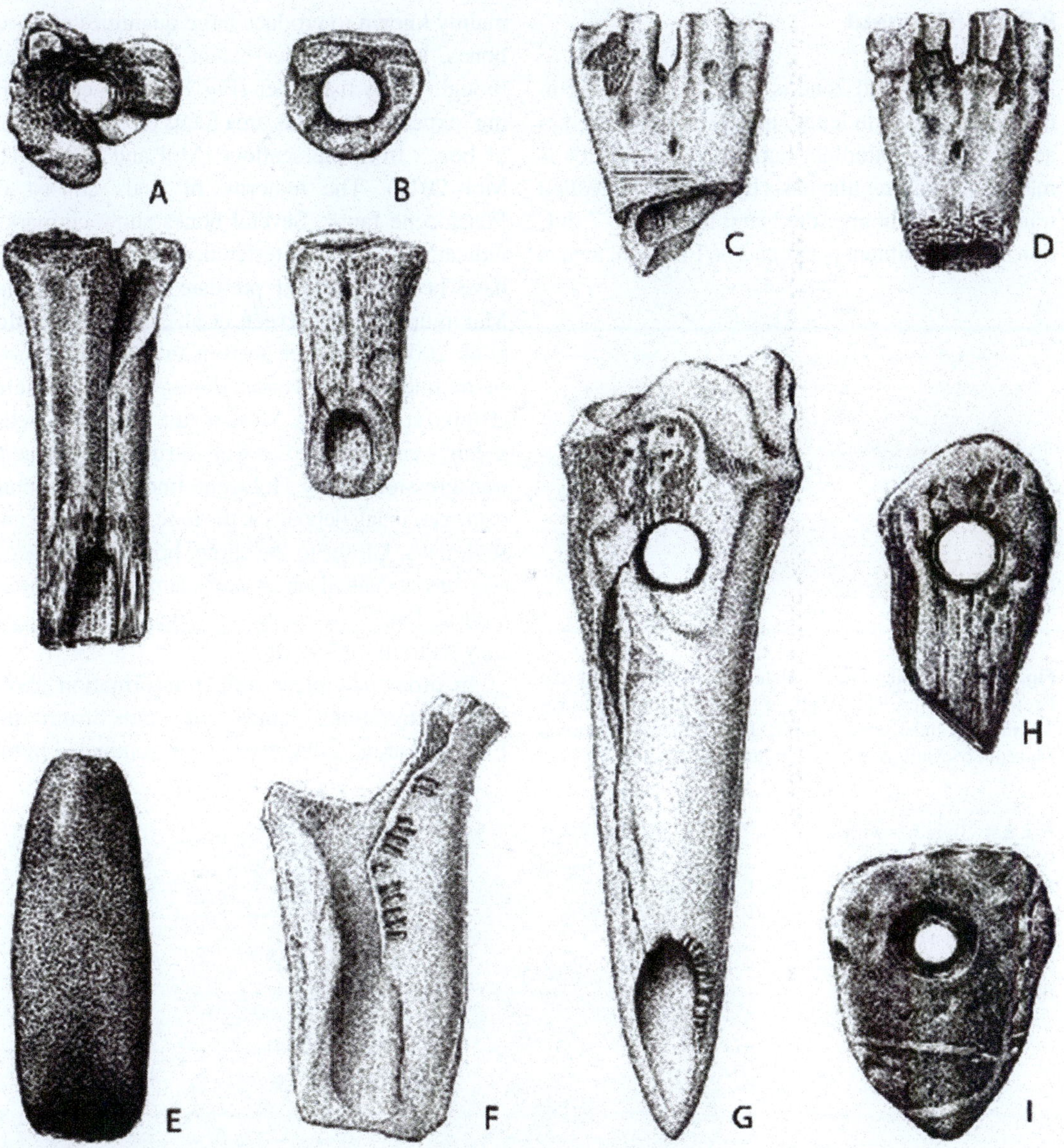

Fig. 8.2 Mesolithic implements from Brown Bank (**a–b**, **f**): socketed adzes; (**c–d**): distal parts of bones with saw and cutmarks; (**e**): ground stone axe (Neolithic?); (**g**): shaft-hole pick; (**h**): perforated antler mattock; (**i**): perforated stone 'wedge'. Drawing by JNJ Caspers, from Louwe Kooijmans (1985)

then permanently flooded. An obvious question is what these objects are doing in such a truly marine environment. Although lost cargo may be an explanation, it is also probable that the axes were ritually deposited, a practice that is well documented for the Neolithic in wetland areas on land (Wentink 2006). The extension of this practice into the marine environment suggests that the sea was as much part of the 'landscape' as were drylands. As such, the location is probably not accidental. Recent work by Gaffney et al. (2009, 145–146) has indicated that during the early Neolithic, the highest parts of the Brown Bank may have been exposed as shallow islands or banks at low tide (Van de Noort 2011, 143–144). Depending on the degree to which islands existed, occupation or ritual deposition at these submerging stepping stones should be taken into account.

8.3.2 Eurogeul

Another major find zone is the Eurogeul. Dug in the 1970s, 57 km in length and 23 m deep, it gives access to the Rotterdam harbour. The corridor is maintained on a regular basis by means of so-called trailing suction hopper dredgers; it yields 5–7 million tons of sediment per year. The Eurogeul area is mainly known to produce large quantities of fossil bones, and to a lesser extent wood and stones, brought up by fishermen (Fig. 8.4). Targeted 'fishing' expeditions in this area focus on the collection of bones from the seafloor (Mol and Post 2010; Mol 2012). The majority of finds consist of Pleistocene fauna. Several bones show cutmarks, indicating human interference. Mammal bones have been dated and pre-date the Last Glacial Maximum (Glimmerveen et al. 2004, 2006; Mol et al. 2006). However, there is evidence that most of the finds were reworked during the Late Glacial (Hijma et al. 2012). Most of the Holocene sediments were reworked as well during the Holocene transgression. Early Holocene finds may lie concentrated in lag deposits at the base of the Holocene sediments. Closer to the shore, however, the geology has remained intact and mammal remains as well as Late Palaeolithic and Mesolithic artefacts may be found in situ (ibid.).

In order to collect further information about the stratigraphic context of finds from the Eurogeul area, research has been conducted in the

Fig. 8.3 Two Neolithic polished flint axes from the Brown Bank area. The length of the axe blades measures 32 cm (top) and 19 cm (bottom). Photo courtesy National Museum of Antiquities, Leiden, the Netherlands

Fig. 8.4 Together with large quantities of fish, fishermen bring up Pleistocene bones in their nets on a daily basis. Photo by Rob Buiter

sand extraction zone for land reclamation in the Rotterdam harbour, the so-called Maasvlakte 2 (Kuitems et al. 2015). This extraction zone is situated next to the Eurogeul. The project involved intensive geological research by means of core drilling and geophysical techniques (seismics, side-scan sonar), as well as targeted 'fishing' trips in the sand extraction zone to collect bones, molluscs, and archaeological objects using a beam trawler. In addition, systematic surveys were conducted on the artificial beach of Maasvlakte 2 by means of manual and mechanical (Mega Beach Cleaner system) collection of objects. The geological investigations and offshore collection of finds made it possible to obtain better insight into the stratigraphical context of finds, but also delivered new insight into Late Pleistocene sea-level fluctuations. Swathes of sand extraction—following lithological layers—were followed by trawl collection of objects, which permitted the development of a coarse biostratigraphy. Based on this sequence and OSL dates of sediment sampled from the high-resolution cores, it could be established that the area saw a marine transgression phase in the early Weichselian (MIS 5d–a), instead of a lowstand as was generally believed. To some extent, this might explain the Weichselian dates obtained on bone remains of beluga and other marine mammal species (Mol et al. 2006).

8.3.3 The Zeeland Ridges: Middeldiep, Westkapelle, Roompot and Onrust

Further south, an area off the Zeeland coast has become known as a Palaeolithic 'treasure trove'. Finds derive from waste heaps at wharves where shells are brought ashore. Since private fossil collectors regularly check the origin of shells, and the dredging areas are relatively restricted in space (often 1 or 2 km in length and several tens of metres wide), a number of 'sites' have been identified in this part of the Southern Bight over the years. All sites are part of the so-called Zeeland Ridges area, a complex of SW–NE oriented sandy ridges, situated parallel to the Dutch coast. These consist of (often re-worked) sediment with material dating to the Early Pleistocene, Eemian, Weichselian and Holocene (Hublin et al. 2009; Amkreutz et al. 2010). The first find zone includes the neighbouring sites of 'Roompot' and 'Onrust', situated in an elongated gully with a depth of 20 m (Louwe Kooijmans 1970/1971; Verhart 2001). These locations yielded several Mesolithic stone tools and waste products, as well as worked pieces of bone and antler, including a bone awl (Verhart 2001). Comparable finds are known closer to shore in a relatively shallow (−7 m) flat area near Westkapelle. Another find zone is the Middeldiep area, part of which is known as the 'Steenbanken' (Stone Banks), adjacent to a deep gully, known as Middeldiep (−27 m). The area has yielded a number of Middle Palaeolithic flint implements, including two hand-axes (Verhart 2001) (Fig. 8.5) and several Levallois flakes (Verhart 2004).

Of particular importance is the find of a Neanderthal frontal bone (Hublin et al. 2009) from the Middeldiep. The skull fragment (Fig. 8.6) was discovered among the remnant cat-

Fig. 8.5 Small Middle Palaeolithic handaxe from the Zeeland Ridges. Photo by Jan Glimmerveen

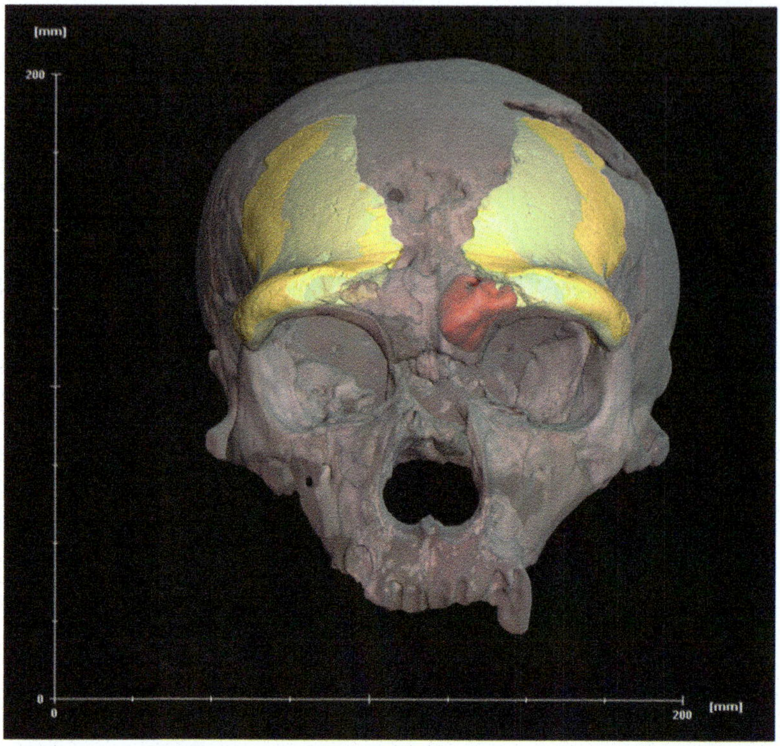

Fig. 8.6 3D scan of the Neanderthal frontal bone (right on the picture) projected on the La Chapelle-aux-Saints (France) Neanderthal skull. The red zone indicates the location of the cavity that was identified as a lesion resulting from an epidermoid cyst. The frontal bone fragment has been mirrored to the other side of the skull. Courtesy J-J Hublin, Max Planck Institute for Evolutionary Anthropology, Leipzig

egory of finds of the collector and identified years after it had been picked up as a potential Neanderthal brow-ridge. Physical and chemical analyses were conducted at the Max Planck Institute in Leipzig. There was insufficient collagen left to produce a radiocarbon date. However, an elaborate analysis of the skull fragment's morphology indicated that it is 3000 times more likely that the fragment belongs to a Neanderthal than to an anatomically modern human. Additional analysis indicated that it probably belonged to a young adult male individual. Furthermore, carbon and nitrogen isotope analysis indicated that the individual probably had a predominantly carnivorous diet, typical for Neanderthals. Finally, the bone showed a cavity behind the brow ridge which was identified as a lesion resulting from an epidermoid cyst. This palaeopathology had never been documented before in Neanderthals.

Finding the first Dutch Neanderthal is of interest (Amkreutz et al. 2010). Unlike the surrounding countries of Germany, Belgium and Britain, the Netherlands has no caves, which are known to provide good contexts for the preservation of Neanderthal remains. The piece therefore forms a welcome addition to the information in the Netherlands from this period, which is otherwise largely based on lithics. The find is, however, even more important since it derives from the North Sea. It is indicative of the enormous potential of the Pleistocene and Holocene archaeological archive offshore from our coast. As such it is also a symbol of the current large increase of this archive through unmonitored industrial activities such as mining for aggregates, fishing, construction of cable infrastructure and other offshore industries.

A recent geological study made it possible to correlate offshore stratigraphic units in the Middeldiep area with onshore deposits dated between 50,000 and 30,000 BP (Busschers et al. 2007; Hijma et al. 2012). This would place the flint tools and Neanderthal fossil at the end of the Middle Palaeolithic, which is in line with the morphological features of the Neanderthal frontal bone (Hublin et al. 2009). The Zeeland Ridges area may currently be designated as one of the most interesting zones where prospects for finding well preserved sites are good.

8.3.4 Maasvlakte-Europoort

When the Port of Rotterdam was extended in the early 1970s, land reclamation was based on the extraction of mudspoil from various locations in the Rotterdam harbour area (Verhart 1988). At the pristine surface and along the new shores of this reclaimed land, Maasvlakte-Europoort, private collectors found fossil bones, as well as bone artefacts, mostly barbed points (Louwe Kooijmans 1970/1971). Over the years, over 500 points (Fig. 8.7) have been reported, which are of Preboreal to Boreal age (Verhart 1988). From the conditions of discovery, it is clear that the Maasvlakte-Europoort collection cannot be considered as an assemblage with high integrity. However, in consideration of information on the origin of the sand used for land reclamation, all material has to come from a rather restricted area. Verhart (1988) has suggested that the concentration of points in the area is related to the specialised exploitation of a near-coastal lagoon, e.g. spearing of aquatic mammals and fish. However, the stratigraphic origin of these finds always remained obscure. New construction works in the Port of Rotterdam subsequently provided opportunities to collect information about the possible context of these materials (see below).

8.3.5 Coastal Areas and Beaches with Coastal Reinforcement

Along the Dutch coast, several locations are known where prehistoric artefacts are regularly found. In general, these are locations where sand

Fig. 8.7 Mesolithic bone points from Maasvlakte-Europoort. The largest point, third from left, is 95 mm in length. Photo courtesy National Museum of Antiquities, Leiden, the Netherlands

is deposited for coastal reinforcement; the sand is dredged up from various, nearby offshore locations (usually at distances up to 10 km) and transported by pipelines or sprayed on the beach (Verhart 1995, 2004; Amkreutz et al. 2017). Clearly, these find spots are of limited archaeological value regarding their geological context, yet they also provide further clues to the potential of the North Sea. A well-known site is the beach at 'Hoek van Holland'. This location has yielded several barbed points as well as a shaft-hole antler axe. At Scheveningen, a shaft-hole antler axe was found amidst sand coming from 30 km offshore. At Monster, flint nodules, some worked flint, barbed points and bones were found amidst sand dredged from a depth of around 26 m at 5–7 km off the coast (Verhart 1995). In both cases an Early Mesolithic date for the finds is most likely. Additional finds have been reported recently from these locations, as well as from a number of other locations along the Dutch coast, e.g. near Kijkduin. Other finds derive from Colijnsplaat and include several pieces of worked antler of Mesolithic and Neolithic age (Louwe Kooijmans 1970/1971). Occasionally, Palaeolithic artefacts are found as well, as at Petten in North Holland. One of the most important sites is the 'Zandmotor' ('Sand engine'). This is a large artificial beach situated in front of Kijkduin and Ter Heijde and intended as a natural method of beach replenishment. In total 21.5 million cubic meters of sand extracted off the coast were deposited here over an area of 128 hectares. This area, apart from many finds of fossil fauna, has now yielded many hundreds of lithic artefacts, including tools of Palaeolithic and Mesolithic date, tools and waste products of bone and antler, pieces of jewellery and human bone dated to the Mesolithic.

8.4 Rotterdam-Yangtze Harbour: Investigating an Underwater Site

Cooperation between the Port of Rotterdam, local and national authorities, and scientists has permitted for the first time an integration with the construction workflow of systematic geoarchaeological surveying and an 'excavation' (Weerts et al. 2012). As Early Holocene land surfaces were expected to be present at depths between 17 and 22 m below sea level, it was impossible to follow the usual land-based surveying strategies. Geological information indicated the presence of one or more Late-Glacial river dunes in the Yangtze harbour basin. A model of the submerged and sediment-covered land surface was constructed on the basis of additional sediment cores and seismic data (Vos et al. 2010). Based on the knowledge of the use of river dunes by prehistoric people in the Central Netherlands river area, two areas of high archaeological potential were selected for further investigation. This led to the discovery of a Mesolithic site at a depth of 20 m below sea level (Moree and Sier 2015).

In the autumn of 2011 a small-scale 'excavation' was conducted by means of a special grab-sampler that permitted exact horizontal and vertical positioning from a pontoon (Fig. 8.8). Each grab sample filled two big bags with sediment that was subsequently wet-sieved through a 10 mm and 2 mm mesh. The archaeological material found consists of thousands of fragments of bone, flint artefacts and charred plant remains (Fig. 8.9). Although the spatial resolution of the excavation is not ideal, the results are of immense value for our understanding of human adaptations in a dynamic landscape. The combined information from the excavation and a series of high-quality piston cores, which were sampled for high-resolution palaeoenvironmental analysis, has permitted a very good insight into the hunter-gatherer occupation history of the dune, in the context of landscape change. Starting in the second half of the Preboreal, activities were conducted on an inland floodplain landscape. The dune continued to be used until the first half of the Early Atlantic period, when sea-level rise led to its flooding (Fig. 8.10). The results from this research currently provide the earliest evidence for Mesolithic coastal activity in this sector of the North Sea (Peeters et al. 2015).

Fig. 8.8 Use of specially designed grab sampler for excavation at the Maasvlakte excavation showing a large bag being filled with sediment sampled from a sand layer at 18 m below sea level. Photo by Dimitri Schiltmans, BOOR

Fig. 8.9 Residue from one of the large bags being checked for the presence of archaeological material, such as knapped flint, burnt bone and charred plant remains. Photo by Dimitri Schiltmans, BOOR

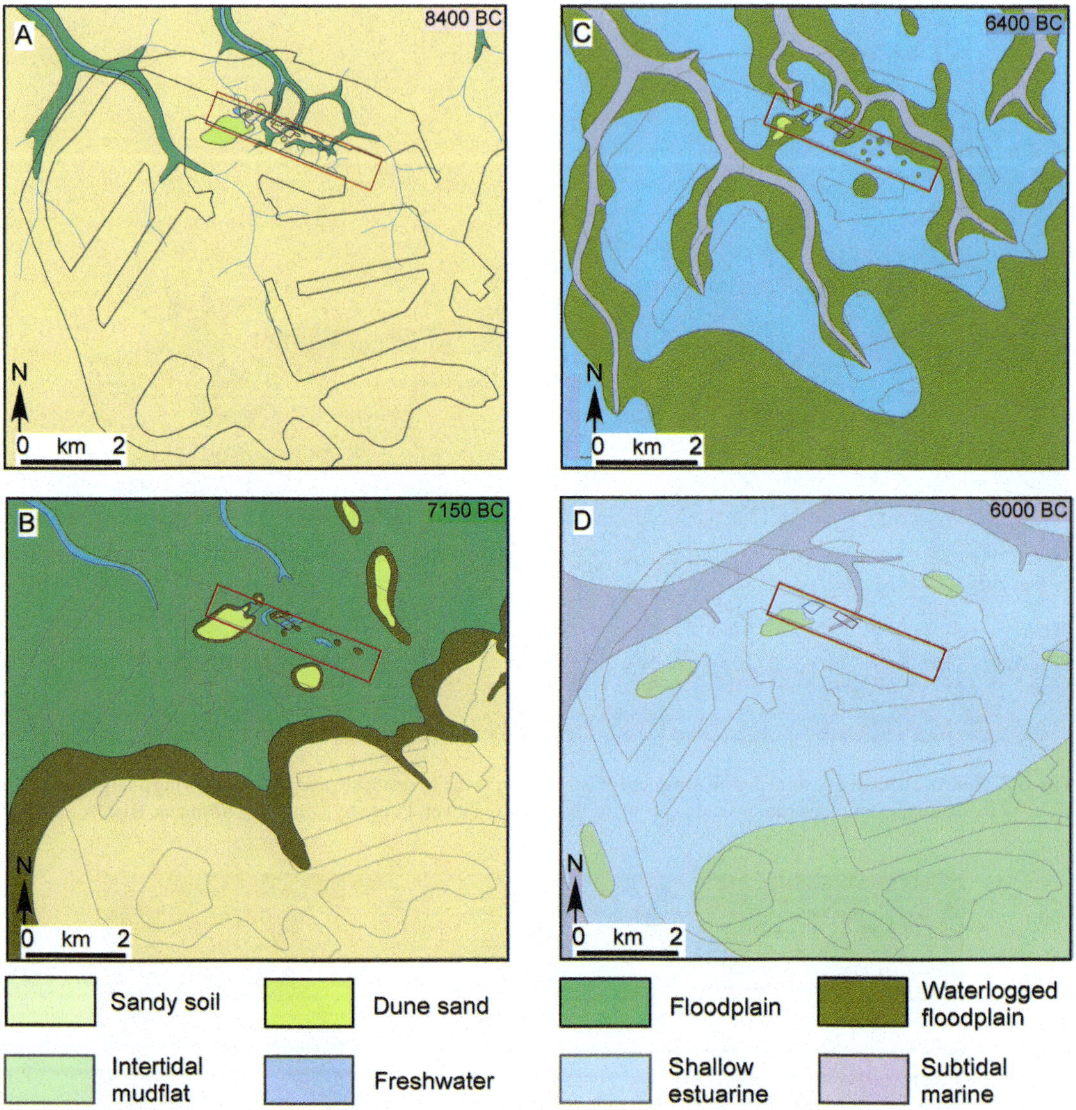

Fig. 8.10 Schematic representation of palaeogeographic development and flooding of the Yangtze Harbour area. (**a**), 8400 cal BC: the investigated site (within the red rectangle) is located at the fringes of the floodplain in the dry hinterland; (**b**), 7150 cal BC: the investigated site is located on the floodplain, which became part of an extended wetland zone; (**c**), 6400 cal BC: the investigated site is located on the floodplain, which became part of a tidal area; (**d**), 6000 cal BC: the investigated site transformed into a mudflat in a subtidal zone in an estuarine area. From Moree and Sier (2015)

8.5 Diets and Cultural Traditions from Submerged Finds

An important criticism of the Dutch situation is that most of the finds are in secondary contexts, either hauled up in fishing nets or more often deposited ashore by the gravel industry, sand replenishment for beach reinforcements or other infrastructural projects. The available contextual information is often limited to a certain area offshore where the sediments were extracted. While this criticism is correct, it fails to address two points. The first is that the quantity and quality of many of the finds indicate that there is much potential for the presence of intact sites. The excavations in the Yangtze harbour also substantiate this potential. While subsurface excavation is costly, more intensive prospection may provide a better context for the coarse find-locations (e.g., Hijma et al. 2012). Secondly, the criticism does not acknowledge the fact that the finds themselves also represent a hitherto little explored potential. The many finds of bone and antler points from various areas on the coast are gradually revealing patterns in style and execution that may be chronological or cultural (Amkreutz and Spithoven 2019). This potential is even more distinctly demonstrated in an analysis of 33 human Mesolithic bones that were fished from the North Sea or deposited ashore by the aggregates industry. The research by Van der Plicht et al. (2016) focused on the isotopic content of these bones and measured the composition of the stable isotope ratios of carbon and nitrogen, $\delta^{13}C$ and $\delta^{15}N$. These are informative on the trophic chain and the dominant type of food consumed, as they differ according to the trophic level of the consumer and the degree to which food intake includes aquatic resources. The raised levels of the Mesolithic bones clearly point to a dominant contribution of freshwater resources (Fig. 8.11). Unfortunately, the reservoir effect hampers the exact calibration of the radiocarbon dates of the bones, but their Mesolithic age and relative chronological position with regard to carbon and nitrogen levels demonstrates a clear trend. It appears that over time Mesolithic diet became increasingly aquatic. Apart from fish, this may also have included waterfowl and species such as otter and beaver. This 'Doggerland-menu' of course relates to the fact that the low-lying basin of the North Sea gradually drowned after the Last Ice Age (between 9500 and 6000 cal BC). Previously, it was often assumed that this submergence forced people out of the North Sea Basin or that they adapted to a diet characterised by marine resources. The current research, however, changes this idea. The importance of freshwater sources indicates that the inhabitants made the most of the changing environment and of the developing wetlands that expanded around them in the delta areas of the Meuse, Rhine and Thames before the area finally flooded. This is not strange as these freshwater wetland areas are among the richest in (food) resources worldwide (Nicholas 2007). Later Mesolithic sites in the Rhine-Meuse delta of the Netherlands confirm this focus on aquatic resources (Louwe Kooijmans 2003; Smits et al. 2010), but the research particularly underlines the fact that flooding should not always be regarded as a catastrophe (see Leary 2009) and that the Mesolithic inhabitants demonstrated a strong flex-

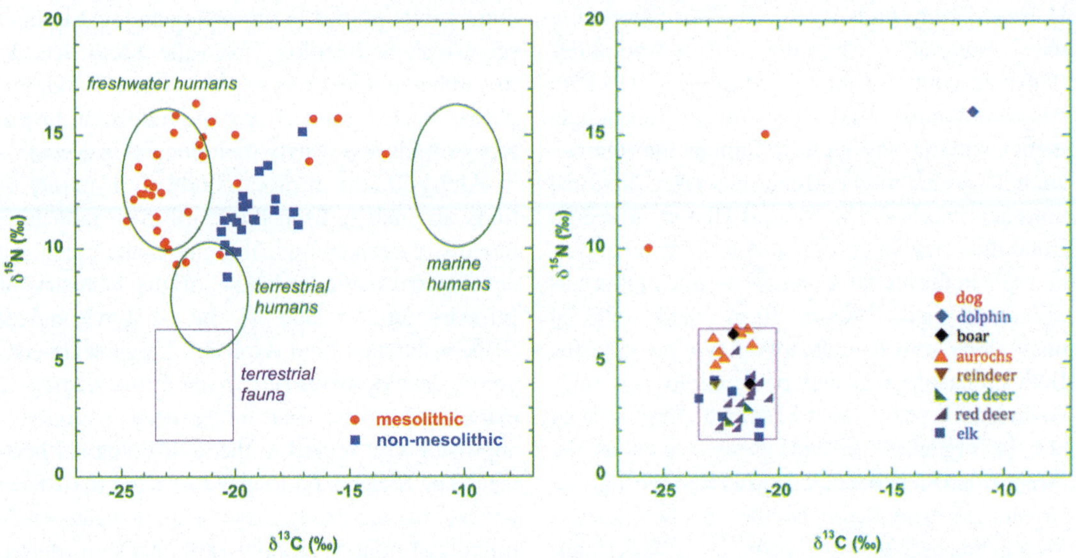

Fig. 8.11 Comparison between stable isotope data of human remains (left) and terrestrial mammals (right) from the North Sea. From Van der Plicht et al. (2016)

ibility in changing their lifeways, traditions and diet (Amkreutz 2013). This is of relevance for further studies into the Mesolithic, but perhaps also has some repercussions for current debates regarding rising water levels and climate change. Finally, it underlines the fact that isolated finds with little contextual information may nevertheless form an important contribution to our knowledge, especially when they come from a largely unexplored and vast prehistoric landscape in front of our coast. Recent information that Mesolithic bone material from the North Sea is particularly suitable for aDNA analysis further underlines this perspective (see Bailey et al., Chap. 10, this volume).

8.6 Collections and Developments in Heritage Management

The responsibility for the offshore archaeological heritage lies with the Cultural Heritage Agency (Rijksdienst voor het Cultureel Erfgoed, RCE). However, legislation concerned with archaeological heritage is restricted to the waters falling under national jurisdiction, and (in strict terms) applies only to items made by humans. Although there is an obligation to report finds from these waters to the authorities, this is often not done. However, stray finds are more often reported to the National Museum of Antiquities (Rijksmuseum van Oudheden Leiden). Palaeontological finds (or other non-human items) without any sign of human interference are not covered by the Monuments Act. National museums, such as the Natural History Museum Rotterdam and Naturalis (Leiden), play an important role in the registration and storage of this category of finds. Apart from these official institutions, private collectors play a pivotal role. Both archaeological and palaeontological finds are mostly kept by individuals, who may, or may not, be organised in workgroups, such as the 'Werkgroep Pleistocene Zoogdieren' (Workgroup Pleistocene Mammals) or the 'Archeologische Werkgemeenschap voor Nederland' (Archaeological Work Community for the Netherlands).

This situation poses some problems, as it is difficult to get an up-to-date overview of prehistoric finds and sites from the Dutch part of the North Sea (also see Verhart 1995). Even more important is the problem of getting access to the wharves where thousands of fossils and artefacts are brought ashore on a yearly basis. Both finds from waters within, and beyond national jurisdiction, are at risk of disappearing from view without ever having been spotted. In addition, as legislation only applies to artefacts, other materials, such as mammal bones, fossil tree trunks or even lumps of peat, which provide a lot of information relevant for the reconstruction of the (environmental) past, are artificially separated from human-made items (see also Van Kolfschoten 2006). This is in marked contrast to the situation of terrestrial archaeology. Fortunately, recent developments demonstrate a change of attitude in this respect, for instance in the context of the Port of Rotterdam. Additionally, in 2016, a group of professional and amateur researchers started to collaborate in the informal 'Doggerland Research Group' (Werkgroep Steentijd Noordzee). They have compiled a list of c. 100 amateurs with archaeological artefacts from the North Sea in their collections. Their aim is to investigate new finds, document sites and collaborate with different government bodies and commercial parties in protecting and researching prehistoric archaeology from the North Sea. As the archaeological record as we know it today is clearly biased, joint efforts to produce an inventory of finds and clarify their context is crucial.

Alongside the efforts to build an inventory of finds originating from the North Sea, indicative models of archaeological significance are being developed on the initiative of the Ministry of Infrastructure and the Environment (Rijkswaterstaat) and the RCE. Large-scale economic developments (e.g. wind farms, aggregate extraction) in the near future need a proactive approach with regard to the archaeological heritage. The focus is no longer restricted to historic wrecks; the prehistoric heritage in its palaeoenvironmental context is now explicitly considered. One model (Fig. 8.12) covers the Dutch sector of the continental shelf and indicates the possible

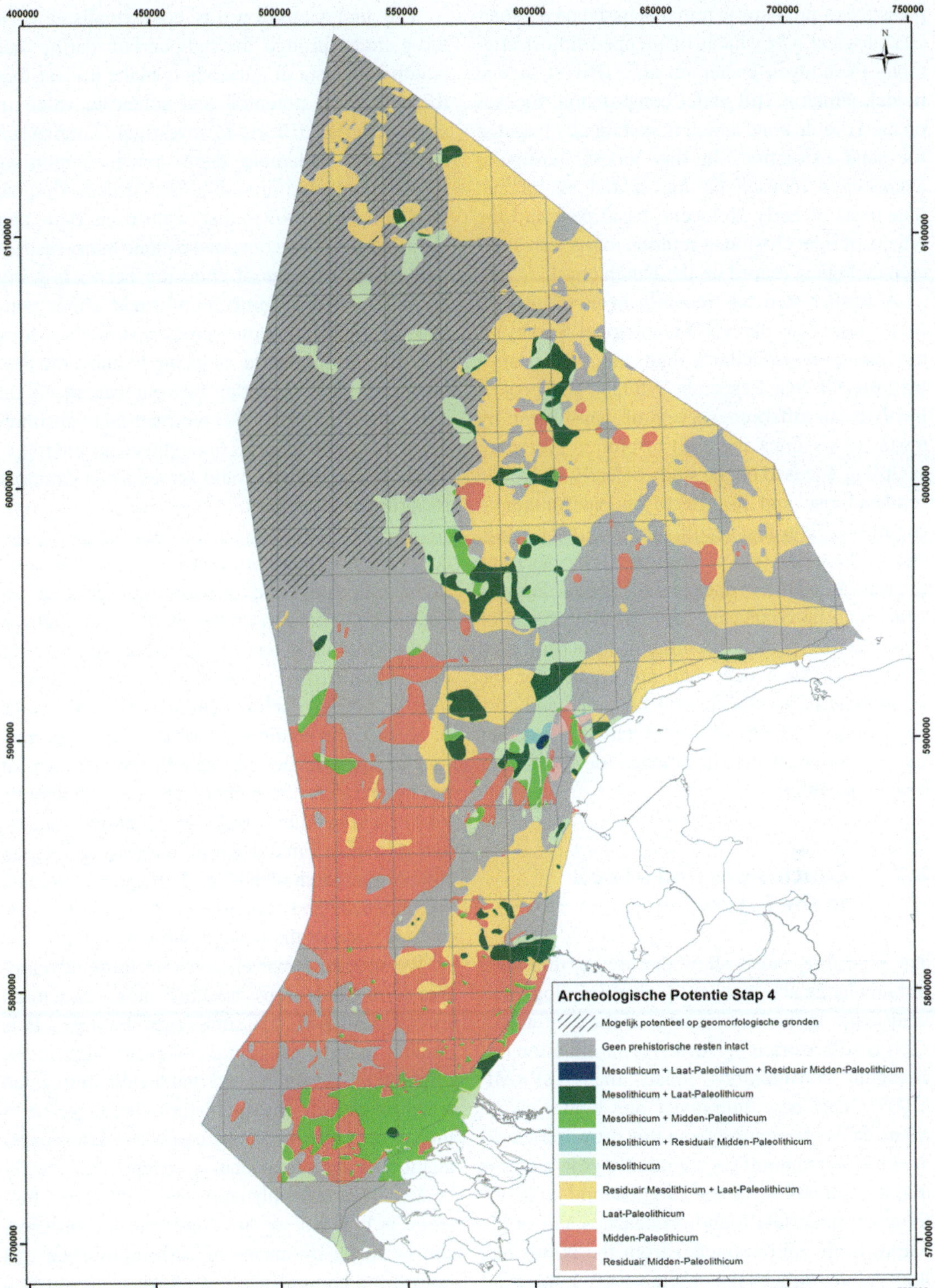

Fig. 8.12 Indicative map of archaeological potential of the Dutch sector of the North Sea. Different areas have potential for archaeology of different periods ranging from Middle Palaeolithic to Mesolithic. Colour codes refer to different archaeological periods or combinations of periods. From Vonhögen-Peeters et al. (2016)

presence of prehistoric remains based on a geoarchaeological assessment of mapped lithostratigraphy (Vonhögen-Peeters et al. 2016). A second model, which is still under construction, focuses on the 12-mile zone which is specifically targeted for sand extraction. In this model, particular attention is focused on the spatial extent and intactness of early Holocene basal peat and the effects of tidal erosion in relation to the terrestrial archaeological record on the Dutch coast.

A further step has recently been taken in a joint project by the TNO-Geological Survey of the Netherlands/Deltares and the Netherlands Institute for Sea Research (NIOZ). The project involves an offshore survey of basal peat by means of seismics and high-resolution vibrocore sampling, focused on the study of Early Holocene sea-level rise and its effects on environmental conditions. The first results are very promising, due to the collection of well-preserved sections of peat, as well as (unexpected) gyttja deposits which can be connected to the initial effects of sea-level rise (personal communication Mr. Freek Busschers). Such new data are crucial for the assessment of previously collected data and core descriptions which mostly focused on other issues than palaeoenvironmental developments and archaeology.

8.7 Conclusions and a Look to the Future

The preceding pages show that the current state of knowledge about submerged archaeology and landscapes in the Dutch sector of the continental shelf is still restricted. However, the situation has changed considerably since the NSPRMF (2009), and new initiatives are being taken. Many finds originating from this drowned world are known, but numbers are rapidly growing now that the richness of archaeological and palaeontological materials found on beaches and originating from sand extracted from the North Sea floor is attracting considerable attention from the media and private collectors, as well as from scientists.

The first steps towards contextualisation of these finds through the radiocarbon dating and isotope analysis of palaeodiets make it clear that there is great potential and scientific value in such material. Efforts to investigate submerged sites at considerable depth under conditions where diving is impossible (Rotterdam Yangtze Harbour) has shown that significant data and information can be obtained despite many restrictions. It is a matter of choosing between doing nothing and remaining uninformed about contexts of finds and human-landscape relationships in a submerging world, or trying to make the best out of it, while accepting that one cannot obtain the resolution/control that we are used to onshore. The experience in terms of engineering and logistics is highly valuable and serves as an example for projects elsewhere.

An important spin-off from the Yangtze project is the compilation of an inventory of collections and finds; the opportunities created for private collectors to collect bones and artefacts on the artificial beach of Maasvlakte 2 and have these identified by palaeontologists and archaeologists triggered a lot of enthusiasm. In the wake of all this, the significance of submerged prehistory and landscapes has caught the attention of policy and decision-makers who have to operate within a dynamic context of economic interest and valuation. Finally, we are witnessing targeted offshore research which has direct archaeological significance, and which goes beyond the submerged parts of the Dutch continental shelf.

We should, however, consider these developments as first steps towards more structured investigations of this submerged heritage. It is evident that the economic stakes are high in the offshore environment, and that legally embedded safeguarding of the prehistoric heritage, be it through excavation or in situ preservation is unrealistic as long as expectations cannot sufficiently be underpinned with empirical data. The challenge is big, but we are confident that progress can be made by means of both established and innovative approaches, but above all through a willingness to experiment and deviate from established conventions.

References

Amkreutz LWSW (2013) Persistent traditions. A long-term perspective on communities in the process of Neolithisation in the Lower Rhine Area (5500–2500 cal BC). Sidestone Press, Leiden

Amkreutz, LWSW, Spithoven M (2019) Hunting beneath the waves. Bone and antler points from North Sea Doggerland in front of the Dutch Coast. Schleswig

Amkreutz LWSW, Glimmerveen JC, Hublin J-J, Roebroeks W, Anthonis L (2010) Een Neanderthaler uit de Noordzee. Westerheem 59(2):50–59

Amkreutz L, Niekus M, Schiltmans D, Smit B (2017) Meer dan bijvangst! De prehistorische archeologie van de Noordzee. Cranium 34(1):34–47

Busschers FS, Kasse C, Van Balen RT, Vandenberghe J, Cohen KM, Weerts HTJ, Wallinga J, Johns C, Cleveringa P, Bunnik FPM (2007) Late Pleistocene evolution of the Rhine-Meuse system in the southern North Sea basin: imprints of climate change, sea-level oscillation and glacio-isostacy. Quat Sci Rev 26:3216–3248

Cohen K, Gibbard P, Weerts H (2014) North Sea palaeogeographical reconstructions for the past 1 Ma. Neth J Geosci 93(1/2):7–30

Cohen KM, Westley K, Erkens G, Hijma MP, Weerts HTJ (2017) The North Sea. In: Flemming NC, Harff J, Moura D, Burgess A, Bailey G (eds) Submerged landscapes of the European continental shelf. Quaternary paleoenvironments. Wiley-Blackwell, Chichester, pp 147–186

Farr RH, Momber G, Satchell J, Flemming N (2017) Paleolandscapes of the Celtic Sea and the channel/La Manche. In: Flemming NC, Harff J, Moura D, Burgess A, Bailey G (eds) Submerged landscapes of the European continental shelf. Quaternary paleoenvironments. Wiley-Blackwell, Chichester, pp 211–241

Gaffney V, Fitch S, Smith D (2009) Europe's lost world, the rediscovery of Doggerland, CBA research report 160. Council for British Archaeology, York

Glimmerveen J (2007) Fish and stones. N Sea Prehist Newsl 1:9–10

Glimmerveen J, Mol D, Post K, Reumer JWF, Van der Plicht H, De Vos J, Van Geel B, Van Reenen G, Pals JP (2004) The North Sea project: the first palaeontological, palynological, and archaeological results. In: Flemming NC (ed) Submarine prehistoric archaeology of the North Sea: research priorities and collaboration with industry, CBA Research Report 141. Council for British Archaeology, York, pp 43–52

Glimmerveen J, Mol D, Van der Plicht J (2006) The Pleistocene reindeer of the North Sea — initial palaeontological data and archaeological remarks. Quat Int 142/143:242–246

Hijma M (2009) From river valley to estuary. The early-mid Holocene transgression of the Rhine-Meuse valley, the Netherlands. Utrecht, Netherlands Geographical Studies 389

Hijma M, Cohen KM, Roebroeks W, Westerhoff WE, Busschers FS (2012) Pleistocene Rhine-Thames landscapes: geological background for hominin occupation of the southern North Sea region. J Quat Sci 27:17–39

Hublin J-J, Weston D, Gunz P, Richards M, Roebroeks W, Glimmerveen J, Anthonis L (2009) Out of the North Sea: the Zeeland Ridges Neandertal. J Hum Evol 57:777–785

Jelgersma S (1979) Sea-level changes in the North Sea Basin. In: Oele E, Schuttenheim RTE, Wiggers AJ (eds) The quaternary history of the North Sea. Acta Universitatis Upsaliensis, Uppsala, pp 233–248

Kuitems M, Van Kolfschoten T, Busschers F, De Loecker D (2015) Geoarchaeological and palaeontological research in the Maasvlakte 2 sand extraction zone and on the artificially created Maasvlakte 2 beach: a synthesis. In: Moree JM, Sier MM (eds) Interdisciplinary research programme Maasvlakte 2, Rotterdam. BOOR, Rotterdam, pp 351–398

Leary J (2009) Perceptions of and responses to the Holocene flooding of the North Sea lowlands. Oxf J Archaeol 28(3):227–237

Louwe Kooijmans LP (1970/1971) Mesolithic bone and antler implements from the North Sea and from the Netherlands. Berichten Rijksdienst voor het Oudheidkundig Bodemonderzoek 20/21:27–73

Louwe Kooijmans LP (1974) The Rhine/Meuse delta. Four studies on its prehistoric occupation and Holocene geology. Oudheidkundige Mededelingen van het Rijksmuseum van Oudheden te Leiden, Leiden, pp 52–54

Louwe Kooijmans LP (1985) Sporen in het land. De Nederlandse Delta in de prehistorie, Meulenhof, Amsterdam

Louwe Kooijmans LP (2003) The Hardinxveld sites in the Rhine/Meuse Delta, the Netherlands, 5500-4500 cal BC. In: Larsson L, Kindgren H, Knutsson K, Loeffler A, Åkerlund A (eds) Mesolithic on the move. Papers presented at the sixth international conference on the Mesolithic in Europe, Stockholm 2000. Oxbow, Oxford, pp 608–624

Maarleveld ThJ (1984) 1983: derde jaar archeologie in Nederlandse wateren. Ministerie van Welzijn, Volksgezondheid en Cultuur, Rijswijk

Maarleveld TJ, Peeters JHM (2004) Can we manage? In: Flemming NC (ed) Submarine prehistoric archaeology of the North Sea: research priorities and collaboration with industry, *CBA research report 141*. Council for British Archaeology, York, pp 102–112

Mol D (2012) Two days of trawling in the Eurogully: did trawler OD7 find a mammoth graveyard? Deposits Mag 30:13–15

Mol D, Post K (2010) Gericht korren op de Noordzee voor de zoogdierpaleontologie: een historisch overzicht van de uitgevoerde expedities. Cranium 27:14–28

Mol D, Post K, Reumer JWF, Van der Plicht J, De Vos J, Van Geel B, Van Reenen G, Pals J-P, Glimmerveen J (2006) The Eurogeul — first report of the palaeontological, palynological and archaeological investigations of this part of the North Sea. Quat Int 142/143:178–185

Mol D, De Vos J, Van Geel B, Glimmerveen J, Van der Plicht J, Post K (2008) Mammoeten, neushoorns en andere dieren van de Noordzeebodem. Kleine encyclopedie van het leven in het Pleistoceen, Diemen

Moree JM, Sier MM (2015) Twenty meters deep! The Mesolithic period at the Yangtze Harbour site – Rotterdam Maasvlakte, the Netherlands. Early Holocene landscape development and habitation. In: Moree JM, Sier MM (eds) Interdisciplinary research programme Maasvlakte 2, Rotterdam. BOOR, Rotterdam, pp 7–350

Nicholas GP (2007) Prehistoric hunter-gatherers in wetland environments: theoretical issues, economic organisation and resource management strategies. In: Lillie M, Ellis S (eds) Wetland archaeology and environments. Regional issues, global perspectives. Oxbow, Oxford, pp 46–64

Peeters H (2011) How wet can it get? Approaches to submerged prehistoric sites and landscapes on the Dutch Continental Shelf. In: Benjamin J, Bonsall C, Pickard C, Fischer A (eds) Submerged prehistory. Oxbow, Oxford, pp 55–64

Peeters JHM, Cohen KM (2014) North Sea submerged landscapes and prehistory. Geology, prehistoric archaeology and research potential of the southern North Sea. Neth J Geosci 93(1/2):3–6

Peeters JHM, Momber G (2014) The southern North Sea and the human occupation of Northwest Europe after the last glacial maximum. Neth J Geosci 93(1/2):55–70

Peeters H, Flemming N, Murphy P (eds) (2009) North Sea prehistory research and management framework. NSPRMF, Amersfoort

Peeters JHM, Brinkhuizen DC, Cohen KM, Kooistra LI, Kubiak-Martens L, Moree JM, Niekus MJLT, Schiltmans DEA, Verbaas A, Verbruggen F, Vos PC, Zeiler JT (2015) Synthesis. In: Moree JM, Sier MM (eds) Interdisciplinary research programme Maasvlakte 2, Rotterdam. BOOR, Rotterdam, pp 287–318

Roebroeks W (2014) Terra Incognita: the Palaeolithic record of northwestern Europe and the information potential of the southern North Sea. Neth J Geosci 93(1/2):43–54

Smits E, Millard AR, Nowell G, Pearson DG (2010) Isotopic investigation of diet and residential mobility in the Neolithic of the Lower Rhine Basin. Eur J Archaeol 13(1):5–31

Van de Noort R (2011) North Sea archaeologies. A maritime biography 10,000BC – AD 1500. Oxford University Press, Oxford

Van der Plicht J, Amkreutz L, Niekus MJLT, Peeters H, Smit B (2016) Surf 'n Turf in Doggerland: dating, stable isotopes and diet of Mesolithic human remains from the southern North Sea. J Archaeol Sci Rep 10:110–118

Van Kolfschoten T (2006) Bones and stones get lost. The Dutch Palaeolithic heritage in danger. In: Rensink E, Peeters H (eds) Preserving the early past. Investigation, selection and preservation of Palaeolithic and Mesolithic sites and landscapes. Amersfoort, Nederlandse Archeologische Rapporten 31, pp 77–86

Verhart LBM (1988) Mesolithic barbed points and other implements from Europoort, The Netherlands. Oudheidkundige Mededelingen Rijksmuseum voor Oudheden Leiden 68:145–194

Verhart LBM (1995) Fishing for the Mesolithic. The North Sea: a submerged prehistoric landscape. In: Fischer A (ed) Man and sea in the Mesolithic. Oxbow, Oxford, pp 291–302

Verhart LBM (2001) De zee neemt, de zee geeft. Spectaculaire vondsten uit de Noordzee Westerheem 50:102–107

Verhart LBM (2004) The implications of prehistoric finds on and off the Dutch coast. In: Flemming NC (ed) Submarine prehistoric archaeology of the North Sea: research priorities and collaboration with industry, CBA research report 141. Council for British Archaeology, York, pp 57–61

Vonhögen-Peeters LM, Van Heteren S, Peeters JHM (2016) Indicatief model van het archeologisch potentieel van de Noordzeebodem. Deltares, Utrecht

Vos P, Van den Berg M, Bunnik F, Cremer H, De Kleine M, Maljers D, Mesdag C (2010) Geoarcheologisch vooronderzoek ten behoeve van het Yangtzehavenproject: inventariserend veldonderzoek, fase 2. Deltares, Utrecht

Weerts H, Otte A, Smit B, Vos P, Schiltmans D, Waldus W, Borst W (2012) Finding a needle in a haystack by using knowledge of Mesolithic human adaptation in a drowning delta. eTopoi J Anc Stud, special volume 3:17–24

Wentink K (2006) Ceci n'est pas une hache. Neolithic depositions in the northern Netherlands, Leiden

9 Belgium: Prehistoric and Protohistoric Archaeology in the Intertidal and Subtidal Zones of the North Sea

Marnix Pieters, Tine Missiaen, Maikel De Clercq,
Ine Demerre, and Sven Van Haelst

Abstract
The evidence for submerged prehistoric remains and landscapes in the Belgian sector of the North Sea is scarce. The majority of finds dredged and trawled from the seafloor consists of palaeontological material, notably mammalian bone remains, dating from the Early and Late Pleistocene, and possibly the Holocene. Scarce palaeontological and archaeological finds come from the intertidal zone. The available data are rather limited and strikingly sparse compared to neighbouring countries, yet it is steadily growing since the SeArch project. Recent investigations of an area with exceptionally high concentrations of Late Pleistocene and Palaeogene bone finds in the 'Scheur' illustrates this development and offers a promising avenue for future research. Growing awareness among a broader public is similarly resulting in an increase in reported chance finds, mainly from beaches. In due course, these growing numbers will hopefully allow quantitative support for certain hypotheses.

Keywords
Palaeontological · Submerged · Protohistoric · Prehistoric · Palaeolandscape · Raversijde

9.1 Introduction

Compared to neighbouring countries, the general attitude in Belgium towards archaeological and palaeontological finds in the North Sea has long been marked by a genuine lack of scientific interest. The sparse archaeological research in the intertidal and subtidal zones of the Belgian part of the North Sea was often very fragmented and focused largely on wreck sites. With the recent SeArch project (Missiaen et al. 2017) and the new national law on underwater cultural heritage (see also later), this attitude has now changed completely. Together with better insights into submerged palaeolandscapes, this has resulted in an increasing awareness of the importance of (reporting) archaeological finds (see also Pieters et al. 2018 for a short overview).

So far, relatively few fossil bones have been reported from the subtidal zone (Fig. 9.1). This is

M. Pieters (✉) · I. Demerre
Agentschap Onroerend Erfgoed, Brussels, Belgium
e-mail: marnix.pieters@vlaanderen.be; ine.demerre@vlaanderen.be

T. Missiaen · S. Van Haelst
Flanders Marine Institute (VLIZ), Ostend, Belgium
e-mail: tine.missiaen@vliz.be; sven.vanhaelst@vliz.be

M. De Clercq
Renard Centre of Marine Geology, Ghent University, Ghent, Belgium
e-mail: Maikel.DeClercq@UGent.be

G. Bailey et al. (eds.), *The Archaeology of Europe's Drowned Landscapes*, Coastal Research Library 35, https://doi.org/10.1007/978-3-030-37367-2_9

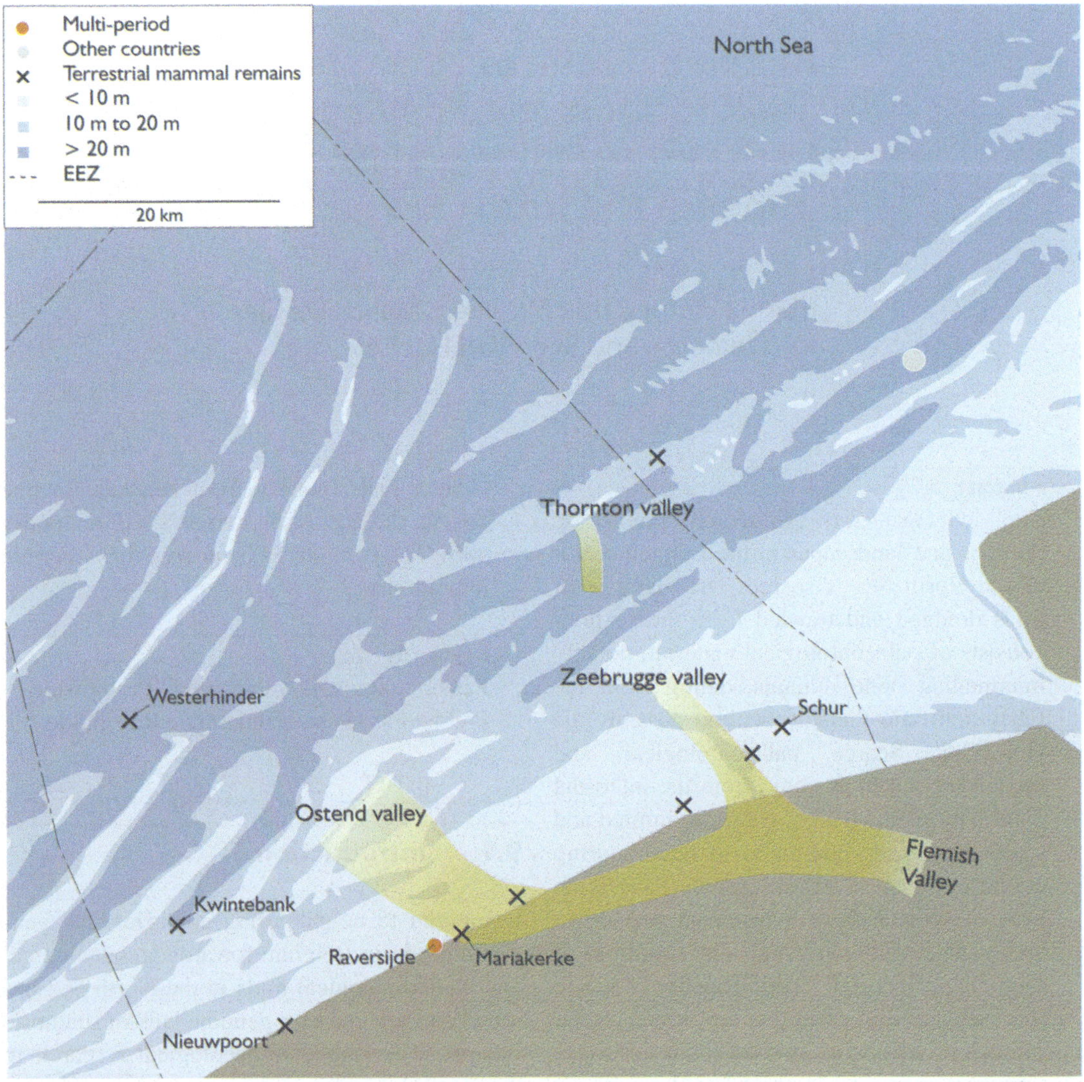

Fig. 9.1 Map with the locations of pre- and protohistoric archaeological evidence in the intertidal and subtidal zones of the Belgian part of the North Sea. Drawing by Moritz Mennenga

in stark contrast to the Netherlands where numerous tools and fossils have been reported (Van Kolfschoten and van Essen 2004; Mol et al. 2006; Hublin et al. 2009; Hijma et al. 2012; Peeters and Amkreutz, Chap. 8, this volume). The few mammal bone fragments that have been fished up largely date from the Last Glacial-Interglacial cycle (c. 120,000–12,000 years BP) (Demerre 2008; Vermeersch et al. 2015). Prehistoric artefacts have only been found in the intertidal zone. Since the late nineteenth century, numerous flint artefacts have been reported from beaches, probably because of the exposure of artefacts by erosion at the shore edge (Pieters et al. 2010). Archaeological investigation of the beach at Raversijde in the 1970s and 1980s resulted in numerous flints and a possible Bronze or Iron Age wooden paddle-like object (Pieters et al. 2010).

Roman artefacts recovered from the seafloor in the subtidal zone are quite scarce and mainly consist of ceramics. It is believed that most of these finds are related to Roman shipwrecks (Pieters et al. 2010). Artefacts from Medieval times mainly involve isolated finds from various locations. In contrast to the Roman finds, all the Medieval artefacts were found close to the coast (< 10 km), which suggests a relation to coastal settlements and/or coastal navigation (Pieters et al. 2010).

Two prominent Roman period settlements are known from the intertidal zone (Wenduine and Raversijde). The latter has been the focus of a large-scale archaeological research project on land carried out by the Flanders Heritage Agency and the province of West Flanders (Pieters et al. 2006). Roman artefacts were often found in peat layers. Through time these peat layers were occasionally exposed (e.g., after major storms), but beach reconstruction during the past two decades in combination with the construction of groynes in the 1970s has resulted in the permanent burial of the peat layers under a thick (1–2 m) sand cover.

Five important Medieval settlements are known from the intertidal zone (Raversijde, Ostend, Bredene, Wenduine and Heist). Four of these have been identified as fishing settlements (the only exception being Ostend, which became a town). Beach finds in Ostend mainly involved potsherds, but other artefacts were also recovered from the harbour mouth (Pieters et al. 2010). In Raversijde the remains of a late Medieval fishing settlement were found (Pieters 1993, Pieters et al. 2013).

In this paper, we will focus on three specific finds from distinct time periods and different environmental settings:

- A wooden peel or paddle found in situ in a peat layer exposed in the intertidal zone in the 1960s and 1970s at Raversijde.
- Lithic artefacts and Iron Age pottery found as stray finds in the intertidal zone during 'field-walking' activities by archaeologists, local historians and amateur researchers in the late nineteenth and early twentieth centuries.
- Fossil remains of Pleistocene/Early Holocene terrestrial mammals found during fishing and dredging activities in the subtidal part of the North Sea but also accidentally found in the intertidal zone (on the beach at ebb tide) by casual passers-by.

9.2 Archaeological Finds

9.2.1 Wooden Peel or Paddle

This oak artefact (Fig. 9.2) was found in 1970 in situ on the beach of Raversijde (Fig. 9.1) in the so-called Subboreal 'surface peat', a well-described and well-known geological unit on the Belgian coastal plain (Baeteman 2007) that extends into the present-day North Sea. The growth of this peat layer has been dated by radiocarbon analysis in the area of the find spot between 4395 ± 30 BP and 2635 ± 35 BP (Deforce and Bastiaens 2013).

The artefact is 112 cm long, shows traces of wear and charring on the upper part of the blade and is perforated by boring organisms, probably the American piddock (*Petricola pholadiformis*) (Pieters et al. 2010, pp. 184–186). To provide for long-term conservation, it was impregnated with PEG (polyethylene glycol), and no attempt was ever made to obtain a radiocarbon date until 2015. The attempt in 2015 was successful in eliminating completely the contamination of the PEG, which is soluble in water (Van Strydonck and Van Bos 1998), and produced a date of 2573 ± 29 BP, a result very close to the end of the peat growth at Raversijde. This result dates the object in cultural terms to the Bronze Age–Iron Age transition in the area of present-day Belgium. No function can be attributed with absolute certainty to this artefact. Its shovel-like shape suggests that it could be a baker's peel (used for removing bread from the oven) or a paddle. The most plausible interpretation would be a 'peel', but a combination of both cannot be excluded—perhaps a peel recycled as a paddle.

The main conclusion from the analysis of this find must be to consider the surface peat in the coastal plain also as a potential archaeological unit, although this possibility has not so far been investigated further. Another possible argument to strengthen the archaeological context of the surface peat is the presence of numerous carbonized

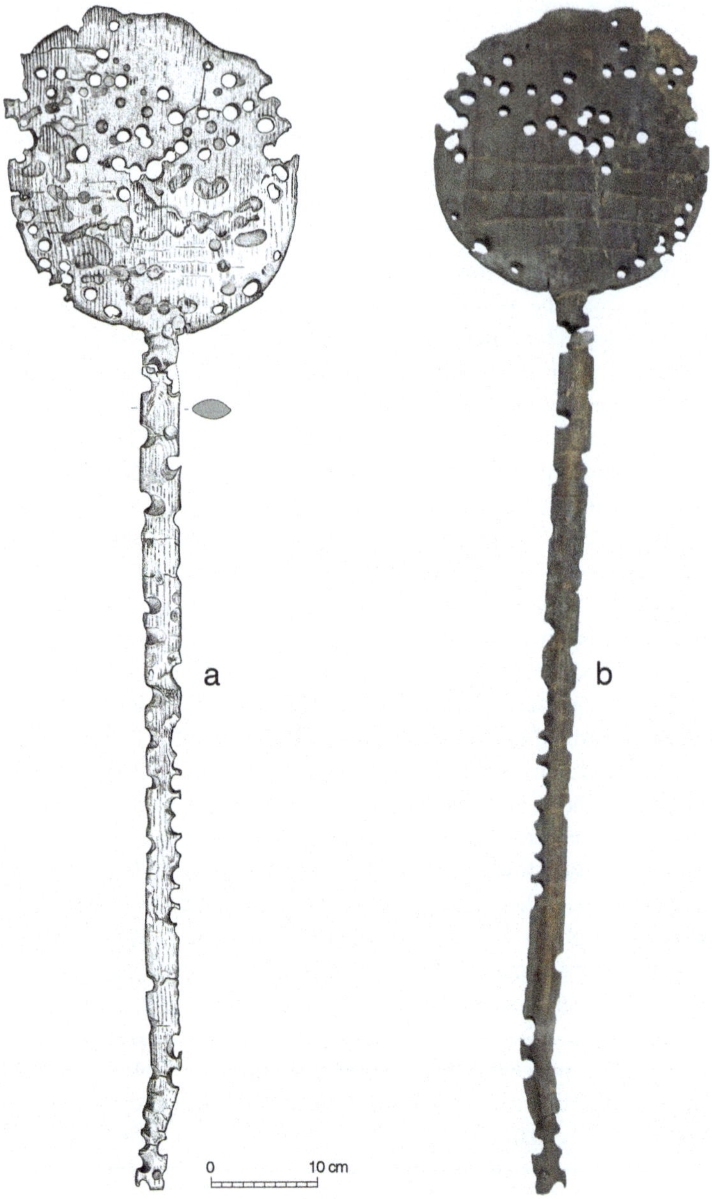

Fig. 9.2 Wooden paddle or peel found by Etienne Cools and Agnès Mortier in the Subboreal peat layer exposed on the beach of Raversijde. The peat was visible until the 1970s. Illustration by Marc Van Meenen and Hans Denis, Flanders Heritage Agency

microscopic plant remains noticed during a pollen study in the same peat layer at Raversijde (Deforce and Bastiaens 2013). The presence of carbonized remains itself is not a guarantee of human activity in the area but deserves archaeological attention. This said, it should also be realized that this Subboreal peat layer was exploited intensively in Roman and Medieval times as a source of fuel (Pieters 2000) and as a consequence not much of it is preserved for future research.

9.2.2 Lithic Artefacts and Iron Age Pottery Stray Finds

The beach of Raversijde is well known among archaeologists for its intertidal Roman and Medieval archaeological structures and finds (Zeebroek et al. 2012). Less well-known is the fact that this beach also produced, according to Chocqueel (1950), some 262 flint artefacts of which nearly 100 have recently been re-examined by an expert in prehistoric lithic artefacts (De Bie 2013). Two so-called Tjongerian points (Fig. 9.3: 1–2) in the 'Chocqueel-assemblage' almost certainly date to the Late Upper Palaeolithic (14,000–12,000 BP) and can be attributed to the Federmesser grouping. The other diagnostic tools in the available assemblage place a second occupation phase in the Neolithic (6500–5000 BP): the polished nature of the pieces and the large scraper suggest middle to late Neolithic (Fig. 9.3: 4–7).

These artefacts were all gathered in the first half of the twentieth century and no additional information is available as to their spatial distribution on the beach, their possible context or any related cultural features. In view of their age (late Palaeolithic and middle to late Neolithic) their reworking from foreshore deposits older than and underlying those surfacing at the beach in the first half of the twentieth century (Fig. 9.4) seems plausible.

Apart from this quantitatively and qualitatively important assemblage of prehistoric artefacts from Raversijde beach, a few flint artefacts were found in the late nineteenth and the first half of the twentieth centuries on almost every beach east of Nieuwpoort (Pieters et al. 2010). On the other hand, there are no records of such finds from the beaches to the west of Nieuwpoort. This remarkable distribution pattern seems at first glance related to processes of erosion and sedimentation: to the west of the river IJzer, the Belgian coast is building up gradually while to the east of the river IJzer the coast is gradually retreating (De Moor and Pissart 1992). This dynamic could explain the above-mentioned distribution pattern.

Neither the objects themselves from these other beaches nor any drawings or photographs of them are still available, which hampers a reinterpretation based only on the textual information. However, the fact that most beaches along the Belgian coast are known as find spots of at least a few flint artefacts emphasizes the archaeological value of the unusual assemblage of artefacts from Raversijde. When looking at the geology of the coastal zone, Raversijde is situated at the edge of a large buried palaeovalley (the so-called Ostend Valley), recently described in more detail (Mathys 2009; De Clercq et al. 2016). The presence of this Late Pleistocene buried palaeovalley makes the washing ashore of artefacts from underlying older deposits plausible.

Between the Stone Age and the Roman period, only a few sherds of Iron Age pottery have been found on the beaches of Bredene and Heist (Pieters et al. 2010).

9.3 Palaeontological Terrestrial Mammal Remains

In this section, we present an overview of locations with remains from fossil terrestrial and marine fauna that have been reported with certainty as coming from the intertidal or subtidal zone. This is largely based on intensive inventory work done by the Flanders Heritage Agency (Demerre 2008) and on an unpublished report from the SeArch project (Vermeersch et al. 2015). Eight find locations are reported: three from the intertidal zone and five from the subtidal.

9.3.1 Intertidal Find Spots

From the beach of Mariakerke (Ostend), there is a fossil tooth from a walrus (*Odobenus rosmarus*) probably of Weichselian age. The tooth belongs to the collection of the local museum De Plate in Ostend. The second intertidal find spot, the beach east of the harbour of Ostend, produced six bone finds, including fragments from a Weichselian mammoth and reindeer. On the beach at Blankenberge, a humerus fragment of a young aurochs was found (Fig. 9.5). Specialists such as A. Ervynck (Belgium) and M. Pathou-Mathis (France) are convinced that this bone fragment belongs to a young aurochs and could be interpreted as a human artefact, as the humerus was broken at exactly the right spot to extract the marrow. Due to unclarified reasons, the radiocarbon analysis did not produce a reliable result despite two successive trials. Another bone frag-

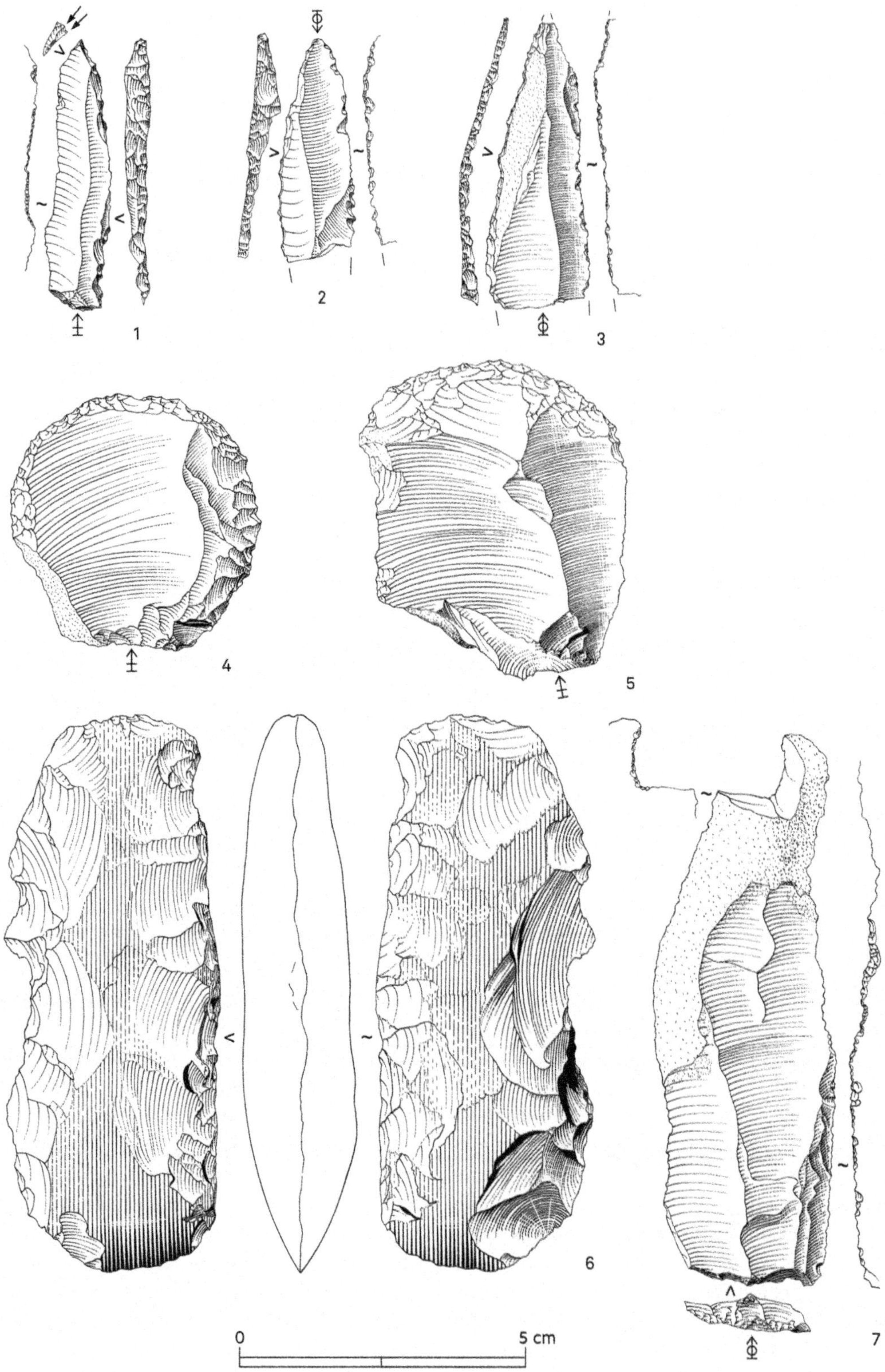

Fig. 9.3 Lithic artefacts found by A. Chocqueel at the beach of Raversijde. Illustration by Marc Van Meenen, Flanders Heritage Agency

Fig. 9.4 Aerial picture of the beach of Raversijde in the early 1970s. The Subboreal peat eroding at the surface is clearly visible. Photo courtesy of Etienne Cools and Agnès Mortier

ment from an aurochs was uncovered in the harbour channel of Blankenberge, close to the above find, and has been submitted for radiocarbon dating.

9.3.2 Offshore Find Spots

A. Gautier in two papers on fossil walrus bones also mentions the find of a mammoth skull (*Elephas primigenius* or *Mammuthus primigenius*) of Late Pleistocene (Weichselian) age dredged from the North Sea near the Westhinder Bank. The skull was of particular interest because of its association with puparia of the subarctic blowfly (*Protophormia terraenovae*) (Gautier 1995; Gautier 1998). The skull itself is actually missing.

T. van Kolfschoten and H. Van Essen mention Early Pleistocene fossil remains from Tiglian deposits (2.45–1.8 million years) found close to the Thornton Bank. This sandbank extends from Belgian waters into the Dutch part of the North Sea, suggesting these finds may originate from the Dutch sector (Van Kolfschoten and Van Essen 2004) as no deposits of this age are found in the Belgian part of the North Sea (Mathys 2009). The authors mention *Anancus arvernensis*, *Mammuthus meriodionalis*, *Eucladoceros* sp. and *Equus* sp. D. Mol (1991) also mentions a molar from *Anancus arvernensis* found next to the Thornton Bank.

A sixth find spot is situated in the 'Scheur', a navigation channel towards the Westerscheldt northeast of the harbour of Zeebrugge, close to the so-called Zeebrugge wreck-site (Vandenberghe 1997). According to oral information by the people responsible for the salvage of this wreck, remains of a woolly rhinoceros (*Coelodonta antiquitatis*) were found in the immediate vicinity of the wreck. In this case also, the palaeontological remains themselves are missing. A small fraction of the fossil collection of the Ostend museum 'De Plate' consists of several palaeontological remains dredged from the

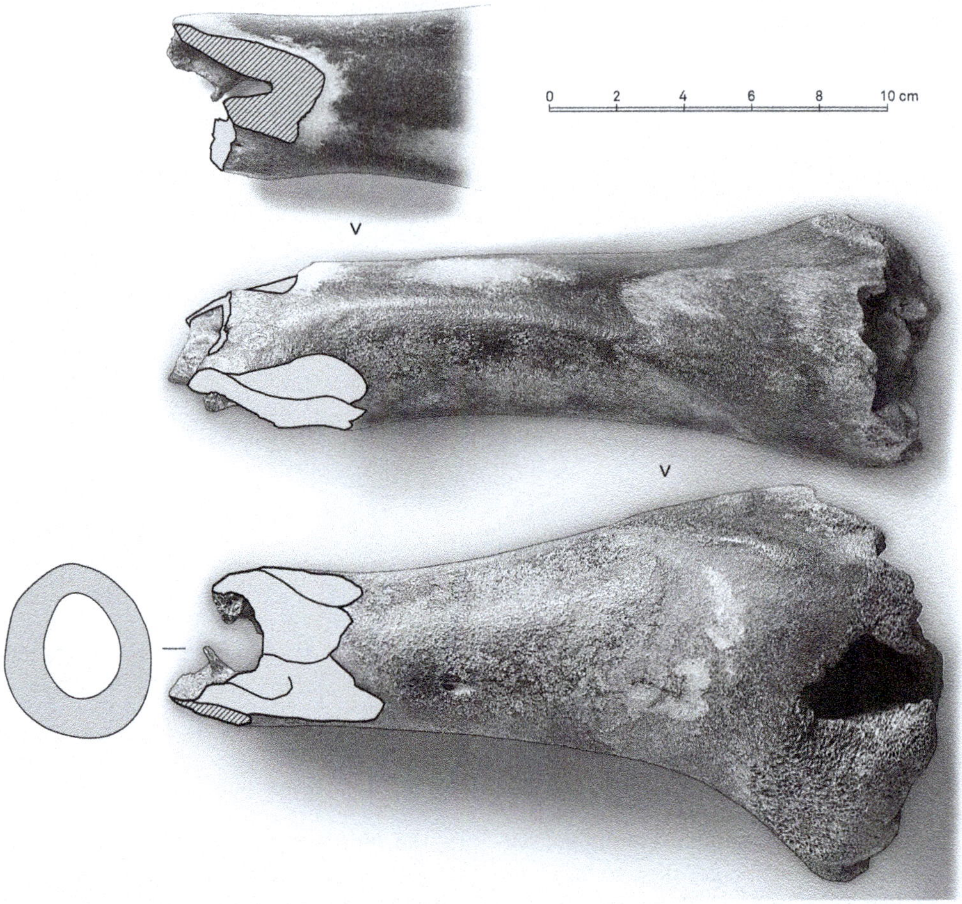

Fig. 9.5 Humerus fragment of a young aurochs. Illustration by Marc Van Meenen and Kris Vandevorst, Flanders Heritage Agency

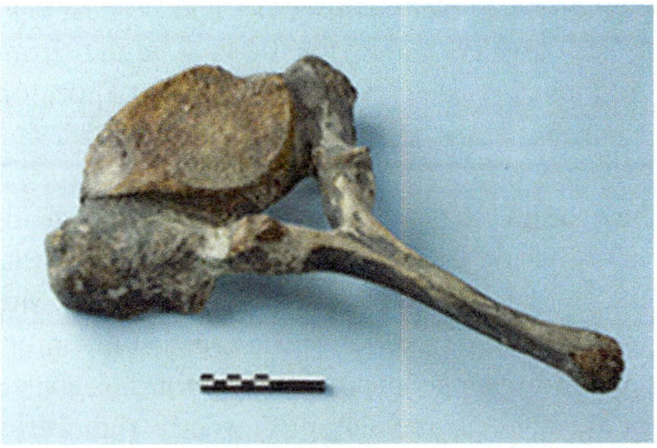

Fig. 9.6 Thoracic vertebra from a mammoth (*Mammuthus primigenius*) or elephant (*Elephas antiquus*) found in the Zeebrugge harbour channel. Photo by Hans Denis, Flanders Heritage Agency

navigation channel leading to the harbour of Zeebrugge. These finds consist of, among others, the thoracic vertebra of a mammoth (*Mammuthus primigenius*) or an elephant (*Elephas antiquus*) (Fig. 9.6). Whether these finds are connected with the find reported from the 'Scheur' is still under debate.

In 2008 a red deer antler was found during sand extraction activities on the Kwinte Bank. There are no tool marks on the antler and it seems to have been shed naturally (Wessex Archaeology 2008). It probably dates from the Early Holocene. Two palaeontological remains in the collection of the 'Aquarium of Ostend', but at present missing, are on a picture labelled as found respectively near the coast of Nieuwpoort and near the coast of Zeebrugge. Again, the latter find points to the same area as the navigation channels of the harbour of Zeebrugge and the 'Scheur'. The find from the coast of Nieuwpoort represents a separate locality.

With regard to age, the finds from the Thornton Bank area are early Pleistocene while the others are late Pleistocene (Eemian or Weichselian) or Early Holocene in age.

9.3.3 The 'Scheur' Case Study

Important new information has recently been obtained at the 'Scheur'. Due to maintenance dredging of the navigation channel, the upper, Holocene layers were removed, thereby exposing older Late Pleistocene and even underlying Palaeogene strata. Dutch fishing activities in the area have uncovered large quantities of palaeontological bone material over the past decades. This only recently came to the attention of Belgian scientists. As a result, two targeted surveys were carried out with RV Simon Stevin from the Flanders Marine Institute in July 2017. This produced more finds and made clear that remains from two fairly distinct geological periods are present in that location: the Pleistocene—mainly remains of walrus (*Odobenus rosmarus*) but also including wild horse, aurochs and red deer—and the Eocene/Early Oligocene (rare whale remains, *Archaeoceti*) (Post et al. 2017). Based on the current finds, the site appears to be one of the largest and southernmost find locations of walrus bone material in the world (K. Post and D. Mol, personal communication, 2017). The bone material is most likely preserved in situ, derived from layers exposed on the seafloor. Determination and dating of the fossils are still ongoing at this moment. Further research (very-high resolution bathymetry and seismics, and shallow coring) will be done in the near future (2018) in order to gain a better understanding of chronology and palaeolandscape evolution.

9.4 The Palaeolandscape Context

If we compare the different find locations with a recent reconstruction of the base of the Quaternary sequence (Top-Palaeogene surface) (Mathys 2009, De Clercq et al. 2016), a correlation between geology and geomorphology (e.g., buried palaeovalleys) and palaeontological material seems plausible, although difficult to prove due to the limited precision of the find locations and the lack of geological context. The beaches of Mariakerke and east of Ostend, the Kwinte Bank and the Westhinder area are all situated in or near the Ostend Valley, whereas the find spot related to Zeebrugge (according to three different descriptions) is situated in the so-called Zeebrugge Valley. The beach of Blankenberge seems to be close to another buried palaeovalley, the so-called Coastal Valley, which is in fact the upstream section of the Ostend Valley (Mathys 2009; De Clercq et al. 2016; Heyse and Demoulin 2018). The find spot near the Thornton Bank may well be related to the buried Thornton Valley, and finally the find spot near Nieuwpoort can be associated with an ancient IJzer Valley. These data are of course not enough for a statistically valid analysis but the coincidence at this level is striking.

In a paper on Pleistocene faunal remains from terrestrial Belgium, Gautier (1985) distinguishes three types of palaeontological accumulations, and this provides additional useful information for understanding the subtidal and intertidal discoveries. One of Gautier's types is particularly relevant, namely faunal remains found as 'concentrations in the Flemish Valley'. The latter is a buried palaeovalley situated, broadly speaking, to the north of Ghent, which connects to the downstream Coastal-Ostend Valleys. The characteristics of these faunal remains found in concentrations, and their association with the presence of fly puparia, points to a specific taphonomic history that seems also to have been at play in the offshore Ostend Valley, notably the skull with puparia dredged up from the Westhinder Bank, slightly downstream of this valley.

According to Gautier (1985) and Gautier and Schuman (1973), the presence of fly puparia (*Protophormia terraenovae*) in the skulls of large Pleistocene mammals, and their concentrated numbers, can be explained according to the following hypothetical sequence of events. Large mammals gathered in the lowland river valleys during winter, where some of them died due to the harsh conditions. In spring the frozen cadavers thawed and were washed downstream by peak spring meltwater discharge after the blowflies had already laid their eggs. Somewhere downstream the cadavers settled in certain areas favourable for burial and good preservation.

The link between the North Sea area and the onshore Flemish Valley makes it interesting to look at the setting of known Middle Palaeolithic archaeological sites on land in relation to the offshore topography (Crombé and Van Der Haegen 1994). Most of the known Middle Palaeolithic finds in Flanders come from the interfluves, but in the southern part of the Flemish Valley, more precisely in the area of Ghent, three sites have been located clearly within the valley infill itself. These sites were detected during development work and sand extraction between 75 and 40 years ago. Unfortunately, precise data are lacking on their lithostratigraphical context or their relationship to the palaeontological remains also reported from these locations. Consequently, no firm conclusions can be drawn on their geographical and geological setting, as they just reflect the locations of infrastructural works near the edges of the Flemish Valley. Nevertheless, the fact that Middle Palaeolithic sites are present in the Flemish Valley, regardless of whether they were in situ or reworked, provides insights into burial and preservation conditions that can plausibly be extended downstream to the Coastal-Ostend Valleys on the Belgian coastal plain. In particular, because of higher water tables, preservation conditions in valley infills are in general expected to be much better than for sites in other geomorphological settings.

The scarcity of reported bone finds from the Belgian part of the North Sea compared to the numerous finds reported from Dutch waters is absolutely striking (e.g. Louwe Kooijmans 1971). This is probably partly due to the scientific neglect of these finds in Belgium, notwithstanding the fact that fishermen regularly reported their finds to museums during the second half of the twentieth century. Another reason could be the different geological setting whereby the Belgian sector is situated in an area dominated by erosion rather than sedimentation, as is the case in the Dutch sector of the North Sea (Hijma et al. 2012).

9.5 Future Potential

The specific finds discussed above cover all that is presently known about pre- and/or protohistorical archaeological and palaeontological finds in the Belgian part of the North Sea including the present-day adjoining beaches. The available data are, on the one hand, rather limited and notably scarce compared to neighbouring countries, but on the other hand they are steadily growing since the SeArch project. The recent investigations (summer of 2017) of an area with exceptionally high concentrations of Late Pleistocene and Palaeogene bone finds in the 'Scheur' is a good illustration of these developments and offers a promising avenue for future research. The growing awareness of the general public is similarly leading to an increase in reported chance finds, mainly from beaches on the modern shoreline. In due course these growing numbers will hopefully make it possible to test more effectively current hypotheses.

Underwater contexts, and wetland conditions more generally, offer better preservation conditions for organic materials, such as the wooden peel or paddle preserved in peat, and the blow fly puparia preserved in a mammoth skull, and this preservation potential will greatly extend the possibilities for new research compared to similar archaeological sites on land and above groundwater levels.

At present, new research in Belgium, focusing on the Pleistocene palaeolandscapes of the Belgian part of the North Sea and their archaeological/palaeontological potential, is in progress (De Clercq 2018). The results so far, which are visualized in a series of palaeogeographical maps, show that the preserved sediments are much more diverse than has previously been assumed (Mostaert et al. 1989; Mathys 2009). The sediments are best preserved in incised depressions and palaeovalleys cut into the Top-Palaeogene surface. They originate from various depositional environments, ranging from fluvial to marine, and provide unprecedented information about the palaeogeographical evolution of the southern North Sea for the last 160,000 years. High-resolution sedimentary and seismic analysis provides detailed information about the depositional circumstances of these sedimentary bodies.

This PhD study will provide a geological context for the Pleistocene archaeology and palaeontology of the Belgian part of the North Sea (similar to Hijma et al. 2012), and will for the first time allow an assessment of the preservation potential of the various depositional environments (cf. Ward and Larcombe 2008). It clearly shows that integration of geophysical, geological/sedimentological and archaeological/palaeontological disciplines not only forms the key towards major breakthroughs in our knowledge on and understanding of the prehistory of Europe, but it also provides the best opportunity to develop tools, such as predictive maps of archaeological potential and risk, that can serve the needs of commercial-industrial end-users of the sea and the underwater archaeological heritage community.

For a long time, the strong divide between archaeology on land and archaeology at sea and the related misunderstandings about the difficulties of underwater research have had a negative impact on the possibilities for research on underwater cultural heritage, notwithstanding the huge potential. In Belgium this is finally changing, be it slowly. New ground-breaking interdisciplinary research is planned for 2018–2019 in the wider Zeebrugge region. The focus of this study, a collaboration between the Ministry of Public Works, Ghent University and Flanders Marine Institute, is to map the Quaternary palaeolandscape evolution and by linking this to the various fossil finds in the area to gain a better understanding of the archaeological potential of this highly dynamic area, including evidence of human occupation. The study will involve novel techniques such as OSL dating and sedaDNA, and the results will undoubtedly transform our understanding of Palaeolithic/Mesolithic archaeology in Belgium.

9.6 Management of the Underwater Cultural Heritage

Underwater cultural heritage as defined by UNESCO in the Belgian part of the North Sea is the responsibility of the Federal State of Belgium (Federal Governmental Service of Mobility and Transport). Since June 2014, Belgium has established a new active legislation related to underwater cultural heritage in the North Sea. That new legal framework has, among other things, defined a 'receiver of underwater cultural heritage', a new function similar to the receiver of wrecks in the United Kingdom but restricted to underwater cultural heritage. This new function is added onto the Federal portfolio of the governor of West Flanders (the only province in Belgium that borders the North Sea). Governors are civil servants of the Belgian regions (Flanders, Brussels Capital Region and Wallonia). The governor of West Flanders has two protocols of collaboration for the execution of tasks as receiver of underwater cultural heritage, one with the Flanders Marine Institute and the other with the Flanders Heritage Agency. The first deals with duties related to storage and conservation of finds, whereas the second is mainly concerned with scientific aspects related to these new heritage-related tasks.

Underwater cultural heritage as defined by UNESCO and present in the intertidal zone of the Belgian part of the North Sea is the regional responsibility of Flanders, taken care of by the Flanders Heritage Agency. The limit between the two zones of responsibility is the multi-annual mean low water at spring tide.

The Flanders Heritage Agency (Agentschap Onroerend Erfgoed) is a scientific institute of the Region of Flanders and has developed over the past two decades scientific and management skills and expertise in underwater cultural heritage including the intertidal zone. Together with the Flanders Marine Institute, the Flanders Heritage Agency also hosts an online database, <www.maritime-archaeology.be>, on maritime archaeological heritage. Thanks to a new research mandate, the Flanders Marine Institute will be able in the near future to perform, coordinate and organize new and complementary research related to underwater cultural heritage, in close collaboration with academic institutions.

References

Baeteman C (2007) De laat holocene evolutie van de Belgische kustvlakte: Sedimentatie-processen versus zeespiegelschommelingen en Duinkerke transgressies. In: De Kraker AMJ, Borger GJ (eds) Veen-Vis-Zout. Landschappelijke dynamiek in de zuidwestelijke delta van de Lage Landen. Geoarchaeological and Bioarchaeological Studies 8:1–17

Chocqueel A (1950) Les civilisations préhistoriques & anciennes de la Flandre Occidentale d'après l'examen d'objets leur ayant appartenu. Editions du Temple, Brussels

Crombé Ph, Van Der Haegen G (1994) Het Midden-Paleolithicum in Noordwestelijk België, Archeologisch Inventaris Vlaanderen. Buitengewone Reeks 3, Ghent

De Bie M (2013) Vuurstenen artefacten. In: Pieters M, Baeteman C, Bastiaens J, Bollen A, Clogg P, Cooremans B, De Bie M, De Buyser F, Decorte G, Deforce K, De Groote A, Demerre I, Demiddele H, Ervynck A, Gevaert G, Goddeeris T, Lentacker A, Schietecatte L, Vandenbruaene M, Van Neer W, Van Strydonck M, Verhaeghe F, Vince A, Watzeels S, Zeebroek I (eds) Het archeologisch onderzoek in Raversijde (Oostende) in de periode 1992–2005. Vuurstenen artefacten, een Romeinse dijk, een 14de-eeuws muntdepot, een 15de-eeuwse sector van een vissersnederzetting en sporen van een vroeg 17-de-eeuwse en een vroeg 18de-eeuwse belegering van Oostende. Relicta Monografieën 8, Brussels, pp 67–72

De Clercq M (2018) Drowned landscapes of the Belgian continental shelf. Implications for northwest European landscape evolution and preservation potential for submerged heritage. PhD dissertation, Ghent University

De Clercq M, Chademenos V, Van Lancker V, Missiaen T (2016) High-resolution DEM for the Top-Palaeogene surface of the Belgian Continental Shelf. J Maps 12(5):1047–1054

De Moor G, Pissart A (1992) Het Reliëf. In: Denis J (ed) Geografie van België. Brussels, pp 130–215

Deforce K, Bastiaens J (2013) Het paleobotanisch onderzoek van het oppervlakteveen in profiel D. In: Pieters M, Baeteman C, Bastiaens J, Bollen A, Clogg P, Cooremans B, De Bie M, De Buyser F, Decorte G, Deforce K, De Groote A, Demerre I, Demiddele H, Ervynck A, Gevaert G, Goddeeris T, Lentacker A, Schietecatte L, Vandenbruaene M, Van Neer W, Van Strydonck M, Verhaeghe F, Vince A, Watzeels S, Zeebroek I (eds) Het archeologisch onderzoek in Raversijde (Oostende) in de periode 1992-2005. Vuurstenen artefacten, een Romeinse dijk, een 14de-eeuws muntdepot, een 15de-eeuwse sector van een vissersnederzetting en sporen van een vroeg 17-de-eeuwse en een vroeg 18de-eeuwse belegering van Oostende. Relicta Monografieën 8, Brussels, pp 52–55

Demerre I (2008) Maritiem archeologisch erfgoedonderzoek in 2006–2007. Twee jaar registratie en verwerking van het erfgoed in en uit de Noordzee. Vlaams Instituut voor het Onroerend Erfgoed, Brussels

Gautier A (1985) Pleistocene zoogdieren van België en waar ze gevonden worden. Lutra 28:121–123

Gautier A (1995) Bovenpleistocene zoogdieren van Oudenaarde Donk (België), fossiele vliegenpoppen uit de Vlaamse Valleien elders en nog een en ander over de Vlaamse Vallei. Cranium 12(2):73–81

Gautier A (1998) Fossiele walrussen (*Odobenus rosmarus*) aan de Belgische kust. Natuurwetenschappelijk Tijdschrift 78:122–127

Gautier A, Schuman H (1973) Puparia of the Subarctic Black Blowfly Protophormia Terraenovae (Robineau-Desvoidy, 1830) in a skull of a Late Eemian (?) Bison at Zemst, Brabant (Belgium). Palaeogeogr Palaeoclimatol Palaeoecol 14:119–125

Heyse I, Demoulin A (2018) The Flemish valley: response of the Scheldt drainage system to climatic and Glacio-Eustatic oscillations. In: Dumoulin A (ed) Landscapes and landforms of Belgium and Luxembourg. Springer International, Cham, pp 297–311

Hijma MP, Cohen KM, Roebroeks W, Westerhoff WE, Busschers FS (2012) Pleistocene Rhine-Thames landscapes: geological background for hominin occupation of the southern North Sea region. J Quat Sci 27(1):17–39

Hublin J-J, Weston D, Gunz P, Richards M, Roebroeks W, Glimmerveen J, Anthonis L (2009) Out of the North Sea: the Zeeland ridges Neandertal. J Hum Evol 57:777–785

Louwe Kooijmans LP (1971) Mesolithic bone and antler implements from the North Sea and from the Netherlands. Berichten van de Rijksdienst voor het Oudheidkundig Bodemonderzoek 20–21:27–73

Mathys M (2009) The quaternary geological evolution of the Belgian Continental Shelf, Southern North Sea. Unpublished PhD thesis, Ghent University

Missiaen T, Pieters M, Maes F, Kruiver P, De Maeyer P, Seys J (2017) The search project: towards an assessment methodology and sustainable management policy for the archaeological heritage of the North Sea in Belgium. In: Bailey GN, Harff J, Sakellariou D (eds)

Under the sea: archaeology and palaeolandscapes of the continental shelf. Springer, Cham, pp 415–424

Mol D (1991) Het ijstijdlandschap van de zuidelijke Noordzee. Grondboor en Hamer 45:9–13

Mol D, Post K, Reumer JWF, Van der Plicht J, De Vos J, Van Geel B, Van Reenen G, Pals JP, Glimmerveen J (2006) The Eurogeul—first report of the palaeontological, palynological and archaeological investigations of this part of the North Sea. Quat Int 142–143:178–185

Mostaert F, Auffret JP, De Batist M, Henriet J-P, Moons A, Sevens E, Van den Broeke I, Verschuren M (1989) Quaternary shelf deposits and drainage patterns off the French and Belgian coasts. In: Henriet J-P, De Moor G, De Batist M (eds) The quaternary and tertiary geology of the Southern Bight, North Sea. Belgian Geological Survey, Brussels, p 241

Pieters M (1993) Archeologisch onderzoek te Raversijde. Interim Verslag 1992. Archeologie in Vlaanderen II:247–264

Pieters M (2000) The impact of peat extractions on soil parent material in the area of the Medieval Village of Walravenside (Ostend, West-Flanders, Belgium). Pedologie-Themata 8:71–75

Pieters M, Demerre I, Zeebroek I (2006) Dijk uit de Romeinse tijd aangesneden onder het middeleeuwse vissersdorp Walraversijde. In: Bosman A, De Clercq W, Hoevenberg J (eds) Romeinendag, Jaarlijks Belgisch congres voor Romeinse archeologie. Universiteit Gent, pp 93–97

Pieters M, Demerre I, Lenaerts T, Zeebroek I, De Bie M, De Clercq W, Dickinson B, Monsieur P (2010) De Noordzee: een waardevol archief onder water. Meer dan 100 jaar onderzoek van strandvondsten en vondsten uit zee in België: een overzicht. Relicta 6:177–218

Pieters M, Baeteman C, Bastiaens J, Bollen A, Clogg P, Cooremans B, De Bie M, De Buyser F, Decorte G, Deforce K, De Groote A, Demerre I, Demiddele H; Ervynck A, Gevaert G, Goddeeris T, Lentacker A, Schietecatte L, Vandenbruaene M, Van Neer W, Van Strydonck M, Verhaeghe F, Vince A, Watzeels S, Zeebroek I (2013) Het archeologisch onderzoek in Raversijde (Oostende) in de periode 1992–2005. Vuurstenen artefacten, een Romeinse dijk, een 14de-eeuws muntdepot, een 15de-eeuwse sector van een vissersnederzetting en sporen van een vroeg 17-de-eeuwse en een vroeg 18de-eeuwse belegering van Oostende, Relicta Monografieën 8, Brussels

Pieters M, De Clercq W, Demerre I, Missiaen T, Van Haelst S (2018) Belgium – finds from trawl fishing, dredging and beach walking. In: Fischer A, Pedersen L (eds) Oceans of archaeology. Jutland Archaeological Society and Aarhus University Press, Aarhus

Post K, Hoekman A, De Wilde B (2017) Oerwalvissen op de bodem van de Noordzee. Cranium 34(1):48

Van Kolfschoten T, Van Essen H (2004) Palaeozoological heritage from the bottom of the North Sea. In: Flemming NC (ed) Submarine prehistoric archaeology of the North Sea, Research priorities and collaboration with industry CBA Research Report, vol 141. Council for British Archaeology, York, pp 70–80

Van Strydonck M, Van Bos M (1998) 14C-datering van met PEG behandeld nat opgegraven hout. Lunula Archaeologica Protohistorica VI:83–87

Vandenberghe S (1997) Découvertes importantes de la fin du 15ème et le début du 16ème siècle provenant du fond de la mer du Nord en face du port de Zeebrugge. In: De Boe G, Verhaeghe F (eds) Travel technology & organisation in Medieval Europe, Papers of the 'Medieval Europe Brugge 1997' Conference, vol 8. IAP-Rapporten 8, pp 87–90

Vermeersch J, Demerre I, Pieters M, Van Haelst S, De Hauwere N (2015) Paleontologische resten afkomstig uit het Belgische deel van de Noordzee of aangetroffen op de aanpalende stranden. Unpublished report SeArch-project WP 1.5.1A

Ward I, Larcombe P (2008) Determining the preservation rating of *submerged* archaeology in the post-glacial southern North Sea: a first-order geomorphological approach. Environ Archaeol 13:59–83

Wessex Archaeology (2008) Hanson_0171_a: Two sherds of Roman Samian ware and a piece of antler. Unpublished report

Zeebroek I, Demerre I, Deckers P (2012) The site of Raversijde beach. A test site for geophysical research techniques. In: Fenwick V, Poudret-Barré A, Momber G, Demerre I, Zeebroek I, Bowens A, Chatelin C (eds) Archaeological Atlas of the 2 Seas. A cross-border maritime archaeology project, Final report 2009–2012. Hampshire and Wight Trust for Maritime Archaeology, Southampton, pp 32–35

Great Britain: The Intertidal and Underwater Archaeology of Britain's Submerged Landscapes

10

Geoff Bailey, Garry Momber, Martin Bell, Louise Tizzard, Karen Hardy, Andrew Bicket, Lauren Tidbury, Jonathan Benjamin, and Alex Hale

Abstract

The submerged landscapes around Great Britain are extensive and would have offered productive territory for hunting, gathering, exploitation of aquatic and marine resources, and—in the final stages of postglacial sea-level rise—opportunities for agriculture. They would also have provided land connections to continental Europe and opportunities for communication by sea travel along now-submerged palaeocoastlines and river estuaries. Most of the archaeological material has been discovered in intertidal or shallow water conditions, but there are also discoveries in deeper water, with dates ranging from earliest human presence nearly one million years ago up to the establishment of modern sea level. Some later material is present where coastlines have continued to sink in more recent millennia. Intertidal sites are especially well represented

G. Bailey (✉)
Department of Archaeology, University of York, York, UK

College of Humanities, Arts and Social Sciences, Flinders University, Adelaide, SA, Australia
e-mail: geoff.bailey@york.ac.uk

G. Momber
Maritime Archaeology Trust, National Oceanography Centre, Southampton, UK

Department of Archaeology, University of York, York, UK
e-mail: garry.momber@maritimearchaeologytrust.org

M. Bell
Department of Archaeology, School of Human and Environmental Sciences, University of Reading, Reading, UK
e-mail: m.g.bell@reading.ac.uk

L. Tizzard
Wessex Archaeology, Portway House, Salisbury, Wiltshire, UK
e-mail: l.tizzard@wessexarch.co.uk

K. Hardy
Department of Archaeology, University of York, York, UK

ICREA, Barcelona, Catalonia, Spain

Departament de Prehistòria, Facultat de Filosofia i Lletres, Universitat Autònoma de Barcelona, Barcelona, Catalonia, Spain
e-mail: karhardy@icrea.cat

A. Bicket
Wessex Archaeology, Edinburgh, UK

Department of Geography, Loughborough University, Loughborough, UK
e-mail: a.bicket@wessexarch.co.uk

L. Tidbury
Museum of London Archaeology, Mortimer Wheeler House, London, UK
e-mail: LTidbury@mola.org.uk

J. Benjamin
College of Humanities, Arts and Social Sciences, Flinders University, Adelaide, SA, Australia
e-mail: Jonathan.benjamin@flinders.edu.au

A. Hale
Historic Environment Scotland, John Sinclair House, Edinburgh, UK
e-mail: alex.hale@hes.scot

G. Bailey et al. (eds.), *The Archaeology of Europe's Drowned Landscapes*, Coastal Research Library 35, https://doi.org/10.1007/978-3-030-37367-2_10

because of relatively large tidal ranges and shallow offshore gradients on many coastlines. These are often associated with remains of submerged forests, which are periodically exposed at low tide and then covered up again by movements of sand. Some of the most distinctive intertidal finds are the human and animal footprints preserved in intertidal sediments in many locations, especially at Goldcliff East. The earliest, at Happisburgh, are dated between 0.78 and 1 Ma. Fully submerged sites include the Mesolithic site of Bouldnor Cliff with its worked timbers, and the Middle Stone Age artefacts from offshore aggregate Area 240 along with well-preserved ice age fauna and environmental indicators. Pioneering work using oil industry seismic records has produced detailed reconstructions of the submerged landscape, and this is being followed up by new work involving targeted acoustic survey and coring of sediments.

Keywords
Submerged forests · Intertidal · Happisburgh · Bouldnor Cliff · Goldcliff East · Area A240

10.1 Introduction

This chapter concerns Great Britain, a geographical term for the largest island of the British Isles, which includes England, Wales and Scotland, their territorial waters and offshore islands such as the Isle of Wight, the Hebrides, the Orkney Islands and the Shetland Islands (Fig. 10.1). The island of Ireland includes the Republic of Ireland and Northern Ireland (which is politically a part of the United Kingdom), and is treated separately (Westley and Woodman, Chap. 11, this volume).[1]

We summarise the geological and environmental changes of the Quaternary period of relevance to archaeological interpretation of Great Britain's submerged landscapes with reference to the more detailed information in Flemming et al. (2017) and provide an overview of the intertidal and underwater archaeological material and the conditions that have contributed to its preservation and discovery, followed by a more detailed description of individual finds. In a concluding assessment, we consider the wider significance of what is already known and some directions for future research.

10.2 The Geological and Palaeogeographical Context

Great Britain has numerous peninsulas, smaller offshore islands and archipelagos, a long coastline relative to its land area (some 17,800 km of coastline relative to an area of 209,000 km^2) and is surrounded by extensive areas of now-submerged landscape (Bicket and Tizzard 2015; Westley 2017; Cohen et al. 2017; Farr et al. 2017). The central and southern sectors of the North Sea Basin formed the principal area of lowland territory on the exposed continental shelf, and this area has come to be known as Doggerland, after the Dogger Bank, which would have formed a low rise on the coastal plain (Coles 1998; Fig. 10.2).[2] At its maximum extent, the

[1]The United Kingdom is a political unit comprising England and the devolved administrations of Scotland, Wales and Northern Ireland (the latter suspended at the time of writing). The British Isles, or the British and Irish Isles, refer to a geographical entity that comprises the islands of Great Britain and Ireland. The Isle of Man and the Channel Islands (principally Guernsey and Jersey) are Crown Dependencies, self-governing administrations outside the United Kingdom but for which the UK government takes responsibility particularly in international affairs. For convenience, we include here brief reference to archaeological evidence on Jersey, though the island is geographically closer to France than to England (see Billard et al., Chap. 12, this volume). Evidence from Gibraltar, which is a British Overseas Territory, is discussed with evidence from Spain by Arias in Chap. 13, this volume.

[2]Note that the area of exposed land in the North Sea at low sea level partly falls within the Exclusive Economic Zones of Norway, Denmark, Germany, the Netherlands, Belgium and France, as well as the United Kingdom, see Fig. 1.1, this volume, and references to underwater investigations of Doggerland occur in the relevant national chapters in this volume.

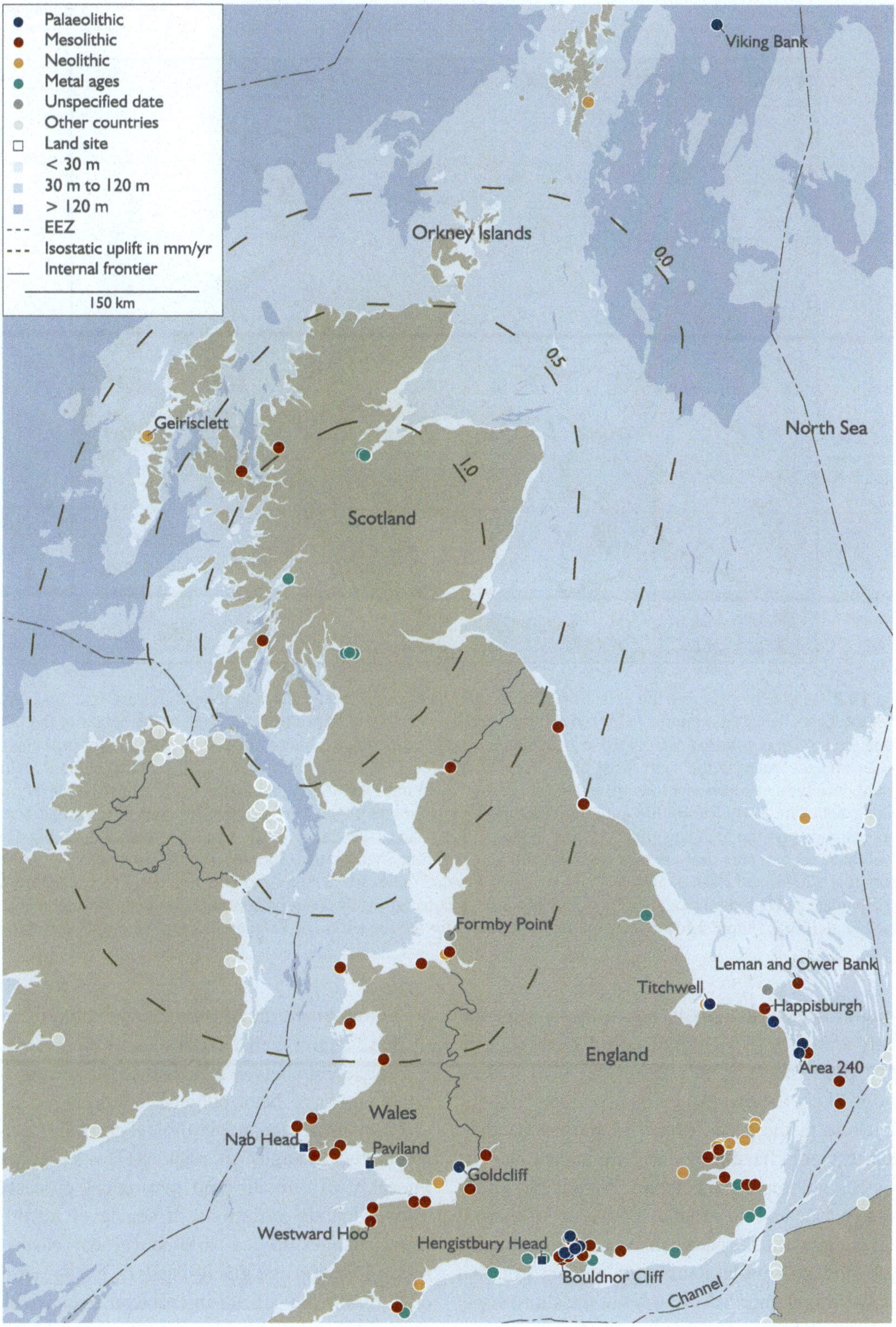

Fig. 10.1 Map showing the location of underwater finds with names of sites and geographical areas mentioned in the text. Within the dotted lines to the North the Earth's crust is rising, while south of it the land has been slightly subsiding since the Stone Age. Site information from the SPLASHCOS Viewer http://splashcos-viewer.eu. Drawing by Moritz Mennenga

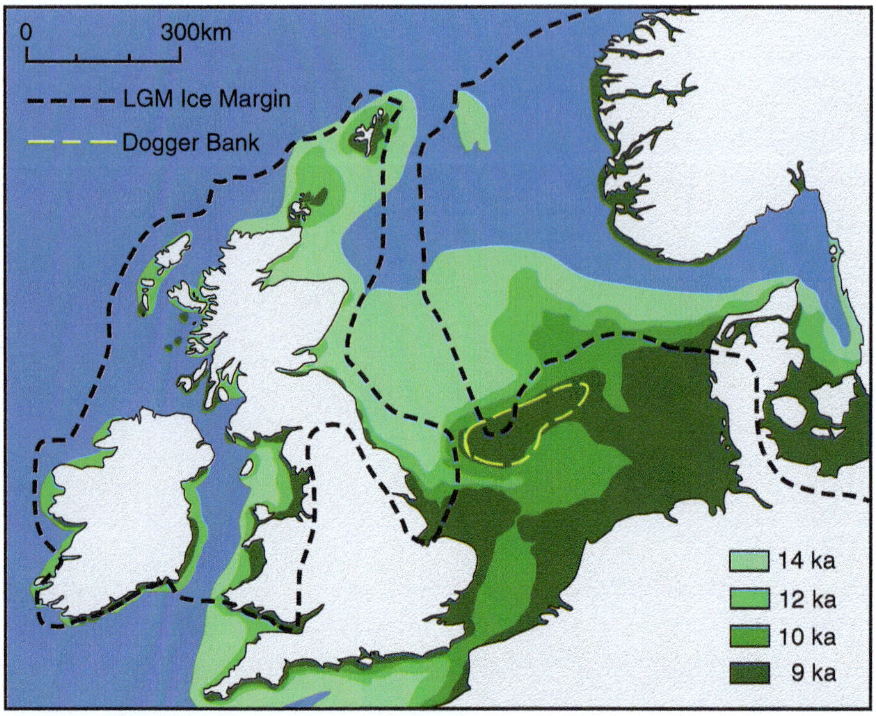

Fig. 10.2 Map demonstrating the extent of land exposed at successive stages during the progressive rise of sea level after the Last Glacial Maximum (LGM). Additional areas of land would have been exposed at the height of the LGM when sea level was at its lowest, but these areas would have been uninhabitable or inaccessible because of proximity to the ice sheets, except to the south-west of the English mainland. Also shown is the area of the Dogger Bank, which would have been exposed as a low hill in a broad coastal plain until late in the process of sea-level rise. Recovery in a fishing net of an Early Mesolithic antler artefact at the eastern end of the Dogger Bank in Danish territorial waters demonstrates that the area was visited when it was still dry land (see Bailey et al., Chap. 3, this volume, Fig. 3.4). Neolithic axes of a later date are discussed in the text. Data from Shennan and Andrews (2000); Shennan et al. (2006); Sturt et al. (2013) and Harff et al. (2017). Dates are calibrated radiocarbon years, but note that dates and coastlines are approximate and based on model predictions only weakly constrained by dated palaeoshorelines features. Lambeck, in Harff et al. (2017, Figure 2.13) gives slightly earlier dates for the separation of Doggerland. Drawing by Geoff Bailey

amount of new land exposed on the continental shelf around Great Britain would have been at least as extensive as the present-day land surface and is likely to have played a particularly significant role in the early history of human occupation. It would have provided a major pathway of contact and migration between Britain and mainland Europe and an extensive territory of coastal plains, low hills, rivers, lakes and marshlands well stocked with large mammals, plant life, fish and wildfowl, marine resources at the shore edge, and opportunities for sea travel along its now-submerged shorelines. Stone Age coastlines and coastal hinterlands should, therefore, be well represented in the underwater archaeological record.

At the same time, the extent and configuration of these submerged territories and the nature of the connection between Britain and the European continent have been subject to highly variable and dynamic palaeogeographical and palaeoenvironmental changes in response to a complex interplay of climatic and geological processes. These include variations in sea level resulting from changes in ice volume (global eustasy), hydro-isostatic and glacio-isostatic changes associated with the glacial-interglacial cycle, which caused episodic subsidence and rebound of the Earth's crust at a regional scale because of the loading and unloading of continental ice sheets, and of water masses in shallow basins, and slow tectonic subsidence of the southern North Sea basin (Westley 2017). These processes have had a major impact not only on the changing configuration of lowland territory, coastlines and land accessible to human settlement, but also on the conditions for the preservation and exposure of submerged landscapes and their associated archaeology.

In the earlier part of the Pleistocene, there was a permanent land connection between Britain and the European mainland across what is now the Dover Strait, and accumulation around the coastal rim of the palaeo-North-Sea of thick fluvial and deltaic sediments that are now submerged. At some time after about half a million years ago, when glacial meltwaters finally broke the land connection between Britain and Europe, Great Britain became isolated from the European mainland during subsequent high sea-level episodes, and access by land was only possible during periods of lower sea level, though not during the glacial maxima, when blanket cover of ice sheets would have deterred settlement (Cohen et al. 2017; Dawson et al. 2017; Farr et al. 2017).

10.3 The Archaeological Context: Submerged Coasts and Marine Resources

The earliest occupation of Britain extends back to nearly one million years (Parfitt et al. 2010), and even from this earliest period coastal areas were clearly an attraction to human populations, including landscapes that are now submerged. Direct evidence for the use of marine resources is absent but some of the earliest sites were certainly very close to their contemporaneous shorelines, notably the group of early sites in SE England: Happisburgh, dated at c. 850,000 years ago and discussed below; the 480,000-year-old site of Boxgrove, where large mammals were trapped and butchered on a coastal plain locally confined by a line of cliffs (Roberts and Parfitt 1999; Pope 2003); and Clacton on the east coast, where the tip of a 400,000 year-old wooden spear was found in fluvial peat deposits exposed by erosion on the foreshore (Allington-Jones 2015). Numerous isolated finds of Lower Palaeolithic handaxes in southern England have also been reported in the National Monuments Record (https://www.pastscape.org.uk/default.aspx, accessed 2018). These have been found on modern beaches where they are clearly eroded from the adjacent coastal cliffs or dredged up from marine sediments in river estuaries.

At the coastal cave of La Cotte de St Brelade on the island of Jersey, occupied periodically from about 238,000 years ago onwards, the remains of butchered mammoths and other large mammals were recovered (see also Billard et al., Chap. 12, this volume, for further information on this region). The La Cotte faunal remains were originally interpreted as the result of hunters driving the large mammals over the cliff edge from above the site, although the possibility of an extensive coastal plain in front of the site was envisaged at lower sea levels with the use of the ravines at the foot of the cliff for trapping animals (Callow 1986; Scott 1986). Recently, more detailed analysis of the submerged landscape has reinforced the significance of the extensive and now-submerged coastal plain as a habitat for large mammals and has identified a dissected geology of ravines and ridges on its inner margin that would have enabled Neanderthal hunters to use the complex topography to engage in ambush-hunting close to the cave (Scott et al. 2014).

Towards the Last Glacial Maximum, when much of northern Britain was covered by ice and the sea was approaching its lowest, the burial of the 'Red Lady' of Paviland in south Wales dated to c. 34,000 cal BP (Jacobi and Higham 2008) close to the southern margin of the ice sheet demonstrates that people were clearly visiting the seashore some distance away to collect marine shells for ornaments. Stable isotope signatures from the human bone remains show that they were also eating marine foods (Richards 2000).

During the Last Glacial Maximum, the British mainland was abandoned, but as the ice retreated, sites linked to late Upper Palaeolithic cultures on the European continent began to appear between about 16,000 and 12,500 years ago. In England, sites are closely linked with the Magdalenian and Federmesser industries of northern France and the Low Countries (Momber and Peeters 2017). To the north, in Scotland, a Hamburgian presence is confirmed at Howburn in the Clyde Valley (Ballin et al. 2010, 2018) while finds of stone artefacts elsewhere in Scotland show typological affinities with other northern European industries such as the Ahrensburgian and the Federmesser (Saville and Ballin 2009; Mithen et al. 2015; Ballin 2017, 2019). Most likely, these British Late-Glacial occurrences were outlying upland camps of hunting territories centred on the submerged plains of the North Sea Basin and the English Channel, as Fischer (2004, p. 35) has surmised for the Hamburgian reindeer-hunting sites of NW Germany. Some are in coastal or near-coastal locations, raising the possibility of an early maritime adaptation and a coastal colonisation of northern Britain from the now-submerged coasts of Doggerland.

Similarly, the late Upper Palaeolithic site of Hengistbury Head on the English south coast, dated at c. 12,500 years ago (Barton 1992), and the early Mesolithic site of Nab Head in Wales, dated at c. 9500 years ago (Tolan-Smith 2008), are located on the present-day coastline and clearly commanded excellent views over low-lying coastal areas and shorelines, and access to their varied resources, all of which are now submerged. Neither site, however, has direct evidence of marine food remains, perhaps because of distance from the contemporaneous shoreline.

Coastal sites with evidence of marine exploitation located above modern sea level on the modern coastline, such as the shell middens of Oronsay in the Western Isles of Scotland (Mellars 1987), are present towards the end of the Mesolithic period after the establishment of modern sea level at about its present position c. 6000 cal BP (Tolan-Smith 2008), and somewhat earlier at about 10,000 cal BP on some of the uplifted coastlines of northern Britain at sites such as Howick (Waddington 2007).

All of this indicates that coastal environments close to the contemporaneous shoreline were a prime attraction to the early inhabitants of Great Britain throughout its 1-million-year sequence, with or without the addition of marine resources. Underwater finds are therefore likely to be of great significance in filling some very large gaps in the Stone Age record, and an increasing number of finds have come to light in the past two decades (Fig. 10.1).

10.4 Conditions of Preservation and Discovery

Two different sets of factors have influenced the current state of knowledge. The first is the variable geomorphological conditions in which material is likely to have been preserved after inundation by sea-level rise and made visible and accessible to observation. The second refers to the nature of the explorations and underwater activities that have resulted in archaeological discoveries, including the impact of offshore industries.

10.4.1 Geomorphology

There is extensive discussion in the recent literature about the potential effects of offshore processes of erosion and sedimentation on the seabed and likely targets for preservation of underwater archaeological deposits around the British Isles (summarised in Flemming et al. 2017). For the present purposes in relation to currently known archaeological discoveries, two general features are of particular relevance: the nature of the intertidal zone; and the effects of tectonic uplift and subsidence.

10.4.1.1 The Intertidal Zone

Because of irregular coastlines, interconnected sea basins and variable seabed bathymetry, the tidal range around the British coast is one of the most variable in Europe and includes some of the highest tidal ranges in the world. Mean spring-tidal range varies from less than 2 m on the English south coast to 14 m in the Severn Estuary (Westley 2017). In general, tidal ranges are highest on parts of the west coast of England, lowest on the south coast of England and the south-west coasts of Scotland, and intermediate elsewhere. When combined with shallow offshore gradients, large tidal ranges can expose very extensive intertidal mudflats at low tide extending for many hundreds of metres out from the modern shoreline. Moreover, the generally gentle water-currents in these shallow areas combined with rapid accumulation of sediments, especially at the mouths of river estuaries, means that archaeological material deposited on land and subsequently inundated by sea-level rise is exposed to minimal disturbance before being covered by protective layers of peat or marine sediment.

10.4.1.2 Submerged Forests

One of the distinctive features of the British coastline, which can be observed on many beaches, both on the open coastline and in more sheltered bays and estuaries, is patches of submerged forest exposed at low tide—a variable mixture of tree stumps, fallen tree trunks and peat sediments (Fulford et al. 1997). Even on more exposed beaches, these old land surfaces can be protected for decades or centuries by accumulations of sand and remain invisible until exposed by storm action and the removal of the overlying sediments. Once they are uncovered, further erosion by waves and marine currents may expose the underlying land surface and reveal archaeological features previously sealed beneath the overlying peat. These submerged forests have been the object of scientific enquiry for over a hundred years (see Reid 1913) and the more extensive deposits have been the focus in recent decades of major research investigations into sea-level change and wetlands archaeology.

These intertidal zones have the advantage that they are easily accessible and can be inspected on foot and followed downslope at extreme low tides and even excavated without the need to go underwater, although the window of opportunity for such investigations may be as little as 1 h in the tidal cycle, and even then, only once every few months for deposits at extreme low water. For the same reason, of course, intertidal deposits are also vulnerable to the destructive effects of human activity along the shoreline including the impact of tourism, construction of roads, buildings and seashore defences, and a variety of recreational, commercial and industrial activities. Many of the features recorded earlier in the last century have disappeared or been destroyed. They are an object lesson in the delicate balance between burial under protective sediments and exposure by erosion that determines the preservation and visibility of so much of the underwater record. They also testify to the surprising richness of a zone that is all too easily assumed to have little surviving prehistoric archaeology because of the supposedly destructive impacts of shallow-water currents and waves in the surf zone. It is no surprise that the majority of the prehistoric underwater finds known from Britain occur in the intertidal zone, or that they include important finds such as wooden artefacts, dugout canoes, wooden structures, and traces of in situ cultural features and activity areas.

10.4.1.3 Uplift and Subsidence

The major process is land uplift in the North (Fig. 10.1) resulting from isostatic adjustment after the melting of the North European ice sheets of the Last Glacial (see also Jöns et al. Chap. 5, and Glorstad et al. Chap. 6, this volume, for discussion of this process). This uplift has been especially marked in parts of central Scotland. Here large areas of the mainland were depressed under the weight of the overlying ice sheet with a corresponding and compensating uplift—the forebulge—around the edge of the ice sheet. Once the ice melted, the land beneath it rebounded and the forebulge subsided. These processes proceeded at different rates and were further compounded by the hydrostatic effects of rising sea level on shallow shelf areas (Dawson et al. 2017). Because of these complex and regionally variable patterns of coastal uplift and subsidence and their interactions with eustatic sea-level rise, some parts of the Scottish coastline have been uplifted exposing Late Glacial and early postglacial shorelines that would otherwise have been submerged by sea-level rise (sometimes followed by renewed submergence), while in other areas, notably around the Outer Hebrides and the Orkney Islands, the land has been sinking, submerging shorelines and coastal landscapes as late in date as 4000 years ago (Fig. 10.3; Dawson et al. 2017).

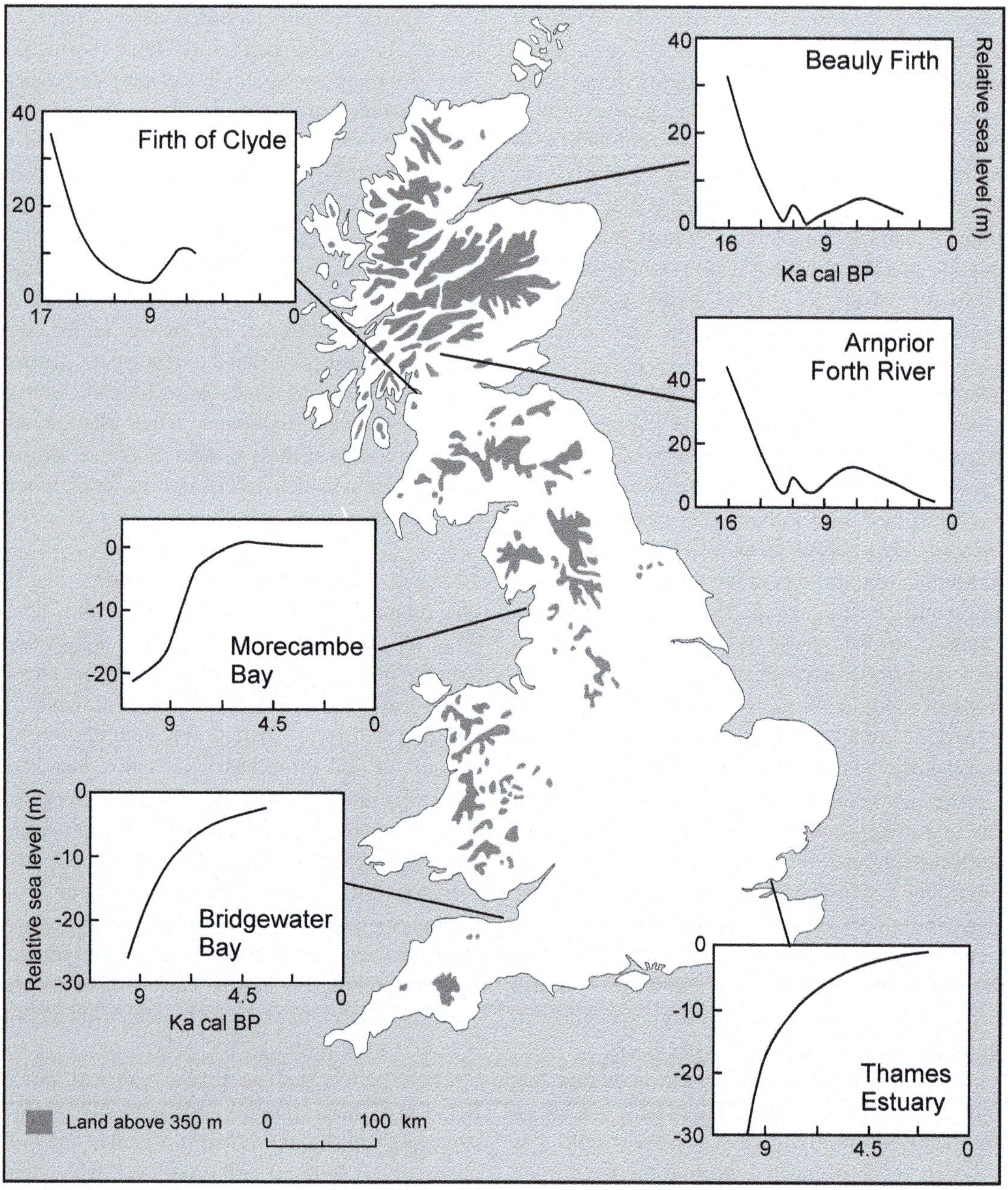

Fig. 10.3 Changes in Late Glacial and Postglacial relative sea level at selected sites in Britain. Sites in the south show a steady rise in sea level following the typical eustatic pattern, whereas the sites in Scotland show a drop in sea-level because isostatic rebound following deglaciation initially outpaced increase in ocean volume. Date scale is in calibrated radiocarbon years, and relative sea level is in metres above or below present sea level. Sea-level data modified from Lambeck (1995). Drawing by Geoff Bailey

The result is that coastal sites of relatively early date, dating back to 10,000 years ago or more, are visible on some of the shorelines of northern Britain (Waddington 2007; Hardy and Wickham-Jones 2009; Mithen et al. 2015; Momber and Peeters 2017). The converse is also the case, that sites of Neolithic date or later may now be submerged, for example around the

Orkney islands, with the potential for recovery of well-preserved organic remains in underwater deposits compared to their dryland equivalents in other regions (Bates et al. 2013).

In the south of England, the land is sinking, in part because of adjustment following the melting of the ice sheets, and in part because of tectonic processes operating on a much longer time scale. This has resulted in many drowned river valleys along the present coastline and the accumulation of thick layers of marine sediments that have preserved earlier land surfaces and archaeological finds, but also hidden them from view. Here, discoveries depend on the re-exposure of earlier sediments: by tidal scour and progressive shoreward erosion; by coastal engineering works such as harbour construction, dredging and drilling of boreholes, by fishing activity, and by activation of submarine channels that have cut through overlying marine sediments to expose earlier terrestrial land surfaces.

10.4.2 Underwater Search and Discovery, and Offshore Industry

Reporting of material along Britain's shorelines by interested members of the public and more systematic scientific investigations of intertidal material exposed on the foreshore have a long history extending well back into the last century (e.g. Warren et al. 1936). Offshore, the main contributing effect has been the chance recovery of material as a by-product of commercial activities such as trawler fishing, most famously the recovery of a Mesolithic harpoon embedded in a lump of peat recovered by the fishing vessel *Colinda* in 1932 (Godwin and Godwin 1933).

In the late 1990s, in response to the multiple threats arising from intensified coastal development and the prospect of future climate and sea-level change, a series of reports on the coastal heritage, including underwater finds, were produced to inform Shoreline Management Plans (Fulford et al. 1997; Wessex Archaeology 2007; Hazell 2008; Murphy 2014). These reports incorporate pre-existing data from the National Monuments Record as well as adding new information.

Offshore, the expansion of industrial exploitation on the seabed in recent decades associated with trawler fishing, oil and gas exploration, sand and aggregate extraction and construction of wind farms has been accompanied by legislation requiring the implementation of archaeological mitigation strategies driven by Environmental Impact Assessments and the need for official marine permits for underwater commercial work (Bicket et al. 2014; Bicket and Tizzard 2015; Sturt et al. 2017; see also Pater, Chap. 26, this volume). Related developments are the existence of a substantial body of existing geological, oceanographic and other marine data that bear indirectly on the archaeological potential of the seabed (Flemming 2004); expansion of specialist underwater units in commercial archaeological companies to service the requirement for Environmental Impact Assessments; a general shift in the intellectual climate of archaeological opinion stimulated by Coles's (1998) paper on Doggerland; and a government tax on offshore and onshore aggregate extraction initiated in 2002. The latter provided substantial funding for archaeological investigations, notably a project that drew on oil company seismic records to reconstruct submerged Stone Age landscapes covering an area of 23,000 km^2 in the North Sea (Gaffney et al. 2007, 2009; Fig. 10.4). Since 2005 the British Marine Aggregate Producers Association (BMAPA) has implemented a protocol for ensuring the reporting of underwater archaeological material brought up during commercial activities (Ransley and Sturt 2013).

10.5 Underwater Finds and the SPLASHCOS Viewer

10.5.1 Sources of Information and Criteria for Inclusion

At the time of writing, the SPLASHCOS viewer contains 258 records of underwater sites or artefacts in Britain, identified mainly from the published literature and from the online data-

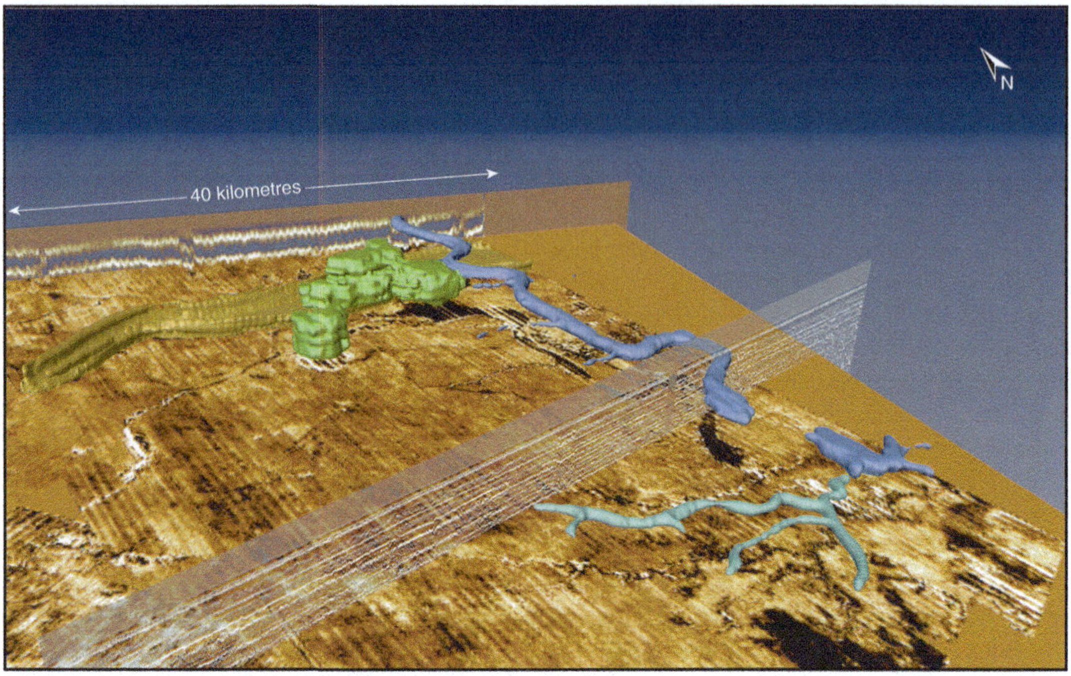

Fig. 10.4 3D reconstruction of an early Holocene meandering river system identified from seismic data beneath the bed of the North Sea. The major channels of the river system are in blue. Green indicates a marshy area. An earlier, Late Glacial, landscape is detectable beneath the Holocene features, and includes a tunnel valley, shown in gold. After Gaffney et al. (2009, figure 5.15), courtesy of Vince Gaffney

bases of archaeological sites maintained by national government agencies and other organisations (Table 10.1).

All these sources have been consulted in compiling the SPLASHCOS record. However, the online sources need to be treated critically. Most include coastal and intertidal finds indiscriminately regardless of whether they are on the landward or the seaward side of the shore margin. Moreover, for intertidal material, the recorded observations rarely pose the question—let alone provide an answer—as to whether the material found has been re-deposited in the intertidal zone because it was eroded out from terrestrial deposits on the landward side of the shore margin, washed up onto the shore from deposits located underwater, or deposited in situ and subsequently exposed by tidal erosion. For our purposes, we include only those sites or archaeological artefacts that were originally deposited on a land surface that was subsequently inundated by relative sea-level rise. Material clearly derived from adjacent cliffs or land surfaces, or where this cannot be ruled out, have been excluded.

Similarly, numerous individual finds of typologically distinctive stone tools such as Palaeolithic handaxes or Mesolithic tranchet axes have been dredged up over the years from marine sediments in the Thames estuary and other similar contexts where it is likely that the material has been displaced by erosion from the adjacent river bank. These too have been excluded. In cases of ambiguity, we have either excluded the material or relied on descriptions in the published literature. For example, an antler harpoon of Mesolithic type found on the shore at Whitby was examined in detail by Mellars (1970), who concluded from its condition that it had originated from a submerged deposit offshore and not from the adjacent shoreline, and this is included in our list.

If we were to include all the material found in the intertidal zone, our total number of find spots could easily be doubled, but for the reasons outlined above this would be a misleading figure. At

Table 10.1 Distribution of underwater and intertidal archaeological material by type of site and archaeological period

Type of site	Palaeo L-M	Palaeo Upper	Pal-Meso	Meso	Meso-Neo	Neo	Neo-Bronze	Bronze	Indet. Stone	Iron	Total N	Total %
In situ cultural features—settlement data	–	–	–	5	–	11	–		–	9	25	*9.6*
Stratified material—multiple finds	3	3	–	12	7	1	2		–	–	28	*10.8*
Unstratified material—multiple finds	1	–	–	19	9	12	–		–	–	41	*15.9*
Single unstratified find	5	5	33	63	15	9	2		21	–	153	*59.3*
Votive deposits	–	–	–	–	–	–	–	11	–	–	11	*4.3*
Total N	**9**	**8**	**33**	**99**	**31**	**33**	**4**	**11**	**21**	**9**	**258**	***99.9***
%	***3.5***	***3.5***	***12.7***	***38.1***	***11.9***	***13.1***	***1.5***	***4.2***	***8.1***	***3.5***		***100.1***
Presence of wood	–	–	–	7	7	13	2		–	9	**38**	***14.7***

L-M: Lower and Middle Palaeolithic; Pal-Meso, Meso-Neo and Neo-Bronze indicates that material is either present from both periods or is not sufficiently diagnostic to assign more precisely to one or the other. Numbers of occurrences with wooden artefacts or evidence of worked wood are also shown. Data are from the SPLASHCOS Viewer at http://splashcos-viewer.eu/, and this in turn is based on a variety of online records. For England, records of archaeological sites are maintained by Historic England, and the database is currently known as the National Record of Historic England, NRHE (formerly the National Monuments Record, NMR) (https://www.pastscape.org.uk/). Historic England also maintains the Intertidal and Coastal Peat Database (https://historicengland.org.uk/research/current/heritage-science/intertidal-peat-database/) and the Rapid Coastal Zone Assessments (e.g. https://historicengland.org.uk/images-books/publications/nwrcza/). In Wales, the equivalent is the National Monuments Record of Wales (NMRW) maintained by The Royal Commission on the Ancient and Historical Monuments of Wales (http://www.coflein.gov.uk/), and in Scotland, Canmore (https://canmore.org.uk/), maintained by Historic Environment Scotland. Note that these are the current names and websites of these government agencies but that both are liable to change on a regular basis. The Archaeological Data Service (ADS) (http://archaeologydataservice.ac.uk/) also provides a permanent archive of some of these records and useful links to other relevant reports such as Wessex Archaeology's (2007) 'Artefacts from the Sea' database. All these sources have been consulted in compiling data for the SPLASHCOS Viewer the same time, it is likely that our reduced list is probably a minimum figure. Where intensive investigation has been carried out, for example along the 6-km stretch of shoreline at Wootton Quarr on the north coast of the Isle of Wight, there are multiple occurrences of stone tools, ceramics, wooden trackways and other cultural features, amounting in total to 180 occurrences covering a time range from the late Mesolithic through to the Medieval period (Tomalin et al. 2012). In effect, the shoreline is virtually a single archaeological site forming a spatio-temporal continuum in a wetlands landscape.

The other issue to note is the upper chronological limit for what we include. The primary interest of this volume is, broadly speaking, the period before the establishment of modern sea level at about 6000–5500 years ago. However, some coastlines in Britain, as elsewhere in Europe, have continued to sink, or have been exposed to landward erosion that has resulted in inundation of former land surfaces and their submergence beneath layers of marine sediment. Some of these later sites are relevant to our interests in demonstrating what happens to archaeological settlements and material remains when they are submerged, and we have included some of these examples.

However, deciding where to place the chronological boundary is necessarily somewhat arbitrary. Many of the estuarine mudflats and saltmarshes on the present-day coastline contain more deeply stratified sequences of sub-surface peat or land surfaces with prehistoric archaeological material that fit our criteria. However, many of these coastal zones continued in use into later prehistoric and historical periods right up to the present-day as coastal land surfaces were progressively transformed into freshwater peatlands and saltmarshes. Indeed, these coastal wetlands contain a rich harvest of material culture especially items made of wood including wooden trackways, fish traps, boats, animal byres and other wooden structures, which demonstrates their continuing attractiveness as sources of food, raw materials and seasonal animal grazing at all periods (Bell 2013). Most of these finds, however, refer to the exploitation of coastal wetlands or saltmarshes in approximately their present position in relation to modern sea level, and do not represent underwater finds in the sense used in this volume. Unless they have demonstrably been inundated by sea-level rise or marine sediments after abandonment, we exclude such finds, though we recognise their significance as evidence of wetland exploitation in more recent millennia. Notable examples that we include are the timber henges off the coast of Norfolk (Brennand and Taylor 2003) and the Iron Age marine crannogs of Scotland (Hale 2004), all of which have been partially inundated since they were in use. We have also included a small number of Bronze Age finds, where the range of artefacts present indicates material that has been deliberately deposited offshore and under water as a votive offering (Samson 2006). Such votive deposits are also common in rivers and lakes and it is possible that Neolithic and earlier artefacts, found on the seabed but excluded from our list because of the ambiguity of their provenance, may fall into the votive category, notably the Neolithic polished axes recovered from the Dogger Bank and discussed below.

10.5.2 Archaeological Overview

Of our 258 records, 153 (59%) are unstratified single finds, while only 53 records (20%) are of multiple finds in stratified context or with in situ cultural features (Table 10.1). In terms of chronology, the material ranges from the Lower Palaeolithic through to the Iron Age. The largest group are Mesolithic finds, comprising 99 records (38%), followed by finds of Late Palaeolithic and/or Mesolithic date and Mesolithic and/or Neolithic—33 (13%) and 31 (12%), respectively. Just five Mesolithic and 11 Neolithic records can be associated with in situ settlement-material, plus the nine Scottish crannogs. However, this may be a considerable underestimate since many finds labelled as 'collective finds of stratified material' may represent in situ cultural features but are too incomplete to judge their context with certainty. Wooden artefacts are preserved at 38 sites (15%), mostly of Mesolithic or Neolithic

Table 10.2 Distribution of underwater and intertidal archaeological material by type of site and depth of recovery

	Intertidal		Offshore	Offshore	Offshore	Total	
Type of site	Indet	SF	≤ 12 m	18–35 m	140–145 m	N	%
In situ cultural features—settlement data	19	8	1	–	–	**28**	*10.9*
Stratified material—multiple finds	7	15	–	2	–	**24**	*9.3*
Stratified material—single find	–	–	–	–	1	**1**	*0.4*
Unstratified material—multiple finds	3	27	1	1	–	**32**	*12.4*
Single unstratified find	87	3	67	5	–	**162**	*62.8*
Votive finds	–	–	10	1	–	**11**	*4.3*
Total	**116**	**53**	**79**	**9**	**1**	**258**	*100.1*
%	*45*	*20.5*	*30.6*	*3.5*	*0.4*	*100*	

Intertidal refers to material observed at low tide, offshore to finds recovered by dredging or coring. SF: Submerged forest. Indet: intertidal deposits not certainly associated with submerged forests, may include peat deposits. Data from http://splashcos-viewer.eu/

age. These statistics with their dominance of Mesolithic and Neolithic artefacts reflect the ease of discovery of material that was inundated only towards the end of postglacial sea-level rise and is therefore exposed in the present-day intertidal zone or in shallow water down to depths of about 10 m.

A further breakdown of the records shows that the majority, 169 (66%), refer to material in the intertidal zone (Table 10.2). In addition, approximately half these intertidal records refer to finds in or beneath submerged forests and this category includes the best preserved of the prehistoric sites including most of the examples of in situ cultural features or settlement data. A further 79 (31%) refer to underwater finds in a depth range down to about 12 m, with just nine finds (4%)—though a very important and informative group—in the 18–35 m depth range, and one outlier at 140–145 m, namely the artefact recovered from coring on the Viking-Bergen bank and associated with what are interpreted as beach deposits.

10.6 Sites in the Intertidal Zone

Many records in this category are individual artefacts or finds of material for which only fleeting observations were possible at the time of discovery, where questions of artefact descriptions, stratigraphic relationships and chronology remain unresolved. This is especially the case for submerged forests, which were often exposed by storms for a limited period, accessible only at low tide and then covered over again by shifting sediments. Much of the archaeological material is fragmentary, of uncertain provenance or date, or poorly reported. Some records suggest finds of major potential significance which remain tantalising in their incompleteness. Thurlestone Sands near Torbay in south Devon is a good example. Here submerged forests were observed after exposure by a hurricane-force gale in 1866, and again in the same area of the beach after a storm in 1923 (https://www.pastscape.org.uk/hob.aspx?hob_id=441056). The finds reported in 1923 included tree stumps, fallen tree trunks, three of which were recovered from the subsoil beneath the forest layer, a Mesolithic tranchet axe and a perforated mace head. One of the trunks was removed, cleaned and found to have been shaped in the form of a dugout canoe with evidence of tool marks. The find was not conserved and has since been lost.

A similar record comes from Hartlepool Bay, NE England, where submerged peats exposed at various points along the foreshore are reported to have yielded a Neolithic human skeleton, Mesolithic flints, a red deer skeleton with a wooden spear embedded in the rib cage, wooden stakes, wattle hurdling and a possible fish weir, with radiocarbon dates indicating a time range of 6000 to 5500 cal BP for at least part of this material (Trechmann 1936; Buglass 1994; Waugham et al. 2005). Isolated finds of human material have been reported elsewhere including a human skeleton from Liverpool Bay beneath peats radiocarbon-dated between 7500 and 7000 cal BP (Huddart et al. 1999).

An important find at Titchwell on the Norfolk coast is a Late Glacial long-blade industry found in situ on a palaeosol exposed at extreme low water 500 m offshore from the modern beach. The old land surface had been exposed by erosion of an overlying peat horizon dated at 9000 cal BP (Wymer and Robins 1994). The industry is of Creswellian type, dated by association with finds elsewhere at 13,000 to 12,000 cal BP. Over 600 artefacts were found in situ along with burnt flints, charcoal and three bovid teeth. A nearby channel probably indicates a contemporaneous stream at a time when sea level was some 40 m below present and the nearest coastline more than 100 km distant.

Other notable finds are the Mesolithic shell midden stratified beneath a submerged forest and later exposed at low tide at Westward Ho! in SW England (Balaam et al. 1987), and extensive traces of Mesolithic and Neolithic settlements and activity areas with flint tools and wooden artefacts in the Severn Estuary (Bell 2007, and below), at Wootton Quarr on the Isle of Wight (Tomalin et al. 2012), and in the estuaries of the Rivers Thames, Blackwater and Crouch on the Essex coast (Wilkinson and Murphy 1986, 1995; Sidell and Haughey 2007).

On the Essex coast, important Neolithic settlements such as the Stumble and Rolls Farm were occupied when sea level was slightly lower than present and were subsequently inundated and covered by marine sediments and later by coastal peat. As Wilkinson and Murphy (1995, p. x, p. 3) have observed, the original land surface sealed beneath the marine sediment preserves extensive spatial distributions of various features such as pits, postholes and other cultural debris that would have little chance of surviving the effects of plough damage to Neolithic sites further inland, where the material has been reduced to amorphous lithic scatters. On the other hand, wooden artefacts are not preserved at these sites because they were not inundated until sometime after abandonment. Interestingly, the peat deposits that later accumulated above these Neolithic layers do contain organic remains of Bronze Age and later date, but these represent the use of a wetland landscape that was not subsequently inundated by further sea-level rise.

Other late-dated examples are the submerged Bronze Age terrestrial landscape recorded at Langstone Harbour on the south coast with urn burials and burnt mounds (Allen and Gardiner 2000), Early Bronze Age circular wooden henges c. 10 m in diameter of possible ritual significance at Holme-next-the-Sea on the Norfolk coast (Brennand and Taylor 2003), and Bronze Age barrows and a possible henge on the Humber River in the northeast of England (Van de Noort 2004).

In Scotland, a chambered stone cairn at Geirisclett on the island of North Uist in the North-west has its base at high water mark in the intertidal zone (Benjamin et al. 2014, Fig. 6), and a series of marine crannogs in the Beauly Firth (in the north-east) and the Firth of Clyde (in the south-west) are shoreline settlements that have been inundated by relative sea-level rise (Fig. 10.5; Hale 2004; Jacobsson et al. 2017).

One of the difficulties of systematic work in the intertidal zone is the brief windows of opportunity for detailed work allowed at low tide, and the likelihood of progressive and rapid destruction of what is visible by continued erosion. Nevertheless, important large-scale, multidisciplinary projects have addressed this challenge and generated important results with a wealth of features including numerous remains of wooden artefacts, especially on the Severn Estuary, at Wootton Quarr on the Isle of Wight, and on the Essex Coast, as detailed above. These projects have generally taken a holistic approach to the palaeoenvironmental and archaeological examination of these wetland landscapes through time and across the boundary between dry land, coastal saltmarshes and the intertidal zone, occasionally extending to offshore and underwater investigations as at Wootton Quarr.

Fig. 10.5 Aerial photograph of Dumbuck crannog, dated at c. 2500 cal BP, showing the remains of a circular timber platform 35 m in diameter in the intertidal zone of the Firth of Clyde, near Glasgow. The platform of alder wood is surrounded by piles of oak, with a stone breakwater and timber-lined dock. Excavations in the late nineteenth century recovered querns, worked wood, and a 12 m-long log boat. This is one of four crannogs in the Firth of Clyde with another four in the Beauly Firth and one on the West coast. Note that crannogs are also present in Scottish freshwater lake-edge settings, but these are not further discussed here. © Crown Copyright Historic Environment Scotland SC970218.

10.7 Intertidal Footprints

In a number of modern intertidal locations, fossilised human and animal footprints have been preserved and exposed. They were impressed in fine-grained sediments on mudflats that originally existed along ancient shorelines. Typically, these sediments build up rapidly as a sequence of thin, stratified layers, representing seasonal alternations between fine-grained sediments accumulated during summer and coarse-grained sediments in winter. Rapid sediment-accumulation and subsequent natural consolidation ensures preservation of delicate features such as footprints, which are then exposed again in the modern intertidal zone by shoreward erosion, often as a series of stepped platforms with overlying layers of laminated sediments progressively peeled away to reveal earlier material. Most examples are from estuarine mudflats and cannot have been far from their contemporaneous coastlines. In such environments, the firmness of the sediment, gentle water conditions and rapid accumulation of sediment would have provided the best conditions for the preservation of surface imprints. The footprints themselves give details of age, sex, stature and physical condition, and the combination of footprints in trails suggests different types of activities.

The oldest examples are at Happisburgh, dated at 0.78–1 million years (Ashton et al. 2014). They represent the oldest footprints outside Africa, and some of the earliest evidence for human occupation in Britain (Fig. 10.6). They were formed on the intertidal mudflat of a river estuary. Of 155 outlines, 49 were clear enough to indicate the direction of travel, and 12 were complete enough for detailed analysis, indicating at least five individuals including adults and children (Fig. 10.7).

The most extensive exposures of such finds are of much later date (broadly Mesolithic or Neolithic), at Formby Point (Huddart et al. 1999) with over 145 footprints (Fig. 10.8) and Goldcliff East, with over 270 human footprints, discussed in more detail below (Bell 2007). Many child footprints are present in both cases. In both areas, there are also many footprints of large animals such as deer (Fig. 10.9), aurochs and wolf or dog, and shore birds such as crane and oystercatcher. The Goldcliff footprints are dated as early as 7500 cal BP. At Formby, there are two stratigraphically separated groups of prints. One is dated c. 5750 cal BP, and can, thus, be associated with Neolithic activity in the area, while the earlier set could be Mesolithic. At this latter site, one arthritic individual and a pregnant woman have been inferred. The trails suggest slower-paced activities by the women and children such as shellfish gathering, and a longer stride length for the men and imprints suggesting running after

Fig. 10.6 General view of laminated silts in the Happisburgh intertidal zone with footprints of early humans. After Ashton et al. 2014

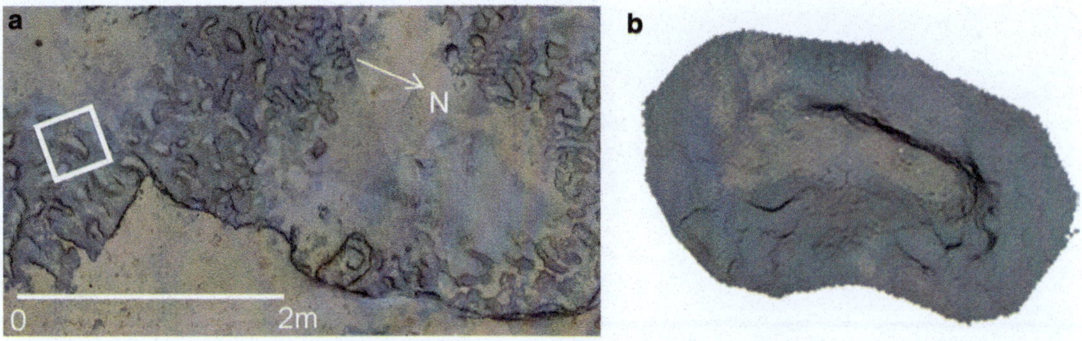

Fig. 10.7 (**a**) Plan view of the Happisburgh footprints. White rectangle indicates location of individual footprint; (**b**) close up of individual footprint. After Ashton et al. 2014. Photo of 3 (**b**) by Sarah Duffy (University of York), courtesy of Nick Ashton

Fig. 10.8 Print in 6–4000-year-old mudflats at Formby Point of a human foot. Image courtesy of Silvia Gonzalez and David Huddart

Fig. 10.9 Prints of red deer hooves at Formby. Note the difference between the deeply impressed prehistoric hoof prints and the modern boot prints in the superficial sediments. Image courtesy of Silvia Gonzalez and David Huddart

hunted game. Another large group of Mesolithic human and animal footprints dated at about 7300 cal BP has recently been reported at Low Hauxley (Eadie and Waddington 2013).

10.7.1 Goldcliff East

Goldcliff East was a bedrock island with settlements on its flanks which were slowly and progressively engulfed by the sea and buried during the period 7500 to 6800 cal BP. The finds are in the intertidal zone of the Severn River estuary where archaeological investigations have revealed traces of a succession of settlements and activity areas (Bell 2007). The most striking evidence is in the form of 270 human footprints that criss-cross the ancient mudflats and have been preserved within intercalated layers of sediment (Fig. 10.10; Scales 2007).

Among the human footprints, 21 individuals have been identified, providing tangible evidence for interactions between group members and patterns of movement between activity areas. Of these, four were adults (over 14 years old), seven were 'sub-adults' (11–14), two were older children (7–11), while the majority of the foot tracks (67%) were made by eight young (3–6-year-old) children. The best preserved occurred in fine-grained sediments indicating activity at the height

Fig. 10.10 General view of footprint tracks at Goldcliff East, Site C. Photo by Eddie Sacre

of summer. Some footprints form trails indicating the movements of a group of people across the mudflats (Figs. 10.11 and 10.12). This high proportion of child footprints highlights the activities of group members who are usually invisible in the archaeological record and are here seen to be fully engaged with the life of the community from an early age.

There are also 149 recorded mammal tracks, mostly of red deer, but including a small number of roe deer, aurochs and a canid, probably a dog. Over 70 bird prints include cranes, grey herons, gulls, oystercatchers and terns.

The artefacts include lithics and a small number of pieces of worked antler and bone. Worked wooden artefacts include cut pieces of roundwood (diameter < 30 mm) and other fragments that may represent pointed stakes, seven probable hand-held tools of uncertain function and possible fragments of fish weirs (Brunning 2007).

Charred seeds, fruits and nuts, faunal remains of mammals and fish, and pollen and insect evidence of environmental conditions and diet are all well preserved in the anaerobic sediments. These finds indicate exploitation in most seasons of the year but especially in late summer and autumn, and provide evidence for the use of fire to manipulate the local vegetation, probably to increase the abundance of woodland-edge plant resources

10.8 Underwater Sites

10.8.1 Bouldnor Cliff: A Submerged Mesolithic Site

This is at present the most intensively explored underwater Stone Age site in Britain, located in southern England offshore of the northern coastline of the Isle of Wight. Here traces of human activity in a former inland environment are preserved in and below a peat layer dated to c. 8100 cal BP with Mesolithic stone tools showing affinities with France and a large collection (some 80 pieces) of worked wood (Figs. 10.13 and 10.14). An unusual find demonstrating the excellent conditions of preservation is a piece of twine or string made from plant fibres (Fig. 10.15).

The site lies 12 m below mean sea level and is associated with a drowned forest. The submerged landscape is a 1 km-long bench eroding out from near the foot of a cliff. Five distinct locations of archaeological remains have been identified so far, associated with a variety of environmental features including streams, sand bars, woodland and fen. Above the archaeological layer, there are estuarine silts that filled the pre-existing valley during the later stages of sea-level rise (Momber et al. 2011; Momber and Peeters 2017).

The wooden material includes oak timbers worked by tangential splitting, a technique not previously recorded in Britain earlier than the Neolithic period some 2500 years later (Taylor 2011). Some of these timbers appear to represent the edge of a platform or collapsed structure.

Extraction of ancient DNA from sediments (sedaDNA) in a sealed context indicates the pres-

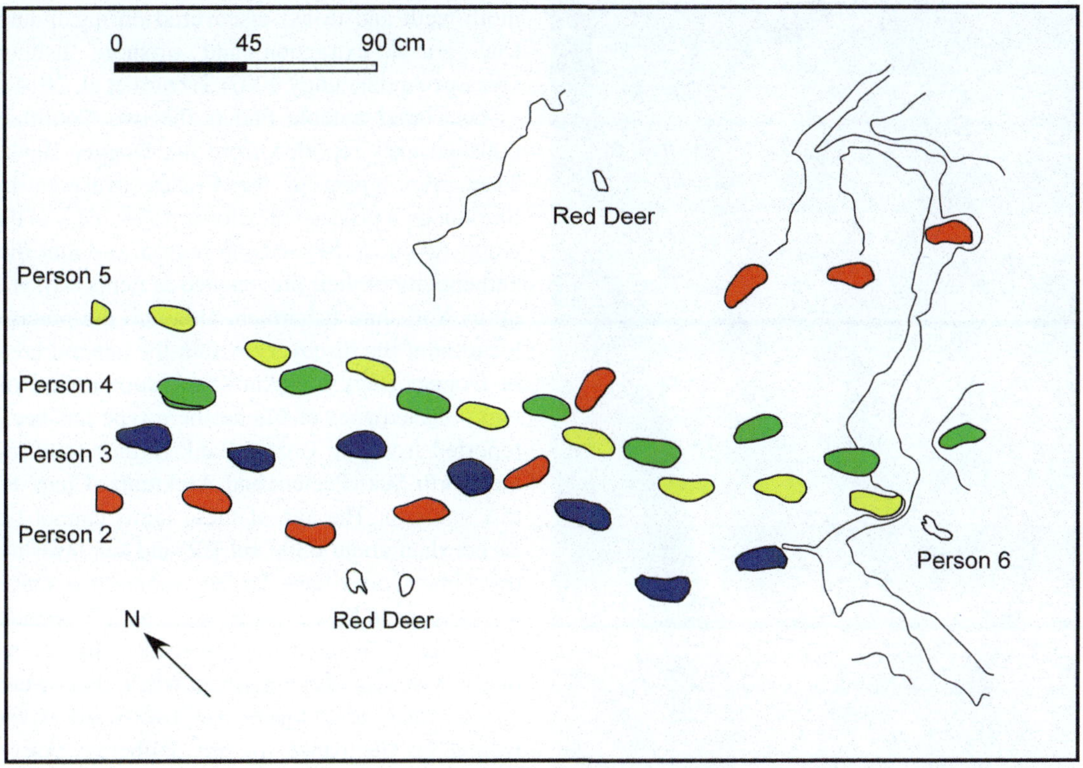

Fig. 10.11 Goldcliff East, Site E. Plan of the trails of Persons 2–5. The footprint of Person 6 is on a lower lamination. Also shown are hoof prints of red deer. After Scales 2007

ence of cultivated einkorn wheat 2000 years before agriculture is conventionally supposed to have reached Britain (Smith et al. 2015). If confirmed, this will reinforce the evidence from the wooden artefacts for the precocious arrival of 'Neolithic' traits at a time when sea levels were lower than present, prime territory in coastal locations that are now submerged was still available, and connections with the European mainland were still easily facilitated along coastlines that connected southern England with northern France (see also Anderson-Whymark et al. 2015).

10.8.2 Deep-Water Finds in the North Sea

Isolated finds have been recovered during trawler-fishing and industrial activities in the British sector of the North Sea. One of the most famous is the barbed antler harpoon from the Leman and Ower Banks (Fig. 10.16). When brought up in the net of a trawler boat in 1932, this excellently preserved item was fixed in a lump of peat, which was originally interpreted as an item lost while fishing along the edge of a now-submerged lake (Godwin and Godwin 1933). The peat was originally dated at 7500 cal BP (8422±170) (Godwin and Willis 1959), but direct AMS (Accelerator Mass Spectrometry) dating of the specimen itself later returned a date of 13,500 cal BP (11,740±150) (Bonsall and Smith 1989; Hedges et al. 1990). A similar find of a harpoon, though less well known, was brought up in c. 1950 in a fishing net 2 km offshore of the Norfolk coast near Trimingham.

Other finds include a worked flint recovered from coring of the seabed at a water depth of −143 m during oil exploration on the Viking Bank in the northern sector of the North Sea (Long et al. 1986), and worked flint and remains of fauna (predominantly mammal bones, mam-

Fig. 10.12 Well-preserved footprint of Person 6 at Goldcliff East, Site E. Photo by Eddie Sacre

moth teeth and tusks) discovered during industrial gravel extraction and offshore fishing (Wessex Archaeology 2007; Tizzard et al. 2014).

One other notable find is the two Neolithic polished axes reported from the Dogger Bank. They were given to the Craven Museum in Skipton by a private collector in 1934 and Van de Noort (2014, p. 59) sees no reason to doubt the authenticity of their provenance as items brought up by Yorkshire fishermen, although the precise location of the discovery within the general area of Dogger Bank is unknown. A similar find of two polished axes of Michelsberg type has been reported from the Brown Bank further south in the North Sea (Peeters and Amkreutz, Chap. 8, this volume). The age of these items cannot be earlier than about 6300 cal BP, and sea level by that time would have largely submerged these locations or at best exposed them as low islands at low tide. It is unclear whether these finds represent visits to these islands when they were above water, are objects lost overboard when fishing on the banks, or are deliberate votive offerings deposited in the sea in the manner more commonly recognised in the Bronze Age (see above). At any rate, they are evidence for sea journeys over considerable distances in the southern North Sea at this time.

Fig. 10.13 Worked flints recovered from submerged peat deposits at Bouldnor Cliff. Scale in centimetres. Photo by Maritime Archaeology Trust

Fig. 10.14 Piece of worked timber from Bouldnor Cliff, showing the freshly exposed condition of the wood with deep cut marks, immediately after removal from the surrounding peat deposits. Photo by Maritime Archaeology Trust

10.8.3 Area 240: Palaeolithic Artefacts in Their Palaeoenvironmental Context

This site was first identified when flint artefacts and faunal remains were found in aggregate stockpiles at a harbour in the Netherlands. Further monitoring (Fig. 10.17) established that the artefacts were dredged from a specific locale within marine aggregate license Area 240, 11 km off the English coast, at depths of 18 to 35 m. Geophysical survey and coring were used to obtain further information on the sediments and landforms associated with the artefacts and their date. Additional samples of bulk sediment were removed from the seabed using a combination of a remotely operated underwater camera for site selection, a beam trawl and a clamshell grab. This resulted in the recovery of additional flint artefacts and faunal remains and established the stratigraphic relationship between the archaeological finds and the dated sediment sequence (Tizzard et al. 2011, 2014, 2015; Bicket et al. 2014; see also Pater, Chap. 26, this volume).

More than 120 specimens, including handaxes (Fig. 10.18) and Early Middle Palaeolithic Levallois flakes, have now been recovered from this locality. Most of the artefacts are associated with stratified floodplain deposits of a river estuary dated to 250,000–200,000 BP, an age that is consistent with well-dated, technologically similar artefact assemblages found elsewhere in Britain and Continental Europe. Around 20% of the assemblage exhibits Levallois technology further supporting a post-300,000 date for some or all the material. Some of the faunal material (e.g. Fig. 10.19) can be associated with earlier channel deposits dating back to more than 500,000 BP, and a smaller group of artefacts occurs in later sediment infills dating to 40,000–30,000 BP.

This early Palaeolithic assemblage has survived multiple cycles of sea-level change during which the area has been exposed and submerged several times. The results of the archaeological fieldwork indicate that many of the Stone Age items are in their original place of deposition. In addition, they demonstrate that it is possible to recover archaeological material and contextual environmental information from deeply submerged sites far off the present coast. This is the only such site identified in the North Sea to date and represents one of the oldest submerged prehistoric sites known anywhere in the world. The material highlights the possibility of further discoveries especially linked to inundated Quaternary floodplains and river systems, which are common around northwest European coasts.

Fig. 10.15 Close-up of a piece of string from Bouldnor Cliff made by twisting together of plant fibres. It is about 10 cm in length, with a diameter of 1–2 mm. The fibres could be from nettle, wild clematis, blackberry or honeysuckle. Materials were identified by David Robinson. Photo by Maritime Archaeology Trust

Fig. 10.17 Archaeologist Andrew Bicket holding a handaxe recovered during monitoring work carried out on behalf of Hanson Aggregates Marine Limited on board one of their vessels. ©Wessex Archaeology

Fig. 10.16 The barbed point made of antler recovered in 1932 by the Colinda fishing vessel. Length of specimen is 219 mm. Found at Leman and Ower Banks. © Norwich Castle Museum and Art Gallery. Reproduced by permission

Fig. 10.18 Bifacially worked handaxe from A240. This is one of over 30 examples of this characteristic artefact type from the Area 240 locality dated to 250–200,000 years ago. Many, like this one, have sharp edges and are in mint condition, indicating that they have remained undisturbed in their original position. Scale in centimetres. © Wessex Archaeology

Fig. 10.19 Tooth of a woolly mammoth recovered from the A240 site. Many fossilised remains of terrestrial mammals have been recovered, including bones of horse, bison and reindeer, most of which are dated to about 500,000 years ago. Scale in centimetres. © Wessex Archaeology

10.9 Discussion

The British underwater material covers a wide range of ages, water depths and types of material evidence, sometimes with excellent conditions of preservation. Intertidal finds dominate numerically and some of the best evidence is associated with or sealed beneath submerged forests or peat deposits that are exposed at low tide, in some cases at considerable depth and distance offshore. These include settlement and activity areas, often with preservation of worked wood, and human and animal footprints. Some of these intertidal finds, exposed only at extreme low spring tides, are further offshore, at a greater depth below mean sea level, and of earlier date, than many of the famous, permanently submerged underwater sites of the Baltic (See Bailey et al. Chap. 3, this volume; Nilsson et al., Chap. 4, this volume; Jöns et al., Chap. 5, this volume) and suggest an untapped potential for new research.

In considering this material as a whole, two sorts of questions come to the fore. The first is the nature of the strategies that can be adopted to achieve purposeful investigation and discovery of new underwater archaeological material, especially in more deeply submerged locations. The second question is what sorts of new information might be obtained from underwater investigation that cannot be found on land, and what new themes such information opens up for investigation into the organisation and development of earlier prehistoric societies. The two types of questions are closely linked; without the intellectual justification for new field investigations, there would be little motivation to undertake them and little appetite to develop new methods and marshal the necessary resources. The currently available evidence provides significant pointers as to the direction of future investigations.

10.9.1 Research Strategies

Taking first the question of how to systematise the search for new information, it is important to emphasise that all the finds reported here have been discovered by chance exposures or dredged up by fishing and industrial activity. Most are in the intertidal zone or in very shallow water, where the age of the material is relatively young (mostly Holocene or Late Glacial). This is also the zone where the chances of discovery are highest, but where the risks of destruction are also very high. The fact that so much evidence, some of it very old as in the case of the Happisburgh footprints, has been exposed in such a geologically active zone over many centuries and even millennia is notable. It demonstrates that archaeological evidence is not necessarily doomed to destruction in the early stages of marine inundation, when the destructive forces of wave action in the surf zone and shallow-water currents are at their greatest. As always in dealing with underwater preservation, local factors are paramount, including minor variations in topography, shoreline configuration and exposure to wind and waves. Accumulation of marine sediments or layers of peat that provide protective cover is clearly a key factor that can contribute to preservation of old land surfaces, regardless of the erosive power of waves and currents. Discovery, of course, depends on intermittent exposure of the underlying deposits by natural erosion or human activities. Almost by definition, such sites when first discovered are already in the process of being exposed to destructive forces. If allowed to continue unabated, continued erosion will lead to further destruction and loss unless the material is covered up again by further accumulation of protective sediments.

The challenge then is to work out how to search for new archaeological material, rather than relying on chance discoveries, and especially how to deal with the challenge of discovering archaeological finds in more deeply submerged areas. New research is currently evolving along two separate lines of enquiry. One is to develop improved methods of mapping and remote sensing that can give detail to the nature of the now-submerged terrestrial landscape and identify target features worth investigating more closely with remotely operated vehicles and cameras, coring and diving. A good example of new work along these lines is the new mapping of Doggerland (Gaffney et al. 2017), in effect an extension of the earlier landscape reconstructions pioneered by the Gaffney team using seismic records from the hydrocarbon industry, but with acoustic techniques more appropriate to the shallower deposits of the seabed that inform on the recently submerged human landscape, followed up by coring of sediments in target areas identified by the mapping. Whether this will identify archaeological materials remains to be seen but the recovery of sedaDNA from sealed terrestrial contexts will add new information on the terrestrial plant and animal life present before submergence.

The other strategy is to work from known finds recovered by chance exposure, trawler-fishing or other industrial activity, and use these to target areas worth more detailed investigation. The success of such an approach has already been indicated by the A240 finds discussed above. Bynoe et al. (2016) draw attention to the detailed archive of records that can help to pinpoint the location of faunal remains recovered from the seabed by trawler fishing. As the A240 discoveries show, if materials disturbed on the seabed and recovered during the course of industrial operations can be traced back to their original position, this may indicate more deeply embedded material in stratified sediments with opportunities for dating, palaeoenvironmental analysis and preservation of organic materials. There is no doubt that large quantities of faunal remains and artefacts are preserved on the seabed of the North Sea, and every reason to suppose that more awaits discovery, with new opportunities for systematic retrieval.

Both strategies are currently being focussed on deeper areas of the submerged shelf and hence on landscapes that extend further back in time and to environmental settings quite different from those currently exposed near the present-day shoreline. Potentially, these offer a much greater payoff in terms of new information. But they also pose greater logistical challenges and costs than working in shallow water. Equally, there is no reason why similar strategies should not be applied in shallower water, with simpler technical requirements and the prospect of recovering significant new data. The existing evidence from intertidal submerged forests and peat deposits indicates that these increase in age with increasing distance and depth from the modern shoreline and that this trend continues offshore. It should be possible through acoustic survey and diver inspection to follow these features into deeper water and to search for channels or other features that have exposed earlier terrestrial deposits beneath the overlying marine sediment, as is the case at Bouldnor Cliff. All the indications are that extensive areas of submerged forest and associated land surfaces exist at greater depth and may be in a less disturbed condition than those known from the present-day intertidal zone. The feasibility of underwater exploration to extend investigations of known intertidal features to the offshore zone has already been demonstrated in the Wootton Quarr project (Edwards and Dix 2012).

Many of the intertidal finds reported in this chapter were found low in the intertidal zone and were investigated on foot during the very short windows of opportunity offered by extreme low tides. An alternative strategy in such circumstances would be to investigate these deposits when covered by the sea at high tides, using well-established techniques of diver inspection, survey and excavation. Prediction of target areas for closer investigation using the Danish fishing site model, suitably adapted to English coastal geomorphology and tidal conditions, is another possibility (Hall 2014).

10.9.2 Research Themes

Turning to substantive themes of early development that are brought uniquely into focus by the underwater evidence, four stand out. The first is the quality of preservation of organic materials and other evidence that would not normally be recovered, or only rarely so, from terrestrial deposits, and their potential to push back in time technological and economic innovations traditionally associated with the Neolithic or later periods. Wooden artefacts are especially well represented and indicate economically significant activities such as building structures, fish traps, landing stages for water craft, and sometimes remains of the boats themselves. Most of the British finds are relatively late in date (Neolithic or Bronze Age) and perhaps occasion relatively little surprise given conventional expectations about the nature of technological progress. But more deeply submerged sites take the story further back in time. At Bouldnor Cliff, for example, there is evidence of worked timbers revealing techniques of wood working and the presence of cultivated cereals some 2000 years earlier than the earliest evidence currently known from sites on land.

It should be emphasised, however, that submergence by marine inundation does not guarantee the survival of wood or other organic materials. If the archaeological site and its artefacts had been exposed to subaerial weathering for long enough before being covered by marine sediments, then organic materials such as wood would have disappeared. Rapid inundation and protective cover of anaerobic sediments is the key to the survival of such materials. Conversely, even without organic preservation, other features such as pits and postholes may be preserved under a protective cover of marine sediments, as noted above on the Essex coast.

A second theme is the deeper history of economic adaptations and especially those associated with the exploitation of marine resources that might be revealed at coastal sites that are now deeply submerged. Conventional archaeological evidence in the form of food remains of marine mollusc shells or bones of fish and sea mammals is not yet known from underwater sites, except for occasional hints from sites on the present-day coast or in the intertidal zone, mainly because the underwater settlement sites where such remains might be expected have not yet been discovered. Artefacts that can be uniquely associated with exploitation of aquatic resources, such as the Leman and Ower harpoon, are rare and at best ambiguous with respect to their function. The Westward Ho! shell midden provides a hint that such deposits can survive inundation and erosion at the shore edge, and other food remains including plant material should have a good chance of preservation in submerged deposits. There is certainly every reason to expect that marine resources should have left evidence of their consumption from the very earliest period of occupation in Britain. Shellfish in the intertidal zone, fish and crustacea in natural rock pools and traps, and sea mammals and sea birds that come ashore for breeding or are beached by storms would all be accessible along shorelines from the earliest period by our earliest hominin ancestors without the need for specialised equipment or boats. Proxy evidence from stable isotope signatures in human bone, as in the Paviland example discussed earlier, provides another potential line of investigation where human skeletal remains are present.

A third theme is the use of coastlines and travel by boat as pathways of communication and population dispersal. The evidence from Norway demonstrates how rapidly people can move into ice-free coastal zones with the aid of boats and a combination of hunting on land and at sea (Bjerck 2008; Glorstad et al., Chap. 6, this volume). Similarities in some of the earliest flaked stone spearheads found in Scotland with those in northern Germany and southern Scandinavia hint at just such a pattern of dispersal and economic adaptation around the now-submerged coastlines of northern Doggerland, perhaps as early as 14,000 years ago. Certainly, by the time that northern coastlines and their coastal archaeological sites become visible from about 10,000 years ago onwards, the distribution of sites around the northern and western coasts of Britain suggests a ready facility for inshore sea-travel and coastwise communication (Waddington 2015). Similarities between Mesolithic artefact types in southern England and northern France similarly suggest patterns of communication around the coastlines that connected the two regions before sea-level rise finally severed the land connection between them.

A final theme of topical interest is the impact of sea-level rise on societies living in the affected coastal regions, and the socio-economic, demographic and psychological impact of such changes (Leary 2015). According to one interpretation of the British evidence, the appearance of sedentary or semi-sedentary coastal settlements with evidence of marine exploitation in northern Britain may represent a response to the loss of territory by sea-level rise (Waddington 2007, 2015). However, such a view presupposes that similar settlements were absent on earlier and now-submerged coastlines, and that their appearance on modern shorelines at a later date is a response to the pressures imposed by loss of pre-existing resources, rather than simply a matter of differential preservation or visibility. It also relies on the questionable assumption that people in earlier periods avoided coastlines and their attendant ecological and economic advantages and preferred to concentrate on terrestrial resources until forced to change by loss of existing land.

Much depends on the pace and scale of land loss in relation to pre-existing population densities. Did the loss of large areas of land happen rapidly enough to stimulate or compel the adoption of new patterns of settlement and new economic and social arrangements? Or was the progressive abandonment of former territory slow enough to be accommodated by orderly retreat without major dislocation of existing social and economic arrangements? Some progress towards answering these questions could be achieved by simulations based on simple demographic assumptions and assessments of land area lost with progressive sea-level rise (see, for example, Williams et al. 2018 for an analysis of a similar set of changes around the Australian coastline).

To what extent was loss of land compensated by new advantages and opportunities, for example the increased productivity of fishing grounds in the shallow seas created by inundation of the continental shelf, or the opening up of new subsistence opportunities in the hinterland by climatic warming? What impact did changes in the coastal zone have on patterns of exploitation, site distributions and settlement in the British hinterlands, and can changes in the late Upper Palaeolithic and early Mesolithic archaeological record in these hinterlands be attributed in any way to social impacts resulting from loss of territory on the retreating coastline? The fact that Neolithic cultural traits, if not actual Neolithic people expanding from continental Europe, were present offshore and in now-submerged lowlands, raises the further question of how far sea-level rise is implicated in the expansion of agricultural economies and the interactions between indigenous and incoming populations. Other climatic and environmental changes may well have been involved in addition to sea-level rise, so that disentangling the relative influence of these different variables is likely to be a challenging proposition. If the evidence currently available is scarcely sufficient to move the debate about the human consequences of sea-level rise much beyond speculation, simply to pose these sorts of questions should stimulate new ideas, new lines of enquiry and new observations. Submerged landscapes, as Sturt et al. (2017) have noted, are good to think with, and the benefits of such a perspective ought to stimulate new investigations not only of the underwater archaeological record, but the archaeological record on land, and ultimately the relationship between the two of them.

10.10 Management of the Underwater Cultural Heritage

The underwater and on-land heritage is protected by legislation at UK Government level, in particular the National Heritage Act of 1983, as amended in 2002 to include underwater archaeology in UK territorial waters. The 2002 National Heritage Act provides additional protection of the underwater cultural heritage through the planning regime. Where underwater cultural heritage is identified through Environmental Impact Assessment, marine licences can be withheld unless appropriate archaeological mitigation is put in place. This may include an exclusion zone, monitoring, survey or excavation (Bicket et al. 2014; see also Pater, Chap. 26, this volume).

Implementation is devolved to separate Executive Agencies in England, Scotland and Wales. The Historic Buildings and Monuments Commission for England, also known as English Heritage, was established in 1983. In 2015, the responsibility for maintaining historic buildings and collections was hived off as a separate charity retaining the title of English Heritage, and responsibility for research and maintaining records of archaeological sites was established under the new title of Historic England (HE), which maintains the National Record of Historic England (formerly known as the National Monuments Record). HE is currently answerable to the UK Government Department for Culture, Media and Sport. The Scottish equivalent is Historic Environment Scotland (since 2014 combining the functions of Historic Scotland and the Royal Commission on the Ancient and Historical Monuments of Scotland). Historic Environment Records are accessed through Canmore (https://canmore.org.uk/), and HES has been accredited as a Marine Environmental Data & Information Network (MEDIN) Data Archive Centre (DAC) (https://www.oceannet.org/about/data-archive-centres), as a repository for marine historic environment data since 2015. There is a similar division of labour in Wales between CADW, responsible for the historic environment and answerable to the Welsh Government Department of Economy, Science and Transport, and The Royal Commission on the Ancient and Historical Monuments of Wales (RCAHMW), responsible for research and maintenance of the National Monuments Record of Wales.

Acknowledgements All authors have contributed to the compilation of the site record in the SPLASHCOS Viewer on which this chapter is based. GB analysed the site records and wrote the text, with contributions from GM, MB, LTz, AB and AH. We thank Nic Flemming and Fraser Sturt for a critical reading of the text and helpful comments.

References

Allen MJ, Gardiner L (2000) Our changing coast: A survey of the intertidal archaeology of Langstone Harbour, Hampshire. Council for British Archaeology, York

Allington-Jones L (2015) The Clacton spear: the last 100 years. Archaeol J 172(2):273–296

Anderson-Whymark H, Garrow D, Sturt F (2015) Microliths and maritime mobility: a continental European-style Late Mesolithic flint assemblage from the Isles of Scilly. Antiquity 89(346):954–971

Ashton N, Lewis SG, De Groote I, Duffy SM, Bates M, Bates R, Hoare P, Lewis M, Parfitt SA, Peglar S, Williams C, Stringer C (2014) Hominin footprints from Early Pleistocene deposits at Happisburgh, UK. PLoS One 9(2):e88329. https://doi.org/10.1371/journal.pone.0088329

Balaam ND, Bell MG, David AEU, Levitan B, Macphail RI, Robinson M, Scaife RG (1987) Prehistoric and Roman-British sites at Westward Ho!, Devon: archaeological and palaeoenvironmental surveys 1983–1984. In: Balaam ND, Levitan B, Straker V (eds) Studies in palaeoeconomy and environment in Southwest England, British archaeological reports British series 181. B. A. R., Oxford, pp 163–264

Ballin TB (2017) Rising waters and processes of diversification and unification in material culture: the flooding of Doggerland and its effect on north-west European prehistoric populations between ca. 13 000 and 1500 cal BC. J Quat Sci 32(2):329–339

Ballin TB (2019) Identification of Scottish Late Upper Palaeolithic industries by detailed technological analysis. Mesolithic Miscellany 27(1)

Ballin TB, Saville A, Tipping R, Ward T (2010) An Upper Palaeolithic flint and chert assemblage from Howburn Farm, South Lanarkshire, Scotland: first results. Oxf J Archaeol 29(4):323–360

Ballin TB, Saville A, Tipping R, Ward T, Housley R, Verrill L, Bradley M, Wilson C, Lincoln P, Macleod A (2018) Reindeer hunters at Howburn Farm, South Lanarkshire. Archaeopress, Oxford

Barton RNE (1992) Hengistbury Head, Dorset, Volume 2: The late upper palaeolithic and early mesolithic sites, Monograph 34. Oxford University Committee for Archaeology, Oxford

Bates MR, Nayling N, Bates R, Dawson S, Huws D, Wickham-Jones C (2013) A multi-disciplinary approach to the archaeological investigation of a bedrock-dominated shallow-marine landscape: an example from the Bay of Firth, Orkney, UK. Int J Naut Archaeol 42(1):24–43

Bell M (2007) Prehistoric coastal communities: the Mesolithic in Western Britain, CBA research report 149. Council for British Archaeology, York

Bell M (2013) Intertidal survey and excavation. In: Menotti F, O'Sullivan A (eds) The Oxford handbook of wetland archaeology. Oxford University Press, Oxford, pp 467–481

Benjamin J, Bicket A, Hale A, Anderson D (2014) A multi-disciplinary approach to researching the inter-tidal and marine archaeology in the Outer Hebrides, Scotland. J Island Coast Archaeol 9:400–424

Bicket A, Tizzard L (2015) A review of the submerged prehistory and palaeolandscapes of the British Isles. Proc Geol Assoc 126(6):643–663. https://doi.org/10.1016/j.pgeola.2015.08.009

Bicket A, Firth A, Tizzard L, Benjamin J (2014) Heritage management and submerged prehistory in the United Kingdom. In: Evans AM, Flatman JC, Flemming NC (eds) Prehistoric archaeology on the continental shelf: A global review. Springer, New York, pp 213–232. https://doi.org/10.1007/978-1-4614-9635-9_12

Bjerck HB (2008) Norwegian Mesolithic trends: a review. In: Bailey G, Spikins P (eds) Mesolithic Europe. Cambridge University Press, Cambridge, pp 60–106

Bonsall C, Smith C (1989) Late Palaeolithic and Mesolithic bone and antler artefacts from Britain: first reaction to accelerator dates. Mesolithic Miscellany 10:33–38

Brennand M, Taylor M (2003) The survey and excavation of a Bronze Age timber circle at Holme-next-the-Sea, Norfolk, 1998–9. Proc Prehist Soc 69:1–84

Brunning R (2007) Mesolithic worked wood. In: Bell M (ed) Prehistoric coastal communities: the Mesolithic in western Britain, CBA research report 149. Council for British Archaeology, York, pp 125–130

Buglass J (1994) A Neolithic fish weir in Cleveland? Nautical Archaeology Newsletter, Autumn 4

Bynoe R, Dix JK, Sturt F (2016) Of mammoths and other monsters: historic approaches to the submerged Palaeolithic. Antiquity 90(352):857–875. https://doi.org/10.15184/aqy.2016.129

Callow P (1986) Pleistocene landscapes and the Palaeolithic economy. In: Callow P, Cornford JM (eds) La Cotte de St. Brelade 1961–1978: excavations by C.B.M. McBurney. Geo Books, Norwich, pp 365–376

Cohen KM, Westley K, Erkens G, Hijma MP, Weerts HJT (2017) The North Sea. In: Flemming NC, Harff J, Moura D, Burgess A, Bailey GN (eds) Submerged landscapes of the European continental shelf: quaternary paleoenvironments. Wiley, Chichester, pp 147–186

Coles BJ (1998) Doggerland: a speculative survey. Proc Prehist Soc 64:45–81

Dawson S, Bates R, Wickham-Jones C, Dawson A (2017) Northern North Sea and Atlantic Northwest approaches. In: Flemming NC, Harff J, Moura D, Burgess A, Bailey GN (eds) Submerged landscapes of the European continental shelf: quaternary paleoenvironments. Wiley, Chichester, pp 187–209

Eadie G, Waddington C (2013) Rescue recording of an eroding inter-tidal peat bed at Low Hauxley, Northumberland (6109). Archaeological research services report 2013/17 http://www.archaeological-researchservices.com/projects/Low%20Hauxley%20Footprints%20&%20Peat%20Deposit.pdf

Edwards R, Dix J (2012) High resolution geophysical investigation of the nearshore environment of the Wootton-Quarr coast. In: Tomalin DJ, Loader RD, Scaife RG (eds) Coastal archaeology in a dynamic environment: a Solent case study, BAR British series 568. Archaeopress, Oxford, pp 88–104

Farr RH, Momber G, Satchell J, Flemming NC (2017) Paleolandscapes of the Celtic Sea and the Channel/La Manche. In: Flemming NC, Harff J, Moura D, Burgess A, Bailey GN (eds) Submerged landscapes of the European continental shelf: quaternary paleoenvironments. Wiley, Chichester, pp 211–239

Fischer A (2004) Submerged Stone Age – Danish examples and North Sea potential. In: Flemming NC (ed) Submarine prehistoric archaeology of the North Sea. Research priorities and collaboration with industry, CBA research report 141. Council for British Archaeology, York, pp 23–36

Flemming NC (ed) (2004) Submarine prehistoric archaeology of the North Sea. Research priorities and collaboration with industry, CBA research report 141. Council for British Archaeology, York

Flemming NC, Harff J, Moura D, Burgess A, Bailey GN (eds) (2017) Submerged landscapes of the European continental shelf: quaternary paleoenvironments. Wiley, Chichester

Fulford M, Champion T, Long A (1997) England's coastal heritage: a survey of English Heritage and the RCHME, Archaeological report no. 15. English Heritage, London

Gaffney V, Thomson K, Fitch S (eds) (2007) Mapping Doggerland: the Mesolithic landscapes of the southern North Sea. Archaeopress, Oxford

Gaffney V, Fitch S, Smith D (eds) (2009) Europe's Lost World: the rediscovery of Doggerland, CBA research report 160. Council for British Archaeology, York

Gaffney V, Allaby R, Bates R, Bates M, Ch'ng E, Fitch S, Garwood P, Momber G, Murgatroyd P, Pallen M, Ramsey E, Smith D, Smith O (2017) Doggerland and the Lost Frontiers Project (2015–2020). In: Bailey GN, Harff J, Sakellariou D (eds) Under the sea: archaeology and palaeolandscapes of the continental shelf. Springer, Cham, pp 305–319

Godwin H, Godwin M (1933) British Maglemose harpoon sites. Antiquity 7:36–48

Godwin H, Willis EH (1959) Cambridge University natural radiocarbon measurements I. Am J Sci Radiocarbon Supplement 1:63–75

Hale A (2004) Scottish marine crannogs, British archaeological reports British series 369. Archaeopress, Oxford

Hall K (2014) Locating potential Mesolithic fish sites in Britain using predictive modelling: applying the 'fishing site model' to British conditions. In: Foulds FWF, Drinkall HC, Perri AR, Clinnick DTG, Walker JWP (eds) Wild things: recent advances in Palaeolithic and Mesolithic research. Oxbow, Oxford, pp 278–291

Hardy K, Wickham-Jones C (2009) Mesolithic and later sites around the Inner Sound, Scotland: Scotland's First Settlers project 1998–2004. Scottish Archaeology Internet Report 31

Harff J, Flemming NC, Groh A, Hünicke B, Lericolais G, Mesched M, Rosentau A, Sakellariou D, Uścinowicz S, Zhang W, Zorita E (2017) Sea level and climate. In: Flemming NC, Harff J, Moura D, Burgess A, Bailey GN (eds) Submerged landscapes of the European continental shelf: quaternary paleoenvironments. Wiley, Chichester, pp 21–49

Hazell ZJ (2008) Offshore and intertidal peat deposits, England – a resource assessment and development of a database. Environ Archaeol 13(2):101–110

Hedges REM, Housley RA, Law IA, Bronk CR (1990) Radiocarbon dates from the Oxford AMS system: archaeometry datelist 10. Archaeometry 32(1):101–108

Huddart D, Roberts G, Gonzalez S (1999) Holocene human and animal footprints and their relationship with coastal environmental change, Formby Point, NW England. Quat Int 55:29–41

Jacobi RM, Higham TFG (2008) The 'Red Lady' ages gracefully: new ultrafiltration AMS determinations from Paviland. J Hum Evol 55(5):898–907

Jacobsson P, Hale A, Hamilton D, Cook G (2017) Radiocarbon wiggle-match dating in the intertidal zone. J Island Coast Archaeol. https://doi.org/10.1080/15564894.2017.1383323

Lambeck K (1995) Late Devensian and Holocene shorelines of the British Isles and North Sea from models of glacio-hydro-isostatic rebound. J Geol Soc Lond 152:437–448

Leary J (2015) The remembered land: surviving sea-level rise after the Last Ice Age. Bloomsbury, London

Long D, Wickham-Jones CR, Ruckley NA (1986) A flint artefact from the northern North Sea. In: Roe DA (ed) Studies in the Upper Palaeolithic of Britain and Northwest Europe, British archaeological reports international series 296. B. A. R, Oxford, pp 55–62

Mellars PA (1970) An antler harpoon-head of 'Obanian' affinities from Whitburn, County Durham. Archaeologia Aeliana 4th Series 48:337–346

Mellars PA (1987) Excavations on Oronsay: prehistoric human ecology on a small island. Edinburgh University Press, Edinburgh

Mithen S, Wicks K, Pine A, Riede F, Lane C, Banerjea R, Cullen V, Gittins M, Pankhurst N (2015) A Lateglacial archaeological site in the far north-west of Europe at Rbha Port an t-Seilich, Isle of Islay, western Scotland: Ahrensburgian-style artefacts, absolute dating and geoarchaeology. J Quat Sci 30(5):396–416

Momber G, Peeters H (2017) Postglacial human dispersal and submerged landscapes in North-west Europe. In: Bailey GN, Harff J, Sakellariou D (eds) Under the sea: archaeology and palaeolandscapes of the continental shelf. Springer, Cham, pp 321–334

Momber G, Tomalin D, Scaife R, Satchell J, Gillespie J (eds) (2011) Mesolithic occupation at Bouldnor Cliff and the submerged prehistoric landscapes of the Solent, CBA research report 164. Council for British Archaeology, York

Murphy P (2014) The English coast: a history and a prospect. Continuum, London

Parfitt SA, Ashton NM, Lewis SG et al (2010) Early Pleistocene human occupation at the edge of the Boreal zone in Northwest Europe. Nature 466:299–233. https://doi.org/10.1038/nature09117

Pope M (2003) Placing Boxgrove into its prehistoric landscape. Archaeol Int 7:13–16. https://doi.org/10.5334/ai.0705

Ransley J, Sturt F (eds) (2013) People and the sea: a maritime archaeological research agenda for England, CBA research report 171. Council for British Archaeology, York

Reid C (1913) Submerged forests. Cambridge University Press, Cambridge

Richards MP (2000) Human and faunal stable isotope analyses from Goat's Hole and Foxhole Caves, Gower. In: Aldhouse-Green S (ed) Paviland Cave and the "Red Lady": a definitive report. Western Academic and Specialist Press, Bristol, pp 71–75

Roberts MB, Parfitt SA (eds) (1999) Boxgrove: A Middle Pleistocene hominid site at Eartham Quarry, Boxgrove, West Sussex. English Heritage, London

Samson AVM (2006) Offshore finds from the Bronze Age in north-western Europe: the shipwreck scenario revisited. Oxf J Archaeol 25(4):371–388

Saville A, Ballin TB (2009) Upper Palaeolithic evidence from Kilmelfort Cave, Argyll: a re-evaluation of the lithic assemblage. Proc Soc Antiqu Scotl 139:9–45

Scales R (2007) Footprint-tracks of people and animals. In: Bell M (ed) Prehistoric coastal communities: the Mesolithic in Western Britain, CBA research report 149. Council for British Archaeology, York, pp 139–159

Scott K (1986) The bone assemblages of layers 3 and 6. In: Callow P, Cornford JM (eds) La Cotte de St. Brelade 1961–1978: excavations by C.B.M. McBurney. Geo Books, Norwich, pp 159–183

Scott B, Bates M, Bates R, Conneller C, Pope M, Shaw A, Smith G (2014) A new view from La Cotte de St Brelade, Jersey. Antiquity 88:13–29

Shennan I, Andrews J (eds) (2000) Holocene land-ocean interaction and environmental change around the North Sea. Geological Society, London. Special Publication 166

Shennan I, Bradley S, Milne G, Brooks A, Bassett S, Hamilton S (2006) Relative sea-level changes, glacial isostatic modelling and ice-sheet reconstructions from the British Isles since the Last Glacial Maximum. J Quat Sci 21:585–599

Sidell J, Haughey F (eds) (2007) Neolithic archaeology in the intertidal zone. Oxbow, Oxford

Smith O, Momber G, Bates R, Garwood P, Fitch S, Pallen M, Gaffney V, Allaby RG (2015) Sedimentary DNA from a submerged site provides evidence of wheat in the British Isles 8000 years before present. Science 347(6225):998–1001. https://doi.org/10.1126/science.1261278

Sturt F, Garrow D, Bradley S (2013) New models of North West European Holocene palaeogeography and inundation. J Archaeol Sci 40:3963–3976

Sturt F, Dix J, Grant MJ (2017) The history of industry-linked research in English waters: lessons for the future. In: Bailey GN, Harff J, Sakellariou D (eds) Under the sea: archaeology and palaeolandscapes of the continental shelf. Springer, Cham, pp 425–436

Taylor M (2011) Waterlogged wood. In: Momber G, Tomalin D, Scaife R, Satchell J, Gillespie J (eds) Mesolithic occupation at Bouldnor Cliff and the submerged prehistoric landscapes of the Solent, CBA research report 164. Council for British Archaeology, York, pp 84–89

Tizzard L, Baggaley PA, Firth AJ (2011) Seabed prehistory: investigating palaeolandsurfaces with Palaeolithic remains from the southern North Sea. In: Benjamin J, Bonsall C, Pickard C, Fischer A (eds) Submerged prehistory. Oxbow, Oxford, pp 65–74

Tizzard L, Bicket AR, Benjamin J, de Loecker D (2014) A Middle Palaeolithic site in the southern North Sea: investigating the archaeology and palaeogeography of Area 240. J Quat Sci 29:698–710. https://doi.org/10.1002/jqs.2743

Tizzard L, Bicket AR, de Loecker D (2015) Seabed prehistory: investigating the palaeogeography and Early Middle Palaeolithic archaeology in the southern North Sea, Wessex archaeology report 35. Wessex Archaeology, Salisbury

Tolan-Smith C (2008) Mesolithic Britain. In: Bailey G, Spikins P (eds) Mesolithic Europe. Cambridge University Press, Cambridge, pp 132–157

Tomalin DJ, Loader RD, Scaife RG (2012) Coastal archaeology in a dynamic environment: A Solent case study, BAR British series 568. Archaeopress, Oxford

Trechmann CT (1936) Mesolithic flints from the submerged forest at West Hartlepool. Proc Prehist Soc 2(2):161–168

Van de Noort R (2004) The Humber wetlands: the archaeology of a dynamic landscape. Windgather Press, Bollington

Van de Noort R (2014) North Sea archaeologies: a maritime biography, 10,000 BC – AD 1500. Oxford University Press, Oxford

Waddington C (ed) (2007) Mesolithic settlement in the North Sea Basin: a case study from Howick, North-east England. Oxbow, Oxford

Waddington C (2015) A case for secondary Mesolithic colonisation of Britain following rapid inundation of the North Sea Plain. In: Ashton N, Harris C (eds) No stone unturned: papers in honour of Roger Jacobi, Lithic studies occasional paper, vol 9. Lithic Studies Society, London, pp 221–232

Warren SH, Piggott S, Clark JGD, Burkitt MC, Godwin H, Godwin ME (1936) Archaeology of the submerged land-surface of the Essex Coast. Proc Prehist Soc 2:178–210

Waugham M, Davidson M, Innes J, Tooley M (2005) Archaeology and environment of submerged landscapes in Hartlepool Bay, England, Tees archaeology monograph series, vol 2. Tees Archaeology, Hartlepool

Wessex Archaeology (2007) Artefacts from the sea. Archaeology Data Service, York. https://doi.org/10.5284/1000260. Accessed July 2014

Westley K (2017) The northern shelf. In: Flemming NC, Harff J, Moura D, Burgess A, Bailey GN (eds) Submerged landscapes of the European continental shelf: quaternary paleoenvironments. Wiley, Chichester, pp 135–145

Wilkinson TJ, Murphy P (1986) Archaeological survey of an intertidal zone: The submerged landscape of the Essex coast, England. J Field Archaeol 13(2):177–194

Wilkinson TJ, Murphy PL (1995) The archaeology of the Essex coast, vol. 1: The Hullbridge survey. East Anglian archaeology 71

Williams AN, Ulm S, Sapienza T, Lewis S, Turney CSM (2018) Sea-level change and demography during the last glacial termination and early Holocene across the Australian continent. Quat Sci Rev 182:144–154

Wymer JJ, Robins PA (1994) A long blade flint industry beneath Boreal peat at Titchwell, Norfolk. Norfolk Archaeol 42:13–37

Ireland: Submerged Prehistoric Sites and Landscapes

11

Kieran Westley and Peter Woodman

Abstract
Evidence of Ireland's drowned landscapes and settlements presently comprises 50 sites spread across the entire island. These comprise mainly intertidal find spots or small collections of flint artefacts. A handful of fully subtidal sites are known, generally from near-shore regions and consisting, with one exception, of isolated single finds. Evidence of organic remains is also sparse, with the exception of Mesolithic and Neolithic wooden fish traps buried in estuarine sediments under Dublin. The relatively small number of sites is probably due to lack of research as much as taphonomic issues, and thus the current evidence hints at the potential archaeological record which may be found underwater. Such evidence could contribute to knowledge of the coastal adaptations and seafaring abilities of Ireland's earliest inhabitants. Nonetheless, taphonomic considerations, specifically relating to Ireland's history of glaciation, sea-level change and also modern oceanographic conditions likely limit the preservation of submerged landscapes and their associated archaeology. Realistically, the Irish shelf is likely characterised by pockets of preservation, which makes detection and study of submerged landscapes difficult but not impossible. A range of potential routes of investigation are identifiable, including site-scale archaeological survey, landscape-scale seabed mapping, archival research and community engagement.

Keywords
Submerged prehistoric sites · Intertidal archaeology · Holocene sea-level rise · Submerged forests · Fish traps · Maritime adaptation

11.1 Introduction

The island of Ireland lies to the west of Great Britain, separated from it by the Irish Sea. At present, it is politically divided between Northern Ireland (part of the United Kingdom) and the Republic of Ireland and, as such, its historic environment is recorded and managed under two separate jurisdictions. However, in this chapter, the submerged archaeological record of the entire island will be considered as a whole, for the simple reason that this political boundary did not exist in prehistory. The aim of this chapter is to review the extant evidence for submerged archae-

K. Westley (✉)
School of Geography and Environmental Sciences, Ulster University, Coleraine, Northern Ireland, UK
e-mail: kl.westley@ulster.ac.uk

P. Woodman (Deceased)
Department of Archaeology, University College Cork, Cork City, Republic of Ireland

G. Bailey et al. (eds.), *The Archaeology of Europe's Drowned Landscapes*, Coastal Research Library 35, https://doi.org/10.1007/978-3-030-37367-2_11

ological landscapes around Ireland, discuss any patterns visible in the data, identify the potential contribution of these landscapes to Irish archaeology and, finally, comment on potential avenues of future investigation. The timeframe under consideration focuses on the Irish Mesolithic and the early Neolithic (c. 10,500–5500 cal BP) because, as demonstrated below, most submerged landscape evidence from Ireland likely dates to this interval.

Before proceeding to the evidence, a brief review of the archaeological, sea-level and palaeogeographic background is needed. This will not only set the scene for readers unfamiliar with Irish prehistory but also identify issues in the extant evidence base which, as will become apparent later, are directly relevant to the study of submerged prehistoric landscapes. More detailed recent overviews of the Irish Mesolithic can be found in Woodman (2015) and Warren (2017), while a similarly recent overview covering submerged Quaternary palaeolandscapes and taphonomic issues can be found in Westley and Edwards (2017).

11.2 Archaeological Background

Ireland's prehistoric record bears some similarities with neighbouring Britain but also some major differences. One key distinction is the lack of Palaeolithic evidence compared to Britain, which was occupied as far back as c. 0.78–0.99 million years ago and is evidenced by well-preserved in situ sites as well as extensive secondary context assemblages (Pettitt and White 2012; Bailey et al., Chap. 10, this volume). The Irish evidence so far consists of four isolated lithic finds (two handaxes and two flakes) of Lower to Middle Palaeolithic appearance. However, the provenance of the handaxes is uncertain while the flakes are derived from reworked contexts (Woodman 2015; Warren 2017). Most recently, a cut-marked bear patella from Alice and Gwendoline Cave (Co. Clare) was radiocarbon dated to 12,810–12,590 cal BP along with a second cut-marked bear vertebra dated to 11,080–10,400 cal BP (Dowd and Carden 2016). The former presents the strongest available evidence for, at the very least, a human incursion into Ireland during the final Palaeolithic.

Even if the possibility of Palaeolithic occupation, or at least pioneer incursion(s), cannot be ruled out, the earliest incontrovertible evidence of full-scale human colonisation and settlement comes during the Mesolithic. This is the site of Mount Sandel, an occupation site situated close to the north coast and dated to c. 9800 cal BP (Bayliss and Woodman 2009; Woodman 2015). By comparison, Britain has Late Glacial hunter-gatherers present throughout the GI-1 interstadial (c. 14,600–12,900 cal BP; Pettitt and White 2012), and even after near-total depopulation in the first half of the cold GS-1/Younger Dryas stadial (c. 12,900–11,500 cal BP), humans had returned from c. 12,000 cal BP onwards in the form of Epi-Ahrensburgian hunters (Pettitt and White 2012; Warren 2017).

The reasons for the missing Irish Palaeolithic remain unclear. For the pre-Last Glacial Maximum (LGM), this could be partly explained by the scouring effect of multiple glaciations and the fact that Ireland's location may have been too peripheral for anything other than sporadic pioneer expeditions. However, for the Late Glacial and Mesolithic, there is still a delayed occupation relative to Britain which cannot be explained by purely taphonomic factors and could relate to social and ecological factors, not least the 'impoverished' nature of the Irish fauna (see Woodman 2015 and Warren 2017 for a fuller account of these issues).

Ireland's earliest occupants likely arrived from Britain, bringing with them similar toolkits and strategies. In Britain, the early Mesolithic (c. 11,500 to 10–10,500 cal BP) is typified by forms such as obliquely blunted points, broad blade microliths and slender uniserial bone/antler barbed points. By c. 10,500–10,000 cal BP, these assemblages gave way to different forms, characteristically narrow blade or geometric microliths and flat, squat barbed bone/antler points (the later Mesolithic; Tolan-Smith 2009; Warren 2017). The first thousand years of the Irish Mesolithic (c. 9800–8800 cal BP) show some similarities with British later Mesolithic assemblages (as expected given the colonists' likely origins), most evidently in the form of geometric microliths, but also differences, such as distinctive small core and flake axes not found in contemporary British sites. Once established in Ireland, the trajectory of the Irish Mesolithic departs further. From c. 8800 to 8600 cal BP, tool assemblages become dominated by a 'macrolithic' industry of large blades, flakes and distinctive butt-trimmed forms which have no British counterpart, except on the Isle of Man. Bone artefacts also differ as the barbed points of the British Mesolithic are not present in Ireland; instead, slender points without barbs were used (Woodman 2012, 2015). Effectively, this means that the British Late Mesolithic can be regarded as equivalent to the Irish Earlier Mesolithic, while the Irish Later Mesolithic has no parallels in mainland Britain (Warren 2015a). For the rest of this chapter, unless otherwise stated, the terms 'Earlier' and 'Later Mesolithic' will be used by reference to the Irish typochronology.

As elsewhere in north-west Europe, the Mesolithic was characterised by mobile populations hunting and gathering terrestrial, lacustrine, riverine and, to some extent, marine resources (note, however, questions regarding the level of coastal adaptation which will be discussed later). This way of life was then replaced by the Neolithic with its new agricultural practices, burial traditions, built structures and material culture. Dating of the key transition interval has been refined to 5700–5570 cal BP and comprised a 40–100 year 'boom' in Neolithic settlement evidenced by the rapid spread of distinct rectangular houses, appearance of domestic cereals (wheat and barley) and adoption of long-term fixed plot agriculture (Whitehouse et al. 2014). However, there are hints of earlier Neolithic-like activity in the form of burials and also domestic cattle bones from the final Mesolithic site of Ferriter's Cove south-west Ireland) dated to 6400–6220 cal BP. Whether this is indicative of an earlier colonisation event or contact between indigenous Mesolithic populations and incoming Neolithic ones is still unclear (Garrow and Sturt 2011; Whitehouse et al. 2014; Warren 2017).

11.3 Sea-Level and Palaeogeographic Change

Like much of north-west Europe, Late Pleistocene and Holocene relative sea-level (RSL) change around Ireland was driven by the interplay between global glacio-eustatic ocean volume change and local- to regional-scale isostatic response to ice loading during the LGM and unloading thereafter. The resulting impact on RSL change was particularly complex as Ireland is located in a transitional zone between uplifting formerly glaciated and collapsing forebulge areas (Edwards and Craven 2017; Westley and Edwards 2017).

At its maximum extent (c. 27,000–23,000 cal BP), ice entirely covered Ireland and its continental shelf as far west and south as the Atlantic shelf break and Celtic Sea shelf edge, respectively. Thickness, duration and timing of maximum extent of ice cover all varied spatially (Clark et al. 2012; Ballantyne and O Cofaigh 2017). Where ice was thickest and longest lasting, isostatic depression and subsequent postglacial rebound were greatest. This is exemplified by Ireland's north-east corner and was further enhanced by proximity to the regional centre of ice loading in Fenno-Scandinavia. Moving south and south-west across the island, isostatic rebound diminished with distance from the centre of ice loading (Westley and Edwards 2017).

This resulted in RSL histories which vary spatially and are often non-monotonic, i.e., characterised by variations in the rate of RSL change and, in some places, complete oscillations. Reconstructing these patterns has been hindered by an uneven distribution of accurate Sea-Level Index Points (SLIP), particularly for the Late Glacial and intervals during which RSL fell below present (Westley and Edwards 2017). Consequently, emphasis has been placed on modelling isostatic rebound and its effect on RSL using Glacio-Isostatic Adjustment (GIA) models constrained by the available field evidence. There are, however, still uncertainties for the earliest Holocene and Late Glacial to be addressed, and different models also deviate from one another—

generally in terms of magnitude rather than pattern—depending on differences in the input ice loading histories (Edwards and Craven 2017). There is also ongoing debate regarding the reconciliation of field data with model results, particularly for the Late Glacial in north-east Ireland (see McCabe et al. 2007; McCabe 2008; Edwards et al. 2008). Consequently, RSL reconstructions for Ireland are still a work-in-progress, though the general pattern of change—certainly for the Holocene—is well established. One clear feature arising from all models and field evidence is that an RSL lowstand occurred across the entire Irish shelf during the Late Glacial and Early Holocene, i.e., coincident with the earliest occupation and colonisation of Ireland. However, its timing and magnitude varied around the island due to differential isostatic rebound (Fig. 11.1).

In the extreme south and south-west, RSL rose from a lowstand of c. −50 to − 90 m (precise values vary depending on the GIA model used; Edwards and Craven 2017) and reached modern sea level in the Late Holocene (Brooks et al. 2008; Bradley et al. 2011; Kuchar et al. 2012). RSL rise was not constant, with models indicating variations in the rate of RSL rise and sometimes oscillations, though of insufficient magnitude to bring RSL above present level. The lack of evidence for RSL above present reflects reduced ice loading and hence the greater influence of glacio-eustasy.

Moving north and east, the influence of isostasy increases. Upon deglaciation, the resulting impact on sea level was a generalised pattern of 'rise-fall-rise' (Edwards and Craven 2017). This consisted of initially higher-than-present RSL,

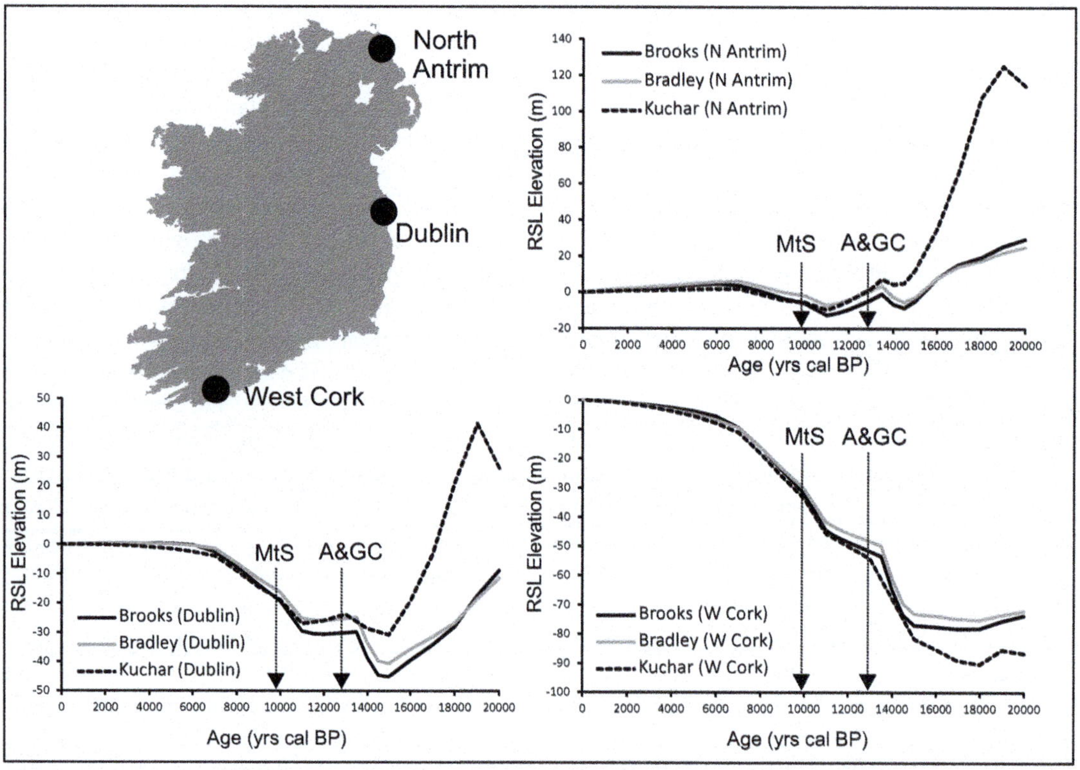

Fig. 11.1 RSL curves derived from the Brooks et al. (2008), Bradley et al. (2011) and Kuchar et al. (2012: min ice model) GIA models. Selected curves highlight the variation between the centre of ice loading in the north-east (North Antrim) and reduced ice loading in the south-west (West Cork). The RSL curve from Dublin presents an intermediate example between the two extremes. Note that despite the range of inter-model variation, they all agree on an RSL lowstand around the time of Ireland's earliest occupation, indicated by dates for Mount Sandel (MtS) and the Alice and Gwendoline Cave bear patella (A&GC). Data courtesy of Robin Edwards and Joseph Kuchar

followed by RSL fall to a lowstand below present, and finally another RSL rise to the present. Again, the magnitude and timings of the various lowstands and highstands are location- and model-dependent. In the extreme north-east, where isostatic recovery was greatest, the final RSL rise resulted in an additional minor mid-Holocene highstand of a few metres above present which, due to continued rebound, gave way to RSL falling to its present level. Zones in between the extreme north-east and south-west saw variations on these patterns. In general, the main trends were a reduction in initial Late Glacial highstand elevation, an increasingly long and deep lowstand and absence of the mid-Holocene highstand as one moves south and west (Edwards and Craven 2017; Westley and Edwards 2017).

Palaeogeographic reconstructions have been produced by correcting Digital Elevation Models (DEMs) of the land and seabed with GIA-modelled histories of RSL change (Fig. 11.2; Edwards and Brooks 2008; Brooks et al. 2011; Sturt et al. 2013). Again, there are minor differences between reconstructions depending on the GIA model used and the resolution of the input bathymetric/topographic surface (Edwards and Craven 2017). However, all recent reconstructions agree that Ireland was cut off from mainland Britain from a relatively early date (c. 16,000 cal BP [Edwards and Brooks 2008], 15,000 cal BP [Brooks et al. 2011]) and was an island by the time of its earliest settlement. Though the reconstructions are coarse and do not take into account bathymetric variation resulting from sedimentation or erosion, they also suggest that most of the larger bays and inlets were not created till after c. 12,000–13,000 cal BP, and most appear to have formed between c. 10,000 and 8000 cal BP (Edwards and Brooks 2008; Brooks et al. 2011; Sturt et al. 2013).

11.4 Overview of Submerged Archaeological Sites

A database of submerged archaeological sites around Ireland, comprising all known instances of subtidal or intertidal assemblages with prehistoric evidence, has been compiled for the SPLASHCOS project (Bailey and Sakellariou 2012). Sites were identified from the published and grey literature, existing gazetteers or Historic Environment Records and the authors' personal knowledge. There are currently 50 sites recorded in the database, of which 11 are fully subtidal, 38 are intertidal and 1 is deeply buried below sea level under reclaimed land (Fig. 11.3). Given the cut-off date of 5500 cal BP specified for the initial database compilation, the recorded sites are mainly Mesolithic, with a handful of early Neolithic entries. Where possible, chronology has been based on radiocarbon dates or, if lacking, diagnostic artefacts. Sites without either radiocarbon dates or diagnostic finds (17 in total) are assumed to predate 5500 cal BP and are conservatively placed in a broad prehistoric period since a Mesolithic/Neolithic attribution is not confirmed but cannot be ruled out. Details of chronological assignation for each site can be found in the database hosted on the SPLASHCOS Viewer (http://www.splashcos-viewer.eu).

The spatial distribution of sites is heavily biased to the north and north-east coasts (35 sites versus 15 for the rest of Ireland). This mirrors the distribution of the terrestrial Irish Mesolithic. Reasons for the dense concentration in the north-east are argued to be a product of a long history of research and extensive lithic evidence resulting from readily available flint sources as much as former occupation preferences (Woodman 2015). Interestingly, this pattern also mirrors the RSL history described above, with the majority of submerged sites located where the lowstand was shallowest and few in the south and west where the lowstand was deepest. The implications of this will be discussed later.

The following subsections will describe in more detail the submerged archaeological record, splitting it according to its current location, i.e., subtidal versus intertidal assemblages. Two other broad categories of evidence not included in the database, but which pertain to submerged landscapes, are also examined. These are intertidal and submerged organic deposits lacking archaeological finds and artefact-bearing raised beach deposits currently situated above present sea level.

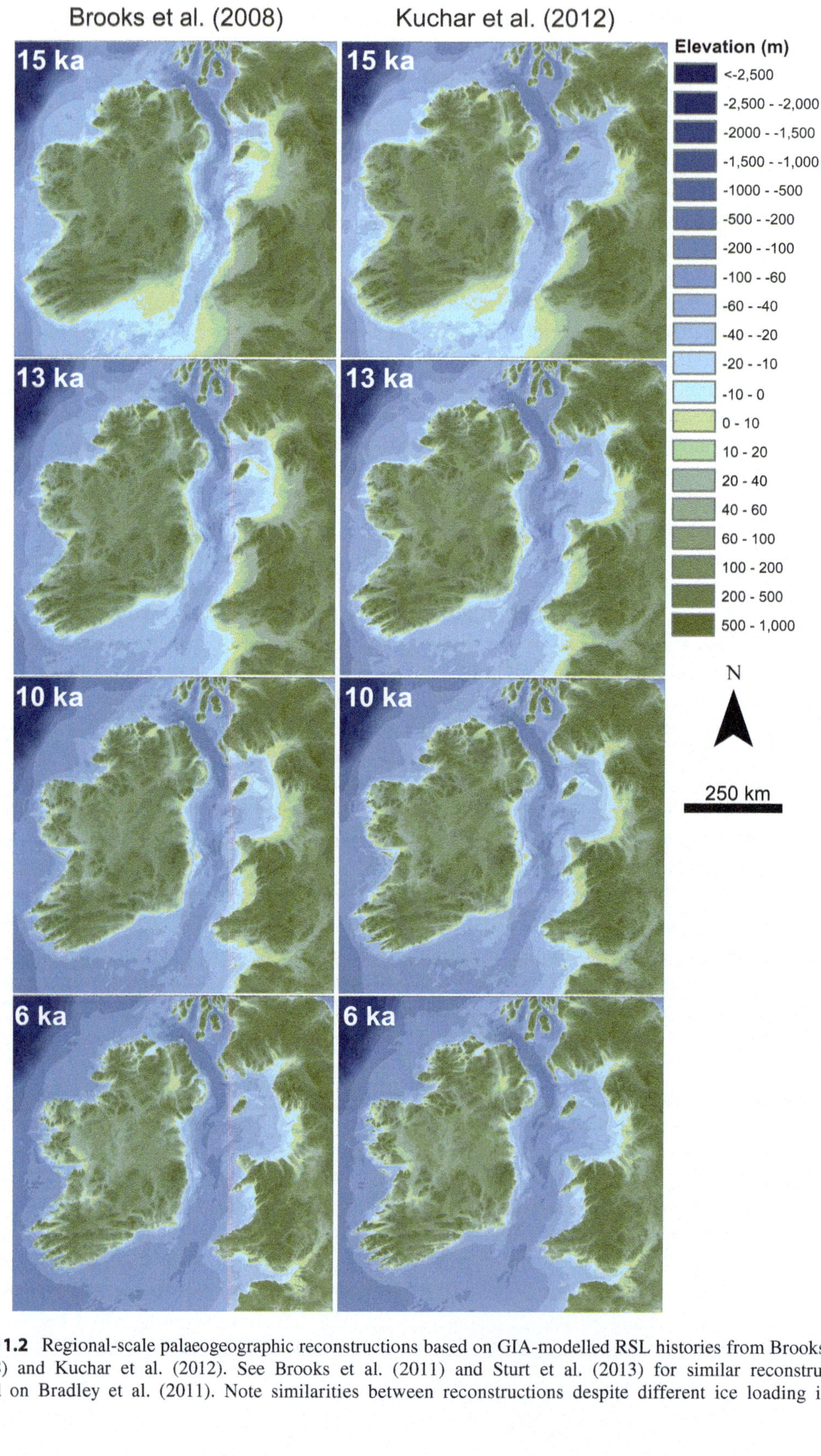

Fig. 11.2 Regional-scale palaeogeographic reconstructions based on GIA-modelled RSL histories from Brooks et al. (2008) and Kuchar et al. (2012). See Brooks et al. (2011) and Sturt et al. (2013) for similar reconstructions based on Bradley et al. (2011). Note similarities between reconstructions despite different ice loading inputs.

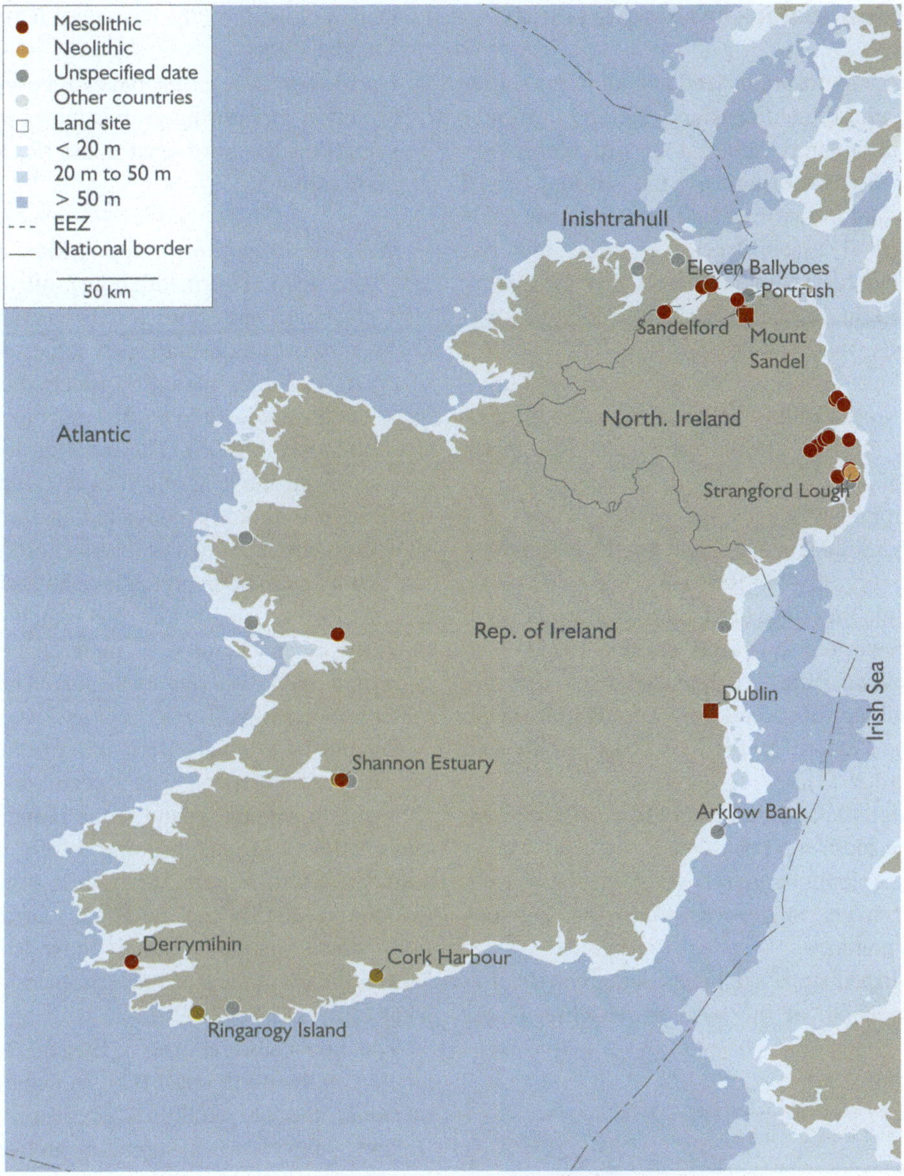

Fig. 11.3 Overview map showing location of all known subtidal and intertidal archaeological sites predating 5500 cal BP from the island of Ireland. Site information from the SPLASHCOS Viewer http://splaschos-viewer.eu. Drawing by Moritz Mennenga

Fig. 11.2 (continued) Timesteps roughly bracket the post-LGM recolonisation of Britain (onset of the Late Glacial interstadial, c. 15,000 cal BP), potential pioneer incursions into Ireland (Alice and Gwendoline Cave bear patella, c. 13,000 cal BP), Mesolithic settlement of Ireland (Mount Sandel, c. 10,000 cal BP) and the end of the Irish Mesolithic (c. 6000 cal BP). Reconstructions should be considered only as first approximations as they are not corrected for palaeotidal changes or bathymetric variations resulting from sediment deposition or erosion. The underlying Digital Elevation Model is the GEBCO (2014) global 30 arc-second grid. RSL data courtesy of Robin Edwards, Tony Brooks and Joseph Kuchar

11.4.1 Subtidal Archaeological Finds

Despite the exposure of land available for settlement around the entire island, Ireland currently has no completely subtidal in situ prehistoric sites. At the time of writing, the subtidal record consists of 11 identified find spots of stray finds or reworked assemblages of lithic material which have been found either by dredging or by divers. Eight of these comprise only isolated stray finds:

1. Two worked flints dredged from the Arklow Bank (10 km off the east coast; Campbell 2003).
2. A ground stone point trawled from near Derrymihin/Minane Island on the south-west coast (Fig. 11.4; O'Riordain 1953).
3. A single flint flake dredged from Larne Lough on the north-east coast (Common 1933).
4. A few possible struck flakes from a larger assemblage of natural gravel dredged from the Boyne Estuary on the east coast (Brady and Pollard 2013).
5. A single flake found by a diver off Glandore on the south-west coast.
6. A possible flint core recovered in a lobster pot off Renvyle on the west coast (M. Gibbons pers. comm.)
7. A polished stone axe head found by a diver at 18 m depth at the base of Muglins Rock, Dalkey Sound, on the east coast (Cooney and Mandal 1998).
8. Up to four struck flint flakes recovered from pollution monitoring grab samples in Belfast Lough on the north-east coast (R. McNeary pers. comm.)

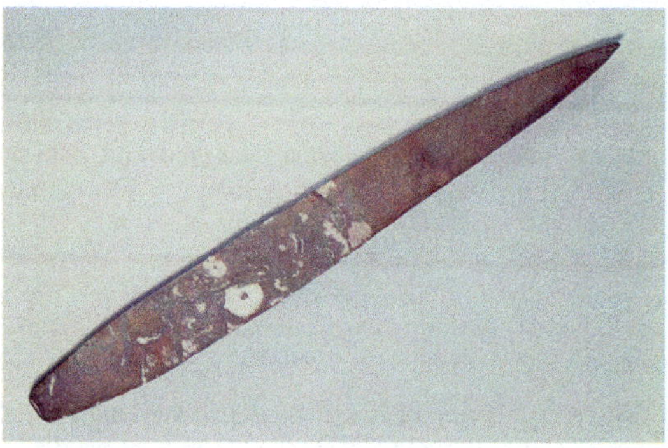

Fig. 11.4 Ground stone point (40 cm long) dredged off Derrymihin, south-west Ireland, possibly a Moynagh point of Later Mesolithic date. The surface discolorations probably result from growth of marine organisms. Photo: National University of Ireland, Galway

With the exception of the Arklow Bank (c. 10–20 m deep, 10 km offshore), all the above sites cluster in nearshore waters, particularly in sheltered bays and estuaries. A larger collection of up to 45 lithics including flakes, blades and axes was also recovered by divers from the River Corrib (Driscoll 2006). The reported location is c. 3 km from the sea, and it is uncertain if this material was directly deposited in the river or submerged by higher river levels caused by RSL rise downstream or changing lake levels in Lough Corrib further upstream. Nevertheless, it is included as an example of material potentially deposited in an estuarine location and later inundated because of RSL rise.

Diagnostic artefacts are lacking from most of the above find spots. The exceptions are the Derrymihin/Minane Island stone point which is classified as a Moynagh Point, a form generally associated with Later Mesolithic assemblages (Woodman 2015), and the River Corrib collection which includes some diagnostic Later Mesolithic items such as butt-trimmed forms and a bar form (Driscoll 2006).

The final subtidal site is Eleven Ballyboes, located on the north coast (Co. Donegal). This is presently the only subtidal prehistoric site in Ireland subject to systematic archaeological investigation (Westley 2015). The location comprises two small (< c. 50 m wide) bays with sandy pocket beaches. Investigation focused here after large numbers of lithics (c. 1600 to date) were discovered strewn across the shore by a group of amateur collectors (McNaught 1998). Follow-up work confirmed that these were Earlier Mesolithic in age, based on the presence of diagnostic narrow blades, platform cores and distinctive flake axes (Fig. 11.5; Costa et al. 2001). In the westernmost bay, systematic survey and test excavation recovered c. 120 lithics from underwater, including buried items as well as examples lying

Fig. 11.5 Eleven Ballyboes (Co. Donegal) views across the (**a**) western and (**b**) eastern bay and beach. (**c**) Representative sample of lithic finds from the intertidal zone including blades, flakes, cores and axes. Photos by K. Westley

loose on the seabed (Fig. 11.6). Some are fresh and unpatinated and contrast with their stained and water-rolled counterparts recovered from the beach, thus indicating that the source deposit was located underwater. Excavation identified this source as a gravel deposit located c. 30 m offshore in c. 2 m water depth from which artefacts are periodically washed out onto the adjacent beach and seabed. However, it also confirmed that this deposit has already been reworked by wave and tidal processes, and consequently, while the finds can be localised to a relatively small area, they are no longer in situ (Westley 2015).

By contrast, the easternmost bay has only produced three undiagnostic and water-rolled subtidal flints. However, it was found to contain a submerged peat layer at roughly the same depth as the find-bearing gravel in the western bay (c. <2 m depth). Organic preservation is excellent, with large chunks of wood, *sphagnum* moss and hazelnuts preserved in situ (Fig. 11.7).

Fig. 11.6 Subtidal Mesolithic flints from Eleven Ballyboes (Co. Donegal). (**a**) Diver with flint blade recovered from the seabed. (**b**) Representative sample of lithic finds from below low water. Photos by W. Forsythe, K. Westley

Fig. 11.7 Submerged peat from the eastern bay, Eleven Ballyboes. (**a**) Seaward edge of the peat lay visibly breaking up into large chunks. (**b**) Exposed surface of the peat layer showing preserved wood. Photos by K. Westley

Radiocarbon dates place the peat between c. 9400 and 8700 cal BP which is in accordance with the Earlier Mesolithic attribution of the lithic finds. No archaeological finds were made from the peat, and it is unclear if this relates to a true lack of evidence or just the limited sampling strategy (Westley 2015). In a European context, Eleven Ballyboes is not especially significant, being relatively small and with no in situ finds yet identified. In an Irish context, it has a somewhat greater importance in that it gives an indication of the nature of the archaeological record potentially located offshore and highlights the localised taphonomic processes which characterise large parts of the Irish shelf. In this case, for example, two adjacent bays show very different stratigraphy and preservation conditions.

11.4.2 Intertidal Archaeological Assemblages

Thirty-eight intertidal locations, mainly located along the north and north-east coast, have also produced archaeological evidence consisting mainly of flint tools washed up onto the upper part of the intertidal shore. The two largest concentrations are around Strangford and Belfast Loughs, both sea loughs created during the early Holocene and where historically many collectors have been active (Woodman 1978; McErlean et al. 2002).

Good examples from these loughs where material has been found close to low-water mark are Sydenham Station (Belfast Lough) and Big Stone Bay (Strangford Lough). The former consists of a selection of struck flints (of Later Mesolithic appearance) recovered from an eroded intertidal surface in the late nineteenth century (Patterson 1892; Woodman 1978). Though the original report described associated faunal remains, these were later radiocarbon dated and found to be much younger, <2000 cal BP (Woodman et al. 1997). Unfortunately, the location is now reclaimed and buried below Belfast City Airport. The latter site also consists of a lithic scatter, probably of Later Mesolithic age, found on intertidal muds (Fig. 11.8). Underlying this is a peat deposit dated to c. 6800–6600 cal BP though whether the lithics and peat are associated is not confirmed (Woodman 1978). Unlike Sydenham, recent visits (2014) show that the site still exists, and relatively fresh lithic material is still present on the intertidal mud, along with traces of the peat. Sites like these could be taken as hints of the archaeological record potentially located further offshore, as demonstrated at Eleven Ballyboes (see above and Westley 2015). In other cases, it is possible that archaeological material in the intertidal zone has washed down from eroding deposits above high water or may be a remnant lag from eroded sites on the foreshore, and these possibilities would require field assessment to confirm.

Fig. 11.8 (**a**) View looking across the intertidal zone at Big Stone Bay and Strangford Lough at low water. (**b**) Later Mesolithic finds collected from intertidal mud surface at Big Stone Bay. Photos by K. Westley

Fig. 11.9 Bann Estuary, Co. Derry. (**a**) Sandelford site exposed during extreme low water. This site is located below Mount Sandel—the earliest dated occupation site in Ireland. (**b**) Mesolithic flints recovered from the intertidal zone at Sandelford. Photos by R. McConkey

A few sites are also located in upper estuarine locations, most notably at Sandelford in the Bann Estuary adjacent to Mount Sandel (Fig. 11.9). Here, a ridge protruding partly across the river is exposed at low water and has produced a collection of several hundred lithics similar to those from Mount Sandel itself, i.e., Earlier Mesolithic (Woodman 1978, 2015). Smaller quantities of lithic material (tens of finds) have also come from the intertidal portions of the River Foyle at Ballynagard and Culmore.

The evidence described so far is overwhelmingly lithic. However, at one location on the beach at Glynn (Larne Lough), a portion of a pointed bone implement was found. Although it has not been radiocarbon dated, it is similar to items interpreted as the tips of fishing spears or eel rakes which have been dredged from the non-tidal part of the River Bann upstream of Mount Sandel (the Cutts locality). Dates on these items consistently place them to between c. 9000 and 7500 cal BP (Woodman 2015). The location of the Glynn find spot may be significant in that nearby raised beach deposits contain both Earlier and Later Mesolithic lithics apparently washed up from further offshore (see below). Along with a worked wooden plank from the Shannon Estuary dating to c. 6700 cal BP, these remain the only known organic artefacts of Mesolithic age from exposed intertidal landscapes (O'Sullivan 2001). Note though that organic remains from buried landscapes are known (see below for discussion of the North Wall Quay fish traps).

As RSL stabilised round the northern coast of Ireland as late as 6000 cal BP, one might expect that much of the material recovered from beaches would date to the Later Mesolithic. However, some of the recovered assemblages contain elements such as distinctive flake axes and core axes which belong to the first thousand years of the Irish Mesolithic. Thus, the pattern is suggestive of activity around a number of the outer estuaries and sea loughs during the Later Mesolithic and perhaps also in the Earlier Mesolithic. Given that RSL was lower, this was not directly on the shore but on higher ground overlooking coastal and estuarine floodplains.

Intertidal collections are largely absent outside the north and north-east coasts barring material from Connemara in the west of Ireland (generally stone axes, probably of Later Mesolithic affinity), which has reportedly been washed up during recent winter storms and, in one instance, may be associated with an intertidal peat deposit (M. Gibbons, pers. comm.). Whether this relates to a true absence of evidence or simply a lack of recognition of the relevant evidence is presently uncertain.

Aside from the Mesolithic evidence, finds dating to the early Neolithic have also been made in intertidal localities. This includes organic artefacts found in the intertidal zone of the Shannon

Fig. 11.10 Neolithic finds from the intertidal zone. (**a**) The Lag: Megalithic tomb in the intertidal zone, Co. Cork. (**b**) Logboat in intertidal mud, Strangford Lough, Co. Down. Photos by P. Walsh; McErlean et al. (2002)

Estuary, such as human remains and a woven basket, both dating to around 5700–5400 cal BP, while an intertidal scatter of lithics, bone and organic artefacts at Carrigdirty, also in the Shannon Estuary, has been interpreted as a possible Early Neolithic occupation site (O'Sullivan 2001). A logboat of similar age (5500–5000 cal BP) was also found in intertidal mud in Strangford Lough (Fig. 11.10b; McErlean et al. 2002). The Neolithic is also represented by two megalithic tombs on the south-west coast at Cork Harbour and Ringarogy Island. Both were presumably built on dryland but are now situated in the intertidal zone (Fig. 11.10a; Power 1994; Shee Twohig 1995). Their location in the far south-west fits the GIA-modelled pattern of RSL change in that this area was the last to attain modern sea level.

11.4.3 Raised Beach Archaeological Finds

The term 'raised beach' refers to palaeoshoreline features, generally consisting of sands and gravels sometimes with marine shells, found on land above present sea level. They represent former coastal, intertidal or nearshore environments left high and dry by RSL fall. In north-west Europe, the principal mechanism behind this is glacio-isostatic uplift, and hence raised beaches concentrate where the LGM ice was thickest and longest lasting (e.g., Scotland, Scandinavia). In Ireland, post-LGM raised beaches are found principally around the north-east coast owing to its greater degree of isostatic recovery and include Late Pleistocene and mid-Holocene examples (Edwards and Craven 2017; Westley and Edwards 2017; see previous section; Fig. 11.1). In particular, the mid-Holocene raised beach is frequently associated with water-rolled flint artefacts which derive from coastal sites, primarily of Later Mesolithic age. Notable examples include sites at Curran Point (Movius et al. 1953/54), Port of Larne (Woodman 2012) and Cushendun (Movius et al. 1940/41).

Raised beach sites, though not currently submerged, have been included in this review because they represent archaeological material which was once inundated and therefore subject to coastal and marine processes. As such, they give an impression of the likely nature of the archaeological evidence which might be found offshore. At the Curran Point, for instance, thousands of water-rolled lithics have been recovered, concentrated mainly in storm beach shingle but also including fresher or unrolled implements from underlying sands and gravels representing former foreshore sediments (Movius et al.

1953/54). In other cases, raised beach deposits may also bury older Earlier Mesolithic deposits. At Cushendun, fine-grained sediments (the lower lagoon silts) are interpreted as a sheltered estuarine environment (Movius et al. 1940/41) and contain flint implements and organic remains dating to c. 8000–9000 cal BP. These occur beneath c. 7 – 8 m of stratified gravels, silts and sand which are variously interpreted as relating to marine transgression, coastal barrier migration and possibly breaching (Roe and Swindles 2008). The overlying sediments also contain horizons of lithics (including Later Mesolithic to Neolithic forms) that tend to be rolled and patinated in contrast to the Earlier Mesolithic finds which are regarded as in situ or, at least, not transported very far (Movius et al. 1940/41; Woodman 1978). Finds of a similar age were also made at the Port of Larne site where two chipping floors, flint implements and a series of pits and hollows—some containing organic remains such as fish bones, hazelnuts and charcoal—were found under several metres of beach shingle. Radiocarbon dates indicate multiple phases of activity ranging from c. 9200 to 7500 cal BP (Woodman 2012).

At worst, raised beach sites indicate that although archaeological deposits on the wave-swept Irish coast are not obliterated by marine transgression, much of the evidence will be heavily reworked (e.g., the storm beach deposits of the Curran Point or the Upper Gravels at Cushendun). At best, they show that better preservation is possible and that significantly variable taphonomic conditions can occur in even a single locality. This ranges from variation between rolled and near-fresh lithics in the Curran Point sands and gravels to the fine-grained lower lagoon silts at Cushendun and the intact chipping floors with faunal and fish remains preserved under beach shingle at the Port of Larne. This is encouraging for the preservation of similar evidence offshore. Moreover, the presence of material within a raised beach could hint that a source deposit is buried offshore in a similar fashion to the intertidal material discussed above. This appears to be the case at Glynn (Larne Lough) where Earlier Mesolithic flints have been found in raised beach sands and gravels apparently washed in from a source deposit submerged in a small adjacent bay (Woodman 1977). The assumption of course is that the locale was persistently attractive to past humans over a long period of time including through an interval of RSL rise.

11.4.4 Intertidal, Submerged and Buried Organic Remains

The most widespread evidence indicative of prehistoric landscape preservation on the Irish seafloor comes from the intertidal zone. Dotted around the Irish coast are numerous peat and forest remains exposed at low tide (Charlesworth 1963; Mitchell 1976). Many sites are found in low-energy sheltered bays and estuaries, such as Strangford Lough and the Shannon Estuary (O'Sullivan 2001; McErlean et al. 2002). However, there are also sites located on more exposed and higher-energy coastlines, such as Portrush on the north coast (Fig. 11.11; Wilson et al. 2011), and various localities on the Atlantic-facing west coast, which are periodically exposed after beach stripping by storms (Williams and Doyle 2014; O'Connell and Molloy 2017; M. Gibbons pers. comm.).

At a handful of sites, peats have been found below the low-water mark, either brought up by dredgers and cores or reported by divers. Known sites include Portrush West Bay (Westley et al. 2014), the Bann Estuary (Carter 1982), Allihies (Breen 1993), Lough Swilly (O'Raw 2003) and Eleven Ballyboes (see above; Westley 2015). These are a strong indication that remnants of the past landscape can be preserved in fully subtidal environments. Radiocarbon dates from these intertidal and subtidal deposits give ages from as early as 13,500 cal BP right up to 5000 cal BP (Brooks and Edwards 2006).

A few locations have also produced faunal remains. For example, Roddans Port on the north-east coast has produced reindeer (Singh 1963; Morrison et al. 1965), red deer have been found in the intertidal portions of the Shannon Estuary (O'Sullivan 2001) and a portion of red

Fig. 11.11 Bed of intertidal peat exposed on the beach at Portrush (West Strand), Co. Antrim. This particular site has been radiocarbon dated to between 7400 and 6500 cal BP. Photo by K. Westley

deer antler was washed up on a beach at Ventry, Co. Kerry (Woodman et al. 1997). At least two subtidal localities have also produced large Pleistocene mammal remains, namely, an undated mammoth tusk dredged from Waterford Harbour (Monaghan 2017) and a possible mammoth humerus dredged from Galway Bay (Adams 1883). Note, however, that identification of the latter has not been verified, and Savage (1966 p. 123) states that it is 'more likely to be that of a circus elephant than a mammoth'. These are the only presently known large mammal remains from submerged Irish landscapes, and the paucity of finds is a marked contrast to the extensive palaeontological collections from the North Sea (Cohen et al. 2017; Bailey et al., Chap. 10, this volume; Peeters and Amkreutz, Chap. 8, this volume).

Additional organic evidence from below present sea level (though not necessarily underwater) has come from deep excavations or borehole sampling on reclaimed land. Particularly well-documented land surfaces include an extensive peat sequence buried as deep as −15 m below present sea level around Belfast Lough and radiocarbon-dated to c. 11,100–9500 cal BP (Manning 1970; Carter 1982; Brooks and Edwards 2006). Even older evidence comes from the south-west of Ireland (Cork Harbour) in the form of series of estuarine or intertidal organic clay silts at c. −17–35 m depth. These are particularly significant as they date from a temperate stage preceding the Last Interglacial (MIS 5), potentially the MIS 11 or MIS 9 interglacials (Dowling et al. 1998; Coxon et al. 2017).

Similar deposits can also survive offshore. For example, Plets et al. (2015) sampled buried fine-grained organic-rich estuarine/intertidal sediments dated to c. 13,300–13,400 cal BP from the inshore part of Bantry Bay (Co. Cork). Using seismic profiles, they also identified an even more deeply buried deposit stratified under glacial material and, from its acoustic characteristics, suggested it was deposited in a low-energy estuarine or lacustrine setting (Plets et al. 2015). Tentative correlations were made with a freshwater organic-rich silt identified in boreholes from the same area by Stillman (1968). Though originally regarded as Allerød/GI-1 in age (c. 13–14 ka cal BP), this assignation was based solely on pollen zonation, and it may be that it actually represents an earlier interglacial, as in Cork Harbour (Plets et al. 2015).

Most buried organic deposits have produced only palaeoecological or, in a few cases, palaeontological evidence, such as red deer, boar and giant deer (*Megaloceros*) excavated from Belfast Harbour (Praeger 1892; Savage 1966). However, their archaeological potential is well illustrated by the discovery of Later Mesolithic wooden fish traps during construction work at Spencer Dock, North Wall Quay, Dublin (Fig. 11.12). These were found buried under reclaimed land at a depth of c. 6.3 m below present sea level on what was once the bank of the Liffey Estuary or a tidal island. They consist of a

Fig. 11.12 Prehistoric fish traps found buried 6.3 m below present sea level at North Wall Quay, Dublin. (**a**) Neolithic weir (5600–5400 cal BP) comprising wattle panels supported by large stakes and bound by hazel withies. (**b**) Detail of horizontal panel from a wattle trap (8040–7840 cal BP). (**c**) Detail of C-shaped trap (8030–7790 cal BP) showing vertical rods and binding withies (scale intervals are 20 cm). Photos by M. McQuade

complex of up to five well-preserved partly broken traps made of hazel stakes, wattle fencing and baskets. Identified traps included up to three ebb weirs, a basket trap and a C-shaped trap. All appear to have been designed to catch fish on the falling tide. Radiocarbon dates place them between c. 8100 and 7700 cal BP (McQuade and O'Donnell 2007). A Neolithic fish trap dated to c. 5600–5400 cal BP was also found at a higher level at this site (McQuade 2008).

11.5 Discussion

The above clearly demonstrates that fragments of submerged landscape and associated archaeology are known from all around Ireland. Nevertheless, it is also apparent that Ireland lacks the extensive palaeolandscape and palaeontological records of the North Sea (Cohen et al. 2017; Peeters and Amkreutz, Chap. 8, this volume) or the unparalleled archaeology of the Baltic (Fischer 2011; Lübke et al. 2011; Bailey et al., Chap. 3, this volume; Nilsson et al., Chap. 4, this volume; Jöns et al., Chap. 5, this volume).

This leads to three key questions. Firstly, why has relatively little submerged landscape evidence been found in Ireland? Secondly, even if more evidence remains to be discovered, would it usefully contribute to our understanding of early Irish prehistory? Thirdly, if deemed worthy of investigation, what are the most effective strategies of investigation?

11.5.1 The Lack of Evidence

It is not worth trying to find a single overarching reason for the paucity of Irish submerged landscape evidence. The reality is that a number of factors likely contribute and which one(s) is/are the most important cannot be determined at present.

Taphonomy has undoubtedly played a part (Westley and Edwards 2017). Much of the Irish coastline is typified by high-energy conditions, particularly the Atlantic-facing south, west and north coasts. Storm impacts can be considerable, and there are many examples of beach stripping and coastal erosion (e.g., Kandrot et al. 2016). Even in the more sheltered Irish Sea, there are zones of strong tidal currents, particularly at the narrow straits of the North Channel and St. George's Channel (Evans 1995; Atkins 1997). Offshore, high-seabed sediment mobility in response to wave and tidal forcing is well documented by mobile bedforms (e.g., Evans et al. 2015). From this, the expectation would be that preservation of prehistoric sites and palaeolandscapes is poor. While likely true for many areas, this is not universal. There is sufficient complexity in the Irish coastline, with numerous bays, inlets and—particularly on the west coast—islands, to create variation in energy levels and taphonomic conditions. Moreover, the above evidence shows that palaeolandscape fragments can be preserved even in high-energy environments, such as the intertidal peats on the wave-swept north and west coasts, as well as in lower-energy inlets and estuaries. It is more likely that pockets of preservation rather than total absence will be the norm. Thus, it can be reasonably argued that the lack of fully subtidal evidence may stem from factors other than destruction by rising sea levels.

One additional mitigating factor is the extent of land exposed during the RSL lowstand. The depth of the lowstand, coupled with the bathymetry of Ireland's shelf, means that maximal subaerial exposure may never have been more than c. 30 km out from the present coastline (Fig. 11.2; Brooks et al. 2011). This contrasts with Britain where the shallow North Sea and English Channel provided a vast landscape open to prehistoric settlement. Even within the Irish Sea itself, bathymetry is such that the largest expanses of land were on the British side of the central channel (Brooks et al. 2011; Sturt et al. 2013). Simply put, the larger expanse of these areas means that the likelihood of settlement, and consequently the preservation of archaeological evidence and the chance of discovery, is greater.

Another reason could relate to the history of research. It has long been recognised that prehistoric sea levels were lower—largely due to intertidal forests and deeply buried peats (e.g., Praeger

1891; Stewart 1911). However, the earliest antiquarians and archaeologists were concerned primarily with the raised beach sites of the north-east (e.g., Coffey and Praeger 1904/05; Movius et al. 1953/54). It was not until the 1990s that efforts were made to systematically undertake maritime archaeology in Ireland and, even then, attention focused on shipwrecks (Breen and Forsythe 2001). The exceptions were the major intertidal surveys of Strangford Lough and the Shannon Estuary (O'Sullivan 2001; McErlean et al. 2002), but even these projects, which did identify intertidal archaeological and palaeolandscape evidence, were multi-period surveys with submerged palaeolandscapes forming part of a body of work. It is only within the last decade that a concerted effort has been made to investigate Irish submerged landscapes (Bell et al. 2006; Westley et al. 2011a, b, 2014; Westley 2015). Even this though has focused almost exclusively on the north coast, and the rest of Ireland has never been fully investigated beyond palaeoenvironmental investigations of a handful of intertidal peats (e.g., Whitehouse et al. 2008; Simpson 2008; Williams and Doyle 2014; O'Connell and Molloy 2017).

A final, more speculative, reason could include the nature of commercial activities on the Irish shelf. The two main industries responsible for the vast majority of chance finds (mainly palaeontological, but also including some archaeological remains) around the British Isles are offshore aggregates extraction and trawling. Aggregates extraction has produced a number of recent finds, not least the Area 240 handaxes off Great Yarmouth (Tizzard et al. 2011; Bailey et al., Chap. 10, this volume). In Irish waters, aggregate deposits suitable for extraction have been identified in the Western Irish Sea, but extensive commercial extraction has not taken place (Highley et al. 2007; O'Mahony et al. 2008). Trawling, particularly beam trawling for flatfish, was historically, and remains, the main source of finds in the Dutch sector of the North Sea (Van Kolfschoten and Laban 1995; Mol et al. 2006; Peeters and Amkreutz, Chap. 8, this volume). The same was true of the British sector until the decline of the fishing industry (Bynoe et al. 2016). One possibility, yet to be verified, is that differences in the Irish fisheries are responsible for the lack of finds. For example, large sectors of the northern Irish Sea fishery focus on the Dublin Bay prawn (Gibson 2011). This species is concentrated in deep water (>30 m) muddy substrates which were probably not exposed during the RSL lowstand and therefore not a likely source of prehistoric finds. Other parts of the Irish shelf, however, are trawled, and it remains to be determined if the lack of finds is a product of different gear (e.g., Irish fishermen primarily use otter trawls rather than beam trawls; Davie and Lordan 2011; see also Maarleveld, Chap. 26, this volume), a lack of recognition of recovered archaeological/palaeontological finds or a true lack of evidence. All the above are speculative and need further research to confirm but suggest legitimate reasons as to why submerged prehistoric evidence is lacking, as well as equally valid reasons to suspect that evidence exists but remains undiscovered.

11.5.2 Contribution to Knowledge

The oft-cited reasons for investigating submerged archaeological landscapes—namely, their ability to provide unique evidence (in some cases including organics preserved by waterlogging) pertaining to coastal and marine adaptations, seafaring and past human responses to sea-level change (Bailey and Flemming 2008; Bailey 2014)—apply also to the Irish situation.

11.5.2.1 Seafaring

That Ireland was apparently not connected to Britain from at least 15,000 cal BP onwards (Brooks et al. 2011) implies that its Mesolithic settlement was accomplished by people with the ability to make seagoing journeys. This is substantiated by Earlier Mesolithic flints from Inishtrahull (Woodman 2015). This small island is located c. 7 km north of Malin Head (Ireland's most northerly point) and is separated by sufficiently deep water to have remained cut off during the RSL lowstand. Once established in Ireland, the seafaring tradition was maintained, as evidenced by the Later Mesolithic occupation of Rathlin Island, like Inishtrahull cut off by several kilometres of deep water (Quinn et al. 2012), and also Later Mesolithic-type material from the Isle of Man and Neolithic contacts across the Irish Sea (Garrow and Sturt 2011).

The circumstantial evidence for seafaring reveals a number of things about Ireland's early colonists. Firstly, the occupation of Inishtrahull and Rathlin suggests an ability to cross the exposed, often treacherous waters of the Atlantic seaboard as well as coping with the more sheltered waters of the Irish Sea. Secondly, boats carried more than just people. Transport of animals may also have played a major role both in enabling permanent Mesolithic settlement in Ireland and later in the Neolithic transition to agriculture. Wild boar in particular is not present in Late Glacial faunal assemblages, and their later appearance in the early Holocene coincident with Mesolithic settlement is suggestive of a deliberate introduction (Montgomery et al. 2014; Warren et al. 2014; Monaghan 2017). This could have been a response to the absence of large game such as red deer, aurochs and elk, which were available in Britain but not Ireland (Woodman 2015; Warren 2017). Later in the Mesolithic and Neolithic, the appearance of new domestic species (e.g., cattle appear by c. 6300 cal BP) and also red deer from c. 5000–4500 cal BP onwards are clear signs of seaborne transport of animals (Montgomery et al. 2014; Monaghan 2017).

Less clear are the boats and the routes used to reach Ireland. Hundreds of Irish logboats have been found and range in date from the Mesolithic to the post-Medieval (Fry 2000). However, the vast majority are from inland waterways, and clearly seagoing crafts are not known from either the Mesolithic or Neolithic. In terms of Mesolithic crossings, Woodman (2015) has suggested three potential routes to Ireland, firstly from Scotland across the North Channel, secondly across the Isle of Man Basin and finally from Wales to south-east Ireland. Note though that these are direct journeys across the Irish Sea and do not account for oceanographic conditions of the time (e.g., tidal currents and dominant wind directions). It may be that short- and long-distance 'coastal tramping' along a north–south axis constrained by tidal currents, as envisaged by Garrow and Sturt (2011) for the Neolithic, may be a more appropriate analogue.

There are also questions regarding the nature and frequency of contact between Ireland, Britain and the outlying islands. In the Neolithic, there must have been seafaring networks across the Irish Sea, suggested by the movement of people, animals, objects and ideas and exemplified by the transition to agriculture, the Neolithic axe trade (including distinctive porcellanite axes sourced from north-east Ireland) and similarities in monument and pottery types (Garrow and Sturt 2011; Cooney et al. 2012). The situation in the Mesolithic is less clear. Individual finds such as the Ferriter's Cove cow bones indicate some contact by the final Mesolithic, but in general, the supporting evidence for maritime networks is rare. One of the central questions in the Irish Mesolithic has been the transition to a very different technique of stone working after c. 8800–8600 cal BP which is not paralleled in Britain but is found on the Isle of Man. The absence of British equivalents can be interpreted in two ways. Firstly, as an indication of isolation from Britain and very limited maritime networks. This, however, is at odds with the evidence for seafaring described above, including the Isle of Man material. Alternatively, it might be taken as evidence of deliberate cultural differentiation in response to high levels of maritime connectivity (Garrow and Sturt 2011; Cooney et al. 2012; Bell and Warren 2013; Warren 2015a, 2017).

What was taking place in the Late Glacial, to enable brief exploratory journeys, as suggested by the Alice and Gwendoline Cave patella (Dowd and Carden 2016), is also uncertain. Stable isotope evidence for the presence of marine resource-using populations at Kendricks Cave, North Wales (Richards et al. 2005), hints at a marine adaptation in the Irish Sea region but how this translates to seafaring is not clear. Recent discoveries of Ahrensburgian-type material (latter half of GS-1 to earliest Holocene) have been made on the Scottish island of Islay (visible from Northern Ireland on clear days). Despite slightly lower RSL, boats were necessary to reach Islay (Mithen et al. 2015), supporting the premise that Late Glacial populations were capable of seagoing or at least island-hopping voyages. An ability to cross even greater distances to reach Ireland would provide greater substantiation of the maritime abilities of these hunter-gatherers. Indeed, that such abilities were present is perhaps unsurprising in light of the extensive but indirect evidence from Scandinavia (the majority of Fosna/Hensbacka sites are located on palaeoshorelines, outer coasts and islands from c. 11,700–11,500 cal BP onwards in western Sweden and Norway) indicative of boat use and suggestive of seal hunting (Schmitt 2015; Schmitt and Svedhage 2015; Bjerck 2017) and the hypothesised links between Scandinavian Fosna/Hensbacka and Scottish groups across the Doggerland region during the GS-1/Holocene transition (Warren 2015a; Ballin and Bjerck 2016).

The ephemeral nature of seafaring and its attendant evidence makes it hard to address these questions. Nevertheless, certainly for the earlier parts of the period of study, the lack of information from contemporaneous shorelines where boats were built, launched and made landfall makes it very difficult to get definitive answers.

11.5.2.2 Coastal Adaptation and Initial Colonisation

One of the long-standing unresolved questions of Irish archaeology is the delay in human occupation relative to mainland Britain. While the pre-LGM absence is probably partly explicable by taphonomy and Ireland's peripheral location, the Late Glacial and Mesolithic delay is more uncertain, certainly in light of the observation (see above) that sea crossings were probably not an obstacle (Warren 2015a).

If the Alice and Gwendoline Cave patella (c. 12,800–12,6000 cal BP) is secure evidence of a pioneer incursion into Ireland, then Late Glacial populations were present in Britain at least 2000 years earlier. Reasons postulated for the delay centre on the impoverished nature of the Late Glacial Irish fauna, which lacked large herbivores other than giant deer (*Megaloceros giganteus*), small numbers of reindeer (but only towards the end of the GI-1 interstadial) and possibly red deer (Montgomery et al. 2014; Monaghan 2017). Crucially, horse, a key resource for British Late Glacial hunters, was absent from Ireland (Woodman 2015; Warren 2015a, 2017).

If one assumes that the British Isles were largely depopulated during the GS-1/Younger Dryas cold snap, then Ahrensburgian hunters had returned by c. 12,000 cal BP, and British Early Mesolithic populations were in place by c. 11,500 cal BP. Once again, there is a delay of c. 1500–2000 years until the colonisation of Ireland, represented by Mount Sandel. Pioneer incursion cannot be ruled out, as implied by the second cut-marked bear bone from Alice and Gwendoline Cave (11,000–10,400 cal BP), but full settlement does not appear to have taken place until after 10,000 cal BP. Again, this delay is generally viewed as product of Ireland's distinctive ecology, with no large ungulates surviving the Younger Dryas (Montgomery et al. 2014; Woodman 2015; Warren 2017).

The absence of large game may have been a deterrent to land-based hunters, both in the Late Glacial and Early Holocene. Therefore, it may be that deliberate importation of species such as wild pig by boat was needed to enable full settlement (Warren 2015a, 2017). However, if incoming populations had some degree of coastal adaptation, as suggested by the circumstantial evidence for seafaring, and the capacity to construct fish traps such as found in the Liffey Estuary (McQuade and O'Donnell 2007), the absence of large game might be expected to be less of a deterrent. The caveat is that we simply do not know what the marine and coastal ecology of the Irish coastline was like in the Late Glacial or aftermath of the GS-1/Younger Dryas and thus what resources were actually available (Woodman 2015). Linked to this are questions of when migratory species such as salmon or eels began to establish themselves in Irish rivers and estuaries. Effectively, while the technical challenge of seafaring may have been overcome, it is an open question as to whether the coastal environment was attractive or supportive of permanent long-term settlement until later in the Holocene.

With the gaps in the archaeological and marine palaeoecological records on both sides of the Irish Sea, it is difficult to see how this question can be resolved. There is no certainty that submerged sites can provide the necessary evidence, but without coastal evidence dating to the Late Glacial and Earlier Mesolithic, we cannot even begin reconstructing the ecological affordances of the coastline and inshore waters.

11.5.2.3 Coastal Change and Human Response

The presence of a maritime adaptation during the permanent settlement of Ireland is substantiated by saltwater fish remains at a handful of sites, chiefly Mount Sandel and Port of Larne for the Earlier Mesolithic and Ferriter's Cove, Baylet and Rockmarshall for the Later Mesolithic (Warren 2015b; Woodman 2015). The latter three sites also have associated shell middens. However, the extant evidence for this maritime adaptation is still sparse; fewer than ten sites have substantial quantities of saltwater fish remains, and there are no shell middens with confirmed dates to the Earlier Mesolithic (Milner and Woodman 2007). Moreover, only one coastal site has preserved fishing gear (North Wall Quay fish traps), though basket fish traps are also known from the inland lake site of Clowanstown. Fish remains were not found at North Wall Quay, but it is likely that exploited species included estuarine (e.g., herring, whiting, bass, sole, flounder, among others) and migratory ones (e.g., eel, salmon) (McQuade and O'Donnell 2007). Estuarine exploitation of migratory species is also suggested by the Cutts bone points (Warren 2015b; Woodman 2015). Marine mammal remains are also rare, with seal found only at Dalkey Island (Warren 2015b). Stable isotopes on human bone from Ferriter's Cove and a dog bone from Dalkey Island indicate a heavily marine-based diet, but on the other hand, human bones from elsewhere such as Killuragh Cave (inland), Rockmarshall (coastal) and Loughan (inland riverine) indicate a more balanced marine-terrestrial diet (Warren 2015b; Woodman 2015).

Notwithstanding the need to be wary of ascribing a single dietary adaptation to the whole island and entire Mesolithic (Woodman 2015), this lack and the spatio-temporal distribution of evidence raises two issues. Firstly, is it a product of submergence or erosion of coastal sites by rising sea level? Secondly, could rising sea level during the early Holocene have led to changing practices during the Mesolithic as coastal palaeolandscapes themselves changed in response?

Regarding the first issue, there are spatio-temporal disparities in the distribution of evidence across Ireland which suggest this is a possibility. For instance, in south-west Ireland, Ferriter's Cove was for many years the only site with evidence of coastal use (Woodman et al. 1999; Milner and Woodman 2007). This has since been supplemented by recent excavation of a final Mesolithic shell midden at Fanore (Lynch 2013). While this is undoubtedly a product of historical factors, such as a heavy focus on the north-east (Woodman 1978, 2015), it also mirrors the predicted deepening of the RSL lowstand from north-east to south-west, in that the greatest concentration of coastal sites has been found where the least land has been inundated (the north-east) and vice versa (i.e., south-west). That Ferriter's Cove and Fanore date to the final Mesolithic could be taken as an indication that earlier coastal evidence from the south-west is concentrated underwater. In contrast, the earliest dated sites with saltwater fish remains are Port of Larne and Mount Sandel, both located in the north-east relatively close to the contemporaneous shoreline because the lowstand was shallower. The absence of Earlier Mesolithic shell middens, which might be expected to have been created near the coast, supports this in that RSL was universally lower in the early Holocene and thus the relevant shorelines are now underwater.

The second issue is presently more speculative and harder to assess, particularly in light of the missing coastal parts of the record which make it hard to directly compare different time intervals. Nonetheless, we must consider that the early inhabitants of Ireland were living at in time of rapid RSL rise. Therefore, what might have been the response to flooding of river valleys to create extended inlets and embayments? Might the creation of larger areas of shallows and wetlands have increased the resources of shallow inshore waters and hence attracted increased coastal exploitation made possible by use of standing gear like the Dublin fish traps (McQuade and O'Donnell 2007)? Alternatively, could there have been an increased focus on topographic prominences, such as headlands, which were raised above rising sea levels and storm waves (e.g., Westley et al. 2014)? The likelihood is that different coastal landscapes had different resources and were used in different ways. However, without accurate reconstructions of what specific areas looked like or offered in resource terms, it becomes difficult to understand how regional strategies varied over time.

11.6 Future Research

If the argument that submerged landscapes can usefully contribute to Irish prehistory is accepted, then how can they be investigated? Given the rather limited archaeological and palaeoenvironmental evidence base, the key aim at this stage should be to enhance it.

With this in mind, the first issue to consider is which areas have the best prospects for finding submerged evidence. The likelihood is that submerged sites, if preserved, would be found at a range of depths around Ireland due to the varying pattern of RSL rise. In the north-east, extant models and raised beach evidence suggest that only the Late Glacial and Earlier Mesolithic coastal landscapes are missing and would be located in relatively shallow (<−20 to −30 m) water. As one moves south and west, the timespan of potential sites will increase, and earlier sites will be submerged at progressively greater depths (e.g., down to −40 to −60 m). This is most obvious for the south coast where most traces of early settlement will lie below sea level as shown by the existence of Neolithic megalithic tombs on the intertidal shore. Given the high-energy conditions that prevail across much of Ireland's coastline, preservation of submerged sites is most likely within sheltered bays, inlets and sea loughs where the extant intertidal record is most commonly preserved. Also, at times of lower sea level, these would have formed river valleys and estuarine locations, which seem to have been favoured places for settlement, judging from the extant intertidal and on-land record. Exploitation of such environments is also attested to by the North Wall quay fish traps (McQuade and O'Donnell 2007), but as shown by this site, archaeological material in such environments could be deeply buried as well as submerged.

The second consideration is how to go about conducting the investigation. There are four potential routes by which future work can proceed. These are not mutually exclusive, and the likelihood is that certain avenues of investigation will be more fruitful in some areas than others:

1. Start with known intertidal sites (both archaeological and palaeoenvironmental) to identify sources of archaeological material (e.g., washed up or eroded down) and determine if these extend into nearshore waters. The value of intensive work on Mesolithic intertidal sites has been demonstrated elsewhere, for example, in the Severn Estuary (Bell 2007), while working from onshore to offshore has already given us the first glimpses of Ireland's submerged prehistoric record as at Eleven Ballyboes (Westley 2015).
2. Use seabed mapping data to firm up evidence of RSL and coastal change and in turn produce more accurate palaeolandscape reconstructions that can go beyond shelf-scale depictions (Fig. 11.2). The ultimate aim would be to use these to guide archaeological investigation in deeper water and also to provide the proper landscape context needed to address some of the questions raised above. Fortunately, Ireland has one of the largest seabed mapping programs in the world (INFOMAR; Dorschel et al. 2011), and the resulting data have already been used in submerged landscape investigation (e.g., Westley et al. 2011a, b, 2014; Plets et al. 2015). However, these data have yet to be used to

their full potential, with areas outside the north coast largely un-investigated for archaeological purposes and a focus on seabed bathymetry rather than buried stratigraphy (e.g., Westley et al. 2011a). It is certainly notable that the Irish Sea has not been subject to the same level of geological reconstruction and palaeolandscape investigation as the North Sea (Gaffney et al. 2007; Cohen et al. 2017) where such work has allowed major advances in understanding of the submerged archaeological record.

3. Conduct archival research on potential submerged evidence located in museum and private collections, as done for the North Sea by Bynoe et al. (2016). This has yet to be done in Ireland. In fact, it is genuinely unclear as to whether submerged evidence, particularly palaeontological finds, actually exists in Irish museum collections. It is possible that the paucity of material is genuinely reflective of the absence of evidence but needs further research to confirm.
4. Finally, make efforts to raise awareness among users of the seabed, particularly the commercial fishing industry and sports divers, to ensure that chance finds are recognised, recovered and properly recorded. Again, similar initiatives have been successful elsewhere, most notably the Protocols for Archaeological Discoveries for the fishing and renewables industry in British waters (Wessex Archaeology 2013; The Crown Estate 2014) and the OHCCMAPP community engagement project in the Outer Hebrides (Benjamin et al. 2014).

11.7 Conclusion

The extant evidence clearly demonstrates the existence of prehistoric landscapes on Ireland's coast, seabed and intertidal zone, in some instances with archaeological material in a range of preservation states from reworked to in situ. It is also clear that these landscapes comprise the coastal zones potentially occupied during Ireland's earliest colonisation and settlement and, as such, could contribute to questions of seafaring and coastal adaptation that cannot be addressed from the terrestrial data alone.

Nonetheless, we must be realistic. Research to date has only scratched the surface and hinted at the full extent and nature of the offshore record. This is clearly illustrated by the fact that the majority of evidence is intertidal rather than subtidal, and of the latter, no definitely in situ material other than the buried North Wall Quay fish traps has been found. We should also consider that Ireland's unique and complex pattern of glaciation, and of sea-level and palaeogeographic change, when coupled with present-day oceanographic conditions, produces a taphonomic situation that is not well suited to extensive archaeological preservation. Simply put, Irish submerged landscapes are not like those of the Baltic or the North Sea. The archaeological record is much shorter, concentrated within the Holocene, with perhaps brief windows into the Late Glacial. Extensive submerged landscapes are likely to be much rarer, given limited shelf exposure, glaciation and high-energy conditions. A situation where fragments of landscape or pockets of preservation characterise the Irish shelf is much more likely. Moreover, there is no guarantee that submerged sites will conclusively resolve the aforementioned questions on seafaring, coastal adaptation and human responses. With the best will in the world, we should accept that the idealised submerged site with fine-grained in situ preservation of seafaring and fishing activities may prove elusive.

On the other hand, by avoiding the offshore landscape, we also ignore a large chunk of Ireland's prehistory and, arguably, will make little headway in dealing with the above questions. We should also not underestimate the knowledge to be gained from looking at the environmental aspects of the submerged record. Improved knowledge of sea-level change and reconstruction of local- to regional-scale palaeogeography and environmental change can place the known archaeological record in a more accurate landscape context and also address some of the nuances surrounding seafaring and coastal adaptation, particularly in relation to ecological affordances and palaeoceanographic conditions. Encouragingly, there are identifiable routes of investigation, based on the limited work done so far in Ireland, as well as approaches adopted from elsewhere. Particularly advantageous in the Irish context is the ready availability of vast quantities of seabed mapping data which have yet to be fully examined from an archaeological perspective. Ultimately though, the investigation of submerged landscapes is not something that can be done overnight. The realistic aspiration would be that, in the coming years, solid foundations can be put in place that will aid understanding and management and, in the long term, facilitate greater discovery of Ireland's submerged prehistory.

11.8 Management of the Underwater Cultural Heritage

The responsible authority for underwater cultural heritage including submerged landscapes lies with two organisations. In the Republic of Ireland, this is the Underwater Archaeology Unit of the National Monuments Service which is itself part of the Department of Culture, Heritage and the Gaeltacht (DCHG) (http://archaeology.ie/, accessed December 2018). In Northern Ireland, it is the Historic Environment Division of the Department for Communities (DfC) (https://www.communities-ni.gov.uk/articles/dfc-marine-historic-environment, accessed December 2018).

Acknowledgements Peter Woodman sadly and unexpectedly passed away when this chapter was near completion. It would have been impossible to write without his unparalleled knowledge of Irish prehistory. Numerous other colleagues also deserve thanks for their assistance with research, fieldwork or simply putting up with long-running discussions on submerged landscapes. These are Rory Quinn, Ruth Plets, Colin Breen, Wes Forsythe, Rory McNeary, Fraser Sturt, Justin Dix, Robin Edwards, Trevor Bell and Brian McNaught. Finally, thanks also to Graeme Warren and two anonymous reviewers for their insightful comments which improved the original draft.

References

Adams AL (1883) Report on the history of Irish fossil mammals. Proceedings of the Royal Irish Academy. Science 3:89–100

Atkins WS (1997) Wind and water. In: Barne JH, Robson CF, Kaznowska SS, Doody JP, Davidson NC, Buck AL (eds) Coasts and seas of the United Kingdom. Region 17 Northern Ireland. Joint Nature Conservation Committee, Peterborough, pp 28–32

Bailey G (2014) New developments in submerged prehistoric archaeology: an overview. In: Evans A, Flatman JC, Flemming NC (eds) Prehistoric archaeology on the continental shelf. A global review. Springer, Dordrecht, pp 291–300

Bailey G, Flemming N (2008) Archaeology of the continental shelf: marine resources, submerged landscapes and underwater archaeology. Quat Sci Rev 27:2153–2165

Bailey G, Sakellariou D (2012) SPLASHCOS: submerged prehistoric archaeology and landscapes of the continental shelf. Antiquity project gallery. http://antiquity.ac.uk/projgall/sakellariou334/

Ballantyne CK, Ó Cofaigh C (2017) The last Irish ice sheet: extent and chronology. In: Coxon P, McCarron S, Mitchell F (eds) Advances in Irish Quaternary studies. Atlantis Press, Paris, pp 101–149

Ballin T, Bjerck H (2016) Lost and found twice: discussion of an early post-glacial single-edged tanged point from Brodgar on Orkney, Scotland. J Lithic Stud 3:1–20

Bayliss A, Woodman PC (2009) A new Bayesian chronology for Mesolithic occupation at Mount Sandel, Northern Ireland. Proc Prehist Soc 75:101–123

Bell M (2007) Prehistoric coastal communities. The Mesolithic in Western Britain, CBA research report 149. Council for British Archaeology, York

Bell M, Warren G (2013) The Mesolithic. In: Ransley J, Sturt F, Dix J, Adams J, Blue L (eds) People and the sea: England's maritime and marine historic environment resource assessment and research agenda, CBA research report, vol 171. Council for British Archaeology, York, pp 30–49

Bell T, O'Sullivan A, Quinn R (2006) Discovering ancient landscapes under the sea. Archaeol Irel 20:12–17

Benjamin J, Bicket A, Anderson D, Hale A (2014) A multi-disciplinary approach to researching the intertidal and marine archaeology in the Outer Hebrides, Scotland. J I Coast Archaeol 9:400–424

Bjerck H (2017) Settlements and seafaring: reflections on the integration of boats and settlements among marine foragers in Early Mesolithic Norway and the Yámana of Tierra del Fuego. J I Coast Archaeol 12:276–299

Bradley SL, Milne GA, Shennan I, Edwards RJ (2011) An improved Glacio isostatic adjustment model for the British Isles. J Quat Sci 26:541–552

Brady N, Pollard E (2013) Archaeology and development in a maritime context, highlighting Drogheda and the

Boyne estuary as a case study. In: Jameson JJ, Eogan J (eds) Training and practice for modern day archaeologists. Springer, New York, pp 237–269

Breen C (1993) The 'Pearl', Tran a Fearla, Allihies. Excavations Bulletin 1993: Summary accounts of archaeological excavations in Ireland. http://www.excavations.ie

Breen C, Forsythe W (2001) Management and protection of the maritime cultural resource in Ireland. Coast Manag 29:41–51

Brooks AJ, Edwards R (2006) The development of a sea-level database for Ireland. Irish J Earth Sci 24:13–27

Brooks AJ, Bradley SL, Edwards RJ, Milne GA, Horton BP, Shennan I (2008) Post-glacial relative sea-level observations from Ireland and their role in glacial rebound modelling. J Quat Sci 23:175–192

Brooks AJ, Bradley SL, Edwards RJ, Goodwyn N (2011) The paleogeography of Northwest Europe during the last 20,000 years. J Maps 2011:573–587

Bynoe R, Dix JK, Sturt F (2016) Of mammoths and other monsters: historic approaches to the submerged Palaeolithic. Antiquity 90:857–875

Campbell K (2003) Arklow Bank Wind Park. Offshore monitoring of dredging. Excavations Bulletin 2003: Summary accounts of archaeological excavations in Ireland. http://www.excavations.ie

Carter RWG (1982) Sea-level changes in Northern Ireland. Proc Geol Assoc 93:7–23

Charlesworth JK (1963) Historical geology of Ireland. Oliver and Boyd, Edinburgh

Clark CD, Hughes ALC, Greenwood SL, Jordan C, Sejrup HP (2012) Pattern and timing of retreat of the last British-Irish Ice Sheet. Quat Sci Rev 44:112–146

Coffey G, Praeger RL (1904/5) The Antrim Raised Beach: a contribution to the Neolithic history of the North of Ireland. Proc R Ir Acad Section C: Archaeol Celtic Stud 25:143–200

Cohen K, Westley K, Erkens G, Hijma MP, Weerts HJT (2017) The North Sea. In: Flemming NC, Harff J, Moura D, Burgess A, Bailey GN (eds) Submerged landscapes of the European continental shelf: Quaternary paleoenvironments. Wiley Blackwell, Chichester, pp 147–186

Common R (1933) Unusual Larne flint flake. Irish Nat J 4:139–140

Cooney G, Mandal S (1998) The Irish stone axe project: monograph 1. Wordwell, Dublin

Cooney G, Mandal S, O'Keefe E, Warren G (2012) Rathlin in early prehistory. In: Forsythe W, McConkey R (eds) Rathlin Island. An archaeological survey of a maritime landscape. TSO, Belfast, pp 46–85

Costa LJ, Sternke F, Woodman PC (2001) The analysis of a lithic assemblage from Eleven Ballyboes, County Donegal. Ulst J Archaeol 60:1–8

Coxon P, Mitchell F, von Engelbrechten S, Vaughan L (2017) Interglacial sequences. In: Coxon P, McCarron S, Mitchell F (eds) Advances in Irish Quaternary studies. Atlantis Press, Paris, pp 43–66

Davie S, Lordan C (2011) Examining changes in Irish fishing practices in response to the cold long-term plan. ICES J Mar Sci 68:1638–1646

Dorschel B, Wheeler AJ, Monteys X, Verbruggen K (2011) Atlas of the deep-water seabed, 1st edn. Springer, Dordrecht

Dowd M, Carden R (2016) First evidence of a Late Upper Palaeolithic human presence in Ireland. Quat Sci Rev 139:158–163

Dowling LA, Sejrup HP, Coxon P, Heijnis H (1998) Palynology, aminostratigraphy and U-series dating of marine Gortian interglacial sediments in Cork Harbour, southern Ireland. Quat Sci Rev 17:945–962

Driscoll K (2006) The early prehistory in the west of Ireland: investigations into the social archaeology of the Mesolithic, west of the Shannon, Ireland. Unpublished MLitt Thesis, National University of Ireland, Galway

Edwards RJ, Brooks AJ (2008) The island of Ireland: drowning the myth of an Irish land-bridge? In: Davenport J, Sleeman D, Woodman P (eds) Mind the pap: postglacial colonisation of Ireland. Special Supplement to the Irish Naturalists' Journal, vol 29, pp 19–34

Edwards RJ, Craven K (2017) Relative Sea-level change around the Irish coast. In: Coxon P, McCarron S, Mitchell F (eds) Advances in Irish Quaternary studies. Atlantis Press, Paris, pp 181–215

Edwards RJ, Brooks AJ, Shennan I, Milne G, Bradley S (2008) Reply: postglacial relative sea-level observations from Ireland and their role in glacial rebound modelling. J Quat Sci 23:821–825

Evans CDR (1995) Wind and water. In: Barne JH, Robson CF, Kaznowksa SS, Doody JP (eds) Coasts and seas of the United Kingdom. Region 12 Wales: Margam to Little Orme. Joint Nature Conservation Committee, Peterborough, pp 29–33

Evans W, Benetti S, Sachetti F, Jackson DWT, Dunlop P, Monteys X (2015) Bedforms on the northwest Irish shelf: indication of modern active sediment transport and over printing of paleo-glacial sedimentary deposits. J Maps 11:561–574

Fischer A (2011) Stone age on the continental shelf: an eroding resource. In: Benjamin J, Bonsall C, Pickard C, Fischer A (eds) Submerged prehistory. Oxbow, Oxford, pp 298–312

Fry M (2000) Coiti: Logboats from Northern Ireland. Greystone Press, Antrim

Gaffney V, Thomson K, Fitch S (2007) Mapping Doggerland: the Mesolithic landscapes of the southern North Sea. Archaeopress, Oxford

Garrow D, Sturt F (2011) Grey waters bright with Neolithic Argonauts? Maritime connections and the Mesolithic-Neolithic transition within the 'western seaways' of Britain, c. 5000-3500 BC. Antiquity 85:59–72

GEBCO (2014) The GEBCO_2014 Grid, version 20150318. www.gebco.net

Gibson C (ed) (2011) Northern Ireland. State of the Seas Report. Agri-Food and Biosciences Institute and Northern Ireland Environment Agency, Belfast

Highley D, Hetherington L, Brown T, Harrison D, Jenkins G (2007) The strategic importance of the marine aggregates industry to the UK. British Geological Survey Research Report OR/07/019

Kandrot S, Farrell E, Devoy R (2016) The morphological response of foredunes at a breached barrier system to winter 2013/14 storms on the south-west coast of Ireland. Earth Surf Process Landf 41:2123–2136

Kuchar J, Milne G, Hubbard A, Patton H, Bradley S, Shennan I, Edwards RJ (2012) Evaluation of a numerical model of the British–Irish ice sheet using relative sea-level data: implications for the interpretation of trimline observations. J Quat Sci 27:597–605

Lübke H, Schmölckem U, Tauber F (2011) Mesolithic hunter-fishers in a changing world: a case study of submerged sites on the Jäckelberg, Wismar Bay, north-eastern Germany. In: Benjamin J, Bonsall C, Pickard C, Fischer A (eds) Submerged prehistory. Oxbow, Oxford, pp 21–37

Lynch M (2013) Prehistoric coastal midden sites in the west. In: Higgins J, Conneely A, Gibbons M (eds) Irish maritime heritage. Proceedings of the 3rd Galway international heritage conference. Crow's Rock Press, Galway, pp 1–10

Manning P (1970) Geology of Belfast and the Lagan valley. Her Majesty's Stationary Office, Belfast

McCabe AM (2008) Comment: postglacial relative sea-level observations from Ireland and their role in glacial rebound modelling. J Quat Sci 23:817–820

McCabe AM, Cooper JAG, Kelley JT (2007) Relative Sea-level changes from NE Ireland during the last glacial termination. J Geol Soc Lond 164:1059–1063

McErlean T, McConkey R, Forsythe W (2002) Strangford Lough: an archaeological survey of the maritime cultural landscape. Blackstaff Press, Belfast

McNaught B (1998) Early Mesolithic site discovered in Donegal. Donegal Annual 50:64–65

McQuade M (2008) Gone fishin. Archaeol Irel 22:8–11

McQuade M, O'Donnell L (2007) Late Mesolithic fish traps from the Liffey estuary, Dublin, Ireland. Antiquity 81:569–584

Milner N, Woodman PC (2007) Deconstructing the myths of Irish shell middens. In: Milner N, Craig O, Bailey G (eds) Shell middens in Atlantic Europe. Oxbow, Oxford, pp 101–110

Mitchell GF (1976) The Irish landscape. Collins, London

Mithen S, Wicks K, Pirie A, Riede F, Lane C, Banerjea R, Cullen V, Gittens M, Pankhurst N (2015) A Lateglacial archaeological site in the far north-west of Europe at Rubha Port an t-Seilich, isle of Islay, western Scotland: Ahrensburgian-style artefacts, absolute dating and geoarchaeology. J Quat Sci 30:396–416

Mol D, Post K, Reumer J, van der Plicht J, de Vos J, van Geel B, van Reenen G, Pals JP, Glimmerveen J (2006) The Eurogeul – first report of the palaeontological, palynological and archaeological investigations of this part of the North Sea. Quat Int 142–143:178–185

Monaghan NT (2017) Irish Quaternary vertebrates. In: Coxon P, McCarron S, Mitchell F (eds) Advances in Irish Quaternary studies. Atlantis Press, Paris, pp 255–291

Montgomery WI, Provan J, McCabe AM, Yalden DW (2014) Origin of British and Irish mammals: disparate post-glacial colonisation and species introductions. Quat Sci Rev 98:144–165

Morrison M, Stephens N, Jope M, Dance S, Anderson F, Round F, Godwin H, Willis E (1965) A submerged late-Quaternary deposit at Roddans Port on the north-east coast of Ireland. Philos Trans R Soc Lond, Ser B 249:221–255

Movius HL, Jessen K, McMillan N, Richardson F (1940/41) An early post-glacial archaeological site at Cushendun, County Antrim. Proc R Ir Acad Section C: Archaeol Celtic Stud Hist Linguist Lit 46:1–84

Movius HL, Jessen K, Benninghoff WS, McMillan N (1953/54) Curran Point, Larne, County Antrim: the type site of the Irish Mesolithic. Proc R Ir Acad Section C: Archaeol Celtic Stud Hist Linguist Lit 56:1–195

O'Connell M, Molloy K (2017) Mid- and late-Holocene environmental change in Western Ireland: new evidence from coastal peats and fossil timbers with particular reference to relative sea-level change. The Holocene 27(12):1825–1845

O'Mahony C, Sutton G, McMahon T, O'Cinneide M, Nixon E (2008) Policy report – issues and recommendations for the development and regulation of marine aggregate extraction in the Irish Sea. Marine Environment and Health Series 32

O'Raw J (2003) The Saldahna disaster in the maritime cultural landscape of Lough Swilly. Unpublished MSc Thesis, Ulster University, Coleraine

O'Riordain SP (1953/54) Lough Gur excavations: Neolithic and Bronze age houses on Knockadoon. Proc R Ir Acad Section C: Archaeol Celtic Stud Hist Linguist Lit 56:297–459

O'Sullivan A (2001) Foragers, farmers and fishers in a coastal landscape: an intertidal archaeological survey of the Shannon estuary. Royal Irish Academy, Dublin

Patterson WH (1892) On a newly discovered site for worked flints in the County of Down. J R Soc Antiqu Ireland 2:154–155

Pettitt P, White M (2012) The British Palaeolithic. Human societies at the edge of the Pleistocene world. Routledge, Abingdon

Plets RMK, Callard SL, Cooper JAG, Long AJ, Quinn RJ, Belkap DF, Edwards RJ, Jackson DWT, Kelley JT, Long D, Milne GA, Monteys X (2015) Late Quaternary evolution and sea-level history of a glaciated marine embayment, Bantry Bay, SW Ireland. Mar Geol 369:251–272

Power D (1994) Archaeological inventory of County Cork, vol 2 (East and South Cork). Stationary Office, Dublin

Praeger RL (1891) A report on the estuarine clays of the North-East of Ireland. Proc R Ir Acad 2:212–289

Praeger RL (1892) The Irish Post-Glacial estuarine deposits. Ir Nat 1:138–141

Quinn R, Lafferty B, Westley K, Plets R (2012) The natural environment. In: Forsythe W, McConkey R (eds) Rathlin Island. An archaeological survey of a maritime landscape. TSO, Belfast, pp 20–44

Richards MP, Jacobi R, Cook J, Pettitt PB, Stringer CB (2005) Isotope evidence for the intensive use of marine foods by Late Upper Palaeolithic humans. J Hum Evol 49:390–394

Roe HM, Swindles GT (2008) Post-glacial evolution of the Lower Dun Valley, Co. Antrim: a re-examination of the Movius (1940) sections at Cushendun. In: Whitehouse NJ, Roe HM, McCarron S, Knight J (eds) North of Ireland field guide. Quaternary Research Association, London, pp 117–127

Savage RJG (1966) Irish Pleistocene mammals. Irish Nat J 15:117–130

Schmitt L (2015) The continental connection: a coastal excursion in time, geography and culture. Nor Archaeol Rev 48:21–26

Schmitt L, Svedhage K (2015) Chronological aspects of the Hensbacka – a group of hunter-gatherers/fishers on the west coast of Sweden during the Pleistocene/Holocene transition: an example of early coastal colonization. Dan J Archaeol 4:1–7

Shee Twohig ES (1995) An inter-tidal passage tomb at 'the Lag', Ringarogy Island, Co. Cork. Archaeol Irel 9:7–9

Simpson D (2008) Greyabbey Bay, Strangford Lough: marine, non-marine and terrestrial micro- and macrofossils. In: Whitehouse NJ, Roe HM, McCarron S, Knight J (eds) North of Ireland field guide. Quaternary Research Association, London, pp 155–167

Singh G (1963) Pollen-analysis of a deposit at Roddans Port, Co. Down, N. Ireland, bearing reindeer antler fragments. Grana Palynol 4:466–474

Stewart SA (1911) The latest fluctuations of the sea-level on our own coasts. Ir Nat 20:17–27

Stillman CJ (1968) The post glacial change in sea level in South-Western Ireland: new evidence from freshwater deposits on the floor of Bantry Bay. Sci Proc R Dublin Soc A3:125–127

Sturt F, Garrow D, Bradley S (2013) New models of North West European Holocene palaeogeography and inundation. J Archaeol Sci 40:3963–3976

The Crown Estate (2014) Protocol for archaeological discoveries: offshore renewables projects. Wessex Archaeology, Salisbury, on behalf of The Crown Estate

Tizzard L, Baggaley P, Firth A (2011) Seabed prehistory: investigating palaeolandsurfaces with Palaeolithic remains from the southern North Sea. In: Benjamin J, Bonsall C, Pickard C, Fischer A (eds) Submerged prehistory. Oxbow, Oxford, pp 65–74

Tolan-Smith C (2009) Mesolithic Britain. In: Bailey G, Spikins P (eds) Mesolithic Europe. Cambridge University Press, Cambridge, pp 132–157

Van Kolfschoten T, Laban C (1995) Pleistocene terrestrial mammal faunas from the North Sea area. Med Rijks Geol Dienst 52:135–151

Warren G (2015a) Britain and Ireland inside Mesolithic Europe. In: Anderson-Whymark H, Garrow D, Sturt F (eds) Continental connections: exploring cross-channel relationships from the Mesolithic to the Iron age. Oxbow, Oxford, pp 43–58

Warren G (2015b) 'Mere food gatherers they, parasites upon nature…': food and drink in the Mesolithic of Ireland. Proc R Ir Acad 115C:1–26

Warren G (2017) The human colonisation of Ireland in Northwest European context. In: Coxon P, McCarron S, Mitchell F (eds) Advances in Irish Quaternary studies. Atlantis Press, Paris, pp 293–316

Warren G, Davis S, McClatchie M, Sands R (2014) The potential role of humans in structuring the wooded landscapes of Mesolithic Ireland: a review of data and discussion of approaches. Veg Hist Archaeobotany 23:629–646

Wessex Archaeology (2013) Fishing industry protocol for archaeological discoveries final report. Wessex Archaeology, Salisbury

Westley K (2015) Submerged archaeological landscape investigation, Eleven Ballyboes, Republic of Ireland. Int J Naut Archaeol 44:243–257

Westley K, Edwards R (2017) Irish Sea and Atlantic Margin. In: Flemming NC, Harff J, Moura D, Burgess A, Bailey GN (eds) Submerged landscapes of the European continental shelf: Quaternary paleoenvironments. Wiley Blackwell, Chichester, pp 241–280

Westley K, Quinn R, Forsythe W, Plets R, Bell T, Benetti S, McGrath F, Robinson R (2011a) Mapping submerged landscapes using multibeam bathymetric data: a case study from the north coast of Ireland. Int J Naut Archaeol 40:99–112

Westley K, Bell T, Plets R, Quinn R (2011b) Investigating submerged archaeological landscapes — a research strategy illustrated with case studies from Ireland and Newfoundland, Canada. In: Benjamin J, Bonsall C, Pickard C, Fischer A (eds) Submerged prehistory. Oxbow, Oxford, pp 129–144

Westley K, Plets R, Quinn R (2014) Holocene palaeogeographic reconstructions of the Ramore Head area, Northern Ireland, using geophysical and geotechnical data: palaeo-landscape mapping and archaeological implications. Geoarchaeol Int J 29:411–430

Whitehouse NJ, Watson J, Turney C (2008) Roddans Port Late Glacial site. In: Whitehouse NJ, Roe HM, McCarron S, Knight J (eds) North of Ireland field guide. Quaternary Research Association, London, pp 168–173

Whitehouse NJ, Schulting R, McClatchie M, Barratt P, McLaughlin TR, Bogaard A, Colledge S, Marchant R, Gaffrey J, Bunting MJ (2014) Neolithic agriculture on the European western frontier: the boom and bust of early farming in Ireland. J Archaeol Sci 51:181–205

Williams DM, Doyle E (2014) Dates from drowned mid-Holocene landscapes off the Central Western Irish seaboard. Irish J Earth Sci 32:23–27

Wilson P, Westley K, Plets R, Dempster M (2011) Radiocarbon dates from the inter-tidal peat bed at Portrush, County Antrim. Ir Geogr 44:323–329
Woodman PC (1977) A narrow-blade Mesolithic site at Glynn, County Antrim. Ulst J Archaeol 40:12–20
Woodman PC (1978) The Mesolithic in Ireland. Archaeopress, Oxford
Woodman PC (2012) Making yourself at home on an island: the first 1000 years (+?) of the Irish Mesolithic. Proc Prehist Soc 78:1–34
Woodman PC (2015) Ireland's first settlers. Time and the Mesolithic. Oxbow, Oxford
Woodman PC, McCarthy M, Monaghan N (1997) The Irish Quaternary Fauna project. Quat Sci Rev 16:129–159
Woodman PC, Anderson E, Finlay N (1999) Excavations at Ferriter's Cove 1983–1995. Last foragers/first farmers in the Dingle peninsula. Wordwell, Dublin

France: Submerged Prehistory on Atlantic and Mediterranean Coasts

12

Cyrille Billard, Marie-Yvane Daire and Chloé Martin, with contributions by Yves Billaud, Catherine Bizien-Jaglin, Antoine Chancerel, Dominique Cliquet, Nathalie Fourment, Henri Gandois, Briagell Huet, Marine Laforge, Loïc Langouët, Luc Laporte, Jean-Marc Large, Frédéric Leroy, Elías López-Romero, Léopold Maurel, Jean-Laurent Monnier, Pierre Régaldo, Anne Ropars, Pierre Stéphan, and Luc Vallin

Abstract

The French coastline accounts for many submerged remains of prehistoric sites along all its coastlines. Undisturbed sites including intertidal deposits containing artefacts give a total of 180 find spots. Each geographical region has its own specific features: submerged peats and fish weirs along the Channel and Atlantic coasts, faunal assemblages from the Channel and the North Sea, underwater sea caves and karsts along parts of the Mediterranean coast and occupations dated to later prehistory on the Mediterranean and in the south-western coastal lagoons. Coastal archaeology successfully confirms the high potential and good preservation of submerged sites, particularly organic artefacts such as wooden objects, features related to fishing and faunal assemblages. New types of sites and research issues have emerged over the past decade, particularly Palaeolithic sites, Neolithic axe deposits in association with megalithic complexes and fishing structures reflecting a widespread activity along the coasts of the Channel and the Atlantic Ocean. The implementation of a scientific programme relating to submerged sites is all the more urgent since offshore industrial projects are rapidly increasing along the French coasts.

Keywords

Megaliths · Fish weirs · Cosquer Cave · Middle Palaeolithic sites · Pleistocene fauna · Karst caves · Intertidal archaeology

C. Billard (✉) · D. Cliquet
DRAC-SRA Normandie, Caen, France

UMR 6566 Centre de Recherche en Archéologie, Archéosciences, Histoire, Caen, France
e-mail: cyrille.billard@culture.gouv.fr; dominique.cliquet@culture.gouv.fr

M.-Y. Daire
CNRS, UMR 6566 Centre de Recherche en Archéologie, Archéosciences, Histoire, Rennes, France

Association Manche Atlantique pour la Recherche Archéologique dans les Îles, Rennes, France
e-mail: marie-yvane.daire@univ-rennes1.fr

G. Bailey et al. (eds.), *The Archaeology of Europe's Drowned Landscapes*, Coastal Research Library 35, https://doi.org/10.1007/978-3-030-37367-2_12

12.1 Introduction

The French coastline extends for 5834 km and faces four marine basins with highly varied coastal and offshore conditions: the southern edge of the North Sea, La Manche or the (English) Channel, the Atlantic and the Mediterranean. Extensive offshore areas of submerged landscape were exposed at the Last Glacial Maximum, especially in the north-west (Farr et al. 2017). These submerged landscapes are of particular interest because they were adjacent to some of the most important concentrations of Palaeolithic and Mesolithic sites in Europe, and their emergence and subsequent inundation are likely to have played an important role in changing patterns of settlement and demography in France and between France and the adjacent regions of Britain and northern Spain (see Bailey et al., Chap. 10, this volume; Arias, Chap. 13, this volume). Little is presently known about the archaeology of deeper areas of this extensive submerged landscape apart from isolated materials, mostly animal bones, dredged up during offshore fishing and industrial activity. Systematic investigation of prehistoric remains is largely confined to shallower water and the intertidal zone.

Despite its variability in tidal range, climate and forms of coastal relief, France harbours many such remains of prehistoric sites inundated by sea-level rise (Fig. 12.1). Interest in this theme began in the nineteenth century with

C. Martin
CNRS, UMR 6566 Centre de Recherche en Archéologie, Archéosciences, Histoire, Rennes, France

Association Manche Atlantique pour la Recherche Archéologique dans les Îles, Rennes, France

Centre Régional d'Archéologie d'Alet, Rennes, France

Y. Billaud · F. Leroy
Département des Recherches Archéologiques Subaquatiques, Marseille, France
Sous-Marines, Marseille, France
e-mail: yves.billaud@culture.gouv.fr; frederic.leroy@culture.gouv.fr

C. Bizien-Jaglin
Centre Régional d'Archéologie d'Alet, Rennes, France

A. Chancerel
Musée National de Préhistoire, Les Eyzies, France
e-mail: antoine.chancerel@culture.gouv.fr

N. Fourment · L. Maurel · P. Régaldo
DRAC-SRA Grande Aquitaine, Bordeaux, France
e-mail: nathalie.fourment@culture.gouv.fr; leopold.maurel@culture.gouv.fr; pierre.regaldo@culture.gouv.fr

H. Gandois
UMR 6566 Centre de Recherche en Archéologie, Archéosciences, Histoire, Caen, France

UMR 8215, Université Paris 1, Panthéon, Sorbonne, Rennes, France

B. Huet
UMR 6566 Centre de Recherche en Archéologie, Archéosciences, Histoire, Caen, France

M. Laforge
UMR 6566 Centre de Recherche en Archéologie, Archéosciences, Histoire, Caen, France

ÉVEHA, Vezin-Le-Coquet, Rennes, France
e-mail: marine.laforge@eveha.fr

L. Langouët
Association Manche Atlantique pour la Recherche Archéologique dans les Îles, Rennes, France

Centre Régional d'Archéologie d'Alet, Rennes, France

L. Laporte · J.-L. Monnier
CNRS, UMR 6566 Centre de Recherche en Archéologie, Archéosciences, Histoire, Rennes, France
e-mail: luc.laporte@univ-rennes1.fr; jean-laurent.monnier@univ-rennes1.fr

J.-M. Large
Groupe vendéen d'études préhistoriques, membre associé UMR 6566 Centre de Recherche en Archéologie, Archéosciences, Histoire, Rennes, France

E. López-Romero
Université de Bordeaux, Bordeaux, France
e-mail: elias.lopez-romero@u-bordeaux-montaigne.fr

A. Ropars
DRAC-SRA Normandie, Caen, France
e-mail: anne.ropars@culture.gouv.fr

P. Stéphan
UMR 6554 LETG-Brest Géomer CNRS – Institut Universitaire Européen de la Mer, Brest, France

L. Vallin
DRAC-SRA Hauts-de-France, Lille, France
e-mail: luc.vallin@culture.gouv.fr

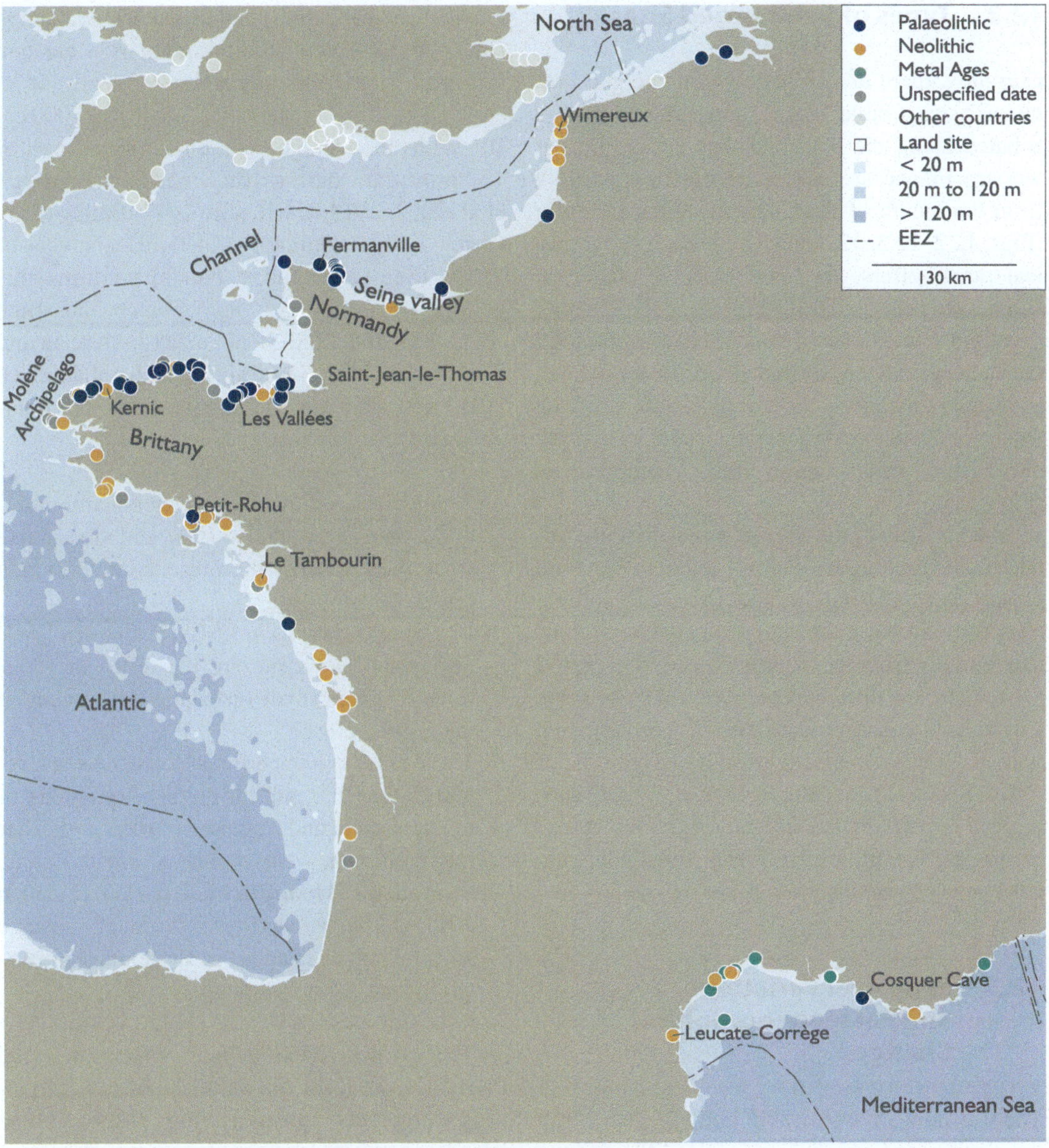

Fig. 12.1 Map of prehistoric submerged sites off the French coasts. Site information from the SPLASHCOS Viewer http://splashcos-viewer.eu. Drawing by Moritz Mennenga

observations of megalithic monuments below high-tide level. Here, we focus on sites with undisturbed material earlier than 5500 cal BP, including deposits in the intertidal zone that contain artefacts and fauna, Bronze Age sites and some multiperiod sites that have been submerged. Sites on the landward side of the shoreline that are in the process of being eroded by wave action are intentionally excluded.

For the presentation of the sites, we divide these into two main groups, the first associated with Palaeolithic sites and Pleistocene faunas and the second with Neolithic and later sites and monuments. Within each group, we present the material in broadly geographical order, working from north to south along the Channel and Atlantic coasts and from west to east along the Mediterranean.

12.2 Types of Submerged Sites

Currently, there are 180 known submerged prehistoric sites (Table 12.1). These are widely distributed along the coastline, but the regions of western France are best represented, especially from the mouth of the Charente to the Cotentin (Figs. 12.2, 12.3, 12.4 and 12.5). Each geographical area offers its own specific features: submerged peats and fish weirs along the Channel and the Atlantic coasts, faunal assemblages along the Channel and the North Sea, occupations dated to later prehistory in the coastal lagoons of the south-west and the Mediterranean and submarine caves and karsts along the Mediterranean (Fig. 12.6).

The majority of the sites (82) are of indeterminate prehistoric age. Of those that can be more closely dated, Palaeolithic sites (42) and Neolithic sites (40) are the main categories with a significant number from the Middle Palaeolithic period (16). As for the different categories of sites, stone fish weirs of indeterminate date (57) predominate, with just 3 wooden examples of Late Neolithic or Early Bronze Age date, followed by 26 submerged or partially submerged megalithic monuments of Neolithic age. The remainder are cultural layers of uncertain function.

12.3 Sea-Level Variations and Palaeogeographical Changes

There is now a considerable body of data about the pattern and chronology of late Pleistocene sea-level change on the French coastline and its relationship to global sea-level changes, based on evidence from raised beaches, offshore sediments, submerged peats and the sedimentary infills of palaeo-valleys, especially from Normandy and Brittany in the north-west and the Golfe de Gascogne in the south-west (Coutard 2003; Toucanne 2008; Laforge 2012; Sorrel et al. 2012; Stéphan and Goslin 2014; Stéphan et al. 2015).

As elsewhere, the general pattern of change since the Last Glacial Maximum is a eustatic sea-level rise from a maximum depth of c. −125 m at c. 17,000 cal BP to −5 m at c. 7000 cal BP with an average rate of sea-level rise of 12 mm/year during this period (Lambeck et al. 2002, 2014; Harff et al. 2017; Bailey et al., Chap. 1, this volume). A detailed analysis of SLIPs (Sea-Level Index Points) confirms this general trend along the French coastline (Figs. 12.7 and 12.8; Stéphan 2012; Stéphan and Goslin 2014; Goslin et al. 2015; Stéphan et al. 2015) and supplies additional details that demonstrate the following:

- Sea-level rise slowed down significantly after 7000 years ago, with a slow and steady rise since then towards the present sea-level at a rate of ≤1 mm/year.
- Regional variations in RSL (relative sea level) during the Holocene are the result of variations in glacio-hydro-isostatic adjustment of the Earth's crust.
- There is no evidence for a Holocene sea level higher than the present or for alternations of transgressive and regressive phases. Earlier interpretations to this effect (e.g., Morzadec-Kerfourn 1974; Ters 1986; Stéphan 2011) failed to take into account isostatic adjustments or changes in shoreline sedimentation and geomorphology.

The Molène archipelago exemplifies the impact of sea-level rise in terms of coastal palaeogeographical changes (Fig. 12.9). These islands constitute the emerged part of a large shallow-water rocky platform that was attached to the mainland at the Last Glacial Maximum. With the gradual rise of sea level, this was transformed at first into a single, large island and then into a variety of smaller islands and islets, removing from view almost of traces of human activity on this submerged landscape apart from a number of stone fish traps (Dréano et al. 2013; Gandois et al. 2013a, b; Stéphan et al. 2013; Pailler et al. 2014).

Table 12.1 Proportions of French submerged sites by type of site and chronology

Type of site	Prehistoric	Palaeolithic	Middle Palaeolithic	Neolithic	Early Neolithic	Late Neolithic/Early Bronze	Total	
							N	%
Wooden fish weirs	0	0	0	0	0	3	**3**	***1.7***
Stone fish weirs	57	0	0	0	0	0	**57**	***31.7***
Marsh/fauna	0	0	0	5	0	0	**5**	***2.8***
Caves	2	1	0	0	0	0	**3**	***1.7***
Settlements	0	1	0	3	0	0	**4**	***2.2***
Megalithic monuments	0	0	0	26	0	0	**26**	***14.4***
Cultural layers of indeterminate function	8	40	16	6	1	11	**82**	***45.6***
Total	**67**	**42**	**16**	**40**	**1**	**14**	**180**	**100.1**
%	***37.2***	***23.3***	***8.9***	*22.2*	***0.6***	*7.8*		***100***

Data from the SPLASHCOS Viewer at http://splashcos-viewer.eu

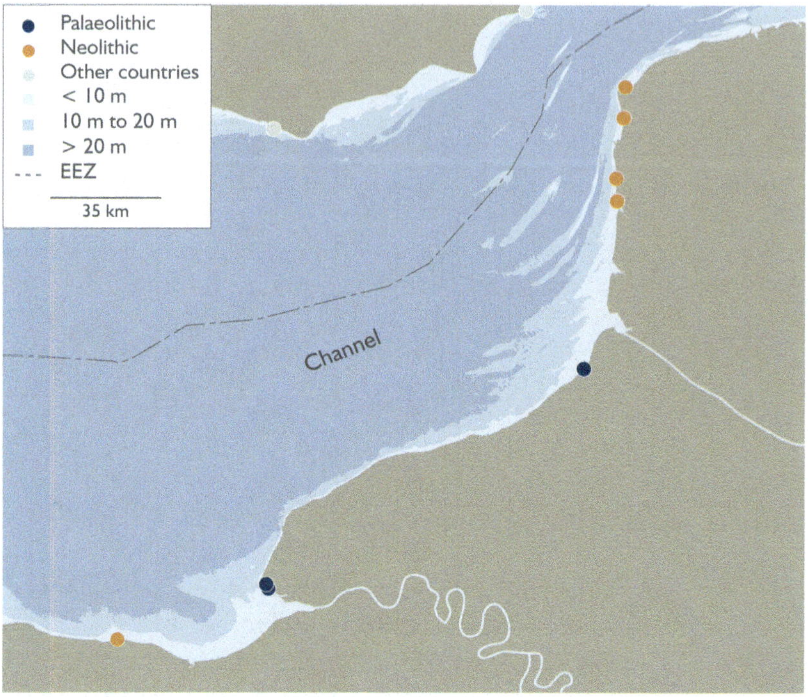

Fig. 12.2 Map of prehistoric submerged sites off the Channel and North Sea coasts of France. Site information from the SPLASHCOS Viewer http://splashcos-viewer.eu. Drawing by Moritz Mennenga

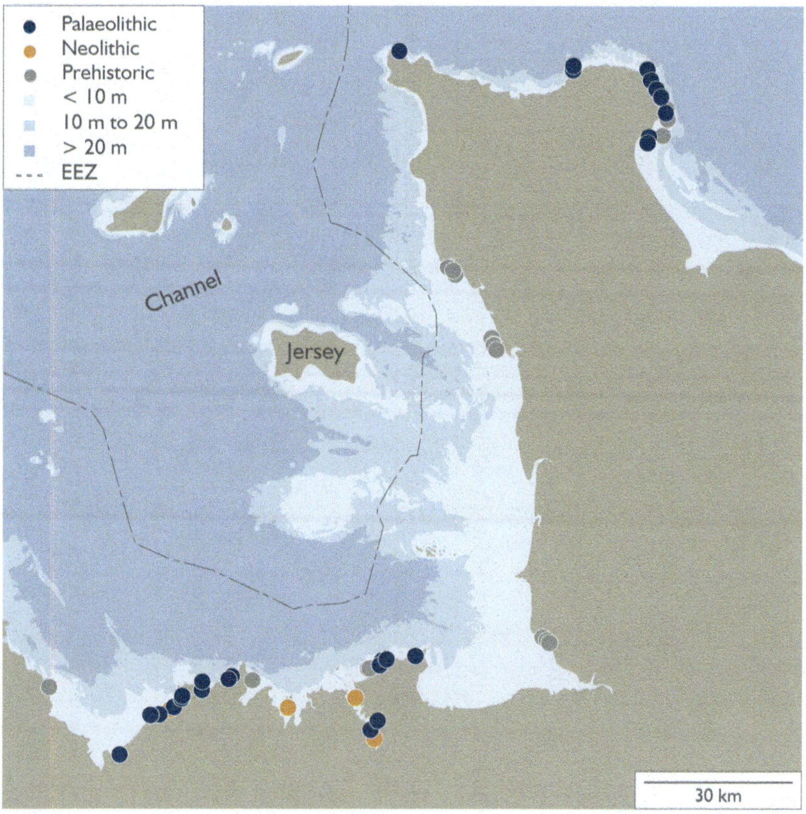

Fig. 12.3 Map of prehistoric submerged sites off the Cotentin coast and the Norman-Breton Gulf. Site information from the SPLASHCOS Viewer http://splashcos-viewer.eu. Drawing by Moritz Mennenga

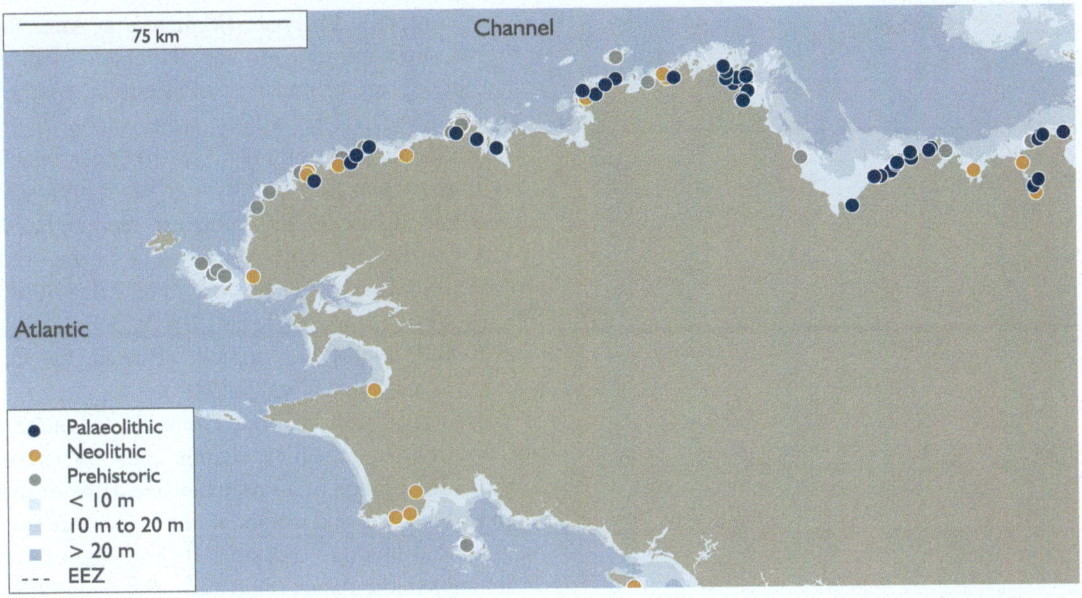

Fig. 12.4 Map of prehistoric submerged sites off the Brittany coast. Site information from the SPLASHCOS Viewer http://splashcos-viewer.eu. Drawing by Moritz Mennenga

12.4 Sites with Palaeolithic Artefacts and Pleistocene Fauna

12.4.1 The Northern Armorican Massif (Golfe Normand-Breton)

There are numerous Lower and Middle Palaeolithic finds along the shorelines of this region associated with Pleistocene deposits, and Middle Palaeolithic sites are especially common along the north coast of Brittany and Cotentin (Fig. 12.3; Coutard 2003; Coutard and Cliquet 2005; Monnier 2006; Lefort et al. 2011; Monnier et al. 2011). All are associated with Pleistocene deposits of various types, including raised beaches, loess and periglacial deposits, and many of the artefacts are made on flint derived from Cretaceous formations that are now submerged. The beach along this coastline is backed by a marine cliff which has been eroded by marine action and reveals stratified Pleistocene deposits in section. Some of these sites are associated with earlier periods of high sea level, particularly MIS 5e (Eemian), others with the Last Glacial Period (MIS 4 and 3). These deposits are geologically in situ, but it is sometimes difficult to distinguish those that are at the base of the cliff from those that are being exposed in the intertidal zone by erosion of the overlying beach deposits (Monnier et al. 2011). We present below examples of the latter type, which clearly belong to a land surface that was subsequently inundated by sea-level rise.

There are several dozens of sites that are geologically in situ in the intertidal zone (Giot et al. 1973; Monnier 1986; Huet 2002, 2006; Laforge 2012), notably the site of Les Gastines (Ille-et-Vilaine), which has undergone a systematic excavation (Monnier 1988). Mention should also be made of some famous sites excavated at the foot of cliffs or near ancient submarine caves and associated with now-submerged landscapes, notably Piégu (Côtes-d'Armor), La Cotte de Saint-Brelade in Jersey (see also Bailey et al., Chap. 10, this volume) Mont-Dol (Ille-et-Vilaine), Karreg-ar-Yellan (Côtes-d'Armor) and Saint-Germain-des-Vaux/Port-Racine (Manche).

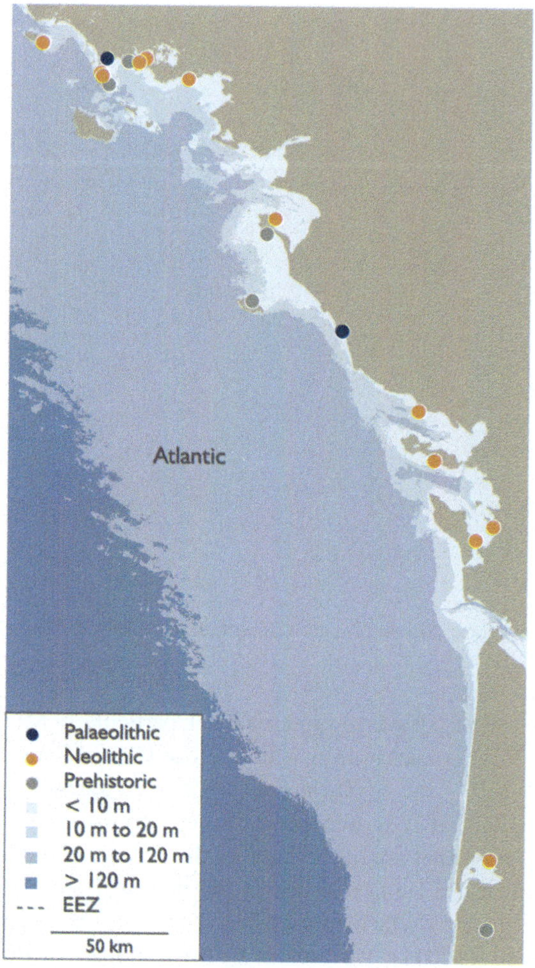

Fig. 12.5 Map of prehistoric submerged sites off the Aquitaine coast. Site information from the SPLASHCOS Viewer http://splashcos-viewer.eu. Drawing by Moritz Mennenga

In Gatteville-Le-Phare (Manche), at Anse du Figard, flint implements are associated with a marine deposit attributed to the Eemian (Michel et al. 1982). The lithics were made from Cenomanian flint coming from a barrier beach and are characterised by a predominantly Levallois debitage with notches, denticulates, side scrapers and some pebble tools. In the tidal creek of Portbail (Manche), a Middle Palaeolithic flint industry is associated with a hard sand-clay level and underlying gravels affected by ice action (Laisné and Vilgrain 2008).

Chausey, offshore from Granville (Manche), is an archipelago of 52 islands and 365 islets. With a tidal range of 14 m, extensive areas of the intertidal zone and some of the islets become exposed at low tide, and several shore areas have yielded Middle Palaeolithic tools (Pruvost 2006).

In the Normandy region, some other Palaeolithic industries have been reported, notably at Agon-Coutainville (Manche), Asnelles, Meuvaines, Ver-sur-Mer and between Villers-sur-Mer and Houlgate (Calvados). Further south, in the Charente, particularly at the mouth of the Seudre and Charente Rivers, there are many submerged remains, the oldest dating back to the Middle Palaeolithic, some near ancient terraces (Fridman 1957), others in the Marennes-Oléron basin (Maufras 1874; Hinout and Rolland 1960) and around the Islands of Aix and Madame (Facon 1967; Néraudeau and Bourgeois 1994).

Fig. 12.6 Map of prehistoric submerged sites off the Mediterranean coast of France. Site information from the SPLASHCOS Viewer http://splashcos-viewer.eu. Drawing by Moritz Mennenga

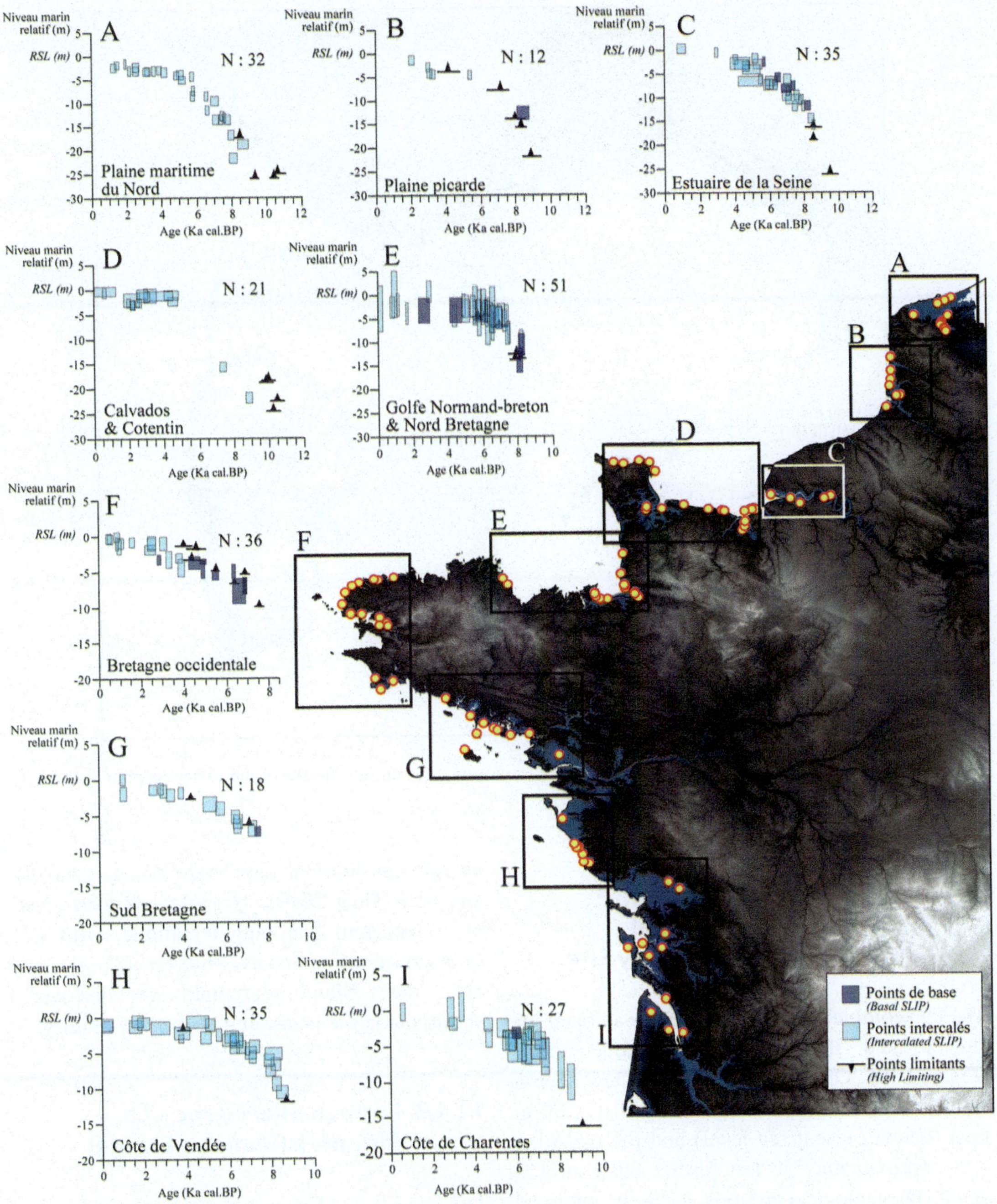

Fig. 12.7 Sea-level variations along French Channel-Atlantic coasts during the Holocene. After Stéphan and Goslin (2014)

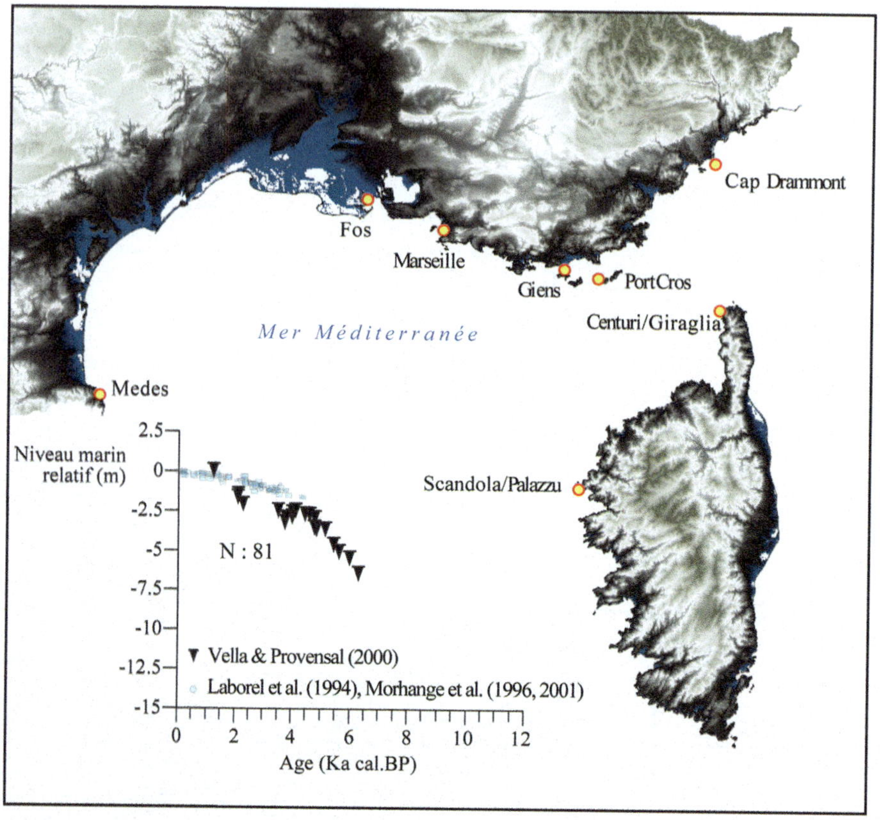

Fig. 12.8 Sea-level variations along the French Mediterranean coast during the Holocene. After Stéphan and Goslin (2014)

12.4.2 Pléneuf-Val-André 'Les Vallées' (Côtes-d'Armor, France): A Neanderthal Workshop and Butchery Site

The Palaeolithic site of Les Vallées is an open-air site located in the intertidal zone today (Fig. 12.10). Originally, it was located near a small stream at a time when sea level was 60 m bpsl (below present sea level) and the coastline was approximately 60 km further out (Laforge 2012). On geological grounds, the archaeological material is associated with a cold period earlier than the Last Interglacial, most likely the Saalian glaciation (MIS 6), confirmed by ESR and U/Th dates on bone giving an average age of 164 ± 13 ka (Bahain et al. 2012).

The material indicates a workshop linked with butchery activities and yielded some 2500 artefacts made on flint and quartz with numerous faunal remains dominated by horse, bison or aurochs and deer (Huet 2010). The artefacts were made by a recurrent knapping technique, with side scrapers of various types, Mousterian points and denticulates. Mammoth remains were discovered at this site at the beginning of the last century.

12.4.3 Fermanville/Biéroc – La Mondrée (Manche-France)

The site of La Mondrée in Fermanville lies at a depth of about 20 m bpsl and was first investigated during the 1970s, when it yielded an abundant Middle Palaeolithic assemblage of c. 2500 worked pieces (Fig. 12.11; Scuvée and Vérague 1988), with subsequent recovery of an equine tooth and two other bone fragments, one probably from an aurochs. New investigations from 2000 to 2002 by diver survey and again in 2010 were

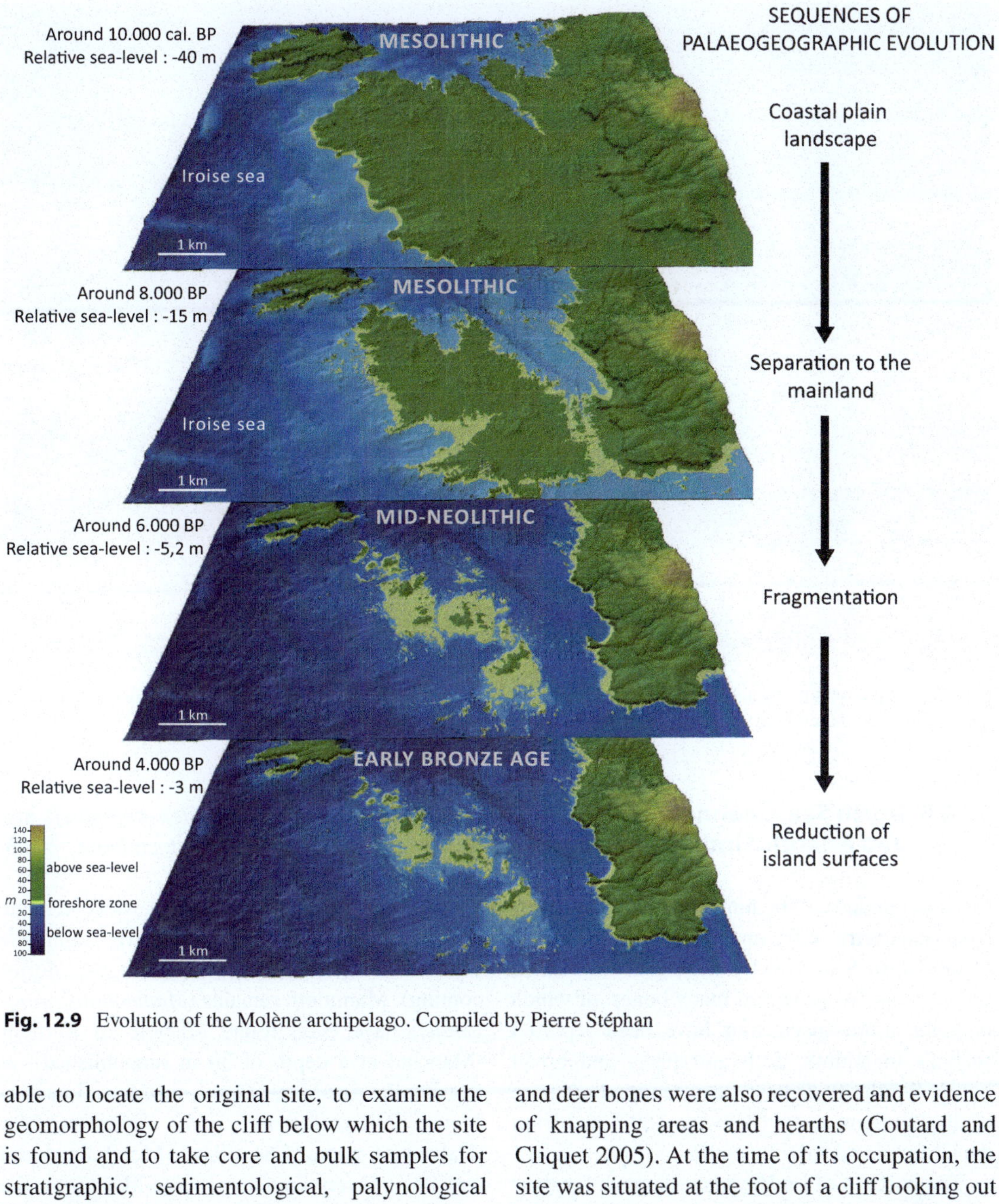

Fig. 12.9 Evolution of the Molène archipelago. Compiled by Pierre Stéphan

able to locate the original site, to examine the geomorphology of the cliff below which the site is found and to take core and bulk samples for stratigraphic, sedimentological, palynological and geochronological (OSL) analysis (Cliquet et al. 2011).

The 2002 survey revealed at least two occupation levels in situ. Pollen data indicate a date of MIS 5a, and OSL dating produced consistent results of c. 70 ka, corresponding to the boundary between MIS 5 and MIS 4 (Mercier et al. 2013). The stone assemblage is a Mousterian industry with scrapers and notched pieces; bovid and deer bones were also recovered and evidence of knapping areas and hearths (Coutard and Cliquet 2005). At the time of its occupation, the site was situated at the foot of a cliff looking out over a narrow coastal plain at the mouth of a small valley that gives access to an extensive plateau in the hinterland. The shoreline was perhaps no more than a few kilometres. This location would have provided shelter, large nodules of Cretaceous flint exposed during periods when sea level was lower than present and easy access to animal herds and to marine resources at the shore edge.

Fig. 12.10 The Middle Palaeolithic site of Les Vallées during excavation: (**a**) faunal remains and quartz flakes; (**b**) quartz cores and flakes. Photos by Briagell Huet

12.4.4 North Sea, Channel and Atlantic Finds

Since the middle of the nineteenth century, geologists have retrieved mammal bones from the bed of the North Sea. Trawler fishing in the Pas de Calais Strait has produced many bones, of which only the more spectacular have been reported such as mammoth teeth and tusks and bovid bones. Similar finds have been reported offshore of the Boulogne coast (Sauvage 1899) and from the Channel and Atlantic coasts.

The beaches of Le Havre are famous for their large Palaeolithic assemblages associated with fauna (Babeau 1906; Romain 1915). At −2 m ODN (Ordnance Datum Newlyn; equivalent to mean tide level), the flint industry of the Station Romain includes bifaces and flakes linked to the MIS 7 interglacial. A rich fauna testifies to temperate environments (wild boar, roe deer, straight-tusked elephant [*Palaeoloxodon antiquus]*) and glacial environments (horse, mammoth, bison, reindeer).

There are numerous finds of Pleistocene fauna off the Normandy coast, particularly along the edges of the Seine palaeo-valley (Auguste, forthcoming). Mammoth remains (*Mammuthus primigenius*) have been found offshore of Barfleur (Manche) at a depth of 30 m (unpublished); at Courseulles-sur-Mer, Graye-sur-Mer, Luc-sur-Mer, Port-en-Bessin and Villers-sur-Mer (Calvados); and at Cauville-sur-Mer, Saint-Jouin and Saint-Valéry-en-Caux (Seine-Maritime), the latter at a depth of 20 fathoms (Mercier 1934). Offshore of Barfleur, a horse tooth has been trawled at a depth of 20 m (unpublished). At Saint-Aubin-sur-Mer, on the Câtel beach (Calvados), mammoth and rhinoceros bones have been reported. At Saint-Côme-de-Fresné/Asnelles Belle-Plage (Calvados), the Eemian beach is cov-

Fig. 12.11 The Fermanville La Mondrée site. Bones of aurochs (top) and deer (middle) picked up from the eroding surface of the site and fragments of an equid tooth (bottom) found in situ. Their deteriorated surfaces are due to biological decomposition. Photo by Dominique Cliquet

ered by peat and green clay of glacial age, in which faunal remains have been recovered as well as some flint tools (Guillaume 1935, 1952).

On the beaches of Brétignolles-sur-Mer (Vendée) particularly at Le Lac and Le Marais Girard, *Elephas antiquus* remains were found beneath a peat deposit lying on an Eemian conglomerate, apparently associated with some small pebble tools (Fig. 12.12; Labrude et al. 2000; Pouit and Viaud 2002). Full details have not been published, apart from a pollen analysis (Bernard and Visset 1992), which suggests an interglacial age, most probably MIS 5 e (Eemian).

12.4.5 Ault-Onival (Somme)

The site of the Ault-Onival beach, on the south shoreline of the Somme estuary, was discovered at the beginning of the twentieth century, and more recently an important assemblage of flint cores, Levallois flakes and proboscidian bones was collected from low in the intertidal zone (Fig. 12.13; Agache 1968; Auguste, 2009; Antoine et al. 2011). The site is dated to the early Pleniglacial or the beginning of the middle Pleniglacial (MIS 4 or the MIS 3–4 transition), based on sedimentary features and an ESR/Uth date (55 ± 10 ka) on a horse tooth, and this is

Fig. 12.12 La Parée beach in Brétignolles-sur-Mer (Vendée): *Elephas antiquus* remains in situ

Fig. 12.13 Surveys by R. Agache on the Middle Palaeolithic site of Ault-Onival (Somme). Photo by DRAC Hauts de France-SRA

consistent with the palaeontological data, mainly horse and *Mammuthus primigenius* (Auguste 2009; Antoine et al. 2011). The technological features of the industry and the position of the site indicate flint knapping activities near a source of raw material (Perpère 1986, 1999). The sedimentary origin of the deposit, as well as its situation under an ancient sea cliff created by coastal erosion, offers parallels with the section of the Sangatte fossil cliff, where dates of MIS 7 demonstrate that the Strait of Dover was open as a sea channel by that time (Antoine 1989; Balescu et al. 1992).

12.4.6 South-West France

No Pleistocene or Palaeolithic finds have been recovered along the coast of Aquitaine, but recent palaeoenvironmental research and archaeological survey in the hinterland raise questions about the environments available on the now-submerged coastal shelf. In the Landes triangle of the Aquitaine region of SW France, there existed an arid 'desert' during the Last Glacial Maximum (MIS 2), where coversands extended over a large area of low relief in the hinterland, exposed to winds and with little surface water, creating an inhospitable environment for human settlement, and a cultural barrier between the Pyrenees and Cantabria to the south and the Charente and the Périgord to the east and the north (Bertran et al. 2013). New studies of lithic technology and raw materials suggest that, despite this barrier, circulation of materials and movements of people persisted between the adjacent regions and that this may have taken place along the coastline of the submerged shelf, which extends for 50–100 km offshore from the present coastline. This hypothesis highlights the potential human significance of the submerged shelf and the need for offshore palaeoenvironmental studies, and ultimately the search for archaeological remains, to further explore this potential.

12.4.7 The Mediterranean Coastal Karst

The French Mediterranean coastline includes the following karstic massifs, from west to east (Fig. 12.6; Billaud 2017):

- Perpignan to Narbonne: the end of the Corbières massif and the massif of La Clape.
- Sète to Montpellier: the Gardiole mountain and the border of the Cévennes.
- East of Marseilles to Toulon: the Couronne Massif and the Calanques.
- Antibes to the Italian frontier: the Préalpes of Grasse and of Nice.

The cave openings in these karstic environments have trapped sediments containing a variety of information about Quaternary human occupation and environment particularly over the last glacio-eustatic cycle (Billaud et al. 2014; Billaud 2017).

Underwater archaeological studies of these submerged karstic openings are still few in number. Apart from the recent operations in the Cosquer Cave (described below), there is older research conducted from 1968 to 1976 by DRASSM (Département des Recherches Archéologiques Sub-aquatiques et Sous-Marines), mainly focussed on the Calanques Marseillaises area and the Villefranche Bay further east (Bonifay 1970, 1973; Bonifay and Courtin 1998). In the Calanques, this research identified a dozen caves, but only four have been investigated in detail. One of them contains bone breccias, and Palaeolithic artefacts were discovered at another (Bonifay and Courtin 1998).

In the Couronne Massif, few cavities are mentioned apart from the Caves of the Mona au Rouet and the three Cap Méjean caves (Bonifay and Courtin 1998, fig. 2), which are openings at a depth of 15–20 m bpsl. Two of these are complex networks only partly explored. One contains cemented deposits which have yielded microfauna (Courtin 1978, p. 738). In the Var department, cavities are also mentioned east of La Ciotat (Courtin 1978). They are situated at a depth of 10–15 m bpsl and contain cemented deposits.

Information about the archaeology of these underwater karsts is uneven because, apart from the Cosquer Cave, they have often been explored for speleological or hydrological purposes rather than archaeological. Nevertheless, the available information indicates the presence of Pleistocene sedimentary sequences and remains of human activity and fauna, suggesting considerable potential for future research.

12.4.8 The Cosquer Cave

In 1991, the discovery of the Cosquer Cave, located on the coast near Cassis east of Marseille, revealed the only Palaeolithic painted cave known to this day east of the River Rhone. This resulted in renewed interest in the detailed study of regional karst and in the reconstruction of sea-level rise after the last glaciation. The Cosquer Cave has been visited by divers since 1984, but its archaeological significance was only formally recorded in 1991 (Clottes et al. 2005; Vanrell 2010). This cavity is presently accessible to divers by an opening at a depth of 37 m bpsl which leads upwards through a narrow gallery to a partially flooded chamber (Fig. 12.14). The original entrance was closed by sea-level rise after the Last Glacial Maximum. However, the water level in the cave is slightly lower than that of the Mediterranean Sea. The emerged part of the cave has provided a rich collection of art, with a total of 188 animal figures (11 species identified, some of which are marine species, including 9 seals, 4 fish and 3 auks), 71 negative handprints and nearly 300 symbolic representations (Fig. 12.15). There are numerous traces of past human activity—handprints, charcoal, hearths and broken concretions. Radiocarbon dates place the occupation between 29,000 and 17,500 BP (Early Gravettian to Epigravettian).

12.5 Later Prehistoric Finds

12.5.1 Intertidal Settlements and Peat Deposits

Ancient peat deposits are exposed at low tide along many shorelines and extend to the subtidal zone. They bear witness to the extension of back-barrier marshes before the end of postglacial sea-level rise.

On the shorelines of Picardie and Boulonnais, low, sandy beaches alternate with rocky headlands, with offshore gravel barriers or bay-mouth bars. On the sandy beaches, peat deposits exposed by marine erosion provide radiocarbon dates between 5700 BP, as in the Canche tidal creek in

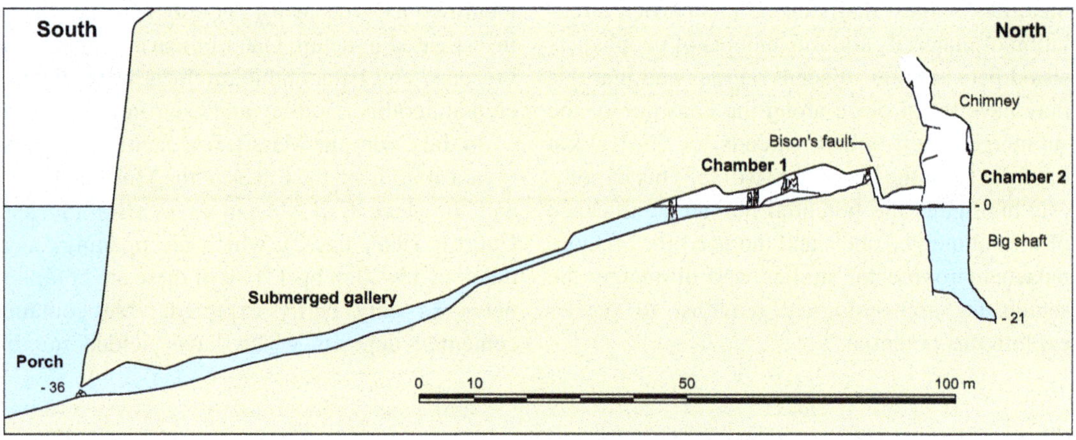

Fig. 12.14 Longitudinal cross-section through Cosquer Cave. Drawing by Yves Billaud

Fig. 12.15 Les Petits Chevaux, Cosquer Cave, operation DRASSM. Photo by Antoine Chené, CNRS

Camiers, and 3600 BP, as at Wimereux (Mariette 1971), sometimes associated with lithic artefacts as at Wissant's Bay (Dutertre 1924, 1925; Briquet 1930), and especially at Tardinghen (Mariette 1971, 1976b; Vallin and Masson 1993).

At Wimereux, the top of the eroded peat yielded flint implements and traces of a hearth (Briquet 1906, 1930; Dutertre 1925, 1930, 1933, 1934; Agache 1968; Mariette 1971). Also, a shell midden was recorded on the edge of the peat where it was covered by humic sands (Munaut and Gilot 1977), and the level beneath provided some sherds attributed to the Middle and Late Neolithic (Mariette 1971, 1976a; Ters 1973).

Similar deposits are present at Hardelot, about 20 km further south (Mariette 1971; Ters 1973; Agache 1968; Bostyn and Vallin 1986), where a Late Neolithic site has been discovered around −2.5 m ODN. Remains of a rectangular building consisting of wooden poles dug in the peat were observed in the nineteenth century (Ovion 1890).

In Brittany, several Neolithic occupations have been identified in submerged peats: a deer antler industry at La Ville-Berneuf (in Pléneuf-Val-André) (Milon and Giot 1949); the remains of postholes, ditches and varied implement dated to 4000–2500 cal BC at Curnic in Guissény (Giot et al. 1965); and Neolithic polished axes, flint tools and ceramics at the mouth of the Quillimadec River (Kerlouan).

On the right bank of the mouth of the Rance River, the remains of a probable settlement discovered at the limit of the present lowest tides yielded two lines of wooden stakes 50 m apart (Langouët 1995; Hénaff et al. 1998), and one of the stakes is dated to 3327–2929 cal BC (Ly-8830), attributable to the Recent Neolithic.

On the Vendée coastline, numerous peat deposits have been observed since the early twentieth century (Welsch 1911, 1917), especially at Brétignolles-sur-Mer, where Early Neolithic material has been recorded at La Parée, −5 and − 6 m ODN, dated at 5207–4458 cal BC, and at La Pointe du Grouin du Cou dated at 5700–4850 cal BC (Joussaume et al. 1986).

The Charente coastline has a different morphology, with a string of islands close offshore, and a continental shelf deeply incised by ancient valleys (Laporte 1998, 2009). There are rare discoveries of the 5th or early 4th millennium (cal. BC), notably the Early Neolithic remains at the Bois-en-Ré beach (Pautreau and Robert 1980) and a Middle Neolithic shell midden (Rousseau et al. 2001). Remains belonging to the later 4th and 3rd millennium are more abundant, reflecting the stabilisation of the shoreline at its present position. Submerged peats on the west coast of Oléron Island facing the mainland have produced many Late Neolithic finds (Welsch 1917; Burnez and Morel 1965; Gabet 1973; Laporte and Glausinger 1986; Laporte et al. 1998; Laporte 2009). At La Pointe d'Ors, there are the remains of a probable domestic stone building within archaeological levels accessible only at low tide (Luguet 1867; Pineau 1891; Boisselier 1893; Dulioust 1958; Gabet 1968; Rouvreau 1968; Rouvreau and Gomez 1973; Joussaume 1981; Cassen 1987; Laporte et al. 1995). These deposits have yielded some wooden artefacts, including examples of stone axes in wooden handles, and similar artefacts are reported on the opposite side of the same bay at Piédemont (Gabet and David 1975).

12.5.2 Megalithic Sites

12.5.2.1 Neolithic Tombs

Several passage graves are located at about mean tide level on the Channel coast. These are typically dated to about 5100 cal BP but may have continued in use until around 4000 cal BP (L'Helgouach 1965; Giot 1958, 1968; Giot and Morzadec 1992). The position of these monuments ranges between 0 m ODN (i.e., +5,27 m above low tide) and − 7 m ODN. Because of the tidal range, these monuments are regularly submerged, in some cases at every high tide, as with the Kernic passage grave at Plouescat (Fig. 12.16). This monument is probably one of the most famous on the northern coast of Brittany. Classified as a 'monument historique', it was excavated in 1983 and 1984 because of the effects of shoreline erosion and was then restored (Lecerf 1985).

On the shore, to the south of Coalen Island (Lanmodez, Côtes-d'Armor), there is a damaged passage grave, which has lost its capstones (Fig. 12.16). Open to the south, the monument is reached during higher tides. It is 7.2 m long and 1.4 m wide and still has 3 standing stones on the west side and 6 on the east. To the north, an isolated half block constitutes perhaps one of the

Fig. 12.16 Partially submerged passage graves on the seashore. Left: Kernic, Plouescat (Finistère); Right: Coalen Island in Lanmodez (Côtes d'Armor). Photos by Marie-Yvane Daire

sides of a terminal chamber ('cella'), and scattered blocks may be the remains of the surrounding structure (Giot 1971).

In the channel between Rovarc'h and Garec'h-Vari islands (Penvenan, Côtes-d'Armor), a collapsed dolmen stands on the shore of Pellinec. In the 1930s, a flagstone measuring 1.80 × 1.60 m and 5 supports lying on the ground could still be seen (Mazères and de Bellaing 1936). In the same commune, the channel of Saint-Gildas island (Penvenan) presents on its south side an eroded and collapsed dolmen, which was still visible at the beginning of the 1930s (Mazères and de Bellaing 1936). In the 1980s, only some scattered blocks remained (Daire, personal observation).

On the north bank of Tresseny Bay, there is a probable passage grave at Lerret in Kerlouan, submerged at each high tide, though much damaged. The monument as a whole, the northern side of which is well preserved, is 23 m long (L'Helgouach 1965; Sparfel and Pailler dir. 2009, pp. 100–101). At Le Conquet on the northern point of Finistère, around forty flagstones indicating a likely collapsed passage grave are visible at low tide on the beach of Les Blancs Sablons (Sparfel and Pailler dir. 2009, p. 175). Four of them are oriented in the same direction, forming a passage with an internal width of 1.7 m.

East of Noirmoutier Island (Vendée), the stone called 'La Table' is the only remnant of a dolmen (Large et al. 2015).

12.5.2.2 Standing Stones and Megalithic Complexes

In several places in Brittany, submerged standing stones mark the seashore and are important amongst other reasons because they are evidence of Holocene marine transgression. In many cases, these stones appear in isolation—all that is left of the original monument complex—but sometimes sufficient numbers are present to indicate a circular layout.

At Côte des Abers (Finistère), several megalithic monuments on the seashore include the 2.5-m high-standing stone of Saint-Cava Bay in Plouguerneau, called Men Ozac'h (Devoir 1912). Before being moved, it was submerged for half the tidal cycle with the base lying 2 m above present mean sea level (Fig. 12.17). This standing stone was associated with the remains of a rectangular enclosure of smaller blocks.

On the south coast of Finistère, the standing stone of Penglaouïc (Loctudy), in the Pont-L'Abbé river, is partly submerged at high tide. It is one of a series of megaliths placed at the mouth of waterways or streams, notably at Lesconil (Fig. 12.17), Léchiagat and Lespurit en Plovan.

Other examples are the large menhir currently lying on the ground on the south of Groix Island in Locmaria Bay, some dozens of metres offshore from the present coastline (López-Romero et al. 2008), and in Brittany, the standing stones of La Pointe du Chevet at Saint-Jacut-de-la-Mer (Côtes-d'Armor) (Langouët 2003) and

Fig. 12.17 Left: Menhirs on the Finistère coast. Left: Men Ozarc'h at Plouguerneau. Right: Lesconil. Photos courtesy of Archives du Labo Archéosciences, UMR 6566 CreAAH, Rennes

Trébeurden, located between Grande Island and Milliau Island (Marchat and Le Brozec 1991).

12.5.2.3 Stone Rows and Megalithic Enclosures

Several partly submerged standing stones have recently been discovered and studied in Morbihan, especially on the seashore of the Quiberon Peninsula, such as the Kerdual alignment, preserved between Carnac and La Trinité-sur-Mer, and the Kerpenhir site (Locmariaquer). Highlighted by Z. Le Rouzic (1965), a set of submerged blocks has long been seen extending into the sea at low tide, a prolongation of the Le Moulin alignments (Saint-Pierre-Quiberon). These include more than a dozen parallel lines of standing stones about 400 m long, associated with a megalithic enclosure (2 large stelae are still visible) and a burial mound. Mapping of the rocky Kerbougnec platform also led to the discovery of a subtidal structure comprising 150 partly submerged monoliths (Cassen et al. 2011).

At Le Petit-Rohu in Quiberon, two pairs of polished axe blades made from Alpine rock embedded in the mud of the foreshore were discovered by holidaymakers in August 2007 at Porh Fetan (Fig. 12.18; Cassen et al. 2008). Further to this discovery, excavations revealed animal footprints, plough marks and a set of 45 collapsed flagstones, which appear to have formed a NNE–SSW alignment. Some of these are visible on the shore at low spring tide, and others are permanently submerged in 5–6 m water. There are numerous other stone rows of similar type in the same area, such as Kerbougnec, which are now submerged in more than 5 m of water. The environmental context shows that the axe blades associated with the Petit-Rohu monoliths were buried in wetlands behind a sandy barrier at the foot of a prominent rocky outcrop. The coastline would have been about 500 m distant at

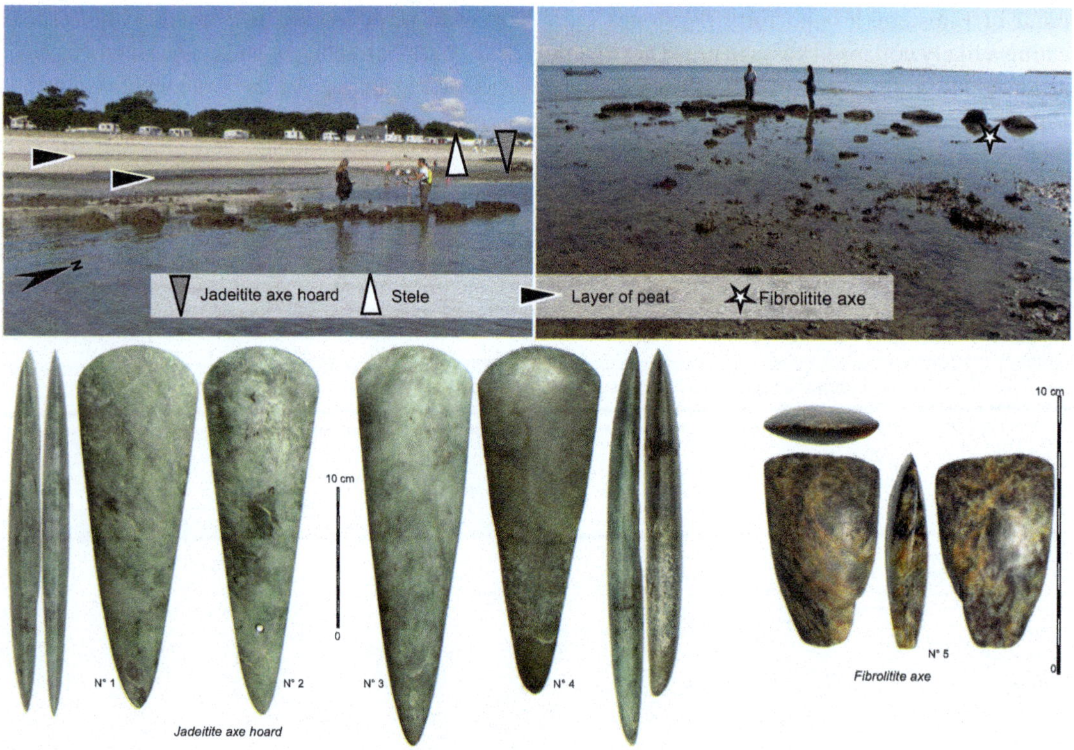

Fig. 12.18 Rows of stelae discovered at low tide on the intertidal platform of Le Rohu in Saint-Pierre-Quiberon and a hoard of jadeite axes. After Cassen et al. (2010)

the time when these artefacts were deposited, estimated on archaeological grounds as the mid-5th millennium cal BC (Cassen et al. 2010).

At the heart of the Morbihan Gulf is the famous double megalithic enclosure of Er Lannic, long recognised as evidence of a Holocene marine transgression (Closmadeuc 1882). Erected around 6000 cal BP, this consists of 49 menhirs, of which 33 are covered at high tide (Gouezin 1998). They form two adjacent horseshoe-shaped enclosures, with a menhir, larger than the others, standing at the point where they meet (Fig. 12.19).

Finally, a cromlech (circle of standing stones) occurs on the Chausey archipelago (Manche). The archipelago was separated from the mainland about 5000–4000 cal BC by a shallow (c. 20-m deep) sea crossing of 14 km. The tidal range in this area is 15 m, and at low tide near the Oeillet inlet, there is a rocky seashore with the cromlech located at around −0.2 m ODN (Fig. 12.19; Chancerel et al. 1994, 1998). This circle (a little flattened) is composed of large granite boulders, of which only five are still in their vertical position. Inside the circle, there is an oval-shaped stone-lined hearth 1.25 m long. Around 600 flint artefacts recovered from the site confirm a Neolithic date. There are also several stone dikes related to the cromlech, which could be contemporaneous with it.

12.5.3 Coastal Lagoons

12.5.3.1 The Atlantic Coast

The coastal lagoons here are well known for their submerged finds, especially from the proto-historic period in the Sanguinet Lake (Maurin 1998). However, the Aquitaine region also offers evidence of several prehistoric sites around the Arcachon Basin and the ancient mouth of the Leyre River. The Betey site in Andernos-Les-Bains is characterised by numerous discoveries made during pedestrian surveys on the shoreline, but without stratigraphic or chronological context (Roussot-Larroque 1997). The subsequent study of the artefacts has indicated four main occupation phases: Mesolithic and Early, Recent and Late Neolithic.

Excavations associated with new harbour work in the area (carried out jointly by INRAP, DRASSM and Aquitaine SRA) have produced stratified and radiocarbon-dated evidence of Neolithic sherds and lithics as well as sherds of Early Bronze Age II/Middle Bronze Age I, although this material has been subjected to rede-position by fluvial action.

12.5.3.2 The Mediterranean

A number of submerged Neolithic finds have been found in coastal lagoons (Leroy 2001). One of the principal sites is Leucate-Corrège discovered by dredging at a depth of −5 m, which includes Early Neolithic material including fish and bird bone, net weights and bone tools (Guilaine et al. 1984). Middle and Late Neolithic settlements have also been recovered at Montpenèdre in Marseillan, in the Thau lagoon, Moutaras at La Grande Motte and Frontignan 'Les Dunes' at depths ranging from 0 to −2 m (Montjardin and Cablat 2004) and at Mèze 'La Conque'. Several Bronze Age sites are present in these coastal contexts, especially in the Thau lagoon (Leroy et al. 2003).

12.5.4 Submerged Fish Weirs

Fish weirs have been systematically studied in Brittany since 2006 (Daire and Langouët 2008; Langouët and Daire 2009; Daire and Langouët 2010; Daire et al. 2013; Gandois and Stéphan 2015; Gandois et al. 2018), and similar inventories have begun in the Departments of Manche (Normandy) and Vendée (Pays de Loire) since 2009. To date, in Brittany, more than 750 fish weirs have been identified and characterised by twenty or so variables, such as the type of weir, length and level in relation to tides (Table 12.2). Most, of course, relate to the present position of the shoreline and are relatively late in date.

The oldest wooden fish weirs are located in the Bay of Mont-Saint-Michel at the mouth of the Rance River and are radiocarbon-dated to the

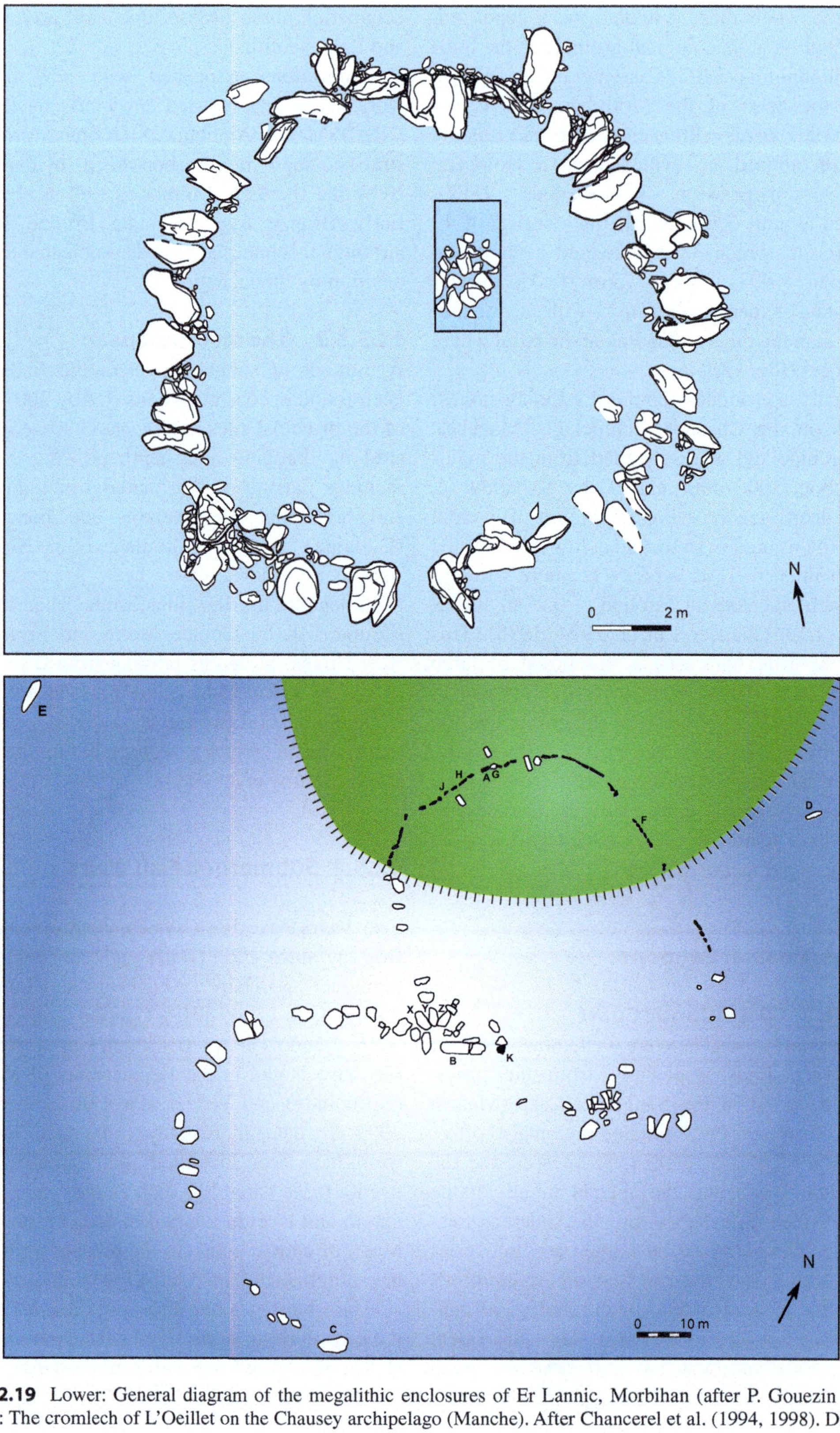

Fig. 12.19 Lower: General diagram of the megalithic enclosures of Er Lannic, Morbihan (after P. Gouezin 1998). Upper: The cromlech of L'Oeillet on the Chausey archipelago (Manche). After Chancerel et al. (1994, 1998). Drawing by G Lannuzel

Table 12.2 Inventory of fish weirs of all periods in Brittany, from the database 'Pêcheries d'Estran de Bretagne', 2017

	Côtes-d'Armor	Finistère	Morbihan	Ille-et-Vilaine	Brittany Total
Fish weirs of all types and periods	304	246	121	84	**755**
Wooden fish weirs of all periods	45	3	14	60	**122**
Stone fish weirs of all periods	239	226	82	23	**570**
Stone fish weirs below the level of the lowest spring tide (% of all stone fish weirs in parentheses)	5 (*2.1*)	66 (*29.2*)	10 (*12.2*)	2 (*8.7*)	**83 (*14.6*)**

This database is in progress and is not yet published. It includes additional finds of fish weirs, listed in this table, but these are not included in Table 12.1

Fig. 12.20 Footprints at Saint-Jean-le-Thomas (Manche). They are associated with the remains of a wooden fish weir (not visible in the image) of Late Neolithic or Early Bronze Age. The scale is 1 m. Photo by Cyrille Billard

Late Neolithic (Hénaff et al. 1998). Early Bronze Age examples are known at Saint-Jean-le-Thomas and Champeaux (Manche) (Fig. 12.20; Billard et al. 2013a, b; all radiocarbon dates in Billard and Bernard dir. 2016).

In Brittany, the search for ancient stone fish weirs, principally conducted from aerial photos of the shore at low tide, has led to the identification of fish weirs under 3–4 m of water. Similar submerged examples have also been observed along the coasts of Vendée and Normandy, four sites and 11 sites, respectively (Fig. 12.21; Large et al. 2009; Billard et al. 2016).

Dating of stone fish weirs is a problem, but a broad dating scheme is possible by relating the position of the weirs to present-day tidal level with reference to the Holocene sea-level curve of the region (Daire and Langouët 2010, 2011, Daire et al. 2013). Those that are permanently submerged must have been built when relative sea level was lower than present and can be attributed to the Mesolithic or Neolithic period. In

Fig. 12.21 Submerged complex of fish weirs at Le Tambourin, Noirmoutier Island (Vendée). Photo by Patrice Birocheau, 2009

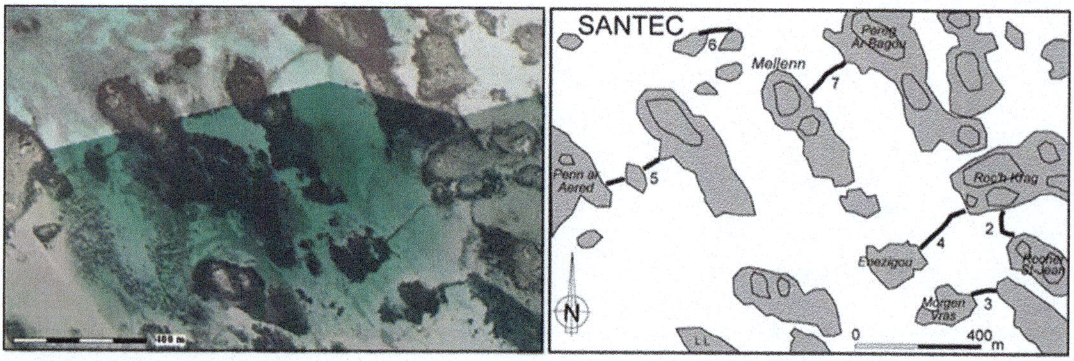

Fig. 12.22 Three submerged fish weirs at Santec (Finistère). They are built across a submarine channel delimited by the rocks marked Melenn et Pereg ar Bagou and form a sequence at progressively shallower depths (7 is deepest, 3 is shallowest), illustrating the adaptation of the fish weir builders to progressive sea level rise. Photo by Ortholittorale 2000; drawing by Loïc Langouët

Brittany, 83 of these have been identified, 14.6% of all known stone fish weirs in the region (Table 12.2). The majority are located on the channel coasts of Finistère, Côtes d'Armor and Ille et Vilaine, with far fewer on the Atlantic coastlines. The oldest, judging by their depth at c. 5 m below the spring tide level, are at Lag Karreg (Brignogan) and Telleskog (Kerlouan) in Finistère.

In some areas, 'horizontal stratigraphies' can be observed, that is, shifts in the position of fish weirs to accommodate the rise in sea level. At Santec, three fish weirs are located at progressively shallower depths from −6.8 m to −2.6 m along a submerged valley channelled between rocky outcrops (Fig. 12.22; Roué and Le Goff 2008, p. 105). In the same way, at Saint-Jean-de-la-Rivière (Manche), six stone fish weirs form a

Fig. 12.23 Submerged fish weir complex at L'Épine, Noirmoutier Island (Vendée). Photo by Patrice Birocheau, 2009

sequence of successively shallower structures across a marine channel, the lowest being fully subtidal and the highest being located close to the neap tide level.

It is probable that many more submerged stone fish weirs remain to be discovered on other coastlines, and surveys are only just beginning, as, for example, at L'Acheneau (Vendée), where the fish weir is submerged at −3 to −9 m (Fig. 12.23; Ydier and De Maisonneuve 2012), and in the Molène archipelago (Gandois et al. 2013a, b).

12.6 Scientific Challenges and Future Prospects

Prehistoric occupations have left significant traces in the intertidal zone and on the seabed, but until recently in France, as in many European countries, maritime archaeology has principally involved the study of ships and maritime trade. However, archaeological investigations along the coastline have successfully confirmed the high potential and good preservation of submerged sites, including organic artefacts such as wooden objects, features related to fishing and faunal assemblages. New types of sites and new problems have emerged in the past decade. The potential for the investigation of submerged Palaeolithic sites is highlighted by the Fermanville 'La Mondrée' site and the famous Cosquer Cave. The study of megalithic complexes needs to be extended offshore to obtain a more complete picture, and fish weirs are producing an unexpectedly abundant and informative indicator of widespread and intensive activity along the Atlantic and Channel coastlines. In many areas of western France, submerged coastal sites represent one of the principal ways to obtain tree ring series for the construction of dendrochronological reference curves.

Little work has been carried out so far on high-resolution mapping of submerged landscapes, and this should be a priority for the future. Bathymetric lidar, in particular, offers interesting prospects for surveys in shallow water. Other priorities are the integration of palaeoenvironmental studies with archaeological survey, particularly in relation to the Pleistocene, and the training of divers who combine diving qualifications with a knowledge of prehistoric archaeology. The implementation of a genuine scientific programme relating to all these issues is all the more urgent in that offshore industrial projects, including wind farms, the quarrying of marine aggregates and the expansion of port areas, are rapidly increasing along the French coastline.

12.7 Management of the Underwater Cultural Heritage

The underwater heritage has been the responsibility of DRASSM (Département des Recherches Archéologiques Subaquatiques et Sous-marines) since 1966. This is a national government agency under the ministry in charge of culture and is based at Marseille. It is responsible for the implementation of legislation relating to all aspects of maritime cultural heritage including prehistoric, archaeological and historical information, with particular reference to the intertidal area and the seabed within the French exclusive economic zone, especially since the ratification by France in 2013 of the 2001 UNESCO Convention. DRASSM registers sites and directs the study and excavation of submarine archaeological sites in France and overseas. It includes freshwater contexts within its remit and is concerned with the evaluation and study of sites as varied as shipwrecks, prehistoric lacustrine dwellings, river installations and submerged caves.

References

Agache R (1968) Informations archéologiques, circonscription de Nord et Picardie. Gallia-Préhistoire 9(2):267–309

Antoine P (1989) Stratigraphie des formations pléistocènes de Sangatte (Pas-de-Calais), d'après les premiers travaux du tunnel sous la Manche. Bulletin de l'Association Française pour l'Etude du. Quaternaire 37:5–17

Antoine P, Auguste P, Bahain JJ, Louguet S (2011) Datation et reconstitution paléoenvironnementale d'un site paléolithique moyen submergé en Manche Est: Ault-Onival (Somme, France). Quaternaire 22(3):221–233

Auguste P (2009) Evolution des peuplements mammaliens en Europe du Nord-Ouest durant le Pléistocène moyen et supérieur. Le cas de la France septentrionale. Quaternaire 20(4):527–550

Babeau L (1906) La préhistoire et la plage sous-marine du Havre. Bulletin de la Société Géologique de Normandie 26:37–45

Bahain JJ, Flaguières C, Laurent M, Shao Q, Dolo JM, Garcia T, Douville E, Frank N, Monnier JL, Hallegouet B, Laforge M, Huet B, Auguste P, Serre F, Gangnepain J (2012) ESR and ESR/U-series dating study of the Middle Paleolithic site of Piégu (Brittany, France) – a multi-material approach. Quat Geochronol 10:424–429

Balescu S, Packman SC, Wintle AG, Grün R (1992) Thermoluminescence dating of the middle Pleistocene raised beach of Sangatte (Northern France). Quat Res 37(3):390–396

Bernard J, Visset L (1992) Une tourbière interglaciaire sur la plage de la Parée à Brétignolles (Vendée – France): première approche pollenanalytique. Quaternaire 3:16–22

Bertran P, Sitzia L, Banks WE, Bateman MD, Demars PY, Hernandez M, Lenoir M, Mercier M, Prodeo F (2013) The Landes de Gascogne (Southwest France): periglacial desert and cultural frontier during the Palaeolithic. J Archaeol Sci 40(5):2274–2285

Billard C, Bernard V (2016) Pêcheries de Normandie: Archéologie et histoire des pêcheries littorales du département de la Manche. Presses Universitaires de Rennes, collection Archéologie et Culture, Rennes, 717 p

Billard C, Bernard V, Bouffigny A, Clavel B, Couturier Y, Jaouen G, Le Digol Y, Quevillon S, Ropars A (2013a) Techniques et modes d'exploitation des pêcheries sur le littoral normand (France) : un essai de bilan de 10 années de travaux archéologiques. In: Daire MY, Dupont C, Baudry A, Billard C, Large JM, Lespez L, Normand E, Scarre C (eds) Ancient maritime communities and the relationship between people and environment along the European Atlantic coasts.

Proceedings of the HOMER 2011 conference, Vannes, 28 septembre-1er octobre 2011, British archaeological reports international series 2570. Archaeopress, Oxford, pp 139–149

Billard C, Bernard V, Quevillon S (2013b) Potentiel et méthodes d'approche des pêcheries littorales de Basse-Normandie (France). In: Daire MY, Dupont C, Baudry A, Billard C, Large JM, Lespez L, Normand E, Scarre C (eds) Ancient maritime communities and the relationship between people and environment along the European Atlantic coasts. Proceedings of the HOMER 2011 conference, Vannes, 28 septembre-1er octobre 2011, British archaeological reports international series 2570. Archaeopress, Oxford, pp 3–11

Billard C, Daire MY, Martin C, Billaud Y, Bizien-Jaglin C, Chancerel A, Cliquet D, Fourment N, Gandois H, Huet B, Laforge M, Langouët L, Laporte L, Large JM, Leroy F, López-Romero E, Monnier JL, Ropars A, Régaldo P, Stéphan P, Vallin L (2016) Un premier inventaire des sites préhistoriques submergés au large des côtes françaises. Bulletin de l'Association Manche Atlantique pour la Recherche Archéologique dans les Iles 29:5–49

Billaud Y (2017) Submerged karst structures of the French Mediterranean coast: an assessment. In: Flemming NC, Harff J, Moura D, Burgess A, Bailey GN (eds) Submerged landscapes of the European continental shelf: Quaternary Paleoenvironments. Wiley, Chichester, pp 333–340

Billaud Y, Chazaly B, Olive M, Vanrell L (2014) Acquisition 3D et documentation multiscalaire de la grotte Cosquer: une réponse aux difficultés d'accès et à une submersion inéluctable? Kartologia 64:7–16

Boisselier A (1893) Le palet de Gargantua et les oscillations du niveau de la mer. Revue des Sciences Naturelles de l'Ouest 3:366–367

Bonifay E (1970) Antiquités archéologiques sous-marines. Gallia Préhistoire 13(2):585–592

Bonifay E (1973) Circonscription des antiquités préhistoriques sous-marines. Gallia Préhistoire 16(2):525–533

Bonifay E, Courtin J (1998) Les remplissages des grottes immergées de la région de Marseille. In: Camps G (ed) L'homme préhistorique et la mer, 120e congrès du commité des travaux historiques et scientifiques, Aix-en-Provence, oct. 1995. Editions du Commité des Travaux Historiques et Scientifques, Paris, pp 11–29

Bostyn F, Vallin L (1986) L'outillage préhistorique en os du Nord-Pas-de-Calais. Inventaire et aspects techniques. Gallia Préhistoire 29(1):193–215

Briquet A (1906) Notes sur quelques formations quaternaires du littoral du Pas-de-Calais. Annales de la Société Géologique du Nord 35:211–236

Briquet A (1930) Le littoral du nord de la France et son évolution morphologique. Editions A. Colin, Paris

Burnez C, Morel J (1965) Contribution à la connaissance de la céramique préhistorique de la Saintonge. Bulletin de Société Prehistorique Française 62/63:555–566

Cassen S (1987) Le Centre-Ouest de la France au IVe millénaire, British archaeological reports international series 342. Archaeopress, Oxford

Cassen S, Boujot C, Errera MD, Marguerie D, Menier D, Pailler Y, Pétrequin P, Poirier S, Veyrat E, Vigier E (2008) Discovery of an underwater deposit of Neolithic polished axeheads and a submerged stone alignment at Petit Rohu near Saint-Pierre-Quiberon (Morbihan, France). Antiquity 82, Project Gallery. http://www.antiquity.ac.uk/projgall/cassen316/

Cassen S, Boujot C, Errera MD, Menier D, Pailler Y, Pétrequin P, Marguerie D, Veyrat E, Vigier E, Poirier S, Dagneau C, Degez D, Lorho T, Neveu-Derotrie H, Obeltz C, Scalliet F, Sparfel Y (2010) Un dépôt sous-marin de lames polies néolithiques en jadéitite et sillimanite et un ouvrage de stèles submergé sur la plage dite du Petit Rohu près Saint-Pierre-Quiberon (Morbihan). Bulletin de Société Prehistorique Française 107:53–84

Cassen S, Baltzer A, Lorin A, Fournier J, Sellier D (2011) Submarine Neolithic stone rows near Carnac (Morbihan, France): preliminary results from acoustic and underwater survey. In: Benjamin J, Bonsall C, Pickard C, Fisher A (eds) Submerged prehistory. Underwater archaeology and the future of submerged European prehistory. Oxbow, Oxford, pp 99–110

Chancerel A, Hérard B, Ghesquière E, Desloges J (1994) Archipel de Chausey. Rapport de prospection 1994. DRAC service régional de l'archéologie, Caen

Chancerel A, Dinard F, Desloges J, Ghesquière E, Hérard B (1998) Archipel de Chausey (Granville – Manche). Rapport de prospection. Campagne de 1996. DRAC service régional de l'archéologie, Caen

Cliquet D, Coutard S, Clet-Pellerin M, Allix J, Tessier B, Lelong F, Baltzer A, Mear Y, Poizot E, Auguste P, Alix P, Olive J, Guesnon J (2011) The Middle Palaeolithic underwater site of La Mondrée, Normandy, France. In: Benjamin J, Bonsall C, Pickard C, Fisher A (eds) Submerged prehistory. Underwater archaeology and the future of submerged European prehistory. Oxbow, Oxford, pp 111–128

Closmadeuc G (1882) Le cromlech d'Er-Lannic et le Golfe du Morbihan, à l'époque dite celtique. Bulletin de la Société Polymathique du Morbihan 1882:8–24

Clottes J, Courtin J, Vanrell L (2005) Cosquer redécouvert. Seuil, Paris

Courtin J (1978) Recherches Archéologiques sous-marines. Gallia Préhistoire 21(2):735–746

Coutard S (2003) Formations quaternaires en bordure d'une mer épicontinentale, la Manche Tectonique, eustatisme, climat et occupations humaines – Exemple du Val de Saire (Normandie, France). Unpublished PhD thesis, Université Caen Normandie, Caen

Coutard S, Cliquet D (2005) Chronostratigraphie des formations pléistocènes et peuplement paléolithique en contexte littoral (Val de Saire, Normandie). Bulletin de Société Prehistorique Française 102:477–499

Daire MY, Langouët L (eds) (2008) Les pêcheries de Bretagne; archéologie et Histoire des pêcheries d'estran. Co-édition CeRAA-AMARAI, Les Dossiers du Centre Régional d'Archéologie d'Alet, AE

Daire, MY, Langouët, MY (2010) Les anciens pièges à poissons des côtes de Bretagne, un patrimoine au rythme des marées. Co-édition CeRAA-AMARAI, Les Dossiers du Centre Régional d'Archéologie d'Alet, AG

Daire MY, Langouët L (2011) Dater les anciennes pêcheries par les niveaux marins: approches méthodologiques et perspectives géo-archéologiques: le Bas-Léon, nord Finistère, Bretagne. Norois 220:3–25

Daire MY, Langouët L, Maheo M (2013) Une archéologie au rythme des marées: actualité des recherches sur les anciens barrages de pêcheries maritimes en Bretagne (France). In: Daire MY, Dupont C, Baudry A, Billard C, Large JM, Lespez L, Normand E, Scarre C (eds) Ancient maritime communities and the relationship between people and environment along the European Atlantic coasts. Proceedings of the HOMER 2011 conference, Vannes, British archaeological reports international series 2570. Archaeopress, Oxford, pp 151–162

Devoir A (1912) Témoins mégalithiques des variations des lignes de rivages armoricains. Bulletin de la Société Archéologique du Finistère 39:220–239

Dréano Y, Gandois H, Pailler Y (2013) L'exploitation des poissons dans l'archipel de Molène (Finistère, France) du Néolithique récent à l'Âge du Bronze ancien. In: Daire MY, Dupont C, Baudry A, Billard C, Large JM, Lespez L, Normand E, Scarre C (eds) Ancient maritime communities and the relationship between people and environment along the European Atlantic coasts. Proceedings of the HOMER 2011 conference, Vannes, British archaeological reports international series 2570. Archaeopress, Oxford, pp 447–458

Dulioust JP (1958) La station d'Ors. Bulletin de la Société Préhistorique Française 55(5/6):262–263

Dutertre AP (1924) Observations sur les formations quaternaires et récentes du Boulonnais. Annales du Musée géologique du Boulonnais 1(4):3–13

Dutertre AP (1925) Notice géologique sur la Pointe aux Oies et les abords de la station zoologique de Wimereux. Travaux de la Station Zoologique de Wimereux 9:66–88

Dutertre AP (1930) Trouvailles dans la tourbe submergée de la Pointe-aux-Oies, près Wimereux (Boulonnais). Annales de la Société Géologique du Nord 55:19–21

Dutertre AP (1933) L'industrie et la faune de la tourbe submergée du rivage du Boulonnais (Note préliminaire). Bulletin de la Société Préhistorique Française 30(11):587–588

Dutertre AP (1934) Notice archéologique sur la tourbe submergée de la Pointe-aux-Oies à Wimereux (Pas-de-Calais). Bulletin de la Société Préhistorique Française 31(7/8):352–359

Facon R (1967) Présence du Moustérien moyen, type La Quina à l'Île Madame (Charente-Maritime). Bulletin de la Société Préhistorique Française 64:275–278

Farr RH, Momber G, Satchell J, Flemming NC (2017) Paleolandscapes of the Celtic Sea and the Channel/La Manche. In: Flemming NC, Harff J, Moura D, Burgess A, Bailey GN (eds) Submerged landscapes of the European continental shelf: Quaternary Paleoenvironments. Wiley, Chichester, pp 211–239

Fridman R (1957) Généralités sur des phénomènes périglaciaires würmiens sur le littoral et les îles de la Charente-maritime. Bulletin de la Société Géographique de Rochefort 7:585–596

Gabet C (1968) Habitat néolithique d'Ors. Bulletin de la Société de Géographie de Rochefort, 2e série, 4

Gabet C (1973) Le Banc de Tourbe sur l'estran de la baie de La Perroche (île d'Oléron). Bulletin de la Société Archéologique et Historique de la Charente Maritime 25:297–307

Gabet C, David P (1975) Le gisement de Piédemont (Commune de Port-des-Barques). Rapport préliminaire. Roccafortis 2e série, III 6:177–193

Gandois H, Stéphan P (2015) Les barrages de pêcheries de l'archipel de Molène (mer d'Iroise, Finistère). Bulletin de l'Association Manche Atlantique pour la Recherche Archéologique dans les Îles 28:45–76

Gandois H (dir.), Stéphan P, Cuisnier D, Gladu Y, Lallement F, Priol H (2013a) Rapport sur les prospections sous-marines et la zone d'estran en mer d'Iroise, Opération n°OA-1746. DRASSM, SRA Bretagne

Gandois H, Pailler Y, Stéphan Y, Nicolas C (2013b) L'érosion marine et ses effets sur les vestiges archéologiques en mer d'Iroise: exemple de l'impact de la tempête de mars 2008 sur l'île Kemenez et ses Ledenez (Le Conquet, Finistère, France). In: Daire MY, Dupont C, Baudry A, Billard C, Large JM, Lespez L, Normand E, Scarre C (eds) Ancient maritime communities and the relationship between people and environment along the European Atlantic coasts. Proceedings of the HOMER 2011 conference, Vannes, British archaeological reports international series 2570. Archaeopress, Oxford, pp 99–110

Gandois H, Stéphan P, Cuisnier D, Hulot O, Ehrhold A, Paul M, Le Dantec N, Franzetti M (2018) The stone tidal fish weirs of the Molène archipelago (Iroise Sea, Brittany, Western France): a long-term tradition starting with early megalithic constructions. Int J Naut Archaeol 47(1):5–27

Giot PR (1958) Kernic, Plouescat. Gallia Préhistoire 1(1):129

Giot PR (1968) La Bretagne au péril des mers holocènes. In: La Préhistoire, problèmes et tendances. CNRS, Paris, pp 203–208

Giot PR (1971) Lanmodez, île Coalen. Gallia Préhistoire 14:341

Giot PR, Morzadec H (1992) Des dolmens à couloirs au péril des mers actuelles. Revue Archéologique de l'Ouest 9:57–66

Giot PR, Hallegouët B, Briard G (1965) Le site du Curnic en Guissény (Finistère). Annales de Bretagne 72(1):49–70

Giot PR, Hallegouët B, Monnier JL (1973) Le Paléolithique ancien du Pays de Léon. Finistère L'Anthropologie 77:497–518

Goslin J, Van Vliet-Lanoë B, Spada G, Bradley S, Tarasov L, Neill S, Suanez S (2015) A new Holocene relative sea-level curve for western Brittany (France): insights on isostatic dynamics along the Atlantic coasts of North-Western Europe. Quat Sci Rev 129:341–365

Gouezin P (1998) Le site mégalithique d'Er Lannic. Editions de l'Association Archéo Douar Mor, Vannes

Guilaine J, Freises A, Montjardin R (1984) Leucate-Corrège, habitat noyé du Néolithique ancien cardial. Archives d'Ecologie Préhistorique. Centre d'anthropologie des sociétés rurales, Toulouse, musée Paul Valéry, Sète

Guillaume L (1935) Observations sur les dépôts quaternaires de Saint-Côme-de-Fresné et Asnelles-Belle-Plage (Calvados). Bulletin de la Société Linéenne de Normandie 8(8):63–72

Guillaume L (1952) Les formations quaternaires de la plage du débarquement britannique de Saint-Côme de Fresné-Asnelles Belle Plage (Calvados). In: Shackleton RM (ed) Part XIII. Proceedings of section M. Other subjects, international geological congress, report of the eighteenth session, Great Britain, 1948, London, pp 105–113

Harff J, Flemming NC, Groh A, Hünicke B, Lericolais G, Meschede M, Rosentau A, Sakellariou D, Uścinowicz S, Zhang W, Zorita E (2017) Sea level and climate. In: Flemming NC, Harff J, Moura D, Burgess A, Bailey GN (eds) Submerged landscapes of the European continental shelf: quaternary paleoenvironments. John Wiley & Sons, Chichester, pp11–49

Hénaff X, Langouët L, Meury JL (1998) Des structures néolithiques en bois dans l'anse du Rosais à Saint-Malo (35), suite. Bulletin de l'Association Manche Atlantique pour la Recherche Archéologique dans les Îles 11:5–11

Hinout J, Rolland C (1960) Gisements côtiers de Saint-Trojan. Bulletin de la Société Préhistorique Française 57(5/6):27

Huet B (2002) Une industrie à composante lithologique mixte: le gisement paléolithique moyen de Goaréva (île de Bréhat, Côtes-d'Armor). Bulletin de la Société Préhistorique Française 99:699–716

Huet B (2006) De l'influence des matières premières lithiques sur les comportements techno-économiques au Paléolithique moyen: l'exemple du Massif armoricain (France). Unpublished PhD thesis Université Rennes 1, Rennes

Huet B (ed) (2010) Le site paléolithique moyen des Vallées (Pléneuf-Val-André, Côtes-d'Armor) Rapport de fouille archéologique programmée, DRASSM et SRA Bretagne

Joussaume R (1981) Le Néolithique de l'Aunis et du Poitou occidental dans son cadre atlantique, Rennes, Travaux du Laboratoire d'Anthropologie-Préhistoire-Protohistoire et Quaternaire Armoricains. CNRS, Paris

Joussaume R, Boiral M, Ters M (1986) Sites préhistoriques submergés à La Tranche-sur-Mer (Vendée). Bulletin de la Société Préhistorique Française 83(11/12):423–435

Labrude C, Large JM, Mangematin J (2000) Le Marais Girard à Brétignolles-sur-Mer (Vendée): une nouvelle approche du site. Bulletin du Groupe Vendéen d'Etudes Préhistoriques 36:13–23

Laforge M (2012) Le cadre chronostratigraphique des peuplements pléistocènes de l'Ouest de la France. Eustatisme, changements climatiques et occupations humaines. Unpublished PhD thesis Université de Rennes 1, Rennes

Laisné G, Vilgrain G (2008) Quand Néandertal occupait le littoral de Portbail (Manche). Rapport PCR "Les Premiers Hommes en Normandie", septième année

Lambeck K, Esat TM, Potter EK (2002) Links between climate and sea-levels for the past three million years. Nature 419:199–206

Lambeck K, Rouby H, Purcell A, Sun Y, Sambridge M (2014) Sea level and global ice volumes from the last glacial maximum to the Holocene. Proc Natl Acad Sci 111:15296–15303

Langouët L (1995) Des structures néolithiques en bois dans l'anse du Rosais à Saint-Malo (35). Bulletin de l'Association Manche Atlantique pour la Recherche Archéologique dans les Îles 8:27–32

Langouët L (2003) Les menhirs de Saint-Jacut-de-la-Mer. Bulletin de l'Association Manche Atlantique pour la Recherche Archéologique dans les Îles 16:61–65

Langouët L, Daire MY (2009) Ancient maritime fish-traps of Brittany (France): a reappraisal of the relationship between human and coastal environment during the Holocene. J Marit Archaeol 4(2):131–148

Laporte L (ed) (1998) L'estuaire de la Charente de la Protohistoire au Moyen Age. Document d'Archéologie Française. Editions de la Maison des Sciences de l'Homme, Paris

Laporte L (ed) (2009) Des premiers paysans aux premiers métallurgistes dur la façade atlantique de la France (3500–2000 av. JC). Mémoire XXXIII, Editions de l'Association des Publications Chauvinoises

Laporte L, Glausinger R (1986) Le site de L'Ecuissière à Dolus (île d'Oléron, Charente-Maritime). Groupe Vendéen d'Etudes Préhistoriques, La Roche-sur-Yon, pp 77–88

Laporte L, Regibus JP, Regibus J, Ricou C, Rodien-Laire C (1995) Inventaire archéologique de l'île d'Oléron (17). Bulletin de l'Association Manche Atlantique pour la Recherche Archéologique dans les Îles 8:5–20

Laporte L, Cros JP, Fontugne M, Gebhardt A, Gruet Y, Marguerie D, Oberlin C (1998) Les occupations néolithiques de la côte occidentale de l'île d'Oléron. In: Camps G (ed) L'homme préhistorique et la mer, Actes du 120e Congrès National des Sociétés Savantes, Aix-en-Provence, 1995. Congrès National des Sociétés Historiques et Scientifiques, Paris, pp 197–238

Large JM, Birocheau P, Corson S, Cousseau F, Large C, Tortuyaux JP (2009) Une archéologie des pêcheries d'estran. L'anse aux Moines et la Pointe du Vieux Moulin au Château-d'Olonne en Vendée Bulletin du Groupe Vendéen d'Etudes Préhistoriques 45:4–45

Large JM, Corson S, Magne J (2015) Le mégalithe de la Table sur le Plateau de la Vendette (Noirmoutier-en L'Île, Vendée). Bulletin du Groupe Vendéen d'Etudes Préhistoriques 51:33–41

Le Rouzic Z (1965) Les monuments mégalithiques de Carnac et de Locmariaquer. Leur destination – Leur âge. Imprimerie Armoricaine, Quimper

Lecerf Y (1985) L'allée couverte de Kernic à Plouescat (Finistère). Bulletin de la Société Archéologique du Finistère 114:17–34

Lefort JP, Hallegouët B, Monnier JL (2011) La mer d'Iroise: une singularité dans l'approvisionnement en silex des hommes du Paléolithique. Revue Archéologique de l'Ouest 28:7–18

Leroy F (2001) Sites lagunaires du Languedoc au Néolithique et à l'Âge du bronze. In: L'Helgouach J, Briard J (eds) Systèmes fluviaux, estuaires et implantations humaines de la Préhistoire aux grandes invasions, actes du 124e congrès national des sociétés historiques et scientifiques, Nantes, 1999. Congrès National des Sociétés Historiques et Scientifiques, Paris, pp 229–239

Leroy F, Bouby L, Guibal F (2003) Les gisements protohistoriques de l'étang de Thau. In: Actes des quatrièmes rencontres méridionales de Préhistoire récente, Nîmes 2000, pp 369–376

L'Helgouach, J (1965) Les Sépultures mégalithiques en Armorique (dolmens à couloirs et allées couvertes). Unpublished PhD thesis, Université de Rennes 1, Rennes

Lopez-Romero E, Langouët L, Bihan S (2008) Nouveaux mégalithes sur l'Ile de Groix. Bulletin de l'Association Manche Atlantique pour la Recherche Archéologique dans les Îles 21:5–14

Luguet H (1867) Note sur des recherches faites dans l'île d'Oléron. Recueil de la Commission des Arts et Monuments Historiques de la Charente-Inférieure 2:301–304

Marchat A, Le Brozec M (1991) Les mégalithes de l'arrondissement de Lannion. Collection Patrimoine archéologique de Bretagne, Institut Culturel de Bretagne et Travaux du Laboratoire d'Anthropologie-Préhistoire de Rennes, Rennes

Mariette H (1971) L'Archéologie des dépôts flandriens du Boulonnais. Quaternaria 14:137–150

Mariette H (1976a) Le Néolithique de la Pointe-aux-Oies (Wimereux, Pas-de-Calais). In: Livret-guide de l'excursion A10 du IXe Congrès de l'UISPP (Nice, 1976): Nord-Ouest de la France (Bassin de la Seine, Bassin de la Somme et Nord), pp 172–173

Mariette H (1976b) Flandrien et archéologie de Wissant (Pas-de-Calais). In: Livret-guide de l'excursion A10 du IXe Congrès de l'UISPP (Nice, 1976): Nord-Ouest de la France (Bassin de la Seine, Bassin de la Somme et Nord), pp 180–182

Maufras E (1874) Du préhistorique de la Charente-Inférieure (deux stations néolithiques à l'île d'Oléron). Association Française pour l'Avancement des Sciences, Congrès de Lille, pp 590–593

Maurin B (1998) 3000 ans sous les eaux: vingt ans de fouilles archéologiques dans le lac de Sanguinet, Landes. Editions Gaia, Paris

Mazères R, de Bellaing E (1936) Port-Blanc en Penvénan. Aperçu géologique et préhistorique. Bulletin de la Société d'Emulation des Côtes-du-Nord LXVIII:195–213

Mercier J (1934) Observations sur les gisements d'*Elephas primigenius* Blum. en Basse-Normandie. Bulletin de la Société Linnéenne de Normandie, 8ème série 6:7–11

Mercier N, Cliquet D, Olive J, Poudret-Barré A, Momber G, Coutard S, Clet M (2013) Approche du gisement moyen sous-marin de la Mondrée à Fermanville (Manche) et évaluation de son potentiel en matière de datation par la méthode de l'OSL. In: Daire MY, Dupont C, Baudry A, Billard C, Large JM, Lespez L, Normand E, Scarre C (eds) Ancient maritime communities and the relationship between people and environment along the European Atlantic coasts. Proceedings of the HOMER 2011 conference, Vannes, British archaeological reports international series 2570. Archaeopress, Oxford, pp 13–22

Michel D, Coutard JP, Helluin M, Lautridou JP, Ozouf JC, Pellerin J (1982) Contribution à l'étude du Paléolithique inférieur et moyen de la région de Barfleur. Le Paléolithique inférieur de Gatteville-Phare (Manche). Bulletin de la Société Préhistorique Française 79(10/12):319–329

Milon Y, Giot PR (1949) IVe Circonscription, informations archéologiques. Gallia Préhistoire 7(2):252–256

Monnier JL (1986) Le gisement paléolithique moyen de Nantois, Pléneuf (Côtes-du-Nord). Bulletin de la Société Préhistorique Française 83:146–150

Monnier JL (1988) Le gisement paléolithique moyen des Gastines (Ille-et-Vilaine). Etude géologique et archéologique. Revue archéologique de l'Ouest 5:11–33

Monnier JL (2006) Les premiers peuplements de l'Ouest de la France. Cadre chronostratigraphique et paléoenvironnemental. Bulletin du Musée d'Anthropologie et Préhistoire de Monaco 46:3–20

Monnier JL, Huet B, Laforge M (2011) Application of sedimentological analysis to correlation of eroded layers under beaches with local and regional Pleistocene stratigraphy. A contribution to geological dating of Palaeolithic sites on the northern coast of Brittany (France). Quat Int 231:78–94

Montjardin R, Cablat A (2004) Le gisement Néolithique ancien des Dunes à Frontignan (Hérault). In: Dartevelle H (ed) Auvergne et Midi. Actes des Vème Rencontres Méridionales de Préhistoire Récente (Clermont-Ferrand, Novembre 2002). Préhistoire du Sud-Ouest, supplément 9:295–312

Morzadec-Kerfourn MT (1974) Variation de la ligne de rivage armoricaine au Quaternaire. Analyses polliniques de dépôts organiques littoraux. Mémoires de la Société Géologique et Minéralogique de Bretagne 17

Munaut AV, Gilot E (1977) Recherches palynologiques et datations 14C dans les régions côtières du Nord de la France. I. Phases transgressives et stabilisations dunaires flandriennes dans l'estuaire de la Canche (Pas-de-Calais). Bulletin de l'Association Française pour l'Etude du Quaternaire 3:17–25

Néraudeau D, Bourgeois D (1994) Le Paléolithique inférieur et moyen de l'île Madame (17). Bulletin de l'Association Manche Atlantique pour la Recherche Archéologique dans les Îles 7:5–14

Ovion Dr (1890) Procès-verbal de la séance du 5 septembre 1888. Bulletin de la Société Académique de l'Arrondissement de Boulogne-sur-Mer 5:372–373

Pailler Y, Stéphan P, Gandois H, Nicolas C, Sparfel Y, Tresset A, Donnart K, Dréano Y, Fichaut B, Suanez S, Dupont C, Audouard L, Marcoux N, Mougne C, Salanova L, Sellami F, Dietsch-Sellami MF (2014) Landscape evolution and human settlement in the Iroise Sea (Finistère, Brittany, France) during the Neolithic and bronze age. Proceedings of the Prehistoric Society 8:105–139

Pautreau JP, Robert PP (1980) Le foyer néolithique ancien des Gouillauds, au Bois, Île de Ré (Charente-Maritime). Bulletin de la Société Préhistorique Française 77(4):123–128

Perpère M (1986) Apport de la typométrie à la définition des éclats Levallois: l'exemple d'Ault. Bulletin de la Société Préhistorique Française 83(4):115–118

Perpère M (1999) Le débitage Levallois d'Ault. Anthropologie 103(3):343–376

Pineau E (1891) Nouvelle contribution à la géographie préhistorique des côtes occidentales de France. Bulletin de la Société Française pour l'Avancement des Sciences 1891:266–267

Pouit D, Viaud JM (2002) Des éléphants antiques (Palaeoloxodon antiquus) à Brétignolles-sur-Mer (Vendée, France): histoire et conséquences des découvertes. Le Naturaliste Vendéen 2:35–59

Pruvost P (2006) Le Paléolithique moyen des Iles Chausey (Manche). Rapport d'activité 2005 du PCR "Les Premiers Hommes en Normandie", pp 27–35

Romain G (1915) La station paléolithique sous-marine dans la plage du Havre. 43e session de l'Association Française pour l'Avancement des Sciences, Le Havre 1914, pp 611–619

Roué D, Le Goff JC (2008) Barrages à poissons et pêcheries à Santec (Nord-Finistère). In: Daire MY, Langouët L (eds) Les pêcheries de Bretagne: archéologie et histoire des pêcheries d'estran, Coéd. Ce.R.A.A.-A.M.A.R.A.A.I., Les Dossiers du Centre Régional d'Archéologie d'Alet, AE, pp 97–111

Rousseau J, Robert PP, Cantin N, Dupont C (2001) Exploitation du territoire insulaire au Néolithique moyen d'après la fouille d'un amas coquillier. Les Gouillauds à Bois-plage en Ré (Charente-Maritime, France). Revue d'Archéométrie 25:225–242

Roussot-Larroque J (1997) La "station sous-marine" du Bétey à Andernos (Gironde) d'après la collection Ferrier. Fédération Historique du Sud-Ouest, Bordeaux, pp 53–71

Rouvreau M (1968) Les fouilles de la Pointe d'Ors. Archéologie Pontoise

Rouvreau M, Gomez J (1973) Les occupations post-néolithiques de la station d'Ors (île d'Oléron). Bulletin de la Société d'Archéologie et d'Histoire de l'Aunis 25:37–44

Sauvage H-E (1899) Le mammouth de la partie sud de la Mer du Nord. Bulletin de la Société académique de l'arrondissement de Boulogne-sur-mer 5(1891–1899):408–416

Scuvée F, Vérague J (1988) Le gisement sous marin du Paléolithique moyen de l'anse de la Mondrée à Fermanville (Manche). LITTUS-C.E.H.P, Cherbourg

Sorrel P, Debret M, Billeaud I, Jaccard SL, Mc Manus JF, Tessier B (2012) Persistent non-solar forcing of Holocene storm dynamics in coastal sedimentary archives. Nat Geosci 5:892–896

Sparfel Y, Pailler Y dir. with contributions of Chaigneau C, Chauris L, Fichaut B, Gouletquer P, Stéphan P, Suanez S, Tanguy B (2009) Les mégalithes de l'arrondissement de Brest, inventaire et essai de synthèse. Editions CeRAA et Institut Culturel de Bretagne, Daint-Malo

Stéphan P (2011) Colmatage sédimentaire des marais maritimes et variations relatives du niveau marin au cours des 6000 dernières années en rade de Brest (Finistère). Norois 220:9–37

Stéphan P (2012) Une régression marine au Bronze final en Bretagne ? Ou le reflet de changements morpho-sédimentaires importants à la côte. In: M. Melin & C. Mougne (dir.), L'homme, ses ressources et son environnement dans le Nord-ouest de la France à l'âge du Bronze: actualité de la recherche, Actes du séminaire archéologique de l'ouest du 22 mars 2012. Editions de Géosciences Rennes, hors-série 7, pp 131–139

Stéphan P, Goslin J (2014) Evolution du niveau marin relatif à l'Holocène le long des côtes françaises de l'Atlantique et de la Manche: réactualisation des données par la méthode des "sea-level index points". Quaternaire 25(4):295–312

Stéphan P, Pailler Y, Tresset A, Gandois H (2013) Changements paléogéographiques de l'archipel de Molène (Finistère, Bretagne, France): Implications sur les peuplements humains du Néolithique à l'Age du Bronze. In: Daire MY, Dupont C, Baudry A, Billard C, Large JM, Lespez L, Normand E, Scarre C (eds) Ancient maritime communities and the relationship between people and environment along the European Atlantic coasts. Proceedings of the HOMER 2011 conference, Vannes, British archaeological reports international series 2570. Archaeopress, Oxford, pp 647–660

Stéphan P, Goslin J, Pailler Y, Manceau R, Suanez S, Van Vliet-Lanoë B, Hénaff A, Delacourt C (2015) Holocene salt-marsh sedimentary infillings and relative sea-level changes in West Brittany (France)

from foraminifera-based transfer functions. Boreas 44(1):153–177
Ters M (1973) Les variations du niveau marin depuis 10.000 ans, le long du littoral atlantique français. In: Le Quaternaire, géodynamique, stratigraphie, environnement. Travaux français récents. 9e Congrès INQUA, Christchurch. Supplement. Bulletin de l'Association Française pour l'Etude du Quaternaire:114–115
Ters M (1986) Variations in Holocene Sea level on the French Atlantic coast and their climatic significance. In: Rampino MR, Sanders JE, Newman WS, Königsson LK (eds) Climate: history, periodicity and predictability. Van Nostrand Reinhold, New York, pp 204–237
Toucanne S (2008) Reconstruction des transferts sédimentaires en provenance du système glaciaire de mer d'Irlande et du paléofleuve Manche au cours des derniers cycles climatiques. Unpublished PhD thesis Université Bordeaux 1, Bordeaux
Vallin L, Masson B (1993) Tardinghem, dunes du Chatelet. Bilan Scientifique Régional du Nord-Pas-de-Calais 80
Vanrell L (2010) La grotte Cosquer (Marseille, Bouches-du-Rhône). In: Archéologie des rivages méditerranéens: 50 ans de recherches: actes du colloque d'Arles (Bouches-du-Rhône) 28-29-30 octobre 2009. Errance, Paris, pp 31–36
Welsch J (1911) La tourbe littorale du Croisic (Loire-Inférieure) et les dépôts analogues de l'Ouest de la France. Bulletin de la Société Scientifique et Naturelle de l'Ouest de la France 3:201–221
Welsch J (1917) Les lignites du littoral et les forêts submergées de l'Ouest de la France. Anthropologie 38:201–234
Ydier V, De Maisonneuve B (2012) Compte-rendu de découverte de bien culturel maritime: août 2012 – Pêcherie de l'Acheneau, l'Épine, Vendée. Rapport du Cercle de Recherche sur l'Histoire et le Patrimoine de la Vendée, St-Gilles-Croix-de-Vie

Spain: Underwater Exploration on a Narrow Continental Shelf

13

Pablo Arias

Abstract

In spite of Spain's long coastline (nearly 8000 km) and its well-established tradition in underwater archaeology, the prehistoric settlement of the continental shelf is practically unknown with very few finds. Underwater research has focused on naval archaeology and, until very recently, no attempt had been made to look for prehistoric underwater sites. In the past decade, new research projects have been launched to explore selected areas on the Cantabrian shelf and offshore of Gibraltar. This chapter summarises the currently available evidence of submerged prehistoric archaeology and the preliminary results of these new projects.

Keywords

Continental shelf · Submerged landscapes · Neanderthal · Mousterian · Upper Palaeolithic · Mesolithic · Neolithic · Bronze Age · Votive deposits

13.1 Introduction[1]

Occupying roughly five sixths of a large peninsula, Spain has a long coastline of some 7905 km. This includes Atlantic and Mediterranean-facing shores—1075 km in the Bay of Biscay, 1761 km facing the Atlantic Ocean and 2058 km along the Mediterranean Sea, plus 1428 km in the Balearic Islands and 1583 km in the Canary Islands. Moreover, some of the most fertile regions and largest concentrations of population are in coastal regions, so Spain can be defined as a maritime country. From another point of view, the Iberian Peninsula, located at a geographic crossroads of Europe, was densely populated throughout prehistory and possesses some of the most important Palaeolithic sites in Europe, notably the caves of northern and south-eastern Spain and Gibraltar, a very high density of Mesolithic coastal sites and important cultures of the Neolithic, Bronze and Iron Ages. This suggests considerable potential for research on the prehistoric settlement of the continental shelf.

As a largely coastal country, it is not surprising that Spain enjoys a rich tradition in the field

[1] The Iberian Peninsula comprises the sovereign states of Spain, Portugal and Andorra and the British Overseas Territory of Gibraltar. In terms of its Stone Age prehistory, it forms a single geographical entity. For convenience, the evidence from Portugal is considered separately in Chapter 14, and that from Gibraltar is included in this chapter.

P. Arias (✉)
Instituto Internacional de Investigaciones Prehistóricas de Cantabria, Universidad de Cantabria-Gobierno de Cantabria-Santander, Santander, Spain
e-mail: pablo.arias@unican.es

G. Bailey et al. (eds.), *The Archaeology of Europe's Drowned Landscapes*, Coastal Research Library 35, https://doi.org/10.1007/978-3-030-37367-2_13

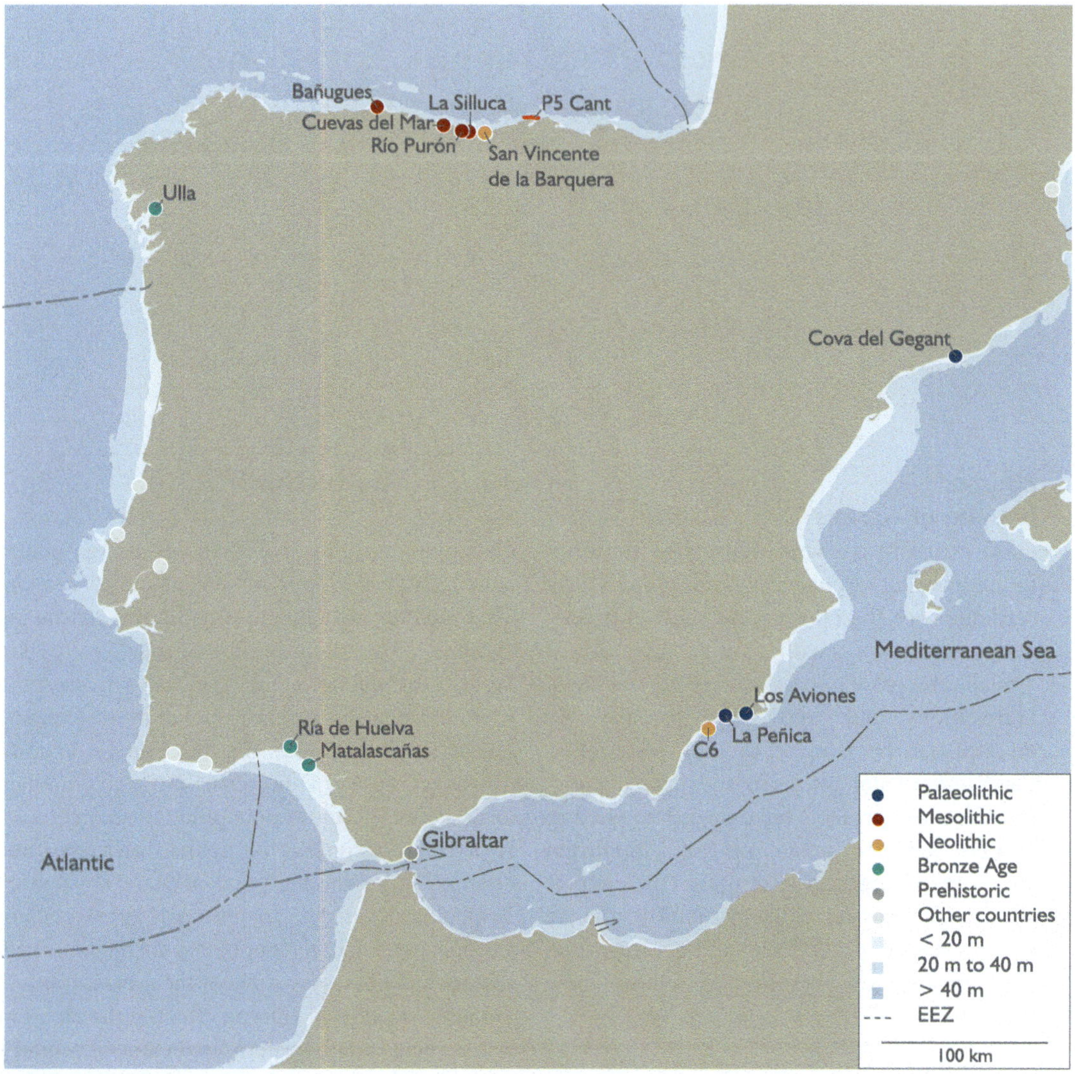

Fig. 13.1 Location of sites and research projects mentioned in the text. Site information from the SPLASHCOS Viewer http://splashcos-viewer.eu. Drawing by Moritz Mennenga

of underwater archaeology (Blánquez and Martínez Maganto 1993; Nieto 2009; Nieto and Bethencourt 2014). This has focused on naval history, with hundreds of shipwrecks and other remains of trade and warfare from the Phoenician, Greek and Roman periods and from the Modern Age, when the Spanish Crown became one of the world's main naval powers, all of which are well represented in the recently refurbished National Underwater Archaeology Museum (*Museo Nacional de Arqueología Subacuática*-ARQUA) in Cartagena. There is a small but significant number of underwater finds ranging from the Palaeolithic to the Bronze Age, mostly discovered by chance (Fig. 13.1), and recent projects dedicated to underwater survey of the continental shelf, which suggests considerable potential for future discoveries. This chapter summarises data on known prehistoric underwater sites, the results of recent offshore surveys and the potential for future research.

13.2 Conditions of Preservation and Discovery

One of the deterrents to underwater investigation on many coastlines, especially those in the Bay of Biscay associated with the large concentrations of Palaeolithic sites in the Cantabrian region, is that sea conditions are challenging for underwater archaeological exploration and potentially unfavourable for the preservation of prehistoric remains, due to strong currents, large waves and a predominance of erosional processes over sedimentation (Moura et al. 2017). Moreover, administrative responsibility for the continental shelf is divided between a number of regional and central government agencies, leading to considerable bureaucratic complexities in the organisation of underwater work aimed at the search for submerged landscapes, and these can act as a further disincentive. Although Spain is not formally a federal state, the regions (autonomous communities) hold considerable legislative and administrative power in numerous fields. One of them is culture, and particularly archaeology, which has been fully devolved to the regions (Martínez Díaz 1992, 1993). However, the management of the continental shelf is the exclusive responsibility of the central government. This duality causes considerable bureaucracy and also hinders the rational planning of research and protection of the heritage, as conflicts of authority and demarcation disputes between the different administrations can easily occur.[2] In fact, despite a few attempts (directed mainly towards naval archaeology), a national management plan for underwater archaeology has still not been devised.

Another factor that has to some extent deterred interest and investigation in offshore prehistoric material is the fact that the Spanish continental shelf is quite narrow (Fig. 13.1); apart from Galicia in the north-west and sections of the Mediterranean coast, relatively little territory has been submerged by postglacial sea-level rise, especially in comparison with northern Europe, weakening the incentive to conduct underwater investigation on the assumption that it would make little difference to current understandings of Palaeolithic and Mesolithic patterns of settlement and adaptation.

This is counterbalanced by a series of positive factors. In the first place, on many coastlines, there is a high density of Palaeolithic and Mesolithic sites. This is especially true of Cantabrian Spain, which displays the highest concentration of coastal Palaeolithic sites in Europe (Straus 1992, 2018; Fano 2004). The density of Mesolithic sites is even higher: in eastern Asturias, around 130 sites have been recorded along only 50 km of coast (see Fano 2004, Arias and Fano 2009). On the other hand, the continental shelf is particularly narrow, averaging 15 km, so the hypothesis can be proposed that the recorded Palaeolithic sites on land are just a part of a larger coastal settlement system which is not too extensive to be explored efficiently but which may include important evidence of sites and activities that are not represented on land in the present-day coastal region. In fact, there is indirect evidence showing that Palaeolithic groups in Cantabrian Spain and elsewhere maintained regular contact with the sea. Even at the maximum lowering of sea level at the Last Glacial Maximum, the shoreline was within quite easy reach of coastal caves that are on or close to the present-day coastline. Some of these coastal caves have stratified deposits with cultural material that reach back to the time of the Last Glacial Maximum or earlier, and it is clear from the faunal remains in these earlier deposits that the now submerged plain was a very significant area of lowland territory for the large mammals that formed an important part of the Palaeolithic subsistence economy and an important part of the territory used by the occupants of these cave sites.

The submerged coastline was also a source of marine foods and close enough to the present-day

[2] As an example, the P5 Cant Project, described below, required permission from three state organisations (Harbour Master's Office, Port of Santander and Civil Guard), two regional government organisations (Directorate General of Culture and Directorate General of Fishing and Food), a hospital (belonging to another department of the regional government) because of safety regulations associated with diving and the Santander Town Corporation (for permission to place a geolocation aerial on land).

shoreline that some of these resources were brought back to caves on the modern coast for processing and consumption. Mollusc shells, in particular, are present in these Palaeolithic deposits, and sometimes quantities of fish bone, for example, at La Riera Cave in Asturias (Straus and Clark 1986), Nerja and Bajondillo in Andalusia (Morales et al. 1998; Cortés-Sánchez et al. 2008, 2011) and Gorham's Cave and Vanguard Cave on the Gibraltar Peninsula (Finlayson et al. 2006; Stringer et al. 2008). Occasionally, remains of marine mammals such as seals or cetaceans are also recorded (Corchón et al. 2008; Álvarez-Fernández 2011; Álvarez-Fernández et al. 2014; Morales et al. in press). The quantities of marine food remains in these Palaeolithic deposits are not as great as those present in later deposits, such as the Asturian shell middens, many of which fill the mouths of small caves and are stratified above deposits with Palaeolithic cultural remains (González Morales 1982). However, that may simply reflect the greater distance between these sites and the contemporaneous palaeo-shorelines of the Palaeolithic and early Mesolithic period and the likelihood that only a portion of the marine catch was carried inland while the rest was processed and consumed closer to the coastline and left its archaeological mark on sites that are now submerged (Bailey and Craighead 2003).

Other evidence of contact with the coast during periods of lower sea level is the existence of depictions of marine animals in Palaeolithic art, such as the representations of fish on the walls of El Pindal (Alcalde del Río et al. 1911; Morales and Roselló 1984–1985) and Altxerri (Fig. 13.2; Altuna 1997) or portable art objects, notably the

Fig. 13.2 Rock engravings representing Pleuronectiform fish (probably flounder—*Platichthys flesus*) on the walls of the cave of Altxerri (Basque country). (**a**) Photo of the engravings. (**b**) The same image with the outlines of the fish highlighted and a scale added. Photo by J. Altura and X. Otero; interpretation after Ruiz 2014

Fig. 13.3 Representation of a bison and a whale on a sperm whale tooth from layer VIIIb (middle Magdalenian) of the cave of Las Caldas (Asturias). Scale in cm. Photo of the original (upper row) and digitally enhanced interpretation of the photo (lower) after Rivero 2015

representation of a cetacean on a sperm whale tooth from the cave of Las Caldas in Asturias (Fig. 13.3; Rivero 2015; Corchón 2017). Moreover, isotopic values demonstrate marine diet for some individuals (unpublished results from Cueva de Los Canes in Asturias). All of these suggest that the submerged areas of continental shelf were densely occupied and that the narrowness of the submerged shelf is, therefore, an incentive to more detailed underwater exploration rather than a deterrent.

A third advantage of working offshore, and this is well demonstrated on the Cantabrian coast, is that the region is rich in karst geology, which in some cases continues beneath the sea, and these are good environments for human settlement and for the preservation of archaeological remains. Similar comments apply to other coastlines, particularly in south-east Spain and Gibraltar. The presence of numerous occupied caves and rockshelters in these regions of karst geology raises the prospect of discovering submerged cliff lines with caves where terrestrial deposits may have been protected from the destructive effects of sea-level rise (see also Canals et al. 2017).

Mention should also be made of inland lake basins. These offer good examples of underwater preservation. The most important of these is Lake Banyoles (Catalonia). Here, a large-scale research project at the early Neolithic site of La Draga has documented a settlement of one of the earliest farming communities in the western Mediterranean, with magnificent preservation of organic remains such as wooden artefacts and plant fibres, dating between c. 5300 and 5000 cal BC (Bosch et al. 2000, 2011, 2012; Piqué et al. 2015). Other examples are the Mesolithic sites of Albufera de Anna in Valencia (Aparicio 1975) and some megalithic monuments flooded by reservoirs in Extremadura (Matamoros et al. 2013–2015).

13.3 Archaeological Overview

13.3.1 Palaeolithic and Mesolithic Finds

One interesting case, because of its potential, is La Peñica or Punta del Faro, a poorly known site that was found in 1980 when the port of Puerto de Mazarrón in Murcia was being dredged. Here, at 3.5 m depth, beneath a layer with Roman material, was a Palaeolithic deposit overlying a fossil beach. The artefacts appear to have been eroded from a nearby rockshelter and are of Middle Palaeolithic affinity, but no further details are available (Montes 1985, 1989). Also in the same region, there is the cave of Los Aviones near Cartagena, where a Mousterian deposit is being eroded from the cave by wave action (Montes 1985, 1989, 1991).

Other cases are known of totally or partially flooded caves in which strata or at least prehistoric materials have been documented. Of particular interest is the cave site of Cova del Gegant near Sitges, south of Barcelona (Daura et al. 2005, 2010). The cave is partly above modern sea level but is only accessible by sea or from above through a narrow, vertical chamber. The deposits inside the cave have been partially eroded by wave action, but some sediments are slightly above sea level and have preserved a stratigraphy with a time span of c. 100–40 ka, with remains of Pleistocene fauna, Mousterian artefacts, a Neanderthal mandible with a uranium series date of 52.3 ± 2.3 ka and some isolated Neanderthal teeth.

We can also mention some Mesolithic sites on the coast of the Bay of Biscay such as La Silluca, Cuevas del Mar and Río Purón (Llanes), which consist of Asturian shell middens located in caves which are currently flooded at high tide (González Morales 1982). A similar case on the Mediterranean coast is Cave C-6 near Cape Cope (Águilas, Murcia), where late Neolithic remains have been found (San Nicolás and Martínez 1987).

More difficult to interpret are the prehistoric remains that have appeared on beaches, of which there are numerous Mesolithic examples, for example, at Bañugues and other sites in the Cape Peñas area of Asturias (Blas et al. 1978) and in the Mongrí area of Gerona. It is difficult to determine whether these represent materials eroded out from submerged coastal sites and redeposited on the shore or more simply reflect the concentration of Mesolithic activities along the present-day shoreline.

Mention should also be made of the remains of Pleistocene fauna that have been occasionally dredged up by fishing boats at different points on the Spanish continental shelf, such as the proboscidean teeth from Catalonia mentioned by Canals et al. (2017, pp. 323–324). However, unless it can be clearly established that they can be related to human activity, they should be regarded as palaeontological deposits, as is the case with the above examples, which include species such as the mastodon which were extinct before the arrival of hominins in the Iberian Peninsula.

13.3.2 Neolithic and Bronze Age Votive Deposits

A number of finds fit into this category, although most are accidental discoveries resulting from commercial fishing or dredging. The only known find associated with systematic archaeological activity is a Neolithic polished axe recovered in an underwater survey in the port of San Vicente de la Barquera in Cantabria (Casado et al. 1992). A copper axe was found in the same area, at Puente de la Maza (Jorge 1953; Arias et al. 2005).

The most significant case in an offshore context of this kind is a late Bronze Age carp's tongue sword, picked up in a fisherman's net two miles off the coast between Matalascañas and Mazagón, Huelva (Meijide 1988, p. 113).

Other finds were discovered in estuaries or ports by dredging and therefore have limited or non-existent contextual information. The most significant of these is the Ría de Huelva hoard, found in a large estuary formed by the confluence of the Rivers Tinto and Odiel. Here, in 1923, a dredger found one of the most significant late Bronze Age assemblages in Europe, a fundamental reference for the transition from Late Bronze Age 2 to 3, the Blackmoor/Braud/Huelva phase according to the seriation proposed by Brandherm (2007; see also Burgess and O'Connor 2008 and Brandherm and Burgess 2008). It comprises 400 objects, including 88 swords or sword fragments, 29 daggers, 85 spearheads, 49 ferrules and arrowheads, 4 helmets and 9 elbow fibulae, buttons, fasteners, pins, torc fragments and harness elements. For many years, this assemblage was attributed to a Bronze Age shipwreck (Almagro 1940, 1958). However, a later reappraisal (Ruiz-Gálvez 1995) suggests that it is an example of a

Bronze Age hoard (Bradley 1990). From this point of view, the Huelva material may be interpreted as a votive offering deposited in an estuary in the late tenth century cal BC.

A similar example occurs in the estuary of the River Ulla in Galicia, where dredging recovered a large group of Late Bronze Age objects: at least two rapiers, five swords, two palstaves and a spearhead (Ruiz-Gálvez 1982; Peña 1985). It is possible that we can add to this list some swords recovered in the large area, currently emerged, corresponding to the former estuary of the River Guadalquivir.

13.4 New Projects

Two exploratory underwater projects have been initiated in recent years, one in Cantabria and the other in Gibraltar, both involving a combination of underwater mapping and diver exploration.

Starting in the north, the project 'El Paleolítico Sumergido: Paisajes y Poblamiento Prehistóricos en la Plataforma Continental Cantábrica' (P5 Cant) (The Submerged Palaeolithic: Prehistoric Landscape and Human Settlement on the Cantabrian Continental Shelf) was carried out between 2015 and 2017.

As noted earlier, there are challenges to underwater exploration in this area because of variable sea conditions, frequent poor weather and the likelihood of erosion of underwater material by strong currents, but also advantages because of the high density of known sites on land that provide clues to target areas for underwater survey.

The aim of the P5 Cant Project was to test for the presence of submerged archaeology, and three areas were selected for surveying according to the following criteria:

- Shallow waters (<30 m).
- Existence of karst massifs, particularly where the caves faced inland (towards the south).
- Proximity to significant concentrations of on-land Palaeolithic sites.
- Association with rapid transgression phases or with former estuaries and bays.
- Absence of large sand banks.

Areas considered most promising according to these criteria were around the karst massif of Mt. Buciero in Santoña, the vicinity of the Cape Mayor and the Santander Bay and the Cuchía area between the palaeovalleys of the Rivers Besaya and Pas. With the means available for the project, only one of these could be explored intensively, and the Cape Mayor-Santander Bay sector was chosen because of its optimal conditions. Surveying focused on two areas with detailed bathymetric survey using a multibeam echosounder (Reson SeaBat 7125), RTK (Real-Time Kinematic) techniques and the production of digital elevation models (Fig. 13.4).

Using this bathymetry, two underwater survey campaigns were carried out, one of them more extensive, with a 'Triggerfish' model ROV of the 'Deep Ocean Engineering' company, directed by Xabier Guinda of the Hydraulic Institute at the University of Cantabria, and the other by diving, led by the archaeologist Pablo Saiz Silió and his team. These explorations succeeded in locating a series of places with potential for the conservation of Palaeolithic or Mesolithic remains: caves (Fig. 13.5), rockshelters and other karst forms (dolines, canyons), and some anomalous concentrations of mollusc shells. In addition, numerous sediment and rock samples were collected and, together with the bathymetric information, this has identified the continuation under the sea of geological and geomorphological structures known on land (Fig. 13.6; Sánchez Carro et al. 2017).

To sum up, the P5 Cant Project was able to produce worthwhile results in the reconstruction of the landscape and geological structure and the location of numerous sites with potential for future research. No less important has been the refinement of the methodology and the formation and training of a highly motivated interdisciplinary team with which the topic can be addressed in the future on realistic foundations.

In Gibraltar, a programme of underwater exploration was initiated in 2005, as a pilot project involving collaboration between the University of York, the Hampshire and Wight Trust for Maritime Archaeology and the Gibraltar Museum with a follow-up season in 2008 including exca-

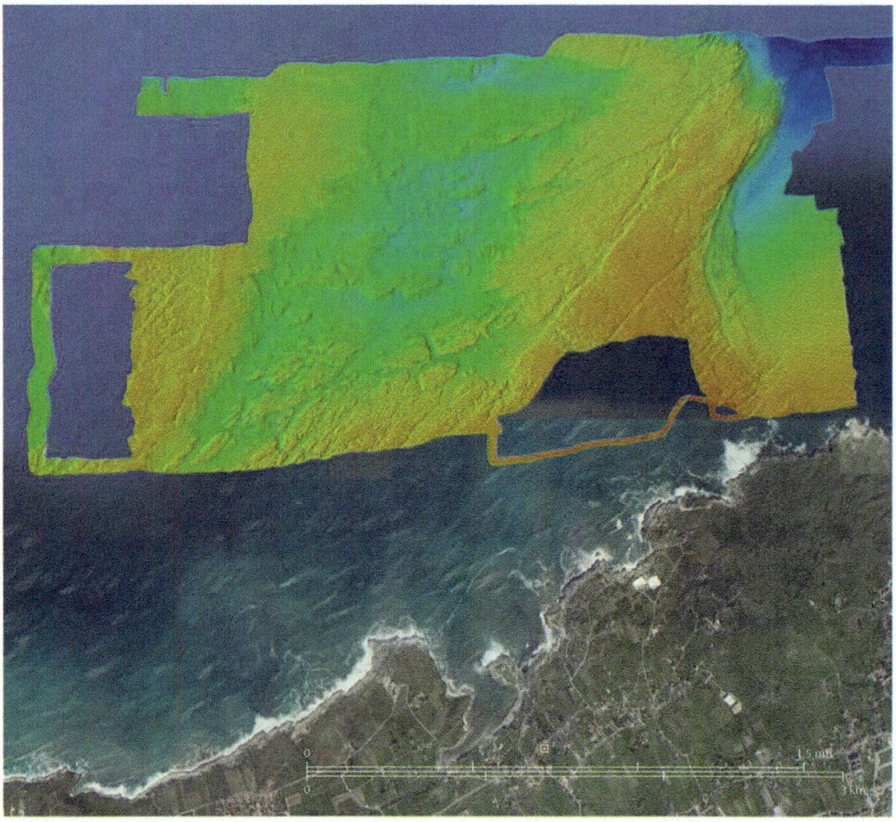

Fig. 13.4 High-resolution bathymetry of La Vaca area, northwest of Santander. Image by V. Bruschi, based on I. Galpasoro

Fig. 13.5 Exploring a submerged cave between the islets of Mouro and La Corbera, near Santander. Photo by P. Saiz Silió

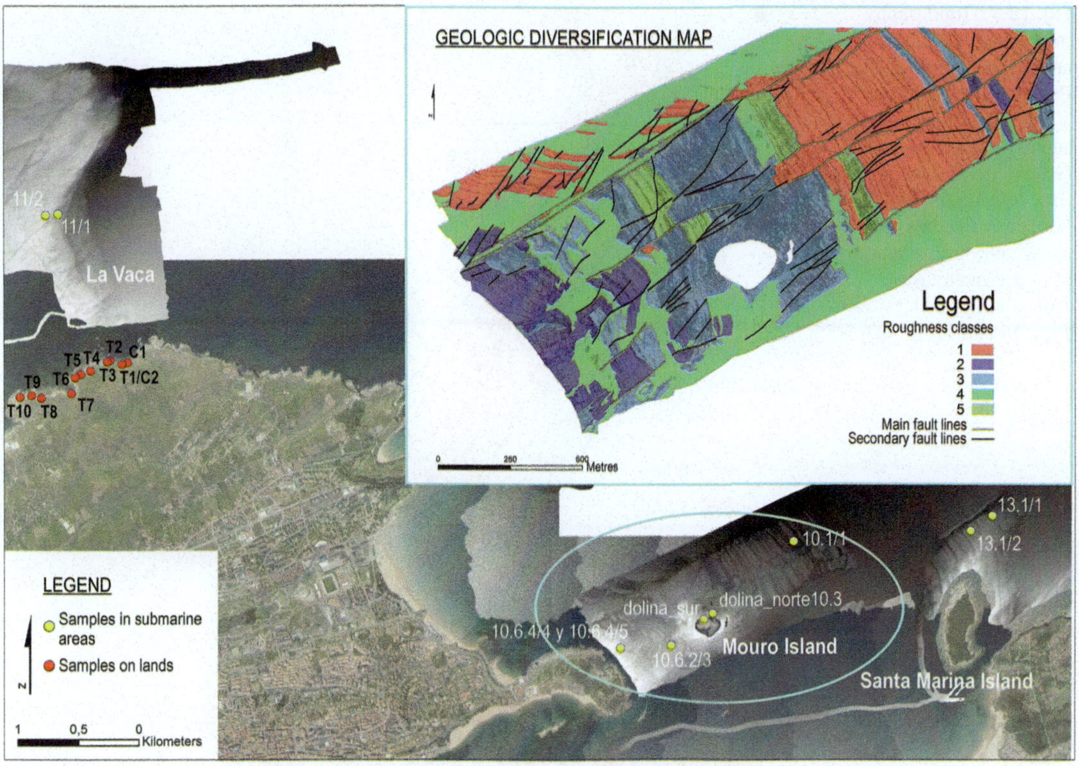

Fig. 13.6 Sampling programme for analysis of the underwater geology in the Cape Mayor-Santander Bay region. Sampling area and rugosity map after an ArcGis analysis showing the alternation of lithology and the continuity of fractures (corresponding to the blue circle in map). After Sánchez Carro et al. 2017

vation of an underwater cave (Bailey 2011). Gibraltar has many coastal caves including Gorham's Cave, Vanguard Cave, Devil's Tower and Forbes Quarry, all of which are famous for their Neanderthal fossils and archaeological sequences (Stringer et al. 2000; Finlayson et al. 2006). The submerged shelf around the Gibraltar Rock forms a relatively compact area, no wider than about 4–5 km, suitable for a detailed underwater landscape survey. Underwater work included landscape survey of most of the Gibraltar shelf using multibeam, sub-bottom profiling and side scan, diver mapping of the caves and their immediate surroundings and diver inspection of ridges and pinnacles on the eastern side of the shelf (Fig. 13.7). A submerged reef, known as Vladi's Reef lies about 200 m offshore and extends over a distance of c. 2 km. A cliff on the north side of the ridge contains a number of cave openings at its base at a depth of about 18–22 m below present sea level, formed by marine solution during an earlier period of high sea level. The two largest of these are clearly large enough for human occupation: Finlayson cave measuring 12 m wide, 4.5 m deep and 1.8 m high from the cave roof to the top of the infill at its entrance and Momber Cave 3.3 m wide, 3.6 m deep and 1.5 m high. These caves appeared on preliminary inspection to have a cave fill, and their aspect, facing away from the open sea and towards the mainland shoreline, suggested the possibility of discovering terrestrial deposits that had been protected from wave erosion during sea-level rise. Preliminary excavations were undertaken at Momber Cave, selected because it was judged to have a better prospect of preserving terrestrial deposits, and a trench 2 m1 m × 0.6 m deep was excavated in the mouth of the cave (Fig. 13.8). The deposits comprise marine cobbles and sands, but further excavation was impeded by closely

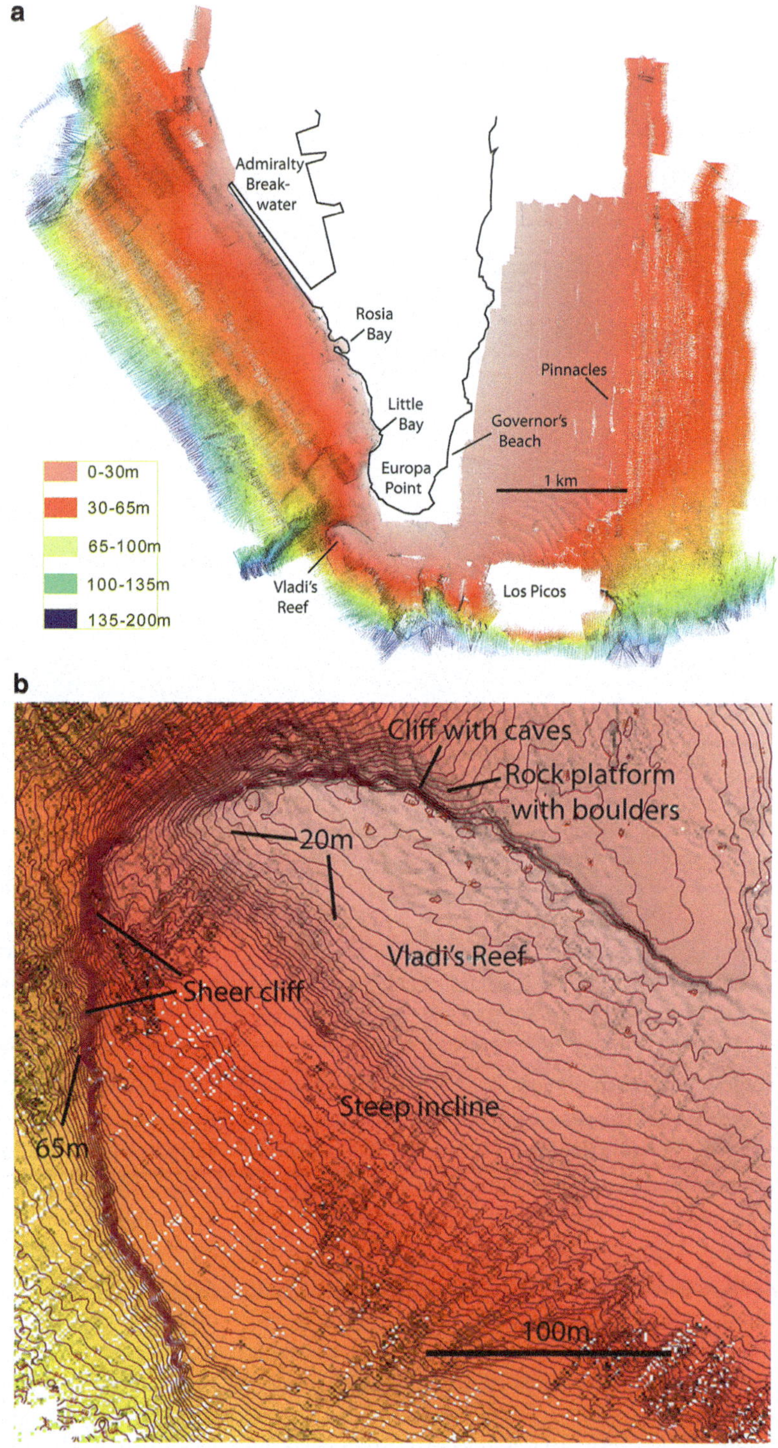

Fig. 13.7 Topography of the continental shelf around the Gibraltar Peninsula, showing Vladi's Reef and other submerged features. (**a**) General depth contours. (**b**) Close-up of Vladi's Reef showing detailed bathymetry and location of underwater caves. After Bailey 2011, Figs. 10 and 11, p. 323

packed boulders at the base of the trench. Major engineering solutions will be required to discover whether terrestrial deposits are preserved beneath these boulders or beneath the boulder field immediately in front of the cave entrance, and no further work has been undertaken. This research

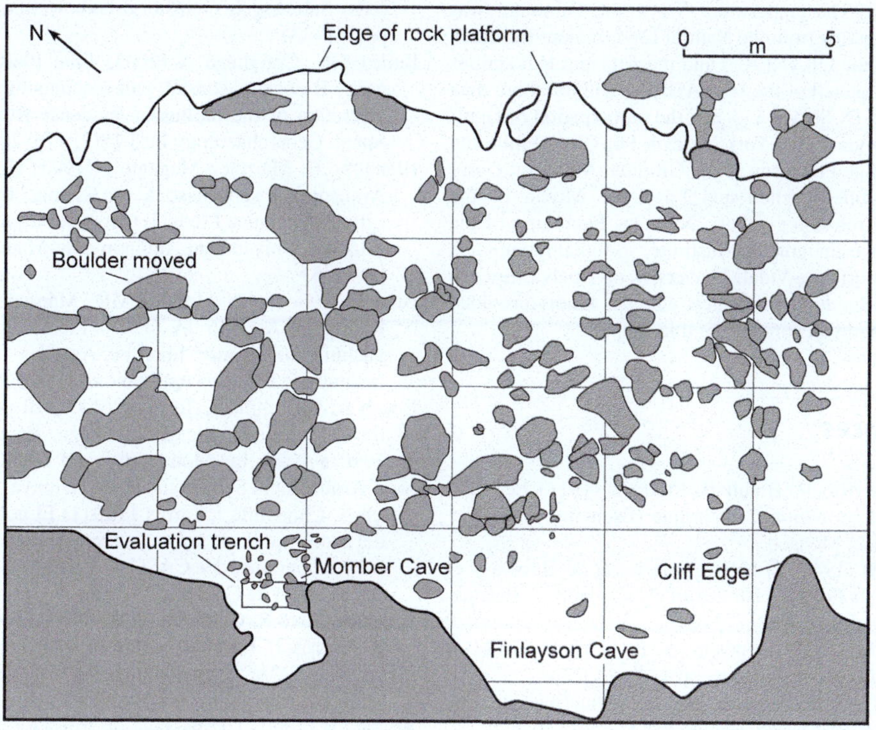

Fig. 13.8 Map of the submerged features on the north side of Vladi's Reef after 2005 survey and excavation. After Bailey 2011, Fig. 12, p. 324

constituted pioneering work on excavation in deep submerged caves, providing experience in the varied technical challenges of this kind of archaeological investigation.

13.5 Concluding Remarks

The Spanish continental shelf holds great potential for archaeological exploration aimed at finding and studying prehistoric sites, particularly Upper Palaeolithic deposits. The survey projects carried out in Cantabria and Gibraltar demonstrate both the challenges and the opportunities of systematic underwater survey. In both cases, the results have provided a better understanding of localised topographic features exposed at lower sea levels and targets suitable for more detailed archaeological investigation and excavation. It is to be hoped that these will be continued as other sectors of the coast are explored. Only in that way will it be possible to approach a fuller understanding of coastal occupation by Pleistocene and early Holocene communities and analyse the use of the littoral environment by hunter-gatherer societies during these earlier periods on a more realistic basis.

Acknowledgements The P5 Cant Project was funded by 'Cantabria Explora', a programme of the Parliament of Cantabria, and included the collaboration of researchers of the Universities of Cantabria (Raúl Medina, Sonia Castanedo, Xabier Guinda, Miguel A. Sánchez Carro, Viola Bruschi, Luis C. Teira, Carlos Duarte, Sara Núñez and Sheila Abada), Barcelona (Miquel Canals, Galderic Lastras, David Amblàs and Xavier Rayo), Salamanca (Esteban Álvarez-Fernández), Burgos (Eneko Iriarte) and La Rioja (Miguel Ángel Fano), the Basque Country (Alejandro Cearreta) and University College Dublin (Frank McDermott); the National Council for Scientific Research-CSIC (José Antonio López); the Spanish Oceanography Institute-IEO (Jesús Rivera); the Prehistory and Archaeology Museum of Cantabria-MUPAC (Roberto Ontañón); and Wessex Archaeology (Inés López-Dóriga), with the technical support of the underwater archaeologist Pablo Saiz Silió for the sampling and diving exploration and of Ibon Galpasoro, of Azti Tecnalia, for the multibeam bathymetry. The underwater work on Gibraltar was funded by grants from the Natural Environment Research Council of the UK (NERC) and the Arts and Humanities Research Council of the UK (AHRC), with offshore data supplied by IX Survey Ltd. and the participation of Geoff Bailey, University of York; Darren Fa, Clive Finlayson and Geraldine Finlayson of the Gibraltar Museum; Garry Momber, Julie Satchell and Lawrence Moran of the Maritime Archaeology Trust; Nicholas Flemming of the National Oceanographic Institute, Southampton; and Joaquín Rodríguez-Vidal, University of Huelva. Special thanks are due to Geoff Bailey, who has kindly provided information on this project for this chapter.

References

Alcalde del Río H, Breuil H, Sierra L (1911) Les cavernes de la région Cantabrique (Espagne), Imp. Vve A. Chêne, Monaco

Almagro M (1940) El hallazgo de la ría de Huelva y el final de la Edad del Bronce en el Occidente de Europa. Ampurias 2:85–143

Almagro M (1958) Depósito de la la ría de Huelva. Inventaria Archaeologica, pp 1–4

Altuna J (1997) Ekain y Altxerri: Dos santuarios paleolíticos en el País Vasco. Haranburu, San Sebastián

Álvarez-FernÁndez E (2011) Humans and marine resource interaction reappraised: archaeofauna remains during the late Pleistocene and Holocene in Cantabrian Spain. J Anthropol Archaeol 30(3):327–343

Álvarez-FernÁndez E, Carriol RP, JordÁ JF, Aura JE, Avezuela B, Badal E, Carrión Y, García-Guinea J, Maestro A, Morales JV, Pérez G, Pérez-Ripoll M, Rodrigo MJ, Scarff JE, Villalba MP, Wood R (2014) Occurrence of whale barnacles in Nerja cave (MÁlaga, southern Spain): indirect evidence of whale consumption by humans in the upper Magdalenian. Quat Int 337:163–169

Aparicio J (1975) Los yacimientos prehistóricos de la Albufera de Anna (Valencia). In: XIII Congreso Nacional de Arqueología, Secretaría de los Congresos Arqueológicos Nacionales, Zaragoza, pp 191–198

Arias P, Fano MÁ (2009) ¿Mesolítico Geométrico o Mesolítico con geométricos? El caso de la región Cantábrica. In: Utrilla P, Montes L (eds) El Mesolítico Geométrico en la Península Ibérica. Universidad de Zaragoza, Zaragoza, pp 69–91

Arias P, Ontañón R, Polanco JA, Setién J, Rovira S, Montero I (2005) Aproximación a la tecnología metalúrgica durante el Calcolítico y la Edad del Bronce de Cantabria. In: González Morales MR, Solórzano J (eds) II Encuentro de Historia de Cantabria. Servicio de Publicaciones de la Universidad de Cantabria-Parlamento de Cantabria, Santander, pp 53–77

Bailey GN (2011) Continental shelf archaeology: where next? In: Benjamin J, Bonsall C, Pickard C, Fischer A (eds) Submerged prehistory. Oxbow Books, Oxford, pp 311–331

Bailey GN, Craighead A (2003) Late Pleistocene and early Holocene coastal palaeoeconomies: a reconsideration of the molluscan evidence from Northern Spain. Geoarchaeology Int J 18(2):175–204

Blánquez JJ, Martínez Maganto J (1993) Historia de la Arqueología subacuástica en España. In: Martínez Lillo S, Blánquez J (eds) II Curso de Arqueología subacuática. Universidad Autónoma de Madrid, Madrid, pp 15–53

Blas MÁ, González Morales MR, Márquez Uría MC, Rodríguez Asensio JA (1978) Picos asturienses de yacimientos al aire libre en Asturias. Boletín del Instituto de Estudios Asturianos 93–94:335–356

Bosch À, Chinchilla J, Tarrús J (2000) El poblat lacustre del neolític antic de la Draga. Excavacions de 1990 a 1998. In: Monografies del CASC 2, Centre d'Arqueologia Subaquàtica de Catalunya, Girona

Bosch À, Chinchilla J, Tarrús J (2011) El poblat lacustre del neolític antic de la Draga. Excavacions 2000–2005. In: Monografies del CASC 9, Centre d'Arqueologia Subaquàtica de Catalunya, Girona

Bosch À, Buxó R, Chinchilla J, Palomo A, Piqué R, Saña M, Tarrús J, Terradas X (2012) El jaciment neolític lacustre de la Draga (Banyoles, Pla de l'Estany). Ajuntament de Banyoles, Banyoles

Bradley R (1990) The passage of arms. An archaeological analysis of prehistoric hoards and votive deposits. Cambridge University Press, Cambridge

Brandherm D (2007) Las espadas del Bronce Final en la Península Ibérica y Baleares. Prähistorische Bronzefunde, IV, 16. Franz Steiner, Stuttgart

Brandherm D, Burgess C (2008) Carp's-tongue problems. In: Verse F, Knoche B, Graefe J, Hohlbein M, Schierhold K, Siemann C, Uckelmann M, Woltermann G (eds) Durch die Zeiten, Festschrift für Albrecht Jockenhövel zum 65. Geburstag. Marie Leidorf, Rahden/Westfalen, pp 133–168

Burgess C, O'Connor B (2008) Iberia, the Atlantic Bronze Age and the Mediterranean. In: Celestino S, Rafel N, Armada X-L (eds) Contacto cultural entre el Mediterráneo y el Atlántico (siglos XII–VIII anr). La precolonización a debate. Madrid: CSIC, Escuela Española de Historia y Arqueología en Roma, Serie Arqueológica 11:41–58

Canals M, Cacho I, Carozza L, Casamor JL, Lastras G, Sanchez-Vidal A (2017) The Western Mediterranean Sea. In: Flemming NC, Harff J, Moura D, Burgess A, Bailey GN (eds) Submerged landscapes of the European continental shelf: quaternary paleoenvironments. Wiley, Chichester, pp 301–331

Casado JL, García-Castrillo G, Gómez Vega B, Sarabia P (1992) Arqueología subacuática en Cantabria. Un patrimonio secular en peligro, Sociedad Estatal para la Ejecución de programas del Quinto Centenario-Diputación Regional de Cantabria-Puerto de Santander-Universidad de Cantabria, Santander

Corchón MS (ed) (2017) La cueva de las Caldas (Priorio, Oviedo): Ocupaciones magdalenienses en el valle del Nalón. Universidad de Salamanca, Salamanca

Corchón MS, Mateos A, Álvarez E, Peñalver E, Delclòs X, van der Made J (2008) Ressources complémentaires et mobilité dans le Magdalénien cantabrique. Nouvelles données sur les mammifères marins, les crustacés, les mollusques et les roches organogènes de la Grotte de Las Caldas (Asturies, Espagne). Anthropologie 112:284–327

Cortés M, Morales A, Simón MD, Lozano MC, Vera JL, Finlayson C, Rodríguez J, Delgado A, Jiménez FJ, Martínez F, Martínez MA, Pascual AJ, Bergadà MM, Gibaja JF, Riquelme JA, López JA, Rodrigo M, Sakai S, Sugisaki S, Finlayson G, Fa DA, Bicho NF (2011) Earliest known use of marine resources by Neanderthals. PLoS One 6(9):e24026

Cortés-Sánchez M, Morales-Muñiz A, Simón Vallejo MD, Bergadà-Zapata MM, Delgado-Huertas A, López-García P, López-Sáez JA, Lozano-Francisco MC, Riquelme-Cantal JA, Roselló-Izquierdo E, Sánchez-Marco A, Vera-Peláez JL (2008) Palaeoenvironmental and cultural dynamics of the coast of Málaga (Andalusia, Spain) during the Upper Pleistocene and early Holocene. Quat Sci Rev 27(23–24):2176–2193

Daura J, Sanz M, Pike AWG, Subirà ME, Quam R, Fullola JM, Arsuaga JL (2005) A Neandertal mandible from the Cova del Gegant (Sitges, Barcelona, Spain). J Hum Evol 49(1):56–70

Daura J, Sanz M, Pike AWG, Subirà ME, Fornós JJ, Fullola JM, Julia R, Zilhão J (2010) Stratigraphic context and direct dating of the Neandertal mandible from Cova del Gegant (Sitges, Barcelona). J Hum Evol 59(1):109–122

Fano MÁ (2004) Las sociedades del Paleolítico en la región Cantábrica. Diputación Foral de Bizkaia, Bilbao

Finlayson C, Giles Pacheco F, Rodríguez Vidal J, Fa DA, Gutierrez López M, Santiago Pérez A, Finlayson G, Allue E, Baena Preysler J, Cáceres I, Carrión J, Fernández Jalvo Y, Gleed-Owen CP, Jimenez Espejo FJ, López P, López Sáez JA, Riquelme Cantal JA, Sánchez Marco A, Giles Guzman F, Brown K, Fuentes N, Valarino CA, Villalpando A, Stringer CB, Martinez Ruiz F, Sakamoto T (2006) Late survival of Neanderthals at the southernmost extreme of Europe. Nature 443:850–853

González Morales MR (1982) El Asturiense y otras culturas locales. La explotación de las áreas litorales de la región cantábrica en los tiempos epipaleolíticos. Ministerio de Cultura, Santander

Jorge M (1953) Hacia una sistematización de la Edad del Bronce en la actual provincia de Santander. Altamira 1, 2, 3:242–282

Martínez Díaz B (1992) Legislación en torno a la protección del patrimonio arqueológico subacuático en España. Cuadernos de Arqueología marítima 1:13–26

Martínez Díaz B (1993) La protección del Patrimonio arqueológico subacuático. In: Martínez Lillo S, Blánquez J (eds) II Curso de Arqueología subacuática. Universidad Autónoma de Madrid, Madrid, pp 121–143

Matamoros P, Carrascosa B, Cerrillo E (2013–2015) La catalogación del Patrimonio arqueológico subacuático en la cuenca extremeña del Tajo. Primeros pasos hacia su salvaguarda Arché Publicación del Instituto Universitario de Restauración del Patrimonio de la Universitat Politècnica de València, 8–10:137–144

Meijide G (1988) Las espadas del Bronce Final en la Península Ibérica, Servicio de Publicación e Intercambio Científico da Universidade de Santiago de. Compostela, Santiago de Compostela

Montes R (1985) El ciclo transgresión-regresión y hundimientos costeros en el sureste español. Su influencia en asentamientos pleistocenos. In: VI Congreso Internacional de Arqueología Submarina, Cartagena, 1982. Subdirección General de Arqueología y Etnografía, Madrid, pp 99–110

Montes R (1989) La grotte "de Los Aviones" Cartagena (Espagne). Bulletin de la Société préhistorique française 86:40–44

Montes R (1991) La cueva de los Aviones. Un yacimiento del Paleolítico Medio (Cartagena, España). In: Excavaciones y prospecciones en la región de Murcia 1985–1986. Consejería de Cultura, Educación y Turismo, Murcia, pp 35–58

Morales A, Roselló E (1984–1985) Algunas consideraciones de índole zoológica en torno al pez representado en la cueva de El Pindal. Ars Praehistorica 3–4:247–251

Morales A, Roselló E, Hernández F (1998) Late Upper Palaeolithic subsistence strategies in southern Iberia: tardiglacial faunas from Cueva de Nerja (Málaga, Spain). Eur J Archaeol 1(1):9–50

Morales JV, Pérez-Ripoll M, Jordá JF, Álvarez-Fernández E, Maestro A, Aura, JE (in press) Mediterranean monk seal hunting in the regional Final Magdalenian of Southern Iberia. A study of the Cueva de Nerja site. Quaternary International, Málaga

Moura D, Gomes A, Horta J (2017) The Iberian Atlantic margin. In: Flemming NC, Harff J, Moura D, Burgess A, Bailey GN (eds) Submerged landscapes of the European continental shelf. Quaternary paleoenvironments. Wiley-Blackwell, Oxford, pp 281–300

Nieto X (2009) La aqueología subacuática en España. In: Nieto X, Cau MÁ (eds) Arqueologia Nàutica Mediterrània. CASC, Girona, pp 17–25

Nieto X, Bethencourt M (2014) Arqueología subacuática española. Actas del I Congreso de Arqueología náutica y subacuática Española, Cartagena, 14, 15 y 16 de marzo de 2013. Universidad de Cádiz, Cádiz

Peña A (1985) Neue bronzeliche Funde in Museo Provincial de Pontevedra. Madrider Mitteilungen 26:22–28

Piqué R, Palomo A, Terradas X, Tarrús J, Bosch À, Chinchilla J, Bodganovic I, López O, Saña M (2015) Characterizing prehistoric archery: technical and functional analyses of the Neolithic bows from La Draga (NE Iberian Peninsula). J Archaeol Sci 55:166–173

Rivero O (2015) Art mobilier des chasseurs magdaléniens à la façade atlantique. Etudes et Recherches Archéologiques de l'Université de Liège, Liège

Ruiz A (2014) Entre el Cantábrico y los Pirineos: El conjunto de Altxerri en el contexto de la actividad gráfica magdaleniense. Nadir, Santander

Ruiz-Gálvez ML (1982) Nueva espada dragada en el río Ulla. Armas arrojadas a las aguas Ampurias 36:181–183

Ruiz-Gálvez ML (1995) Ritos de paso y puntos de paso. La Ría de Huelva en el mundo del Bronce Final Europeo. In: Complutum Extra 5, Servicio de Publicaciones de la Universidad Complutense, Madrid

San Nicolás M, Martínez C (1987) Prospecciones arqueológicas en la cueva C-6, cabo Cope, Águilas. In: Excavaciones y prospecciones arqueológicas. Servicio Regional de Patrimonio Histórico, Murcia, pp 10–16

Sánchez Carro MÁ, Bruschi V, Saiz Silió P, Iriarte E, Galpasoro I, Arias P (2017) El Paleolítico sumergido: batimetría y caracterización geológica del fondo marino en zonas próximas a Santander para la localización de yacimientos arqueológicos. Geo-Temas 17:35–38

Straus LG (1992) Iberia before the Iberians. The stone age prehistory of Cantabrian Spain. University of New Mexico Press, Albuquerque

Straus LG (2018) The upper Paleolithic of Iberia. Trab Prehist 75(1):9–51

Straus LG, Clark GA (eds) (1986) La Riera Cave. Stone Age hunter-gatherer adaptations in northern Spain. Arizona State University, Tempe

Stringer CB, Barton RNE, Finlayson JC (eds) (2000) Neanderthals on the edge: papers from a conference marking the 150th anniversary of the Forbes' quarry discovery, Gibraltar. Oxbow, Oxford

Stringer CB, Finlayson JC, Barton RNE, Fernandez-Jalvo Y, Caceres I, Sabin RC, Rhodes EJ, Currant AP, Rodriguez-Vidal J, Giles-Pacheco F, Riquelme-Cantal JA (2008) Neanderthal exploitation of marine mammals in Gibraltar. Proc Natl Acad Sci U S A 105:14319–14324

Portugal: Intertidal Archaeology and Submerged Coastal Landscapes

14

Nuno Bicho, Leandro Infantini, and João Marreiros

Abstract

People have lived along the Portuguese coast at least since Marine Isotope Stage (MIS) 5, over 100,000 years ago. While there are many known sites and archaeological evidence of human settlement on land, there are only a handful of known submerged or near-submerged locations, and these are all in the present-day intertidal zone. Throughout the period, diet was based on both terrestrial and marine mammals as well as fish and shellfish. This dietary regime was already in existence when Neanderthals populated the region. It appears that marine foods started to be used more frequently when the first anatomically modern humans arrived, some 33,000 years ago. Among the Mesolithic people who lived along the estuaries of the major rivers, marine food frequently composed c. 50% of the diet. With the introduction of agriculture, coastal resources became less important, although some submerged sites of Neolithic date testify to habitation directly at the seashore. Despite the current absence of known underwater prehistoric sites, apart from those in the intertidal zone, recent studies of offshore topography have identified a number of areas with good potential for archaeological prospection, including cliff lines with underwater caves and drowned river valleys.

Keywords

Submerged · Underwater · Muge · Shell middens · Mesolithic · Palaeolithic · Neolithic

14.1 Introduction

Portugal is a small country on the southwestern limit of the European continent. With an area slightly over 92,250 km^2, it has a coastline that exceeds 1000 km in extension. The long stretches of ocean-exposed beaches are intersected by several major estuaries, as well as extensive coastal lagoons bordered by barrier islands. While there are many important rivers that reach the Atlantic coast, namely, the Douro, with its mouth in the city of Porto, or the Guadiana that constitutes the border between Portugal and Spain in the south, there are two very large estuaries, the Tagus and the Sado, forming large bodies of water many square kilometres in area, with depths from 40 to 60 m and > 100 km in length. In addition, the Portuguese coast has two important barrier-island complexes, the Ria de Aveiro and Ria de Formosa in Faro, respectively, 45 and 60 km long. The continental shelf exposed during periods of lowest sea level is relatively narrow—c. 20–50 km wide—with several deep canyons rep-

N. Bicho (✉) · L. Infantini · J. Marreiros
ICArEHB, Universidade do Algarve, Faro, Portugal
e-mail: nbicho@ualg.pt; jmmarreiros@ualg.pt

G. Bailey et al. (eds.), *The Archaeology of Europe's Drowned Landscapes*, Coastal Research Library 35, https://doi.org/10.1007/978-3-030-37367-2_14

resenting ancient river channels that would have brought the coastline of the Last Glacial Maximum to within a kilometre or two of the modern coastline (Fig. 14.1; Moura et al. 2017).

Today, during the summer months mostly around the southern coast, there is a very important upwelling system (Fiuza 1983; Loureiro et al. 2005). This system has produced a highly rich and diverse biomass that has strongly moulded the modern marine and coastal resources of the Portuguese economy, including the highly productive fish and shellfish industries. The Portuguese upwelling, strong though it is today, was three to seven times higher during the last 100,000 years of the Pleistocene (Abrantes 1990, 2000), making the Atlantic coast an extremely rich and abundant food resource. Archaeological evidence, examples of which are presented below, testifies that the rich marine fauna was frequently exploited by humans from the Middle Palaeolithic period onwards and was likely fundamental for the development of the first human societies in SW Europe (Bicho and Haws 2008).

Fig. 14.1 Map of Portugal showing distribution of sites. Site information from the SPLASHCOS Viewer http://splashcos-viewer.eu. Drawing by Moritz Mennenga

At present, we have identified only six prehistoric underwater sites—these include fully submerged as well as intertidal locations; three are dated to the Middle Palaeolithic, one to the Mesolithic and two to Neolithic occurrences. These sites were originally open-air locations, and those found in the intertidal zone are especially liable to destruction by wave action. Four have been partially excavated, and three are still part of ongoing projects in central and southern Portugal.

14.2 Research History and Archaeological Context

In 1879, Nery Delgado, a member of the National Portuguese Geologic Commission, excavated the cave site of Furninha (Fig. 14.1), located on the Atlantic limestone cliff on the Peniche Peninsula in central Portugal (Bicho and Cardoso 2010). The cave entrance located some 5 m above the modern sea level, with a 9-m-thick sequence, gave one of the earliest Portuguese examples of marine resource exploitation during the Middle and Upper Palaeolithic (Delgado 1884). Since then, other sites located usually within a few metres of the modern sea level have been found and have been excavated along the Portuguese Atlantic coast. Probably the most notable case is the cave of Figueira Brava (Antunes 2000a, 2000b), a Mousterian site some 30 km south of the Tagus estuary where remains of marine species such as dolphin and penguin were found as part of the diet. The cave entrance is just a few metres above modern sea level and might have been opened during MIS 5e.

This trend of a component of marine and aquatic resources as part of the human diet is also present in many other sites, some further from the coast. This is the case at Vale Boi (Bicho 2004; Bicho and Haws 2008; Bicho 2015) in Algarve near the Cape of St. Vincent, with collection of shellfish, among other marine resources, starting at least some 33,000 years ago, and the cave of Picareiro (Bicho and Haws 2008), an inland site some 70 km north of Lisbon with coastal fish and shellfish remains starting during the Gravettian 28,000 years ago. These include clupeidae species (possibly shad or sardine) as well as cockles, clams and mussels (Bicho et al. 2003). In addition, proxy evidence is present at the Middle Palaeolithic site of Mira Nascente, near Nazaré (Benedetti et al. 2009; Haws et al. 2011). This open-air site in a coastal sand dune is dated to c. 40 to 42,000 cal BP. Sedimentology indicates that the site was located on the edge of a lake very close to the shoreline. Use-wear analysis suggests that the extremely well-preserved flint assemblage was likely used for fish processing (Haws et al. 2011).

During the Last Glacial Maximum, c. 25,000–20,000 cal BP, sea level was as much as c. 130 m lower than today (Lambeck et al. 2014), and the habitable land extended as much as 10–50 km out from the modern coastline. No trace of Palaeolithic habitation in this deeply drowned landscape is currently available. There are, most likely, two main reasons for this: the lack of systematic surveys and a generally highly erosive coastline due to the combined effect of waves, currents and tides.

Another relevant element in the history of coastal archaeological research in Portugal is the Early and Middle Holocene timeframe. During the Early Holocene, the Portuguese coast south of the Nazaré canyon was the location for a large number of coastal Epipalaeolithic and Mesolithic sites both in Portuguese Estremadura (Sousa 2004; Araújo 2009, 2012; Bicho et al. 2010; Araújo et al. 2014; Sousa and Soares 2016) and southern Portugal (Carvalho 2008; Valente 2008; Bicho 2009). Estuarine Mesolithic shell middens have been known since 1863 in the Muge area in the Tagus valley, some 60 km north-east of Lisbon. The earliest occupations are dated to before 6000 cal BC and are thought to be the consequence of major coastal changes (Bicho et al. 2010) linked to the changing coastal morphology, rapid sea-level rise and a dramatic drop in coastal marine biomass and the so-called 8.2 ka cold event (c. 8200 cal BP or c. 6200 cal BC) (Von Grafenstein et al. 1998). These locations are represented by some of Europe's largest shell middens with layers of marine shells up to 5 m thick. These accumulated next to a contemporaneous inlet in Muge and in Sado around 8000 years ago (Umbelino 2006; Bicho et al. 2010; Peyroteo Stjerna 2016). Stable isotope measurements of bones of humans buried in the shell mounds (Fig. 14.2) suggest that ca. 25–70% of the protein diet of these individuals derived from seafood (e.g., Umbelino 2006; Bicho et al. 2010; Peyroteo Stjerna 2016).

The emergence of agricultural societies in central and southern Portugal, with a process of acculturation and genetic mixing of the local Mesolithic communities with new immigrants from the East, took place some 7500 years ago (Bicho et al. 2017). Neolithic people, marked by Cardial ceramics, arrived in the western Algarve and the Portuguese Estremadura from the Mediterranean by sea (Carvalho 2008).

14.3 Sites

The above cases can be used as proxies for the use and exploitation of coastal environments during the Palaeolithic and Mesolithic on the Portuguese Atlantic coast, but unfortunately, there are very few cases of underwater prehistoric occurrences. Despite its near-1000-km-long Atlantic coastline, with its estuarine and lagoon coastal settings, Portugal has only six underwater archaeological sites and two examples of submerged forests.

Praia Rei Cortiço (Fig. 14.3). This is a Middle Palaeolithic coastal site located near Peniche (Haws et al. 2011; Minckley et al. 2015), where pollen and sediment evidence indicate that the site was originally located in a coastal

Fig. 14.2 Mesolithic human burial in Cabeço da Amoreira, Muge. The isotopic information indicates that this skeleton belongs to a person whose diet was composed of c. 50% marine resources. Photo by Nuno Bicho

Fig. 14.3 Excavation at Praia Rei Cortiço during low tide. The archaeologists are standing just below the horizon that represents Neanderthal coastal occupation during the MIS 5 c. 115,000 years ago. Most of the site was washed away during a high tide in 2014. Photo by Jonathan Haws

lagoon environment with a sea level similar to that of today (Minckley et al. 2015). Lithic artefacts are abundant and are typical of a Mousterian occupation. Cores follow a centripetal reduction sequence with common discoidal cores and very rare Levallois examples. Raw materials include quartzite, quartz and rare chert elements.

Praia do Pedrogão. This is also a Middle Palaeolithic coastal site (Aubry et al. 2005; Benedetti et al. 2009) located in the county of Leiria. The site is on the present shore, just on the higher section of the intertidal zone and, thus, during spring tides is subject to strong erosional forces. The lithic assemblage is composed of quartzite and quartz artefacts, with rare retouched stone tools. The technology is marked by the common presence of Levallois reduction sequences producing flakes.

Cabeço da Amoreira, in Muge. In 2011, a deposit was found below the modern local water table and associated with a prehistoric estuarine shoreline. The known sequence (testing has not yet reached the base of the cultural deposits) covers the period from at least Early Neolithic up to Iron Age horizons (Bicho et al. 2017) and is characterized by layers with very high organic content. The lithic assemblages are limited, while ceramics are very common, with some decorated examples.

The São Julião Mesolithic shell midden. Located north of Lisbon, near the city of Mafra, this is a shell midden complex, with at least four loci, of which at least one is within reach of wave action during very high spring tides in the estuary of the small Falcão River (Sousa 2004; Sousa et al. 2016). These shell middens have a limited number of very simple lithic assemblages, composed mostly of flakes. On the other hand, the sites have many hearths associated with diverse mollusc species including bivalves and gastropods.

The Middle Palaeolithic cave of Ibn Ammar. Located in the estuary of the Arade River (Fig. 14.4), near Portimão, this cave has many lower chambers that are inundated at high tide during spring tides, but the floor of the largest chamber is permanently underwater to a depth of at least 50 cm. The human occupations (Mousterian, Early Neolithic and later) have been located only in the upper chambers, where it is possible to excavate (Bicho 2004; Carvalho 2008). The prehistoric artefacts found inside the cave include Mousterian centripetal discoidal cores and flakes made on quartz, quartzite and flint (Bicho 2004) and early Neolithic ceramics (Carvalho 2008).

The Neolithic site of Praia do Forte Novo. Located in coastal Algarve, this site was found during an extreme low water stand for a few days in 1999, and since then, it has been covered by sand and is usually underwater (Fig. 14.5) (Rocha 2003; Rocha and Barros 1999/2000). Thousands of pottery fragments have been recovered as well as hearths and features that may represent partly eroded remains of saltpans and huts.

Maceda beach submerged forest. Here, petrified macrofossil wood remains have been found near the city of Porto and dated to MIS 3–2 (Granja and Carvalho 1995; Granja et al. 2008).

Barreiro submerged forest. This is north of Lisbon and dated to the Early Holocene (García-Amorena et al. 2007).

All these sites are located within the intertidal zone, although the latter three are at the extreme lower limit and are exposed only in those rare moments of extremely low spring tides. Both Middle Palaeolithic sites, Praia Rei Cortiço and Praia do Pedrogão, are at the upper limit and are only reached infrequently by wave action.

14.4 Prospects for Preservation and Discovery of Underwater Sites

Until recently and following the idea of a major adaptation in late Pleistocene and Early Holocene times (Binford 1968; Flannery 1969), it was thought that marine resources were important in the economy and diet of the human populations

Fig. 14.4 View from Ibn Ammar entrance towards the estuary during low tide, when shellfish resources are available as well as fresh water coming out from the various cave springs (seen on the left side of the photo, where the wall is located). Photo by Nuno Bicho

Fig. 14.5 The dark organic layer seen in the centre of the photo is rich in remains of coastal habitation. This c. 7000-year-old site, located in the intertidal zone, is shown here on a rare occasion of surface sand erosion due to wave and current action. Praia do Forte Novo. Photo by Leonor Rocha

only from the Late Upper Palaeolithic onwards in Portugal. Data coming from different sites in central and southern Portugal, as well as a general change in paradigm in world archaeology (Erlandson 2001; Bailey and Milner 2002; Bailey 2004), have shifted the climate of opinion towards the view that marine resources and coastal adaptations were in use much earlier than previously thought (Bicho and Haws 2008). One of the key factors masking the archaeological evidence is the fact that many sites dating to the Palaeolithic and even to the Mesolithic are now submerged on the continental platform, down to 150 m, due not only to general sea-level rise (Dias et al. 2000) but also to tectonic episodes during the last 120,000 years (Bicho and Haws 2008).

The submergence of sites as deep as 150 m on a platform that extends, at that depth, in some places as much as 50 km offshore poses difficult challenges for underwater prehistoric research, more so in the open Atlantic Portuguese coastal setting. However, in certain areas, there is steep bathymetry that brings potential areas of survey, testing and excavation much closer to the modern coast (Infantini et al. 2012). That is the case, for example, with the limestone cliff areas of the Setubal Peninsula or the western Algarve and the Nazaré and Portimão canyons. Not only is there a high potential for submerged caves, but also for underwater relict valleys and cliffs dating to the late Pleistocene. A perfect example is that of the Armação de Pêra Bay, where an MIS 3 barrier and cliff was found at a depth below sea level of some 20 m (Infantini 2012; Infantini et al. 2012). Directly to the north and on the top of the cliff, there is a large submerged platform. Bathymetric analysis shows a palaeolandscape that includes a barrier or coastal lagoon and dune field (Fig. 14.6), slowly modified and submerged during the Late Upper Palaeolithic and Early Holocene (Infantini 2012).

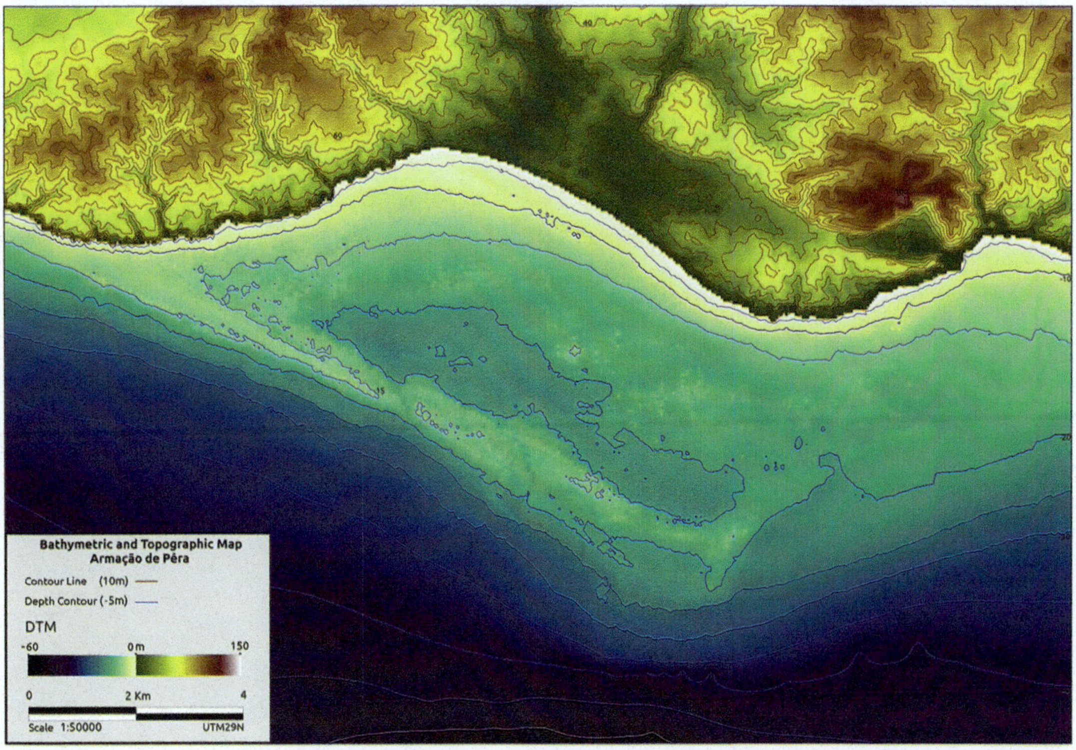

Fig. 14.6 A drowned Stone Age landscape some 20 m below present-day average sea level in Armação de Pêra Bay. The c. 12-km-long reef, representing a former peninsula, may have protected traces of human settlement from the erosive powers of waves from the open sea, making the area an obvious candidate for future underwater reconnaissance. Image by Leandro Infantini

In addition to these specific regions with a steep bathymetry, there are two main coastal regions, one extending from the Nazaré canyon to the Sado estuary and the other comprising the southern coast between Cape St. Vincent and the Mondego estuary (Moura et al. 2017). Sediment deposition is higher in these regions, providing potentially moderate to good conditions for underwater preservation. Relict beach sediments in both areas are common, probably dating to the Last Interglacial, where there is a high potential for finding Middle Palaeolithic occupations, both open-air and cave sites. This potential is present in the coastal limestone around the Tagus and Sado estuaries as well as on the Sagres coast (Fig. 14.7) where many underwater caves are known, including at least two with dry sections inside the caves.

Palaeolagoon and coastal lake and island barrier contexts known in the Óbidos region, between the Tagus and Sado north of Sines and in the Faro area, also have very high potential, following detailed and systematic coastal and underwater surveys, for finding prehistoric settlements, similar to those described above. In many cases, there are similar geomorphological settings to those of the large Tagus and Sado river basins, but there are much smaller basins such as the Mira, Odeceixe, Aljezur, Carrapateira and Portimão, among many others, that may provide particularly good areas for finding either underwater or intertidal archaeological sites.

Many other areas need to be scrutinized, topographically reconstructed, tested and, in specific cases, excavated. A multidisciplinary research program at the University of Algarve is now under development to study the prehistoric human occupation of coastal cave settings in southern Portugal.

Fig. 14.7 General view of the southwestern coast of Sagres, with limestone cliffs 70–100-m high where caves are very common. Photo by Nuno Bicho

14.5 Management of the Underwater Cultural Heritage

In the mid-1980s, the director of the National Museum of Archaeology, Francisco Alves, started to develop the field of underwater archaeology. His work eventually led to an autonomous department, the Centro Nacional de Arqueologia Náutica e Subaquática, in the Portuguese Archaeological Institute (Instituto Português de Arqueologia, IPA), the institution that supervised archaeological activity in Portugal from 1997 to 2006. IPA was replaced by IGESPAR (Instituto de Gestão do Património Arquitectónico e Arqueológico), and this in turn was replaced by the Direção Geral do Património Cultural (DGPC) in 2012. Within this governmental institution, there is a specific department that oversees the underwater archaeological heritage, the Divisão de Arqueologia Náutica e Subaquática. Presently. DGPC is the only institution in charge of the underwater archaeological heritage.

The Portuguese archaeological heritage is listed and available to the general public in an online database known initially as the Endovélico and very recently changed to the 'Portal do Arqueólogo'. Here, one can search for sites, absolute dates, site locations, periods as well as excavators (http://arqueologia.patrimoniocultural.pt/index.php?sid=sitios).

Acknowledgements We thank COST Action TD0902 SPLASHCOS for funding and particularly the Chair and Vice-Chair of the Action, Geoff Bailey and Dimitri Sakellariou. We also thank Fundação para a Ciência e Tecnologia for funding through the projects PTDC/EPH-ARQ/4998/2012 and PTDC/HAR-ARQ/27833/2017.

References

Abrantes F (1990) Increased upwelling off Portugal during the last deglaciation: diatom evidence. Mar Micropaleontol 17:285–310

Abrantes F (2000) 200 000 yr diatom records from Atlantic upwelling sites reveal maximum productivity during LGM and a shift in phytoplankton community structure at 185 000 yr. Earth Planet Sci Lett 176:7–16

Antunes M (2000a) The Pleistocene fauna from Gruta da Figueira Brava: a synthesis. Memórias da Academia das Ciências de Lisboa Classe de Ciências XXVIII:259–282

Antunes M (2000b) Gruta da Figueira Brava: Pleistocene marine mammals. Memórias da Academia das Ciências de Lisboa Classe de Ciências XXXVIII:245–258

Araújo A (2009) Hunter-gatherers adaptations during Pleistocene/Holocene transition in Portugal: data and explanatory models. In: McCartan S, Schulting R, Warren G, Woodman P (eds) Mesolithic horizons. Papers Presented at the Seventh International Conference on the Mesolithic in Europe Belfast, vol 2005. Oxbow, Oxford, pp 533–540

Araújo A (2012) Une histoire des premières communautés mésolithiques au Portugal. Unpublished PhD Thesis, Université de Paris 1, Panthéon, Sorbonne

Araújo A, Moreno-García M, Gabriel S (2014) O concheiro de Toledo no contexto do Mesolítico inicial do litoral da Estremadura. Rev Port Arqueol 17:5–34

Aubry T, Ribeiro J, Anglucci D (2005) Testemunhos da ocupação pelo Homem de Neanderthal: O sítio da Praia do Pedrogão. In: Carvalho S (ed) Habitantes e habitats: Pré e proto-História na Bacia do Lis. Câmara Municipal de Leiria, Leiria, pp 56–66

Bailey G (2004) World prehistory from the margins: the role of coastlines in human evolution. Journal of Interdisciplinary Studies in History and Archaeology 1:39–50

Bailey G, Milner N (2002) Coastal hunter-gatherers and social evolution: marginal or central? Before farming: The Archaeology of Old World Hunter-Gatherers 3(4):1–15

Benedetti M, Haws J, Funk D, Daniels M, Hespe P, Bicho N, Minckley T, Ellwood B, Forman S (2009) Late Pleistocene raised beaches of coastal Estremadura, Central Portugal. Quat Sci Rev 28:3437–3447

Bicho N (2004) As comunidades humanas de caçadores-recolectores do Algarve Ocidental – perspectiva ecológica. In: Ferro AA, Tavares MJ, Cardoso J (eds) Evolução Geohistórica do Litoral Português e fenómenos correlativos. Universidade Aberta, Lisbon, pp 359–396

Bicho N (2009) On the edge: early Holocene adaptations in Southwestern Iberia. J Anthropol Res 65(2):185–206

Bicho N (2015) Ecological niches of the Iberian Peninsula: a comparative analysis of European coastal adaptations. In: Sanz N (ed) Human origin sites and the World Heritage Convention in Eurasia. UNESCO – World Heritage Papers, pp 108–126

Bicho N, Cardoso J (2010) Paleolithic occupations and lithic assemblages from Furninha Cave, Peniche (Portugal). Zephyrus 66:17–37

Bicho N, Haws J (2008) At the land's end: marine resources and the importance of fluctuations in the coast line in the prehistoric hunter-gatherer economy of Portugal. Quat Sci Rev 27:2166–2175

Bicho N, Haws J, Hockett B, Markova A, Belcher W (2003) Paleoecologia e ocupação humana da Lapa do

Picareiro: resultados preliminares. Rev Port Arqueol 6(2):49–81
Bicho N, Umbelino C, Detry C, Pereira T (2010) The emergence of the Muge Mesolithic shell middens (central Portugal) and the 8200 cal yr BP cold event. Journal of Island and Coastal Archaeology 5:86–104
Bicho N, Cascalheira J, Gonçalves C, Umbelino C, García Rivero D, André L (2017) Resilience, replacement and acculturation in the Mesolithic/Neolithic transition: the case of Muge, central Portugal. Quat Int 446:31–42
Binford L (1968) Post-Pleistocene adaptations. In: Binford S, Binford L (eds) New perspectives in archeology. Aldine, Chicago, pp 313–341
Carvalho A (2008) A neolitização do Portugal Meridional. Os exemplos do Maciço Calcário estremenho e do Algarve Ocidental. Universidade do Algarve, Faro
Delgado JFN (1884) La grotte de Furninha a Peniche. Congrès International d'Anthropologie et d'Archéologie Préhistoriques. In: Compte-Rendu de la neuvième Session à Lisbonne (1880). Typographie de l'Académie Royale des Sciences, Lisboa, pp 207–278
Dias JA, Boski T, Rodrigues A, Magalhães F (2000) Coast line evolution in Portugal since the Last Glacial Maximum until present – a synthesis. Mar Geol 170:177–186
Erlandson JM (2001) The archaeology of aquatic adaptations: paradigms for a new millennium. J Archaeol Res 9:287–350
Fiuza A (1983) Upwelling patterns off Portugal. In: Suess E, Thiede J (eds) Coastal upwelling: its sedimentary record. Plenum, New York, pp 85–98
Flannery KV (1969) Origins and ecological effects of early domestication in Iran and the Near East. In: Ucko PJ, Dimbleby GW (eds) The domestication and exploitation of plants and animals. Aldine, Chicago, pp 73–100
García-Amorena I, Gómez Manzaneque F, Rubiales J, Granja H, Soares de Carvalho G, Morla C (2007) The Late Quaternary coastal forests of western Iberia: a study of their macroremains. Palaeogeogr Palaeoclimatol Palaeoecol 254:448–461
Granja HM, Carvalho GS (1995) Sea-level changes during the Pleistocene-Holocene in the NW coastal zone of Portugal. Terra Nova 7:60–67
Granja HM, De Groot TAM, Costa AL (2008) Evidence for Pleistocene wet aeolian dune and interdune accumulation, S. Pedro da Maceda, north-west Portugal. Sedimentology 55:1203–1226
Haws J, Funk CL, Benedetti M, Bicho N, Daniels M, Minkley T, Denniston R, Jeraj M, Gibaja J, Hockett B (2011) Paleolithic seascapes along the west coast of Portugal. In: Bicho N, Haws J, Davis L (eds) Trekking the shore: changing coastlines and the antiquity of coastal settlement. Springer, New York, pp 203–246
Infantini L (2012) Paisagem pré-histórica submersa da Baía de Armação de Pêra. Unpublished MA thesis, Universidade do Algarve, Faro
Infantini L, Moura D, Bicho N (2012) Utilização de ferramenta SIG para o estudo da morfologia submersa da Baía de Armação de Pêra (Algarve). In: Campar A, Bettencourt A, Moura D, Monteiro-Rodrigues S, Caetano-Alves M (eds) Environmental change and human interaction along the Western Atlantic edge. Associaçao Portuguesa para o Estudo do Quaternario, Coimbra, pp 227–242
Lambeck K, Rouby H, Purcell A, Sun Y, Sambridge M (2014) Sea level and global ice volumes from the Last Glacial Maximum to the Holocene. Proc Natl Acad Sci 111(43):15296–15303
Loureiro S, Newton A, Icely J (2005) Microplankton composition, production and upwelling dynamics in Sagres (SW Portugal) during the summer of 2001. Sci Mar 69:323–341
Minckley T, Haws J, Benedetti M, Brewer S, Forman S (2015) Last interglacial vegetation and climate history from the Portuguese coast. J Quat Sci 30:59–69
Moura D, Gomes A, Horta J (2017) The Iberian Atlantic margin. In: Flemming NC, Harff J, Moura D, Burgess A, Bailey GN (eds) Submerged landscapes of the European continental shelf: quaternary paleoenvironments. Wiley, Chichester, pp 281–300
Peyroteo Stjerna R (2016) On death in the Mesolithic, or the mortuary practices of the last hunter-gatherers of the south-western Iberian Peninsula, 7th–6th millennium BCE. Uppsala University, Occasional Papers in Archaeology 60, Uppsala
Rocha L (2003) Intervenções de emergência realizados pela extensão de Silves do Instituto Português de Arqueologia. Xelb 4:161–174
Rocha L, Barros P (1999/2000) Escavações de emergência no povoado da Praia do Forte Novo; Quarteira (Loulé). Al'-ulya 7:19–25
Sousa AC (ed) (2004) O núcleo C do concheiro de S. Julião (Carvoeira, Mafra): a escavação de emergência de 1999. Câmara Municipal, Cadernos de Arqueologia de Mafra 2, Mafra
Sousa AC, Soares AM (2016) Continuity or discontinuity? Aquatic exploitation in the Portuguese Estremadura during the Atlantic period: São Julião and Magoito shell middens as case studies. In: Dupont C, Marchand G (eds) Archéologie des chasseurs-cueilleurs maritimes. De la fonction des habitats à l'organisation de l'espace littoral. Actes de la Séance de la Societé Préhistorique Française, Rennes 10–11 Avril, vol 2014, pp 191–212
Sousa AC, Miranda M, Soares AM (2016) O Concheiro de São Julião (Carvoeira, Mafra): as intervenções de 2007 e 2014, novos dados e novas leituras. Rev Port Arqueol 19:11–26
Umbelino C (2006) Outros sabores do passado: as análises de oligoelementos e de isótopos estáveis na reconstituição da dieta das comunidades humanas do Mesolítico Final e do Neolítico Final-Calcolítico do território português. Unpublished PhD thesis, Universidade de Coimbra, Coimbra

Valente M (2008) As últimas sociedades de caçadores-recolectores no Centro e Sul de Portugal (10.000–6.000 anos BP): aproveitamento dos recursos animais. Unpublished PhD thesis, Universidade do Algarve, Faro

Von Grafenstein U, Erlenkeuser H, Müller J, Jouzel J, Johnsen S (1998) The cold event 8200 years ago documented in oxygen isotope records of precipitation in Europe and Greenland. Clim Dyn 14:73–81

Part III

The Mediterranean and the Black Sea

The Mediterranean and the Black Sea: Introduction

15

Nena Galanidou and Geoff Bailey

Abstract

The chapters in this section cover a very large area and two major marine basins, reflecting the somewhat scattered distribution of underwater finds and the patchy record of underwater research. Nevertheless, there are concentrations of finds with a detail and quality of evidence to match the best that has been found around the coastlines of north-west Europe. The evidence also includes types of sites and preservation of material items that have no equivalent elsewhere, including submerged village settlements with timber-built platforms and dwelling structures, and settlements with remains of stone-built dwellings, burials, street plans, water wells and other features. Most of these sites are Neolithic or Bronze Age in date, but there are also earlier finds extending back as far as the Middle Palaeolithic. The types of sites represented also include underwater caves containing deposits with palaeontological and archaeological remains. Sea-level change presents a complex story in this region, especially around the coastlines of southern Italy and the Aegean and the Black Sea, and intensive research has focussed on this issue, with implications for the changing palaeogeography of coastlines and islands and the possibility of sea crossings and maritime connections. This introductory overview brings together results from the chapters in this section under four main themes: underwater caves, sea crossings, mapping of submerged landscapes and predictive modelling of underwater site locations and underwater settlements. New and active research is under way in some regions, including the extension of mapping and site survey to more deeply submerged areas of the continental shelf. These results suggest very considerable potential for new discoveries and the need to extend underwater research to other coastal states that have so far made little or no contribution to the prehistoric archaeology of submerged landscapes.

Keywords

Underwater caves · Palaeolithic · Neolithic · Bronze Age · Submerged landscapes · Predictive modelling

N. Galanidou (✉)
Department of History and Archaeology, University of Crete, Rethymno, Greece
e-mail: galanidou@uoc.gr

G. Bailey
Department of Archaeology, University of York, York, UK

College of Humanities, Arts and Social Sciences, Flinders University, Adelaide, SA, Australia
e-mail: geoff.bailey@york.ac.uk

G. Bailey et al. (eds.), *The Archaeology of Europe's Drowned Landscapes*, Coastal Research Library 35, https://doi.org/10.1007/978-3-030-37367-2_15

15.1 Introduction

Of all the regional sections, this one covers the largest area and at the same time has the smallest number of underwater finds, with 138 recorded finds (Appendix I). Known sites are relatively few or exist as isolated discoveries, and some countries have yet to yield any finds at all, although those included here, namely, Malta (Gambin, Chap. 17, this volume) and Ukraine (Kadurin et al., Chap. 21, this volume), clearly have promising conditions for new discoveries. As indicated in the Introduction (Bailey et al., Chap. 1, this volume), this paucity of finds reflects in a general way both differences in coastal geology and geomorphology and different intellectual interests and histories of underwater investigation. Nevertheless, the known underwater finds extend from as early as the Middle Palaeolithic in Croatia (Radić Rossi et al., Chap. 18, this volume) and Israel (Galili et al., Chap. 23, this volume) to the Bronze Age or later, especially in Italy (Castagnino Berlinghieri et al., Chap. 16, this volume), Croatia (Radić Rossi et al., Chap. 18, this volume), Greece (Galanidou et al., Chap. 19, this volume) and Bulgaria (Peev et al., Chap. 20, this volume). They represent a variety of site types including cave deposits, palaeontological finds, votive deposits and village settlements with details of settlement layout and features such as dwelling structures of timber and stone, pits and burials. In two areas, Bulgaria (Peev et al., Chap. 20, this volume) and Israel (Galili et al., Chap. 23, this volume), the concentration of underwater sites along limited stretches of coastline and the range of finds and quality of preservation are comparable to the underwater settlements of the western Baltic in Denmark and Germany (Bailey et al., Chap. 3, this volume; Jöns et al., Chap. 5, this volume).

It is broadly true to say that investigation of submerged Stone Age landscapes throughout these marine basins has been overshadowed by an interest in shipwrecks and submerged remains of shoreline infrastructure such as harbours, fish tanks, ship sheds and settlements relating to the Bronze Age, Iron Age and classical antiquity. In part, this reflects the fact that the Mediterranean and the Black Sea already from an early period were coming under the influence of maritime exploration, trade and colonisation associated with the development and expansion of Neolithic and Bronze Age societies at a time when Stone Age societies in NW Europe were hunting and fishing along the now-submerged coastlines of the North Sea and the Baltic or expanding into the newly deglaciated regions further north. It also reflects the fact that some coastal regions, notably in southern Italy and Greece, have continued to sink after stabilisation of eustatic sea level because of their tectonic history, with the partial or total submergence of coastal settlements and harbours of later periods.

Another disincentive to underwater exploration for submerged Stone Age sites and landscapes noted by Arias (Chap. 13, this volume) is the relative narrowness of the continental shelf along many sections of the Mediterranean coastline— $\leq$ 5–10 km. Here, the amount of land exposed at lowest sea level was relatively small, and its periodic exposure and inundation by changes in sea level had less dramatic effects than on shallower continental shelves elsewhere. Narrow shelves also encourage the view, not necessarily justified, that sites such as caves situated on the present-day coast are close enough to provide a sufficient window into the use of the submerged landscape and its palaeoshorelines without the need for underwater exploration. Because of the geology and topography of the Mediterranean coast, coastal caves of this type are common. They include some of the most important and best-known Palaeolithic sequences in the Mediterranean such as Gorham's Cave in Gibraltar (Arias, Chap. 13, this volume), the Monte Circeo caves of Italy (Castagnino Berlinghieri et al., Chap. 16, this volume), Crvena Stijena in Montenegro on the east Adriatic coast (Whallon 2018), the caves of the Mani Peninsula and Franchthi Cave in Greece (Galanidou et al., Chap. 19, this volume), Ksar Akil in Lebanon (Tixier 1974), the Mount Carmel Caves in Israel (Galili et al., Chap. 23, this volume) and the Haua Fteah in Cyrenaica, Libya (McBurney 1967; Barker et al. 2007). Similar examples are present on Atlantic coastlines, notably in northern Spain

(Arias, Chap. 13, this volume) and on the island of Jersey (Bailey et al., Chap. 10, this volume).

These sites, not surprisingly given their deep stratified sequences and abundant remains, have attracted considerable attention and resources in their investigation and play an important role in providing insights into the use of the adjacent territory that is now submerged. However, as the chapters in this section make clear, these on-land caves are part of a continuum that extends to submerged caves offshore and below present sea level. Moreover, on-land caves on the modern coastline need to be complemented by underwater investigation of the adjacent submerged landscape, the ways in which that landscape and its resources were modified by sea-level change, and how these changes in their turn affected the site catchment of the caves located on the modern coast, their varying attractiveness for human occupation at different periods and the nature of the food remains and artefacts deposited within them. Moreover, submerged cave sites, even those quite close to the present coastline, may give evidence of human activities not represented in the deposits of caves on land.

Notwithstanding these disincentives to underwater exploration, it is worth noting that the preconditions that have encouraged underwater investigations in north-west Europe, namely, exposure of artefacts and other features in the intertidal or shallow water zone, are clearly present on some Mediterranean and Black Sea coastlines. These examples include stone tools eroded out from underwater deposits or visible in pedestrian surveys along the modern shoreline (Arias, Chap. 13, this volume; Radić Rossi et al., Chap. 18, this volume); soft sediments and peats in shallow bays where culture layers and artefacts are often brought to light by commercial activities such as dredging, notably in France, Croatia and Bulgaria (Billard et al., Chap. 12, this volume; Radić Rossi et al., Chap. 18, this volume; Peev et al., Chap. 20, this volume); and remains of stone structures easily visible in shallow water in Italy, Greece and Israel (Castagnino Berlinghieri et al., Chap. 16, this volume; Galanidou et al., Chap. 19, this volume; Galili et al., Chap. 23, this volume). There is also a strong scientific tradition of investigating sea-level change, a theme increasingly focussed on the human impact of such changes (Benjamin et al. 2017).

In at least three cases, underwater investigation motivated primarily by the search for shipwrecks has resulted in the discovery of important underlying prehistoric deposits with cultural material, notably at Cala Tramontana on the island of Pantelleria (Abelli et al. 2016, p. 97; Castagnino Berlinghieri et al., Chap. 16, this volume), Zambratija in Croatia (Benjamin et al., 2011, p. 194; Radić Rossi et al., Chap. 18, this volume) and Urdoviza on the Bulgarian coast (Angelova and Draganov 2003, p. 12; Peev et al., Chap. 20, this volume).

Closer examination of the Mediterranean and Black Sea underwater sites reveals some similarities with the more abundant material in the other marine basins, as well as contrasts. We highlight four themes: underwater caves, sea crossings, mapping and predictive modelling and underwater settlements.

15.2 Underwater Caves

Caves with an opening at sea level and with preservation of deposits with evidence of human occupation when sea level was lower than present have been recognised as a potential window into the submerged landscape since Blanc's (1940) observations at Grotta Palinuro in Italy, where he observed cemented deposits containing bone fragments attached to the cave walls extending below modern sea level. Since that time, thousands of underwater caves have been explored by divers in the karstic geology around the northern coastline of the Mediterranean. However, reports of underwater deposits are rare, and those with archaeological remains are even rarer. Flemming and Antonioli (2017) offer a comprehensive overview and suggest that the rarity of archaeological remains may be due to a variety of causes. These include removal of deposits by marine erosion; the preservation of deposits by cementation or beneath a protective overburden of marine sediments, rockfalls and biogenic accretions, posing major logistical obstacles to access and excavation; and perhaps also the lack of archaeological expertise and an eye for anthropogenic indicators amongst the divers who have conducted underwater exploration. As Billard et al. (Chap. 7, this volume) note in relation to the coastal karst of the French Mediterranean, most of these underwater investigations have been carried out for speleological or hydrological purposes rather than archaeological ones.

Caves that are empty of deposits or lacking archaeological evidence of use are common in surveys on land. Here, caves with archaeological deposits are usually shallow openings with specific conditions of size, aspect, elevation above the valley floor, location with respect to topography and animal movements, and geomorphological conditions conducive to the accumulation and retention of sediments. These sites are usually only a small percentage of the total available in the limestone regions of Europe. So it is not surprising that many underwater caves lack human evidence. The critical question is whether cave deposits with remains of terrestrial deposits and human activities could have survived the destructive processes of marine inundation.

Examples from the following chapters include both caves with openings at or close to modern sea level, which cannot be accessed today except from the sea, and also fully submerged caves. Examples of the former are Cova del Gegant in Spain, with Mousterian artefacts and Neanderthal remains (Arias, Chap. 13, this volume); Grotta Verde and Grotta dei Cervi on the island of Sardinia, Italy, the former with Neolithic burials and the latter a palaeontological deposit (Castagnino Berlinghieri et al., Chap. 16, this volume); the Vamos Cave on Crete with palaeontological deposits but no evidence of human activity; and the Kalamakia Cave on the Mani Peninsula of the Greek mainland with Middle Palaeolithic occupations (Galanidou et al., Chap. 19, this volume).

Fully submerged caves include an important group of sites in the Couronne and Calanques massifs of the French Mediterranean coast, some with breccias containing animal bones and stone artefacts, and the exceptional case of the Cosquer Cave (Billard et al., Chap. 12, this volume). Here, an inner chamber contains painted wall art and remains of hearths dated to the LGM (Last Glacial Maximum). This inner chamber is above present sea level but can only be reached along a narrow upward sloping corridor 150 m from a cave mouth that is now 37 m below present sea level. Also in this category are submerged caves off the island of Corfu at a depth of 40 m, with stone artefacts on the seabed nearby that may have been eroded from deposits within the caves and off Agios Petros in the northern Sporades (Galanidou et al., Chap. 19, this volume).

As far as preservation is concerned, deposits have remained in place either because they are high enough to be clear of wave action even at highest sea level, with evidence that earlier and deeper deposits have been washed out during marine transgression, or because of cementation that has helped to resist wave attack. The presence of faunal remains in submerged cave deposits on Sardinia and Crete shows that cave deposits can be preserved after submergence. However, artefacts are absent in both cases either because the caves were not suitable for human occupation or perhaps because humans were not present on these islands during the Pleistocene, a point that we return to below.

All these indicators suggest considerable further potential for the archaeological investigation of underwater caves, especially those where local conditions may have resulted in the preservation of deposits. Landward-facing caves protected from the full force of waves are one possibility, and it is notable that the Vamos cave in Crete falls into this category. Preservation of terrestrial deposits protected beneath a later overburden of rockfalls and marine sediments offers another possibility. But the logistical problems of investigating these deposits should not be underestimated. Preliminary excavation into the uppermost deposits of a cave in a landward-facing cliff on Vladi's Reef, and at a depth of 19 m, just offshore of the Gibraltar coastline, is reported by Arias et al. (Chap. 13, this volume). However, further work has been deferred because of the resources and engineering solutions needed to remove rockfalls that block the cave entrance.

15.3 Sea Crossings

The Mediterranean has four large islands that have always been detached from their adjacent mainland throughout the Pleistocene sea-level cycle: Cyprus (Ammerman, Chap. 22, this volume), Crete (Galanidou et al., Chap. 19, this volume), Sardinia and Corsica—a single land mass at lower sea levels—and Sicily, though the latter was briefly connected to the Italian mainland at the LGM (Castagnino Berlinghieri et al., Chap. 16, this volume). They are well known for their unusual mammalian fauna of now-extinct animals, which show the evolution in isolation of species that have no mainland analogues, most famously the pygmy elephant and pygmy hippo (Davis 1967, Athanassiou et al. 2015 and references therein). There is also a scattering of smaller islands, notably in the Aegean, some of which, such as the Melos group, coalesced to form a larger land mass when sea levels were lower.

Although it was once assumed that these islands were not colonised until the expansion of Neolithic farmers and that the extinction of the endemic fauna was the result of that human invasion, it is now widely accepted that sea crossings to some of the islands in the Aegean and eastern Mediterranean were undertaken by Epipalaeolithic or Mesolithic seafarers as early as 13,000 years ago, well before the arrival of Neolithic farmers (Laskaris et al. 2011; Galanidou et al., Chap. 19, this volume; Ammerman, Chap. 22, this volume). The further question arises as to whether these large islands, apparently with sufficient resources to sustain resident human populations of hunter-gatherers, were reached at a much earlier date in the Pleistocene, with all that this implies about the seafaring capabilities of Palaeolithic populations (Galanidou 2009, 2014). Yet it is also important to emphasise that sea crossings include both organised voyages and serendipitous crossings or accidental drifting (Dennell et al. 2014); in the latter case, the archaeological imprint would be minimal if any (Papoulia 2016, p. 43).

The clear evidence in Australasia that short sea crossings took place as early as 800,000 years ago, and in a sustained fashion over longer distances from at least as early as 60,000 years ago (Hiscock 2008; Clarkson et al. 2017), certainly encourages the search for similar evidence in the Mediterranean. Indeed, repeated claims have been made for a Palaeolithic presence on the large Mediterranean islands, most recently the finds from south-western Crete (Runnels 2014 and references therein). The early Palaeolithic status of these finds has been questioned and remains contested because of doubts about the human origin of the material in question or about its stratigraphic provenance and dating (Broodbank 2014; Galanidou 2014; Phoca-Cosmetatou and Rabett 2014).

The investigation of submerged landscapes discussed in the following chapters does not at present offer any decisive resolution of these uncertainties, but the focus on sea-level change and palaeoshoreline reconstruction certainly raises some relevant questions and intriguing clues.

The first and most important question is the changing palaeogeography of island connections with changes in sea level. Perhaps the best evidence of islands that were never connected to their adjacent mainland but were nevertheless occupied, or at any rate visited, during the Palaeolithic comes from the Ionian coast of Greece, where lithic assemblages with Middle and Upper Palaeolithic affinities have been found on the offshore islands of Zakynthos, Kephallinia and Atokos (Ferentinos et al. 2012; Galanidou 2018; Galanidou et al., Chap. 19, this volume). The sea crossings required would always have been quite short—less than 7–5 km (Papoulia 2017, p. 82; Galanidou et al. in prep), and it is possible that the limited number of artefacts attributed to the Palaeolithic are nothing more than the result of serendipitous crossings to these Ionian islands (Galanidou 2018; Papoulia 2018).

In Italy, there is the intriguing underwater find of Cala Tramontana on the island of Pantelleria (Castagnino Berlinghieri et al., Chap. 16, this volume). The island was never closer than 20 km to the nearest mainland, even at the lowest sea level, so that sea crossings over a significant distance would always have been required to reach it. It was important during the Neolithic period as a major source of obsidian, but Cala Tramontana appears to be of earlier date. The deposits with the artefacts are at 18–21 m below present sea level. They are not directly dated but fall within the period 9600 to 7700 cal BP according to the local sea-level curve, and their expedient manufacture is consistent with short visits to the island. Whether the visits were specifically to collect obsidian and take it away for use elsewhere is not clear, but the evidence raises the intriguing question of whether obsidian artefacts sourced to Pantelleria are present in pre-Neolithic archaeological deposits elsewhere. In any case, the Cala Tramontana find is relatively late in date and casts no light on the possibility of earlier sea crossings.

In Italy, detailed studies of sea-level change and palaeogeographic reconstructions offer additional insights (Castagnino Berlinghieri et al., Chap. 16, this volume). Sardinia was separated from the Italian mainland even at lowest sea level during the LGM by a sea crossing of 12 km and has no certain evidence of human presence before the Mesolithic period at 8700 cal BP. Stone tools of earlier data have been claimed but not substantiated. The island of Pianosa in the Tuscan Sea was connected to the mainland during the LGM, and the presence of Upper Palaeolithic industries shows human entry and occupation in that period. Subsequently, with sea-level rise, the island was cut off by a sea channel 10 km wide and was apparently abandoned, only to be reoccupied after a long interval during the Neolithic period.

Sicily is especially interesting because the Strait of Messina between Sicily and the Italian mainland is only 4 km wide, at most, but is notorious for its treacherous currents. Detailed modelling of changes in channel geometry and tidal velocities in response to sea-level change shows that these treacherous conditions would have persisted at lower sea levels but that a land connection would have existed for about 10,000 years during the LGM. It is during this period that abundant evidence of human occupation in the form of Upper Palaeolithic industries appears in Sicily along with the entry of a modern fauna including red deer and wild ass (Castagnino Berlinghieri et al., Chap. 16, this volume).

One interesting consequence of the LGM land connection between mainland Italy and Sicily is that low sea levels at this time would have opened up a land corridor extending further south from Sicily to incorporate Malta in an elongated peninsula, representing an attractive thoroughfare for the dispersal of mammals and a cul-de-sac that would have facilitated their trapping by human hunters. Gambin (Chap. 17, this volume) notes that there are pollen indicators suggesting the presence of agriculture on Malta at a date when sea levels were lower than present and before the earliest dated archaeological sites on the modern coast. Extensive mapping of the submerged landscape has also identified features that might hold promise for archaeological exploration in search of earlier sites, and that possibility represents a high priority for future research.

The implication of the above examples from western Greece and Italy is that short sea crossings, of the order of 5 km or less, especially in sheltered waters, were certainly achievable at least as early as the Middle Palaeolithic but that longer distances in the 5–10 km range or more were marginal for Palaeolithic populations and that unequivocal evidence for such crossings is not apparent until the very end of the Pleistocene or the early to mid-Holocene period (Epipalaeolithic, Mesolithic and Neolithic). Similar considerations apply to Crete where mapping of palaeogeographic changes taking account of eustatic and tectonic effects indicates that Crete could have been reached periodically during Pleistocene low sea-level stands by island-hopping across relatively narrow sea channels from mainland Greece or Turkey (Sakellariou and Galanidou 2016, 2017; Galanidou et al., Chap. 19, this volume).

However, it would be unwise to overgeneralise about Palaeolithic seafaring capabilities from these limited examples. Here, as in all other aspects of underwater research, local factors are likely to be of paramount importance and highly variable. The Italian examples refer to islands that were either too small or too distant to encourage regular visitation or were isolated by short and narrow straits where funnelling of sea currents would naturally create more dangerous conditions for sea crossings. Equally, the presence of a Pleistocene fauna on these islands, including animals with good swimming abilities such as deer and mammoth, suggests that sea channels were periodically narrow enough to be swum across; and if these animals could make the crossing, then we should allow the same possibility to humans. At any rate, the current evidence offers little encouragement for the idea of organised sea crossings over long distances from the North African coastline to southern Europe during the Pleistocene. These would have required sea journeys of at least 50 km, with the exception of the 11-km-wide Strait of Gibraltar, but even in the latter case, with substantial Palaeolithic caves on both sides of the Strait, decisive evidence of Palaeolithic sea crossings has yet to be identified (Bailey et al. 2008). The only other comment relevant to this discussion is that since the period of greatest interest is one when sea levels were substantially lower than present, there is a strong likelihood that coastal sites with decisive archaeological evidence are now submerged, and this in its turn should reinforce the incentive for underwater exploration.

15.4 Mapping and Predictive Modelling

Key to the above discussion of sea crossings is the mapping of palaeoshorelines and palaeo-oceanographic conditions. The importance of mapping is not confined only to reconstructing palaeocoastlines but also extends to reconstructions of the environment, topography and resources that would have been available on the submerged landscape and ultimately to prediction of target areas worth searching for archaeological remains. Many examples are presented in this section, ranging from large-scale mapping of broad areas to more localised investigations. An example of the former is the palaeogeographic mapping of the Aegean Basin (Galanidou et al., Chap. 19, this volume; Figs. 19.2 and 19.3), which shows how dramatically the proportion of land and sea has changed on Pleistocene timescales.

Examples from other chapters refer to regional or more localised investigations involving offshore survey and comprising varying combinations of acoustic survey, diving and coring in marine basins adjacent to concentrations of known archaeological sites on land that can provide a basis for predicting the locations of underwater sites. These include geophysical survey combined with diving and excavation of an underwater cave on the narrow shelf adjacent to the Neanderthal sites of Gorham's Cave and Vanguard Cave in Gibraltar and geophysical survey, landscape reconstruction and diving offshore of northern Spain with its many Palaeolithic cave sequences (Arias, Chap. 13, this volume); geophysical survey, diving and coring in the Bay of Kiladha in southern Greece opposite the Franchthi Cave and geophysical survey and landscape reconstruction in the Inner Ionian Sea and the Kalloni Gulf of Lesbos in the Aegean, offshore of islands with significant Middle Palaeolithic and Lower Palaeolithic finds, respectively (Galanidou et al., Chap. 19, this volume); predictive site location and offshore geophysical survey on the NW Black Sea shelf of the Ukraine offshore of the abundant record of Upper Palaeolithic sites on the adjacent land (Kadurin et al., Chap. 21, this volume); and Cyprus, where the association of stone tools with aeolianite ridges on the modern coast has led to diver survey for similar features underwater.

The most ambitious of these projects is described in the chapter on Ukraine where the chance recovery of flint artefacts in sediment cores from earlier geological research has encouraged a new investigation with archaeological objectives (Kadurin et al., Chap. 21, this volume). The new research involved the analysis of the association between archaeological sites on land and topographic features, geology and raw material resources, an exercise facilitated by the availability of a large sample of Upper Palaeolithic sites on the tributaries of the large rivers that drain into the north-west sector of the Black Sea. The results were then used to identify likely target areas on the submerged shelf, using a combination of geological mapping and geophysical survey to assess conditions of preservation and chances of site discovery. Target areas with the appropriate combination of features were further explored, but limitations of funds and equipment prevented further investigation, and identified the need for a programme of sediment coring as an important requirement of future research.

Two of the localised studies summarised above provide instructive examples of successful predictive modelling. Underwater research in Kiladha Bay was carried out in the 1980s as a pioneering exercise in geophysical survey to reconstruct features of the submerged topography, combined with hand coring in shallow water 2–3 km offshore at a depth of c. 5 m. Two cores revealed the presence of Neolithic pottery, charred plant fragments, fish vertebrae and mollusc shells, indicating the presence of a former site on the banks of a palaeo-river. New investigations are under way to build on these early investigations (Gifford 1983; Galanidou et al., Chap. 19, this volume). The work in Cyprus began as a search for early lithic material in support of a pre-Neolithic occupation, stimulated by an interest in early seafaring (Ammerman, Chap. 22, this volume). The discovery of aeolianite ridges on the coast with surface lithics of Epipalaeolithic type belonging to a period when sea level was significantly lower than present led to a successful search for similar features underwater some 130 m offshore at a depth of 12 m.

These two examples provide an interesting case of proceeding in small steps from the known archaeology on land to likely targets nearby in shallow water. In both cases, new work is under way or planned to extend this predictive process to the search for earlier sites in deeper water.

15.5 Underwater Villages

Although the total number of underwater sites in this region is quite small, the number of submerged settlements (including in situ cultural layers) is relatively large, 76 out of 138 finds (55%), with notable concentrations in Bulgaria and Israel (Appendix 1). These underwater settlements are broadly of two types: sites associated with shallow coastal lagoons and anaerobic sediments with timber remains, often discovered by dredging of sediments to keep open channels for sea traffic, and sites with remains of stone structures visible at the shore edge or in shallow water. Examples of the former are Leucate-Corrège on the French Mediterranean (Billard et al., Chap. 12, this volume), Zambratija in Croatia (Radić Rossi et al., Chap. 18, this volume) and an important group of sites on the Black Sea coast of Bulgaria (Peev et al., Chap. 20, this volume). Examples of the latter are partially submerged settlements in Greece of Neolithic or Bronze Age date, notably the site of Pavlopetri (Galanidou et al., Chap. 19) and another important group of Neolithic sites on the Carmel coast of Israel (Galili et al., Chap. 23, this volume).

The Bulgarian sites have been only partially excavated and published, but it is clear that they represent substantial villages of Eneolithic and Early Bronze Age date, c. 6500–4500 cal BP, with extensive wooden structures now submerged at depths down to about −7 m. These comprise wooden platforms that appear to have been built to provide a dry and stable surface on low-lying ground liable to flooding and remains of dwellings. There are large quantities of ceramics, stone and bone artefacts, animal bones and evidence of subsistence economies that combined cereal cultivation and domestic livestock with hunting of wild animals such as wildfowl, fishing and sea mammal hunting.

The Black Sea has a complex and much-contested pattern of sea-level change following the LGM because of its isolation from the world ocean, with claims for a dramatic flood that would have drowned large areas of land as global sea-level rise overtopped the Bosphorus sill that separates the Black Sea from the Sea of Marmara and the Mediterranean. The Bulgarian underwater sites are later in date than this postulated flood event and are not related to it, but they provide some insight into the conditions of preservation associated with underwater sites in the region, as well as evidence of ongoing complexities of relative sea-level change at quite a late date, in which glacio-hydro-isostatic warping of the Earth's crust is implicated, with the possible addition of a tectonic component. The existing sites offer considerable potential for new research, and new underwater investigations are under way involving excavation and more extensive mapping and coring of the submerged landscape in deeper water.

The sites on the Carmel coast in Israel represent one of the best studied groups of underwater sites in the Mediterranean and Black Sea basins, with investigation and excavation over a long period. As in Bulgaria, these underwater sites appear to fill a gap in the archaeological sequence, indicating the importance of underwater investigations in identifying the missing coastal component of site distributions and settlement patterns in a wider region. The Israeli sites range in age from about 9000 to 6800 cal BP and in depth from the intertidal zone to −12 m. They are justly famous for their extraordinary range and quality of evidence, which includes stone structures, pits, courtyards, stone and wood-lined water wells, stone-lined graves, a megalithic ritual structure and a wide variety of well-preserved organic materials including wood, remains of plant foods, basketry and bone. The economy combined fishing, cereal cultivation, domestic livestock and hunting of wild animals, and the settlements attest to a high degree of residential stability founded on this mixed marine-terrestrial subsistence base until rising sea level finally forced their abandonment. The large sample of well-preserved human skeletons recovered from underwater burials has facilitated insights into demographic patterns and disease.

15.6 Conclusions

Despite the large area covered in this section and the somewhat scattered and patchy distribution of underwater sites, the material already discovered includes a considerable range and variety of sites, many with a high quality of preservation, and new research is moving forward in several regions, with indications of the potential for many new discoveries.

At the same time, there are still large gaps in geographical coverage. Turkey remains largely a blank where 'underwater archaeology of prehistoric landscapes is not even at an incipient stage' (Özdoğan 2011, p. 226) and has been much overshadowed by work on shipwrecks. What few finds have been discovered are mostly submerged or partially submerged remains of Neolithic and Bronze Age settlements around the Sea of Marmara in the north-west and are well summarised by Özdoğan (2011) and Öniz (2018). With coastlines facing the eastern Mediterranean, the Aegean and the Black Sea and a variety of coastal and offshore topographies including rocky coastlines with important Palaeolithic coastal cave sequences, such as Yarimburgaz and Okuzini (Yalçinkaya et al. 2002; Clark Howell et al. 2010), the coastal regions of Turkey occupy a key geographical position between the Near East and Europe. There is every reason to suppose that they hide underwater palaeolandscapes and archaeological material comparable to the discoveries in the neighbouring regions of Greece and Bulgaria. Elsewhere, the Adriatic coastline of Italy has scarcely been explored for its underwater potential, and the African shores of the Mediterranean are a complete blank.

The arrangements in place for the management of the underwater cultural heritage are quite variable. Arias (Chap. 13, this volume) draws attention to the bureaucratic complexities of securing permission for offshore survey in Spain requiring the involvement of seven different organisations and governmental agencies with overlapping interests. More generally, arrangements for incorporating the underwater cultural heritage in monitoring offshore industrial work and establishing collaborations with offshore industrial companies appear to be less well developed than in NW Europe, although the potentially destructive pressures of coastal and offshore construction work and commercial activities are every bit as intensive as in other regions, and this is a major concern throughout the Mediterranean and Black Sea basins.

There are, thus, considerable challenges confronting underwater research in this region, as well as growth areas of new and energetic research activity, with opportunities for new discoveries and the potential to illuminate in new ways some of the most important themes of human development and the growth of civilisations in a region that plays a key role in the larger picture of world prehistory.

References

Abelli L, Agosto MV, Casalbore D, Romagnoli C, Bosman A, Antonioli F, Pierdomenico M, Sposato A, Chiocci FL (2016) Marine geological and archaeological evidence of a possible pre-Neolithic site in Pantelleria Island, Central Mediterranean Sea. In: Harff J, Bailey G, Lüth F (eds) Geology and archaeology: submerged landscapes of the continental shelf, vol 411. Geological Society, London, pp 97–110

Angelova H, Draganov V (2003) Underwater archaeological excavations of submerged Late Eneolithic and Early Bronze Age settlements in Kiten and Sozopol (South Bulgarian Black Sea coast). In: Angelova H (ed) Honorem Mihaili Lazarov. Thracia Pontica 6, vol 2. Centre for Underwater Archaeology, Sofia, pp 9–22

Athanassiou A, Herridge V, Reese D, Iliopoulos G, Roussiakis S, Misopoulou V, Tsiolakis E, Theodorou G (2015) Cranial evidence for the presence of a second endemic elephant species on Cyprus. Quat Int 379:47–57

Bailey G, Carrión JS, Fa DA, Finlayson C, Finlayson G, Rodríguez-Vidal J (eds) (2008) The coastal shelf of the Mediterranean and beyond: corridor and refugium for human populations in the Pleistocene. Quat Sci Rev 27(23–24):2095

Barker G, Hunt C, Reynolds T (2007) The Haua Fteah, Cyrenaica (Northeast Libya): renewed investigations of the cave and its landscape. Libyan Stud 38:93–114

Benjamin J, Bekić L, Komšo D, Koncani Uhač I, Bonsall C (2011) Investigating the submerged prehistory of the eastern Adriatic: progress and prospects. In: Benjamin J, Bonsall C, Pickard C, Fischer A (eds) Submerged prehistory. Oxbow, Oxford, pp 193–206

Benjamin J, Rovere A, Fontana A, Furlani S, Vacchi M, Inglis RH, Galili E, Antonioli F, Sivan D, Miko S, Mourtzas N, Felia I, Meredith-Williams M, Goodman-Tchernov B, Kolaiti E, Anzidei M, Gehrels R (2017) Late Quaternary sea-level changes and early human societies in the central and eastern Mediterranean Basin: an interdisciplinary review. Quat Int 449:29–57

Blanc AC (1940) Industrie musteriane e paleolitiche superiore nelle dune fossile e nelle grotte litorannee del Capo Palinuro. Rendiconti della Reale Academia d'Italia 10, 7(1):602–615

Broodbank C (2014) So... what? Does the paradigm currently want to budge so much? J Mediterr Archaeol 27(2):267–272

Clark Howell F, Arsebuk G, Kuhn SL, Ozbasaran M, Stiner MC (eds) (2010) Culture and biology at a crossroads – the middle Pleistocene record of Yarimburgaz Cave (Thrace, Turkey). Ege Yayinlari

Clarkson C, Jacobs Z, Marwick B, Fullagar R, Wallis L, Smith M, Roberts RG, Hayes E, Lowe K, Carah X et al (2017) Human occupation of northern Australia by 65,000 years ago. Nature 54(7663):306–310. https://doi.org/10.1038/nature22968

Davis SJM (1967) The archaeology of animals. Batsford, London

Dennell RW, Louys J, O'Regan HJ, Wilkinson DM (2014) The origins and persistence of Homo floresiensis on Flores: biogeographical and ecological perspectives. Quat Sci Rev 96:98–107. https://doi.org/10.1016/j.quascirev.2013.06.031

Ferentinos G, Gkioni M, Geraga M, Papatheodorou G (2012) Early seafaring activity in the southern Ionian Islands, Mediterranean Sea. J Archaeol Sci 39:2167–2176. https://doi.org/10.1016/j.jas.2012.01.032

Flemming N, Antonioli F (2017) Prehistoric archaeology, palaeontology, and climate change indicators from caves submerged by change of sea level. In: Campbell PB (ed) The archaeology of underwater caves. Highfield Press, Southampton, pp 23–38

Galanidou N (2009) Foragers or farmers? Aegean prehistoric research and the beginnings of the Cretan prehistory thread. In: Kopaka K (ed) Aegean prehistoric research in the beginning of the 21st century. Crete University Press, Herakleio, pp 13–42

Galanidou N (2014) Archaic hominins on Crete: fact or fiction? J Mediterr Archaeol 27(2):260–267

Galanidou N, Sakellariou D, Zavitsanou A, Papoulia C, Rousakis G, Alexopoulos I (n.d.in prep) Neanderthal sea crossings in the Inner Ionian Sea Archipelago: Palaeolithic archaeology and palaeogeographic reconstruction during Late Quaternary low sea-level periods

Galanidou N (2018) Parting the waters: Palaeolithic archaeology in the central Ionian Sea. Journal of Greek Archaeology 3:1–22

Gifford J (1983) Core sampling of a Holocene marine sedimentary sequence and underlying Neolithic cultural material off Franchthi Cave, Greece. In: Masters PM, Flemming NC (eds) Quaternary coastlines and marine archaeology. Academic, London/New York, pp 269–281

Hiscock P (2008) Archaeology of ancient Australia. Routledge, London

McBurney CBM (1967) The Haua Fteah (Cyrenaica) and the Stone Age of the South-East Mediterranean. Cambridge University Press, London

Öniz H (2018) Turkey–Gateway between Asia and Europe. In: Fischer A, Pedersen L (eds) Oceans of archaeology, Jutland Archaeological Society Publications vol 101. Jutland Archaeological Society, Højbjerg, pp 100–107

Papoulia C (2016) Late Pleistocene to early Holocene sea-crossings in the Aegean: direct, indirect and controversial evidence. In: Ghilardi M (ed) Géoarchéologie des Îles de Méditerranée. CNRS Editions, Paris, pp 33–46

Papoulia C (2017) Seaward dispersals to the NE Mediterranean islands in the Pleistocene. The lithic evidence in retrospect. Quat Int 431:64–87. https://doi.org/10.1016/j.quaint.2016.02.019

Papoulia C (2018) Pleistocene sea-crossings and submerged terrestrial routes. A view from the Inner Ionian Archipelago. PhD Dissertation, University of Crete

Phoca-Cosmetatou N, Rabett RJ (2014) Reflections on Pleistocene island occupation. J Mediterr Archaeol 27(2):255–278

Runnels C (2014) Early Palaeolithic on the Greek Islands? J Mediterr Archaeol 27(2):211–230

Sakellariou D, Galanidou N (2016) Pleistocene submerged landscapes and Palaeolithic archaeology in the tectonically active Aegean region. In: Harff J, Bailey G, Lüth F (eds) Geology and archaeology: submerged landscapes of the continental shelf, vol 411. Geological Society London Special Publications, London, pp 145–178. https://doi.org/10.1144/SP411.9

Sakellariou D, Galanidou N (2017) Aegean Pleistocene landscapes above and below sea-level: Palaeogeographic reconstruction and hominin dispersals. In: Bailey GN, Harff J, Sakellariou D (eds) Under the sea: archaeology and palaeolandscapes of the continental shelf. Springer, Cham, pp 335–359

Tixier J (1974) Fouille à Ksar'Aqil, Liban (1969–1974). Paléorient 2:187–192

Whallon R (ed) (2018) Crvena Stijena in cultural and ecological context: interdisciplinary research in Montenegro. Montenegrin Academy of Sciences, Podgorica

Yalçinkaya I, Otte M, Kozlowski JK, Bar-Yosef O (2002) La grotte d' Öküzini: Évolution du Paléolithique final du sud-ouest de l'Anatolie. Études et Recherches Archéologiques de l'Université de Liège 96. Liège, Belgium

Laskaris N, Sampson A, Mavridis F, Liritzis I (2011) Late Pleistocene/early Holocene seafaring in the Aegean: new obsidian hydration dates with the SIMS-SS method. J Archaeol Sci 38(9):2475–2479

Özdoğan M (2011) Submerged sites and drowned topographies along the Anatolian coasts: an overview. In: Benjamin J, Bonsall C, Pickard C, Fischer A (eds) Submerged prehistory. Oxbow, Oxford, pp 219–229

Italy: The Archaeology of Palaeoshorelines, Coastal Caves and Seafaring Connections

16

Elena Flavia Castagnino Berlinghieri,
Fabrizio Antonioli, and Geoff Bailey

Watching a coast as it slips by the ship is like thinking about an enigma. There it is before you—smiling, frowning, inviting, grand, mean, insipid, or savage, and always mute with an air of whispering. Come and find out.
—Joseph Conrad, Heart of Darkness

Abstract

Italy has very few currently known finds of underwater archaeology from the Stone Age, despite a substantial history of geological and geoarchaeological investigation into sea-level change and archaeological investigations of underwater remains of coastal settlements and shipwrecks from later periods. Nevertheless, there are traces of prehistoric cultural activity in submerged coastal caves, open-air sites with stone tools and potsherds that have survived inundation, and Bronze Age stone-built structures that have been partially submerged by recent changes in relative sea level and give some measure of the potentially destructive effects of marine erosion. The tradition of geoarchaeological investigation into late Pleistocene and early Holocene sea-level change has provided detailed maps of palaeoshorelines that provide interesting insights into the likelihood of early sea crossings and connectivity between the Italian mainland and its major offshore islands and potential targets for the discovery of new archaeological finds. Moreover, maritime activities and coastal settlement have played an important role throughout Italian prehistory, including consumption of marine foods such as molluscs at the shore edge, sea travel to exploit the resources of offshore islands, agricultural dispersal, and the obsidian trade. These developments were under way at a time when sea levels were mostly lower than the present, and most of the relevant archaeological evidence is likely to

E. F. Castagnino Berlinghieri (✉)
Dipartimento di Scienze Biologiche, Geologiche e Ambientali, Università di Catania, Catania, Italy

Soprintendenza per i Beni Culturali e Ambientali di Siracusa, Sicily, Italy

F. Antonioli
Italian National Agency for New Technologies, Energy and Sustainable Economic Development (ENEA), Rome, Italy
e-mail: fabrizio.antonioli@enea.it

G. Bailey
Department of Archaeology, University of York, York, UK

College of Humanities, Arts and Social Sciences, Flinders University, Adelaide, SA, Australia
e-mail: geoff.bailey@york.ac.uk

G. Bailey et al. (eds.), *The Archaeology of Europe's Drowned Landscapes*, Coastal Research Library 35, https://doi.org/10.1007/978-3-030-37367-2_16

be on palaeoshorelines that are now submerged, emphasising the need for new underwater investigations to fill this important gap in the archaeological record.

Keywords
Submerged caves · Palaeoshorelines · Offshore islands · Obsidian trade · Seafaring · Neolithic · Bronze Age

16.1 Introduction

Italy comprises the Italian Peninsula, the two largest islands in the Mediterranean, namely, Sicily and Sardinia, and over 300 smaller islands within its territorial waters, with a total coastline of c. 8000 km. It faces four different seas: the Adriatic to the east between Italy and Croatia, the Ionian Sea to the south-east between Italy and Greece, the Tyrrhenian Sea to the west, and the Ligurian Sea to the north-west, and no point is further than 120 km from the nearest sea coast (Fig. 16.1). Its coastlines are also highly variable in terms of their present-day morphology and the width and character of the offshore shelf (Antonioli et al. 2017).

Yet, despite Italy's strongly maritime character, the probable importance of coastlines and sea crossings during prehistory, and the sensitivity of its coastal and island palaeogeography to the impact of relative sea-level change, underwater investigations have mainly concentrated on the remains of Classical and later periods such as shipwrecks and remains of harbours, cities or other shoreline structures submerged by tectonic movements and on the search for sea-level indicators. Only eight submerged prehistoric sites are currently known: one palaeontological find of late Pleistocene age but not certainly associated with artefacts or human activity, two Neolithic sites, and five Bronze Age structures. In this chapter, we examine the archaeological and palaeogeographical significance of sea-level change, describe the underwater material that is currently known, and consider future directions of research.

16.2 Palaeogeographical and Archaeological Context

In broad terms, the lowering of sea level at the LGM (Last Glacial Maximum) had its main effect on the eastern side of the Peninsula, creating an extensive area of land in what is now the northern Adriatic basin between northern Italy and Croatia (Fig. 16.1). On the western side of Italy, the continental shelf exposed at low sea level was much narrower, but here sea-level change would have had important effects in joining together or isolating offshore islands including the important land masses of Sardinia and Sicily. Sardinia and Corsica formed a single, large land island at lowest sea level. Though it was not connected to the Italian mainland during the low sea-level stands of the late Pleistocene, it was separated only by a relatively narrow sea channel. It was also large enough to sustain a mammalian fauna that had evolved in isolation since earlier in the Pleistocene (Antonioli et al. 2011; Palombo et al. 2017).

Similarly, Malta would have formed a southern peninsula of Sicily (see Gambin, Chap. 17, volume), and the narrow channel between Sicily and southern Italy—the Strait of Messina, famous in Classical Mythology for its treacherous currents flowing between the twin hazards of Scylla and Charybdis—would have been connected to the mainland, but only intermittently at lowest sea-level stands (Antonioli et al. 2016). The distance to North Africa would also have been reduced at lowest sea-level stands, though sea crossings of at least 40–50 km would have been necessary, even at the Last Glacial Maximum.

Because of its complex geological history, the regional response to global sea-level changes has been highly variable in different parts of Italy. Tectonic uplift or subsidence plays a particularly prominent role in southern Italy and Sicily where plate boundaries associated with the collision of the African and the Eurasian Plates are associ-

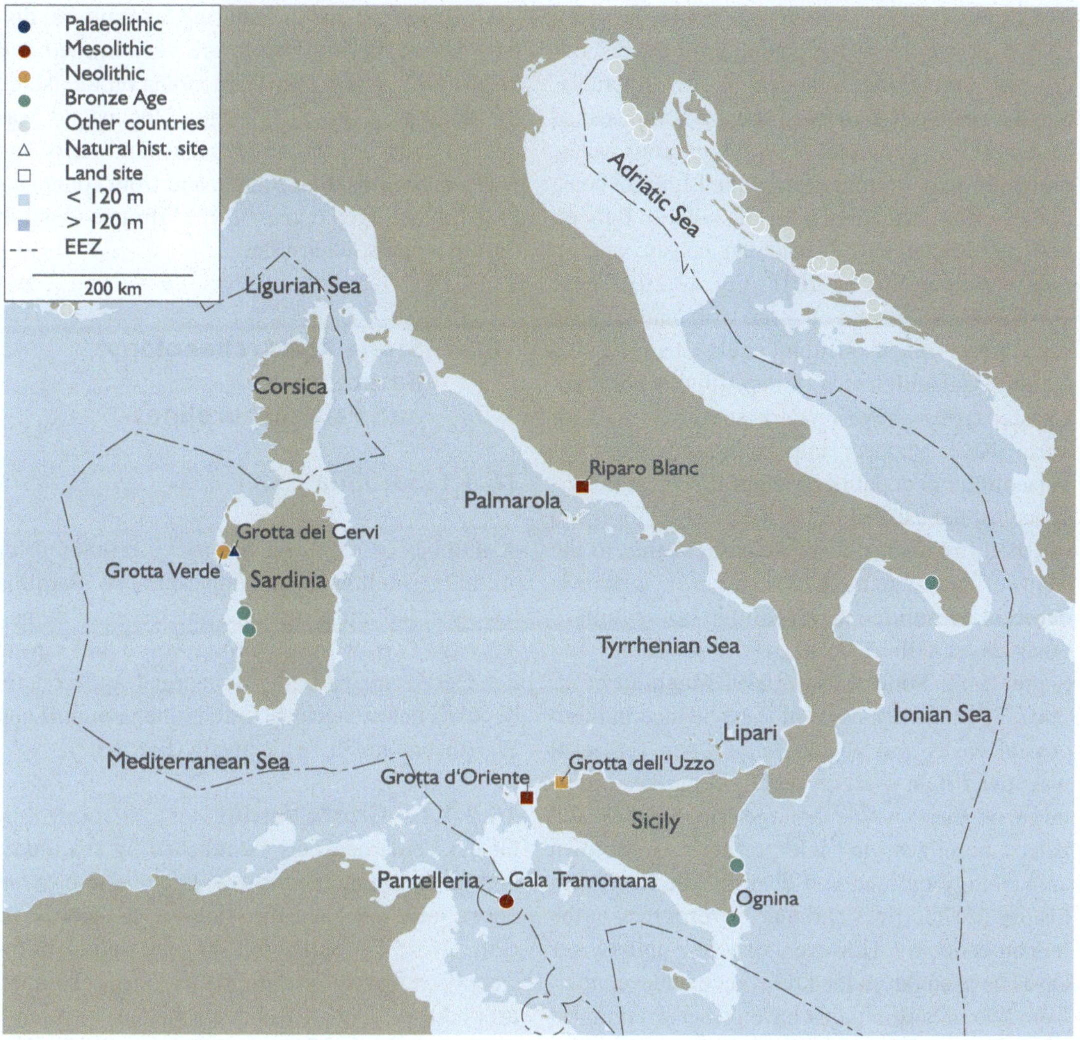

Fig. 16.1 Map of Italy showing distribution of underwater sites, key names and other places mentioned in the text. Site information from the SPLASHCOS Viewer http://splashcos-viewer.eu. Drawing by Moritz Mennenga

ated with intensive seismic and volcanic activity. These have resulted in considerable vertical movement of coastlines, compounded by regional variations in glacio-hydro-isostatic adjustments of the Earth's crust. Consequently, detailed geological studies have been devoted to unravelling the complex interplay between eustatic sea-level changes and isostatic and tectonic movements, making use of natural features such as beach deposits and cultural remains from later periods such as Roman fish tanks and including underwater features as well as those on land (Lambeck et al. 2004; Antonioli et al. 2006, 2011; Auriemma and Solinas 2009; Lambeck et al. 2011; Rovere et al. 2011a, b; Benjamin et al. 2017). These studies have important implications for palaeogeographic reconstructions of shoreline changes, the connectivity between the Italian Peninsula and its offshore islands, and the impact of sea-level change on the archaeological record. They play an important role in providing accurate maps of palaeoshorelines with which to investigate archaeological problems.

Despite the few underwater finds currently known, nevertheless, the potential for the discovery of prehistoric remains is considerable. Sections of the Italian coastline are composed of limestone karst complexes with numerous caves, particularly in Sardinia and on the Ligurian coast in the north-west, along parts of southern Italy on both its Ionian and Tyrrhenian coasts, and in Sicily (Antonioli et al. 2017, Fig. 13.20). These karst geologies extend offshore with many underwater caves (see also Billard et al., Chap. 12, this volume; Galanidou et al., Chap. 19, this volume). Coastal caves above modern sea level in these areas have yielded important sequences of Palaeolithic occupation, human fossils, and evidence for the use of marine molluscs as food and for shell ornaments, in some cases dating to the Mousterian as far back as 115 ka, or with evidence of Mesolithic and Neolithic occupation in other cases (Blanc 1939, 1958/61; Blanc and Segre 1953; Stiner 1994, 1999; Mannino et al. 2007, 2011, 2012). Most of the evidence in these coastal caves was accumulated when sea level was lower than present, highlighting the importance of these now-submerged landscapes and palaeoshorelines in interpreting the known archaeology on land and charting the long-term history of shorelines and marine resources in the human economy. However, very few underwater caves have so far yielded relevant evidence, apart from two in Sardinia that have preserved remains of artefacts or fauna and other examples of potential importance discussed below.

Other issues that focus on the reconstruction of palaeoshorelines and underwater investigations are the development of trade in obsidian and the spread of early farming, both of which were under way during the Neolithic period at a time when relative sea level was still lower than the present. The obsidian trade is of special interest because the major sources of obsidian in the central Mediterranean are the Italian islands of Sardinia, Palmarola, Lipari, and Pantelleria (Castagnino Berlinghieri 2011), and obsidian from these sources was traded widely into mainland Italy and beyond and to North Africa (Tykot 1996). However, underwater finds that relate to these developments in seafaring, agricultural dispersal, and maritime trade are very limited and known only from Cala Tramontana on the island of Pantelleria, notable because the island was never connected to any adjacent land mass and was a major obsidian source, and from some partially submerged sites of the Bronze Age in Sardinia and southern Italy.

16.3 Stone Age Archaeology: Karst Caves and Palaeoshorelines

16.3.1 Sardinia

A number of caves on the rocky promontories around a small bay near Alghero in NW Sardinia have yielded faunal remains and traces of human activity. Two of these, Grotta Verde and Grotta dei Cervi, are partially submerged and contain deposits below sea level with human remains and Pleistocene fauna, respectively (Fig. 16.2).

16.3.1.1 Grotta Verde

Grotta Verde can only be reached by sea today, and its main entrance is a wide opening c. 90 m above sea level in the side of a steep cliff (Fig. 16.3; Lo Schiavo 1986). A tunnel drops steeply down to a partially flooded chamber, around the edges of which are niches between 8 and 10 m bpsl (below present sea level) with bone remains of human burials, Neolithic pottery of impressed cardial type, and some bones of domestic animals, mainly ovicaprid with fewer bones of pig and cattle (Fig. 16.4; Tanda 1980, 1987a, b; Atzeni 1981; Antonioli et al. 1996; Wilkens 2012). The ceramics indicate a date broadly between the 9th and 7th millennium BP, and the whole assemblage is consistent with ritual activity associated with burial rites. At this period, sea level was c. 20 m bpsl. Molluscs shells (*Patella caerulea* and *Phorcus turbinatus*) collected as food at the nearby Cala Vergine cave (on the east shore of Punta del Quadro; Fig. 16.2) and a human incisor from the Medusa Cave (both caves are above present sea level) have both been

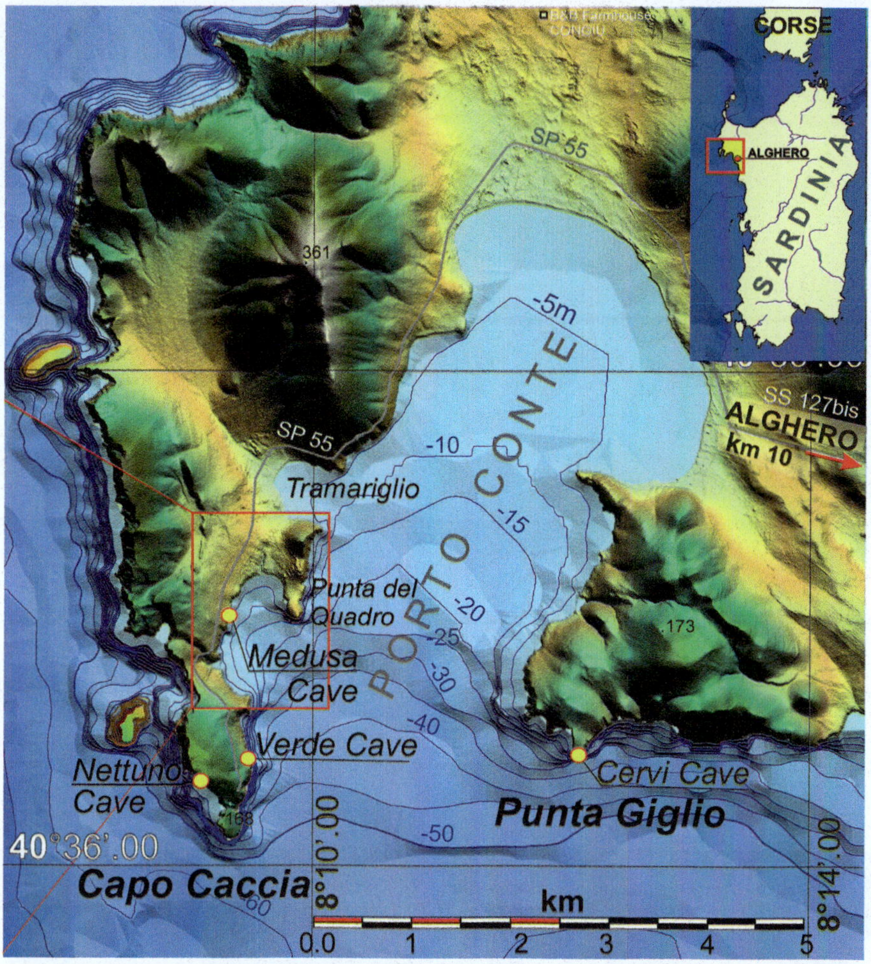

Fig. 16.2 Map of NW Sardinia showing the topography and bathymetry around the bay of Porto Conte and the location of Grotta Verde and Grotta dei Cervi. After Palombo et al. (2017, Fig. 1, p. 118). Reproduced with permission

directly radiocarbon-dated with similar readings of c. 7.4 ka cal BP (Palombo et al. 2017), of similar date to the Grotta Verde remains.

16.3.1.2 Grotta dei Cervi

The Grotta dei Cervi is a cave opening at present sea level and contains a partially submerged bone breccia with remains of a Pleistocene deer, *Praemegaceros cazioti* (Fig. 16.5), a member of an endemic Pleistocene fauna that has persisted on the island since at least the Middle Pleistocene and is well represented in other fossiliferous cave deposits on Sardinia (Palombo et al. 2017). The Grotta dei Cervi deposit was formed when sea level was substantially lower than present, with a date on a deer tooth of 11.8 ka cal BP. The cave was subsequently partly eroded and submerged. When sea level was lower, the cave could have been approached up a gently inclined slope from the surrounding coastal plain, and the animal appears to have died there of natural causes. There are tooth marks on the bones caused by rodents or other deer (who derive calcium by chewing bones) but no evidence of cut marks to suggest human intervention (Palombo et al. 2017).

These caves are of particular relevance to discussions about earliest human entry into Sardinia and the possibility of Pleistocene sea crossings.

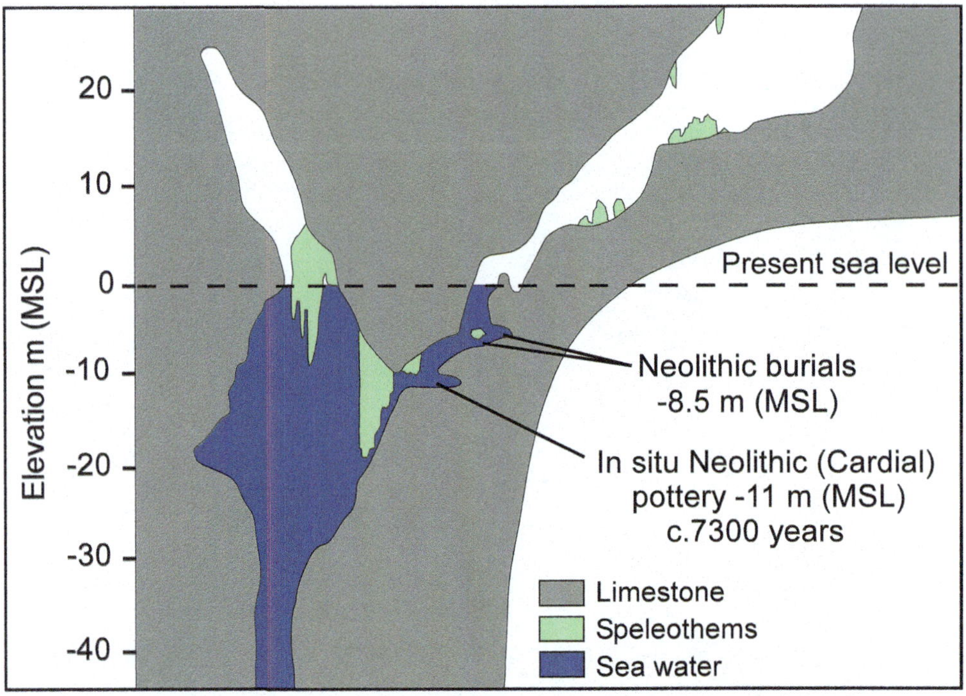

Fig. 16.3 Cross section of the Grotta Verde showing the location of the Neolithic necropolis. After Antonioli et al. (1996) and Palombo et al. (2017, Fig. 6, p. 123). Reproduced with permission

Fig. 16.4 Neolithic pottery from the Grotta Verde. (**a**) Potsherds with cardial impression. (**b**) Globular vessels with pierced lugs. The vessel on the right has cardial-impressed decoration of parallel wavy lines on the collar. Photos courtesy of Museo Nazionale "Giovanni Antonio Sanna", Soprintendenza per i Beni Archeologici di Sassari e Nuoro

Recent research on sea-level markers demonstrates that the Earth's crust in this region has been quite stable over the past 125 kyr (Antonioli et al. 2011) and that palaeoshorelines can be reconstructed with some confidence. These show that, at the time of the Last Glacial Maximum, Sardinia and Corsica were combined as a single land mass. However, movement from the Italian mainland would have required a sea crossing of 12 km across a narrow strait (Fig. 16.6). There are claims for earlier Stone Age finds in Sardinia, but these remain controversial and are not universally accepted because of doubts about stratigraphic integrity and chronology (Klein Hofmeijer et al. 1989; Mussi 2002; see also Dawson 2013). The implication of the evidence from the Alghero

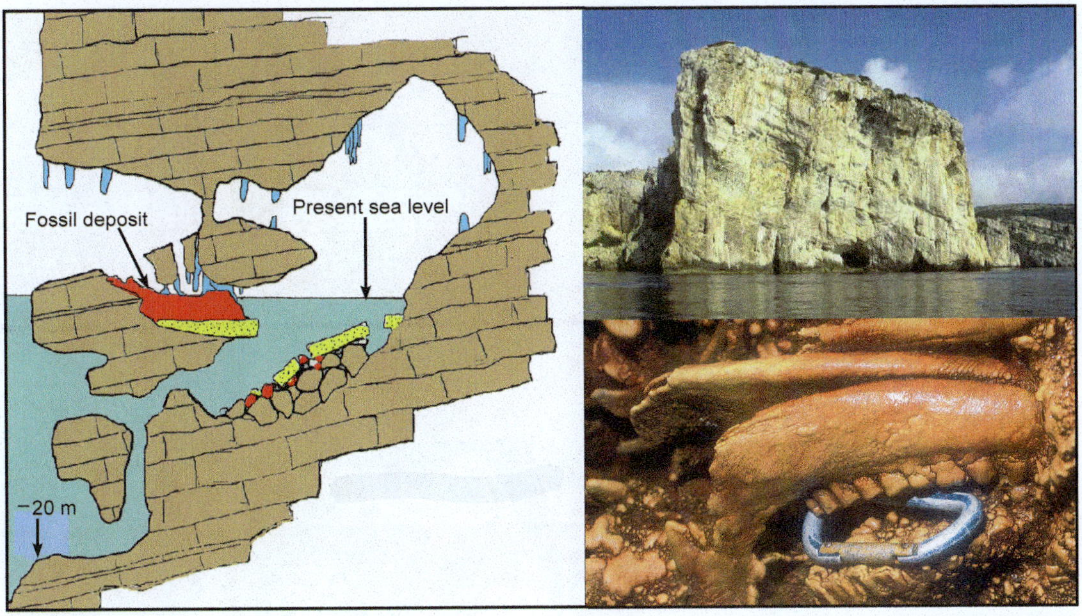

Fig. 16.5 The Grotta dei Cervi showing a cross section of the site and its deposits, and a mandible of *Praemegaceros cazioti* in situ in the breccia deposit. After Palombo et al. (2017, Fig. 6, p. 123). Reproduced with permission

caves is that humans were not present during the Pleistocene and that the sea crossing was a deterrent to human entry until the Mesolithic period, when the first unequivocal archaeological evidence appears (Floris et al. 2012; Palombo et al. 2017, p. 118). However, the question of the earliest date of human entry remains unresolved, and the search for underwater finds may help to produce new and relevant discoveries.

16.3.2 The Tyrrhenian Coast

On the opposite side of the strait from Corsica-Sardinia, reconstruction of palaeoshorelines shows that the Tuscan islands of Elba and Pianosa formed a single land mass attached to the Italian mainland at the Last Glacial Maximum and for some millennia afterwards (Fig. 16.7). Epigravettian (late Upper Palaeolithic) material on Pianosa is consistent with access by land at that time (Ducci and Perazzi 1998, 2000). Subsequently, the island was cut off by a sea crossing of c. 10 km and appears to have been abandoned, judging by the absence of Mesolithic material, only to be reoccupied during the Neolithic period, when seafaring is widely in evidence in the wider Mediterranean. Here too the implication is that sea crossings of 10 km or more were a deterrent to sea crossings until quite a late period in the prehistory of the region. But, as in Sardinia, the possibility that new archaeological sites including underwater remains will be discovered to fill some of the gaps in the archaeological record should be recognised.

Further south, in the Monte Circeo massif on the Latium shoreline, the rockshelter of Riparo Blanc contains Mesolithic occupation with evidence of food remains dominated by shells of marine molluscs and echinoderms, some fish bones, and rare terrestrial fauna (Colonese et al. 2011). This rockshelter is above sea level, but a lower cavity (Cala dell'Alabastro), accessible only from underwater, exists offshore and has yet to be explored (Fig. 16.8). Moreover, the tectonic stability of the area over the past 125,000 years (Ferranti et al. 2006) means that palaeoshorelines at successive periods can be reconstructed with some confidence (Fig. 16.9).

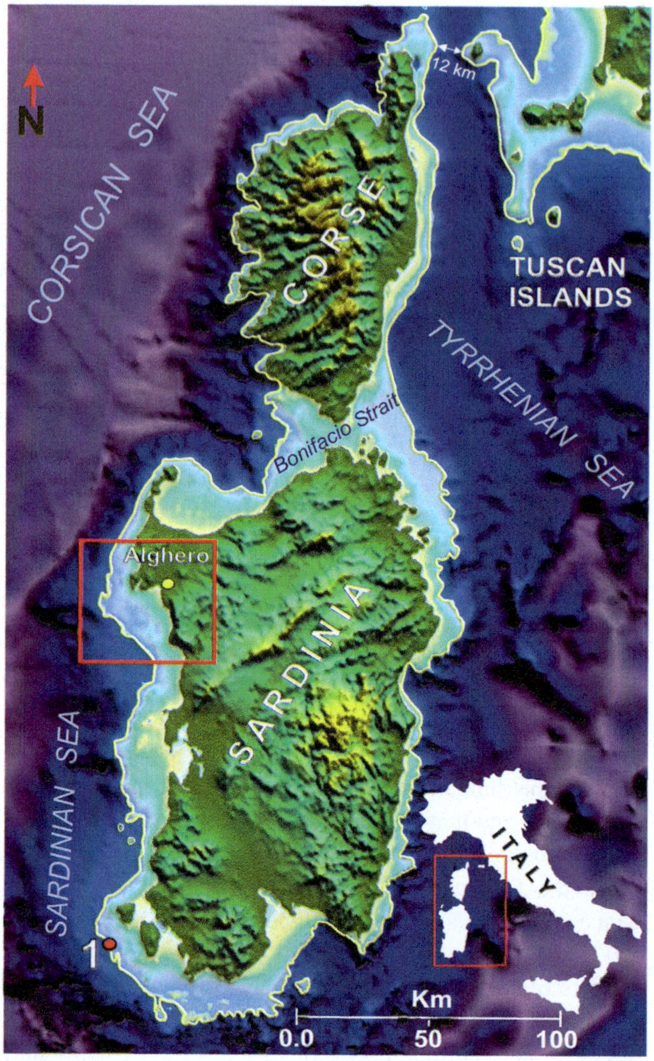

Fig. 16.6 Map showing the coastal palaeogeography of Sardinia and Corsica at the Last Glacial Maximum and the sea crossing to the Italian mainland. After Palombo et al. (2017, Fig. 9, p. 125). Reproduced with permission

16.3.3 Sicily

Along the northern coast of Sicily and the offshore Ègadi Islands to the west, there are onshore coastal cave occupations dating from about 16,000 cal BP onwards, with late Upper Palaeolithic (Final Epigravettian), Mesolithic, and Neolithic levels, notably at Grotta d' Oriente on the island of Favignana (mainly Upper Palaeolithic and Mesolithic at a time when the island was still attached to Sicily) and at Grotta dell'Uzzo on Sicily (mainly Mesolithic and Neolithic) (Tusa 1996, 1999). These sites show progressive isolation and abandonment of the islands with sea-level rise and variable emphasis on consumption of molluscs or other marine foods, their importance increasing through time as sea level rose close to the present level (Compagnoni 1991; Bonfiglio and Piperno 1996; Mannino et al. 2007, 2011, 2012). From the point of view of underwater finds, the main interest of this region is the possibility of finding submerged caves with underwater archaeological evidence. The cave of Ficarella just below Grotta dell'Uzzo is one such possibility and would have overlooked an inlet now at −26 m, which would have been available during the late Mesolithic and early Neolithic (Fig. 16.10).

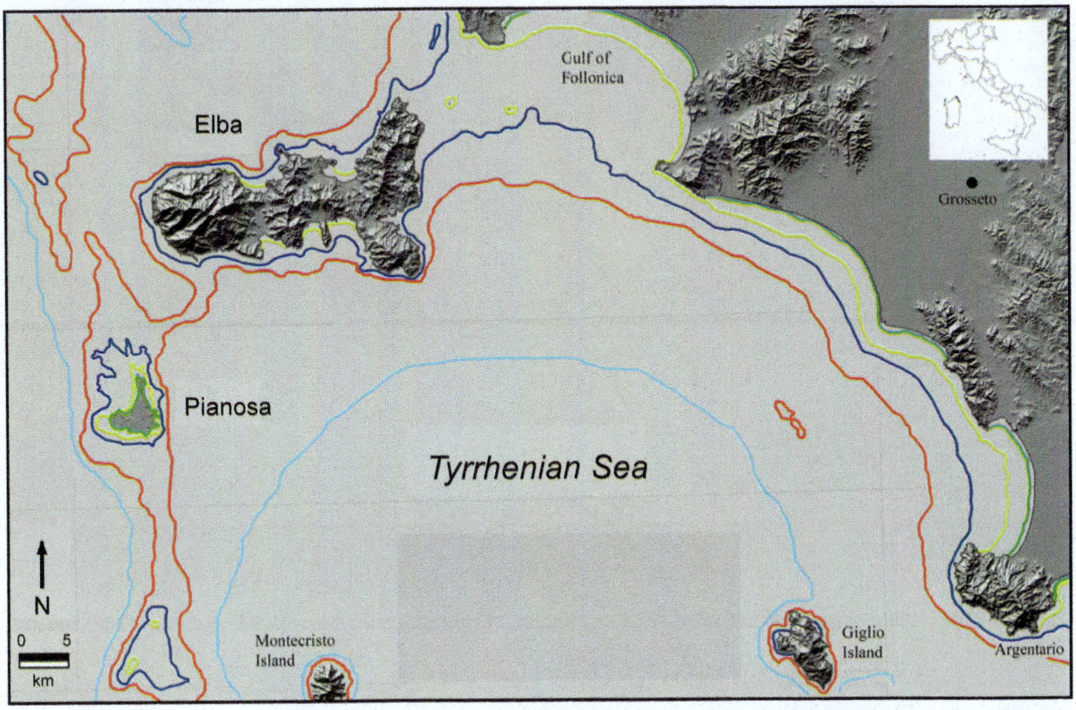

Fig. 16.7 The coastline of northern Lazio showing the relationship between the Italian mainland and the islands of Elba and Pianosa at different sea levels. Shoreline positions are based on bathymetry from GEBCO, adjusted for glacio-hydro-isostatic effects according to the model of Lambeck et al. (2011). Dark blue: Mesolithic coastline (8 ka cal BP). Red: Upper Palaeolithic coastline (11 ka cal BP). Light blue: LGM coastline (22 ka cal BP)

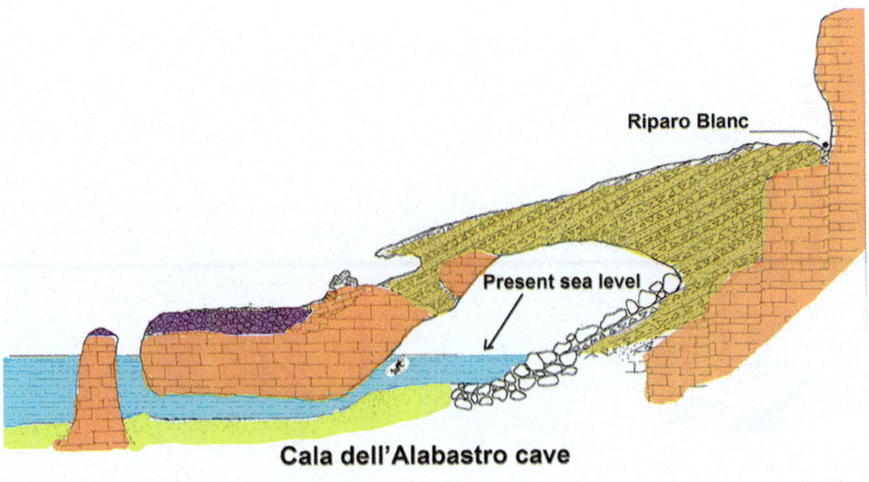

Fig. 16.8 Cross section of the shoreline on Monte Circeo showing the location of the Riparo Blanc rockshelter above present sea level, which has archaeological deposits, and the partially submerged cave of Cala dell'Alabastro. After Ferranti and Antonioli (1994)

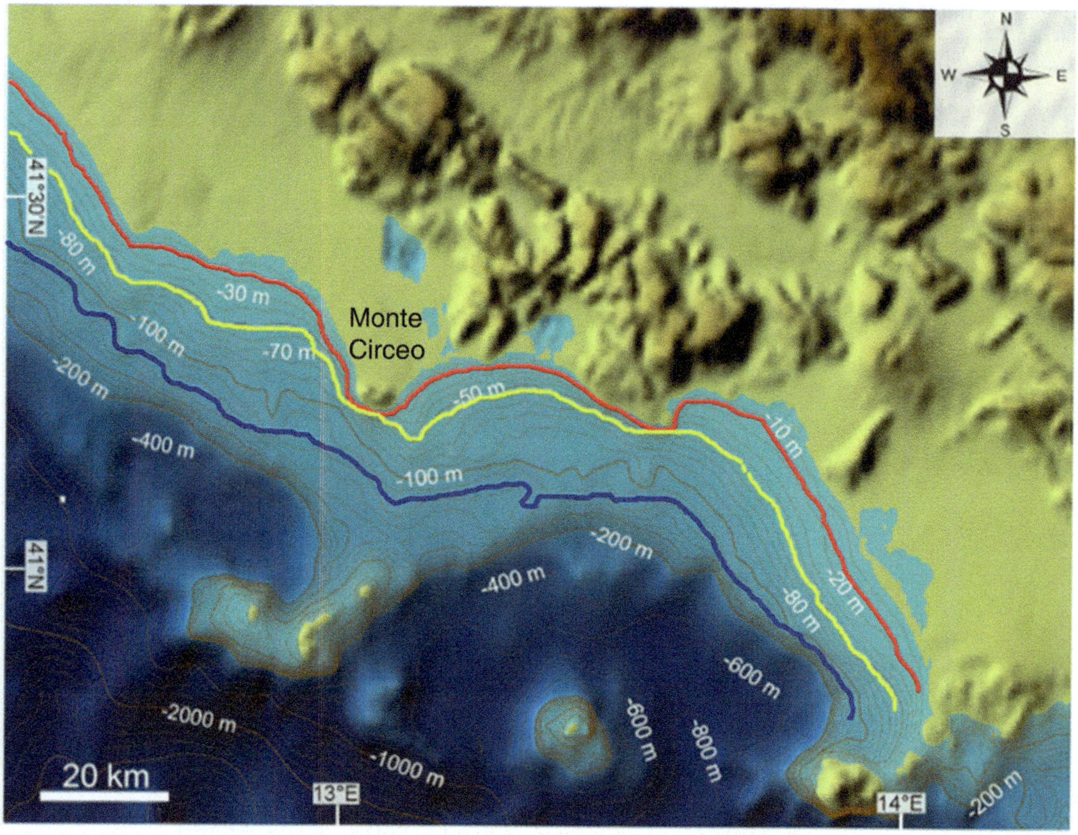

Fig. 16.9 The coastline of central Lazio showing palaeoshorelines for the LGM at 22 ka (dark blue line), the Mesolithic at 11 ka (yellow line), and the Neolithic at 8 ka (red line). Location of shorelines is based on GEBCO bathymetry adjusted for the Lambeck et al. (2011) model

Fig. 16.10 Grotta dell'Uzzo, marked by the red arrow on the right, looking south, with the position of the submerged cave of Ficarella shown by the red arrow on the left. See Antonioli (1997) for further detail. Photo by Fabrizio Antonioli

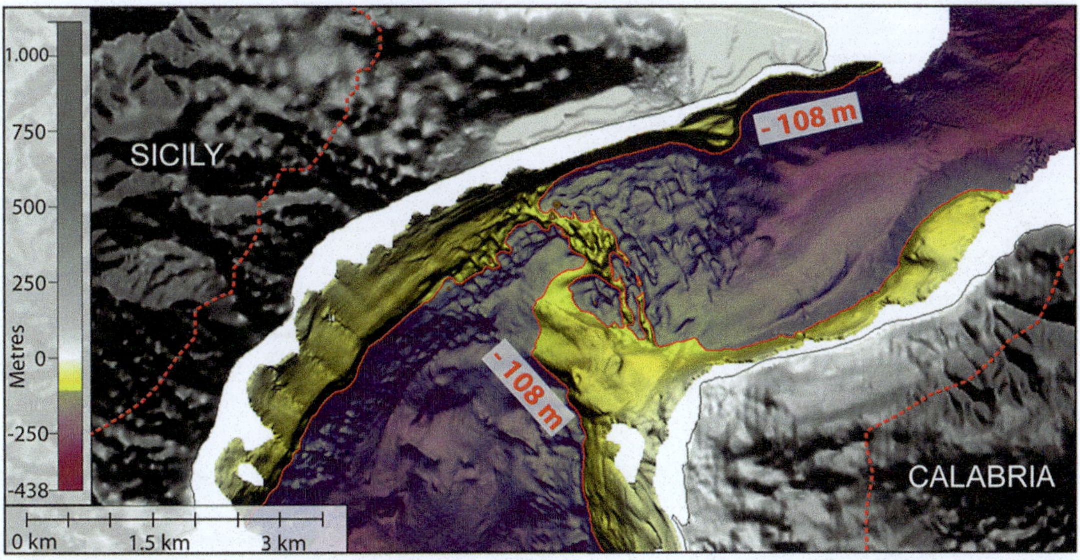

Fig. 16.11 Reconstruction of the palaeogeography of the Strait of Messina showing the land bridge that would have existed at the lowest sea-level stand of the LGM (at −108 m, shown by the continuous red line). The modern coastline is shown by a grey line, and the dotted red line indicates the coastline during the high sea-level stand of MIS 5.5. After Antonioli et al. (2016, Fig. 7)

The submerged sill in the Strait of Messina is located today at a minimum depth of 81 m below present sea level and represents the only connection between Sicily and mainland Italy (and thus Europe) during the last lowstand, when the sea level locally stood at about −126 m. Detailed analysis combining bathymetry, underwater geological investigations, modelling of eustatic sea-level change, and evidence of tectonic movements demonstrates the emergence of the sill as a narrow land bridge for a period of about 1500 years between 21.5 and 20 ka cal BP (Fig. 16.11; Antonioli et al. 2016). If allowance is made for the accumulation of a 20 m thickness of sediments on the seabed which were subsequently eroded away, and hence for a shallower sill than today, the time interval for a dry land connection between mainland Italy and Sicily would have been c. 27–17 ka cal BP. Despite the presence of an ancient Pleistocene fauna of dwarf elephants and hippo on Sicily, and claims for Lower and Middle Palaeolithic stone tools, radiocarbon dating of stratified palaeontological and archaeological data shows that the earliest reliable evidence for human presence only appears during and after this time interval of 27–17 ka, along with the introduction of a modern fauna including red deer (*Cervus elaphus*) and wild ass (*Equus hydruntinus*) (Antonioli et al. 2016). The shortest sea crossing today is only 4 km at present sea level and is notoriously dangerous for seafaring, memorably symbolised in the famous myth of Scylla and Charybdis. When sea level was lower, the crossing distance would have been even shorter, but palaeotidal modelling demonstrates turbulent currents of 17 knots, c. 20 miles per hour (Antonioli et al. 2016, pp 121–124), indicating even more hazardous sea conditions than today—presumably sufficient to deter any sea crossing in the Palaeolithic period.

16.3.4 Pantelleria

16.3.4.1 Cala Tramontana

Pantelleria is a small island lying 20 km southwest of Sicily on the shortest route between Tunisia and Sicily and is famous as a major source of obsidian exploited during the Neolithic and later from about 7000 years onwards. The submerged site of Cala Tramontana was discovered during underwater

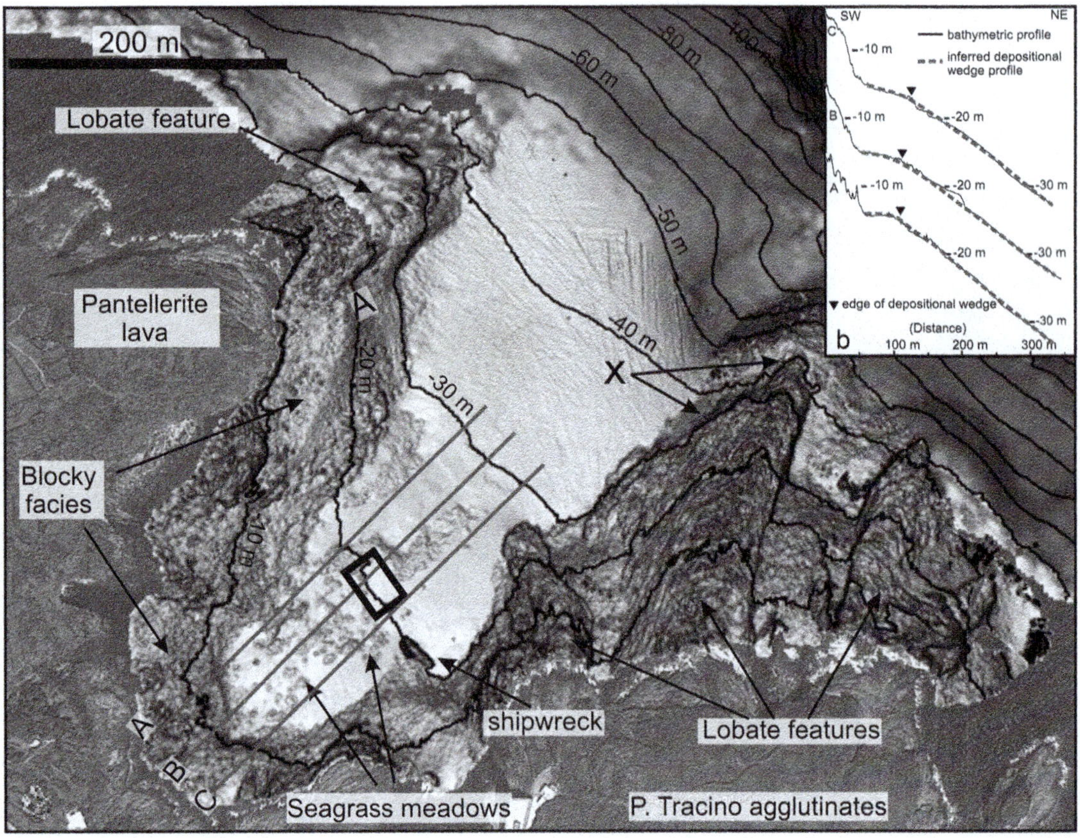

Fig. 16.12 Relief map of Cala Tramontana showing the bathymetry of the bay and the location of archaeological and geomorphological features. The black box marks the area of the prehistoric finds. After Abelli et al. (2016, Fig. 4). Reproduced with permission

investigation of a shipwreck. High-resolution bathymetry and scuba diving resulted in the discovery of a submerged assemblage of 770 artefacts at a depth of 18–21 m, mostly made on locally available flint but including some made on local obsidian (Fig. 16.12; Abelli et al. 2016). The assemblage is based on a simple core and flake technology consistent with expedient exploitation of local raw materials during short-lived visits to the island (Fig. 16.13). Reconstruction of the now-submerged palaeoshoreline shows that conditions were quite different from the present-day coastline, offering a shallow, protected bay suitable for access by boat, whereas the modern shoreline consists of steep cliffs. Dating is only possible by indirect means, using the eustatic sea-level curve, modified to take account of estimates of vertical tectonic movement related to volcanic activity, and gives a date between 9600 and 7700 years ago. This is the earliest date for human presence on the island and clear evidence for a pre-Neolithic sea crossing.

16.4 Bronze Age Finds

16.4.1 Sardinia

16.4.1.1 Nuraghe of Funtana Meiga and Marceddi Lagoon

Sardinia is well known for its many thousands of Bronze Age circular stone towers, or Nuraghe. Most are on dry land, but two examples are

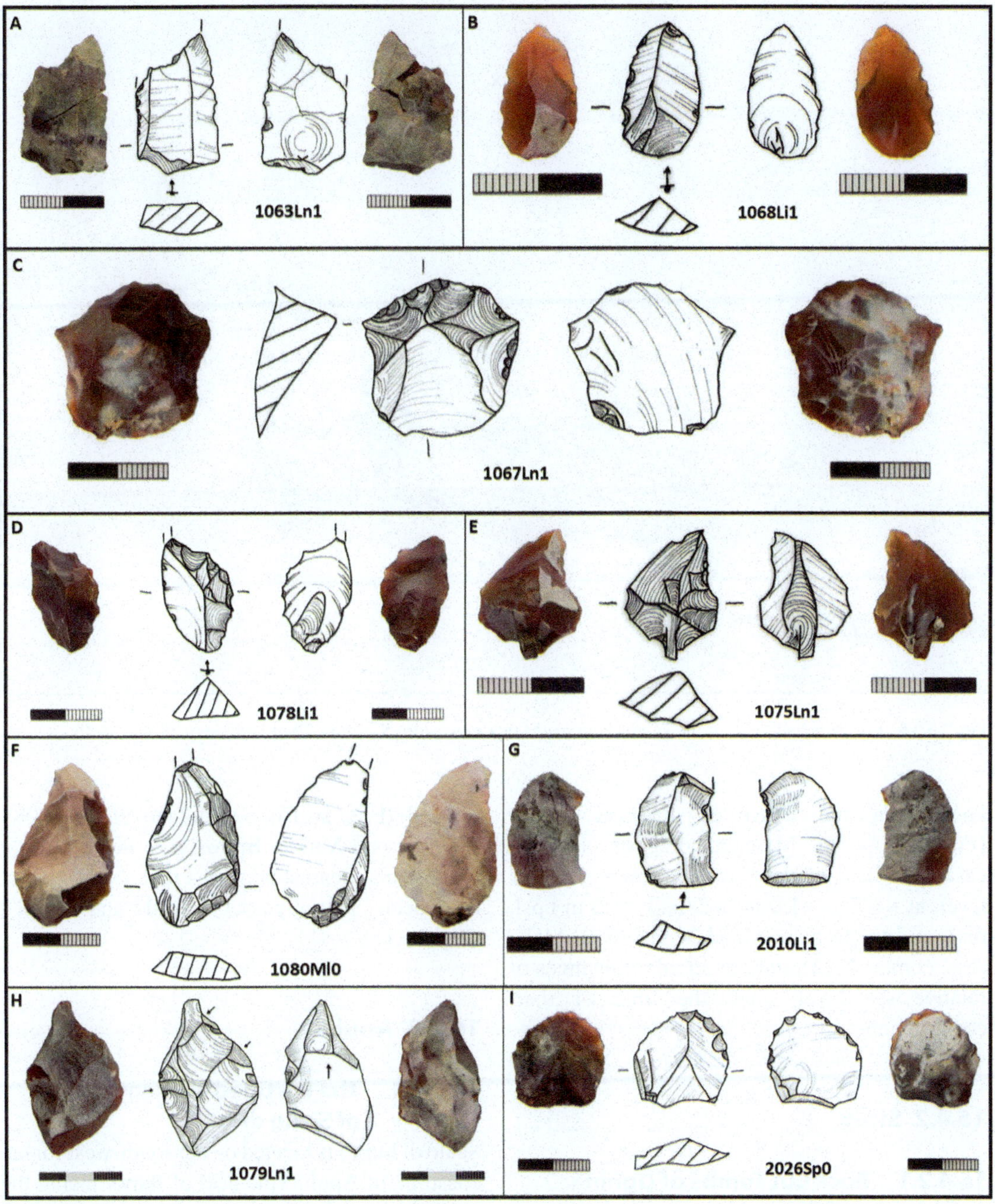

Fig. 16.13 A selection of prehistoric stone tools from the submerged site of Cala Tramontana. After Abelli et al. (2016, Fig. 5). Reproduced with permission

Fig. 16.14 Location and view of the site of Funtana Meiga in western Sardinia. Photo courtesy of Rita Melis

known on the coast at Funtana Meiga (Fig. 16.14) and Marceddi Lagoon. In both cases, the base of the tower is now below sea level, at −0.5 to −1.0 m bpsl and −1.0 m bpsl, respectively (Depalmas and Melis 2010; Melis pers. comm. 2013), indicating minor changes of relative sea level since the time of their construction.

16.4.2 Sicily

16.4.2.1 Rock Cut Tombs of Ognina and Thapsos

Both sites are in south-east Sicily and are partially submerged, demonstrating minor changes of relative sea level of 1–2 m over the past 3000 years and the effects of marine erosion. Ognina is a rock cut chamber tomb located on a former small peninsula connected to the mainland by a narrow rocky isthmus which is now submerged (Fig. 16.15; Scicchitano et al. 2008). Thapsos is a Middle Bronze Age necropolis on the Magnisi Peninsula, with a rock cut tomb that has been partially submerged and eroded (Fig. 16.16).

16.4.3 Apulia

16.4.3.1 The Fortified Settlement of Scalo di Furno

Scalo di Furno is located on the south-west Ionian coastline of Apulia (the heel of Italy), facing the Ionian Sea. It is a fortified settlement occupied for a period of more than 1000 years, beginning in the early Middle Bronze Age c. 1800 with local, handmade impasto pottery, some Late Helladic IIIA–IIIC sherds probably originating from Greece (or from the Aegean Region), and bronze objects (Lo Porto 1990). Part of the settlement has been submerged, and underwater

Fig. 16.15 The now-submerged entrance to the tomb of Ognina. After Scicchitano et al. (2016)

archaeological survey has revealed the remains of a submerged wall, c. 17 m long by 5 m wide and 1 m high, which represents part of the original fortified wall (Scarano et al. 2008; Scarano 2011; Alfonso et al. 2012). Also present at a depth of c. –3.5 m on the seabed is an area of 2000 m^2 with the remains of a flagstone floor and archaeological deposits containing hundreds of impasto sherds typical of the local Middle Bronze Age and many fragments of animal bone.

16.5 Conclusion

Investigation of Italy's submerged prehistoric archaeology is still at a comparatively early stage, and while there is a well-established tradition of underwater archaeological investigation, it has been largely focussed on the Bronze Age and Classical periods. At present, there are very few underwater finds from the Stone Age, and they are located mainly on the western coast of the Italian Peninsula and its offshore islands. Even here, they are widely scattered, while the Adriatic coastline is a complete blank (see Radić Rossi et al., Chap. 18, this volume). Nevertheless, there is considerable scope for future discoveries, and the finds that are known provide some important pointers for the future.

First, there is a very substantial foundation of geological and geophysical knowledge and expertise about relative sea-level change, the changing palaeogeography of the submerged landscape, and the configuration of palaeoshorelines. These already provide intriguing indicators about the isolation or connectivity of offshore islands resulting from late Pleistocene and early Holocene sea-level change and the possibility of sea crossings during the late Palaeolithic and Mesolithic periods.

These palaeogeographic reconstructions also provide a useful starting point for identifying target areas for more detailed underwater investigation. Underwater coastal caves are an obvious

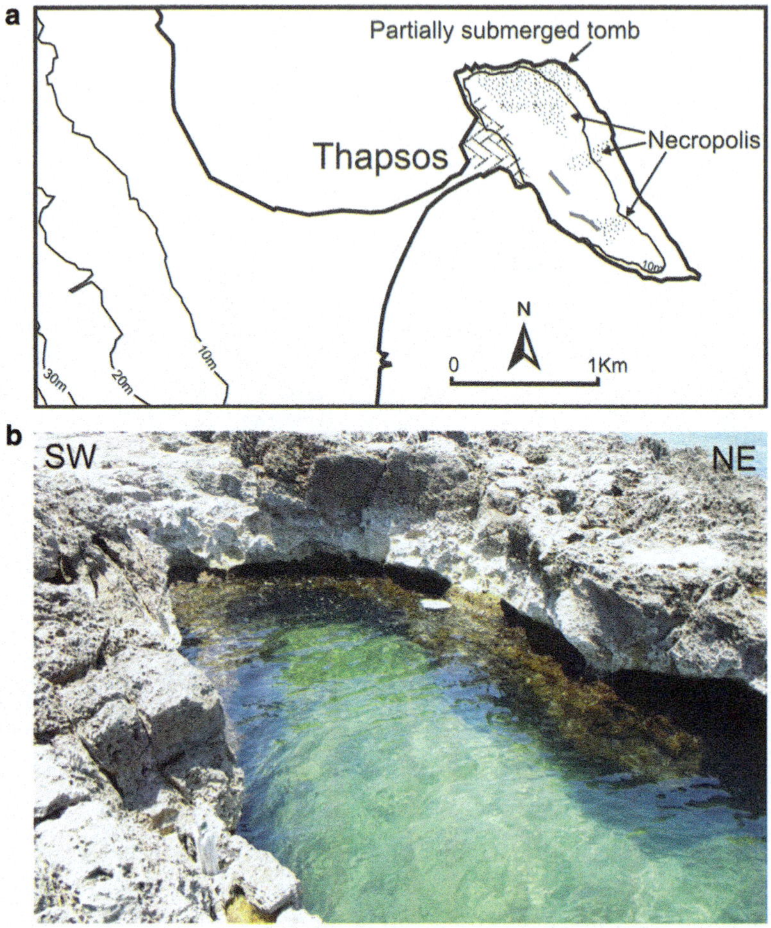

Fig. 16.16 The necropolis of Thapsos showing the location of the partially submerged tomb (**a**). Photo showing the remains of the main sub-circular chamber and a series of niches cut into the bedrock (**b**). After Scicchitano et al. (2008)

target for closer survey. There are, of course, risks that archaeological deposits will have been washed out of coastal caves by wave action during submergence, but the remains in the Sardinian caves of Grotta Verde and Grotta dei Cervi show that sufficient may be preserved to provide valuable information. The presence of onshore coastal caves with archaeological deposits provides another potential starting point in searching for underwater caves as in the example of the Ficarella cave at Grotta dell'Uzzo and the Cala dell'Alabastro at Riparo Blanc.

Another potential target is palaeoshorelines that provided natural harbours and protection for boats. The case of Cala Tramontana shows the importance of safe access for boats and the importance of local conditions in facilitating early visits to offshore islands. It also demonstrates how radically shoreline topography may change, even with relatively small changes in sea level, and that the modern coastal topography may be a very poor guide to conditions in earlier periods.

A second point is that there is considerable expertise in Italian underwater archaeology. Much of that expertise has been devoted to shipwrecks, underwater remains of Etruscan and Roman settlements or shoreline structures, and the search for archaeological or geological markers of submerged palaeoshorelines. However, the case of Cala Tramontana shows that surveys motivated by the investigation of shipwrecks may also reveal prehistoric Stone Age finds on the seabed nearby and that the search for prehistoric material can be efficiently combined with investigations of submerged historical material and sea-level indicators (see also Rovere et al. 2011b).

The Bronze Age finds are of interest in demonstrating how much damage can be caused by marine erosion even to such substantial structures as rock cut tombs, as in the example of Thapsos. But equally, they demonstrate what can survive once submerged, as in the example of the flagstones, potsherds, and animal bones of the Scala di Furno settlement. The same point can be made about the Cala Tramontana material. Both demonstrate that open-air archaeological deposits can survive the potentially damaging effects of marine inundation and be discovered on the seabed. There may be problems of mixing of material from different periods or cases where the underwater finds have been displaced from their primary position by marine currents and redeposited, an issue identified, for example, at Cala Tramontana (Abelli et al. 2016, pp. 107–108). But these problems can, in principle, be solved by appropriately designed geoarchaeological analysis.

Finally, we emphasise that there has been a strong maritime aspect to Italian prehistory for at least the past 100,000 years, with evidence that coastlines were of interest throughout this time span, initially as sources of marine food from at least the Mousterian period, later from at least pre-Neolithic times as starting points for some form of seafaring to reach offshore islands, and from the earliest Neolithic period onwards as a focus for coastal settlement, agricultural dispersal, and trade in obsidian and other goods. The dates for the earliest examples of these activities remain unclear, and the evidence currently available hints at the possibility that some of them may go further back in time than previously thought. Moreover, sea levels were lower than present throughout this time span, certainly well into the Neolithic period, and in some regions until the Bronze Age or later because of isostatic and tectonic movements. It follows that the palaeoshorelines where most of the archaeological evidence for early coastal activity is likely to be concentrated are now under water, a point that should provide a powerful motivation for developing future investigation of Italy's underwater prehistory.

16.6 Management of the Underwater Cultural Heritage

The central government agency responsible for underwater archaeology is the Italian Ministry for Cultural Heritage and Tourism (M.I.B.A.C.T., Ministero per i Beni e le Attività Culturali e del Turismo), while the investigation and management of underwater cultural heritage at the regional level is the responsibility of the local 'Soprintendenze', with the exception of the Autonomous Region of Sicily. The last, in fact, is the only region of Italy governed by the 'Superintendence of the Sea' responsible for the protection, evaluation, and management of the cultural, natural, and underwater heritage of Sicily and its adjacent islands and part of the Department for Cultural Heritage and Sicilian Identity (Regione Siciliana Assessorato Beni Culturali Identità Siciliana). Italy is a party to the UNESCO 2001 Convention on the Protection of the Underwater Cultural Heritage and, as a member of the European Union, is subject to EU regulations about attention to underwater cultural heritage in Environmental Impact Assessments conducted by offshore developers (see Chapters in Part IV of this volume), but there are not yet any examples of new information on submerged landscapes or underwater prehistory recovered in this way.

Acknowledgements We thank Nic Flemming and all the other SPLASHCOS members for their continued cooperation and support, and Sebastiano Tusa, Carmelo Monaco, Giovanni Scicchitano, Rita Melis, Fulvia Lo Schiavo, Margherita Mussi, Flavio Enei, Luisanna Usai, and Daniela Rovina for information about sites which are still under study. Author contributions: EFCB, FA, fieldwork, research, analysis, and writing; GB, research, analysis, and writing.

References

Abelli L, Agosto MV, Casalbore D, Romagnoli C, Bosman A, Antonioli F, Pierdomenico M, Sposato A, Chiocci FL (2016) Marine geological and archaeological evidence of a possible pre-Neolithic site in Pantelleria Island, Central Mediterranean Sea. In: Harff J, Bailey G, Lüth F (eds) Geology and archaeology: submerged

landscapes of the continental shelf. Geological Society, London, Special Publications 411, pp 97–110
Alfonso C, Auriemma R, Scarano T, Mastronuzzi G et al (2012) Ancient coastal landscape of the marine protected area of Porto Cesareo (Lecce, Italy): recent research. Int J Soc Underwater Technol 30(4):207–215
Antonioli F, Ferranti L, Lo Schiavo F (1996) The submerged Neolithic burials of the Grotta Verde at Capo Caccia (Sardinia, Italy) implication for the Holocene sea-level rise. Memorie Descrittive del Servizio Geologico Nazionale 52:329–336
Antonioli F (1997) Problematiche relative alle variazioni recenti del livello del mare e sue interazioni con le comunità preistoriche. In: Tusa S (ed) Prima Sicilia, alle origini della società siciliana. Ediprint, Palermo, pp 146–155
Antonioli F, Ferranti L, Lambeck K, Kershaw S, Verrubbi V, Dai Pra G (2006) Late Pleistocene to Holocene record of changing uplift rates in southern Calabria and eastern Sicily (Southern Italy, Central Mediterranean Sea). Tectonophysics 422:23–40
Antonioli F, Dorefice M, Ducci S, Firmati M, Foresi LM, Graciotti R, Perazzi P, Panatloni M, Principe C (2011) Palaeogeographic reconstruction of northern Tyrrhenian coast using archaeological and geomorphological markers at Pianosa island (Italy). Quater Int 232:31–44
Antonioli G, Lo Presti V, Morticelli MG, Bonfiglio L, Mannino MA, Palombo MR, Sannino G, Ferranti L, Furlani S, Lambeck K, Canese S, Catalano R, Chiocci FL, Mangano G, Scicchitano G, Tonielli R (2016) Timing of the emergence of the Europe–Sicily bridge (40–17 cal BP) and its implications for the spread of modern humans. In: Harff J, Bailey G, Lüth F (eds) Geology and archaeology: submerged landscapes of the continental shelf. Geological Society, London, Special Publications 411, pp 111–144
Antonioli F, Chiocci FL, Anzidei M, Capotondi L, Casalbore D, Magri D, Silenzi S (2017) The central Mediterranean. In: Flemming NC, Harff J, Moura D, Burgess A, Bailey GN (eds) Submerged landscapes of the European continental shelf: quaternary paleoenvironments. John Wiley & Sons, Chichester, pp 341–376
Atzeni E (1981) Aspetti e Sviluppi Culturali del Neolitico e della prima Età dei Metalli in Sardegna. In: Ichnussa. La Sardegna dalle origini all'età classica. Scheiwiller, Milano, pp 19–51
Auriemma R, Solinas E (2009) Archaeological sites as sea level change markers: a review. Quaternary. International 206:134–146
Benjamin J, Rovere A, Fontana A, Furlani S, Vacchi M, Inglis RH, Galili E, Antonioli F, Sivan D, Miko S, Mourtzas N, Felia I, Meredith-Williams M, Goodman-Tchernov B, Kolaiti E, Anzidei M, Gehrels R (2017) Late Quaternary sea-level changes and early human societies in the central and eastern Mediterranean Basin: an interdisciplinary review. Quaternary International 449:29–57
Blanc AC (1939) Il Monte Circeo:le sue grotte paleolitiche ed il suo uomo fossile. Bollettino Reale Società Geografica Italiana 17:485–493
Blanc AC (1958/61) Industria musteriana su calcare e su valve di *Meretrix chione* associatà con fossili di elefante e rinocerante in nuovi giacimenti costieri del Capo de Leuca. Quaternaria 5:308–313
Blanc AC, Segre AG (1953) Le Quaternaire du Monte Circeo. Livret-Guide, IVème Congrès INQUA. Roma, pp. 23-108 Excursion au Mont Circé. INQUA, IV° Congrès International. Roma-Pisa
Bonfiglio L, Piperno M (1996) Early faunal and human populations. In: Leighton R (ed) Early societies in Sicily. New developments in archaeological research. Accordia Specialist Studies on Italy, Volume, vol 5. University of London, Accordia Research Centre, pp 21–30
Castagnino Berlinghieri EF (2011) Subsistence, mechanisms of interaction and human mobility in the Neolithic western Mediterranean: the nature of the occupation of Lipari (Aeolian Islands, Sicily). In: Phoca Cosmetatou P (ed) The first Mediterranean islanders: Initial occupation and survival strategies. Oxford University School of Archaeology, Monograph Series, vol 73, pp 109–139
Colonese AC, Mannino MA, Bar-Yosef Mayer DE, Fa DA, Finlayson JC, Lubell D, Stiner MC (2011) Marine mollusc exploitation in Mediterranean prehistory: an overview. Quater Int 239:86–103
Compagnoni B (1991) La malacofauna del sito mesoneolitico della Grotta dell'Uzzo (Trapani). Rivista di Scienze Preistoriche 43:49–72
Dawson H (2013) Mediterranean voyages: the archaeology of island colonisation and abandonment. Routledge, London
Depalmas A, Melis RT (2010) The Nuragic people: their settlements, economic activities and use of the land, Sardinia, Italy. In: Martini IP, Chesworth W (eds) Landscapes and societies: selected cases. Springer, Dordrecht, pp 167–186
Ducci S, Perazzi P (1998) Una testimonianza del Neolitico a ceramica impressa cardiale nel'ArcipelagoToscano: l'isolotto della Scola presso l'Isola di Pianosa (LI). Atti XIII Congr. UISPP, Forli 1996, vol 3, sez 9, pp 425–430
Ducci S, Perazzi P (2000) I Neolitico antico dell'Arcipelago Toscano. In Tozzi C, Weiss MC (eds) Les premiers peuplements Holocènes de l'aire corso-toscane. Il primo popolamento olocenico dell'area corso-toscana. Progeto Interreg II, Corsica 1997–1999. Pisa, pp 53–56
Ferranti L, Antonioli F (1994) La grotta sommersa ed il Riparo Blanc a san Felice Circeo. Memorie della Società Speleologica Italiana 6:137–142
Ferranti L, Antonioli F, Mauz B, Amorosi A, Dai Prà G, Mastronuzzi G, Monaco C, Orrù P, Pappalardo M, Radtke U, Renda P, Romano P, Sansò P, Verrubbi V (2006) Markers of the last interglacial sea level high

stand along the coast of Italy: tectonic implications. Quater Int 145–146:30–54

Floris R, Melis RT, Mussi M, Palombo MR, Iacumin P, Usai A, Mascia F (2012) La presenza umana nella Sardegna centro occidentale durante l'Olocene antico: il sito di S'Omu e S'Orku (Arbus, VS). Atti della XLIV Riunione Scientifica. La preistoria e la protostoria della Sardegna. Cagliari, Barumini, Sassari 23-28 novembre 2009. Istituto Italiano di Preistoria e Protostoria, Firenze, pp 999–1004

Klein Hofmeijer G, lderliesten C, Van Der Borg K, Houston CM, De Jong AFM, Martini F, Sanges M, Sondaar PY, de Visser JA (1989) Dating of the upper pleistocene lithic industry of Sardinia. Radiocarbon 31(3):986–991

Lambeck K, Antonioli F, Purcell A, Silenzi S (2004) Sea-level change along the Italian coast for the past 10,000 yrs. Quater Sci Rev 23:1567–1598

Lambeck K, Antonioli F, Anzidei M, Ferranti L, Leoni G, Scicchitano G, Silenzi S (2011) Sea level change along the Italian coast during the Holocene and projections for the future. Quater Int 232:250–257

Lo Schiavo F (1986) *Ricerche subacquee* nella Grotta Verde di Alghero. Archeologia Subacquea 3. Supplemento al Bollettino d'Arte 37–38:63–67

Lo Porto FG (1990) Porto Cesareo. Insediamento protostorico di Scala di Furno. In: D'Amicis A (ed) Archeologia dei Messapi. Edipuglia, Bari, pp 221–232

Mussi M (2002) Earliest Italy: an overview of the Italian Paleolithic and Mesolithic. Kluwer Press, NewYork

Mannino M, Thomas KD, Leng MJ, Piperno M, Tusa S, Tagliacozzo A (2007) Marine resources in the Mesolithic and Neolithic at the Grotta dell'Uzzo (Sicily): evidence from isotope analyses of marine shells. Archaeometry 49:117–133

Mannino M, Di Salvo R, Schimmenti V, Di Patti C, Incarbona A, Sineo L, Richards MP (2011) Upper Palaeolithic hunter-gatherer subsistence in Mediterranean coastal environments: an isotopic study of the diets of the earliest directly-dated humans from Sicily. J Archaeol Sci 38:3094–3100

Mannino MA, Catalano G, Talamo S, Mannino G, Di Salvo R et al (2012) Origin and diet of the prehistoric hunter-gatherers on the Mediterranean island of Favignana (Ègadi Islands, Sicily). PLoS ONE 7(11):e49802. https://doi.org/10.1371/journal.pone.0049802

Palombo MR, Antonioli F, Lo Presti V, Mannino MA, Melis RT, Orru P, Stocchi P, Talamo S, Quarta G, Calcagnile L, Deiana G, Altamura S (2017) The late Pleistocene to Holocene palaeogeographic evolution of the Porto Conte area: clues for a better understanding of human colonization of Sardinia and faunal dynamics during the last 30 ka. Quater Int 439:117–140

Rovere A, Vacchi M, Firpo M, Carobene L (2011a) Underwater geomorphology of the rocky coastal tracts between Finale Ligure and Vado Ligure (western Liguria, NW Mediterranean Sea). Quater Int 232:187–200

Rovere A, Antonioli F, Enei F, Giorgi S (2011b) Relative sea level change at the archaeological site of Pyrgi (Santa Severa, Rome) during the last seven millennia. Quater Int 232:82–91

Scarano T, Auriemma R, Mastronuzzi G, Sansò P (2008) L'archeologia del paesaggio costiero e la ricostruzione delle trasformazioni ambientali: gli insediamenti di Torre Santa Sabina e Torre Guaceto (Carovigno, Br). Secondo Simposio Internazionale "Il Monitoraggio Costiero Mediterraneo: Problematiche e Tecniche di Misura", Napoli, 1–6 giugno 2008, CNR-IBIMET, Firenze, pp 391–402

Scarano T (2011) Refuge or dwelling place? The middle Bronze Age fortification walls of Roca (Lecce): spatial and functional analysis of Postern C. Rivista di Scienze Preistoriche 61:95–122

Scicchitano G, Antonioli F, Castagnino Berlinghieri EF, Dutton A, Monaco C (2008) Submerged archaeological sites along the Ionian coast of southeastern Sicily (Italy) and implications for the Holocene relative sea-level change. Quater Res 70:29–36

Scicchitano G, Castagnino Berlinghieri EF, Antonioli F, Spampinato CR, Monaco C (2016) Sacred landscapes and changing sea levels: New interdisciplinary data from the Early Neolithic to the present in south-eastern Sicily. In: Harff J, Bailey G, Lüth F (eds) Geology and archaeology: submerged landscapes of the continental shelf, Special Publications 411. Geological Society, London, pp 233–253

Stiner MC (1994) Honor among thieves: a zooarchaeological study of Neandertal ecology. Princeton University Press, Princeton, NJ

Stiner MC (1999) Palaeolithic mollusc exploitation at Riparo Mochi (Balzi Rossi), Italy: Food and ornaments from the Aurignacian through Epigravettian. Antiquity73:735–54

Tanda G (1980) Il Neolitico antico e medio della Grotta Verde, Alghero. In: Atti della XXII Riunione scientifica nella Sardegna centro-settentrionale, 21-27 Ottobre 1978, IIPP, Firenze, pp 45–94

Tanda G (1987a) Nouveaux éléments pour une définition culturelle des materiaux de la Grotta Verde (Alghero - Sassari, Sardaigne). In : 'Premières communautés paysannes en Méditerranée Occidentale. Actes du Colloque International du C.N.R.S. (Montpellier, 26-29 avril 1983)', Paris 1987, pp 425–431

Tanda G (1987b) Le culture preistoriche. In: Brigaglia M, Castellaccio AA, Contu E, Doneddu GS, GCR F, Lo Schiavo F, Manconi F, Mastino A, Meloni G, Tanda G, Tangheroni M, Turtas R (eds) La Provincia di Sassari: I secoli e la storia. Sassari, Amministrazione Provinciale di Sassari, vol 1987. stampa Amilcare Pizzi, Cinisello B, pp 10–22

Tusa S (1996) From hunter-gatherers to farmers in western Sicily. In: Leighton R (ed) Early societies in Sicily. New developments in archaeological research, Accordia Specialist Studies on Italy, vol 5. Accordia Research Centre, University of London, London, pp 41–56

Tusa S (1999) La Sicilia nella preistoria, 2nd edn. Sellerio editore, Palermo

Tykot RH (1996) Obsidian procurement and distribution in the Central and Western Mediterranean. J Medit Archaeol 9(1):39–82

Wilkens B (2012) Resti faunistici dei livelli neolitici della Grotta Verde di Capo Caccia (Alghero, SS) Atti 6 Convegno Nazionale di Archeozoologia (Orecchiella, 2009), pp 125–129

Malta: Submerged Landscapes and Early Navigation

17

Timmy Gambin

Abstract

The island of Malta is best known for its spectacular stone temples built by Neolithic farmers about 6000 years ago, who traded regularly with Sicily and other areas in the central Mediterranean. At a distance of over 100 km to the nearest continental land mass, it is generally assumed that Malta was uninhabited before the Neolithic period and was then occupied by competent Neolithic seafaring colonists. However, radiocarbon-dated pollen assemblages from fluvial sediments indicate an earlier phase of Maltese prehistory that preceded the period of the temple builders and included agriculture at a time when sea level was c. 6 m lower than the present day. These submerged coasts would have served as landing places for early navigators and offered fertile land nearby, but they have not yet been subject to systematic underwater archaeological survey. There is also the possibility of even earlier habitation at a time when Malta was connected to Sicily by dry land.

Keywords

Neolithic · Seafaring · Megalithic monuments · Submerged landscapes

17.1 Neolithic Sea Connections

The Maltese archipelago lies 96 km and 290 km from the closest points in Sicily and Northern Africa, respectively (Fig. 17.1). The two largest islands, Malta and Gozo, are presently inhabited. There are no mountains, and the highest point is 253 m above present sea level. Geologically, the islands are part of the European mainland, and at the time of the Last Glacial Maximum, some 20,000 years ago, a large land bridge connected them with south-western Sicily.

Currently, the earliest archaeological evidence for human presence in Malta comes from the cave site of Char Dalam in the south of the island, where deposits dated by radiocarbon to c. 7000 cal BP contain pottery with close decorative parallels to the Neolithic habitation at Stentinello in south-eastern Sicily (Trump 2002, p. 46). At Skorba, archaeological remains from the same period were described as a 'substantial wall […] that does not look like the work of pioneer colonists but of a well-established farming community' (Trump 2002, p. 23).

During the late Neolithic period, c. 6100–4500 cal BP, sometimes referred to as the Temple Period, the inhabitants of Malta developed a cul-

T. Gambin (✉)
Department of Classics and Archaeology, Archaeology Centre, University of Malta, Msida, Malta
e-mail: timmy.gambin@um.edu.mt

G. Bailey et al. (eds.), *The Archaeology of Europe's Drowned Landscapes*, Coastal Research Library 35, https://doi.org/10.1007/978-3-030-37367-2_17

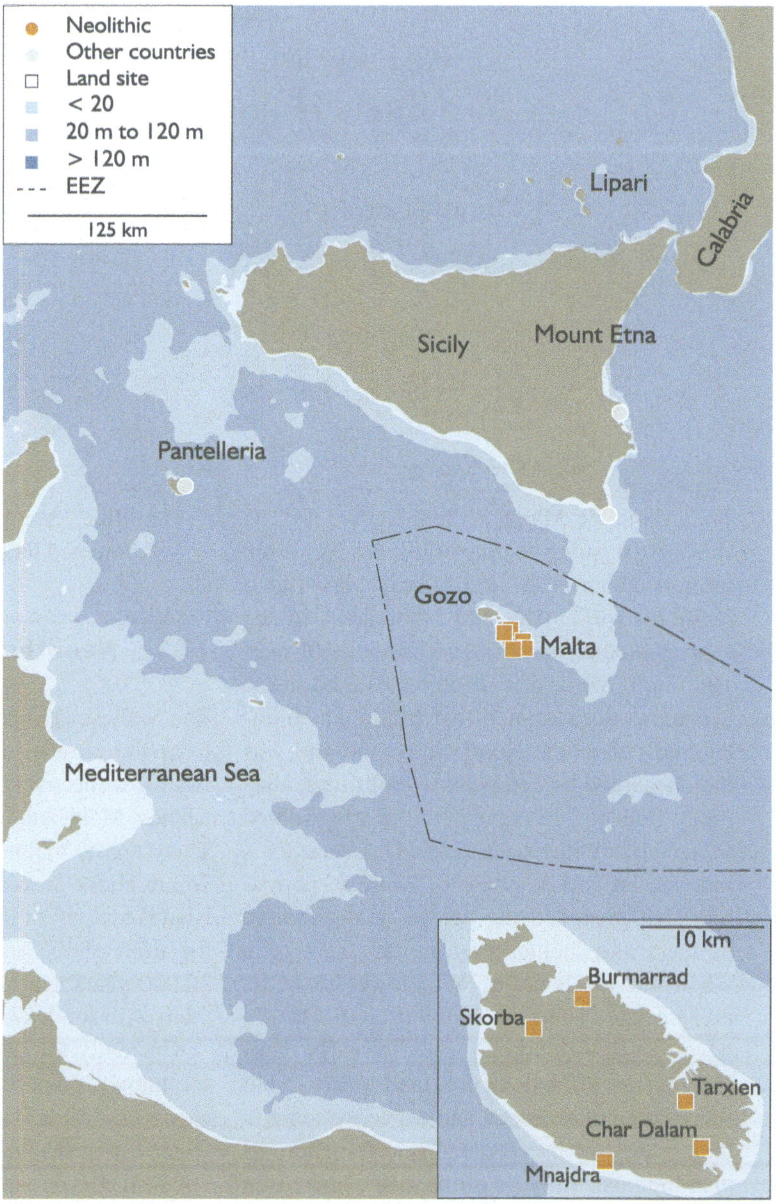

Fig. 17.1 Sites and geographical areas mentioned in the text. The 120 m depth contour is roughly equivalent to the coastline when sea level was at its lowest at the Last Glacial Maximum, while the 20 m contour represents a time somewhat prior to the currently known earliest traces of human activity on Malta. Drawing by Moritz Mennenga

ture that is best known for its spectacular 'temples' built of large worked and decorated stones (Evans 1971; Pace 1996; Ugolini 2012) (Fig. 17.2). At least 26 such megalithic monuments have been discovered, and UNESCO has declared the surviving structures as a World Heritage Site. Many of them are located in prominent positions that overlook natural harbours and have commanding views over the sea. Since the sea level of that period was approximately the same as today, this topographic situation also applies to the period when the monuments were erected and used (Fig. 17.3). Their distribution in the landscape points to the importance of open-sea navigation and to access points leading to and from the sea (Grima 2004).

There is ample evidence for Malta's connectivity with various other parts of the central Mediterranean during the Temple Period. This includes a variety of imported objects and raw

Fig. 17.2 The Mnajdra Temples built close to the cliff edge looking towards the sea. Photo by Alecastorina93, made available under Creative Commons License Attribution-ShareAlike 3.0 Unported (CC BY-SA 3.0). https://creativecommons.org/licenses/by-sa/3.0/deed.en

materials, for instance, obsidian from both Lipari and Pantelleria, at distances of 159 km and 204 km, respectively (see also Castagnino Berlinghieri et al., this volume). Imports of red ochre, flint and pumice are also represented in the archaeological record of this time. Likewise, miniature axes found in Malta are made from various rocks originating from Calabria in the southern Italian mainland and from Mount Etna on Sicily (Trump 2002, p. 38–41). The presence of a few 'exotic' ceramic sherds confirms trade in finished goods originating from Sicily (Trump 2002, p. 211).

17.2 Predecessors

The arrival of agriculture and the demonstrable maritime activities of the subsequent periods took place when sea level was as much as 6 m lower than present (Lambeck et al. 2011; Furlani et al. 2013, 2017). It is therefore possible that some of the earliest farming and navigation activities would have involved the use of coastal areas that are now submerged. The existence of inundated archaeological sites and therefore of significant research potential in the field of submerged prehistory around the Maltese islands can thus be assumed. In fact, pollen analyses of radiocarbon-dated samples from fluvial sediments have recently provided important evidence for agricultural activity in the form of traces of cereals prior to the Temple Period as early as 7200 cal BP (Marriner et al. 2012; Djamali et al. 2013).

To date, research on early Maltese prehistory has focused on the Late Neolithic monuments and other terrestrial sites such as burials (Pace 2004). Due to this emphasis, evidence of possible earlier periods of habitation, which could have been characterized by sites and items of a less spectacular nature, may have been overlooked.

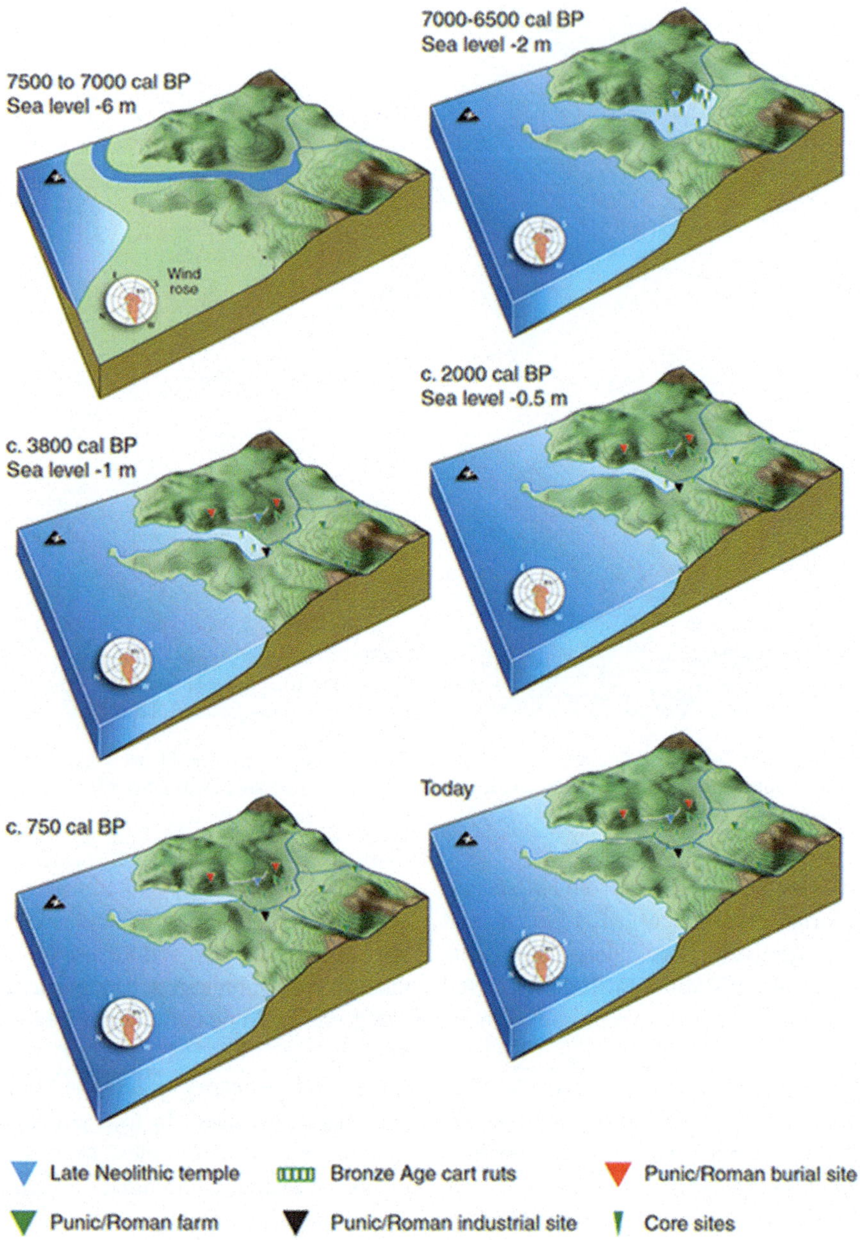

Fig. 17.3 Landscape changes due to sea-level rise and deposition of river sediments at Burmarrad. Traces of intensive land use (farming) have been found in up to 7200-year-old marine sediments below the present river plain. The information on sea-level change and the silting up of the low-lying areas is based on data from the geological core sites shown with green symbols. Drawing by Timmy Gambin with information from the Paleomed Project

Farmers make much stronger imprints in pollen diagrams than do fishermen and hunter-gatherers. This could be one reason for the current lack of traces of preagricultural inhabitants in the pollen diagrams from the islands. Should pre-Neolithic people have inhabited the area, however, there is little doubt that the most attractive foraging grounds would have been in coastal lowlands around large bays and estuaries that are today submerged under many metres of water. To date, very little research on this specific field has been done.

17.3 Potential Research Areas

Malta has a long tradition of underwater archaeology focused on more recent millennia, primarily remains of ancient shipwrecks as well as other remains datable to the historical period (Azzopardi and Gambin 2012). Systematic searches for submerged Stone Age sites have not yet been attempted. However, the reconstructed topography of the submerged coastal zone during this period shows plains and rivers that may have been conducive to human settlement (Fig. 17.3). Today, these areas are covered by thick marine sediments as well as by dense mattes of *Posidonia oceanica* (Fig. 17.4). This cover of *Posidonia* mattes probably has positive effects on the preservation of archaeological deposits. However, at the same time, it reduces the potential outcome of visual archaeological inspection. Given the presence of Upper Palaeolithic sites such as San Corrado, Grotta Giovanni and Canicattini Bagni in south-eastern Sicily (Bietti 1990), it would be reasonable to assume that similar remains may yet be found in submerged landscapes off the northern coasts of Malta and Gozo.

Fig. 17.4 *Posidonia oceanica* vegetation in different stages of development. The thick mass of vegetation to the right, known as a matte, is approximately 6 m in height, concealing any archaeological material that may be present on the seabed. Photo by Timmy Gambin

Other areas may be more profitable in the archaeological search for traces of preagricultural and Early Neolithic habitation – that is, the Stone Age estuarine floodplains. The inner margins of these are present in the modern landscape, but the lower reaches of these palaeoestuaries are located below sea level and covered by very deep marine sediments from later periods (Fig. 17.3). The anaerobic nature of sediments in such geological settings provides the ideal context for the survival of organic materials such as wood and raises the possibility of finding, amongst other artefacts, remains of prehistoric watercraft that may have been abandoned in the estuaries.

An ongoing large-scale offshore remote-sensing project sheds light on the extensive submerged landscape, which existed around the Maltese islands prior to the rise in sea level (Micallef et al. 2013). Although collected primarily for geological purposes, these data have identified ancient rivers, palaeoshorelines, sinkholes and submerged caves, all areas of high potential for future underwater archaeological research, and these provide a valuable basis for the preparation of underwater surveys.

17.4 Management of the Underwater Cultural Heritage

The Superintendence of Cultural Heritage is responsible for the protection, conservation and promotion of cultural heritage, and Heritage Malta for the management of cultural heritage sites. Research and training are undertaken by the University of Malta.

References

Azzopardi E, Gambin T (2012) Archaeology and the sea in the Maltese islands. Midsea Books, Malta

Bietti A (1990) The Late Upper Paleolithic in Italy: an overview. J World Prehist 4(1):95–155

Djamali M, Gambin B, Marriner N, Andrieu-Ponel V, Gambin T, Gandouin E, Lanfranco S, Médail F, Pavon D, Ponel P, Morhange C (2013) Vegetation dynamics during the early to mid-Holocene transition in NW Malta, human impact versus climatic forcing. Veg Hist Archeobotany 22(5):367–380

Evans JD (1971) The prehistoric antiquities of the Maltese islands, a survey. Athlone Press, London

Furlani S, Antonioli F, Biolchi S, Gambin T, Abela R, Anzidei M, Devoto S, Lo Presti V (2013) Late Holocene relative sea level changes in Malta. Quat Int 288:146–157

Furlani S, Antonioli F, Gambin T, Abela R, Ninfo A, Zavagno E, Micallef A, Cucchi F (2017) Marine notches in the Maltese islands (central Mediterranean Sea). Quat Int 439(Part A):158–168

Grima R (2004) The landscape context of megalithic architecture. In: Cilia D (ed) Malta before history. Miranda Publishers, Malta, pp 327–345

Lambeck K, Antonioli F, Anzidei M, Ferranti L, Leoni G, Scicchitano G, Silenzi S (2011) Sea level change along the Italian coast during the Holocene and projections for the future. Quat Int 232:250–257

Marriner N, Gambin T, Djamali M, Morhange C, Spiteri M (2012) Geoarchaeology of the Burmarrad ria and early Holocene human impacts in western Malta. Palaeogeography, Palaeogeogr Palaeoclimatol Palaeoecol 339–341:52–65

Micallef A, Foglini F, Le Bas T, Angeletti L, Maselli V, Pasuto A, Taviani M (2013) The submerged paleolandscape of the Maltese islands: morphology, evolution and relation to Quaternary environmental change. Mar Geol 335:129–147

Pace A (1996) Maltese prehistoric art 5000–2500 BC. Fondazzjoni Patrimonju Malti, Malta

Pace A (2004) The sites. In: Cilia D (ed) Malta before history. The world's oldest freestanding architecture. Miranda Books, Malta, pp 42–211

Trump DH (2002) Malta: prehistory and temples. Midsea Books, Malta

Ugolini LM (2012) Malta: origins of Mediterranean civilization – a re-edition with a foreword and an Introduction by Pessina, A. and Vella, N.C. with an English translation by Scerri L.J. Midsea Books, Malta

Croatia: Submerged Prehistoric Sites in a Karstic Landscape

18

Irena Radić Rossi, Ivor Karavanić, and Valerija Butorac

Abstract

Croatia has a long history of underwater archaeological research, especially of shipwrecks and the history of sea travel and trade in Classical Antiquity, but also including intermittent discoveries of submerged prehistoric archaeology. Most of the prehistoric finds have been discovered by chance because of construction work and development at the shore edge or during underwater investigations of shipwrecks. Eustatic sea-level changes would have exposed very extensive areas of now-submerged landscape, especially in the northern Adriatic, of great importance in the Palaeolithic and early Mesolithic periods. Because of sinking coastlines in more recent millennia, submerged palaeoshorelines and archaeological remains of settlement activity extend as late as the medieval period. In consequence, the chronological range of prehistoric underwater finds extends from the Mousterian period through to the Late Iron Age. Known sites currently number 33 in the SPLASHCOS Viewer with the greatest number belonging to the Neolithic or Bronze Age periods, but ongoing underwater surveys continue to add new sites to the list. Systematic research has intensified in the past decade and demonstrates the presence of in situ culture layers, excellent conditions of preservation including wooden remains in many cases, and the presence of artificial structures of stone and wood possibly built as protection against sea-level rise or as fish traps. Existing discoveries demonstrate the scope for new research and new discoveries and the integration of archaeological investigations with palaeoenvironmental and palaeoclimatic analyses of submerged sediments in lakes and on the seabed. A major challenge for the future is to develop better procedures for the integration of scientific research, commercial and industrial development, and the management and protection of the underwater heritage.

Keywords

Mousterian · Palaeolithic · Mesolithic · Neolithic · Bronze Age · Submerged landscapes · Underwater settlements · Wood

I. R. Rossi (✉)
Department of Archaeology, University of Zadar, Zadar, Croatia
e-mail: irradic@unizd.hr

I. Karavanić
Faculty of Humanities and Social Sciences, Department of Archaeology, University of Zagreb, Zagreb, Croatia

Department of Anthropology, University of Wyoming, Laramie, WY, USA
e-mail: ikaravan@ffzg.hr

V. Butorac
Faculty of Sciences, Department of Geography, University of Zagreb, Zagreb, Croatia
e-mail: vbutorac@geog.pmf.hr

G. Bailey et al. (eds.), *The Archaeology of Europe's Drowned Landscapes*, Coastal Research Library 35, https://doi.org/10.1007/978-3-030-37367-2_18

18.1 Introduction

Underwater archaeological research in Croatia, as in many other Mediterranean countries, focused for decades mainly on Classical Antiquity, thanks to abundant evidence relating to Greek and Roman seafaring (Vrsalović 2011). Submerged prehistoric sites were rarely reported in archaeological bibliographies (e.g., Brusić 1977; Mihovilić 1992, 1995a, b), and it is only during the past decade that increased attention has been paid to this theme, notably an overview of selected finds by Benjamin et al. (2011).

Nevertheless, isolated occurrences have been reported since the nineteenth century, such as the remains of a prehistoric fortified settlement on the island of Korčula in Southern Dalmatia (Vuletić Vukasović and Radić 1887) and a Neolithic axe recovered from the estuary of the Cetina River in Central Dalmatia (Bulić 1898). In the 1960s, temporary drying out of the Nin saltpans for repairs revealed remains of a submerged Early Neolithic settlement with a rich cultural layer at Čvrljevića Ograda (Batović 1965), while partially or fully submerged Late Iron Age burials with contracted skeletons in chambers made of stone slabs were discovered at the entrance to the Nin lagoon at Kraljicina plaža and in the shallow coastal zone of the Ždrijac Peninsula (Brusić 1969, 2002; Glogović 1989). A well-preserved example with stone slabs 1 m high and 'lying at the depth of over 1 m' contained skeletal remains and a bronze needle with a spirally twisted head dated to the eighth century BC Glogović 1989, p. 9).

In the 1970s, a trial survey in South Liburnia (Zadar and Šibenik area) revealed at least four submerged coastal sites with Palaeolithic, Bronze Age, and Iron Age material (Brusić 1977). All the sites—Ričul west of Turanj, Oštarije near Biograd, Sv. Justina in front of Pakoštane, and Stipanac in Lake Prokljan—are on small islands and are partly submerged, with evidence of structures (whether natural or artificial is not clear) that once connected them to the mainland (see Fig. 18.1). This evidence pointed to the likelihood that systematic surveys on both sides of the Pašman Channel would lead to new discoveries both on land (Batović 1988, 1990a) and underwater (Radić Rossi 2011).

During the past decade, several factors have promoted new interest in the detection and interpretation of submerged prehistoric sites. These include more detailed investigation of sea-level change and submerged karstic features (Benac et al. 2004, 2008; Surić et al. 2004, 2005, 2010; Surić 2009; Surić and Juračić 2010; Faivre et al. 2010; Furlani et al. 2011; Marriner et al. 2014; Radić Rossi and Cukrov 2017; Lončar et al. 2017, 2018; Faivre and Butorac 2018; Radić Rossi et al. 2018), leading to increased awareness of the prehistoric archaeological potential of the Croatian seabed (Benjamin and Črešnar 2009; Benjamin et al. 2011) and of the threats to the coastal and underwater heritage posed by construction works and environmental change (Radić Rossi 2008).

Typically, in the past, underwater sites have been located through deepening and reconstruction of harbours and saltpans (Batović 1965; Mihovilić 1992; Radić Rossi 2008), recreational diving (Brusić 2004; Radić Rossi 2009), and pedestrian survey of beaches and intertidal zones (Mihovilić 1995a, b). More recently, new methods have been introduced including the use of satellite imagery to assess larger areas, systematic survey by SCUBA (Koncani Uhač 2009; Bekić et al. 2011; Pešić 2013; Brusić and Parica 2014) or new technologies (Radić Rossi and Boetto 2013; Soura et al. 2013), and the more systematic incorporation of geological, geomorphological, and palaeolandscape research (Marriner et al. 2014; Miko et al. 2016; Ilijanić et al. 2018; Radić Rossi et al. 2018). Excavations have begun in some promising shallow-water areas (Koncani Uhač and Čuka 2016; Čelhar et al. 2017).

This chapter summarises the current state of research, drawing on published sources and the personal experience of the authors. We outline the evidence for changes in palaeogeography and sea level and describe selected evidence from sites that best illustrate the chronological range and the variety and preservation of underwater material, proceeding in chronological order from earliest to latest.

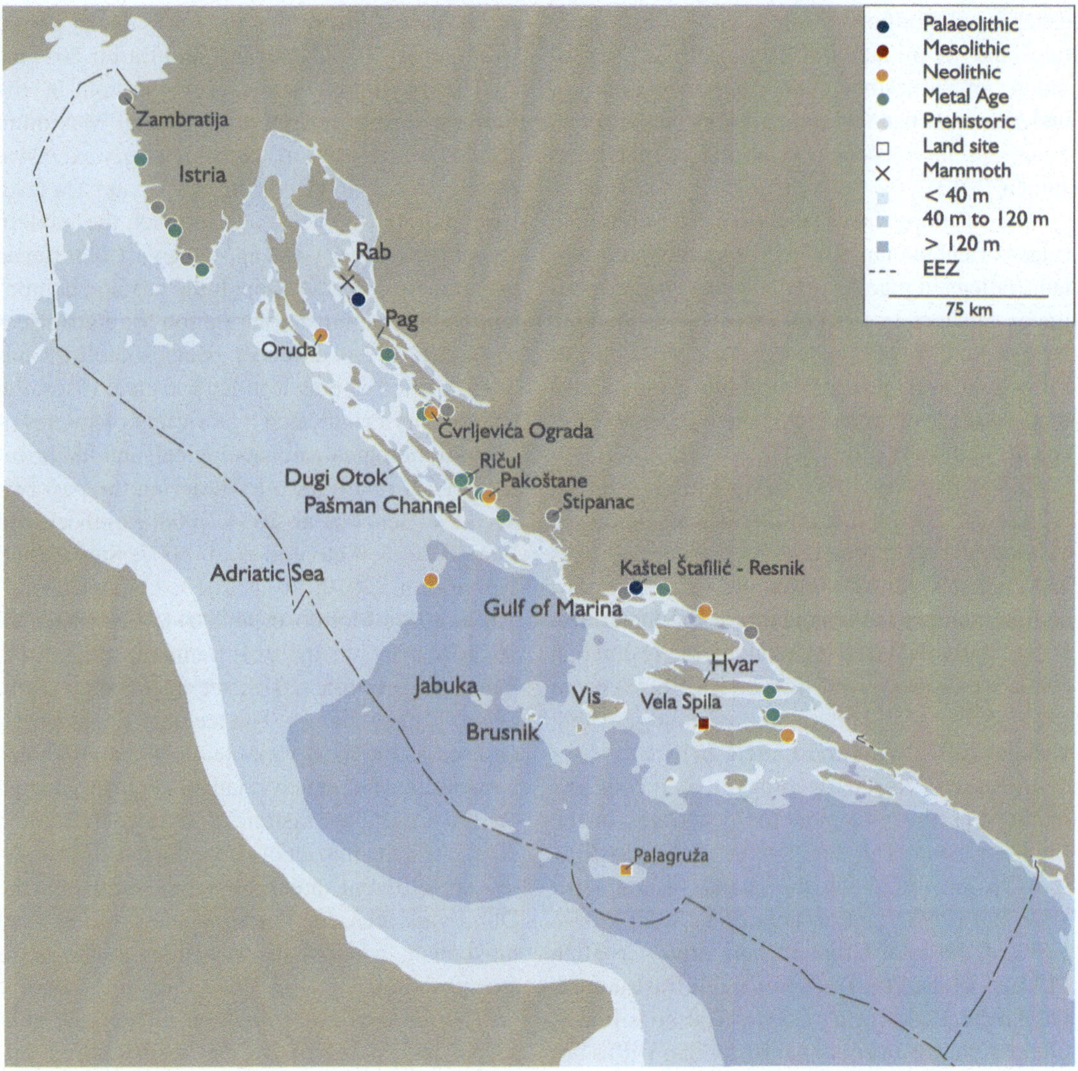

Fig. 18.1 Distribution of underwater sites showing names of key sites and other places mentioned in the text and the boundary of the Exclusive Economic Zone. Site information from the SPLASHCOS Viewer http://splashcos-viewer.eu. Drawing by Moritz Mennenga

18.2 Palaeogeography and Sea-Level Change

The Adriatic Sea is the northernmost basin of the Mediterranean, and the east Adriatic coast, including part of northern Italy, Slovenia, Croatia, Bosnia-Herzegovina, Montenegro, and Albania, is dominated by the maritime sector of the Dinaric Alps. This mountain range originated 240 million years ago (Ma) with shallow-water sedimentation on the Adria microplate, which continued throughout the Mesozoic era (240–65 Ma), resulting in an 8-km-thick succession of dolomites and limestones—the Adriatic Carbonate Platform. Uplift of this platform began towards the end of the Mesozoic with subduction of the Adria microplate beneath the Eurasian Plate and uplift of the Dinaric Alps with major faulting oriented on a NW–SE axis. The main episodes of mountain building continued into the Miocene period (c. 23–5 Ma) accompanied by exceptional faulting, folding, and thrust structures and the initiation of karst processes (dissolution of limestone and the formation of underground channels and caves) (Vlahović et al. 2002, 2005; Mocochain et al. 2009; Pikelj and Juračić 2013).

The resulting landscape is one of mountain ridges oriented on a NW–SE axis with intermontane valleys in parallel, and this topography continues offshore to form an extensive submerged landscape that was exposed during periods of lower sea level, just leaving the hill tops above modern sea level as an archipelago of some 1200 islands and islets (Fig. 18.1).

This pattern of offshore anticlines (ridges), represented by the present-day islands, and synclines (valleys), represented by present-day bays and channels, aligned in parallel with the coast and the main strike orientation of the Dinarides, is especially well developed along the Dalmatian coast, which has given its name to this type of coast in the world literature (Fairbridge 1968; Kelletat 2005; Pikelj and Juračić 2013).

Because of its karst geology, Croatia has one of the most heavily indented coastlines in the Mediterranean. The total length of the coastline is 6176 km, of which the islands account for 4398 km (71%) and an area of c. 3259 km^2—5.8% of the total Croatian land area (Croatian Ministry of the Sea, Transport and Infrastructure; Duplančić Leder et al. 2004). The coastline is predominantly an erosional rocky one with scattered pocket beaches and accumulation of sediment only near river mouths (Bognar et al. 2012). In the north of the Adriatic, the rivers flow through flysch geology susceptible to erosion and have carried large quantities of sediment into the sea. Along the Croatian coast, the rivers carry relatively little sediment, and fresh water also flows underground and emerges as submarine springs, which are abundant along the coast.

The largest palaeokarst depressions lying between present-day islands at depths of −40 m to −90 m contain sedimentological and geomorphological archives of climate and relative sea-level changes and would have favoured the development of shallow freshwater lakes when sea level was low during the Last Glacial period (Miko et al. 2016; Karavanić and Barbir 2017).

The maximum sea-level regression in the northern Adriatic at the Last Glacial Maximum (LGM) was c. −112 m, exposing a very extensive area of coastal lowland (Fig. 18.1), and sea level rose after 18,000 years ago to reach the modern level after c. 6000 years ago (Šegota 1983; Benac and Juračić 1998). Although the eustatic component is the dominant contribution to relative sea-level rise, the position of palaeoshorelines has been affected by both tectonic and glacio-isostatic movements as indicated by a variety of studies of geomorphological, archaeological, and historical markers of relative sea level during the past two decades (Benac et al. 2004, 2008; Lambeck and Purcell 2005; Antonioli et al. 2007; Surić 2009; Faivre et al. 2010, 2011, 2013; Lambeck et al. 2010, 2011; Florido et al. 2011; Furlani et al. 2011, Marriner et al. 2014; Benjamin et al. 2017; Faivre and Butorac 2018; see also Bailey et al., Chap. 1, this volume). The tectonic signal is recognised as the most variable and uncertain component contributing to relative sea-level change on the Croatian coastline (Faivre and Butorac 2018), and the impact of neotectonic activity varies in different areas. In the south, in the Dubrovnik area, the coastline shows no vertical movement, whereas the coastlines in the north are subsiding.

The topography of the now-submerged landscape, with its extensive and well-watered valleys partially enclosed by hill barriers, would have created an attractive environment for hunter-gatherer populations and large mammals during the Last Glacial. As sea level rose, these valleys would have been transformed into brackish lagoons and marshes and ultimately marine basins, with considerable potential for the exploitation of aquatic resources. Moreover, as this landscape became slowly inundated by sea-level rise, archaeological deposits in this hill-valley system would have been relatively well protected by hill ridges and islands from the destructive impact of waves, which can reach a maximum height of 10 m during storms at the present day. This factor, together with relatively low rates of marine sedimentation sufficient to provide a pro-

Fig. 18.2 Remains of *Mammuthus meridionalis adriacus* n. ssp. (a) Diaphysis of left femur, anterior view. (b) Upper third molar, buccal view. The finds were dredged from a depth of 80 m on the seabed between the islands of Rab and Laganj. Palaeonotlogical analysis indicates an early Middle Pleistocene age. Scale in cm. Photos courtesy of the Archives of the Institute for Quaternary Paleontology and Geology, Croatian Academy of Sciences and Arts

tective cover but not so thick as to deeply bury archaeological deposits beyond easy reach, suggests good prospects for the preservation and discovery of underwater archaeological sites. In 1979, a fishing trawler dredged a mammoth molar tooth and femur *(Mammuthus meridionalis adriacus* n. spp) from a shallow depression on the seabed with a fill of quaternary terrestrial sediments, c. 8 km south-west of the island of Pag at a depth of −80 m (Fig. 18.2; Malez and Lenardić-Fabić 1988; Mauch Lenardić 2012). Dated on palaeontological grounds to the early Middle Pleistocene (≤ 750 ka), this find is testament to the preservation potential of the Adriatic seabed.

18.3 Archaeological Context

The archaeological sequence in Croatia extends back to the Lower Palaeolithic period, represented by the site of Šandalja I (Malez 1979; Table 18.1). Open-air finds on the island of Dugi Otok have also been attributed to the Lower Palaeolithic (Batović 1988), but from their similarities with well-provenanced material elsewhere, it is more likely that these belong to the Middle Palaeolithic (Karavanić 2000, 2004; Krile and Vujević 2017). Middle Palaeolithic sites are more common and are best known from the cave sites of Krapina and Vindija in the north-west, famous for their Neanderthal fossils. In recent decades, new cave and open-air sites have been found in the hinterland of the northern Dalmatian coast and open-air sites on some of the neighbouring islands (Vujević 2009; see also Perhoč 2009). Systematic excavations at the cave sites of Mujina pećina and Velika pećina in Kličevica have produced stratified lithic and faunal assemblages dated at 45–39 ka cal. BP, a period that spans the late Neanderthals and the earliest known anatomically modern humans (AMH) in Europe (Karavanić et al. 2008). Open-air sites occur near sources of chert in the hinterland and on the neighbouring islands. They indicate visits to the islands, most likely when connected to the mainland at low sea level and suggest the exploitation of the now-submerged landscape.

Early Upper Palaeolithic sites are rare in the eastern Adriatic (Karavanić 2009; Mihailović 2009), and there appears to be a chronological gap between them and the preceding Middle Palaeolithic. The reasons for this are unclear but may reflect concentrations of settlement on the now-submerged landscape at a time when it was expanding to its maximum extent at the LGM and loss of sites inundated by subsequent sea-level rise, a genuine temporal hiatus between the last Neanderthals and the first anatomically modern humans—perhaps related to the devastation caused by the Campanian Ignimbrite volcanic eruption, low population densities in the region, or simply a lack of research (Karavanić 2009; Papagianni 2009; Papagianni and Morse 2013; Mihailović and Whallon 2017).

Late Upper Palaeolithic (Epigravettian) sites appear in larger numbers after the LGM and are

Table 18.1 Chronological chart showing principal periods and cultures for Croatia

cal BC (approximately)	Period	Culture or Important Sites	
		Istria	Dalmatia
400–0	Late Iron Age	Histrian Culture Phase VI	Liburnian Culture: Phases V-VI Delmatian Culture: Phases IV-V Greek Culture
1,100–400	Early Iron Age	Histrian Culture: Phases I-V	Liburnian Culture: Phases I-IV Delmatian Culture: Phases I-III
1,300–1,100	Late Bronze Age	Urnfield Culture	
1,900–1,300	Middle Bronze Age	Castellieri Culture	Dinara 2
2,300–1,900	Early Bronze Age	Ljubljana-Adriatic	Cetina/Dinara 1
3,000–2,300	Late Eneolithic		Ljubljana-Adriatic
4,000–3,000	Middle Eneolithic Early Eneolithic	Brijuni Culture	Nakovana Culture
4,800–4,000	Late Neolithic	Hvar Culture	Hvar Culture
5,500–4,800	Middle Neolithic	Danilo Culture	Danilo Culture
6,000–5,500	Early Neolithic	Impresso Culture	Vela Luka Culture Impresso Culture
10,000–6,000	Mesolithic	Pupićina peć Šandalja II	Vela Spila, Kopačina Vlakno
30,000–10,000	Upper Palaeolithic (Epigravettian, Gravettian Aurignacian)	Šandalja II	Vlakno
200,000–30,000	Middle Palaeolithic (Mousterian)		Mujina pećina
>200,000	Lower Palaeolithic	Šandalja I	

present on the Adriatic coast and its hinterland and offshore islands, with evidence for movement of raw materials between northern Italy, Istria, and inland Croatia (Malez 1979; Cancellieri 2011; Vukosavljević et al. 2011, 2015; Karavanić et al. 2013, 2015). It is likely that there was extensive activity on a landscape that is now submerged at depths of 60–110 m below present sea level and that many sites present there have been submerged by subsequent sea-level rise (Šegota 1982; Miracle 1995; Whallon 2007). Vlakno Cave on the Island of Dugi Otok has deposits dated to 14.6–14.1 ka cal BP and would have looked out onto this submerged landscape. The presence in the archaeological deposits of marine shells of *Cyclope rustica*, *Columbella rustica*,

Glycymeris, and *Dentalium*, collected for decorative purposes, is evidence of visits to a coastline that would have been more distant than today (Benjamin and Črešnar 2009; Vujević and Parica 2009; Vukosavljević and Karavanić 2015).

The Mesolithic is well represented by numerous sites in the Adriatic region, especially in Istria but also in Dalmatia (Komšo 2008). Marine shells found at many inland Mesolithic sites in northern Italy (e.g., Bus de la Vecia, Pradestel, Riparo Gaban, Romagnano, Vatre di Bambana) indicate contact with an Adriatic coastline that would still have been lower than present (Álvarez Fernández 2003; Komšo 2008; Komšo and Vukosavljević 2011), and by implication the use of the now-submerged landscape and its palaeoshorelines. At the cave site of Vela Spila on Korčula Island, which would have been a peninsula during the Mesolithic period, the Mesolithic layers, dated at c. 9 ka cal BP, contain artefacts made from gabbrodiorite rock sourced to the islands of Vis, Jabuka, and Brusnik, providing evidence of navigation in the open sea over distances of at least 20 km (Radić 2011). Large quantities of fish bones were found in the Mesolithic layers, principally of mackerel (Scombridae) and eel (Congridae and Muraenidae), also dolphin bones (Radić 2011; Rainsford et al. 2014), providing additional evidence for the importance of marine resources.

The transition from Mesolithic to Neolithic at c. 8 ka cal BP involved mainly immigrants from further east bringing new crops and domestic animals. The extent to which the pre-existing hunter-gatherer population was involved in these new developments remains unclear since there is a chronological gap of at least several hundred years between the Mesolithic and the Early Neolithic (Forenbaher and Kaiser 2005; Forenbaher and Miracle 2006; Forenbaher et al. 2013; Moore 2014). What is clear from the distribution of Early Neolithic *impresso* ware is that the pattern of dispersal followed the coastline and that the immigrants and their crops and animals must have travelled by sea, involving significant seafaring abilities including crossings of the Adriatic between Italy and Croatia. This is especially clear from the presence of Early Neolithic pottery and tools made on Italian (Apulian) chert on the island of Palagruža, midway between the coastlines of Italy and Croatia, which must have involved sea crossings of at least 48 km (Korbar et al. 2009; Forenbaher and Kaiser 2005; Forenbaher 2011, 2018a).

Alongside the establishment of farming in areas with suitable soils, this maritime emphasis persisted and intensified in the Middle and Late Neolithic (Danilo and Hvar cultures), respectively (Table 18.1; Težak Gregl 1998), with the use of caves as well as open-air sites in the Danilo culture, the importation of Middle Neolithic ware from southern Italy, and the circulation of obsidian sourced from the islands of Lipari in the Tyrrhenian Sea and Melos in the Aegean (Radić 2011), with obsidian from both sources present on the Palagruža Archipelago (Tykot in Forenbaher 2018a; see also Castagnino Berlinghieri et al., Chap. 16 and Galanidou et al., Chap. 19, this volume).

Given this maritime emphasis, and the fact that palaeoshorelines of this period are now submerged at depths as much as 10 m below present, it must follow that important archaeological evidence, and especially remains of settlements on the shore associated with sea travel, must be sought on the seabed. Similar comments apply to the subsequent periods of the prehistoric sequence, the Late Eneolithic and the Bronze Age, when relative sea level continued to rise, submerging palaeoshorelines of this period at depths of 2–3 m. This was a period that, along with the immigration of new people and developments in social organisation, witnessed an intensification of maritime activities in trade and the use of offshore islands (Majnarić Pandžić 1998; Forenbaher 2018a, b).

Throughout the archaeological sequence described above, submerged landscapes and palaeoshorelines would clearly have been significant in a variety of roles, as areas of fertile territory for hunting and gathering, especially during periods of extreme low sea level in the northern Adriatic, as locations for the exploitation of marine resources, as pathways of communication and colonisation, as submerged territory that may have harboured settlements now lost that would otherwise explain gaps in the archaeological sequence on land, and in the later periods as bases for sea travel and trade. The evidence for these activities must now lie, for the most part, below present sea level.

18.4 Underwater Sites

There are 33 underwater sites currently registered in the SPLASHCOS Viewer (Table 18.2, Fig. 18.1). They range in age and site type from unstratified Middle Palaeolithic stone tools to Iron Age burials. The majority, 17 sites (52%), represent in situ cultural layers, with collections of unstratified material in second place at 11 sites (33%). All sites in these two categories lie in a depth range of 1.5–5.5 m below present sea level, and the majority are Neolithic or Bronze Age in date. One of these, shown as prehistoric in Table 18.1, has both Palaeolithic stone tools and Bronze Age material and is also anomalous in the sense that it is located offshore of the small island of Stipanac in Lake Prokljan, which is a brackish water lake 12 km inland from the coast but connected to the sea via the Krka River.

18.4.1 Palaeolithic

18.4.1.1 Kaštel Štafilić–Resnik

This site is in the Gulf of Kaštela and was found by a local diver c. 300 m offshore at a depth of 4 m and in the immediate vicinity of a well-known Hellenistic and Roman settlement on one side and a presumed site from the Neolithic period on the other (Brusić 2004; Radić Rossi 2009; Kamenjarin and Šuta 2011). Research-led investigation, including underwater excavation, recovered additional artefacts, in total about 100 pieces, of which half are tools. These are of Middle Palaeolithic type including typical Mousterian centripetal cores and side scrapers, with the addition of some possible Upper Palaeolithic tools (Fig. 18.3; Karavanić et al. 2009, 2015; Karavanić 2015; Janković et al. 2011; Barbir 2015). Most of the artefacts are on the seabed surface and have been disturbed by marine currents but probably not moved very far from their original position. The material is similar to the finds from the Mousterian cave site of Mujina pećina, which is c. 8 km inland from Kaštel Štafilić–Resnik. Mujina pećina was used as a seasonal site in autumn and spring, and the two sites could have been used by the same people at different seasons (Miracle 2005; Karavanić et al. 2008; Karavanić et al. 2014a, b). About 150 m west of the Mousterian site, at Stara plaža, Early Neolithic potsherd and numerous wooden stakes have been found in water depths of 0.5–3 m, but a single radiocarbon date on the wood indicating a date in the Classical period puts a question mark over the status of this material and indicates the need for further investigation.

Other Palaeolithic finds are a single Mousterian tool offshore of Povljana on the island of Pag (Batović 1990b; Vujević et al. 2017) and stone tools of possible Mousterian type at a depth of 3 m at Stipanac in Lake Prokljan (Brusić 1977; Malez 1979; Vujević et al. 2017; Karavanić and Barbir 2017) close to a small island which is a rich source of chert raw material (D. Vujević and M. Parica, personal communication 2018).

Table 18.2 Distribution of Croatian underwater finds by site type and broad chronological period

Site type	Pal	Neo	Neo-Bronze	Bronze	Bronze-Iron	Iron	Prehist	Total	
								N	%
Culture layer in situ	–	5	4	7	–	1	–	**17**	***52***
Unstratified collection	2	1	3	2	1	1	1	**11**	***33***
Single find	1	1	–	1	–	–	–	**3**	**9**
Burial	–	–	–	–	–	2	–	**2**	**6**
Total N	3	7	7	10	1	4	1	**33**	***100***
%	***9***	***21***	***21***	***30***	***3***	***12***	***3***		***99***

Data from the SPLASHCOS Viewer at http://splashcos-viewer.eu

Fig. 18.3 Middle Palaeolithic (Mousterian) lithic artefacts from Kaštel Štafilić–Resnik collected from the seabed at a depth of 3–4.5 m. Left: Core. Right: Flakes with natural edge damage. Scale in cm. Photo by Ivor Karavanić

18.4.2 Neolithic and Eneolithic

Underwater finds of Neolithic or Eneolithic age are relatively abundant, and many refer to remains of settlements judging by the presence of pottery, animal bones, and wooden artefacts that may refer to dwelling structures or other wooden structures of unknown purpose.

18.4.2.1 Pakoštane

This site lies c. 100 m offshore between the modern coast (Janice) and the small island of St. Justina (Fig. 18.4) and was discovered during the investigation of a Roman shipwreck (Radić Rossi and Antonioli 2008; Bekić et al. 2011, 2016; Pešić 2012; Radić Rossi et al. 2018; Vujević and Meštrov 2018). The prehistoric finds are at a depth of 4–4.5 m below present sea level. Stone artefacts were scattered over an area of 2500 m^2. A test excavation was carried out in a 2 × 4 m trench, revealing cultural finds in a 1-m-thick layer closely associated with the remains of a wooden structure. The total assemblage including surface and excavated finds comprises over 2000 potsherds of coarse pottery, stone artefacts, animal bones, and worked wood (Fig.18.5). The stone artefacts total 156 pieces of which 42 are worked tools, mostly edge-retouched blades and bladelets. The wood is radiocarbon-dated at 6.6 ka cal BP indicating a Late Neolithic date for the assemblage. On the seaward side of the site is an extensive stone structure which represents a natural beachrock formation indicating a palaeo-shoreline dated at 5.4–5.2 ka cal BP (Eneolithic). The number and variety of artefacts and the occurrence of technical pieces demonstrating in situ manufacture of the flint tools all indicate the presence of a settlement, and the geological context demonstrates that it was located on dry land at the time of occupation.

18.4.2.2 Čvrljevića Ograda

This site was discovered in the inner Nin lagoon during maintenance of the Nin saltpans in 1955, and further investigation found cultural materials over an area of 40 m × 100 m. Rescue excavations identified a 20- to 50-cm-thick cultural layer with potsherds of Early Neolithic *impresso* ware (Fig. 18.6), grindstones made of sandstone, 36 chert tools, marine shells, and bones of domes-

Fig. 18.4 Aerial view of Pakoštane showing the location of the archaeological site (circled in red) and the Pašman Channel in the background. Photo courtesy of Tourist Board Pakoštane

Fig. 18.5 Diver at Pakoštane holding a piece of wood from the top layer of the site. Photo by Fabrizio Antonioli

tic animals (cattle, sheep, and goat) (Batović 1965). The site appears to have been located on the edge of a marine lagoon that subsequently dried out. At the time of occupation, it was several metres above sea level according to Batović (1965) but was later inundated by sea-level rise and is now below present sea level. Relatively little is known in detail about this site, but it is important as one of the few submerged sites of Early Neolithic date and a site that first helped to identify the presence of an Early Neolithic in Croatia and evidence that shorelines of this period had been submerged by subsequent sea-level rise. Two other sites, the Školjić Rock in front of Neviđane on the island of Pašman and Donje Blato–Mlini on the island of Korčula (Vuletić Vukasović and Radić 1887), have recently been attributed to the Neolithic period (Mate Parica personal communication 2018) and await more detailed investigation.

18.4.2.3 Zambratija

Well-preserved remains of a settlement dated to the Early Eneolithic at c. 6 ka cal BP were discovered in the bay of Zambratija in 2008 during the investigation of a sewn plank boat of Late Bronze Age/Iron Age date (c. 3 ka cal BP) and other remains of Roman and Late Medieval data (Koncani Uhač 2009; Benjamin et al. 2011; Koncani Uhač and Čuka 2016; Koncani Uhač et al. 2017).

The site lies in the north-western part of the bay, at a depth of 2.4–3.1 m in a natural depression separated from the outer sea by a natural low limestone ridge. Multibeam survey suggests that the present-day bay was originally subdivided into three smaller bays at the time of prehistoric occupation. The prehistoric settlement remains extend over an area of about 6500 m^2. So far over 100 wooden stakes have been discovered, many of them vertically driven into deposits of peat and clay (Fig. 18.7). They are concentrated in the north-western corner of the depression and have been interpreted as piles to stabilise dwelling structures located on marshy ground close to the original shoreline. Koncani Uhač and Čuka (2016) report the presence in the sediments of aquatic plant remains such as alkali bulrush (*Scirpus maritimus* L.), hairy sedge (*Carex* cf. *hirta L.*), pondweed (*Potamogeton* sp.), and water caltrop (*Trapa natans* L.), and more detailed analysis of the palaeolandscape is ongoing.[1] Other material recovered from the cultural layers comprises pottery, including 300 diagnostic sherds of the Early Eneolithic Nakovana culture, several round spindle whorls, and animal bones.

18.4.2.4 Other Finds

Three other Istrian sites have provided underwater evidence, including pottery and faunal remains, attributed to the Eneolithic period. One is in the shallow bay of Veruda south of Pula where dredging of the harbour recovered two polished axe heads made of volcanic rocks, chipped stone artefacts made of local dark chert, and a shaft-hole axe (Fig. 18.8; Mihovilić 1992, 1995a, b; Benjamin et al. 2011). At the second site on the shore of Cape Gale in Peroj, potsherds, some 500 stone artefacts made on chert, other stone tools, and faunal remains have been found in the intertidal zone to a depth of c. 0.5 m. However, their status as evidence of a submerged settlement is unclear since they could have been eroded from a settlement located on the present-day shoreline. Finally, five pieces of chert, of which one is a probable tool, at a depth of 1.5–2.5 m, indicate a possible underwater site, maybe from the same period, in the bay of Veštar (Benjamin et al. 2011, p. 195).

Elsewhere, Eneolithic material found underwater is limited to chance finds of limited material or uncertain provenance. These include pottery fragments found beneath a c. 200-year-old wooden ship's hull between the islands of Oruda and Palacol at a depth of 5 m (Mihajlović 2011;

[1] The palaeolandscape study is being carried out as a PhD topic by Katarina Jerbić at Flinders University, Australia.

Fig. 18.6 Early Neolithic *impresso* potsherds from Čvrljevića Ograda. Scale in cm. After Batović (1965)

Benjamin et al. 2011) and long chert blades and core fragments disturbed by ship propellers at a depth of 5–6 m at Baška Voda near Makarska (Zubčić 2004; Benjamin and Črešnar 2009; Benjamin et al. 2011). Perhaps the most intriguing find is a submerged structure at Ljubačka Vala 2.6 m below present sea level east of the Gulf of Nin (Parica and Ilkić 2017). There are two embankments made of branches and rocks, the longer extending for 100 m, and this has been interpreted as a fish trap. Radiocarbon dating indicates a Late Eneolithic/Early Bronze Age date.

One other find to mention is a complete globular ceramic vessel of Late Eneolithic/Early Bronze Age date dredged up from a depth of 30 m in the Gulf of Marina (Fig. 18.9; Radić Rossi 2011). The precise location was not recorded, but the pot must presumably represent an object lost overboard during sea travel or perhaps an item lost along with the rest of the cargo in a shipwreck or possibly a votive offering.

18.4.3 The Bronze Age

Submerged evidence notably increases with the Bronze and Iron Ages with a number of apparently well-preserved coastal sites close to the present-day shoreline that have been submerged by later sea-level rise.

Fig. 18.7 View of the Eneolithic settlement area at Zambratija showing oak piles used in the construction of wooden structures. Scale in 20 cm units. After Koncani Uhač and Čuka 2016. Photo courtesy of the authors

18.4.3.1 The Pašman Channel: Ričul and Oštarije

Ričul is located off the north-east shore of a small island in the Pašman Channel (Fig. 18.10). Brusić (1977) first identified a humus layer at a depth of 2–3 m, with abundant sherds of prehistoric pottery, marine shells and animal bones, and an artificial causeway at a depth of 1.5–3 m made of cobbles and massive timbers to provide a connection between the island and the mainland.

New investigations began in 2012 followed by excavations (Čelhar et al. 2017). The new research confirmed the existence of a 125-m-long linear structure connecting the island to the mainland and also identified a lateral embankment built of wooden stakes driven into the original gravel beach and consolidated with cobbles, stone slabs, and potsherds (Fig. 18.11). The top of the embankment now lies at −2.2 m and the seabed at −3–4 m, and the excavators interpreted it as an artificial structure built to protect the settlement from rising sea level. The settlement covers an area of 1 ha, but a large part of it has been destroyed by gravel extraction. Pottery (Fig. 18.12) and radiocarbon dates indicate a Middle Bronze Age date, c. 3.5–3.3 ka cal BP.

Oštarije (Kumentić) is 5 km to the south-east and is near a small island c. 200 m offshore from the mainland near Crvena Luka. The intervening channel is no deeper than 2.5 m and formed a dry connecting causeway during the Bronze Age. The settlement comprises structures with drywall foundations or carved into the rock at depths of 1.5–2 m below sea level, and the pottery is similar to the finds from Ričul (Brusić 1977).

Fig. 18.8 Artefacts from Pula–Veruda. (Upper) Stone shaft-hole axe. (Lower) Selection of blades made on local chert. Scale in cm. Courtesy of the Archaeological Museum of Istria, Pula

Fig. 18.9 Globular vessel recovered from a depth of 30 m in the Gulf of Marina. The vessel is made of fine ware with incised decoration on the cylindrical neck and the upper part of the body. This is in the style of Ljubljana-Adriatic pottery of the Late Eneolithic/Early Bronze Age (Table 18.1; Forenbaher 2018b). Photo by Damil Kalogjera

Fig. 18.10 Aerial view of the Pašman Channel looking north-west, with the position of the underwater Bronze Age Ričul site shown by a red arrow on the right and the underwater Iron Age site of Garmenjak near the island of Pašman on the left. Photo by Ervin Šilić

Fig. 18.11 Concentration of wooden poles at the site of Ričul, interpreted as part of an artificial embankment. Photo by Mate Parica

Judging by the number of sites on both sides of the Pašman Channel including Bronze Age settlements, Iron Age hillforts, and the submerged sites of Pakoštane Ričul and Oštarije, the region was clearly an important focus of settlement throughout later prehistory and increasingly an important seafaring route and a centre for controlling access, movement, and exchange of goods between central and northern Dalmatia. The combination of further work on the submerged sites together with studies of the nearby Vrana Lake (Ilijanić et al. 2018) also offer promising prospects for new palaeoenvironmental investigations.

18.4.3.2 Other Sites

Elsewhere, mention should be made of the Vranjic Peninsula in the Gulf of Kaštela, where rescue excavations salvaged remains of a Bronze Age settlement and a stratigraphic profile showing changes in the coastal environment over a 3000-year period (Radić Rossi 2008; (Arena et al. forthcoming). In the Bay of Šimuni on the island of Pag, remains of a submerged coastal settlement at a depth of 2–3 m have recently been discovered with rows of wooden stakes and numerous potsherds (Bekić 2017). Submerged Bronze Age material has also been found in the anchialine cave of Živa Voda on the island of Hvar (Figs. 18.13, 18.14; Radić Rossi and Cukrov 2017). As on land, so underwater, there is evidence of continuity from the Bronze Age to the Iron Age because of a continuing rise in relative sea level. The most distinctive submerged Iron Age finds so far discovered are the burials of the Liburnian period at Nin and on the island of Murter (Brusić 1969, 2002, 2005).

Fig. 18.12 Selection of Bronze Age pottery at Ričul showing characteristic features. Photos by Mate Parica; drawings by Dario Vujević

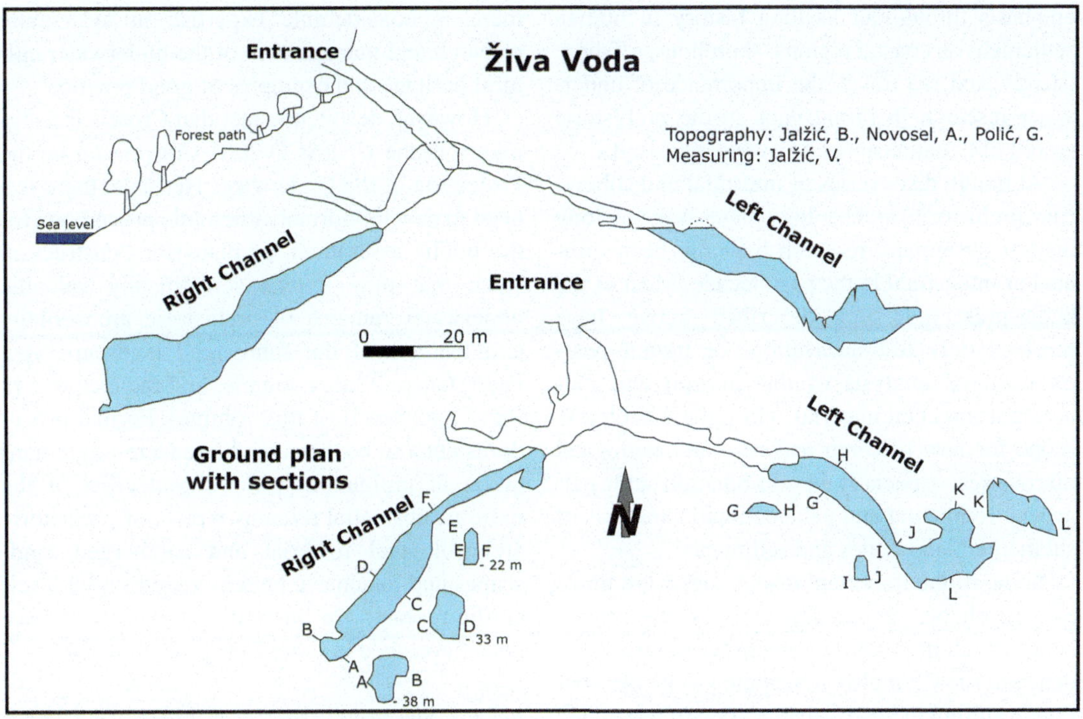

Fig. 18.13 Plan and cross section of the cave of Živa Voda cave on the island of Hvar. This is an anchialine cave, that is, a cave with a submarine connection to the sea and a stratified water column with fresh water at the surface. These are important sources of fresh water. Potsherds of Bronze Age or later periods have been recovered by speleologists from the left channel of this cave system, but it has not been archaeologically researched (Radić Rossi and Cukrov 2017). Drawing by Branko Jalžić, Anđelko Novosel, Gordan Polić, and Vedran Jalžić

Fig. 18.14 Bronze Age ceramic bowl recovered from the left channel of Živa Voda cave on the island of Hvar. The diameter of the bowl is c. 20 cm. Photo by Irena Radić Rossi

18.5 Conclusion: Future Challenges and Management of the Underwater Heritage

The above account highlights the chronological range, variety, and potential significance of submerged prehistory in Croatia. Because of sinking coastlines in recent millennia, added to the effects of postglacial sea-level rise, underwater prehistoric finds range in age from the deeply submerged Middle Pleistocene mammoth remains of Pag and the Mousterian artefacts of Resnik to burials and settlements of the Iron Age. Moreover, because of its mountainous topography and the emphasis throughout its long history of human settlement on coastal regions, shorelines, offshore islands, and sea travel, the importance of underwater research in filling gaps in the prehistoric record of Croatia cannot be overemphasised.

Although discoveries of intertidal and submarine prehistoric archaeology have quite a long history, systematic research has only been initiated or intensified in the past decade. Most of the better-preserved sites described above have received only test excavations or partial post-excavational analysis of the cultural and palaeoenvironmental material. Thus, there is clearly scope for new research and new archaeological discoveries, especially in combination with palaeoenvironmental and palaeoclimatic analysis of submerged landscapes and sediments.

Some of the important discoveries were made as a result of construction work in the coastal zone, which has exposed the archaeological evidence to view but also contributed to its destruction, resulting in hasty salvage operations and the loss of much valuable information. One of the major challenges for the future is to ensure a better integration of policies on heritage protection, commercial and industrial development, and scientific research.

At present, the main government agency responsible for protection and management of the underwater (and terrestrial) cultural heritage and for the issue of permits for field research is the Ministry of Culture of the Republic of Croatia. Implementation is delegated to regional conservation offices in Pula, Rijeka, Gospić, Zadar, Šibenik, Trogir, Split, and Dubrovnik. Financial support is shared between the Ministry of Culture, the Croatian Science Foundation, and the local communities. The principal organisations involved in research and site protection are the Department of Archaeology of the University of Zadar, the Archaeological Museum of Zadar, the UNESCO II Category Centre for Underwater Archaeology in Zadar, the Archaeological Museum of Istria in Pula, the Department of Archaeology of the Faculty of Humanities and Social Sciences, University of Zagreb, and the Department of Underwater Archaeology of the Croatian Conservation Institute in Zagreb. Thus, there is considerable expertise in systematic research and management of the underwater cultural heritage and examples of good practice.

However, despite the fact that Croatia is a signatory to the UNESCO 2001 Convention on the Protection of the Underwater Heritage, there is a need to develop formal systematic procedures for the public assessment of plans for construction work that may impact scientifically valuable underwater remains of prehistoric archaeology and features of the submerged landscape (see Pater, Chap. 27 this volume, and the chapters in Part I and Part II of this volume). Formal procedures need to be developed to ensure implementation of mitigation work or protection of the more insubstantial remains typical of prehistoric archaeological material and submerged landscapes and to achieve better integration with scientific research, in line with practices that have been developed in other European countries.

Acknowledgements Research on underwater Croatian prehistoric sites was mainly funded by the Ministry of Culture of Republic of Croatia and by the Croatian Science Foundation through the research projects IP-09.01/320 (*Late Mousterian in the eastern Adriatic – towards understanding of the late Neanderthals' identity and their demise*) and IP-09-2014-8211 (Archaeology of Adriatic Shipbuilding and Seafaring – AdriaS). We express our gratitude to Alberta Arena, Neven Cukrov, Martina Čelhar, Stašo Forenbaher, Branko Jalžić, Darko Komšo, Ida Koncani Uhač, Jadranka Mauh Lenardić, Marko Meštrov, Mate Parica, Petronije Tasić, and Dario Vujević for their precious help in completing and illustrating this chapter. We are also indebted to Geoff Bailey for his suggestions, comments, and editing of this paper.

References

Antonioli F, Anzidei M, Lambeck K, Auriemma R, Gaddi D, Furlani S, Orrù P, Solinas E, Gaspari A, Karinja S, Kovačić V, Surace V (2007) Sea-level change during the Holocene in Sardinia and in the Northeastern Adriatic (Central Mediterranean Sea) from archaeological and geomorphological data. Quat Sci Rev 26:2463–2486

Arena A, Barbarić V, Radić Rossi I (forthcoming) Vranjic (Kaštela Bay) between the late Early and the Middle Bronze Age (19th–14th century BCE). In: Okolica Kaštelanskog zaljeva u Prošlosti, Proceedings of the Symposium of the Croatian Archaeological Society, Kaštel Lukšić, 2–6 October 2017

Barbir A (2015) Litički materijal s podvodnog srednjopaleolitičkog nalazišta Kaštel Štafilić – Resnik, Graduate thesis. University of Zagreb, Faculty of Humanities and Social Sciences, Zagreb

Batović Š (1965) Neolitski ostaci iz Nina i njihov položaj u okviru neolita na Mediteranu. Diadora 3:5–40

Batović Š (1988) Paleolitički i mezolitički ostaci s Dugog otoka. Poročilo o raziskovanju paleolita, neolita in eneolita v Sloveniji 16:7–54

Batović Š (1990a) Novija istraživanja prapovijesti u biogradskom kraju. In: Batović Š, Obad S, Jakšić N, Došen B, Božulić G (eds) Biograd i njegova okolica u prošlosti. Zavod za povijesne znanosti Filozofskog fakulteta u Zadru, pp 85–171

Batović Š (1990b) Rekognosciranje Otoka Paga u 1989. godini. Obavijesti Hrvatskog arheološkog društva 1:26–32

Bekić L (2017) Šimuni, novo podvodno brončanodobno nalazište u Zadarskoj županiji / Šimuni, New Bronze Age Underwater Site in Zadar County. Diadora 31:41–50

Bekić L, Ilkić M, Brusić Z, Meštrov M, Parica M, Pešić M, Scholz R (2011) New Underwater archaeological research in Pakoštane near Zadar / Nova podvodna arheološka istraživanja u Pakoštanima kod Zadra. Submerged Heritage / Potopljena baština 1:25–31

Bekić L, Pešić M, Scholz R, Meštrov M (2016) Podvodna arheološka istraživanja na prapovijesnom nalazištu Pakoštane – Janice / Underwater archaeological research at the prehistoric site of Pakoštane – Janice. Diadora 29:7–22

Benac Č, Juračić M (1998) Geomorphological indicators of the sea level changes during Upper Pleistocene (Wurm) and Holocene in the Kvarner region (NE Adriatic Sea). Acta Geographica Croatica 33(1):27–45

Benac Č, Juračić M, Bakaran-Petricioli T (2004) Submerged tidal notches in Rijeka Bay NE Adriatic Sea: indicators of relative sea-level change and of recent tectonic movements. Mar Geol 212:21–33

Benac Č, Juračić M, Blašković I (2008) Tidal notches in Vinodol Channel and Bakar bay: Indicators of recent tectonics. Mar Geol 248:151–160

Benjamin J, Črešnar M (2009) Submerged prehistoric site discovery and a proposed application in the Croatian Adriatic. In: Bekić L (ed) Jurišićev zbornik. Hrvatski restauratorski zavod. Međunarodni centar za podvodnu arheologiju u Zadru, Zagreb, pp 52–67

Benjamin J, Bekić L, Komšo D, Koncani Uhač I, Bonsall C (2011) Investigating the submerged prehistory of the Eastern Adriatic: progress and prospects. In: Benjamin J, Bonsall C, Pickard C, Fischer A (eds) Submerged prehistory. Oxbow, Oxford, pp 193–206

Benjamin J, Rovere A, Fontana A, Furlani S, Vacchi M, Inglis RH, Galili E, Antonioli F, Sivan D, Miko S, Mourtzas N, Felia I, Meredith-Williams M, Goodman-Tchernov B, Kolaiti E, Anzidei M, Gehrels R (2017) Late Quaternary sea-level changes and early human societies in the central and eastern Mediterranean Basin: An interdisciplinary review. Quat Int 449:29–57

Bognar A, Faivre S, Buzjak N, Pahernik M, Bočić N (2012) Recent landform evolution in the Dinaric and Pannonian regions of Croatia. In: Lóczy D, Stankoviansky M, Kotarba A (eds) Recent landform evolution – The Carpatho-Balkan-Dinaric Region. Springer, Dordrecht, pp 313–344

Brusić Z (1969) Rezultati podvodnih istraživanja u Ninu. Mornarički glasnik 22:218–222

Brusić Z (1977) Prehistorijski podmorski nalazi na području južne Liburnije. Radovi Centra Jugoslavenske Akademije Znanosti i Umjetnosti u Zadru 24:53–60

Brusić Z (2002) Nekropole liburnskih naselja Nina i Kose kod Ljupča. Histria Antiqua 8:213–242

Brusić Z (2004) Resnik – hidroarheološka istraživanja, katalog izložbe. Muzej grada Kaštela, Kaštela

Brusić Z (2005) Arheološki spomenici otoka Murtera. Murterski godišnjak 2/2004:91–101

Brusić Z, Parica M (2014) Ljubač – liburnsko pristanište. Hrvatski arheološki godišnjak / Croatian Archaeological Yearbook 10/2013:467–468

Bulić F (1898) Ritrovamenti antichi risguardanti la topografia urbana dell'antica Salona. Bulletino di archeologia e storia dalmata 21:157–159

Cancellieri E (2011) From the watershed to the Great Adriatic Plain: an investigation on humans and landscape ecology during the late Upper Paleolithic. The significance of lithic technology. Dissertation Università degli Studi di Ferrara, Ferrara

Čelhar M, Parica M, Ilkić M, Vujević D (2017) A Bronze Age underwater site near the islet of Ričul in northern Dalmatia (Croatia). Skyllis 17(1):21–34

Duplančić Leder T, Ujević T, Čala M (2004) Coastline lengths and areas of islands in the Croatian part of the Adriatic Sea determined from the topographic maps at the scale of 1:25 000. Geoadria 9(1):5–32

Fairbridge RW (1968) The Encyclopedia of geomorphology. Reinhold Book, New York

Faivre S, Butorac V (2018) Recently submerged tidal notches in the wider Makarska area (Central Adriatic, Croatia). Quat Int 494:225–235

Faivre S, Fouache E, Kovačić V, Glušćević S (2010) Geomorphological and archaeological indicators of Croatian shoreline evolution over the last two thousand years. Geo Acta 3:125–133

Faivre S, Fouache E, Ghilardi M, Antoniloli F, Furlani S, Kovačić V (2011) Relative sea level change in western Istria (Croatia) during the last millenium. Quat Int 232:132–143

Faivre S, Bakran-Petricioli T, Horvatinčić N, Sironić A (2013) Distinct phases of relative sea level changes in the central Adriatic during the last 1500 years – influence of climatic variations? Palaeogeogr Palaeoclimatol Palaeoecol 369:163–174

Fernández Á (2003) Shell beads of the last hunter-gatherers and earliest farmers in south-western Europe. Munibe 61:129–138

Florido E, Auriemma R, Faivre S, Radić Rossi I, Antonioli F, Furlani S, Spada G (2011) Istrian and Dalmatian fishtanks as sea-level markers. Quat Int 232:105–113

Forenbaher S (2011) Adriatic offshore islands and long-distance interaction in prehistory. In: Forenbaher S (ed) A connecting sea: maritime interaction in Adriatic prehistory, BAR International Series, vol 2037. Archaeopress, Oxford, pp 73–87

Forenbaher S (2018a) Special place, interesting times: The island of Palagruža and transitional periods in Adriatic prehistory. Archaeopress, Oxford

Forenbaher S (2018b) Ljubljana and Cetina: Pottery styles of the Third Millennium BC in the Eastern Adriatic. Prilozi instituta za arheologiju 35:113–158

Forenbaher S, Kaiser T (2005) Palagruža and the spread of farming in the Adriatic. Opuscula archaeologica 29:7–23

Forenbaher S, Miracle P (2006) The spread of farming in the Eastern Adriatic. Documenta Praehistorica 33:589–609

Forenbaher S, Kaiser T, Miracle P (2013) Dating the East Adriatic Neolithic. Eur J Archaeol 16(4):589–609

Furlani S, Cucchi F, Biolchi S, Odorico R (2011) Notches in the Northern Adriatic Sea: genesis and development. Quat Int 232:158–168

Glogović D (1989) Prilozi poznavanju željeznog doba na Sjevernom Jadranu. Hrvatsko Primorje i otoci. Jugoslavenska akademija znanosti i umjetnosti, Zagreb

Ilijanić N, Miko S, Hasan O, Marković T (2018) Geological, structural-tectonic and hydrogeological characteristics of the Pakoštane area and Lake Vrana. In: Radić Rossi I, Boetto G (eds) Pakoštane – Veli Školj: The Late Roman shipwreck in its geological-geographic and cultural-historical context. University of Zadar & Institute for Maritime Heritage ARS NAUTICA, Zadar, pp 161–177

Janković I, Mihelić S, Karavanić I (2011) Put neandertalaca / The Neandertal Trail. Arheološki muzej u Zagrebu, Zagreb

Kamenjarin I, Šuta I (2011) Antički Sikuli / Ancient Siculi, exhibition catalogue. Museum of the Town of Kaštela, Kaštel Lukšić

Karavanić I (2000) Research on the Middle Palaeolithic in Dalmatia, Croatia. Antiquity 74:777–778

Karavanić I (2004) The Middle Palaeolithic settlement of Croatia. In: Conard NJ (ed) Settlement dynamics of the Middle Palaeolithic and Middle Stone Age, vol II. Kerns Verlag, Tübingen, pp 251–267

Karavanić I (2009) Adriatic coast of Croatia and its hinterland from 50 000 to 25 000 BP. In: Camps M, Szmidt C (eds) The Mediterranean from 50 000 to 25 000 BP: Turning points and new directions. Oxbow Books, Oxford, pp 163–178

Karavanić I (2015) Research on underwater Mousterian: The site of Resnik – Kaštel Štafilić, Dalmatia, Croatia. In: Sázelová S, Novák M, Mizerová A (eds) Forgotten times and spaces: New perspectives in paleoanthropological, paleoetnological and archeological studies. Institute of Archeology of the Czech Academy of Sciences & Masaryk University, Brno, pp 73–79

Karavanić I, Barbir A (2017) Underwater Palaeolithic sites in Croatia. In: 5th Regional Scientific Meeting on Quaternary Geology Dedicated to Geohazards and Final conference of the LoLADRIA project "Submerged Pleistocene landscapes of the Adriatic Sea", 9–10 November 2017, Starigrad-Paklenica, Croatia, Abstracts. Croatian Academy of Science and Arts, Zagreb, p 28

Karavanić I, Miracle PT, Culiberg M, Kurtanjek D, Zupanič J, Golubić V, Paunović M, Mauch Lenardić J, Malez V, Šošić R, Janković I, Smith FH (2008) The Middle Paleolithic from Mujina Pećina, Dalmatia, Croatia. J Field Archaeol 33:259–277

Karavanić I, Zubčić K, Pešić M, Parica M, Šošić Klindžić R (2009) Kaštel Štafilić – podvodno paleolitičko nalazište. Hrvatski arheološki godišnjak / Croatian Archaeological Yearbook 5/2008:549–551

Karavanić I, Vukosavljević N, Šošić Klindžić R, Kurtanjek D, Zupanič J (2013) The lithic and bone industries of the Epigravettian layers from Šandalja II near Pula. Vjesnik za arheologiju i povijest dalmatinsku 106:7–73

Karavanić I, Janković I, Ahern JCM, Smith FH (2014a) Current Research on the Middle Paleolithic Cave, Open-air and Underwater Sites in Dalmatia, Croatia. In: Sázelová S, Hupková A, Morkovsky T (eds) Mikulov Anthropology Meeting 2014, [The Dolní Věstonice Studies 20]. Academy of Sciences of the Czech Republic, Institute of Archeology, Brno and Masaryk University, Department of Anthropology, Brno, pp 31–36

Karavanić I, Vukosavljević N, Šošić Klindžić R, Ahern JCM, Čondić N, Becker R, Zubčić K, Šuta I, Gerometta K, Boschian G (2014b) The Late Mousterian in the eastern Adriatic – towards understanding of late Neanderthals' identity and their demise project: a summary of the 1st year of research. Prilozi Instituta za arheologiju u Zagrebu 31:139–157

Karavanić I, Šošić Klindžić R, Ahern JCM, Čondić N, Janković I, Zubčić K, Smith FH (2015) Recent research on the Croatian Middle/Upper Paleolithic interface in the context of Central and Southeast Europe. In: Harvati K, Roksandić M (eds) Human evolution in the Southern Balkans. Springer, Dordrecht, pp 153–169

Kelletat DH (2005) Dalmatian coasts. In: Schwartz M (ed) Encyclopedia of coastal science, Encyclopedia of Earth Sciences Series. Springer, Dordrecht, pp 355–356

Komšo D (2008) The Mesolithic in Croatia. Opvscula archaeologica 30(2006):55–91

Komšo D, Vukosavljević N (2011) Connecting coast and inland: perforated marine and freshwater snail shells in the Croatian Mesolithic. Quat Int 244:117–125

Koncani Uhač I (2009) Podvodna arheološka istraživanja u uvali Zambratija. Histria Antiqua 17:263–268

Koncani Uhač I, Čuka M (2016) Doprinos poznavanju podmorskog eneolitičkog nalazišta u uvali Zambratija. Histria archaeologica 4:25–74

Koncani Uhač I, Boetto G, Uhač M (2017) Zambratija – prehistoric sewn ship. Archaeological Museum of Istria, Pula

Korbar T, Montanari A, Koch G, Mariani S, DePaolo D, Turchyn AV, Miknić M, Tari V (2009) Geologic reconnaissance of the island of Velika Palagruža (central Adriatic, Croatia). Geologia Croatica 62/2:75–94

Krile I, Vujević D (2017) A contribution to the study of early prehistory of Veli Rat on the island of Dugi Otok. Diadora 31:7–26

Lambeck K, Purcell A (2005) Sea-level change in the Mediterranean Sea since the LGM: model predictions for tectonically stable areas. Quat Sci Rev 24(18–19):1969–1988

Lambeck K, Woodroffe CD, Antonioli F, Anzidei M, Roland Gehrels W, Laborel J, Wright AJ (2010) Paleoenvironmental records, geophysical modeling and reconstruction of sea-level trends and variability on centennial and longer timescales. In: Church JA, Philip L, Woodworth PL, Aarup T, Wilson WS (eds) Understanding sea-level rise and variability. Wiley-Blackwell, Oxford, pp 64–121

Lambeck K, Antonioli F, Anzidei M, Ferranti L, Leoni G, Scicchitano G, Silenzi S (2011) Sea level change along the Italian coast during the Holocene and projections for the future. Quat Int 232:250–257

Lončar N, Bar-Matthews M, Ayalon A, Surić M, Faivre S (2017) Early and mid-Holocene environmental conditions in the eastern Adriatic recorded in speleothems from Mala špilja Cave and Velika špilja cave (Mljet island, Croatia). Acta carsologica 46.2/3:229–249

Lončar N, Bar-Matthews M, Ayalon A, Faivre S, Surić M (2018) Holocene climatic conditions in the eastern Adriatic recorded in stalagmites from Strašna peć Cave (Croatia). Quaternary international. https://doi.org/10.1016/j.quaint.2018.11.006

Majnarić Pandžić N (1998) Brončano i željezno doba. (Dimitrijević S, Težak Gregl T, Majnarić Pandžić N, eds). Prapovijest. Naprijed, Zagreb, pp 161–169

Malez M (1979) Nalazišta paleolitskog i mezolitskog doba u Hrvatskoj. In: Basler Đ (ed) Praistorija Jugoslavenskih Zemalja; Paleolit i Mezolit. Svjetlost, Sarajevo, pp 227–276

Malez M, Lenardić-Fabić J (1988) New subspecies of the southern elephant (*Mammuthus meridionalis adriacus n. sp.*) from the bottom of the Adriatic Sea (Croatia, Yugoslavia). Palaeontol Jugoslav 37:1–36

Marriner N, Morhange C, Faivre S, Flaux C, Vacchi M, Miko D, Dumas V, Boetto G, Radić Rossi I (2014) Post-Roman sea-level changes on Pag Island (Adriatic Sea): Dating Croatia's "enigmatic" coastal notch? Geomorphology 221:83–94

Mauch Lenardić J (2012) Miocene to late Pleistocene proboscideans of Croatia. Quat Int 276–277:120–128

Mihailović D (2009) Upper Paleolithic and Mesolithic chipped stone industries from Crvena Stijena. Prehistoric settlements in caves and rock-shelters of Serbia and Montenegro. Fascicule II. University of Belgrade, Belgrade

Mihailović D, Whallon R (2017) Crvena Stijena revisited: the late Mousterian assemblages. Quat Int 450:36–49

Mihajlović I (2011) Podmorje Kvarnera (Oruda, Unije). Hrvatski arheološki godišnjak/Croatian Archaeological Yearbook 7/2010:480–481

Mihovilić K (1992) Bay of Veruda – Pula. Site of stone objects. Poročilo o Raziskovanju Paleolita, Neolita in Eneolita v Sloveniji 20:87–93

Mihovilić K (1995a) Prapovijesni lokaliteti obalne Istre. Histria Antiqua 1:58–63

Mihovilić K (1995b) Školjić (Funtana) i tragovi prapovijesnih obalnih i otočnih lokaliteta Istre. Histria archaeologica 26:28–57

Miko S, Ilijanić N, Hasan O, Razum I, Durn T, Brunović D, Papatheodorou G, Bakrač K, Hajek Tadesse V, Crmarić R (2016) Late Quaternary evolution of lakes and submerged paleo-karst on the Eastern Adriatic. In: Mandić O, Pavelić D, Kovačić M, Sant K, Andrić N, Hrvatović H (eds) RCMNS Interim Colloquium. Croatian Geological Society, Zagreb, pp 17–18

Miracle PT (1995) Broad-spectrum adaptations re-examined: Hunter-gatherer responses to Late-Glacial environmental changes in the Eastern Adriatic, PhD dissertation, University of Michigan

Miracle PT (2005) Late Mousterian subsistence and cave-use in Dalmatia: the zooarchaeology of Mujina Pećina, Croatia. Int J Osteoarchaeol 15:84–105

Mocochain L, Audra P, Clauzon G, Bellier O, Bigot JY, Parize O, Monteil P (2009) The effect of river dynamics induced by the Messinian Salinity Crisis on karst landscape and caves: example of the Lower Ardèche River (mid Rhone valley). Geomorphology 109(12):46–61

Moore AMT (2014) The spread of farming to the Adriatic: New insights from Dalmatia. In: Ammerman AJ, Davis TW (eds) Island archaeology and the origins of seafaring in the eastern Mediterranean. Proceedings of the Wenner-Gren Workshop held at Reggio Calabria on October 19–21, 2012. Eurasian Prehistory (1–2):155–164

Papagianni D (2009) Mediterranean southeast Europe in the Late Middle and Early Upper Palaeolithic: modern human route to Europe or Neanderthal refugium? In: Camps M, Szmidt C (eds) The Mediterranean from 50 000 to 25 000 BP: turning points and new directions. Oxbow Books, Oxford, pp 115–136

Papagianni D, Morse MA (2013) The Neanderthals rediscovered: how modern science is rewriting their story. Thames & Hudson, New York

Parica M, Ilkić M (2017) Podmorski arheološki nalazi u okolici Ljupča. In: Faričić J, Lenkić J (eds) Župa Ljubač – zrcalo povijesnih i geografskih mijena u sjeverozapadnom dijelu Ravnih Kotara. University of Zadar, Zdar, pp 98–111

Perhoč Z (2009) Sources of chert in Middle Dalmatia: Supplying raw material to prehistoric lithic industries. In: Forenbaher S (ed) A connecting sea: maritime interaction in Adriatic prehistory, BAR International Series 2037. Archaeopress, Oxford, pp 25–45

Pešić M (2012) Underwater archaeological research at the Janice site in Pakoštane / Podvodna arheološka istraživanja nalazišta Janice – Pakoštane. Submerged Heritage / Potopljena baština 2:39–43

Pešić M (2013) Research of the waters of the islet of Babuljaš near Pakoštane/Istraživanje podmorja otočića Babuljaša kod Pakoštana. Submerged Hertiage / Potopljena baština 3:25–30

Pikelj K, Juračić M (2013) Eastern Adriatic Coast (EAC): geomorphology and coastal vulnerability of a karstic coast. J Coast Res 29(4):944–957

Radić D (2011) The beginnings of trans-Adriatic navigation: a view from Vela Spila Cave (Korčula Island). In: Forenbaher S (ed) A connecting sea: maritime interaction in Adriatic prehistory, BAR International Series 2037. Archaeopress, Oxford, pp 13–24

Radić Rossi I (2008) Rescue excavations at Vranjic near Split, Croatia. In: Radić Rossi I, Gaspari A, Pydyn A (eds) Proceedings of the 13th annual meeting of the European Association of Archaeologists, Zadar, Croatia, 18–23 September 2007. Croatian Archaeological Society, Zagreb, pp 151–166

Radić Rossi I (2009) Arheološka baština u podmorju Kaštelanskoga zaljeva. Archaeologia Adriatica 2/2:489–506

Radić Rossi I (2011) Problematika prapovijesnih i antičkih nalazišta u hrvatskom podmorju, diss. University of Zadar, Zadar

Radić Rossi I, Antonioli F (2008) Preliminary considerations on the ancient port of Pakoštane (Croatia) based on archaeological and geomorphologic research. Book of abstracts; 14th Annual Meeting of the European Association of Archaeologists, 16–21 September 2008, Valetta, Malta, pp 96–97

Radić Rossi I, Boetto G (2013) Međunarodni istraživački projekt u uvali Caski na otoku Pagu; Prethodno izvješće o rezultatima istraživačke kampanje 2012. Histria Antiqua 22:377–390

Radić Rossi I, Cukrov N (2017) Archaeological potential of the anchialine caves in Croatia. In: Bailey GN, Harff J, Sakellariou D (eds) Under the sea: archaeology and palaeolandscapes of the continental shelf. Springer, Cham, pp 255–266

Radić Rossi I, Antonioli F, Spada G, Zubčić K, Meštrov M (2018) Sea level and shoreline changes in the Pakoštane area over the past six millennia. In: Radić Rossi I, Boetto G (eds) Pakoštane – Veli Školj: The Late Roman shipwreck in its geological-geographic and cultural-historical context. University of Zadar & Institute for Maritime Heritage ARS NAUTICA, Zadar, pp 229–248

Rainsford C, O'Connor T, Miracle P (2014) Fishing in the Adriatic at the Mesolithic–Neolithic transition: Evidence from Vela Spila, Croatia. Environ Archaeol 19(3):311–320

Šegota T (1982) Razina mora i vertikalno gibanje dna Jadranskog mora od ris-virmskog interglacijala do danas. Geološki vjesnik 35:93–109

Šegota T (1983) Paleogeografske promjene u Jadranskom moru od Virmskog maksimuma do danas. Acta Geographica Croatica 17–18(1):11–14

Soura K, Geraga M, Papatheodorou G, Christodoulou D, Baika K, Kordella S, Fakiris E, Radić Rossi I, Boetto G, Zubčić K (2013) The study of palaeogeographic evolution of Caska Bay (Pag Island, Croatia) the last 10 kyrs using marine remote sensing techniques. In: SPLASHCOS: Under the Sea; Archaeology and Palaeolandscapes, Szczecin, Poland, 23–27 September 2013, Book of Abstracts. COST; European Cooperation in Science and Technology, Szcecin, p 27

Surić M (2009) Rekonstruiranje promjena morske razine na istočnoj obali Jadrana (Hrvatska) – pregled / Reconstructing sea-level changes on the Eastern Adriatic Sea (Croatia) – an overview. Geoadria 14/2:181–199

Surić M, Juračić M (2010) Late Pleistocene-Holocene environmental changes – records from the submerged speleothems along the Eastern Adriatic coast (Croatia). Geologia Croatica 63/2:155–169

Surić M, Juračić M, Horvatinčić N (2004) Comparison of C14 and Th230/U234 dating of speleothems from submarine caves in the Adriatic Sea (Croatia). Acta Carsologica 33(2):239–248

Surić M, Juračić M, Horvatinčić N, Krajcar Bronić I (2005) Late Pleistocene – Holocene sea-level rise and the pattern of coastal karst inundation – records from submerged speleothems along the Eastern Adriatic Coast (Croatia). Mar Geol 214:163–175

Surić M, Lončarić R, Lončar N (2010) Submerged caves of Croatia: distribution, classification and origin. Environ Earth Sci 61(7):1473–1480

Težak Gregl T (1998) Neolitik i eneolitik. (Dimitrijević S, Težak Gregl T, Majnarić Pandžić N, eds). Prapovijest. Naprijed, Zagreb, pp 57–158

Vlahović I, Tišljar J, Velić I, Matičec D (2002) The karst Dinarides are composed of relics of a single Mesozoic platform: facts and consequences. Geologia Croatica 55(2):171–183

Vlahović I, Tišljar J, Velić I, Matičec D (2005) Evolution of the Adriatic Carbonate Platform: Palaeogeography, main events and depositional dynamics. Palaeogeogr Palaeoclimatol Palaeoecol 220:333–360

Vrsalović D (2011) Arheološka istraživanja u podmorju istočnog Jadrana: prilog poznavanju trgovačkih plovnih putova i gospodarskih prilika na Jadranu u antici. Književni krug Split, Archaeological Museum of Split

Vujević D (2009) The relationship between the Middle Palaeolithic sites in the Zadar hinterland and the Zadar Islands. In: Forenbaher S (ed) A connecting sea: Maritime interaction in Adriatic prehistory, BAR International Series 2037. Archaeopress, Oxford, pp 1–12

Vujević D, Parica M (2009) Nakit i umjetnost pećine Vlakno / Jewelery and Art of Vlakno Cave. Archaeologia Adriatica 3(1):23–34

Vujević D, Meštrov M (2018) Prehistoric lithic finds from the Janice site in Pakoštane. In: Radić Rossi I, Boetto G (eds) Pakoštane – Veli Školj: the late Roman shipwreck in its geological-geographic and cultural-historical context. University of Zadar & Institute

for Maritime Heritage ARS NAUTICA, Zadar, pp 199–210
Vujević D, Perhoč Z, Ivančić T (2017) Micro-Mousterian in northern Dalmatia. Quat Int 450:50–67
Vukosavljević N, Karavanić I (2015) Kasnogornjopaleolitički i mezolitički ukrasi od probušenih morskih i slatkovodnih puževa iz špilje Zale. In: Vukosavljević N, Karavanić I (eds) Arheologija špilje Zale: od paleolitičkih lovaca skupljača do rimskih osvajača. Katedra Čakavskog sabora Modruše, Modruš, pp 157–174
Vukosavljević N, Perhoč Z, Čečuk B, Karavanić I (2011) Late Glacial knapped stone industry of Kopačina Cave. Vjesnik za arheologiju i povijest dalmatinsku 104:7–54
Vukosavljević N, Perhoč Z, Karavanić I (2015) Litički skup nalaza od lomljenog kamena iz špilje Zale: kasni gornji paleolitik I mezolitik. In: Vukosavljević N, Karavanić I (eds) Arheologija špilje Zale: od paleolitičkih lovaca skupljača do rimskih osvajača. Katedra Čakavskog sabora Modruše, Modruš, pp 73–118
Vuletić Vukasović V, Radić F (1887) Arheologičke bilježke s putovanja po otoku Korčuli u mjesecu svibnju 1887 god. Vjesnik Hrvatskog arheologičkog družtva 9:104–111
Whallon R (2007) Social territories around the Adriatic in the late Pleistocene. In: Whallon R (ed) Late Paleolithic environments and cultural relations around the Adriatic, BAR International Series 1716. Archaeopress, Oxford, pp 61–65
Zubčić K (2004) Novootkriveni prapovijesni podmorski nalazi u Baškoj vodi. Obavijesti Hrvatskog arheološkog društva 36/1:35–39

Web Sources

Croatian Ministry of the Sea, Transport and Infrastructure. http://www.mppi.hr/default.aspx?id=503. Accessed on 26 Feb 2019
Flanders Marine Institute (2018). IHO Sea Areas, version 3. Available online at http://www.marineregions.org/. https://doi.org/10.14284/323

Greece: Unstable Landscapes and Underwater Archaeology

19

Nena Galanidou, Katerina Dellaporta, and Dimitris Sakellariou

Abstract

The submerged archaeology of Greece extends from the Palaeolithic to the early Byzantine period. It offers valuable information on some of the critical themes of Eurasian prehistory: hominin dispersals, settlement patterns, strategies of survival, population movements and sea voyaging, communication and trade, high-energy destructive events and climate change. This overview focuses on the prehistoric record. It includes partly or fully submerged palaeontological sites as well as archaeological sites. All these are testimonies to the more extensive coastal mosaic of biotopes that were available to prehistoric people prior to c. 4000 cal BP in the Holocene and during the cold and arid periods of the Pleistocene. They show coastal and maritime lifeways in dynamically changing landscapes connecting Asia and Europe. They are now located on the Greek continental shelf due to eustatic and isostatic change as well as the heavy imprint of tectonic activity.

Keywords

Submerged landscapes · Underwater · Caves · Neolithic · Middle Palaeolithic · Early Bronze Age · Tectonics · Kalamakia Cave · Agios Petros · Pavlopetri

19.1 Introduction

Greece is a country of strong maritime traditions with deep historical roots and evidence of earliest human occupation extending back to the Lower Palaeolithic at least 0.5 million years ago (Galanidou et al. 2016a, b). The modern territory of the Greek State includes a large portion of the Aegean Basin to the east and the Ionian Basin to the west, with a total of 3500 offshore islands that represent the high points of a drowned coastal shelf, and 17,000 km of coastline, a quarter of the total Mediterranean coastline. The greater part of this figure is represented by islands (Fig. 19.1). Greece is also a region of extreme, long-term tectonic activity. Because of its position at the convergence of the tectonic plates of Africa, Europe and Anatolia, intense earthquake activity and volcanism are present, resulting in uplift or subsidence of the land surface at both localized and regional scales. These have amplified or

N. Galanidou (✉)
Department of History and Archaeology, University of Crete, Rethymno, Greece
e-mail: galanidou@uoc.gr

K. Dellaporta
Hellenic Ministry of Culture and Sports, Byzantine and Christian Museum, Athens, Greece

D. Sakellariou
Hellenic Centre for Marine Research, Institute of Oceanography, Anavyssos, Greece
e-mail: sakell@hcmr.gr

G. Bailey et al. (eds.), *The Archaeology of Europe's Drowned Landscapes*, Coastal Research Library 35, https://doi.org/10.1007/978-3-030-37367-2_19

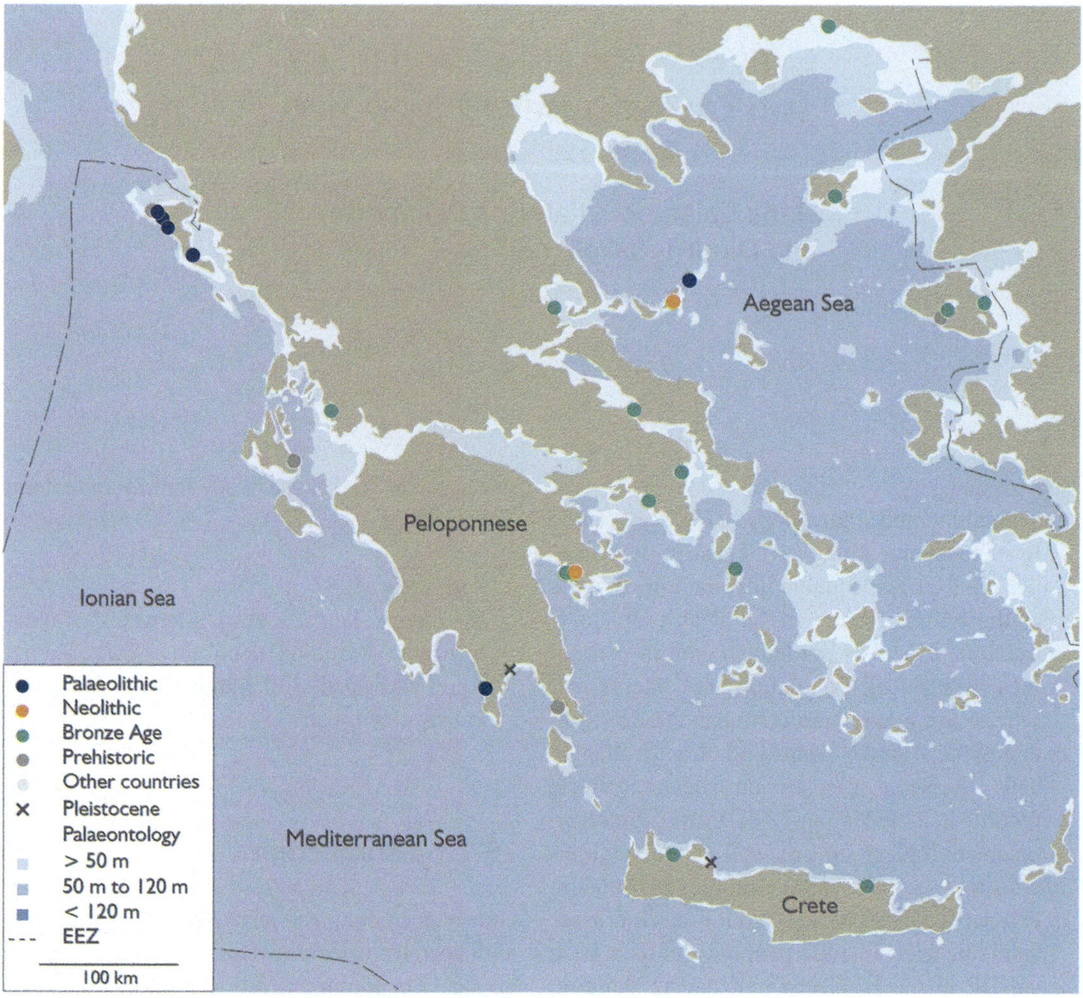

Fig. 19.1 Map of Greece, showing the submerged prehistoric sites. Site information from the SPLASHCOS Viewer http://splashcos-viewer.eu. Drawing by Moritz Mennenga

moderated the effects of eustatic sea-level change associated with the growth and melting of the continental ice sheets, with important consequences for palaeogeographic reconstructions and the position of prehistoric coastlines (Flemming 1968a, b; Lambeck 1995, 1996; Perissoratis and Conispoliatis 2003; Sakellariou and Galanidou 2016, 2017; Sakellariou et al. 2017).

The archaeological record on the coast and underwater includes three distinct categories (excluding shipwrecks):

1. Settlements, structures or karstic features that were in the past on dry land but are now fully inundated
2. Settlements and caves on the present-day coastline that partly extend into shallow waters or are bounded or eroded by the sea
3. Structures for the exploitation of marine resources or maritime activity such as salt-pans, fish farms, harbour installations, lighting/signalling constructions and quarries for the extraction of coastal rocks

The first two categories are the subject of this study. Here we gather information from a literature review, personal communication with the pioneers of underwater and coastal research, namely, Augustus Sordinas and Nic Flemming, as well as our own hands-on experience, to put together an overview of the prehistoric sites and landscapes of the Greek continental shelf up to and including the Early Bronze Age.

The submerged prehistoric sites from Greece recorded in the SPLASHCOS database number 27 entries (Fig. 19.1). These sites are testimony to an extensive mosaic of coastal biotopes that were available prior to about 4000 cal BP. The seabed has yielded partially or totally inundated settlements with architectural features (remains of walls, foundations or stone-built structures), loose finds such as stone tools and ceramics, hoards and cave sites containing bones of extinct Pleistocene animals. Although of pure palaeontological interest, the latter sites are important in showing that terrestrial cave deposits can be preserved during sea-level rise and are not completely washed out and destroyed, and we include them in the present discussion. We also pursue in a systematic way cave sites on the modern coastline that are partially submerged or inaccessible today except from the shoreline or the sea, but with stratified terrestrial deposits containing artefacts, faunal remains and palaeoenvironmental indicators in their dry interiors. However, we exclude the submerged sites dating from the Middle Bronze Age onwards that today lie in the shallow waters of Greece due to coastal subsidence. We organise the material in chronological order, beginning with palaeontological remains and cave sites and proceeding to coastal villages and other finds. As the volume of information is large, the present account aims to highlight the main patterns in the submerged record rather than offer an exhaustive appraisal of individual sites (for full details see http://splashcos-viewer.eu). To capture the current dynamism of continental shelf prehistoric research, we also present recent work on seabed surveying and mapping motivated by explicit archaeological and historical questions.

19.2 Archaeological Context

The extensive submerged landscapes of the Greek territory hold clues to understanding some of the foundational developments of early European history. These include the earliest Pleistocene hominin expansions out of Africa and Asia following terrestrial routes into Europe, the deep history of Palaeolithic sea voyaging, Neolithic farmer diasporas from Anatolia and the Near East, some of which certainly followed coastal pathways and involved sea crossings and island colonisation, and the seaborne trade of the Bronze Age that underpinned Europe's earliest urban civilization, the so-called Minoan, reaching far-flung locations across the eastern Mediterranean Sea.

The Aegean is emerging as a probable basin where land connections opened up between Anatolia and the Balkan Peninsula when sea level was lower than the present, offering extensive and productive, but now submerged, landscapes and routes of dispersal from Early Pleistocene times onwards for animal and hominin populations. The presence of Middle Pleistocene archaeological finds on the island of Lesbos, and Early Pleistocene palaeontological finds there and on other islands in the east Aegean, is indirect evidence of sea-level fluctuations in the early Quaternary, since these islands were likely only accessible via land bridges or short sea crossings made possible during low sea-level stands (Galanidou et al. 2013, 2016a, b). On the island of Lesbos, there are seven Early Pleistocene fossiliferous sites with continental faunas that demonstrate biogeographic integration with Asia (Lyras and van der Geer 2007). Both the Plio-Pleistocene palaeontological record of Greece (e.g. Koufos and Kostopoulos 2016) and the Lower Palaeolithic archaeology of the eastern Aegean Sea (Galanidou et al. 2013, 2016a, b) prompt the articulation of a new scenario for the early penetration and later colonization of Europe via terrestrial pathways now inundated on the Aegean seafloor. Research in marine geoscience and Palaeolithic archaeology is intensifying in the field and the laboratory to map these submerged Aegean landscapes, to identify the most promising areas for underwater archaeological survey and to reconstruct the transformation of landscapes and seascapes during the low sea-level periods of the Pleistocene. The ultimate aim is to obtain a holistic view of northeast Mediterranean palaeogeography—what was land, coast and sea—and assess its impact on the history of hominin dispersals.

The Aegean and Ionian regions are also emerging as a likely focus for initial attempts at sea crossings and early sea voyaging as documented by the colonisation or exploitation of offshore islands, extending back well before the spread of farming to at least 13,000 years ago and possibly earlier (Perlès 1979; Renfrew and Aspinall 1990; Kopaka and Matzanas 2009; Strasser et al. 2010; Ammerman 2014). The site of Maroulas on the island of Kythnos (in the Cyclades) and stratum X at Knossos on Crete offer unequivocal evidence for maritime travels at around 9500 and 9000 years ago, respectively (Evans 1994; Efstratiou 2005; Sampson et al. 2010; Galanidou 2011). Since the earliest points of arrival or departure, and the most favourable terrain for the establishment of settlements, are likely to have been located on or close to the seashore, especially on the smaller islands, much if not most of the relevant evidence for early sea voyaging must now lie underwater.

All the above factors point to the likely significance of the Greek shelf as a key area for investigating Eurasian prehistory.

19.3 Geological and Palaeogeographical Background

Geological processes play an unusually significant role in determining the effects of eustatic changes in sea-level fluctuation on the position of prehistoric coastlines, and the area of land made available during low sea-level stands. This is especially the case in the Aegean Basin, where the crust 'flows' to the south-south-west, confined between the North Anatolian and Kephallinia Faults to the north and north-west and the East Hellenic Trench to the south-east, and overrides the subducting oceanic crust of the Eastern Mediterranean. The deformation is accommodated by dextral and sinistral shearing which creates a complex pattern of localized subsidence along with local uplift (Sakellariou and Tsampouraki-Kraounaki 2018).

The island of Crete is a good example of the complexities associated with local tectonic deformation. The island as a whole is undergoing general long-term uplift but is divided by multi-directional faulting into several different blocks displaying different rates of vertical movement (Pirazzoli et al. 1982; Pirazzoli 1988; Pirazzoli et al. 1996; Stiros 2001; Van Hinsbergen and Meleunkamp 2006; Shaw et al. 2008; Mouslopoulou et al. 2015). The highest uplift rate is in western Crete, which has risen dramatically in historical times, with evidence of uplift by about 7 m in AD 365 as a result of a single powerful earthquake (Shaw et al. 2008), holding out the prospect of discovering early prehistoric coastal evidence that may have been lifted clear of late glacial sea-level rise. These same processes also result in localised subsidence, which is why some Bronze Age sites on Crete are submerged below present sea level.

However, the long-term trend in the Aegean Basin as a whole is subsidence (Lykousis 2009), and this has two important archaeological consequences. First, on the Pleistocene time scale, it means that progressively more extensive areas of land would have been exposed for human exploitation during periods of low sea level as one goes further back in time. According to the reconstructions of Lykousis (2009), during the glacial maxima of MIS (Marine Isotope Stage) 10 and 12, at ~340 ka and 420 ka, respectively, the Aegean Basin would have been an almost continuous land mass connecting Anatolia and the Greek mainland, with some large inland lake basins (Fig. 19.2). Crete would have remained an offshore island but separated by only a very narrow strait at its western end from the Greek mainland (Sakellariou and Galanidou 2016, 2017). These episodes were well within the period when early humans were present in the region and have significant implications for patterns of population

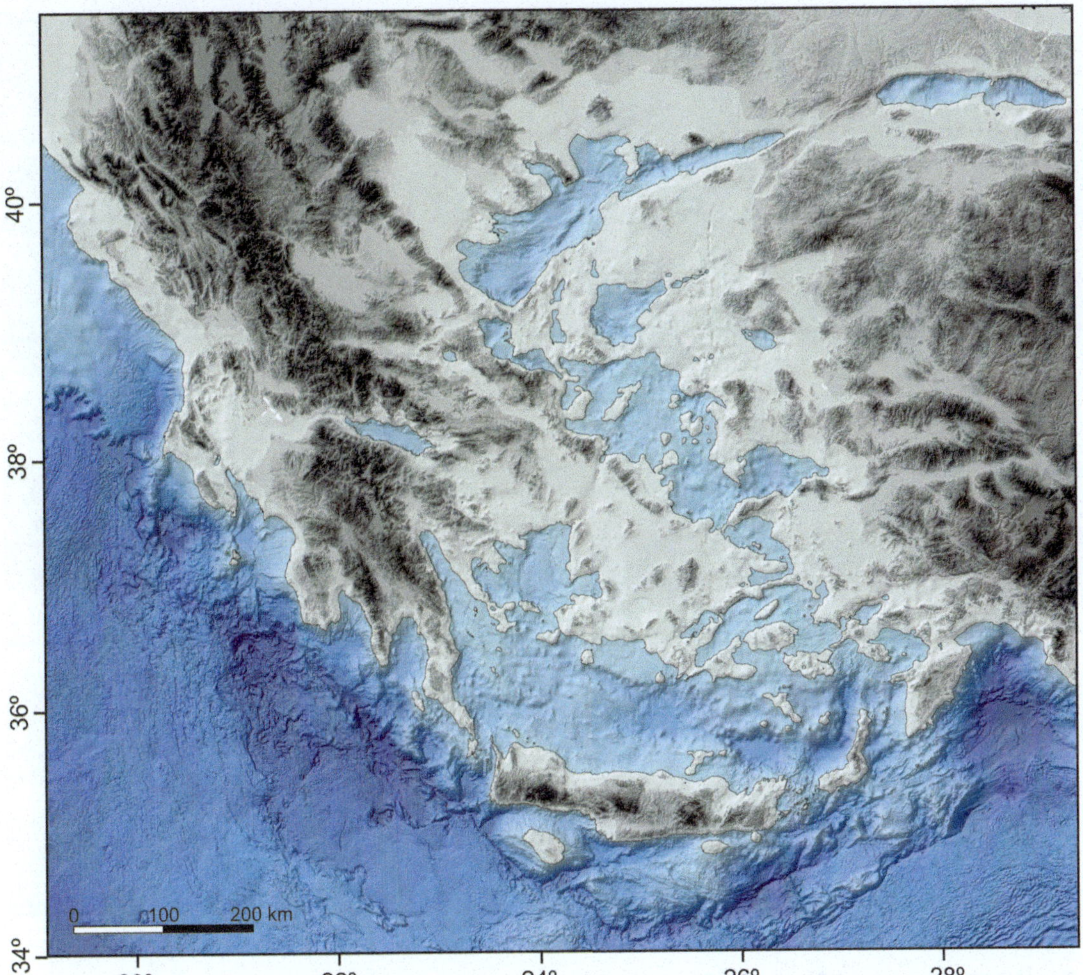

Fig. 19.2 Conceptual palaeogeographic model of the Aegean with areas of land exposed at glacial maxima over the past 500 kyr, modified after Lykousis (2009) and Sakellariou and Galanidou (2016). Land topography is shown in gray (dark gray=high altitudes; light gray=low altitudes). Seafloor topography is shown in blue (dark blue=deep; light blue=shallow). Map created by Dimitris Sakellariou using the latest EMODNET bathymetry (www.emodnet-bathymetry.eu)

dispersal and distribution. At the maximum lowering of sea level of the Last Glacial Maximum (LGM) at 20 ka, some sea crossings would have been shorter, and islands would have coalesced to form mega-islands, notably the Cyclades group (Fig. 19.3).

A second consequence of long-term subsidence of the Aegean Basin is that many coastlines have undergone progressive submergence in recent millennia. This means that archaeological remains of settlements occupied after cessation of eustatic sea-level rise, at around 5000–4000 cal BP (Lambeck et al. 2014, Fig. 4a), are now partially or wholly submerged all over coastal Greece, including material from the Final Neolithic to as recent as the Byzantine period. One such example is the cluster of Middle and Late Minoan harbour sites partially submerged in the intertidal zone all along the rim of central and eastern Crete (e.g. Shaw 1990; Theodoulou 2011; Watrouss 2012) or the submerged sandstone quarry and building at Malia (Guest-Papamanoli

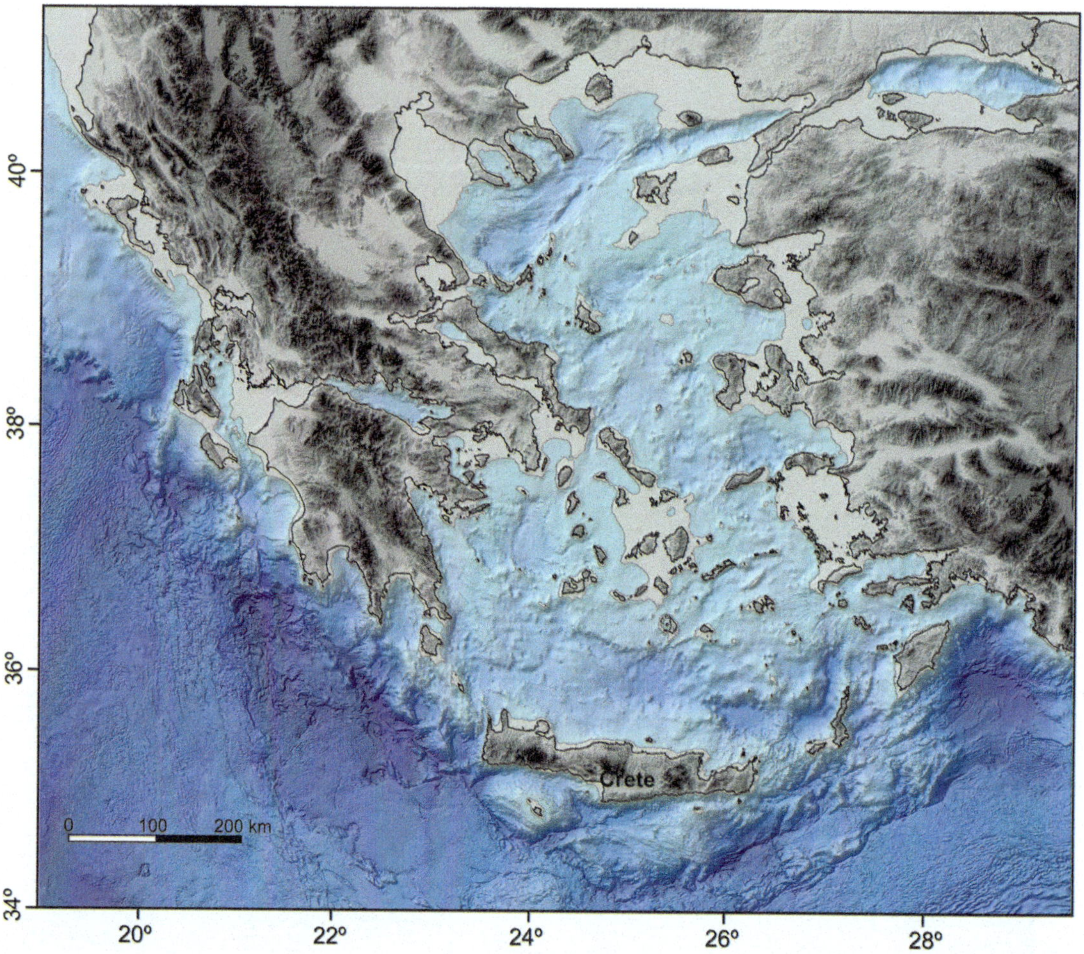

Fig. 19.3 Palaeogeographic reconstruction of the Aegean during the peak of the Last Glacial Maximum, 20 ka BP, based on Lykousis (2009), Kapsimalis et al. (2009), Sakellariou and Galanidou (2016) and references therein. Legend as in Fig. 19.2. Map created by Dimitris Sakellariou using the latest EMODNET bathymetry (www.emodnet-bathymetry.eu)

and Treuil 1979; Guest-Papamanoli 1989). Other later examples are the harbour sites of the Lesbos Pentapolis (Spencer 1995; Theodoulou 2008, 2010, 2017), an inundated Roman villa in the waters of Fourkari, Troizinia near the Ermionis coast (Frost 1977), the submerged antiquities of Samos ranging from harbour installations to a Byzantine hoard (Simossi 1988, 1993; Oekonomides and Drosogianni 1989) and the famous and contested site complex at Kenchreai in the Gulf of Corinth (Scranton et al. 1978). This material is of importance in charting changes of relative sea level and monitoring preservation and recovery patterns despite the potentially destructive impact of waves and currents in the surf zone and in shallow water, but a comprehensive account of the submerged ancient ruins of more recent millennia found in the shallow waters of Greece (e.g. Flemming et al. 1973; Flemming 1978, Table 1; Psychoyos 1988; Baika 2008) lies outside the chronological and thematic scope of this volume.

On the western side of the country, compression and mountain building with a component of strike-slip movement are the main tectonic processes resulting from plate motions, with progressive uplift of the Pindus mountains and submergence of the large basins along the present

Ionian coastline such as the Amvracian Gulf. Changes of eustatic sea level are the main factor affecting the palaeogeography. At the LGM, the islands of Corfu and Lefkas were attached to the mainland and formed high points in an extensive coastal lowland (Fig. 19.2 and 19.3). However, the Ionian islands of Kephallinia and Zakynthos remained separated from the mainland by deep sea channels, and the Middle Palaeolithic and Upper Palaeolithic artefacts found on these islands must have been deposited there by people capable of making short (~ 1–5 km) sea crossings (Ferentinos et al. 2012; Yiannouli 2016, 2017; Papoulia 2017; Zavitsanou et al. 2015).

19.4 Palaeontological Sites

Two caves containing Pleistocene animal bones, but no archaeology, are relevant to the theme of this chapter, for they offer evidence of terrestrial deposits preserved below sea level. Glyfada Cave in the Mani Peninsula has a river running through it and an entrance at 0.5 m apsl (above present sea level) (Fig. 19.4). Of the 10.6 km thus far explored, 1.8 km is under water, the water drowning a palaeokarst at an average depth of 20–25 m. Palaeontological remains of Upper Pleistocene terrestrial and marine mammals, such as deer, hyena, panther, lion, hippopotamus, rodents and seals, as well as birds, were collected from depths ranging between 0.5 and 2.5 m bpsl (below present sea level). They clearly show a cave that was once near the shore and frequented by these animals either as prey or as predators. The largest assemblage of *Hippopotamus amphibius* in Greece, more than 350 whole bones and 900 fragments, comes from the underwater parts of this cave. A radiocarbon age of 31,650 ± 550 BP above the stratigraphic horizon in which most of the bones were collected gives a minimum age (Giannopoulos 2000, p. 402). No synchronous

Fig. 19.4 The Glyfada Cave in south Peloponesse: (**a**) the 'headless panther' chamber of the cave; (**b**) the 'bat' chamber; (**c**) underwater view of the palaeokarst at a depth of −20 m bpsl; (**d**) a hippo bone collected from a depth of −1.5 m bpsl. Photos courtesy of Vasilis Giannopoulos

Fig. 19.5 The submerged Vamos Cave in West Crete: (**a**) reconstruction of the profile (blue denotes air and green denotes water); (**b**) the landscape around the cave today; (**c**) the cave entrance: its floor is at 10 m bpsl whilst the ceiling is at 3.5 m bpsl; (**d**) an *Elephas chaniensis* bone retrieved from a depth of 1.5 m bpsl. Photos courtesy of Vasilis Giannopoulos

Palaeolithic archaeology has been found here, though the data to hand suggest that before 40,000 years ago, entrance to the cave was possible through a large opening.

On the north coast of western Crete lies Vamos Cave, a large part of which is now underwater. Fossil bones of Pleistocene cervids and elephant, *Elephas chaniensis*, were found at depths ranging between 1.5 and 4.5 m bpsl (Fig. 19.5). The latter was a species endemic to Crete that probably went extinct at the end of the Pleistocene. The cave was on dry land during the Last Glacial (Symeonides et al. 2001). The bones were deposited on the cave floor at periods when sea level dropped by at least 10–20 m. At that time, the entrance of the cave was facing the shore.

19.5 Archaeological Overview

19.5.1 Coastal and Underwater Caves

On the Mani Peninsula in the South Peloponnese, the prevailing hydrogeological conditions of the Taygetos mountain chain favour the occurrence of coastal and submarine groundwater discharges and karst development. This results in the presence of submarine springs at Stoupa and Lakonis, in the Messenian and the Laconian Gulfs, respectively, and many caves in the intertidal and subtidal zones (Sakellariou and Galanidou 2016). Off the shores of western Mani, the seabed features at least 12 submarine karstic formations with large concentrations at depths of 25–20 m and 12–8 m bpsl (Bassiakos 1993; Giannopoulos 2000, p. 236). The coastline bordering the Mani Peninsula has numerous caves that were inhabited by Middle Palaeolithic and Upper Palaeolithc hunter-gatherers (Darlas and De Lumley 2002; Darlas and Psathi 2008, 2016). Later, during the Holocene, the coastal caves were used for habitation, burial or ritual activity by Neolithic farmers and herders (Papathanasopoulos 2011).

The contents of three caves situated on the Mani coastline, namely, Kalamakia, Apidima and Lakonis I, indicate a strong Neanderthal presence from the end of the Middle Pleistocene including earlier Neanderthals at Apidima and classic Neanderthals at the other two sites. These three cave sites are the single and only sources of Neanderthal fossils in Greece. The inventory consists of two crania extracted from the Apidima Cave, Apidima 1 and Apidima 2, 14 teeth, cranial and post-cranial fragments from the Kalamakia Cave on the rocky shore of Oitylo bay (not far from Apidima) and a molar from Lakonis I (Harvati et al. 2003, 2010, 2013). Laser ablation U-series dating has returned a minimum age of 160 ka BP for Apidima 2 (Bartsiokas et al. 2017).

19.5.1.1 Kalamakia Cave

Here, recurrent Middle Palaeolithic occupations are dated between 100 and 39 ka BP. The site is 10 m from the water line and 2.3 m apsl, at the foot of a 25-m-high cliff (Darlas and De Lumley 1995, 1998, 2002) (Fig. 19.6a). It forms a gallery 20 m deep, while the mouth is 7 m wide and 8 m high (Fig. 19.6b). The Holocene transgression led to sea levels rising to the present level, a little below that of the cave. Winter waves reach the

Fig. 19.6 The Kalamakia Cave: (**a**) general view from offshore; (**b**) close up of the cave entrance; (**c**) the cave walls showing dense lithophage bore holes up to the full wall height, an indication that during the Pleistocene the cave was under water for long periods; (**d**) broken sea shells found in the intact cave fill demonstrate that the inhabitants made use of marine resources. Photos courtesy of Andreas Darlas

base of the fill, partially eroding it and producing the current condition of the cave. Systematic archaeological excavations conducted on the cave floor show that recurrent Neanderthal occupations took place here. These occupations were not uninterrupted as Pleistocene sea-level changes had dramatic effects on the cave, rendering it alternately a continental and a submarine cave. During the glacial periods, the cave emerged from the sea and was used for Neanderthal habitation, gradually filling with continental sediments and remains of human activities. During interglacial periods the sea rose and flooded it, turning it into an underwater cave and largely stripping it of the previous deposits. Thus, only the deposits of a single cycle, the most recent, were preserved inside the cave each time. Today, therefore, it only contains deposits from the Last Interglacial and the Last Glacial period and almost no evidence of deposits from the preceding phases (Darlas 2012).

The cave walls bear dense lithophage bore holes (Fig. 19.6c) up to their full height, an indication that during the Pleistocene, the cave was underwater for long periods. The >7 m-thick fill consists of a base of marine deposits dated to the Last Interglacial period, covered by continental deposits dated to the Last Glacial period. The marine deposits consist of two layers. At the base, on the bedrock, is a thin layer of limestone with algae (Unit 0), dating from the MIS 5e transgression, at ~125 ka. Above this is a thin layer of beach rock (Unit II) corresponding to MIS 5c—a shell from this level has been dated to 109,000+14,000/−13,000 BP (U/Th). Apart from shells and sand (Fig. 19.6b), this layer contains remains of human activity, including artefacts from the terrestrial fill of the preceding phase, MIS 5d, which was washed out by the transgression of phase MIS 5c. In contrast, the sea level rise of MIS 5a was not high enough to reach the cave, and continental deposits from the Last glacial accumulated to a thickness of almost 7 m (Units III–VI).

19.5.1.2 Lakonis

On the eastern side of the Mani Peninsula 3 km northeast of Gytheio, lies Lakonis, a cluster of five karstic cavities extending over a distance of 200 m along the Sellinitsa coast. Excavations conducted at Lakonis I, a cavern whose roof has collapsed, unveiled a sequence spanning the MP and UP; its early stage is coeval with that of Kalamakia (Panagopoulou et al. 2004). The archaeological inventory consists of 50,000 stone artefacts and faunal remains (including deer, wild boar, large bovids, rhinos and hippos). The lithic industries are made of local volcanic rocks, while exogenous flint varieties are also present, bearing witness to possible movements of groups and provisioning of raw materials from a wide catchment area. A series of hearths completes the picture of a recurring use of the site. Part of the larger karstic complex is now submerged, and the waters around it in the Laconian Gulf have been declared a protected heritage zone partly due to the preservation of Middle Palaeolithic terrestrial deposits on the seabed (see http://listedmonuments.culture.gr/fek.php?ID_FEKYA=9716).

19.5.1.3 The Mani Submerged Landscape

Extrapolation from faunal remains in the caves above present sea level suggests that the now-submerged landscapes fronting the Mani caves must have once comprised a rich mosaic of lagoons and marshlands, grasslands, parklands and woodlands (Elefanti et al. 2008). These lowland biotopes supported diverse communities of plant and animal life and were important in the survival strategies of the Neanderthals and the later *Homo sapiens* groups. To judge by the faunal and marine remains recovered from the sites, marine molluscs (Fig. 19.6d) and birds contributed to subsistence as well as hunting of land mammals, and these activities must have taken place on land that is now submerged (Lebreton et al. 2008; Roger and Darlas 2008a, b). Some of the marine shells were modified to use as tools

(Darlas 2007; Douka and Spinapolice 2012) reinforcing the evidence of a strong relationship to the coast and its resources.

19.5.2 Offshore Caves and Stone Artefacts

A small number of submerged caves completes the inventory of inundated karstic contexts, of which those around Corfu are of particular note. Nic Flemming had a leading role in the exploration of the seabed around Corfu more than 30 years ago, and on the west coast, he located a submerged cave at Linodoros Point at a depth of 8–9 m bpsl (Flemming and Kazianis 1987, p. 3 and 6). A cluster of caves now at depths of 40 m bpsl off the shores of Kyra Panagia, an islet in the Northern Sporades Islands of the Aegean Sea, is also known (Flemming 1983; Efstratiou 2001). The Middle Palaeolithic stone artefacts recovered under water in the same area may originate from archaeological deposits in these caves.

Apart from these cave contexts, loose prehistoric knapped stone artefacts complete the picture. In the Ionian Sea, fishermen from Corfu reportedly trawl up Stone Age artefacts from the seafloor to the north of the island between the islands of Mathraki, Othonoi and Errikousa (Sordinas pers. comm. 2003). In 1987, research on the shore of western Corfu and offshore by Nic Flemming resulted directly from previous work by Augustus Sordinas (1983). The main acoustic surveys have not been published, although the tapes were submitted to the Ephorate of Underwater Antiquities (Flemming pers. comm. 2013). This work brought to light lithic artefacts, cores, flakes, debitage and the occasional tool of unconfirmed Stone Age date on land, scattered among the rocks on the shore. Lithics were found on the seabed at distances ranging from 200 to 700 m, at Agios Georgios and the Diaplo islet, to 2 km away from the nearest shore at Karavi (Flemming and Kazianis 1987, p. 4). A diagnostic Mousterian tool produced by the Levallois technique was recovered about 200 m offshore off the village of Aghios Georgios in SW Corfu and reported by Flemming (pers. comm. 2013)

19.5.3 Neolithic and Bronze Age Sites

19.5.3.1 Agios Petros

On the other side of Greece, the small islet of Agios Petros, in the Northern Sporades (Fig. 19.7), has a Neolithic village, part of which lies submerged to a depth of 10 m bpsl (Flemming 1983, 1985). Excavations of the terrestrial sector conducted by Theocharis in 1969–1971 and Efstratiou in 1981 brought to light houses on the outskirts of the village on a promontory with a natural harbour. The site has returned a single radiocarbon date (BM-2020R) of 5740–5530 cal BC (1σ range). Occupation began towards the end of the Early Neolithic (EN) and beginning of the Middle Neolithic (MN) and lasted for a millennium. A wide range of ceramic vessels, figurines, fishing equipment made of bone, and knapped, polished and ground stone tools were recovered (Efstratiou 1985). The typical early painted and monochrome ceramics recovered suggest affinities with the contemporaneous EN cultural traditions of Thessaly and the Cyclades, with those of Anatolia across the Aegean and with the Balkan Peninsula to the north. A unique tradition, by Greek standards, of painted (red-on-white) ceramics and figurines is recorded on site. The latter shows very characteristic resemblances with Anatolian figurines (Haçilar VI) and one of Balkan type (rod-head figurine). It is estimated that the small village was inhabited by no more than a few dozen people. They lived in rectangular stone houses/structures, of which only a few survive in a bad state of preservation due to weather erosion. Two child burials in bedrock cuttings were also found.

Dimitris Theocharis opened the way to underwater work on the islet, but his work has remained unpublished. In 1981 Nic Flemming conducted new underwater investigation to map the submarine topography and history of the coastline and to examine the presence of archaeological remains (Flemming 1983, 1985). He excavated several 1-m-square quadrats on the seabed (Fig. 19.7a, b) and sampled the sediments for particle-size analysis. Animal bones, lithics and ceramic finds were retrieved from the sea floor.

Fig. 19.7 Underwater investigations and mapping of the seabed offshore of Agios Petros, Kyra Panagia, Northeast Aegean, conducted by Nic Flemming in 1973. Upper: 1-m square grid on the seabed; lower: 1-m-square frame for use in excavation being lowered onto the sea. Photos courtesy of Nic Flemming

The study of the rich faunal assemblages found above and below sea level inform on a diet based on domesticated animals (94.2–97.4% of the assemblage) such as sheep, goat and pig. Birds including the bone of a cormorant and marine fish are also recorded. The different lines of underwater archaeological evidence indicate that marine submergence had a scattering rather than a destructive effect upon the material (Flemming 1985, p. 140). Since EN times, sea level at Agios Petros has risen in the order of 10 m. Two successive episodes of sea-level rise are identifiable, both occurring in the life-span of the Neolithic village, and they were at times rapid enough for deposits to be drowned without prolonged exposure to the sorting effects of wave action (Efstratiou 1985, p. 164). Underwater reconnaissance discovered two submarine springs, and these were probably the main water supply for this small Neolithic community.

19.5.3.2 Pavlopetri

Pavlopetri lies in the shallow waters of the Vatika Bay in the southeast of the Malea Peninsula in the Peloponnese and is perhaps the most impressive and best-known submerged site in the Aegean Basin by virtue of its size, content and preservation quality (Fig. 19.8). A century after its original discovery by geologist Phokion Negris (1904) and more than 40 years after its rediscovery by Nic Flemming (1968a, b) and early exploration by Harding et al. (1969), new offshore work has begun under the scientific direction of Jon Henderson to shed light on this site and its palaeogeographic setting (Henderson et al. 2011) in collaboration with the Ephorate of Underwater Antiquities (EUA) and the Hellenic Centre for Marine Research (HCMR).

The site was inhabited from at least the Early Bronze Age (EBA) through to the end of the Late Bronze Age. It exists on land, on the islet and on the seafloor at a depth of 1–4 m bpsl. The submerged part of the settlement, extending from about 20 m off the Pounda shore to 300 m south, consists of Bronze Age buildings, rooms, courtyards, streets, cist graves and rock-cut tombs found in situ (Fig. 19.9 and 19.10). Henderson et al. (2011, 2013) estimate that it covered an area of approximately 8 ha and had a likely maximum settlement population ranging between 500 and 2000 people (Henderson et al. 2013, p. 247). Sea level has risen by an estimated 4–5 m during the last 5000 years.

Almost 40% of the ceramic component recovered in the 2009 season dates to the EBA. Coupled with the discovery of a monumental trapezoidal building comparable with other EBA structures, all this suggests that, *contra* earlier interpreta-

Fig. 19.8 Aerial view of Pavlopetri submerged remains at Pounta Beach, Viglafia, on the north-western edge of the Bay of Vatika, Peloponnese. Modified after Henderson et al. 2013

Fig. 19.9 Submerged architectural remains at Pavlopetri, with remains of a stone wall showing the outline of a rectangular building. Photo courtesy of Jon Henderson and the Pavlopetri Project

Fig. 19.10 Close-up of the remains of a wall foundation at Pavlopetri. Photo courtesy of Jon Henderson and the Pavlopetri Project

tions, settlement at Pavlopetri may indeed be a typical case of the EBA preference for promontories with immediate access to the sea and its resources of marine foods and opportunities for seaborne communication. Pottery finds at Pavlopetri are suggestive of links with contemporaneous Cycladic, Cretan and north-eastern Aegean cultural traditions (Henderson et al. 2011, 2013).

19.5.3.3 Platiyali

The submerged EBA settlement at Platiyali is situated in the innermost part of a closed 1 km-wide and 2 km-long bay south of Astakos in Aetoloakarnania. The bay lies between the Capes of Carlo Glossa and Steni Gonia and its entrance is protected by the Echinades islets in the Inner Ionian Sea. The underwater investigation in the area took place in 1986 because of planned construction of a ship-breaking yard. To judge from the slight seabed slope, the area where the settlement was established must have been a low coastal plain before it became inundated. Freshwater springs are present today both under water and on the shoreline. Building walls are discernible across a 400-m zone along the shoreline to a distance of 130 m from the shore, and to a depth of at least 5 m bpsl. The settlement area lying on the seabed is estimated to cover at least 5 ha and was submerged at 4.5 ka BP. A remarkable finding is the burial of an infant in a vessel. The settlement layout, the architectural forms, including examples of an Early Helladic megaron at 23 × 11 m, the ceramic shapes and techniques, the stone-knapping technology and use of obsidian and the presence of child-burials in jars, all lead to the conclusion that Platiyali conforms to the general character of the Early Helladic II culture in the Aegean (Dellaporta and Spondylis 1987, 1988; Dellaporta et al. 1988, 1990).

The coastal location of Platiyali and the rather difficult access to it from the land meant that the inhabitants naturally looked to the sea to provide the main route of communication with other cultural centres. It is therefore quite likely that the site was an important port-of-call for the maritime trade in obsidian, acting as a gateway community into north-west Greece. The submergence of the settlement some 4500 years ago was the reason for its preservation to this day.

19.5.3.4 Salanti

Salanti Argolidas is a third EBA settlement, situated in the middle of the Didymon cove, just north of the Franchthi Cave. It extends 400 m on a north-west–south-east axis and is approximately 30 m wide in parallel with the beach and at a depth of 4 m bpsl. The Ephorate of Underwater Antiquities conducted a seabed survey and estimated the settlement area at 8.9 ha. The slight slope of the seabed indicates that the settlement was originally located in a low valley and at some distance from the sea. Visible on the seabed are foundation walls made of rough stones and of two types: a double layer of 50–60 cm thickness with intermediate fill and a single layer of 30 cm thickness. A semi-circular building 4 m in diameter belongs to the latter type. Both types are often found in EBA settlements in the Peloponnese and Central Greece. The inventory of finds includes beak-spouted jugs, domestic coarse ware, open vessels with textured impressions on the upper part and grinding stones and piercers made on volcanic material from the Saronic Gulf. The typology of ceramic finds suggests an EH II date (Blackman 2001, p. 30).

The geographic location and topography of Pavlopetri, Platiyiali and Salanti corresponds to those of many Aegean sites, where occupation begins in the Early Bronze Age and often runs through to the end of this period. The inventory of villages partly submerged or eroded by the sea is long: Manika on Euboea (Sampson and Maroukian 1989), Agia Irini on Kea (Caskey 1962, 1964, 1966, 1971, 1972; Mourtzas 2010), Nea Makri in Attica (Theocharis 1954, 1956; Gkopha 2000; Kourtir and Chalakies on Lesbos (Chatzi 1986; Spencer 1995; Argyri and Theodoulou 2012) and Sissi Spiliada and Psathi on Crete (Hood 1965; Mytilinaiou 1997; Simossi 2003; Theodoulou 2015). At Agia Irini, the occupation began during the Final Neolithic. Built in coastal settings, preferably located on peninsulas or promontories by natural harbours, these sites were excellent departure and arrival points for sea journeys to the wider world. They were usually involved with transport of goods and commerce through maritime routes and are evidence of a widespread coastally oriented way of life. At Agios Kosmas in Attica, an EBA coastal cemetery is submerged in shallow water in the Saronic Gulf (Mylonas 1952, 1959).

19.5.3.5 The Glyfada-Mesi Hoard

In recent years, one of the most dramatic underwater discoveries is the EBA hoard of 110 bronze axes discovered on the seabed of Glyfada-Mesi in Rhodope, Thrace (Simossi 2009), the largest hoard of this period found in Greece or the neighbouring Balkans. It was found at a depth of 3.5 m bpsl and at a distance of 450 m from the nearest shore. A type of double axe with two percussion surfaces is the dominant artefact type, while lath-hammers and single-bladed axes are also present in smaller numbers. The submerged site extends to an area of approximately 10 m^2 (Morgan et al. 2009, p. 67). Embedded in the sea floor below the tools are the bases of two EH vessels and slag with impressions of basket-work is also present. Simossi (2009, p. 100) suggests that the hoard was deliberately concealed on dry land, a recognized practice of the period in sites of the mainland, north and central Aegean, before the sea level rose.

19.6 Mapping the Prehistoric Record of the Greek Shelf

A new direction of research lies in the development of collaborations between archaeologists and oceanographers to map the submerged landscape. Three such projects are currently underway, both using known Stone Age sites on land as a starting point for offshore exploration.

The first project 'Exploring the Submerged Caves and Prehistoric Landscapes of the Inner Ionian Sea Archipelago' is focussed on the coastal and inundated landscape of the semi-enclosed marine area between the western coast of Central Greece (Akarnania), and the islands of Lefkas, Kephallinia and Ithaca. The point of departure is the record of Middle Palaeolithic sites on Lefkas and the islands and islets of the Inner Ionian Sea (Galanidou 2015, 2018; Galanidou et al. 2016b, 2018; Papoulia 2017, 2018). The area was frequented by Neanderthal groups and the palaeogeography of settlement, and the nature of the interconnections between the present-day islands would have been especially sensitive to sea-level change. A seismic reflection survey conducted in the northern part of the archipelago has given new information on the palaeogeographic evolution of the area during the Late Quaternary low sea-level periods of MIS 2, 6 and 8 (Zavitsanou et al. 2015; Zavitsanou 2016). With a single exception, that of Atokos, the islands of the Inner Ionian Sea Archipelago were connected to the mainland during low sea-level stands and easily accessible on foot. The archaeology of Atokos has significant implications for the early history of sea crossings in a semi-enclosed and protected sea where the destination was visible and close at hand (Galanidou 2018; Papoulia 2018). Further out to sea, to the south-west, the islands of Kephallinia and Ithaca have remained separated from the mainland throughout the last 300 kyr.

The second project has a longer history that began in the late 1970s and early 1980s with the search in the Bay of Kiladha in the Western Argolic Gulf for a submerged Neolithic settlement related to the Franchthi Cave occupation. On this occasion, marine geophysical surveying and coring in the bay was conducted (Van Andel et al. 1980; Van Andel 1987; Gifford 1983, 1990). Gifford reported that Neolithic pottery was found in situ in sediments at a water depth of 4.5 m, and 5.5 m below the sediment surface. Two cores were taken through the seabed midway between the Franchthi headland and Koronis Island, offering conclusive evidence that a palaeo-river flowed through this strait, with occupation on the banks of the river.

The search for submerged traces of prehistoric human activity near the cave was resumed in 2012 by the Bay of Kiladha project and was intensified in 2014 with the Terra Submersa expedition, a collaboration between the University of Geneva, the Hellenic Centre for Marine Research, the Greek Ephorate of Underwater Antiquities, PlanetSolar, the Laténium (Switzerland) and the Swiss School of Archaeology in Greece (Beck et al. 2017; see https://www.unige.ch/terrasubmersa/en). Offshore and nearshore surveys were conducted on board two vessels: the Swiss solar-powered PlanetSolar and HCMR's research vessel Alkyon. Detailed swath bathymetry, side-scan sonar imaging and sub-bottom profiling in the wider Gulf and in Koilada Bay, along with offshore drilling in the Koilada Bay offshore of Franchthi Cave in collaboration with the University of Bern, revealed detailed information on palaeoshorelines of the Neolithic period and earlier (MIS 2 and 6). They identified the exposed Neolithic surface and demonstrated the presence of pottery sherds, one dating to the Early to Middle Neolithic period (Sakellariou et al. 2015; Surdez et al. 2018; Beck et al. 2017).

The third project was instigated by archaeological work conducted on Lesbos Island, in the northeast Aegean Sea, which began in 2012 with the investigation of Rodafnidia, Lisvori, the first large-scale Acheulean site in south-east Europe and western Anatolia, dated by relative and p-IRSL methods to the Middle Pleistocene (Galanidou et al. 2013, 2016a). A systematic marine survey in the Kalloni Gulf and off the southern coast of the island has been coordinated by the Department of History and Archaeology of the University of Crete and implemented by the Hellenic Centre for Marine Research in collaboration with the Department of Marine Sciences of the Aegean University with the aim of reconstructing the palaeoenvironment and the palaeoshorelines during the low sea-level periods of the last 500 kyr in the broader area. Processing and interpretation of the seismic profiling data are in progress and are expected to shed light on the palaeogeography of the area and on the terrestrial connection between Lesbos and the west Anatolia.

19.7 Management of the Underwater Cultural Heritage

The protection of antiquities is enshrined in the Constitution of Greece (Article 24), and the government organization legally responsible for the protection of archaeological sites and monuments is the General Directorate of Antiquities and Cultural Heritage under the Ministry of Culture and Sports. Responsibility for underwater archaeology lies with the Ephorate of Underwater Antiquities (EUA), founded in 1976, https://www.culture.gr/en/ministry/SitePages/viewyphresia.aspx?iID=1369. Under the Archaeological Law of 2002 https://www.forin.gr/laws/law/2795/gia-thn-prostastia-twn-arxaiothtwn-kai-en-genei-ths-politistikhs-klhronomias, specific regulations were introduced for the first time to cover the entire spectrum of underwater antiquities with regard to their protection and management as a whole, in keeping with the United Nation's Convention on the Law of the Sea and in the spirit of the UNESCO Convention on the Protection of the Underwater Cultural Heritage, although Greece is not yet a formal signatory to the latter Convention (Dellaporta 2004, 2013, 2014; see also Peeters et al., Chap. 24, this volume; Dromgoole, Chap. 26, this volume). The mission of the EUA is the exploration and protection of the underwater heritage of the Greek seas, rivers and lakes, the discovery, study, conservation and public promotion and understanding of the submerged ancient heritage, including shipwrecks, settlements and harbours, declaration of protected underwater areas and supervision of oceanographic research and works conducted beneath the sea (for a history of EUA and its activities, see Kritzas 1978; Dellaporta 2002; Simossi 2009; Theodoulou 2011).

The Hellenic Centre for Marine Research (HCMR) of the Ministry of Education is a large-scale marine research institute that has contributed significantly to the exploration of the submerged cultural heritage, leading to the discovery and study of numerous deep- and shallow-water archaeological sites (https://www.hcmr.gr/en). It owns and operates three oceanographic vessels, a two-person submersible and four remotely operated vehicles (ROVs), which support the activities of all research sectors and are being used for the exploration of the marine environment and the seafloor. The HCMR has established the new field of Marine Geo-Archaeology with its main objective the multidisciplinary research of submerged antiquities and landscapes (Sakellariou 2010; Sakellariou et al. 2010, 2011). It has introduced the use of state-of-the-art marine techniques in deep and shallow water and archaeological surveys promoting collaboration between marine geoscientists, engineers and archaeologists.

Both institutions have been working to improve knowledge, raise public interest and contribute to the conservation and protection of the underwater cultural heritage in line with the UNESCO Convention, EU declarations and the national law.

Acknowledgements We would like to thank Augustus Sordinas and Nic Flemming for sharing with us their precious knowledge on the archaeology of the seabed of Greece, a knowledge that largely remains unpublished. Hauke Jöns has provided much-appreciated support in assembling the geographic evidence for the Greek submerged material included in the SPLASHCOS Viewer. Moritz Mennenga has been efficient in retrieving information from the data base and producing various versions of the map of the submerged archaeology of Greece at very short notice. Nikos Efstratiou, Evgenia Yiannouli, Andreas Darlas and Thanasis Yiannopoulos shared with us information on their coastal and underwater fieldwork, photos and bibliography. Panos Zervoudakis contributed valuable bibliographic support and editorial skills especially during the finishing stages of this work. They are all warmly acknowledged.

References

Ammerman AJ (2014) Setting our sights on the distant horizon. In: Ammerman AJ, Davis T (eds) Proceedings of the Wenner Gren Foundation workshop on Island archaeology and the origins of seafaring in the eastern Mediterranean, Reggio Calabria, October 19–21, 2012. J Eurasian Prehistory 10 (1–2):203–236

Argyri S, Theodoulou T (2012) Archaeological survey in the Kalloni Gulf Lesbos (2005). Enalia XI Annual, IENAE, Athens, pp 100–109 (in Greek)

Baika K (2008) Archaeological indicators of relative sea-level changes in the Attico-Cycladic massif: preliminary results. Bull Geol Soc Greece 42(2):33–48

Bartsiokas A, Arsuaga JL, Aubert M, Grün R (2017) U-series dating and classification of the Apidima 2 hominin from Mani Peninsula, Southern Greece. J Hum Evol 109:22–29

Bassiakos IE (1993) Dating of fossils from caves and speleothems: evidence from Electron Spin Resonance (E.S.R.) technique, the study of underground karst morphology and the relevant radiometric and geological conditions in speleoenvironments of Dyros, Mani. Unpublished PhD thesis submitted to the University of Athens (in Greek)

Beck J, Sakellariou D, Koutsoumba D (2017) Submerged Neolithic landscapes off Franchthi Cave: the measurements from the Terra Submersa expedition and their implications. In Sarris A, Kalogiropoulou E, Kalayci T, Karimali L (eds) Communities, landscapes, and interaction in Neolithic Greece. Proceedings of the International conference Rethymno 29–30 May 2015, Archaeological series 20, pp 261–268

Blackman D (2001) Archaeology in Greece 2000-2001. Archaeol Rep 47:1–144

Caskey LJ (1962) Excavations in Keos 1966-1970. Hesperia 31:263–283

Caskey LJ (1964) Excavations in Keos 1966-1970. Hesperia 33:314–335

Caskey LJ (1966) Excavations in Keos 1966-1970. Hesperia 35:363–376
Caskey LJ (1971) Investigations in Keos I. Hesperia 40:358–396
Caskey LJ (1972) Investigations in Keos II. Hesperia 40:357–401
Chatzi D (1986) Skamnioud-Kourtir. Archaeologikon Deltion 41:484
Darlas A (2007) Le Moustérien de Grèce à la lumière des récentes recherches. L'Anthropologie 111:346–366
Darlas A (2012) Geomorphological evolution and occupation of the caves of the western coast of Mani during the late Pleistocene and Holocene. In: Zacharias N, Georgakopoulou M, Polykreti K, Fakorellis G, Vakoulis T (eds) Proceedings of the 5th symposium of the Greek Archaeometry Society 2008. Papazisi, Athens, pp 237–253 (in Greek)
Darlas A, De Lumley H (1995) Fouilles franco-helléniques de la grotte de Kalamakia (Aréopolis, Péloponnèse). Bull Corresp Hell 119(2):793–798
Darlas A, De Lumley H (1998) Fouilles franco-helléniques de la grotte de Kalamakia (Aréopolis; Péloponnèse). Bull Corresp Hell 122(2):655–661
Darlas A, De Lumley H (2002) Fouilles franco-helléniques de la grotte de Kalamakia (Aréopolis, Péloponnèse). Bull Corresp Hell 126(2):685–689
Darlas A, Psathi E (2008) Le Paléolithique supérieur dans la péninsule du Mani (Péloponnèse, Grèce). In: Darlas A, Mihailovic D (eds) The Palaeolithic of the Balkans, British archaeological reports international series 1819. Archaeopress, Oxford, pp 51–59
Darlas A, Psathi E (2016) The middle and upper Palaeolithic in the West of Mani Peninsula (Southern Greece). In: Harvati K, Roksandic M (eds) Paleoanthropology of the Balkans and Anatolia: human evolution and its context, Vertebrate paleobiology and paleoanthropology series. Springer, Dordrecht, pp 95–118
Dellaporta C (2002) Underwater cultural heritage in Greece. Problems of protection and promotion. Strumenti per la protezione del patrimonio culturale marino aspetti archeologici. Atti del convegno (8–10 marzo 2001), pp 125–128
Dellaporta KP (2004) Tutela legale e gestione del patrimonio archeologico subacqueo in Grecia. Mediterraneum 4:65–73
Dellaporta KP (2013) Das Archäologische Kulturerbe unter Wasser in Griechenland – Erforschung, Gefährdung und Schutzmaβnahmen. In: Reinfeld M (ed) Archäologie im Mittelmeer. Auf der Suche nach versunkenen Schiffswracks und vergessenen Häfen, Αντικεn Welt. Philipp von Zabern, Darmstadt, pp 73–80
Dellaporta KP (2014) The international and national legislation for the protection of underwater antiquities. In: Choúlia-Kapelonī S (ed) 3rd international conference of experts on the return of cultural property, Athens - ancient Olympia, October 23–26, 2013: proceedings. Archaeological Receipts Fund, Athens, pp 87–92
Dellaporta A, Spondylis I (1987) Un habitat Helladique Ancien II à Platiyali Astakou. In: Tzalas HE (ed) Tropis II. 2nd International symposium on ship construction, Delphi 1987, Proceedings. Hellenic Institute for the Preservation of Nautical Tradition, Athens, pp 27–29
Dellaporta A, Spondylis I (1988) Platiyiali Astakos: a submerged Early Helladic settlement. Proceedings of the first archaeological and historical conference of Aetoloakarnania, Agrinio, 21–23 October 1988. Agrinio, pp 39–46 (in Greek)
Dellaporta A, Spondilis E, Baxevanakis Y (1988) Early Helladic settlement at Platiyiali Astakos (preliminary observations). Anthropologica Analecta 49:7–19 (in Greek)
Dellaporta A, Spondilis E, Baxevanakis Y (1990) Platiyali-Astakos: a submerged Early Helladic site in Akarnania. Enalia 1:44–46
Douka K, Spinapolice EE (2012) Neanderthal shell tool production: evidence from Middle Palaeolithic Italy and Greece. J World Prehistory 12:45–79
Efstratiou N (1985) A Neolithic site in the Northern Sporades. Aegean relationships during the Neolithic of the 5th millennium BC. BAR International series 155. British Archaeological Reports, Oxford
Efstratiou N (2001) The Neolithic settlement of Agios Petros in Kyra-Panagia, Halonnesos and the island habitation patterns in the Aegean – a reconsideration. In: Sampson A (ed) Archaeology in the Northern Sporades. Municipality of Alonessos, Alonessos, pp 231–250 (in Greek)
Efstratiou N (2005) Tracing the story of the first farmers in Greece–a long and winding road. In: Lichter C (ed) How did farming reach Europe? Anatolian-European relations from the second half of the 7th through the first half of the 6th millennium cal BC: proceedings of the international workshop, Istanbul, 20–22 May 2004 (BYZAS 2). Deutsches Archäologisches Institut Abteilung, Istanbul, pp 143–153
Elefanti P, Panagopoulou E, Karkanas P (2008) The transition from the Middle to the Upper Palaeolithic in the southern Balkans: The evidence from the Lakonis I Cave, Greece. Eurasian Prehistory 5:85–95
Evans JD (1994) The early millennia: Continuity and change in a farming settlement. In: Hood S, Evely D, Evely RDG, Hughes-Brock H, Momigliano N (eds) Knossos: a labyrinth of history. Papers presented in honour of Sinclair Hood. British School at Athens, London, pp 1–20
Ferentinos G, Gkioni M, Geraga M, Papatheodorou G (2012) Early seafaring activity in the southern Ionian Islands, Mediterranean Sea. J Archaeol Sci 39:2167–2176. https://doi.org/10.1016/j.jas.2012.01.032
Flemming NC (1968a) Holocene earth movements and Eustatic sea level change in the Peloponnese. Nature 217:1031–1032
Flemming NC (1968b) Mediterranean sea level changes. Sci J 4:51–55
Flemming NC (1978) Holocene Eustatic changes and coastal tectonics in the northeast Mediterranean: implications for models of crustal consumption. Philoso Trans R Soc London 289(1362):405–458

Flemming NC (1983) Preliminary geomorphological survey of an early Neolithic submerged site in the Sporadhes, N. Aegean. In: Masters PM, Flemming NC (eds) Quaternary coastlines and marine archaeology. Academic Press, London, pp 233–268

Flemming NC (1985) Preliminary geomorphological survey of the Neolithic site of Agios Petros. In: Efstratiou N (ed) A Neolithic site in the Northern Sporades. Aegean relationships during the Neolithic of the 5th millennium BC, BAR International series 155. British Archaeological Reports, Oxford, pp 137–141

Flemming NC, Kazianis D (1987) Preliminary report on the Otranto Straits project. Unpublished report submitted to the Department of Antiquities, Underwater Section, Greece

Flemming NC, Czartoryska NMG, Hunter PM (1973) Archaeological evidence for eustatic and tectonic components of relative sea level change in the South Aegean. In: Blackman DJ (ed) Marine archaeology, Colston Papers 23. Butterworth Press, London, pp 1–66

Frost FJ (1977) Phourkari. A villa complex in the Argolid (Greece). Int J Naut Archaeol 6(3):233–238

Galanidou N (2011) Mesolithic cave use in Greece and the mosaic of human communities. J Mediterr Archaeol 24(2):219–242

Galanidou N (2015) Seascape survey on the Inner Ionian Sea Archipelago. In: Carver M, Gaydarska B, Monton S (eds) Field archaeology from around the world. Ideas and approaches. Springer, Cham, pp 101–106

Galanidou N (2018) Parting the waters: Palaeolithic archaeology in the central Ionian Sea. J Greek Archaeol 3:1–22

Galanidou N, Cole J, Iliopoulos G, McNabb J (2013) East meets west: The Middle Pleistocene site of Rodafnidia on Lesvos, Greece. Antiquity Project Gallery 87(336). http://www.antiquity.ac.uk/projgall/galanidou336/

Galanidou N, Athanassas C, Cole J, Iliopoulos G, Katerinopoulos A, Magganas A, McNabb J (2016a) The Acheulian Site at Rodafnidia, Lisvori, on Lesbos, Greece: 2010-2012. In: Harvati K, Roksandic M (eds) Paleoanthropology of the Balkans and Anatolia: human evolution and its context, Vertebrate paleobiology and paleoanthropology series. Springer, Dordrecht, pp 119–138

Galanidou N, Iliopoulos G, Papoulia C (2016b) The Palaeolithic settlement of Lefkas. Archaeological evidence in a palaeogeographic context. J Greek Archaeol 1:1–31

Galanidou N, Vikatou O, Gatsi-Stavropoulou M, Vasilakis A, Iliopoulos G, Staikou V, Veikou M, Forsén J, Morgan C, Vroom J, Papoulia C, Zervoudakis P, Prassas K (2018) The archaeological survey in the Inner Ionian Sea Archipelago. In: Vikatou O, Staikou V, Saranti F (eds) Archaeological work in Aetoloakarnania and Lefkas from prehistoric to Roman times. Proceedings of the 2nd International archaeological and historical conference, December 6-8 2013, Messolonghi. Messolonghi pp 445–473 (in Greek)

Giannopoulos B (2000) Contribution to the study of modern and old environments of the most important Greek caves. Unpublished PhD thesis, University of Athens (in Greek)

Gifford J (1983) Core sampling of a Holocene marine sedimentary sequence and underlying Neolithic cultural material off Franchthi Cave, Greece. In: Masters PM, Flemming NC (eds) Quaternary coastlines and marine archaeology. Academic Press, London, pp 269–282

Gifford JA (1990) Analysis of submarine sediments off Franchthi Cave. In: Wilkinson TJ, Duhon ST (eds) Franchthi Paralia: the sediments, stratigraphy, and offshore investigations. Indiana University Press, Bloomington/Indianapolis, pp 85–116

Gkopha MP (2000) Neolithic Attica (Vol. 13). Archeological Society at Athens (in Greek)

Guest-Papamanoli A (1989) Les carrières de grès dunaire à Malia. Bull Corresp Hell 113(1):113–122

Guest-Papamanoli A, Treuil R (1979) Bâtiment immergé. Bull Corresp Hell 103:668–669

Harding A, Cadogan G, Howell R (1969) Pavlopetri: an underwater Bronze Age town in Laconia. Annu Br Sch Athens 64:113–142

Harvati K, Panagopoulou E, Karkanas P (2003) First Neanderthal remains from Greece: the evidence from Lakonis. J Hum Evol 45:465–473

Harvati K, Stringer C, Karkanas P (2010) Multivariate analysis and classification of the Apidima 2 cranium from Mani, Southern Greece. J Hum Evol 60(2):246–250

Harvati K, Darlas A, Bailey SE, Rein TR, El Zaatari S, Fiorenza L, Kullmer O, Psathi E (2013) New Neanderthal remains from Mani peninsula, Southern Greece: The Kalamakia Middle Paleolithic cave site. J Hum Evol 64(6):1–14

Henderson J, Gallou C, Flemming NC, Spondylis E (2011) The submerged prehistoric town at Pavlopetri in southern Greece. In: Benjamin J, Bonsall CL, Pickard C, Fischer A (eds) Submerged prehistory. Oxbow, Oxford, pp 207–218

Henderson J, Pizarro O, Johnson-Roberson M, Mahon I (2013) Mapping submerged archaeological sites using stereo-vision photogrammetry. Int J Naut Archaeol 42(2):243–256

Hood MSF (1965) Minoan sites in the far west of Crete. The Annual of the British School at Athens 60:99–113

Kapsimalis V, Pavlopoulos K, Panagiotopoulos I, Drakopoulou P, Vandarakis D, Sakellariou D, Anagnostou C (2009) Geoarchaeological challenges in the Cyclades continental shelf (Aegean Sea). Zeitschrift für Geomorphologie, Supplementary Issue 53(1):169-190

Kopaka K, Matzanas C (2009) Palaeolithic industries from the island of Gavdos, near neighbour to Crete in Greece. Antiquity Project Gallery 83(321). http://antiquity.ac.uk/projgall/kopaka321/

Koufos GD, Kostopoulos DS (2016) The Plio-Pleistocene large mammal record of Greece: implications for early human dispersals into Europe. In: Harvati K, Roksandic M (eds) Paleoanthropology of the Balkans and Anatolia: human evolution and its

context, Vertebrate paleobiology and paleoanthropology series. Springer, Dordrecht, pp 269–280
Kritzas C (1978) Underwater archaeology in Greece. In: Rackl HW (ed) Diving into the past. Gutenberg, Athens, pp 414–429 (in Greek)
Lambeck K (1995) Late Pleistocene and Holocene sea-level change in Greece and south-western Turkey: a separation of eustatic, isostatic, and tectonic contributions. Geophys J Int 122:1022–1044. https://doi.org/10.1111/j.1365-246X.1995.tb06853.x
Lambeck K (1996) Sea-level change and shoreline evolution in Aegean Greece since Upper Palaeolithic time. Antiquity 70:588–611
Lambeck K, Rouby H, Purcell A, Sun Y, Sambridge M (2014) Sea level and global ice volumes from the Last Glacial Maximum to the Holocene. Proc Natl Acad Sci 111(43):15296–15303
Lebreton V, Psathi E, Darlas A (2008) Environnement vegetal des Neanderthaleniens de la grotte de Kalamakia (Areopolis, Grèce). In: Darlas A, Mihailovic D (eds) The Palaeolithic of the Balkans, British archaeological reports international series 1819. Archaeopress, Oxford, pp 61–68
Lykousis V (2009) Sea-level changes and shelf break prograding sequences during the last 400ka in the Aegean margins: subsidence rates and palaeogeographic implications. Cont Shelf Res 29(16):2037–2044
Lyras GA, van der Geer AA (2007) The late Pliocene vertebrate fauna of Vatera (Lesvos Island, Greece). Cranium 24(2):11–24
Morgan C, Pitt RK, Whitelaw T (2009) Archaeology in Greece 2008–2009. Archaeol Rep 55:1–101
Mourtzas ND (2010) Sea level changes along the coasts of Kea island and paleogeographical coastal reconstruction of archaeological sites. Bull Geol Soci Greece 43(1):453–463
Mouslopoulou V, Begg J, Nicol A, Oncken O, Prior C (2015) Formation of late quaternary paleoshorelines in Crete, Eastern Mediterranean. Earth Planet Sci Lett 431:294–307
Mylonas GE (1952) Aghios Kosmas excavation 1951. Archaeologiki Efimeris 91:117–134 (in Greek)
Mylonas GE (1959) Aghios Kosmas: an early bronze settlement and cemetery in Attica. Princeton University Press, Princeton
Mytilinaiou DK (1997) Proanaktoriki keramiki apo ti thesi Psathi Kydonias. Kritiki Estia 6:194–236
Negris P (1904) Vestiges antiques submergés. Mitteilungen des Deutschen Archäologischen Instituts, Athenische Abteilung 29:340–363
Oekonomides M, Drosogianni P (1989) A hoard of gold Byzantine coins from Samos. Rev Numismatique 31:145–182
Panagopoulou E, Karkanas P, Tsartsidou G, Kotjabopoulou E, Harvati K, Ntinou M (2004) Late Pleistocene archaeological and fossil human evidence from Lakonis cave, southern Greece. J Field Archaeol 29(3–4):323–349
Papathanasopoulos GA (2011) Neolithic Diro, Alepotrypa Cave, vol I. Melissa and Diro Centre of Neolithic Studies, Athens (in Greek)
Papoulia C (2017) Seaward dispersals to the NE Mediterranean islands in the Pleistocene. The lithic evidence in retrospect. Quat Int 431:64–87
Papoulia C (2018) Pleistocene sea-crossings and submerged terrestrial routes. A view from the Inner Ionian Archipelago. PhD Dissertation, University of Crete
Perissoratis C, Conispoliatis N (2003) The impacts of sea-level changes during latest Pleistocene and Holocene times on the morphology of the Ionian and Aegean seas (SE Alpine Europe). Mar Geol 196:145–156
Perlès C (1979) Des navigateurs méditerranéens il y a10,000 ans. La Recherche 96:82–83
Pirazzoli PA (1988) Sea-level changes and crustal movements in the Hellenic Arc (Greece). The contribution of archaeological and historical data. In: Raban A (ed) Archaeology of coastal changes, British archaeological reports international series 404. Archaeopress, Oxford, pp 157–184
Pirazzoli PA, Thommeret J, Thommeret Y, Laborel J, Montaggioni L (1982) Crustal block movements from Holocene shorelines: Crete and Antikythira (Greece). Tectonophysics 86:27–43
Pirazzoli PA, Laborel J, Stiros SC (1996) Earthquake clustering in the Eastern Mediterranean during historical times. J Geophys Res Solid Earth 101(B3):6083–6097
Psychoyos O (1988) Déplacements de la ligne de rivage et sites archéologiques dans les régions côtières de la Mer Égée, au Néolithique et à l'Âge du Bronze, vol. 62. P. Åströms förlag, Jonsered
Renfrew C, Aspinall A (1990) Aegean obsidian and Franchthi Cave. In: Perlès C (ed) Les industries lithiques taillées de Franchthi (Argolide, Grèce), tome 2: Les industries lithiques du Mésolithique et du Néolithique initial, Excavations at Franchthi Cave, fasc, 5. Indiana University Press, Bloomington/Indianapolis, pp 257–70
Roger T, Darlas A (2008a) Upper Pleistocene bird remains from Kalamakia Cave (Greece). In: Darlas A, Mihailovic D (eds) The Palaeolithic of the Balkans, British archaeological reports international series 1819. Archaeopress, Oxford, pp 69–76
Roger T, Darlas A (2008b) Microvertèbres, paléoenvironnment et paléoclimat de la grotte de Kalamakia (Péloponnèse, Grèce). In: Darlas A, Mihailovic D (eds) The Palaeolithic of the Balkans, British archaeological reports international series 1819. Archaeopress, Oxford, pp 77–84
Sakellariou D (2010) Submerged cultural remains in long-term uplifting regions: examples from the Hellenic Arc. In: Proceedings, INQUA 501 – IGCP 521 Sixth plenary meeting and field trip, Hydrobiological Station of Rhodes, 27 Sept–6 Oct 2010, Rhodes
Sakellariou D, Galanidou N (2016) Pleistocene submerged landscapes and Palaeolithic archaeology in the tectonically active Aegean region. In: Harff J, Bailey G, Lüth F (eds) Geology and archaeology: submerged landscapes of the continental shelf. Geological Society London Special Publications 411, London, pp 145–178. https://doi.org/10.1144/SP411.9
Sakellariou D, Galanidou N (2017) Aegean Pleistocene landscapes above and below sea-level:

Palaeogeographic Reconstruction and hominin dispersals. In: Bailey GN, Harff J, Sakellariou D (eds) Under the sea: archaeology and palaeolandscapes of the continental shelf. Springer, Cham, pp 335–359

Sakellariou D, Tsampouraki-Kraounaki K (2018) Plio-Quaternary extension and strike-slip tectonics in the Aegean. In: Duarte J (ed) Transform plate boundaries and fracture zones. Elsevier, New York

Sakellariou D, Lykousis V, Rousakis G (2010) Reconstructing prehistoric landscapes in tectonically active regions: the Corinth and North Evia prehistoric lakes during LGM. XIX Congress of the Carpathian-Balkan Geological Association/Special Session S30, Thessaloniki, Greece, 23–26 September 2010

Sakellariou D, Bailey G, Flemming NC (2011) "Project Deukalion" discovering Europe's submerged prehistoric landscapes on the seabed. In: DG MARE, Lunchtime conference – submerged prehistoric landscapes, Brussels, 28 June 2010. http://www.splashcos.org/research/dpg

Sakellariou D, Lykousis V, Geraga M, Rousakis G, Soukisian T (2017) Late Pleistocene environmental factors of the Aegean Region (Aegean Sea including the Hellenic Arc) and the identification of potential areas for seabed prehistoric sites and landscapes. In: Flemming NC, Harff J, Moura D, Burgess A, Bailey GN (eds) Submerged landscapes of the European continental shelf: quaternary paleoenvironments. Wiley, Chichester, pp 405–429

Sakellariou D, Beck J, Rousakis G, Georgiou P, Panagiotopoulos I, Morfis I, Tsampouraki-Kraounaki K, Zavitsanou A (2015) Submerged prehistoric landscapes off Franchthi Cave, East Argolic Gulf: Preliminary results. In: 11th Panhellenic symposium on oceanography and fisheries, proceedings. Mytilene, pp 993–996

Sampson AA, Maroukian H (1989) The Coastal topography and archaeology of Manika, an Early Helladic town in West Central Euboea Greece. In: Parish R (ed) Archaeometry. Proceedings of the 25th International symposium Athens 19–23 May 1986. Elsevier, New York, pp 461–467

Sampson AA, Kaczanowska M, Kozłowski JK (2010) The prehistory of the island of Kythnos (Cyclades, Greece) and the Mesolithic settlement at Maroulas. Polish Academy of Arts and Sciences, Kraków

Scranton R, Shaw JW, Ibrahim L (1978) Topography and architecture (Kenchreai I). Brill, Leiden

Shaw JW (1990) Bronze Age Aegean harboursides. In: Hardy DA, Doumas CG, Sakellarakis JA, Warren PM (eds) Thera and the Aegean World III, vol. I, Archaeology. Proceedings of the third international congress, Santorini, Greece, 3–9 September 1989. The Thera Foundation, London, pp 420–436

Shaw B, Ambraseys NN, England PC, Floyd MA, Gorman GJ, Higham TFG, Jackson JA, Nocquet JM, Pain CC, Piggott MD (2008) Eastern Mediterranean tectonics and tsunami hazard inferred from the AD 365 earthquake. Nat Geosci 1:268–276

Simossi A (1988) The ancient port of Samos. Archaeologica Analekta ex Athinon XXI: 111–125 (in Greek)

Simossi A (1993) Le port du guerre de Thasos. In: Lazarov M, Angelova C (eds) Thracia Pontica 6. La Thrace et les sociétés maritimes anciennes. Sozopol, pp 271–286

Simossi A (2003) A Coastal Minoan settlement at Speliada, Seision, Crete. Enalia Annual VII: 57–65 (in Greek)

Simossi A (2009) Ephorate of Underwater Antiquities. 30 years of research activity. Archaiologia 115:95–105 (in Greek)

Sordinas A (1983) Quaternary shorelines in the region of Corfu and adjacent islets, western Greece. In: Masters PM, Flemming NC (eds) Quaternary coastlines and marine archaeology. Academic Press, London/New York, pp 335–344

Spencer N (1995) A gazeteer of archaeological sites in Lesbos, British archaeological reports international series 623. Archaeopress, Oxford

Stiros S (2001) The AD 365 Crete earthquake and possible seismic clustering during the fourth to sixth centuries AD in the Eastern Mediterranean: a review of historical and archaeological data. J Struct Geol 23:545–562

Strasser TF, Panagopoulou E, Runnels CN, Murray PM, Thompson N, Karkanas P, McCoy FW, Wegmann KW (2010) Stone age seafaring in the Mediterranean: evidence from the Plakias region for Lower Palaeolithic and Mesolithic habitation of Crete. Hesperia 79:145–190. https://doi.org/10.2972/hesp.79.2.145

Surdez M, Beck J, Sakellariou D, Vogel H, Birchler EP, Koutsoumba D, Anselmetti FS (2018) Flooding a landscape: impact of Holocene transgression on coastal sedimentology and underwater archaeology in Kiladha Bay (Greece). Swiss J Geosci. Published online 03 May 2018. https://doi.org/10.1007/s00015-018-0309-4

Symeonides NK, Theodorou GE, Giannopoulos VI (2001) New data on *Elephas chaniensis* (Vamos cave, Chania, Crete). In: Cavarretta G, Gioia P, Mussi M, Palombo MR (eds) The world of elephants. Proceedings of the First International Congress. Consiglio Nazionale delle Ricerche, Rome, pp 510–513

Theocharis DR (1954) Excavation of a Neolithic settlement in Nea Makri. PAE 1954:114–122 (in Greek)

Theocharis DR (1956) Nea Makri, Eine grosse Neolithische Siedlung in der nahe von Marathon. Mitteilungen des Deutschen Archäologischen Instituts, Athenische Abteilung 71:1–29

Theodoulou T (2008) The harbour network of ancient Lesbos. First step of an underwater approach. In: Tzalas H (ed) Tropis X, Hydra, vol 28. Hellenic Institute for the Preservation of Nautical Tradition, Athens, pp 93–102

Theodoulou T (2010) Underwater research in the ancient ports of Lesbos. Archaiologia 116:93–102 (in Greek)

Theodoulou T (2011) Underwater archaeological research in Greece – a brief outline. Ariadni 17:13–84 (in Greek)

Theodoulou T (2015) Brief review of underwater and coastal archaeological research in Crete until 2014. Kritika Chronika 35:31–52
Theodoulou T (2017) Recording the harbour network of ancient Lesbos (2007 mission). In: Triantafyllides P (ed), Archaeological work in the Aegean Islands, International conference, Rhodes 27 November – 1 December 2013, vol. B, Mytilene, pp 123–130
Van Andel TH (1987) The adjacent sea. In: Van Andel TH, Sutton SB (eds) Landscape and people of the Franchthi Region, Fascicle 2, Excavations at Franchthi Cave, Greece Indiana University Press, Bloomington/ Indianapolis, pp 31–54
Van Andel TH, Jacobsen TW, Jolly JB, Lianos N (1980) Late Quaternary history of the coastal zone near Franchthi Cave, Southern Argolid, Greece. J Field Archaeol 7:389–402
Van Hinsbergen DJJ, Meleunkamp JE (2006) Neogene supradetachment basin development on Crete (Greece) during exhumation of the South Aegean core complex. Basin Res 18:103–124
Watrous LV (2012) The harbor complex of the Minoan town at Gournia. Am J Archaeol 116(3):521–541
Yiannouli E (2016) A.Sho.Re. 2011-2015, SE Kephallenia in the Ionian Sea: Investigating the geoarchaeology of the coastal zone. In: Photos-Jones E, Bassiakos Y, Filippaki E, Hein A, Karatasios I, Kilikoglou V, Kouloumpi E (eds) Proceedings of the 6th Symposium of the Hellenic Society for Archaeometry, Athens 16–18 May 2013, The Acropolis Museum. British Archaeological Reports IS 2780, Oxford, pp 179–185
Yiannouli E (2017) The work of the University of the Peloponnese in the island of Kefalonia: The research program of the coastal zone (Archaeological Shoreline Research).In: 10th International Panionian conference, 30 April–4 May 2014, Kerkyraika Chronika, Kerkyra, pp 327–340 (in Greek)
Zavitsanou A (2016) Paleogeographical Reconstruction of the Northern Inner Ionian Sea Archipelago during Late Quaternary Low Sea Level Periods. M.Sc. Dissertation, University of Athens (in Greek)
Zavitsanou A, Sakellariou D, Rousakis G, Georgiou P, Galanidou N (2015) Paleogeographic reconstruction of the Sea during Late Pleistocene low sea level stands: preliminary results. In: Proceedings of the 11th Panhellenic symposium on oceanography and fisheries, 13–17 May 2015, Mytilene, Greece. Mytilene, pp 997–1000

20 Bulgaria: Sea-Level Change and Submerged Settlements on the Black Sea

Preslav Peev, R. Helen Farr, Vladimir Slavchev, Michael J. Grant, Jon Adams, and Geoff Bailey

Abstract

The Black Sea is recognised as having great potential for the preservation of submerged prehistoric sites because of the large area of land exposed on the continental shelf at lowest sea levels, especially along its western and north-western coastlines. However, very few have been discovered to date, and those that are known are located in Bulgaria. Because of the complexities associated with the periodic isolation of the Black Sea from the Mediterranean and its reconnection, offshore research has tended to focus on geological and geophysical investigation of inundation history, with unresolved and strongly held disagreements about the timing and rapidity of sea level rise at the end of the Last Glacial and its potential human impact. In Bulgaria, a rich concentration of underwater prehistoric sites has been discovered, thanks to dredging activities earlier in the twentieth century and a long tradition of underwater archaeological investigations going back to the 1970s. These demonstrate the presence of substantial in situ village settlements of Eneolithic and Early Bronze Age date in shallow water (<10 m), with excellent preservation of large ceramic assemblages, wooden structures and faunal and palynological data on palaeodiet and palaeoeconomy. More recently, large-scale marine-geological and archaeological projects have begun, aimed at integrating shallow-water inshore investigations with offshore survey in deeper water.

P. Peev (✉)
Institute of Oceanology, Bulgarian Academy of Sciences, Varna, Bulgaria
e-mail: peev@io-bas.bg

R. H. Farr
The Centre for Maritime Archaeology, University of Southampton, Highfield, Southampton, UK
e-mail: r.h.farr@soton.ac.uk

V. Slavchev
Museum of Archaeology, Varna, Bulgaria

M. J. Grant
Coastal and Offshore Archaeological Research Services (COARS), Ocean and Earth Science, National Oceanography Centre Southampton, University of Southampton, Southampton, UK
e-mail: m.j.grant@soton.ac.uk

J. Adams
The Centre for Maritime Archaeology, University of Southampton, Highfield, Southampton, UK
e-mail: jjra@soton.ac.uk

G. Bailey
Department of Archaeology, University of York, York, UK

College of Humanities, Arts and Social Sciences, Flinders University, Adelaide, SA, Australia
e-mail: geoff.bailey@york.ac.uk

G. Bailey et al. (eds.), *The Archaeology of Europe's Drowned Landscapes*, Coastal Research Library 35, https://doi.org/10.1007/978-3-030-37367-2_20

Keywords

Eneolithic · Early Bronze Age · Sea-level change · Underwater settlements · Flood hypothesis

20.1 Introduction

The Black Sea is an intercontinental enclosed-type marine basin, located between Europe and Asia, bounded by Bulgaria and Romania to the west, Ukraine to the north-west, Russia and Georgia to the north-east and east and Turkey to the south, with a surface area of 416,790 km^2 and a maximum depth of 2212 m. It is bordered by the East European Plain to the north, the Pontic Mountains of the Asia Minor Peninsula to the south, the Danubian Plain and folded morpho-structures of the Balkans to the west and the Caucasus Mountain Chain with the Colchis Lowland to the east (Fig. 20.1; Ivanov and Belokopytov 2013; Kotsev et al. 2017; Yanko-Hombach et al. 2017; Kadurin et al., Chap. 21, this volume).

The Black Sea is a constituent part of the Greater Mediterranean Sea, connected to the Sea of Marmara in the south via the Bosphorus channel, and to the Sea of Azov by the Kerch Strait in the north. However, periodically in the Late Quaternary during periods of low global sea level, the Black Sea was isolated and formed a freshwater or brackish lake (Badertscher et al. 2011). The most recent episode of isolation was during the Last Glacial period, and the question as to when this isolation occurred, how long it persisted, the lake levels during this period and the nature of the reconnection of the lake with the global marine system through the Bosphorus, informs an intense debate about Black Sea inundation and its human impact (for details of this debate see Ryan et al. 1997, 2003; Aksu et al.

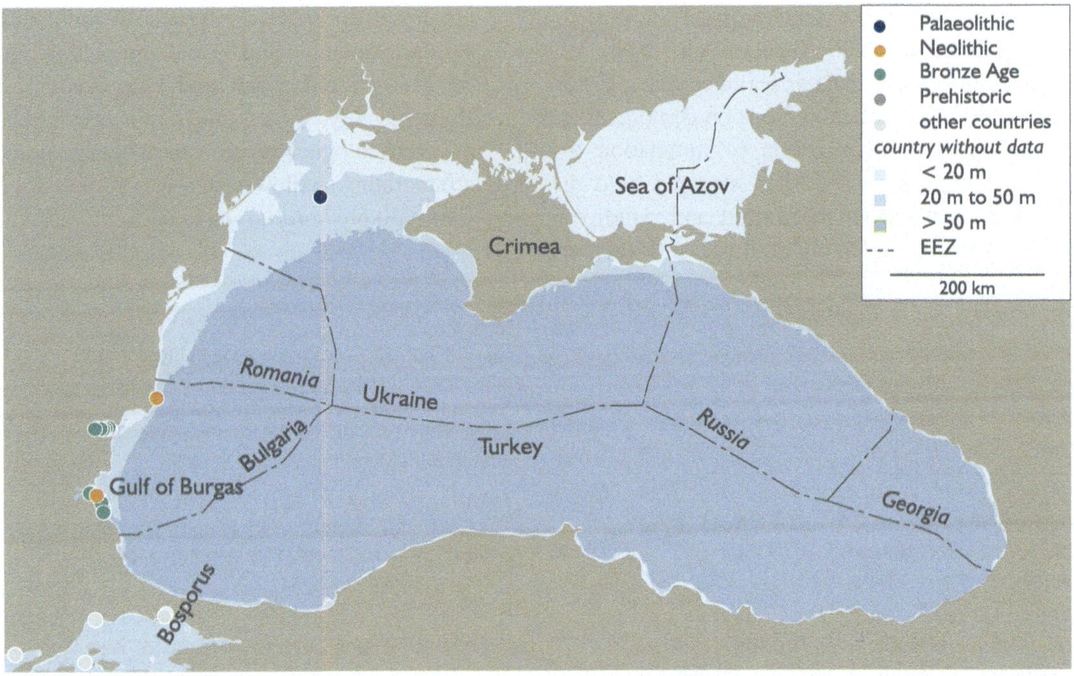

Fig. 20.1 General map of the Black Sea showing distribution of underwater sites, simplified bathymetry and the boundaries of the exclusive economic zones of each state (the boundary between Russia and the Ukraine is in dispute at the time of writing following the annexation of Crimea in 2014). The Palaeolithic site in Ukraine refers to artefacts recovered from a marine core (see Kadurin et al., Chap. 21, this volume). Site information from the SPLASHCOS Viewer http://splashcos-viewer.eu. Drawing by Moritz Mennenga

2002; Hiscott et al. 2007; Ryan 2007; Giosan et al. 2009; Constantinescu et al. 2015; Lericolais 2017; Turney and Brown 2007; Yanko-Hombach et al. 2007a, b, 2011, 2017; Yanchilina et al. 2017; Sturt et al. 2018; Kadurin et al., Chap. 21, this volume).

In terms of its cultural significance, the Black Sea forms a meeting point between Europe and Asia. It encompasses diverse climates and environments from the Mediterranean and the Near East in the south to the Eurasian Steppe in the north. In addition, it drains some of Europe's largest rivers, notably the Danube, Dneiper, Dniester and Southern Bug, rivers that enable connectivity deep into the continent. At various times, the Black Sea has been described as a barrier, frontier or buffer between the 'civilized' and 'Barbarian' worlds (Ivanova 2013, p. 2), at other times as a medium for providing access to resources, ease of movement and avenues for cultural connectivity and trade throughout the region (Draganov 1995, p. 225). The archaeology reflects these varying potentials, with a long sequence of human presence that extends back to the Early Pleistocene at Dmanisi in Georgia at 1.85–1.78 million years ago (Mgeladze et al. 2011). Numerous Middle and Upper Palaeolithic sites are also present, especially in the north and west and in Anatolia to the south (Özdoğan 2011; Kadurin et al., this volume). In terms of later prehistoric periods, the Black Sea region has culturally diverse and complex Neolithic and Chalcolithic sites and was central to the dispersal of agriculture from the Near East and to the development of metallurgy.

Given the extent of the submerged landscapes around the Black Sea, especially in the west and the north, and the complex history of sea-level change, an investigation of their potential to preserve prehistoric sites is compelling. The coast of Bulgaria is of central importance in such an investigation. It has a relatively wide continental shelf, c. 40–50 km in width, which would have exposed substantial areas of habitable territory at low sea-level stands. It is close to the main pathways of cultural connection between Anatolia and Europe via the Sea of Marmara to the south and the Danube River basin to the north. Most importantly, it is the only country with a Black Sea coastline that has produced unquestionable evidence for the preservation of underwater prehistoric settlements associated with submerged palaeoshorelines (for a recent summary, see Peev and Slavchev 2018).

The aim of this chapter is to summarise the issues associated with sea-level change in the Black Sea and their impact on the palaeogeography of the Bulgarian coast and the potential for archaeological preservation, to present the evidence from the known submerged settlements and to consider their wider implications.

20.2 Conditions of Preservation

20.2.1 Sea-Level Change

The Black Sea basin has undergone varying fluctuations in level and nature throughout the Quaternary. At various times in its history, when eustatic sea levels were low, the Black Sea became isolated from the global ocean system (Badertscher et al. 2011; Özdoğan 2011). The precise timing of these periods, the nature of the basin, changes in salinity and lake levels and the subsequent process of transgression have been fiercely debated (Yanko-Hombach et al. 2007a, b, 2011, 2017; Nicholas et al. 2011; Lericolais 2017; Yanchilina et al. 2017). Understanding the chronology, process and rate of coastal inundation is key to our understanding of prehistoric activity on the shelf, and also to the potential for preservation of submerged archaeological sites.

A central and unresolved issue is the level to which the water in the Black Sea dropped during the Last Glacial Maximum (c. 21 ka cal BP) or Early Neoeuxinian in regional terminology. Lowstand estimates range from c. 120 m bpsl[1] (below present sea level) (Alekseev et al. 1986; Filipova-Marinova 2007; Lericolais et al. 2011; Yanchilina et al. 2017) to 45 m bpsl (Yanko-Hombach 2007), while modelling of glacio-hydro-isostatic factors

[1] All depths in this chapter are given as bpsl (below present sea level) or as minus figures (e.g. −120 m) unless otherwise stated.

(GIA/Glacio-Isostatic Adjustment) indicates a low stand of 140–145 m bpsl (Lambeck and Purcell 2005, Fig. 14a). The water level during the low stand at the LGM informs subsequent models for processes of sea-level rise and reconnection with the Mediterranean, and also indicates the extent of the inhabitable land mass that we may now be missing from the archaeological record on the coastal shelves.

Once isolated, the Black Sea became a brackish to saline lake (1–10 ‰) isolated from the Sea of Marmara and the Mediterranean by the Bosphorus sill at c. 40 m bpsl. During this period, the extent of inflow from the Sea of Azov and freshwater run-off through the major continental river systems, including Late Pleistocene glacial meltwater, is a matter of dispute (Yanko-Hombach et al. 2007a, b, 2011, 2017; Soulet et al. 2013), as is the nature and timing of the transgression and the reconnection with the Mediterranean.

Reconnection with the Mediterranean has variously been put at dates ranging from 8.4 ka cal BP (Ryan et al. 2003) to 9.3 ka cal BP (Yanchilina et al. 2017) to 10.3–9.5 ka cal BP (Lambeck et al. 2007) and 9.6–9.2 ka cal BP (Nicholas et al. 2011). One hypothesis—the original 'Flood Hypothesis'— is that lake levels were at a low stand of c. −100 m at about 10,000 years ago shortly before global sea-level rise overtopped the Bosphorus sill, followed by a rapid inundation as the Black Sea was abruptly reconnected to the Mediterranean (Ryan et al. 1997; Peychev and Peev 2006). An alternative flood hypothesis suggests that the rise in the water level of the Black Sea occurred earlier, at about 12 ka cal BP, and had reached the Bosphorus sill before the ingress of the Mediterranean, resulting in a more gradual reconnection (Hiscott et al. 2007). Others again have argued for a gradual reconnection but with greater oscillations of sea level (Pirazzoli, 1996; Balabanov, 2007; Sorokin and Kuprin 2007 Yanko-Hombach et al. 2007a, b). Lericolais (2017) and Yanko-Hombach et al. (2017) provide recent summaries of the various alternatives (see also Kadurin et al., this volume), and Lambeck et al. (2007) and Lambeck and Purcell (2005) provide the most up-to-date modelling of GIA effects in the Black Sea. That the same body of data has produced such significantly different interpretations may seem puzzling. However, the data are drawn from climatology, geology, geophysics, biology and archaeology and were collected by different people at different times for different purposes. Resolution is inevitably varied, and as different research teams privilege different sectors or forms of data, it is perhaps less surprising that there are varied interpretations. Another challenge that has led to this diversity of interpretations is that there have been significant fluctuations in the radiocarbon reservoir effect in the Black Sea during the Late Glacial, resulting in different calibrations of radiocarbon dates derived from shell material (Soulet et al. 2011).

Recently these questions have been the focus of a large-scale project (the Black Sea Maritime Archaeology Project, BSMAP) working on the Bulgarian shelf to collect a new body of integrated data recovered with methodological consistency and to address the nature of transgression after the last glacial maximum. BSMAP has undertaken a geophysical survey of 2000 km^2 of the Bulgarian Shelf and conducted an extensive geological core sampling programme which has recovered 92 cores, amounting to almost 400 m of sediment. At the time of writing, analysis of these cores is well advanced and publication of the results in preparation.

20.2.2 Palaeogeographic Context

The nature, timing and rate of the Holocene transgression of the Black Sea are important in assessing the potential for preservation of submerged prehistoric sites. Clearly different hypotheses result in different models of the extent of submerged coastal shelf, and different narratives of the impact of transgression. However, it is generally agreed that many prehistoric sites are likely to have been located in areas that are now inundated. The archaeological record also supports this, although after initial discoveries on the Bulgarian coast during the 1980s and 1990s, further underwater exploration in the Black Sea has revealed little or nothing in the way of new prehistoric finds (Ballard et al. 2001; Sturt et al. 2018, see also Kadurin et al., Chap. 21, this volume).

The Bulgarian coastline extends for 432 km from Cape Sivriburun in the north (marking the state border with Romania) to the Rezovska River in the south (marking the state border with Turkey). The coastline comprises a rich mosaic of relict and modern land forms, e.g. denudation surfaces, marine terraces, cliffs and ravines formed by erosion, landslides, sandy beaches with dunes, river mouths, inlets and lagoons (Stancheva 2013; Stanchev et al. 2013).

Neotectonic movements, while generally understated, may be relevant to resolving discrepancies between different low-stand elevations across the western shelf (Glebov and Shel'ting 2007; Meriç et al. 2018). The Black Sea Basin originated as a back-arc basin in the early Cretaceous, with extension and thinning of the Earth's crust at a subduction plate boundary. Ongoing deformation similar to that in the Aegean has resulted in long-term subsidence especially around the basin margins with more localised episodes of uplift and subsidence on fault zones, but there is little evidence of geologically recent tectonic movement in the Black Sea coastal zone of SE Bulgaria (Nikishin et al. 2003; Stephenson and Schellart 2010; Özdoğan 2011; Caraivan et al. 2017; Kalafat 2017; see also Galanidou et al., Chap. 19, this volume). Archaeologists have inferred upward tilting of the Bulgarian coast in the north from the different depths of on-land and underwater Late Eneolithic and Early Bronze Age sites in different areas (Draganov 1995, p. 225; Angelova and Draganov 2003, p 10). Orachev (1990) also infers vertical uplift in the north from changes in coastal geomorphology and the partial collapse and submergence of the Thracian city of Byzone on Cape Chiracman during an earthquake in the first-century BC. However, uncertainties about the degree and extent of earth movements and the depth of submerged settlements in relation to contemporaneous sea level preclude further interpretation of tectonic effects.

The bathymetry of the Black Sea (see Yanko-Hombach et al. 2017, Kadurin et al., Chap. 21, this volume) indicates that the broadest shelves are located in the west (Bulgaria and Romania) and the north-west (Ukraine), and it is unsurprising that these areas have been the main focus for underwater research, most of which has been devoted to geological issues. This research has demonstrated the preservation of Quaternary features such as palaeorivers, submerged ridges, potential dune formations and peat horizons (Giosan et al. 2009; Lericolais 2017; Yanchilina et al. 2017; Yanko-Hombach et al. 2017; Kadurin et al., Chap. 21, this volume). It is worth bearing in mind that these are not the only contexts in which submerged sites can be preserved; there is also potential for underwater cave sites, which are more likely to be found along the southern and eastern Black Sea coasts.

In Bulgaria, the average width of the shelf down to the 120 m bathymetric contour is c. 40–50 km giving a maximum area of submerged landscape of as much as 12,380 km^2 (Fig. 20.1). Considering that the land territory of Bulgaria totals c.111,000 km^2, the potential survey area for underwater exploration relating to human presence and utilization of marine resources is quite large.

The Romanian coast to the north is geomorphologically similar to that of Bulgaria, extending for 245 km, with coastal inlets and shallow bays, and is just as likely to contain submerged prehistoric sites as Bulgaria (see Caraivan et al. 2017). The UNESCO report on Romanian underwater cultural heritage (UNESCO 2010a) references the potential for submerged Eneolithic and Neolithic sites and refers to locations at Taşaul and Techirghiol, located only c.50 km north of the famous Bulgarian Neolithic sites of Durankulak and the submerged necropolis site at Shabla. One consideration for the preservation and visibility of submerged prehistoric sites on the Romanian coastal shelf is the likelihood that sites are buried under sediments deposited from the Danube.

The Ukrainian shelf in the north-west is the widest continental shelf, extending out to 220 km and representing 16% of the Black Sea area. This shelf has a long history of geological research, but no submerged prehistoric sites have yet been discovered in this region, apart from isolated flints found in sediment cores during geological surveys (Yanko-Hombach et al. 2017; Kadurin et al., Chap. 21, this volume).

20.3 Archaeological Context

The submerged prehistoric sites so far discovered in the Black Sea are located on the Bulgarian coast. It is likely that this has more to do with a tradition of maritime archaeological research in the region rather than differential preservation. Underwater archaeological fieldwork has taken place in Bulgaria over the past 60 years, beginning with the pioneering work of scholars such as Lyuba Ognenova-Marinova and Mihail Lazarov. An important impetus to the study of prehistoric material was the establishment of the Centre for Underwater Archaeology (CUA) in Sozopol in 1978. This was the first institute to focus on underwater archaeology and the protection and management of underwater cultural heritage in the Black Sea. Many of the investigations of underwater settlements were made in the 1980s and 1990s under the leadership of the late Hristina Angelova, who was Director of the CUA from 1993 to 2016 (Porozhanov 1991, 2004; Ivanov 1993; Draganov 1995, 1998; Angelova and Draganov 2003; Velkovsky et al. 2013).

Palaeolithic and Mesolithic sites in this region are rare. For the Late Palaeolithic, there is the site of Little Cave near the town of Beloslav in the Varna district, with flint tools and associated faunal remains dating to c.14,000–12,000 cal BP (Todorova 1978). Little is known about the Mesolithic generally, apart from the site of Pobiti Kamuni, which contains one of the largest Mesolithic assemblages in south-eastern Europe with more than 12,500 stone artefacts. The finds are unstratified but are attributed on typological grounds to a period of c. 10,000 cal BP (c. 8000 cal BC) (Margos 1972; Todorova 1995). Otherwise, there appears to be a hiatus of more than 1000 years before the appearance of Neolithic farmers in Bulgarian territory.

One suggestion is that the rarity of Mesolithic sites is due to the fact that they were located along palaeocoastlines and palaeorivers that were subsequently inundated by marine transgression during the Holocene (Chapman 1989; Gurova and Bonsall 2014). Certainly, the potential for discovery of Mesolithic sites on the submerged shelf is important, not only for understanding Mesolithic subsistence and settlement but also for understanding the process of Neolithization within the region.

The earliest Neolithic cultures known in Bulgaria appear at about 6500 cal BC and have obvious affinities with the Neolithic in Anatolia (Table 20.1; Boyadziev 1995; Todorova 1995). The distribution of sites is concentrated along the major river systems in central, western and northern Bulgaria, and there is almost no evidence of settlement in the Black Sea coastal region until the Eneolithic (Chalcolithic) period. As for the Mesolithic period, so too for the Neolithic period, there is the question of a possible Neolithic presence on now-submerged territory, which might account for the rarity of Early and Middle Neolithic sites in the coastal region. However, no submerged settlements have yet been found earlier than the Eneolithic period, possibly because sites of earlier date lie at a greater depth than the underwater sites found so far and are buried under a greater thickness of riverine and lacustrine sediments.

The Eneolithic period is marked by the appearance of the Varna culture in the coastal region of

Table 20.1 Chronology of principal subdivisions of the archaeological sequence in Bulgaria from the beginning of the Neolithic to the end of the Early Bronze Age. Note that the boundaries are approximate and that some may be time-transgressive, that relatively few sites have been directly dated by radiocarbon dating and that many sites and culture layers are dated by comparison of their ceramic typology with better dated sites elsewhere. After Boyadziev (1995) and Todorova (1995)

		Date range	
Period	Sub-period	cal BC	Cal BP
Neolithic	Early	6500–5450	8450–7400
	Middle	5450–5150	7400–7100
	Late	5150–4850	7100–6800
Eneolithic	Early	4850–4550	6800–6500
	Middle	4550–4450	6500–6400
	Late	4450–3850	6400–5800
Transitional Period		3850–3200	5800–5150
Bronze Age	Early	3200–2500	5150–4450

NE Bulgaria, famous for its evidence of copper mining, fortified dwellings, palaces and shrines and its large cemeteries, notably at Varna, dated at 4650–4450 cal BC, and Durankulak, with 294 and 1204 graves, respectively (Todorova 1995, Higham et al. 2007). The burial evidence indicates differences of social rank with the individuals of highest status associated with rich grave goods of copper and gold.

The Eneolithic period is followed by the enigmatic Transitional Period, which witnessed the demise of the Varna culture and reduced evidence of settlement along the Bulgarian coast more widely in the centuries between about 3850 and 3200 cal BC (Table 20.1). This reduction in settlement evidence is variously attributed to social collapse or environmental processes such as climate change and flooding of productive farm land by sea-level rise (Todorova 1995). This period was followed, in its turn, by the renewed appearance of settlements after c. 3200 cal BC, associated with the Early Bronze Age.

It is during the Eneolithic and Early Bronze Age periods that evidence of underwater settlements is best represented.

20.4 Underwater Sites

There are 18 underwater sites in the SPLASHCOS Viewer (Fig. 20.2; Table 20.2). At least six sites appear to have in situ cultural deposits, and that figure could be higher. However, relatively few sites have been investigated by systematic excavation, and many have been damaged by construction work or dredging and were investigated under less than ideal conditions, sometimes as rescue operations, and details are not available or easily accessible in many cases. Dating is also often based on typochronological comparisons of pottery types and rarely on direct radiocarbon dates. One site, Shabla, has two human burials, which appear to be the remains of a necropolis, and one (Stomoplo) is of unknown status.

Of these sites, Shabla is on the northern coast of Bulgaria,13 are located in the Varna-Beloslav Lake area (Ivanov 1993) and five along the southern Bulgarian shore (Draganov 1995; Filipova-Marinova et al. 2011). Ten of these have material dating to the later phases of the Late Eneolithic (c 4500–3850 cal BC) (Ivanov 1993; Draganov 1995; Lazarov 1996), and 17 sites contain cultural remains belonging to the Early Bronze Age (3200–2500 cal BC).

All sites are in shallow water at depths ranging from c.3.5 to 9 m. Organic remains of wood have been recovered from six sites, animal bones from four and human bone from the Shabla burials (Table 20.2)

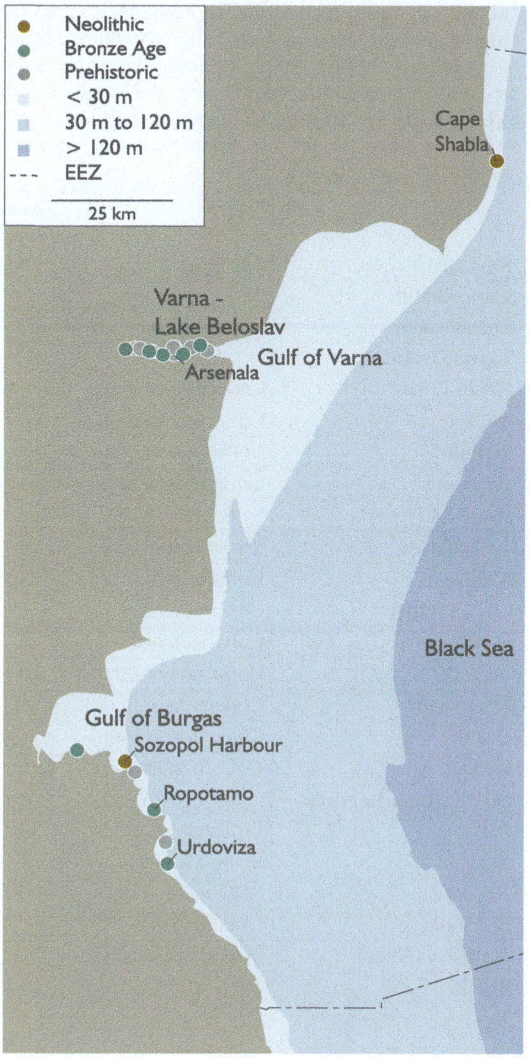

Fig. 20.2 Detailed map of submerged sites on the Bulgarian coast. Site information from the SPLASHCOS Viewer http://splashcos-viewer.eu. Drawing by Moritz Mennenga

Table 20.2 List of underwater archaeological sites, showing chronology, maximum depth bpsl and the presence or absence of wood preservation and animal or human bone. *Eneo* Eneolithic; *EBA* Early Bronze Age. The category of site type 'collection of finds' means that the information about context and recovery methods is not good enough to establish the presence of an in situ cultural layer, but equally that cannot be ruled out. Depth measurements are approximate. The SPLASHCOS Viewer data has 22 entries, 2 separate entries for Shabla (based on the different depths of the two burials) and 3 separate entries for Sozopol, representing deposits in different locations within Sozopol harbour. We have amalgamated these in this table

		Archaeological period		Max depth m	Bone	Wood
Region/site name	Site Type	Eneo	EBA			
Cape Shabla						
Shabla I & II	Human burials	√		3.5–6.5	√ (Human)	
Varna Group						
Hristo Botev	Collection of finds		√	8		
Morfot	Collection of finds	√	√	9		
Topolite	Collection of finds		√	8		
Ezerovo I	Settlement in situ	√	√	8.5		√
Ezerovo II	Settlement in situ	√	√	8.5		√
Arsenala	Settlement in situ	√	√	7	√	√
Ladjata	Collection of finds		√	7.5		
Strashimirovo I	Collection of finds	√	√	5		
Strashimirovo II	Collection of finds		√	4		
Beloslav	Collection of finds	√	√	8		
Povelyanovo	Collection of finds	√	√	6.5		
Baltata	Collection of finds		√	5		
Rodopa	Collection of finds	√	√	6		
South Bulgaria group						
Atiya	Collection of finds		√	9		
Sozopol	Settlement in situ	√	√	5–8	√	√
Ropotamo	Settlement in situ		√	4–5	√	√
Stomoplo	Unknown					
Urdoviza (Kiten)	Settlement in situ		√	6–7	√	√
Totals	**19**	**10**	**17**		**4**	**6**

Although full details are lacking at many sites because of the circumstances of discovery, nevertheless it appears from the recovery of pottery and other indications including remains of wooden structures and abundant artefacts that the sites were substantial settlements. More detailed information including palaeoenvironmental data have been obtained from excavations conducted at Sozopol, Urdoviza and Ropotamo.

20.5 Northern Bulgaria

20.5.1 Cape Shabla

This submerged prehistoric necropolis has been recorded in the area of Cape Shabla (Peychev and Peev 2006; Peev 2008a, 2009). Two burials have been excavated between c. 6.5 m and 3.5 m bpsl and are dated to the Eneolithic period (Fig. 20.3). The necropolis is related to the on-land settlements of Shabla I and/or Shabla II (Todorova 1984). Other than these two burials, the necropolis has not been further investigated.

20.5.2 The Varna-Beloslav Lake Area

Varna Lake is a typical inlet situated in the western part of the Gulf of Varna (Fig. 20.4). It is the largest inlet by volume and depth on the Bulgarian coast with an area of 17 km^2, a maximum depth of 19 m and a water volume of 166 million m^3. It was formed by rising sea levels during the

Fig. 20.3 Underwater photo of prehistoric grave at depth 3.5 m below present Black Sea level off Cape Shabla. Photo by Damyan Damyanov

Fig. 20.4 View of Lake Varna, looking south. Photo courtesy of Museum of Archaeology Varna

Holocene, after c. 7870 cal BP (5870 cal BC) according to Filipova-Marinova et al. (2016). The basin remained a freshwater lake supplied by groundwater and inflowing rivers, with a limited connection to the Black Sea, until it was modified in 1909 to connect with Varna Bay through an artificial channel, after which it became a brackish lagoon. In 1975 a new canal was dug with a depth of 12 m. As a result, the salinity of the water increased to 15–16 ‰ (Peev 2008b). Today, the lake is separated from the sea by the constantly growing Asparuhovo sand spit (Dachev 2003) and is divided into two lakes, Lake Varna and Lake Beloslav, resulting from progressive accumulation of sediment and decreased river flow.

Drilling surveys located 650 m south of the dredged shipping channel have revealed a 51.4 m-deep sequence of Holocene sediments indicating oscillating sea-levels (Popov and Mishev 1974). The sequence is as follows (from the base upwards):

1. A peat layer 0.6 m thick
2. Marine sediments up to 18 m thick containing a fauna dated to the Holocene
3. Subaerial deposits with 1.5 m of peat
4. Sand mixed with gastropod shells, 18 m thick, indicating marine conditions
5. A third peat layer 1 m thick and accumulated clayey sand

This suggests that there were two cycles of marine transgression followed by minor regression. This sequence is of particular interest since, although not fully dated, it indicates a pattern of fluctuations in sea level similar to evidence indicated by stratigraphic sections in underwater archaeological settlements and by pollen sequences, as discussed below.

20.5.3 Arsenala

Along the shorelines of the Varna-Beloslav Lake, the remains of eight submerged settlements from the second stage of the Eneolithic have been found (Fig. 20.2; Ivanov 1993). The earliest discovery was on the north shore of the lake near the village of Strashimirovo, first mentioned in 1921 by H. and K. Shkorpil (1921). However, only one site, Arsenala, has undergone underwater excavations (Fig. 20.5). An area of 25 m^2 was excavated and revealed two cultural layers identified as dating to the Eneolithic and the EBA, respectively. The EBA layer reaches a thickness of 3.5 m. This overlies a sterile layer c. 0.9 m thick, which is rich in marine molluscs. The Eneolithic lies beneath this in a layer c. 0.9 m thick with remains of wooden posts used for house construction, ceramics and animal bones (Ivanov 1987). The Eneolithic layer includes decorated ceramics from the Varna culture (Fig. 20.6), and the EBA layer is notable for cult items including male and female figurines.

The pollen analytical work of Filipova-Marinova et al. (2016) in Lake Varna shows two episodes of deforestation and cereal cultivation with an interval in between of re-afforestation and reduction in indicators of human influence. The first episode of deforestation is dated to 6140–5820 cal BP (4190–3870 cal BC), corresponding to the Late Eneolithic occupation at Arsenala. The second episode begins at 5500 cal BP (3550 cal BC) and corresponds to the middle

Fig. 20.5 Early Bronze Age pottery shown in situ at the submerged site of Arsenala. Scale in 10 cm intervals. Photo courtesy of Museum of Archaeology Varna 1988

of the Transitional Period (Table 20.1), while the interval between lasted for 320 years and is consistent with the occupational hiatus at Arsenala associated with marine sediments.

20.6 Southern Bulgaria

20.6.1 Urdoviza

This site in the bay of Kiten (a name by which the site is also referred to in the literature) was excavated between 1986 and 1989 and again in the 1990s over a total area of 925 m^2 (Draganov 1995; Angelova and Draganov 2003; Stanimirov 2003; Porozhanov 2004). Interestingly, the prehistoric material was discovered by chance during an underwater investigation of a post-Medieval shipwreck, beneath which the prehistoric deposits were revealed. The seabed at this location is 6.3 m bpsl, and the cultural deposit extends from this level down to −7.05 m (Fig. 20.7). Bathymetric and sidescan surveys of the wider surroundings indicate that, when occupied, the settlement was 500–600 m from the nearest shoreline and sea level at that time was at −8 m.

Over 3000 pots or potsherds of typical EBA forms were recovered (Fig. 20.8), many of which are complete vessels, supposedly surviving because the deposits were quickly sealed by marine deposits after inundation and abandonment of the settlement. Other cultural remains are stone and antler artefacts, bone awls and spatulae made from bird bone, clay figurines, a bone amulet, two stone moulds for making bronze axes, a bronze knife and parts of horse bridles.

Excavations also revealed at least 300 timber posts mostly of oak. Some of these were vertical posts driven vertically into the ground and attached to horizontal members by jointing and ropes (Fig. 20.7). Draganov (1995) describes fragments of floor and wall daub, hearths and clay floor coverings, and suggests that the horizontal timbers were designed to support dwelling structures raised slightly above a ground surface that was marshy or liable to flooding. Dendrochronological analysis of 83 of these timbers produced a 285-year long tree ring chronology—one of the longest from Eastern Europe and the Aegean—and evidence of 5 building phases at Urdoviza over a 64-year period of occupation between 4778 and 4715 cal BP (2828–2765 cal BC) (Kuniholm et al. 1998, 2007).

The faunal remains comprise over 7000 specimens and are dominated by wild animals, especially aurochs (*Bos primigenius*), red deer (*Cervus elaphus*), roe deer (*Capreolus capreolus*) boar (*Sus scrofa*) and various bird species. There are also numerous bones of tuna and dolphin, suggesting the use of boats for offshore and deep-water marine exploitation. Domestic animals are cattle and horse, the latter including 40 skulls, 20 of which are intact and arranged in such a way as to suggest some cultic significance along with 5 pairs of aurochs horns apparently arranged in a deliberate pattern. Pollen data show evidence of forest clearance but no indications of cereal agriculture.

Fig. 20.6 Pottery from Arsenala. Photo courtesy of Museum of Archaeology Varna 1988

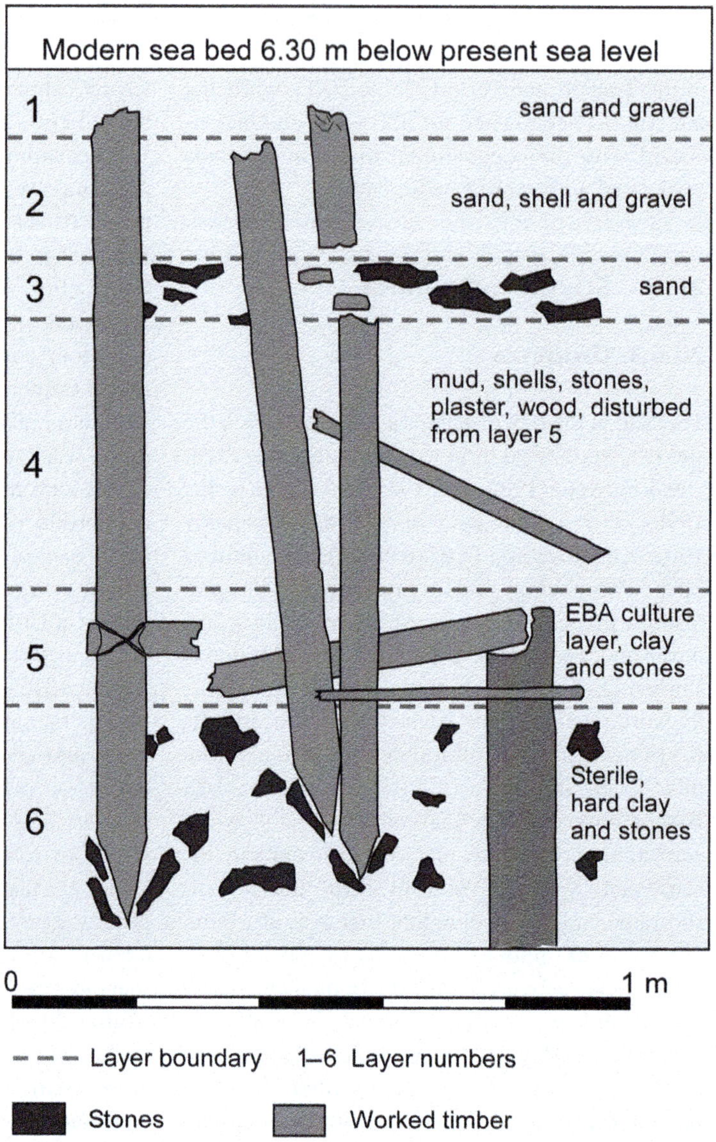

Fig. 20.7 Stratigraphic section through the cultural layer at Urdoviza showing the stratigraphy and some of the timber posts. Drawing by Geoff Bailey, after Angelova and Draganov (2003, Fig. 3, p. 15)

Fig. 20.8 Pottery from the Early Bronze Age submerged settlements of Ezerovo and Urdoviza. Photo courtesy of the Centre for Underwater Archeology, Sozopol and the Museum of Archaeology, Varna

20.6.2 Sozopol

Two closely adjacent areas of submerged settlement were investigated by excavations in Sozopol harbour in the 1980s and the early 1990s, one area with material of Eneolithic date and the other EBA (Draganov 1995, 1998; Angelova and Draganov 2003; Filipova-Marinova et al. 2011). Both produced abundant ceramics, timber posts and faunal remains. The results have been reported in less detail than at Urdoviza. Spassov and Iliev (1994) examined 941 identifiable bone remains indicating a faunal assemblage similar to Urdoviza. Wild animals dominate, especially aurochs, red deer and fallow deer, and domestic animals included a significant proportion of sheep and goat (a difference from Urdoviza) as well as domestic cattle and some domestic pig. Numerous bird bones are present, mostly of species associated with coastal wetlands, numerous bones of two species of dolphins, the bottle-nosed dolphin (*Tursiops truncatus*) and the common dolphin (*Delphinus delphis*), and just two vertebrae of tuna (*Thunnus thunnus*). Although the provenance of some of the bones was uncertain, it is clear that the faunal composition of the two settlements, and especially the emphasis on marine resources, is broadly similar. Pollen analysis and radiocarbon dating of sediment cores from Sozopol harbour shows that both the Eneolithic and EBA periods of settlement were associated with forest clearance and the presence of cultivated wheat and barley and weeds typically associated with cultivation (Filipova-Marinova et al. 2011), marking a difference from Urdoviza where evidence of cereal cultivation is absent

20.6.3 Ropotamo

In conjunction with the offshore coring programme, the Black Sea Maritime Archaeology Project reviewed the available data relating to submerged prehistoric material along the Bulgarian coast (summarised above) with a view to identifying a suitable target for more detailed investigation of the relationship between archaeological material and sea-level change. Work has focused on the mouth of the Ropotamo River. In the 1980s, dredging activities reportedly recovered Eneolithic and EBA ceramics and timber structures (Karayotov 1990). Although little detail was published, there was a clear possibility of locating the original source of the material.

In June 2017, two trenches each 5 m^2 were excavated and revealed a deeply stratified deposit more than 3 m thick with its surface on the seabed at c.2.3 m below present sea level (Fig. 20.9). Below modern, mobile seabed sediments, large quantities of Ottoman ceramics were found above a dense deposit of Byzantine material that included ceramics, tile, mortar and wood, possibly from waterside infrastructure that had subsided into the water due to tectonic activity. Below this was a thick layer of marine sediment in which some Greek ceramic sherds were found. Below this at 2.5 m below the seabed, Bronze Age material was discovered: ceramics identified as EBA material, including a whole vessel (Fig. 20.10), burnt clay from domestic hearths and structural timber. Radiocarbon dating from three of the in situ vertical timber posts suggests construction between 3080 and 2970 cal BC (5050–4920 cal BP). These remains had been covered by marine deposits in which there is a high density of oyster shell, indicating inundation. The relatively uneroded condition of the archaeological material suggests that inundation may have been rapid, creating an environment of shallow flowing water with high levels of light and oxygen suitable for the lodging of oyster spat. Archaeological excavation is ongoing.

Initial results of this new work at Ropotamo are consistent with the earlier findings from Urdoviza and Sozopol for a sea level during the EBA period between c.5 m and 8 m bpsl, with a subsequent rise in sea level, rapidly inundating the settlements and forcing their abandonment.

20.7 Discussion

Both underwater archaeological sites and palaeoenvironmental sequences provide a consistent picture of two phases of settlement on the

Fig. 20.9 Ropotamo Bay. Aerial view of the site overlaid with bathymetry in order to show location of the underwater excavation trenches. Photo by Rodrigo Pacheco-Ruiz, University of Southampton, courtesy of BSMAP

Fig. 20.10 Professor Kroum Batchvarov (University of Connecticut) excavating a complete EBA ceramic bowl at Ropotamo. Note the thick layer of oyster shells immediately above it. Photo by Jon Adams

Bulgarian coast between about c. 4400 and 2500 cal BC, the first associated with the Late Eneolithic period (c. 4450– 3850 cal BC), and the second associated with the EBA (c. 3200–2500 cal BC). There is some variation and margin of error in these date ranges, but all the evidence points to a hiatus in occupation along the Bulgaria coast, with abandonment of settlements and the re-establishment of forested conditions. The length of this interval is debated, and more excavation and dating are needed.

These two periods of occupation witnessed the establishment of settlements along the sheltered shorelines of inlets and river estuaries near the coast, with timber-built platforms and dwellings established on low-lying ground liable to flooding, forest clearance and cereal agriculture. There was some dependency on domestic animals, but also an emphasis on hunting of wild animals and marine resources including tuna fish and dolphins, the latter indicating the probable use of boats. The occupational hiatus seems to have been widespread within the Bulgarian Black Sea coastal region, with evidence for the expansion of forests and abandonment of villages in a number of locations. A variety of causes have been invoked to explain these changes including environmental changes—of sea-level or climate—and social factors (Todorova 1995).

The relationship of these underwater settlements to sea-level change raises two questions. The first is why settlements dated between c.4500 and 2500 cal BC were located on shorelines that are now underwater in association with a sea level as much as 8 m bpsl on the Bulgarian coastline. This is long after a connection between the Black Sea and the world oceans had become established, and a period when eustatic sea level was within a metre or two of the modern level. Lambeck and Purcell (2005) have modelled the effect of glacio-hydro-isostatic movements of the Earth's crust on relative sea level throughout the Mediterranean and the Black Sea. Their models indicate that relative sea level on the coast of Bulgaria at 6000 cal BP (c. 4000 cal BC) was about −5 m (Lambeck and Purcell 2005, Fig. 14f). Allowing for the margins of error in the various measurements involved, this might be sufficient to account for the depth of the underwater settlements. Whether an additional component of vertical tectonic movement resulting from plate motions needs to be involved remains uncertain.

The second question is the apparent evidence for a temporary sea-level oscillation between the Eneolithic and the Early Bronze Age, involving a temporary and short-lived increase in relative sea level followed by a retreat. The evidence of stratigraphic sequences, described above, which show marine sediments sandwiched between terrestrial deposits, provides some support for this fluctuation in height of sea level, or in rate of sea-level variation. How widespread it was, and the causes of it, whether resulting from regional tectonic movements or geomorphological processes of lateral shoreline retreat and advance, remain unclear, and work is ongoing to resolve this problem (compare Filipova-Marinova 2007 and Kislov 2018).

Whatever the causes of this sea-level fluctuation, it offers support for the hypothesis that the gap in occupation between the Eneolithic and the Early Bronze Age was due to a temporary rise in relative sea level that removed low-lying land on the coast that had provided favourable conditions for economic subsistence, including fertile soils for crop cultivation and wetlands rich in bird life.

These submerged settlements demonstrate the excellent quality of evidence that can be preserved as a result of rapid inundation and submergence of archaeological material in anaerobic sediments. They are also testament to the potentially disruptive effects of sea-level rise on pre-existing patterns of social and economic organisation. Both factors have implications for earlier periods. In particular, they raise the question of whether earlier settlements of a similar nature existed along palaeoshorelines that are now submerged, with archaeological settlements protected beneath marine sediments. As noted earlier, the absence of Mesolithic and Neolithic sites on the Bulgarian coast has been attributed to removal of evidence by sea-level rise (Gurova and Bonsall 2014). Given the emphasis on hunting, fowling and marine exploitation even in the Eneolithic and EBA periods, and the importance of forest clearings along the edges of coastlines, rivers and lakes, it is likely that coastal lowlands would have been just as attractive to earlier agricultural and pre-agricultural populations and just as capable of supporting large settlements. The challenge will be to find these earlier sites, particularly if they are deeply buried beneath later marine deposits.

New work is now under way, including the use of predictive modelling and remote sensing in shallow water (Prahov 2013), and underwater excavation in shallow water and geophysical prospection and coring in deeper water offshore by BSMAP. Without such investigations, the nature of the relationship between the underwater Eneolithic and Early Bronze Age settlements and changes in sea level will remain unclear, while questions about the nature of earlier human settlement on the more deeply submerged landscapes of the Black Sea shelf, and the likely impact of more or less rapid sea-level rise in the Late Glacial and Early Postglacial, will remain largely in the realms of speculation.

20.8 Conclusion: Issues of Future Research and Management

The submerged prehistoric settlements of the Bulgarian coast, though relatively late in date, provide a unique insight into the social and economic history of the period and a unique archive of data with which to investigate the interactions between social development and changes in the natural environment. Even in these relatively shallow and accessible waters, there remain large challenges. Only a small fraction of the known underwater settlements has been systematically investigated, and much of the material recovered in earlier excavations has untapped potential for further analysis and renewed underwater investigations of the locations which have previously yielded archaeological discoveries. At the same time, the coastal zone is increasingly an area of intensive commercial and industrial development and economic activity, especially in more shallow, inshore waters. Fishing, gathering of *Rapana* (a large gastropod), hydrocarbon exploration, installation of oil and gas pipelines and underwater cables and the dredging of ports and shipping channels, all have the potential to damage or destroy the submerged cultural heritage. Bulgaria is exemplary in its systematic management of archaeology and coordination of protection with research, conservation and collaboration across a wide range of disciplines and institutions at a national and an international level. Bulgaria is a party to the UNESCO 2001 Convention on the Protection of the Underwater Cultural Heritage. The competent state authority responsible for the protection of the underwater cultural heritage is the Centre for Underwater Archaeology, Sozopol, which falls under the jurisdiction of the Ministry of Culture (UNESCO 2010b) and is responsible for investigation, recording, protection, consultancy, public outreach and training. What is now required are the resources on a scale to match the demands of this new and challenging agenda.

Acknowledgements We thank the following for their support of the research reported here: HERAS Project (Submarine Archaeological Heritage on the Western Black Sea Shelf - HERAS), CBC Program Romania-Bulgaria 2007–2013, the Black Sea Maritime Archaeology Project (BSMAP) supported by the Julia and Hans Rausing Trust, the Centre for Underwater Archaeology Sozopol, the University of Southampton, and the EU through COST Action TD 0902 SPLASHCOS. We also thank John Chapman and Bisserka Gaydarska for provision of information, critical comments on the text and helpful discussion of chronology, and Kurt Lambeck and Dimitris Sakellariou for information on sea-levels. Author contributions: PP, RHF, VS: fieldwork, research, analysis, writing; MJG, JA: fieldwork, research; and GB: research, analysis, writing.

References

Aksu AE, Hiscott RN, Kaminski MA, Mudie PJ, Gillespie H, Abrajano T, Yaşar D (2002) Late Glacial-Holocene palaeoceanography of the Black Sea and Marmara shelf: Stable isotopic, foraminiferal and coccolithic evidence. Marine Geol 190:119–149

Alekseev M, Chistyakov A, Shterbakov F (1986) Chetvertichnaya Geologiya Materikovyih Okrain. Nedra, Moscow (in Russian)

Angelova H, Draganov V (2003) Underwater archaeological excavations of submerged Late Eneolithic and Early Bronze Age settlements in Kiten and Sozopol (South Bulgarian Black Sea coast). In: Angelova H (ed) In Honorem Mihaili Lazarov. Thracia Pontica 6.2:9–22

Badertscher S, Fleitmann D, Cheng H, Edwards RL, Göktürk OM, Zumbühl A, Leuenberger M, Tüysüz O (2011) Pleistocene water intrusions from the Mediterranean and Caspian seas into the Black Sea. Nat Geosci 4:236–239

Balabanov IP (2007) Holocene sea-level changes of the Black Sea. In: Yanko-Hombach V, Gilbert A, Panin N, Dolukhanov P (eds) The Black Sea flood question: Changes in coastline, climate, and human settlement. Springer, Dordrecht, pp 711–730

Ballard RD, Hiebert FT, Coleman DF, Ward C, Smith JS, Willis K, Foley B, Croff K, Major C, Torre F (2001) Deepwater archaeology of the Black Sea: the 2000 season at Sinop, Turkey. Am J Archaeol 105:607–623

Boyadziev YD (1995) Chronology of prehistoric cultures in Bulgaria. In: Bailey D, Panayotov I (eds) Prehistoric Bulgaria. Prehistory Press, Madison, pp 149–191

Caraivan G, Voinea V, Popescu D, Cerchi C (2017) Submerged prehistoric potential of the Romanian Black Sea shelf. In: Flemming NC, Harff J, Moura D, Burgess A, Bailey GN (eds) Submerged landscapes of the European continental shelf: quaternary paleoenvironments. Wiley, Chichester, pp 497–517

Chapman J (1989) Demographic trends in Neothermal South-east Europe. In: Bonsall C (ed) The Mesolithic in Europe. John Donald, Edinburgh, pp 500–515

Constantinescu AM, Toucanne S, Dennielou B, Jorry SJ, Mulder T, Lericolais G (2015) Evolution of the Danube deep-sea fan since the last glacial maximum: new insights into Black Sea water-level fluctuations. Mar Geol 367:50–68

Dachev V (2003) Genesis and evolution of Varna Central Beach. Proc Inst Oceanol 4:74–82 (in Bulgarian)

Draganov V (1995) Submerged coastal settlements from the Final Eneolithic and the Early Bronze Age in the sea around Sozopol and Urdoviza Bay near Kiten. In: Bailey D, Panayotov I (eds) Prehistoric Bulgaria. Prehistory Press, Madison, pp 225–241

Draganov V (1998) James Harvey Gaul and the present state of Eneolithic research in northeastern Bulgaria and Thrace: the areas of the Krivodol-Gumelnita-Karanovo VI and Varna cultures. In: Stefanovich M, Todorova H, Hauptmann H (eds) In the steps of James Harvey Gaul, James Harvey Gaul – in memoriam, vol 1. The James Harvey Gaul Foundation, Sofia, pp 203–221

Filipova-Marinova M (2007) Archaeological and paleontological evidence of climate dynamics, sea-level change, and coastline migration of the Bulgarian sector of the Circum-Pontic region. In: Yanko-Hombach V, Gilbert A, Panin N, Dolukhanov P (eds) The Black Sea flood question: changes in coastline, climate, and human settlement. Springer, Dordrecht, pp 453–482

Filipova-Marinova M, Liviu G, Angelova H, Preisinger A, Pavlov D, Vergiev S (2011) Palaeoecology of submerged prehistoric settlements in Sozopol harbour, Bulgaria. In: Benjamin J, Bonsall C, Pickard C, Fischer A (eds) Submerged prehistory. Oxbow, Oxford, pp 230–244

Filipova-Marinova M, Pavlov D, Giosan L (2016) Multi-proxy records of Holocene palaeoenvironmental changes in the Varna Lake area, western Black Sea coast. Quater Int 401:99–108. https://doi.org/10.1016/j.quaint.2015.05.009

Giosan L, Filip F, Constatinescu S (2009) Was the Black Sea catastrophically flooded in the early Holocene? Quater Sci Rev 28:1–6. https://doi.org/10.1016/j.quascirev.2008.10.012

Glebov AY, Shel'ting SK (2007) Sea-level changes and coastline migrations in the Russian sector of the Black Sea: application to the Noah's Flood hypothesis. In: Yanko-Hombach V, Gilbert A, Panin N, Dolukhanov P (eds) The Black Sea flood question: changes in coastline, climate, and human settlement. Springer, Dordrecht, pp 731–773

Gurova M, Bonsall C (2014) 'Pre-Neolithic' South East Europe: a Bulgarian perspective. Documenta Praehistoria 41:95–109

Higham T, Chapman J, Slavchev V, Gaydarska B, Honch N, Yordanov Y, Dimitrova B (2007) New perspectives on the Varna cemetery (Bulgaria)–AMS dates and social implications. Antiquity 81:640–654

Hiscott RN, Aksu AE, Mudie PJ, Marret F, Abrajano T, Kaminski MA, Evans J, Çakiroğlu AI, Yaşarf D (2007) A gradual drowning of the southwestern Black Sea shelf: Evidence for a progressive rather than abrupt Holocene reconnection with the eastern Mediterranean Sea through the Marmara Sea Gateway. Quater Int 167–168:19–34. https://doi.org/10.1016/j.quaint.2006.11.007

Ivanov I (1987) Podvodni arheologicheski prouchvania na selishte Arsenala. Arheologicheski otkritiya i razkopki prez 1986. Razgrad, 281–283 (in Bulgarian)

Ivanov I (1993) A la question de la localisation et des études des sites submergés dans les lacs de Varna. Pontica 26:19–26

Ivanov VA, Belokopytov VN (2013) Oceanography of the Black Sea. Hydrophysical Institute NAS of Ukraine, Sevastopol

Ivanova M (2013) The Black Sea and the early civilizations of Europe, the Near East and Asia. Cambridge University Press, Cambridge. https://doi.org/10.1017/CBO9781139424738

Kalafat D (2017) Seismicity and tectonics of the Black Sea. Int J Earth Sci Geophy 3:011. https://www.vibgyorpublishers.org/content/international-journal-of-earth-science-and-geophysics/ijesg-3-011.pdf

Karayotov I (1990) The Antique and Medieval port at the mouth of the river Ropotamo. Acta TAB 5:64–66

Kislov A (2018) On the interpretation of century–millennium-scale variations of the Black Sea level during the first quarter of the Holocene. Quater Int 465(A):99–104. https://doi.org/10.1016/j.quaint.2016.09.008

Kotsev I, Stanchev H, Stancheva M, Filipova-Marinova M, Yankova M, Peev P, Peychev V, Krastev A, Palazov A, Pavlov D (2017) Sensitivity mapping and analysis of the Bulgarian Black Sea coastal zone. Total E&P Bulgaria B.V, Sofia

Kuniholm PI, Kromer B, Tarter SL, Griggds CB (1998) An Early Bronze Age settlement at Sozopol, near Burgas, Bulgaria., dated by dendrochronology and radiocarbon. In: Stefanovich M, Todorova H, Hauptmann H (eds) In the steps of James Harvey Gaul, Volume 1: James Harvey Gaul – in memoriam. The James Harvey Gaul Foundation, Sofia, pp 399–409

Kuniholm PI, Newton MW, Kromer B (2007) Dendrochronology of submerged Bulgarian sites. In: Yanko-Hombach V, Gilbert A, Panin N, Dolukhanov P (eds) The Black Sea flood question: changes in coastline, climate, and human settlement. Springer, Dordrecht, pp 483–488

Lambeck K, Purcell A (2005) Sea-level change in the Mediterranean Sea since the LGM: model predictions for tectonically stable areas. Quater Sci Rev 24(18–19):1969–1988

Lambeck J, Sivan D, Purcell A (2007) Timing of the last Mediterranean Sea–Black Sea connection from isostatic models and regional sea-level data. In: Yanko-Hombach V, Gilbert A, Panin N, Dolukhanov P (eds) The Black Sea flood question: changes in coastline, climate, and human settlement. Springer, Dordrecht, pp 797–808

Lazarov M (1996) Sunken settlements along the Western Black Sea in the context of Pontic and Eastern Mediterranean history. History 1:48–61 (in Bulgarian)

Lericolais G (2017) Late Pleistocene environmental factors defining the Black Sea, and submerged landscapes on the western continental shelf. In: Flemming NC, Harff J, Moura D, Burgess A, Bailey GN (eds) Submerged landscapes of the European continental shelf: quaternary paleoenvironments. Wiley, Chichester, pp 479–495

Lericolais G, Guichard F, Morigi C, Popescu I, Bulois C, Gillet H, Ryan WBF (2011) Assessment of the Black Sea water-level fluctuations since the Last Glacial Maximum. In: Buynevich IV, Yanko-Hombach V, Gilbert AS, Martin RE (eds) Geology and geoarchaeology of the Black Sea region: beyond the Flood Hypothesis, Geological Society of America Special Paper 473, pp 33–50

Margos A (1972) Mikroliti ot eneolitni selishta vuv Varnensko. Izvestiya na Narodniya Muzei Varna 8(23):233–235 (in Bulgarian)

Meriç E, Nazik A, Yümün ZU, Büyükmeriç Y, Avşar N, Yildiz A, Sagular EK, Koral H, Gökaşan E (2018) Fauna and flora of drilling and core data from the İznik Lake: The Marmara and the Black Sea connection. Quater Int 486:156–184

Mgeladze A, Lordkipanidze D, Moncel MH, Despriee J, Chagelishvili R, Nioradze M, Nioradze G (2011) Hominin occupations at the Dmanisi site, Georgia, Southern Caucasus: Raw materials and technical behaviours of Europe's first hominins. J Human Evol 60:571–596. https://doi.org/10.1016/j.jhevol.2010.10.008

Nicholas WA, Chivas AR, Murray-Wallace CV, Fink D (2011) Prompt transgression and gradual salinisation of the Black Sea during the early Holocene constrained by amino acid racemization and radiocarbon dating. Quater Sci Rev 30(27–28):3769–3790

Nikishin AM, Korotaev MV, Eshov AV, Brunet M-F (2003) The Black Sea basin: Tectonic history and Neogene–Quaternary rapid subsidence modelling. Sediment Geol 156:149–168

Nikolova L (1994) Data about sea contacts during the Early Bronze Age in South-Eastern Europe (C. 3500/3400 – 2350/2250 B.C.). Thracia Pontica 5:57–86

Orachev A (1990) Contributions to the paleogeography of the coast of Dobroudja. Dobrudzha 7:35–52 (In Bulgarian)

Özdoğan M (2011) Submerged sites and drowned topographies along the Anatolian coasts: an overview. In: Benjamin J, Bonsall C, Pickard C, Fischer A (eds) Submerged prehistory. Oxbow, Oxford, pp 219–229

Peev P (2008a) Underwater sites in the area of cape Shabla (North-East Bulgaria). In: Kostov R, Gaydarska B, Gurova M (eds) Geoarchaeology and Archaeomineralogy. Proceedings of the International Conference 29–30 October – Sofia. St. Ivan Rilski, Sofia, pp 303–304

Peev P (2008b) Physico-geographical description of the Varna bay. In: Varna in the middle ages. Part 1. From the 7th to 10th Century. Zograf, Varna, pp 250–280 (in Bulgarian)

Peev P (2009) The Neolithisation of the Eastern Balkan Peninsula and fluctuations of the Black Sea level. Quater Int 197:87–92. https://doi.org/10.1016/j.quaint.2008.04.019

Peev P, Slavchev V (2018) Bulgaria – burials and wooden settlement structures. In: Fischer A, Pedersen L (eds) Oceans of archaeology. Jutland Archaeological Society and Aarhus University Press, Aarhus, pp 94–99

Peychev V, Peev P (2006) Evolyutzia na bulgarskoto chernomorsko kraibrezhie sled rannia holotzen. Slavena, Varna (in Bulgarian)

Pirazzoli PA (1996) Sea-Level changes: the last 20,000 years. Wiley, Chichester

Popov V, Mishev K (1974) Geomorphology of the Bulgarian Black Sea coast and shelf. BAS, Sofia (in Bulgarian)

Porozhanov K (1991) Le site submergé d'Ourdoviza. Thracia Pontica 4:109–112

Porozhanov K (2004) The Early Bronze Age sunken settlement by Urdoviza (1986-1988 Archaeological Campaigns). In: Slavchev V (ed) Studies in honour of Prof. Dr.hist.Sc. Henrietta Todorova. Dobrudzha 21 (2003), pp 309–322 (in Bulgarian with English Summary)

Prahov N (2013) Archaeological predictive model for late Chalcolithic and Early Bronze Age coastal settlements along the southern Bulgarian sea coast. In: Angelova H, Özdoğan M (eds) International Conference. Where are the sites? Research, protection and management of cultural heritage, 5–8 December 2013, Ahtopol. Organised by Centre for Underwater Archaeology, Sozopol, Bulgaria, pp 117–146

Ryan WBF (2007) Status of the Black Sea flood hypothesis. In: Yanko-Hombach V, Gilbert A, Panin N,

Dolukhanov P (eds) The Black Sea Flood Question: changes in coastline, climate, and human settlement. Springer, Dordrecht, pp 63–88

Ryan WBF, Pitman WC, Major CO, Shimkus K, Moskalenko V, Jones GA, Dimitrov P, Gorür N, Sakinç M, Yüce H (1997) An abrupt drowning of the Black Sea shelf. Mar Geol 138:119–126. https://doi.org/10.1016/S0025-3227(97)00007-8

Ryan WBF, Major CO, Lericolais G, Goldstein SL (2003) Catastrophic flooding of the Black Sea. Ann Rev Earth Planet Sci 31:525–554

Shkorpil H, Shkorpil K (1921) Nakolni postroiki v ezeroto. Izvestiya na Varnenskoto Arheologichesko Druzhestvo 7:79–80 (in Bulgarian)

Sorokin VM, Kuprin PN (2007) On the character of Black Sea level rise during the Holocene. Moscow Univ Geol Bull 62:334–341. https://doi.org/10.3103/S0145875207050055

Soulet G, Ménot G, Garreta V, Rostek F, Zaragosi S, Lericolais G, Bard E (2011) Black Sea "Lake" reservoir age evolution since the Last Glacial — Hydrologic and climatic implications. Earth Planet Sci Lett 308(1–2):245–258. https://doi.org/10.1016/j.epsl.2011.06.002

Soulet G, Ménot G, Bayon G, Rostek F, Ponzevera E, Toucanne S, Lericolais G, Bard E (2013) Abrupt drainage cycles of the Fennoscandian Ice Sheet. Proc Nat Acad Sci USA 110(17):6682–6687 www.pnas.org/cgi/doi/10.1073/pnas.1214676110

Spassov N, Iliev N (1994) Animal remains from the submerged Late Eneolithic - Early Bronze Age settlements in Sozopol (south Bulgarian Black Sea coast). Proc Int Symp Thracia Pontica VI 18-24 September 1994, pp 287–314

Stanchev H, Young R, Stancheva M (2013) Integrating GIS and high-resolution orthophoto images for the development of a geomorphic shoreline classification and risk assessment – a case study of cliff/bluff erosion along the Bulgarian coast. J Coast Conserv 17:719–728. https://doi.org/10.1007/s11852-013-0271-2

Stancheva M (2013) Bulgaria. In: Pranzini E, Williams A (eds) Coastal erosion and protection in Europe. Routledge, Abingdon, pp 378–395

Stanimirov S (2003) The western Black Sea coast in the Eneolithic and Bronze Ages. Athena Rev 3(4) http://www.athenapub.com/AR/12blksea.htm

Stephenson R, Schellart WP (2010) The Black Sea back-arc basin: Insights to its origin from geodynamic models of modern analogues. Geol Soc London Spec Publications 340:11–12. https://doi.org/10.1144/SP340.2

Sturt F, Flemming NC, Carabias D, Jöns H, Adams J (2018) The next frontiers in research on submerged prehistoric sites and landscapes on the continental shelf. Proceedings of the Geologists' Association (2018). doi:https://doi.org/10.1016/j.pgeola.2018.04.008

Todorova H (1978) Prouchvania na neolita I eneolita v Longoza (nachaloto na VI—kraya na V hilyadoletie pr.n.e.). Izvestiya na Narodniya Muzei Varna 14(29):1–9 (in Bulgarian)

Todorova H (1984) Dobrudzha during Prehistoric Age. In: Fol A, Dimitrov S (eds) History of Dobrudzha Tome 1. BAS, Sofia, pp 23–71 (in Bulgarian)

Todorova H (1995) The Neolithic, Eneolithic and Transitional period in Bulgarian prehistory. In: Bailey D, Panayotov I (eds) Prehistoric Bulgaria. Prehistory Press, Madison, pp 79–98

Turney CSM, Brown H (2007) Catastrophic early Holocene sea level rise, human migration and the Neolithic transition in Europe. Quater Sci Rev 26:2036–2041. https://doi.org/10.1016/j.quascirev.2007.07.003

UNESCO (2010a) Romania: National Report on underwater cultural heritage: Report made in the UNESCO Regional Meeting in Istanbul 25–27 October, 2010. http://unesdoc.unesco.org/images/0018/001899/189966E.pdf

UNESCO (2010b) Bulgaria: National Report on underwater cultural heritage: Report made in the UNESCO Regional Meeting in Istanbul 25–27 October, 2010. http://unesdoc.unesco.org/images/0018/001899/189942E.pdf

Velkovsky K, Angelova H, Dragonov V, Prahov N, Dragomir G, Orachev A, Petrov P (2013) Geophysical prospecting and underwater archaeology. Bathymetric and sidescan sonar surveys in the aquatoria of Sozopol-Sinemorets, 2013. In: Angelova H, Özdoğan M (eds) International Conference. Where are the sites? Research, protection and management of cultural heritage, 5–8 December 2013, Ahtopol. Organised by Centre for Underwater Archaeology, Sozopol, Bulgaria, pp 147–154

Yanchilina AG, Ryan WBF, McManus JF, Dimitrov P, Dimitrov D, Slavova K, Filipova-Marinova M (2017) Compilation of geophysical, geochronological, and geochemical evidence indicates a rapid Mediterranean-derived submergence of the Black Sea's shelf and subsequent substantial salinification in the early Holocene. Mar Geol 383:14–34. https://doi.org/10.1016/j.margeo.2016.11.001

Yanko-Hombach V (2007) Controversy over Noah's Flood in the Black Sea: Geological and foraminiferal evidence from the shelf. In: Yanko-Hombach V, Gilbert A, Panin N, Dolukhanov P (eds) The Black Sea flood question: changes in coastline, climate, and human settlement. Springer, Dordrecht, pp 149–203. https://doi.org/10.1007/978-1-4020-5302-3_7

Yanko-Hombach V, Gilbert AS, Dolukhanov P (2007a) Controversy over the great flood hypotheses in the Black Sea in light of geological, paleontological, and archaeological evidence. Quater Int 167:91–113. https://doi.org/10.1016/j.quaint.2006.08.004

Yanko-Hombach V, Gilbert A, Panin N, Dolukhanov P (eds) (2007b) The Black Sea Flood Question: changes in coastline, climate, and human settlement. Springer, Dordrecht

Yanko-Hombach V, Mudie P, Gilbert AS (2011) Was the Black Sea catastrophically flooded during the post-glacial? Geological evidence and archaeological impacts. In: Benjamin J, Bonsall C, Pickard C, Fischer

A (eds) Submerged prehistory. Oxbow, Oxford, pp 45–262
Yanko-Hombach V, Mudie PJ, Kadurin S, Larchenkov E (2014) Holocene marine transgression in the Black Sea: New evidence from the Northwestern Black Sea shelf. Quater Int 345:100–118
Yanko-Hombach V, Schnyukov E, Pasynkov A, Sorokin V, Kuprin P, Maslakov N, Motnenko I, Smyntyna O (2017) Late Pleistocene-Holocene environmental factors defining the Azov-Black Sea basin, and the identification of potential sample areas for seabed prehistoric site prospecting and landscape exploration on the Black Sea continental shelf. In: Flemming NC, Harff J, Moura D, Burgess A, Bailey GN (eds) Submerged landscapes of the European continental shelf: quaternary paleoenvironments. Wiley, Chichester, pp 431–478

21 The Ukraine: In Search of Submerged Late Palaeolithic Sites on the North-Western Black Sea Shelf

Sergey Kadurin, Valentina Yanko-Hombach, and Olena Smyntyna

Abstract

The continental shelf offshore of the Ukraine coast represents one of the largest areas of submerged landscape in the Black Sea, comparable to the extensive shelf areas of Northwest Europe. The general region has a long history of Palaeolithic occupation and numerous archaeological sites associated with the major river systems draining from the north, including many famous Palaeolithic settlements. The submerged landscape exposed during periods of low sea level would have offered an extensive and attractive extra increment of land. Underwater archaeological survey in the Black Sea has so far failed to produce convincing evidence of pre-Holocene sites, but the expectation that the Ukrainian shelf harbours Late Upper Palaeolithic sites is strengthened by the presence of occasional flint artefacts in sediment cores recovered from the shelf area during geological surveys. This chapter develops a predictive model for identifying target areas in the search for Late Palaeolithic sites on this submerged landscape. We focus on the Dniester-Kuyalnik region and analyse the location and distribution of on-land Late Palaeolithic sites in relation to the topographic and palaeoenvironmental features of their local surroundings to identify key determinants of site location. We then use the maps produced by the many hundreds of sediment sequences recovered by drilling and coring on the shelf, many radiometrically dated, to identify similar features on the submerged landscape. We also assess the likelihood of site preservation and visibility under different scenarios—rapid or gradual—of sea-level rise. A preliminary test of the model was attempted with remotely operated vehicles, video and acoustic equipment, but was terminated prematurely by logistic problems, and further investigation awaits improved funding.

Keywords

Sea-level change · Coastline migration · Human adaptation · Palaeolithic · Holocene · Underwater survey

S. Kadurin
Department of Engineering Geology and Hydrogeology, Odessa I.I. Mechnikov National University, Odessa, Ukraine

V. Yanko-Hombach (✉)
Department of Physical and Marine Geology, Odessa I.I. Mechnikov National University, Odessa, Ukraine

Avalon Institute of Applied Science, Winnipeg, MB, Canada
e-mail: valyan@onu.edu.ua; valyan@avalon-institute.org

O. Smyntyna
Department of Archaeology and Ethnology of Ukraine, Odessa I.I. Mechnikov National University, Odessa, Ukraine
e-mail: smyntyna_olena@onu.edu.ua

G. Bailey et al. (eds.), *The Archaeology of Europe's Drowned Landscapes*, Coastal Research Library 35, https://doi.org/10.1007/978-3-030-37367-2_21

21.1 Introduction

The Ukrainian continental shelf exposed as a sub-aerial landscape at the maximum lowering of sea level at the Last Glacial Maximum represents one of the largest areas of now-submerged landscape in the Black Sea (Fig. 21.1), and one of the largest in Europe apart from the North Sea (see also Bailey et al., Chap. 1, Fig. 1.1, this volume; Peev et al., Chap. 20, this volume). Palynological data show that this new territory was a steppe region that could have been exploited extensively by hunter-gatherers migrating in from west to east and from the north-west to the south (Stanko 2007; Yanko-Hombach et al. 2011a, b).

Many Upper Palaeolithic archaeological sites are present in these source regions, extending from Romania through the Ukraine to the Crimea, particularly in the basins of the major rivers draining into the Black Sea from the north such as the Dniester and the Dnieper (Yanko-Hombach et al. 2017, Figure 16.24). Notable examples are Movileni, Lespezi, Mitoc-Malul Galben and Dobrudja in Romania, Molodova in the Middle Dniester region, Liubymovka in the Lower Dnieper in the Ukraine and Siuren 1 and Vishenne 2 in Crimea; (Otte et al. 1997; Demidenko 2000–2001; Chabai 2007). The similarities in stone tool assemblages between these sites suggest a high degree of mobility and connectivity over wide areas.

Moreover, several flint tools have been retrieved from boreholes in various places on the north-western shelf (Stanko 2007, p. 374). In fact, flint pebbles and splinters were mentioned during description of samples recovered from geological survey. However, these descriptions were made by geologists who lacked the knowledge to distinguish between natural pieces of flint and artefacts.

All these factors suggest a high likelihood that the new land exposed by lowered sea level would

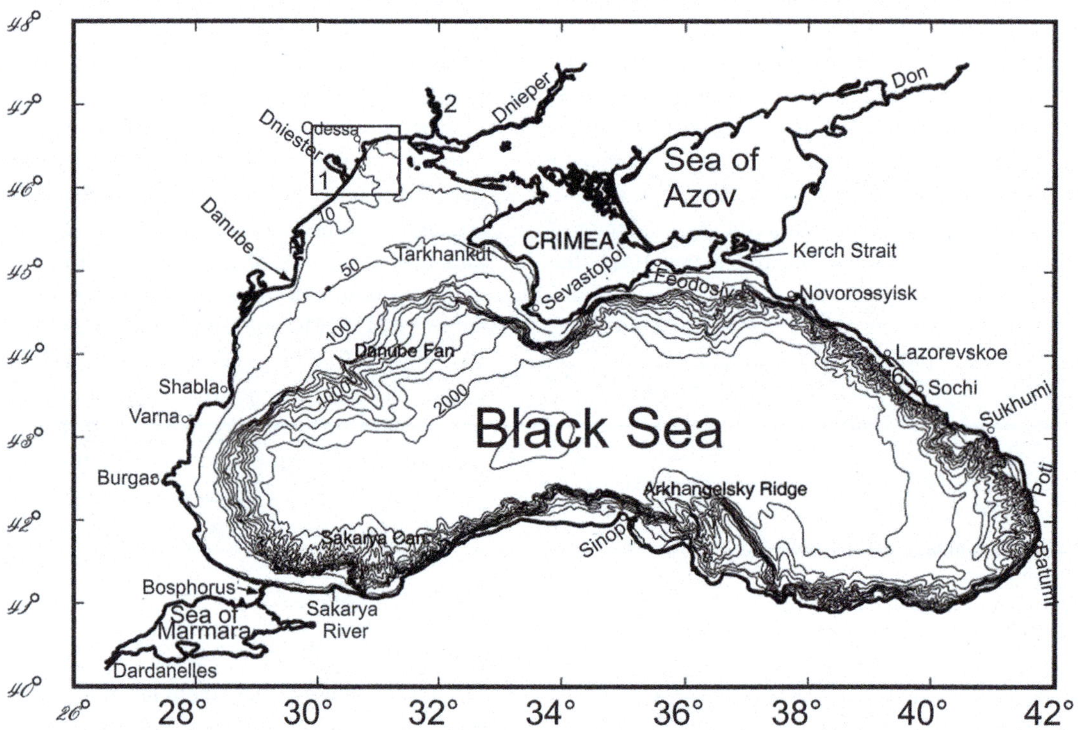

Fig. 21.1 Map of the Black Sea and adjacent regions, showing the extent of the shelf areas, the connections to the Marmara and Aegean seas via the Bosphorus and Dardanelles Straits and the location of the case study area. (*1*) The Dniester Estuary region; (*2*) The Berezan Estuary region. From Yanko-Hombach et al. 2017

have provided a very significant and productive extension of new territory attractive to human settlement. This possibility was explicitly recognised over a decade ago by Stanko (2007) who, in writing about the north-western Black Sea shelf, noted:

> Fluctuations in sea level and the commensurate shrinking and expansion of littoral areas had considerable impact on the settlement pattern of the final Palaeolithic and Mesolithic societies there. (Stanko 2007, p. 372)

> Archaeological surveys targeted at final Palaeolithic and Mesolithic sites on the northwestern Black Sea shelf and along the submerged river valleys might be deemed promising... Such ... survey ... might solve a number of major problems related to the character and chronology of the submergence, migrations, and the interrelationship between prehistoric groups of the Balkans, Central Europe, and Crimea. (Stanko 2007, p. 374)

Until recently, there has been little opportunity to pursue these ideas. Blavatsky (1972) provided the first English language review of underwater archaeological studies in the Black Sea, describing submerged Greco-Roman ruins at water depths of 4–8 m near the Azov Sea entrance and in Taman Bay, near Phanagoria. Similar submerged historical archaeological sites were found in the Bug River Estuary, Olbia Pontica (Kryzhitskiy et al. 1999) in the Dniester Estuary, ancient Tyras (Samoilova 1988) and off the south-eastern Crimean Peninsula (Bolikhovskaya et al. 2004). None of these studies reported any submerged prehistoric settlements.

Dimitrov and Dimitrov (2004) reviewed underwater archaeological studies related to a 'Varna culture' that appeared near the present-day coast of Bulgaria around 5000 BC. Drowned settlements in Lakes Durankulak and Varna were dated to 5270 BC (Dimitrov and Dimitrov 2004, p. 49) by correlation with dated settlements on the lake shore. They claimed that 'Before the Flood [about 7600 years ago], Neolithic people inhabited not only today's coast but also that part of the bottom (called the shelf) which was land' (Dimitrov and Dimitrov 2004, p. 51). So far, however, drowned prehistoric archaeological sites have been found only close to the present Black Sea shore of Cape Shabla north of Varna and in Lake Varna in water less than 10 m deep (Peev 2009). The Shabla site was indirectly dated to the Late Neolithic (Eneolithic) by correlation with the sea-level curve of Peychev and Peev (2006), and submerged settlements in the coastal Varna-Beloslav Lake were indirectly dated to the Late Eneolithic and Early Bronze Age (Peev 2009, 91; see also Peev et al., Chap. 20, this volume).

Coleman and Ballard (2007, p. 677) reviewed evidence for submerged palaeoshorelines in the southern and western Black Sea and their implications for prehistoric inundation. Despite clusters of Neolithic to Bronze Age sites near the present coastline, evidence of prehistoric occupation at water depths greater than 10 m is restricted to one ceramic plate of debatable Neolithic age from −90 m off Varna, and photographs of boulders at −90 m depth off Sinop, possibly related to human habitations along a shoreline inundated during the Neolithic over 8000 years ago (Ballard et al. 2001). In fact, underwater artefacts and shipwrecks recovered to date from this region are of historical age (Ward and Ballard 2004; Ward and Horlings 2008). However, Algan et al. (2009) found archaeological remnants of a Neolithic culture in Istanbul dating between 8400 and 7300 cal BP, indicating shoreline occupation when sea level was 6 m lower than today. These archaic Fikirtepe pottery sherds are immediately overlain by Early Iron Age artefacts, dated by molluscs from 3335 to 3260 cal BP.

Thus, despite decades of searching for submerged prehistoric habitations, there have been no definite finds below a water depth of −10 m, and reports of Neolithic settlements (e.g., Ballard 2001) are based on debatable Early Holocene sea-level estimates (Yanko-Hombach et al. 2011a, b). However, the region is extremely rich in archaeological sites and information relating to the history of sea-level change and the geomorphology and palaeoenvironment of the continental shelf, thanks to systematic studies that have been carried out by several generations of Soviet, Russian and Ukrainian researchers. Discussions have been further intensified during the past decade by a series of multidisciplinary and international research networks involving the increasing participation of western scholars and focussed on Black Sea sea-level change, shoreline dynamics at the Pleistocene-Holocene boundary, their implications for understanding global climate change and their relationship to transformations in human occupation systems, subsistence strategies and stone-tool technologies (Yanko-Hombach and Smyntyna 2009; Yanko-Hombach et al. 2011a, b, 2014; Lericolais 2017; Yanko-Hombach 2017; Yanko-Hombach et al. 2017).[1]

The main goal of this chapter is to identify potential locations for submerged prehistoric sites and landscapes on the north-western Black Sea shelf using the on-land archaeological data as a basis for prediction. Here we:

1. Look for regularities in the spatial distribution of prehistoric sites and artefacts known today in the on-land archaeological record.
2. Examine geomorphological, geological and climatic data to identify the environmental features associated with the locations of these prehistoric sites.
3. Develop a predictive model, based on (1) and (2) for locating submerged prehistoric sites and provide some evaluation of the conditions in which they are likely to have survived the potentially destructive effects of sea-level rise.

21.2 Case Study Area

The Dniester-Kuyalnik interfluve, including both on-shore and underwater areas, was chosen as the case study area (Fig. 21.1). The area is well studied geologically and geomorphologically and has a high density of Late Palaeolithic sites supplemented by geographic coordinates and extensive field studies (Table 21.1). The area is located within the third alluvial terrace of the Dniester River, is crossed by valleys of small rivers (e.g. the Baraboy) and contains the system of valleys associated with the Sukhoy Liman (an ancient estuary) as well as good-quality and well-exposed flint outcrops (Sibirchenko et al. 1983).

[1]See the Acknowledgements for full details of these research networks.

21.3 Materials and Methods

There are about 250 known Late (Upper) Palaeolithic archaeological sites in the south-western part of the Ukraine (Yanko-Hombach et al. 2011a, b, Figure 7). They comprise mainly Aurignacian industries (32–24 ka; Anikovich et al. 2007) and Epigravettian (19–12 ka; Zaliznyak 2014). More than 30 of them were discovered on the present-day on-shore segment of the case study area (Table 21.1).

Each site has GPS coordinates and a short geographic designation enabling one to place it on a geomorphological map (Fig. 21.2). The map is based on information from hundreds of gravity cores and drill holes with lengths up to 5 and 100 m, respectively, many of which were dated by radiometric methods. Descriptions of the main cores and geological maps and reports in digital format are available at http://geoinf.kiev.ua/ and http://geoinf.kiev.ua/wp/kartograma.htm.

In order to increase the probability of finding submerged archaeological sites on the shelf, a method with clearly defined topographic, palaeogeographical and geological search criteria was implemented. These criteria are designed to identify topographic positions that have a good overview of the terrain, nearby freshwater sources and exposures of flint material for stone tool manufacture, a combination of qualities we think would have been influential in site selection.

To further narrow the search, we used GIS software to reconstruct the size of a typical site catchment or site territory based on the distance between site clusters but without taking account of topographic irregularities and barriers (see, e.g. Bailey and Davidson 1983).

21.4 Results

21.4.1 Palaeogeography and Site Prediction

Geomorphologically, the study area includes a watershed plateau, river terraces and slopes, river valleys and hills (Fig. 21.2). The watershed plateau is represented by a Pleistocene succession of

Table 21.1 List of archaeological sites and artefacts with coordinates, geographical location and archaeological age based on Kurten (1969), Sinitsyn (1997, p. 58), Sapozhnikov and Sapozhnikova (2011, table 4, pp. 39–43), Dodonov et al. (2000, p. 307), Djindjian et al. (2006) and Stanko (1999)

Site no	Site name	Type	Latitude	Longitude	Geographic location	Age yrs. ago
1	Cobruciu	Artefact	46.6622	29.7578	Dniester	14,000–12,000
2	Varvara I	Artefact	46.7728	29.9689	Kuchurgan	18,000–12,000
3	Calfa	Artefact	46.9072	29.4558	Dniester	18,000–16,500
4	Bolshaya Akkarzha	Site	46.3425	30.6095	Velykodolynskoe	19,000–17,000
5	Anetovka II	Artefact	47.7039	31.1222	Bakshala	19,100–18,000
6	Anetovka I	Artefact	47.7069	31.1636	Bakshala	19,500–19,100
7	Tatarbunary	Site	45.8572	29.6562	Koghylnik	20,000–12,000
8	Bolshaya Akkarzha II	Site	46.3343	30.6066	Velykodolynskoe	22,000–18,000
9	Bolshaya Akkarzha III	Site	46.3264	30.6040	Velykodolynskoe	22,000–18,000
10	Gribovka	Site	46.1944	30.5576	Baraboy	22,000–20,000
11	Gribovka I	Site	46.2004	30.5498	Baraboy	22,000–20,000
12	Otarik	Site	46.3321	30.3376	Sukhoy Liman	22,000–20,000
13	Skurta	Site	46.3321	30.3376	Sukhoy Liman	22,000–20,000
14	Roxolany	Site	46.1628	30.4736	Dniester	24,000–20,000
15	Caragas	Artefact	46.7881	29.6597	Dniester	24,000–20,000
16	Anetovka XXII-1989	Artefact	47.7183	31.1764	Southern Bug	24,000–20,000
17	Anetovka XXI	Artefact	47.7186	31.1811	Southern Bug	24,000–20,000
18	Baraboy 3	Site	46.2020	30.5765	Baraboy	25,000–22,000
19	Baraboy 2	Site	46.2110	30.5548	Baraboy	25,000–22,000
20	Efymovka	Site	46.3605	30.3454	Dniester	25,000–22,000
21	Efymovka2	Site	46.3516	30.3466	Dniester	25,000–22,000
22	Nikolaevka II	Site	46.3321	30.3376	Dniester	25,000–22,000
23	Peshcera Nordmana	Site	46.4792	30.7345	Black Sea	28,000–26,000
24	Novaya Doina	Site	46.3321	30.3376	Sukhoy Liman	32,000–11,000
25	Olanesti	Artefact	46.4975	29.9675	Dniester	32,000–12,000
26	Krasnaya Kosa	Artefact	46.3475	30.0953	Dnestrovsky Liman	32,000–12,000
27	Kazatskoe	Artefact	46.3533	30.0953	Dnestrovsky Liman	32,000–12,000
28	Turlaki	Artefact	46.1992	30.2331	Dnestrovsky Liman	32,000–12,000
29	Roxolany	Artefact	46.1622	30.4747	Dnestrovsky Liman	32,000–12,000
30	Otaryk	Artefact	46.2061	30.4208	Dnestrovsky Liman	32,000–12,000
31	Aneovka XIII	Artefact	47.7200	31.1764	Bakshala	32,000–24,000
32	Bornatove-zatoka	Site	46.7756	30.4909	Khadzhybey	32,000–26,000
33	Zeleny Khutor	Artefact	46.5006	30.1733	Kuludorova	32,000–28,000
34	Kuludar	Artefact	46.5008	30.2033	Kuludorova Balka	32,000–28,000
35	Tatarka 2	Site	46.4067	30.5907	Dalnik	34,000–22,000

alternating aeolian-deluvial, aeolian and eluvial rocks forming loess loams with buried soils up to 40 m in thickness (Avrametz et al. 2007).

The slopes of the river valleys mainly comprise alluvial sediments accumulated on 12 Quaternary terraces of the rivers flowing into the Black Sea (Veklich and Sirenko 1976; Veklich et al. 1993; Sibirchenko et al. 1983; Ivanov 1987; Shmuratko 2001). One place where all 12 terraces are exposed is the lower reaches of the Dniester River. The deposits of the terraces are basically similar in their lithological characteristics, being represented by sands with pebbles in the lower part and sandy silts and clays, sometimes peat, in the upper part. However, there are some differences in the composition of the heavy mineral fraction caused by differences in the redistribution of eroded sedimentary material. The most significant variations are present within the third terrace, where numerous exposures of abundant, good-quality flint are present (Veklich and Sirenko 1976; Veklich et al. 1993).

Fig. 21.2 Geomorphological map of the pre-Holocene (largely Late Pleistocene) geology, produced at scales ranging from 1:200,000 to 1:50,000. Data from quadrants L-36-XIII and L-36-XIV mapped by Prichernomor SRGE [Black Sea Area State Regional Geologic Enterprise] (www.pgrgp.com.ua; Podoplelov et al. 1975; Karpov et al. 1978; Sibirchenko et al. 1983; Gozhik et al. 1987; Avrametz et al. 2007, see also Yanko-Hombach et al. 2017, Fig. 16.3). The map shows locations of on-land Late Palaeolithic sites and artefacts together with underwater areas that appear promising for locating submerged Late Palaeolithic sites according to the predictive model. See text and Fig. 21.3 for further detail and discussion

Valley bottoms are filled with typical floodplain sediments, such as interbedded sandy silt, clay and peat.

The hills, such as the Dnestrovsky (I) and Katranya (II) Banks, are formed of Upper Pleistocene marine sediments of Karangatian age—MIS 5e, 74–123 ka BP (Yanko-Hombach and Motnenko 2016a, b). Lithologically, they are represented by cemented sandy coquina (limestone with shells) that was exposed during the post-Karangatian regression, passed the first stage of lithification in a coastal environment and formed positive relief forms with a cemented crust over the underlying loose sediments. During the Holocene transgressions, these hills became underwater banks.

All analysed sites are located on the alluvial terraces of secondary rivers or tributaries and form small groups or clusters, each of which includes two to four sites (Fig. 21.3a). The distance between clusters is typically within the range of a 1-day walk (Fig. 21.3b). The distribution of these distances follows a normal probability (Gaussian) curve (Fig. 21.3c) with an average value of 19.49 km and a standard deviation of 6.51 km (95% confidence interval). Moreover, these sites are broadly contemporaneous, suggesting that their distribution reflects a real pattern of prehistoric land use and that the factors that determine the location of the site clusters are concentrations of attractive resources and a distance to the nearest neighbouring cluster that creates a suitably sized territory large enough to support each cluster. It is possible that some sites are missing because they have not been preserved or discovered, but the region has been well studied and the distribution of known sites is clearly not random.

The boundary values (maximum and minimum inter-cluster distances) are 16.74 km and 22.24 km. If these distances are accepted as maximum and minimum spacings between clusters, then a predictive model can be created by drawing concentric circles with these radii around

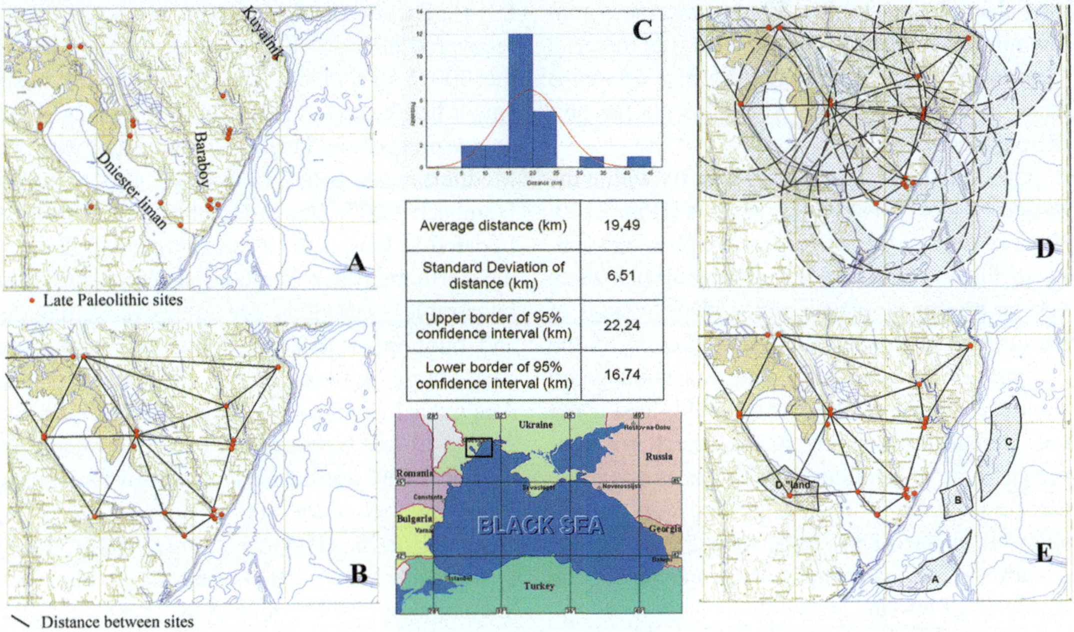

Fig. 21.3 Geostatistical predictive model for identifying target areas to search for submerged Late Palaeolithic sites in the case study area. (**a**) Location of known Late Palaeolithic sites on-land and within the main river valleys today; (**b**) distance between clusters of sites; (**c**) histogram of distances showing normal (Gaussian) type of distribution; (**d**) modelling of concentric catchment circles based on maximum and minimum inter-cluster distances (95% confidence interval); (**e**) areas that lie at the intersections of catchment circles and could therefore be promising targets in the search for submerged Late Palaeolithic sites

each site cluster. The circles then map out a catchment within which known sites are located in the centre and other possible sites would lie within the inner and outer radii corresponding to these values (Fig. 21.3d). Based upon inter-site distances exhibited by known sites, the intersection areas of these catchment circles would be the most promising in the search for new clusters of sites (Fig. 21.3e). Overlain on a geomorphological map of the study area (Fig. 21.2), these intersection areas (A, B, C, D in Figs. 21.2 and 21.3e) show possible locations of submerged archaeological sites that satisfy the search criteria.

21.4.2 Underwater Target Areas

There are three intersection areas (Fig. 21.2) that could be suitable for our purpose. The first one (A) is located 9 km from the present coastline at water depths of 15–20 m bsl (below sea level) in front of the present spit of the Dniester Estuary. Here all search criteria converge. There are small rivers flowing into the large Dniester River forming bluffs and terraces with a good overview of the surrounding terrain. The main elevation point is the Dniesterovsky Bank, which would have been a hill rising above the surrounding plain during Late Palaeolithic time. There are exposed outcrops of good-quality flint material in significant quantity, and according to geostatistical data, the location is at the appropriate distance from the nearest cluster of known sites. Area A seems quite promising in the search for submerged Late Palaeolithic sites.

The second area (B) is located 7 km from the modern coastline opposite Sandzeika village at a water depth of 20–22 m (Fig. 21.2). In the Late Palaeolithic, there was a system of small rivers with elevated bluffs and terraces (terrace III). These rivers in recent times formed the Sukhoy Liman. From the elevated points, there was a good view to the other side of the river valley as well as to the confluence of these tributaries with the Palaeo-Dnieper valley. The river terraces contain numerous outcrops of exposed flint material (Sibirchenko et al. 1983). The distance of area B from other archaeological sites also fits the predictive model.

The third area (C) is located directly within the Palaeo-Dnieper River valley at 25 m (Fig. 21.2). None of the search criteria (except close proximity to flint) is satisfied here. For example, there was no prominent elevated area, the flat sides of the large valley were wetlands (indicated by the presence of peat deposits), and flint nodules are small and poor in quality. As such, this area is not considered as promising in the search for submerged Late Palaeolithic sites. There is one more area D (Fig. 21.2) located presently on the left side of the Palaeo-Dniester River at 2 m asl (above sea level). This area might be promising for the discovery of on-land archaeological sites.

21.5 Discussion

Three areas, A, B and C, described above (Fig. 21.2) seem to be promising for finding submerged Late Palaeolithic archaeological sites at water depths of 20–22 m. Could these sites be present below sea level and if so when were they inundated? To answer this question, the dynamics of sea level, climate and salinity variation during the Late Palaeolithic period should be considered using geological, palynological and palaeontological data together with archaeological evidence of human adaptive strategies. This may give us further clues as to whether we are correct in our assumptions regarding the search for submerged Late Palaeolithic sites.

21.5.1 Sea-Level Change and Palaeoenvironment

In terms of chronology, the Late Palaeolithic corresponds approximately to the following stratigraphic units: the Tarkhankutian (c. 44–32 ka cal BP) and the Lower (c. 32–20 ka cal BP) and Upper Neoeuxinian (c. 20–11.5 ka cal BP) beds (Yanko-Hombach 2007; Yanko-Hombach et al. 2014). The Tarkhankutian beds are occasionally recovered in cores (Fig. 21.4) penetrating below the 30–35 m isobath. The Tarkhankutian basin was a brackish (8–11 psu) body of water, 30 m below present sea level, and connected with the Mediterranean Sea (Nevesskaya 1965; Yanko-Hombach 2007; Yanko-Hombach et al. 2017). Located to the south of the continental glaciers, the northern Black Sea coastal region provided favourable conditions for intensive settlement and hunting, as evidenced by the presence of such large settlements as Sagaidak 1 and Anetovka 1 (Yanko-Hombach et al. 2011a, b).

The Lower Neoeuxinian beds deposited during the LGM are distributed everywhere in the Black Sea below the 100 m isobath (Fig. 21.4) (Kvasov 1975; Fedorov 1977, 1978, 1988; Shcherbakov et al. 1978; Abashin et al. 1982; Shcherbakov 1983; Shnyukov 1985; Svitoch et al. 1998; Kuprin 2002; Kuprin and Sorokin 2007; Yanko-Hombach 2007; Yanko-Hombach et al. 2014, 2017; Harff et al. 2016). During that time, the Black Sea was a brackish lake with salinity around 5 psu and a level at −100 m. The lake was isolated from both the Caspian and Mediterranean seas (Kuprin and Sorokin 2007; Yanko-Hombach 2007; Yanko-Hombach et al. 2011a, b, 2014, 2017). A large portion of the present shelf above the 100 m isobath was exposed and eroded. The north-western shelf was downcut some 40 m to the basement by the Pre-Danube, Pre-Dnieper and Pre-Dniester rivers and covered by subaerial loams (e.g. Shcherbakov et al. 1978; Shcherbakov 1983; Inozemtsev et al. 1984; Fedorov 1988). River mouths were relocated 80–100 km seaward (Gozhik 1984; Shnyukov 1985), where they possessed poorly developed deltas and opened directly into canyons on the continental slope (Fig. 21.5; Yanko-Hombach et al. 2017).

At the LGM, the entire Eastern European platform was covered with periglacial tundra-steppe vegetation (Artyushenko 1970; Artyushenko et al. 1972; Mudie et al. 2014). Extremely harsh climatic conditions caused the displacement of the late Palaeolithic inhabitants from Central and Eastern Europe to the northern Black Sea coastal regions (Yanko-Hombach et al. 2011a, Figure 6).

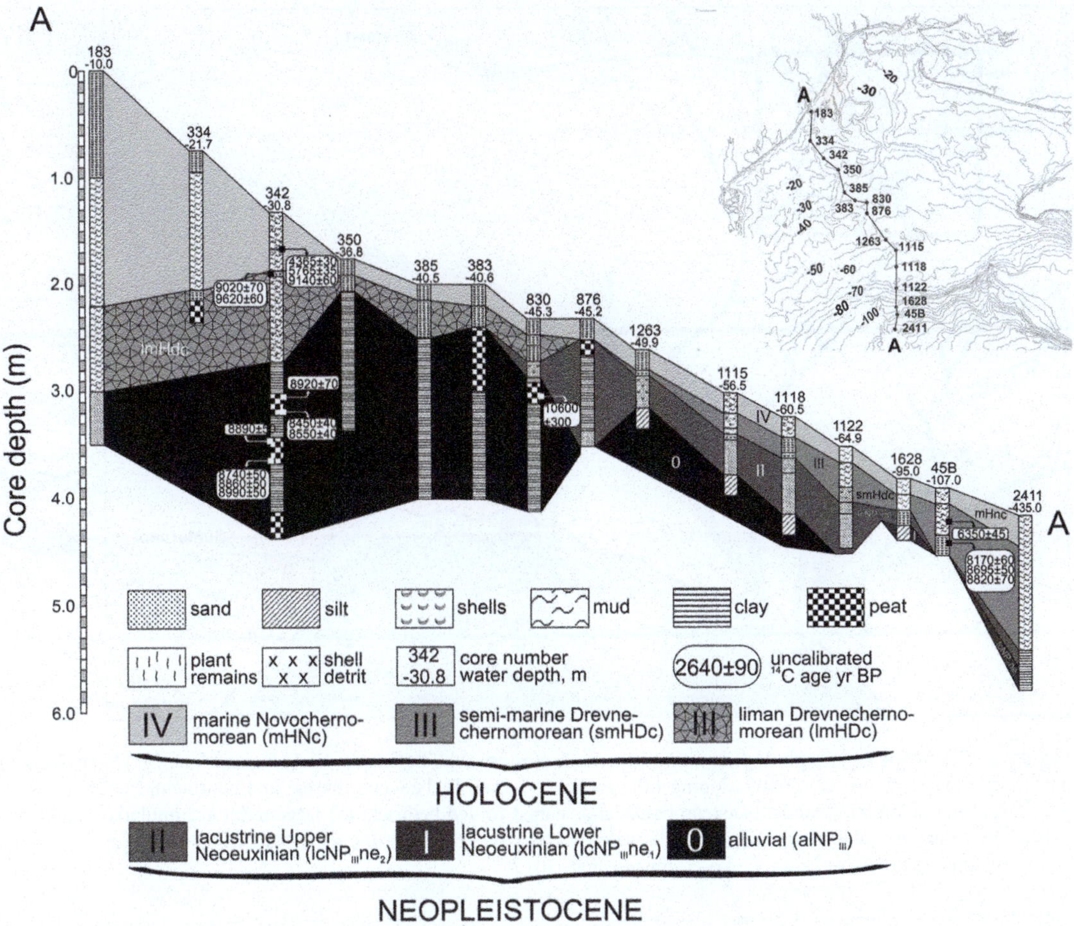

Fig. 21.4 Palaeo-Dniester valley cross-shelf profile, showing sediment lithology and genetic type, depositional setting and stratigraphic ages. Inset shows shelf bathymetry, locations of transect cores and present-day estuaries at the seaward end of the Dnieper, Bug and Dniester rivers. Shell detrit = detrital shell fragments. Profile is based on results of a 1:200,000 marine geological survey. From Yanko-Hombach et al. 2014, 2017

In fact, the Black Sea steppes represent a natural refuge for the populations of the contiguous territories, which is reflected in the mosaic of modern ethnic groups in the region and confirmed by the highest population densities in Eastern Europe (Yanko-Hombach et al. 2011a, Figure 7). In search of protection from the harsh conditions of the periglacial regions and the need to find animals for hunting, bison hunters moved to the south and south-west along the river valleys of the Dnieper, the Southern Bug, the Dniester and the Prut and reached the shore of the Early Neoeuxinian lake. However, it is unlikely that they settled along the immediate coastline. Palynological data from high-resolution marine cores suggest a marshy and mosquito-infested shoreline subject to periodic river flooding (Mudie et al. 2002, 2007). This might have provided good hunting and fishing but poor conditions for the location of settlement because of brackish water and soils prone to salinization and waterlogging. This is consistent with the evidence that prehistoric populations preferred to settle in small river valleys. These marshy conditions may have been inhospitable for permanent settlement, but there is no evidence that prehistoric people would have avoided coastal locations on the Black Sea because of the presence of saltwater (Yanko-Hombach et al. 2014).

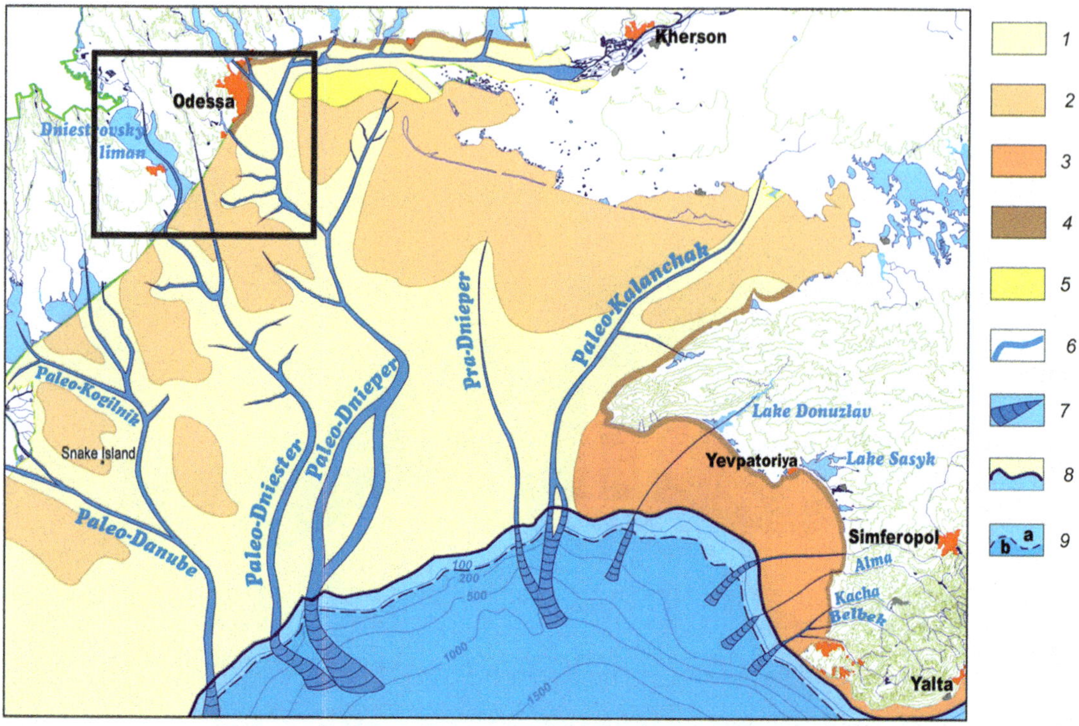

Fig. 21.5 Schematic geomorphological map of the north-western shelf during the Early Neoeuxinian: (*1*) Palaeovalleys and large alluvial plains; (*2*) gently sloping hills formed by Late Pleistocene marine and continental landforms; (*3*) gently sloping piedmont uplands; (*4*) steep banks and seabed folded bedrock; (*5*) large relict accumulative landforms formed by sand; (*6*) palaeoriver bed; (*7*) underwater debris cone; (*8*) coastline of 18 ka BP; (*9*) shelf edge, (a) outer shelf, (b) continental slope. From Yanko-Hombach et al. 2017

These changes of environmental conditions in the north-western Black Sea region resulted in the emergence of a new occupation system based on the collective hunting of bison herds, which involved the short-term concentration of human groups at specialised base camps, designed for implementation of economic activities connected with the preparation for a major hunt and further celebration of its results, and a set of other sites briefly visited during the rest of the year. This occupation strategy is particularly well demonstrated by the key archaeological sites of the region belonging to the Epigravettian, notably Anetivka 2 (on the Southern Bug River). Within walking distance of this site, there are over 30 Late Palaeolithic archaeological sites, all within reach of the Bakshala River, a tributary of the Southern Bug (Stanko 1999, p. 322; Stanko 1997, Figure 23).

The Lower Neoeuxinian beds are often overlapped by sub-aerial loams and later by aquatic sediments with freshwater ostracods *Candona* and *Candoniella* indicating the transformation of bottom conditions from an erosional to a subaquatic accumulative phase at the beginning of the Late Neoeuxinian transgression (Gozhik 1984; Shnyukov 1985).

The Upper Neoeuxinian beds cover the Black Sea floor almost everywhere on the north-western shelf below the 37–39 m isobath (Fig. 21.4; Larchenkov and Kadurin 2011; Mudie et al. 2014; Yanko-Hombach et al. 2014, 2017). Palaeosalinity for the Late Neoeuxinian lake was about 5 psu in shallow areas and could have reached 7 psu and even 12 psu in deeper parts of the basin (Nevesskaya 1965; Mudie et al. 2001; Mudie et al. 2011; Marret et al. 2009; Yanko-Hombach et al. 2014, 2017). Despite a relatively high salinity, Mediterranean species are absent, and Caspian immigrants are abundant.

Prevalence of broad-leaved trees (e.g. *Quercus*, *Carpinus*, *Ulmus*, *Salix*, *Betula*) over conifers (e.g. *Pinus*) and grasses in the spore-pollen diagrams indicates that climate became warmer compared to the Early Neoeuxinian (Pop 1957; Bozilova 1975). During the Late Neoeuxinian transgression, a relatively wide and shallow, sub-horizontal inner shelf with depths of c. 10–15 m was formed. Due to an increase in the base level and consequent increased lateral erosion, the river valleys expanded due to bank scouring, and the accumulation of alluvium intensified. The absence of Mediterranean organisms clearly shows the lack of connection with the Mediterranean Sea. The widespread presence of Caspian species indicates a possible connection with the Caspian Sea. It is also possible that under favourable conditions Caspian species could expand from sheltered habitats in river mouths and lagoons, as was noted by Nevesskaya (1965). The Caspian waters, together with strong freshwater runoff from the land, raised the level of the Late Neoeuxinian lake by 70 m within 3–4000 years. In many places, the Upper Neoeuxinian beds are overlain by peat (Fig. 21.4) with an age of ~11.7 ka cal BP (10 ka radiocarbon years BP) and/or coarse sediments. Their maximum depth of deposition does not exceed 50 m. These peats were formed during the Younger Dryas around 11.7 ka cal BP when the level of the lake dropped to 55 m (Yanko-Hombach et al. 2011b).

It is difficult to reconstruct the lifeways of the Late Palaeolithic human population during this Late Neoeuxinian period due to the almost complete absence of large, long-term settlements, and consequently a scarcity of chronometric dates. For whatever reason, perhaps because of human overexploitation of the northern Black Sea steppe region, population density in this region appears to have sharply decreased, and in the Dniester-Bug interfluve, it became the lowest for the steppe zone within the modern territory of Ukraine (Yanko-Hombach et al. 2011a, Figure 6). As of now, only one ancient occupation—Vladimirovka (layer 3)—can be associated with any degree of certainty to the Late Neoeuxinian, and its technocomplex retains all the features typical of the Anetivka Late Palaeolithic culture (Chernysh 1953).

21.5.2 Conditions of Preservation and Site Discovery

Areas A and B seem never to have been submerged by transgression during the Late Pleistocene because neither the Tarkhankutian nor Late Neoeuxinian transgression exceeded the 30 m and 37 m isobath bsl, respectively. However, the transgressions would have caused an expansion of the river valleys, and this would have initiated meandering and steady infilling by fluvial sediments. In this way, the river valleys would have provided ample accommodation for sedimentary accumulation and buried settlements; i.e. any sites would have a good chance to be preserved.

Today, these areas are below sea level, however, and this means that they were inundated during the Holocene Mediterranean transgression. What are the chances that archaeological sites would have been preserved? There are two basic scenarios proposed for this transgression: progressive (Aksu et al. 2002) or progressive but oscillating (Balabanov 2007; Yanko-Hombach 2007; Yanko-Hombach et al. 2007) and catastrophic (Ryan 2007) or rapid (Nicholas et al. 2011; Lericolais 2017; Yanchilina et al. 2017).

In the case of a gradual transgression, there should be an upward succession of coarse to fine grained sediments. On the inner shelf, alluvial facies should be covered by marshy, estuarine and marine ones. Such a picture can be observed in Core 342 recovered from the Dniester palaeo-valley (Mudie et al. 2014; Yanko-Hombach et al. 2014). Over the course of a progressive transgression, the elevated watershed plateaux would have been prone to coastal abrasion and destroyed by the gradual sea-level rise. But the terraces and their associated protruding bluffs within the meanders of small rivers and tributaries would have been preserved, and these features, according to our data, are present in areas A and

B. In the case of a gradual but oscillating scenario, low-amplitude sea-level fluctuations would be unlikely to destroy thick sequences of sediments or the settlements buried within them.

In the case of a catastrophic or rapid transgression, there would be sharp erosional downcutting in the river valleys, and this would destroy the bluffs and other elevated elements of relief together with the settlements buried within the sediments of previous transgressions. Such a picture of course is an ideal. In reality, it could be different.

In order to verify our predictive model and to improve the search for submerged Late Palaeolithic sites, an expedition 'Coast of Gods' was undertaken in 2012 with the participation of the first author and the use of a seine-fishing vessel 'MBSS 10–06' (Kerch registration port) and a scientific team to search for underwater prehistoric archaeological sites and materials on the Black Sea shelf. We used a remotely controlled submarine 'Sophocles-1' with a 'Towed Bottom Video, TBV-2', an underwater towing system (test model of the National Naval Construction University named after admiral Makarov, Nikolayev city) and sidescan sonar. Unfortunately, due to technical problems and harsh weather conditions, the expedition did not achieve its goal and was postponed pending acquisition of new funding.

21.6 Conclusions

Using on-land archaeological data on the locations of known Late Upper Palaeolithic sites, in a case study area located on the Dniester-Kuyalnik interfluve, we have analysed the main regularities in their spatial distribution. We found that all the sites examined are located on the lower alluvial terraces of smaller, tributary rivers and form small clusters, each of which includes two to four sites. The distribution of distances between the clusters of sites follows a normal (Gaussian) distribution with an average 19.49 km in walking distance between each cluster and its nearest neighbour.

We used geomorphological, geological and climatic data to identify the environmental factors determining the choice of site locations and to predict likely target areas for archaeological survey offshore on the now-submerged landscape. This predictive model identified two likely areas. The first area (A) is 9 km from the present coastline at water depths of 15–20 m in front of the present spit of the Dniester Estuary. Here there would have been small rivers flowing into the large palaeo-Dniester River, forming bluffs and terraces with good overviews of the surrounding terrain. The main elevation point is the Dniesterovsky Bank, which would have been a hill on the exposed land surface during the Late Palaeolithic. The area has outcrops of flint of good quality and quantity on the third alluvial terrace of the Dniester River, and according to our analysis, it is located at an appropriate distance from the nearest cluster of known sites in the present-day landscape. The second area (B) is located offshore from Sandzeika village, 7 km from the coastline at a water depth of 20–22 m on a river terrace of the Sukhoy palaeoriver, a tributary of the palaeo-Dnieper, with a good view to the other side of the river valley as well as towards its confluence with the palaeo-Dnieper valley. The distance of area B from other archaeological sites also fits the predictive model.

In the case of a gradual or gradual-but-oscillating transgression by water from the Mediterranean at the Pleistocene-Holocene boundary, submerged archaeological sites should be preserved by being buried by thick fluvial sediments accumulated during the Late Pleistocene (Tarkhankutian and Late Neoeuxinian) transgressions. In the case of a catastrophic or rapid transgression, a sharp erosional downcutting within the river valleys would have destroyed the bluffs and other features of elevated relief together with any prehistoric settlements buried by sediments accumulated during previous transgressions.

Although the proposed model for finding submerged Late Palaeolithic archaeological sites on the north-western Black Sea shelf was not verified by successful underwater fieldwork, we identify the reasons for this failure not with the predictive model itself, but instead with technological aspects of the fieldwork and the equipment used.

Taking into account the peculiarities of the region under study—depth below present sea level of around 30 m, distance from the present seashore of about 10 km, and most importantly the significant thickness of Holocene sediments—traditional underwater archaeology employing scuba-diving may not be effective for future survey. The only reliable source of information is a programme of underwater drilling, which will require substantial funding and the joint efforts of a multidisciplinary scientific team.

Acknowledgments We are grateful to Prof. Allan Gilbert from Fordham University, USA, for editing the English text and for valuable comments and to Geoff Bailey for his editorial suggestions and further work on the final text. This paper is a contribution to IGCP 521 'Black Sea-Mediterranean Corridor during the last 30 ky: sea level change and human adaptation'; INQUA 0501 'Caspian-Black Sea-Mediterranean Corridor during the last 30 ka: Sea-level change and human adaptive strategies'; IGCP 610 'From the Caspian to Mediterranean: Environmental Change and Human Response during the Quaternary'; the Russian–Ukrainian project No. Ф28/428-2009 'The Northwestern Black Sea Region and Global Climate Change: Environmental Evolution during the last 20 ka and Forecast for the 21st Century' sponsored by the State Fund for Fundamental Research, Ukraine; 'Theories on interaction between the environment and human society in the north-western Black Sea during the Late Pleistocene and Holocene' sponsored by the Ministry of Education and Science of Ukraine (2016–2018); INQUA Focus Group POCAS (SACCOM 1709F) 'Ponto-Caspian stratigraphy and geochronology' (2017–2020); and COST Action TD0902 SPLASHCOS project 'Submerged Prehistoric Archaeology and Landscapes of the Continental Shelf'.

In this paper, we have transliterated Cyrillic letters into the Latin alphabet according to the BGN/PCGN Romanization system for Russian used by Oxford University Press. Exceptions are the names of authors, which we have left in their own preferred transliterations, as well as geographical names as presented most commonly in the majority of English-language papers.

References

Abashin AA, Mel'nik VI, Sidenko OG (1982) Bottom morphology. Characteristics of relief. In: Shnyukov EF (ed) Geology of the Ukrainian shelf. Environment, history and methodology of the study. Naukova Dumka, Kiev, pp 82–88 (in Russian)

Aksu AE, Hiscott RN, Kaminski MA et al (2002) Last glacial-Holocene paleoceanography of the Black Sea and Marmara Sea: stable isotopic, foraminiferal and coccolith evidence. Mar Geol 190:119–149

Algan O, Yalçın MN, Özdoğan M, Yılmaz İ, Sarı E, Kırcı-Elmas E, Ongan D, Bulkan-Yeşiladalı Ö, Yılmaz Y, Karamut İ (2009) A short note on the geoarchaeological significance of the ancient Theodosius Harbour (İstanbul-Turkey). Quat Res 72:457–461

Anikovich M, Anisutkin N, Vishnyatsty L (2007) Uzlovye problemy perekhoda k verkhnemu paleolitu Evrazii [Key points of transition to Late Paleolithic in Eurasia]. In: Platonova N (ed) Trudy Kostenkovsko-Borshcevskoy arheologicheskoy expeditsii v. 5 [Russian academy of science. The works of Kostenkovo-Borcevskiy archaeological expedition], vol 5. Nestor-Istoria, St. Petersburg, pp 203–211 (in Russian)

Artyushenko AT (1970) Rastitelnost' Lesostepi i Stepi Ukrainy v chetvertichnom periode [Vegetation of forest-steppe and steppe Ukraine in quaternary]. Naukova dumka, Kiev

Artyushenko AT, Pashkevich GA, Kareva EB (1972) Razvitie rastitel'nosti yuga Ukrainy v antropogene po dannym sporo-pyl'tsevogo analiza [Development of vegetation of the south of Ukraine in anthropogene]. Bull Comm Study Quat Period 39:82–89 (in Russian)

Avrametz VM, Kakaranza SD, Sibirchenko MG et al (2007) Zvit z provedennya geologichnoi ziomky masshtabu 1:200,000 pivnichno-zakhidnoi chastyny shelfu Chornogo morya v mezhakh arkushiv L-36-XIII. –XIV, XV [report on geological survey 1:200,000 within quadrants L-36-XIII. L-36-XIV, L-36-XV]. Prichonomorske SRGE [Black Sea Area State Regional Geologic Enterprise], Odessa (in Ukrainian)

Bailey GN, Davidson I (1983) Site exploitation territory and topography: two case studies from Palaeolithic Spain. J Archaeol Sci 10:87–115

Balabanov IP (2007) Holocene sea-level changes of the Black Sea. In: Yanko-Hombach V, Gilbert AS, Panin N, Dolukhanov PM (eds) The Black Sea flood question: changes in coastline, climate and human settlement. Springer, Dordrecht, pp 711–730

Ballard RD (2001) Black Sea mysteries. Natl Geogr 199(5):52–69

Ballard RD, Hiebert FT, Coleman DF, Ward C, Smith JS, Willis K, Foley B, Croff K, Major C, Torre F (2001) Deepwater archaeology of the Black Sea: the 2000 season at Sinop, Turkey. Am J Archaeol 105:607–623

Blavatsky VD (1972) Submerged sectors of towns on the Black Sea coast. In: Underwater archaeology, a nascent discipline. UNESCO, Paris, pp 117–119

Bolikhovskaya N, Kaitamba M, Porotov A, Fouache E (2004) Environmental changes of the northeastern Black Sea's coastal region during the middle and late Holocene. In: Scott EM, Alekseev AY, Zaitseva G (eds) Impact of the environment on human migration in Eurasia, NATO science series IV: earth and environmental sciences, vol 42. Springer, Dordrecht, pp 209–223

Bozilova E (1975) Correlation of the vegetational development and climatic changes in the Rila and Pirin mountains in the Late Glacial and Post Glacial time. In: Jordanov D, Bondev I, Kozuharov S, Kuzmanov B, Palamarev E (eds) Problems of Balkan flora and vegetation. Proceedings of the first international symposium on Balkan flora and vegetation, June 7–14, 1973. Varna Pensoft, Sofia, pp 64–71

Chabai VP (2007) The Middle Paleolithic and Early Upper Paleolithic in the northern Black Sea region. In: Yanko-Hombach V, Gilbert AS, Panin N, Dolukhanov PM (eds) The Black Sea flood question: changes in coastline, climate and human settlement. Springer, Dordrecht, pp 279–296

Chernysh FP (1953) Vladimirskaya paleoliticheskaya stoyanka [Vladimirskaya Paleolithic site]. Bull Comm Study Quat Period 17:43–51 (in Russian)

Coleman DF, Ballard RD (2007) Submerged paleoshorelines in the southern and western Black Sea – implications for inundated prehistoric archaeological sites. In: Yanko-Hombach V, Gilbert AS, Panin N, Dolukhanov PM (eds) The Black Sea flood question: changes in coastline, climate and human settlement. Springer, Dordrecht, pp 671–696

Demidenko Yu E (2000–2001) The European Early Aurignacian of Krems-Dufour type industries: a view from Eastern Europe. Préhistoire européenne 16–17:147–162

Dimitrov P, Dimitrov D (2004) The Black Sea, the flood, and the ancient myths. Slavena, Varna

Djindjian F, Sapozhnikov I, Stepanchuk V, Sapozhnikova G (2006) Upper Palaeolithic chronology, cultural facies and economic complexes of the northern Black Sea area. In: Torti JSS, Alcántara AMM, Fullole I, Pericot JM (eds) La Cuenca mediterránea durante el Paleolítico Superior 38.000–10.000 anès. IV Symposium de Prehistoria de la Cueva de Nerja. Fundacion Cueva de Nerja, Nerja, pp 46–59

Dodonov AE, Tchepalyga AL, Mihailescu CD, Zhou LP, Markova AK, Trubikhin VM, Simakova AN, Konikov EG (2000) Last-interglacial records from Central Asia to the northern Black Sea shoreline: stratigraphy and correlation. Geologie en Mijnbouw Netherlands Journal of Geosciences 79(2/3):303–311

Fedorov PV (1977) Pozdnechetvertichnaya istoriya Chernogo moria i razvitie yuzhnykh morey Evropy [Late quaternary history of the Black Sea and southern seas of Europe]. In: Kaplin PA, Shcherbakov FA (eds) Paleogeografiya i otlozheniya pleistotsena yuzhnykh morey SSSR [Paleogeography and deposits of the Pleistocene of the southern seas of the USSR]. Nauka, Moscow, pp 25–32 (in Russian)

Fedorov PV (1978) Pleistotsen Ponto-Kaspiya [The Pleistocene of the Ponto-Caspian]. Nauka, Moscow (in Russian)

Fedorov PV (1988) The problem of changes in the level of the Black Sea during the Pleistocene. Int Geol Rev 30:635–641

Gozhik PF (1984) Istoriya razvitiya limanov [History of the development of the limans]. In: Shnyukov EF (ed) Geologiya shel'fa USSR Limany [Geology of the Ukrainian Shelf Limans]. Naukova Dumka, Kiev, pp 76–80 (in Russian)

Gozhik PF, Karpov VA, Ivanov VG, Sibirchenko MG (1987) Golotsen severo-zapadnoy chasti Chernogo morya [Holocene of the northwestern part of the Black Sea]. Geologicheskiy Institut Ukrainskoy Akademii Nauk [Geological Institute of the Ukrainian Academy of Sciences]. Pre-Print 87-41 (in Russian)

Harff J, Bailey GN, Lüth F (2016) Geology and archaeology: submerged landscapes of the continental shelf: an introduction. In: Harff J, Bailey GN, Lüth F (eds) Geology and archaeology: submerged landscapes of the continental shelf. Geological Society, London, Special Publication 411, pp 1–8

Inozemtsev YI, Lutsiv YK, Sobotovich EV et al (1984) Holocene geochronology and facies complexes of the Pontic area. In: Shnyukov EF (ed) Izuchenie geologicheskoy istorii i protsessov sovremennogo osadkoobrazovaniya Chernogo i Baltiyskogo morey [Study of the geological history and processes of recent sedimentation in the Black and Baltic Seas]. Part 1. Naukova Dumka, Kiev, pp 103–113 (in Russian)

Ivanov VG (ed) (1987) Report on the development of stratigraphic scheme and legend of quaternary sediments of the Black Sea. Report of the Prichernomorskaya Expedition, Odessa (in Russian)

Karpov, VA, Ivanov, VG, Sibirchenko, MG et al. (1978) Otchet o geologicheskom izuchenii s s'emke Odesskogo zaliva Chernogo moria masshtaba 1:50,000 (List 36-51-A, C, G), 1976–1978 [Report of the Prichernomorskaya expedition on geological investigation and survey 1:50,000 within Odessa Bay of the Black Sea], Odessa (in Russian)

Kryzhitskiy SD, Rusyaiva AS, Krapivina VV, Leuipunskaya NA, Skrzhinskaya MV, Anokhin VA (1999) Olbia. Ancient State in the Northern Black Sea Region. Naukova dumka, Kiev (in Russian)

Kuprin PN (2002) Lithology and paleogeography of the Neoeuxinian (Late Pleistocene) stage of the Black Sea. Biulleten' Moskovskogo Obshchestva ispytatelei prirody. Otdel geologicheskiy [Bulletin of Moscow Society of Naturalists, Geological Branch] 77:59–69 (in Russian)

Kuprin PN, Sorokin VM (2007) On the postglacial changes in the level of the Black Sea. In: Yanko-Hombach V, Gilbert AS, Panin N, Dolukhanov PM (eds) The Black Sea flood question: changes in coastline, climate and human settlement. Springer, Dordrecht, pp 205–220

Kurten B (1969) Radiocarbon date for the cave bear remains (*Ursus spelaeus*) from Odessa. Soc Scient Fennica Comm Biolog 31:1–3

Kvasov DD (1975) Pozdnechetvertichnaya istoriya krupnykh ozer i vnutrennikh morey Vostochnoy Evropy [The Late Quaternary history of the large lakes and inland seas of Eastern Europe]. Nauka, Moscow (in Russian)

Larchenkov E, Kadurin S (2011) Paleogeography of the Pontic Lowland and northwestern Black Sea shelf for the past 25 ky. In: Buynevich I, Yanko-Hombach V, Gilbert A, Martin R (eds) Geology and geoarchaeology of the Black Sea region: beyond the flood hypothesis, Geological Society of America Special Paper 473. Geological Society of America, Colorado, pp 71–88

Lericolais G (2017) Late Pleistocene environmental factors defining the Black Sea, and submerged landscapes of the western continental shelf. In: Flemming NC, Harff J, Moura D, Burgess A, Bailey GN (eds) Submerged landscapes of the European continental shelf: quaternary paleoenvironments. Wiley, Chichester, pp 479–496

Marret F, Mudie P, Aksu A, Hiscott RN (2009) A Holocene dinocyst record of a two-step transformation of the Neoeuxinian brackish water lake into the Black Sea. Quat Int 193:72–86

Mudie PJ, Aksu AE, Yaşar D (2001) Late quaternary dinoflagellate cysts from the Black, Marmara and Aegean seas: variations in assemblages, morphology and paleosalinity. Mar Micropaleontol 43:155–178

Mudie PJ, Rochon A, Aksu AE (2002) Pollen stratigraphy of Late Quaternary cores from Marmara Sea: land-sea correlation and paleoclimatic history. Mar Geol 190:233–260

Mudie PJ, Marret F, Aksu AE, Hiscott RN, Gillespie H (2007) Palynological evidence for climatic change, anthropogenic activity and outflow of Black Sea water during the Late Pleistocene and Holocene: centennial-to decadal-scale records from the Black and Marmara Seas. Quat Int 167–168:73–90

Mudie PJ, Leroy SAG, Marret F et al (2011) Nonpollen palynomorphs: indicators of salinity and environmental change in the Caspian–Black Sea–Mediterranean corridor. In: Buynevich I, Yanko-Hombach V, Gilbert A, Martin R (eds) Geology and geoarchaeology of the Black Sea region: beyond the flood hypothesis, Geological Society of America special paper 473. Geological Society of America, Colorado, pp 245–262

Mudie PJ, Yanko-Hombach V, Kadurin S (2014) The Black Sea dating game and Holocene marine transgression. Open J Marine Sci 4:1–7

Nevesskaya LA (1965) Pozdnechetvertichnye dvustvorchatie mollyuski Chernogo moria, ikh sidtematica i ecologia [Late Quaternary Bivalve Molluscs of the Black Sea, their systematics and ecology]. Trudy Paleontologicheskogo Instituta Akademii Nauk SSSR [Proceedings of the Paleontologcal Institute of the USSR Academy of Sciences] 105. Nauka, Moscow, pp 1–39 (in Russian)

Nicholas WA, Chivas AR, Murray-Wallace CV, Fink D (2011) Prompt transgression and gradual salinisation of the Black Sea during the early Holocene constrained by amino acid racemization and radiocarbon dating. Quat Sci Rev 30:3769–3790

Otte M, Noiret P, López Bayón I (1997) Aspects of the Upper Palaeolithic in Central Europe. Préhistoire européenne 11:277–301

Peev PI (2009) The neolithisation of the Eastern Balkan Peninsula and fluctuations of the Black Sea level. Quat Int 197(1–2):87–92

Peychev V, Peev P (2006) Evolution of the Bulgarian Black Sea coast during the Holocene. Slavena, Varna (in Bulgarian)

Podoplelov ON, Karpov VA, Ivanov VG et al. (1973–1975) Otchet o geologicheskoi s'emki shelfa severo-zapadnoy chasti Chernogo moria masshtaba 1:50,000 [Report on geological survey 1:50,000 within north-western part of the Black Sea]. Krymgeologya, Odessa (in Russian)

Pop I (1957) Palinologicheskie issledovaniya v Rumynii i ikh glavneyshie rezultaty [Palynological investigations in Romania and their main results]. Bot J 42(3):72–79

Ryan WBF (2007) Status of the Black Sea flood hypothesis. In: Yanko-Hombach V, Gilbert AS, Panin N, Dolukhanov PM (eds) The Black Sea flood question: changes in coastline, climate and human settlement. Springer, Dordrecht, pp 63–88

Samoilova TL (1988) Tira v VI-Ivv. do n.e. [Tyras in VI – I centuries BC]. Naukova Dumka. Kiev (in Russian)

Sapozhnikov IV, Sapozhnikova GV (2011) Kamennyi vek Severo-Zapadnogo Prichernomoria. Stratum Plus 1:15–331 (in Russian)

Shcherbakov FA (1983) Materikovye okrainy v pozdnem pleistotsene i golotsene [Continental margins in the Late Pleistocene and Holocene]. Nauka, Moscow (in Russian)

Shcherbakov FA, Kuprin PN, Potapova LI et al (1978) Osadkonakoplenie na kontinental'noi okraine Chernogo moria [Sedimentation on the continental shelf of the Black Sea]. Nauka, Moscow (in Russian)

Shmuratko VI (2001) Gravitatsionno-rezonansnaiya ekzotektonika [Gravity-resonance exotectonic]. Astroprint, Odessa (in Russian)

Shnyukov EF (ed) (1985) Geologiya shel'fa USSR: Litologiya [Geology of Ukrainian shelf: lithology]. Naukova Dumka, Kiev (in Russian)

Sibirchenko MG, Karpov VA, Ivanov V et al. (1983) Otchet po izucheniyu litilogicheskogo sostava donnykh otlozheniy shelfa Chernogo morya s zeliyu sostavleniya geologolitologicheskoy karty masshtaba 1:200,000 [Report on Lithological Study of the Black Sea Shelf for the Preparation of Geological-Lithological Map 1:200,000]. Krymgeologia, Odessa (in Russian)

Sinitsyn AA, Praslov ND, Svezhencev YS, Sulerzhitskiy LD (1997) Radiouglerodnaya hronologia verhnego paleolita Vostochnoj Evropy. V. In: Sinitsyn AA, Praslov ND (eds) Radiouglerodnaya hronologia paleolita Vostochnoj Evropy i Severnoy Azii: Problemy i perspektivy. Akadem-Print, Sankt-Peterburg, pp 21–66 (in Russian)

Stanko VN (1997) Nekotorye itogi izucheniya pozdnego paleolita Severo-Zapadnogo Prichernomorya Yuzhnobugskaya gruppa pamyatnikov [Some results of investigations of the Late Paleolithic of North-Western Black Sea region 1 Southern Bug group of sites] In: Stanko VN (ed) Archeologiya i Etnologiya Vostochnoy Evropy. Materialy i issledovaniya [Archaeology and Ethnology of Eastern Europe. Materials and Investigations]. Hermes, Odessa, pp 14–27 (in Russian)

Stanko VN (1999) Anetovka 2 – pozdnepaleoliticheskoe poselenie I svyatilische ochotnikov na bizonov v Severnov Prichernomorie [Anetovka 2 – Late Paleolithic settlement and sacral place of bison hunt-

ers in North Black Sea region]. Stratum Plus 1:322 (in Russian)
Stanko VN (2007) Fluctuations in the level of the Black Sea and Mesolithic settlement of the northern Pontic area. In: Yanko-Hombach V, Gilbert AS, Panin N, Dolukhanov PM (eds) The Black Sea flood question: changes in coastline, climate and human settlement. Springer, Dordrecht, pp 371–386
Svitoch AA, Selivanov AO, Yanina TA (1998) Paleogeographic events of the Ponto-Caspian and Mediterranean in Pleistocene: data on reconstruction and correlation. Moscow State University, Moscow (in Russian)
Veklich MF, Sirenko NA (1976) Pliotsen i pleistotsen levoberezhiya Nizhney Volgi i Krymskoy ravniny [Pliocene and Pleistocene of the left side of the Lower Dnieper and Crimean plain]. Naukova Dumka, Kiev (in Russian)
Veklich MF, Sirenko NA, Matviishina GN (1993) Stratigraficheskaya skhema chetvertichnikh otlozheniy Ukrainy [Stratigraphic scheme of the Ukrainian quaternary deposits]. Naukova Dumka, Kiev (in Russian)
Ward C, Ballard R (2004) Black Sea shipwreck survey 2000. Int J Naut Archaeol 33(1):2–13
Ward C, Horlings R (2008) The remote exploration and archaeological survey of four Byzantine ships in the Black Sea. In: Ballard R (ed) Archaeological oceanography. Princeton University Press, Princeton, pp 148–173
Yanchilina AG, Ryan WBF, McManus F et al (2017) Compilation of geophysical, geochronological, and geochemical evidence indicates a rapid Mediterranean-derived submergence of the Black Sea's shelf and subsequent substantial salinification in the early Holocene. Mar Geol 383:14–34
Yanko-Hombach V (2007) Controversy over Noah's Flood in the Black Sea: geological and foraminiferal evidence from the shelf. In: Yanko-Hombach V, Gilbert AS, Panin N, Dolukhanov PM (eds) The Black Sea flood question: changes in coastline, climate and human settlement. Springer, Dordrecht, pp 149–204
Yanko-Hombach V (2017) Editorial to IGCP 610 special volume of Quaternary International. Quat Int 409:1–7
Yanko-Hombach V, Motnenko I (2016a) The Karangatian epoch in the Neopleistocene history of the Black Sea. EGU General Assembly 2016, 17–22 April, Vienna, Austria, p 3152
Yanko-Hombach V, Motnenko I (2016b) The Karangatian epoch (MIS 5e) in the Black Sea basin. In: Gilbert A, Yanko-Hombach V (eds) Proceedings of IGCP 610 fourth plenary conference and field trip: "from the Caspian to Mediterranean: environmental change and human response during the quaternary", Tbilisi, Georgia, 1–9 October 2016, pp 202–206
Yanko-Hombach V, Smyntyna O (2009) Quaternary history of the Black Sea and adjacent regions: proceedings, IGCP 521-INQUA 0501 plenary meeting and field trip, Odessa, Ukraine. Quat Int 197:1–5
Yanko-Hombach V, Gilbert AS, Panin N, Dolukhanov PM (eds) (2007) The Black Sea flood question: changes in coastline, climate and human settlement. Springer, Dordrecht
Yanko-Hombach V, Smyntyna O, Kadurin SV et al (2011a) Black Sea level changes and adaptation strategies of prehistoric populations during the last 30 ka. Geologiya I mineral'nye resursy Mirovogo Okeana [Geol Miner Resour World Ocean] 2:61–94 (in Russian)
Yanko-Hombach V, Mudie P, Gilbert AS (2011b) Was the Black Sea catastrophically flooded during the post-glacial? Geological evidence and archaeological impacts. In: Benjamin J, Bonsall C, Pickard C, Fischer A (eds) Submerged prehistory. Oxbow Books, Oxford, pp 45–262
Yanko-Hombach V, Mudie PJ, Kadurin S, Larchenkov E (2014) Holocene marine transgression in the Black Sea: new evidence from the northwestern Black Sea shelf. Quat Int 345:100–118
Yanko-Hombach V, Schnyukov E, Pasynkov A, Sorokin V, Kuprin P, Maslakov N, Motnenko I, Smyntyna O (2017) Late Pleistocene-Holocene environmental factors defining the Azov-Black Sea basin, and the identification of potential sample areas for seabed prehistoric site prospecting and landscape exploration on the Black Sea continental shelf. In: Flemming NC, Harff J, Moura D, Burgess A, Bailey GN (eds) Submerged landscapes of the European continental shelf: quaternary paleoenvironments. Wiley, Chichester, pp 431–478
Zaliznyak L (2014) Paleolit Centralnoy Ukrainy [Paleolithic in Central Ukraine]. Ark Dent 3:3–17 (in Ukrainian)

Cyprus: The Submerged Final Palaeolithic of Aspros Dive Site C

22

Albert J. Ammerman

Abstract

Underwater reconnaissance work in front of the early site of Aspros on the west coast of Cyprus has led to the recovery of lithic artefacts that date to the Final Palaeolithic. Survey work on Cyprus was initially focused on the search for pre-Neolithic archaeology on land, resulting in the discovery of sites on formations of aeolianite along the modern coastline. Given their inferred pre-Neolithic age at a time when sea levels were significantly lower than present, their geomorphological setting was used to predict targets for underwater survey offshore, resulting in the discovery of Dive Site C. In the eastern Mediterranean, no other site of this age has been found in a submerged context so far. At the end of the Pleistocene, this place stood at the foot of an aeolianite cliff that once formed part of the north bank of the Aspros River. Detailed analysis of the lithics at Dive Site C has identified diagnostic chipped stone tools in a hyper-microlithic tradition. This chapter sets out the steps that led to the underwater discovery, describes the lithic industries, presents the resulting gains in new knowledge and highlights the contribution that underwater archaeology is making to a better understanding of the Epipalaeolithic on Cyprus. Of particular interest are the close parallels between the reduction technology and the stone tool types at Aspros Dive Cite C and those from the levels of the Öküzini Cave in southern Turkey, which date to the time of the Younger Dryas.

Keywords

Underwater archaeology · Cyprus · Final Palaeolithic · Aspros · Voyaging · Younger Dryas

22.1 Introduction

One of the challenges in the study of voyaging foragers in the Mediterranean is that sea levels were appreciably lower in the time before the Neolithic period. In the case of the site of Aspros on land (Figs. 22.1, and 22.2), which dates to the Late Palaeolithic or the Epipalaeolithic as it is called locally on Cyprus and in the Levant (Ammerman 2013, 2014a; Kaczanowska et al. 2014; see also Knapp 2013), the distance from the shoreline today to the one 12,000 years ago was around 1.4 km (Fig. 22.3). In turn this raises the question of whether what one finds when one does a survey on land provides a good guide to what was happening on the coastline at the end of the Pleistocene. This has been an open question for some time. One way to address this is by con-

A. J. Ammerman (✉)
Department of the Classics, Colgate University, Hamilton, NY, USA
e-mail: aammerman@colgate.edu

G. Bailey et al. (eds.), *The Archaeology of Europe's Drowned Landscapes*, Coastal Research Library 35, https://doi.org/10.1007/978-3-030-37367-2_22

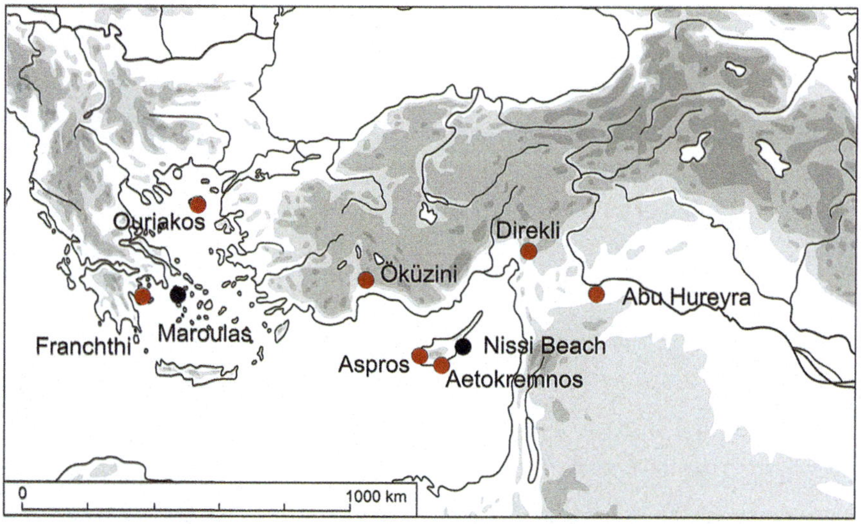

Fig. 22.1 Map showing the location of seven sites that date to the time of the Younger Dryas (c. 10,800–9500 cal BC) in the eastern Mediterranean: Abu Hureyra, Aetokremnos, Aspros, Direkli Cave, Öküzini Cave, Ouriakos and Franchthi (red dots). Black dots give the location of two sites that date to the first half of the ninth millennium cal BC: Nissi Beach and Maroulas

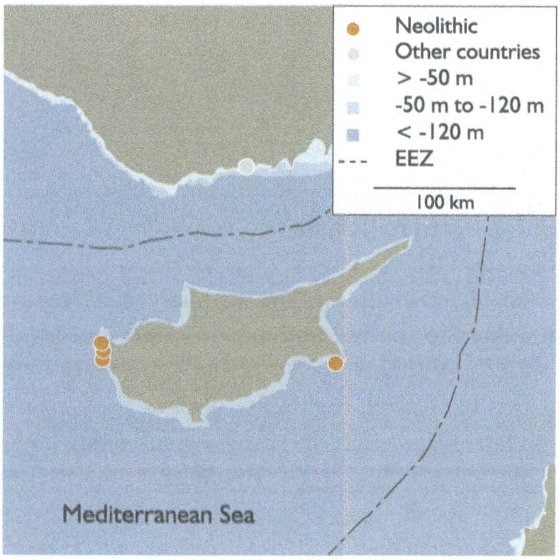

Fig. 22.2 Location of underwater sites on Cyprus. Sites in the west are at Aspros, the site in the east is Nissi Beach. Site information from the SPLASHCOS Viewer http://splashcos-viewer.eu. Drawing by Moritz Mennenga

ducting reconnaissance work on the seafloor. While a good case can be made in the context of method and theory in archaeology for engaging with submerged prehistory (e.g., Bailey 2013 and references therein), archaeologists on Cyprus have been hesitant to attempt this kind of work. There were practical considerations that kept them from getting their feet wet. To begin with, it takes time, effort and money to put together the team and equipment to do underwater archaeology. Moreover, the sizes of the lithics that one is attempting to find are quite small, the pieces of chipped stone are often less than 1 cm in length, and their visibility on the seafloor is even lower than it is on land. For many years it was hard enough for archaeologists working on land to find pre-Neolithic sites on the islands in the eastern Mediterranean (Ammerman 2010). Given the risks involved—the chances of returning home empty-handed—it was understandable that archaeologists on Cyprus held back from taking the plunge. Dive Site C was discovered in July 2007 on the west coast of Cyprus at a depth of 12 m in the water, at a place that once stood on the north bank of the Aspros River when sea level was lower than today (Ammerman et al. 2008, 2011).

It is worth underscoring two things by way of introduction. First, what was undertaken at Dive Site C was not a routine piece of work in the field. None of us had ever worked at a Palaeolithic site

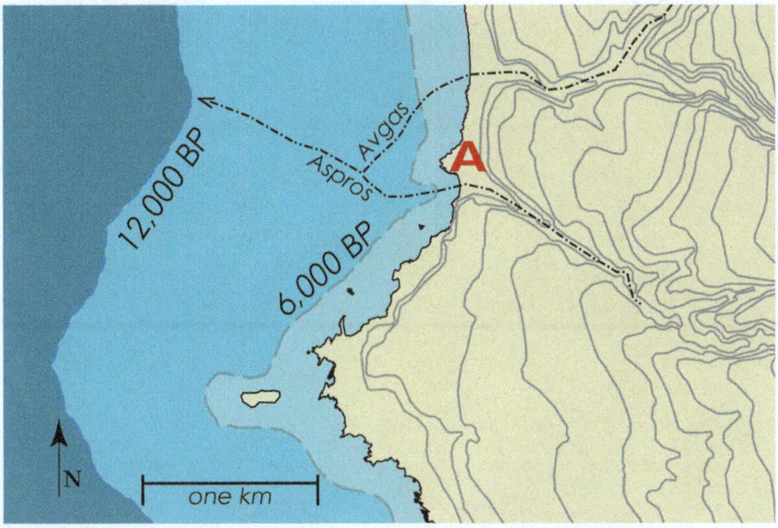

Fig. 22.3 Map showing the site of Aspros on land (A), located above the confluence of the Avgas and Aspros Rivers, with their approximate courses to three shorelines: (1) c. 12,000 years BP (10,000 cal BC) at c.70 m below sea level today; (2) c. 6000 BP (c. 4000 cal BC) at c. 8 m below sea level today, dashed line; and (3) the present day, 0 m, solid line. For land above sea level today, the contour lines are at 20 m intervals

in a submerged context before, and underwater archaeology was not part of my own training. In retrospect, one can see it as the logical thing to do. Perhaps 'pioneering venture' is the best term to use in describing what we were setting out to do. Second, we had great good fortune to find Dive Site C, given the limited time and resources that were available to us. In addition, doing underwater archaeology in a hot summer month led fortuitously to the important discovery of sea salt in shallow basins in the aeolianite bedrock on the coast (Fig. 22.4; Ammerman et al. 2008, Fig. 10; Ammerman 2014a, Fig. 19). As an annually renewing resource that occurs right on the shoreline, salt might be a good reason for foragers based on the mainland to voyage out to Cyprus in the summer months.

On the other hand, we failed to appreciate the full significance of the lithic finds in our two preliminary reports (Ammerman et al. 2008, p 19; Ammerman et al. 2011, p 266–267), and it was only when Janusz Kozlowski and Malgorzata Kaczanowska, leading lithic specialists from Poland, agreed to step in and re-examine all of the lithics found at the sites of Aspros and Nissi Beach including the material from Dive Site C that we were able to appreciate their full significance (Kaczanowska and Kozlowski 2014a), in particular the close parallels between the chipped stone tools and reduction technology at Dive Site C and those found in the upper levels at the Öküzini Cave in southern Turkey (Yalçinkaya et al. 2002).

This chapter outlines the steps that led to underwater work at Aspros in 2007, sets out the predictive strategy used to locate underwater sites, and summarizes the results of underwater reconnaissance with particular attention to the analysis of the lithics at Aspros Dive Sie C (for further detail see Ammerman 2013, 2014a).

22.2 Steps in the Fieldwork on Land

When I first went out to Cyprus in 2003 to find the missing pre-Neolithic sites on the island, the notion of doing underwater archaeology was the last thing that I had in mind. As late as 2003, there was only one reliable pre-Neolithic site on

Fig. 22.4 Sea salt on the shoreline at Aspros. The salt forms when spray from the sea collects in shallow depressions in the aeolianite rock and the water dries out in the sun during the summer months. The sea salt there was a resource that renewed itself each year. Photo by Albert Ammerman

the island of Cyprus, the much-debated rock shelter known as Aetokremnos on the Akrotiri Peninsula (Simmons 1999; Ammermans and Noller 2005; Zazzo et al. 2015). Even as late as 2007, there was in the literature no site of Mesolithic or Palaeolithic age that was known on the island of Crete. In short, the whole question of early voyaging to offshore islands in the Mediterranean Sea—Cyprus, Crete, Corsica, Sardinia and the Balearic Islands—was still up in the air at the time.

On Cyprus we took a new approach—one that focused on the location of sites on or near the coast and the visibility of early sites on the modern land surface (Ammerman et al. 2006). We soon found Nissi Beach at Agia Napa in January of 2004 and then Aspros in the Akamas in December of the same year (Fig. 22.1; Ammerman 2010). By the end of 2005, the reconnaissance work on land had led to the discovery of ten sites on formations of aeolianite on the coasts of Cyprus, all of them with pebble-and-flake-based lithic reduction technologies quite different from the blade-oriented traditions found at Neolithic sites on the island and therefore inferred to be pre-Neolithic in date. This challenged conventional wisdom at the time, namely, that hunter-gatherers were reluctant seafarers in the Mediterranean and that pre-Neolithic sites were difficult to find on its islands (Cherry 1990). We now had good evidence for crossing the sea to the offshore island of Cyprus in the time before the Neolithic period. In light of the heightened mobility of Natufian hunter-foragers in the Levant during the late Epipalaeolithic (Bar-Yosef 1998; Boyd 2006), one could also put forward the working hypothesis that, in response to the cold snap of the Younger Dryas (dated to the period 12,800–11,500 cal BP), voyaging to Cyprus now made it possible for populations in the Levant to cope with a climatic downturn by enlarging even further their territorial range (Broodbank 2006; Ammerman 2010).

The next step in the fieldwork was to place the early sites at Aspros and Nissi Beach into their environmental context. Jay Noller, a soil scientist and coastal geologist, took the lead in this line of investigation. He studied first the aeolianite rock (old lithified sand dunes) and then examined the thin palaeosols that had formed on top of this bedrock during the closing years of the Pleistocene. The palaeosols at both sites were found to have very low productivity in ecological terms. In short, the soils are of little or no use for purposes of agriculture either today or in prehistoric times. Noller also mapped the geomorphology of the area in the vicinity of Aspros. Of particular interest for our present purposes, he also produced a map that shows the confluence of the Avgas and Aspros Rivers with the approximate positions of the shorelines at 6000 and 12,000 years ago (Fig. 22.3; Ammerman et al. 2007, Fig. 3). In the case of the earlier shoreline, it was taken to stand at c. 70 m below modern sea level. This reconstruction stands up quite well today in the context of the slightly lower value for sea level at the end of the Younger Dryas (c. 60 m) recently proposed by Lambeck and co-workers (Lambeck et al. 2014). The distance from the west side of the site of Aspros on land to the shoreline at c. 10,000 cal BC is just over 1.4 km.

The third step was to excavate six small trial trenches at Aspros in 2007 and 2008 (for their locations, see Ammerman et al. 2008, Fig. 4). While the excavations of the palaeosols did yield a certain number of lithics in each case (ones similar to those on the site's surface), none of the trenches produced a good feature or concentration of lithics. The counts of the lithics recovered from each of the first four trenches are given in our third report on early sites on Cyprus (Ammerman et al. 2008, Table 4). In turn this now led to the realization that the trial trenches were all made in a place that was too far from the coast at the time of interest. If we wished to work at sites closer to the shoreline in late Palaeolithic times, we would have to look for them offshore and under water (Ammerman et al. 2011, p 269).

22.3 Underwater Reconnaissance Work at Dive Site C

Alan Blum, a geologist at Cornell University, first drew my attention to the effect of 'case hardening' of the aeolianite bedrock that occurs on the north side of the Aspros River. As shown in Fig. 22.5, the aeolianite has a hard, upper carapace, but it is easy for the lower part of the cliff to become undercut when it is exposed to the elements over time, due to a lesser degree of lithification with depth in this rock. Shallow rock shelters can even form in some places along this cliff, which is 4–5 m high. In addition, the case hardening of the aeolianite in the valley, as mentioned before, creates a more resistant place on the coast, one that has a much better chance of surviving a marine transgression. Accordingly, we chose to focus our underwater searches on the submerged north bank of the palaeochannel, and this was the key decision behind the success of the work at Dive Site C.

Dive Site C is located some 130 m from the present-day coastline (Fig. 22.6). The foot of the aeolianite cliff on the river's north bank stands at about 12 m below sea level today, and the top of the cliff stands some 4 m above the foot of this cliff. Six dives were made in this place during the field season. On the whole, visibility was more limited in the upper part of Dive Site C (the flat area at the top of the cliff). For this reason, less effort was put into coverage of the seabed there.

Fig. 22.5 View of the aeolianite cliff on the south side of the site of Aspros on land with members of the dive team in the foreground. Note the collapse of the overhanging upper part of the aeolianite cliff on the left. No archaeology is observed in the rock shelter to the right since local shepherds have removed all the deposits down to the bedrock over the centuries. Photo by Albert Ammerman

The dives were made using standard SCUBA equipment, and each one lasted for less than an hour. Once a good potential place such as Dive Site C was identified by means of wider coverage, two or three divers working together as a 'lithic recovery' team were then assigned to cover it more intensively (Figs. 22.7, and 22.8), while a supervisor kept an eye on the work in progress and recorded the area. Professional underwater cinematography was also used to document each dive site. Those working on a lithic recovery team were asked to collect everything they encountered, including pieces of very small size, and not to be selective in what they picked up. Later, the materials would be sorted and studied. During a given dive, each person collected his or her own pieces and put them in a separate labelled bag. In terms of methodology, it was of interest to see whether or not there were patterns of difference in recovery from one diver to the next. This turned out to be the case: two of the divers were consistently better at finding the smaller pieces.

22.4 The Study of the Lithics

Only 38 lithic artefacts were identified in the original study of the material from Dive Site C, but the later analysis by Kaczanowska and Kozlowski (2014a) identified a total of 60 lithic artefacts from the same set of material. This came as a complete surprise to us. We were expecting to see some degree of difference to emerge from the re-examination of the material but not such a big one. Kaczanowska and Kozlowski were now able to identify 17 retouched tools, 1 core, 2 core fragments and 4 blade fragments (see Figs. 22.9, and 22.10; Table 22.1). Even for them the study of the lithics at Dive Site C was far from easy given the microlithic character of the material. They also analyzed all the lithic materials recov-

Fig. 22.6 View from the south showing the location of Dive Site C and the landscape of the Akamas to the north. The RIB boat is underlined in red and the site of Aspros on land is just visible on the far right. Photo by Albert Ammerman

ered at the site of Aspros on land (Kaczanowska et al. 2014) as well as those found on land at the site of Nissi Beach (Kaczanowska and Kozlowski 2014b).

One of the striking features of the raw material used for making the lithics at Dive Site C is their diversity. Three basic classes of raw materials are observed: flints and cherts, radiolarites, and coarse-grained quartzite. Macroscopically the flints comprise the most diverse group, dividing into 13 variants on the basis of colour, translucency and inclusions. In all there are 16 different raw materials that were recognized by Kaczanowska and Kozlowski (2014a; see also the last column in Table 22.1). There are two main inferences to draw from this. First, it is fair to say that the hunter-gatherers were not all that selective in the pebbles they chose to work. Second, what one appears to be dealing with at Dive Site C are people who were on the move and who went to sea from time to time as part of their mobile way of life.

For the lithics recovered on land at Aspros, Kozlowski believes that all of the chipped stone there (more than 1700 pieces) dates to the time of the Epipalaeolithic. He finds no evidence for cores or stone tool types associated with either the PPNA or the PPNB on Cyprus. This holds for Dive Site C as well. This result makes good sense since most of the landscape at Aspros consists of formations of aeolianite, which were poor for early forms of agro-pastoralism. On the other hand, there are clear differences between the lithics found at Aspros (on land) and those recovered at Dive Site C. In short, the lithic tradition at the latter can be described as a hyper-microlithic one. The chipped stone pieces there are consistently of small size, including the core

Fig. 22.7 Work in progress at Dive Site C: note the base of the cliff on the left of the diver on the left side of the photograph. Photo by Albert Ammerman

and two core fragments (Fig. 22.9:1–3), and there is a much higher proportion of backed pieces and segments (Fig. 22.10: 1–5). There are interestingly close parallels between the chipped stone tools at Dive Site C and those found in the upper levels of the Ökuzini Cave in Turkey (Kaczanowska and Kozlowski 2014a, p 63). Here it is worth adding that the upper level at Öküzini has been radiocarbon dated to the time of the Younger Dryas (Yalçinkaya et al. 2002). This is why the lithics at Dive Site C can be firmly attributed to the Final Palaeolithic. At the same time, what is found at Dive Site C now offers a new face of the Epipalaeolithic on Cyprus—the assemblages of this age that occur at other early sites on the island all have a more generic character and lack close parallels with lithic traditions on the mainland.

22.5 Discussion

At all the pre-Neolithic sites found on formations of aeolianite on Cyprus, one finds much the same pebble-and-flake-based tradition of lithic reduction technology. While it was definitely rewarding to find the missing pre-Neolithic sites on the island, we initially tended to discount what appeared to be a rough and expedient way of making chipped stone tools. Most of the archaeologists working on our project came from a background in Neolithic studies,

Fig. 22.8 Work in progress at Dive Site C: note that the diver on the left is using a slow hand-waving technique to increase the visibility of small chipped stone tools. Photo by Albert Ammerman

where the lithic traditions are commonly oriented toward the production of blades and not flakes. It took us a number of years to get over this bias and see the lithics at Aspros (through the eyes of Kozlowski) in a more positive light. Thus, far from this being a makeshift and expedient lithic technology, it is better described as a sophisticated and demanding one. For instance, the small segments are quite difficult to make. This reduction technology in combination with the use of the wide range of raw materials observed at Dive Site C makes good sense for mobile hunter-gatherers. For voyaging foragers on the move, this meant that they did not have to stop and search for an outcrop of flint or chert. Instead it was easier to look on the beach or in shallow water for suitable pebbles that were more immediately at hand. Here it is of interest to recall the standardized, or the so-called simple, lithic tradition that was widely used by early voyaging foragers in the case of Australia (Balme and O'Connor 2014). Finally, the best way to reach the island of Cyprus by means of a small boat or watercraft in the summer months (when sea salt could be collected on the shoreline) was by voyaging from the west (Bar-Yosef Mayer et al. 2015), that is, from the south coast of Turkey where the Öküzini Cave is located.

In closing, it is worth remembering that many prehistoric archaeologists who work on Cyprus had serious doubts that anything of interest would come out of the underwater reconnaissance work

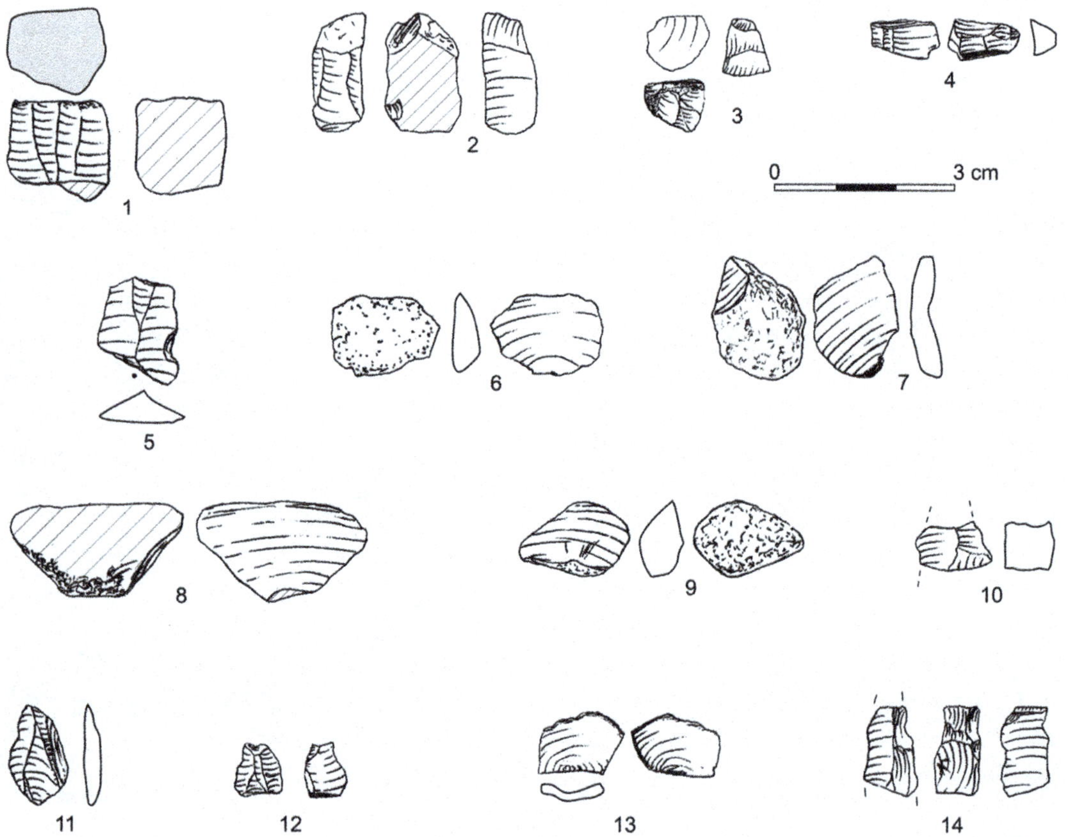

Fig. 22.9 Drawings of the chipped stone artefacts recovered from Dive Site C at Aspros. From Kaczanowska and Kozlowski (2014a: Plate 1). See Table 22.1 for the description of the respective pieces. Drawings by Janusz Kozłowski

that we were planning to do in front of Aspros in 2007. This was fully understandable since we were trying to do something new on the island. In fact, there was the risk that our pioneering venture would fail. Now we know that it is not only possible to do this kind of work on Cyprus but that it can be organized by an archaeologist who had almost no previous experience in the field of underwater archaeology.

The existence of submerged sites of late Palaeolithic age on Cyprus is no longer just an idea. It is a reality. In turn this now changes our perspective on where and on how we should conduct fieldwork in research on voyaging foragers in the eastern Mediterranean (Ammerman 2014b).

We are, of course, just at the very beginning of research on submerged prehistoric sites on Cyprus. If we now look forward to what needs to be done next at Dive Site C, it is useful to consider three lines of work. First, there is the need to return to the site and enlarge in spatial terms the collection of lithics at the foot of the cliff and also in the area at the top of it where four pieces were recovered in 2007 (see the last column in Table 22.1). Here it would be helpful to bring out to Cyprus mechanical sieving equipment of the kind used in Denmark for this

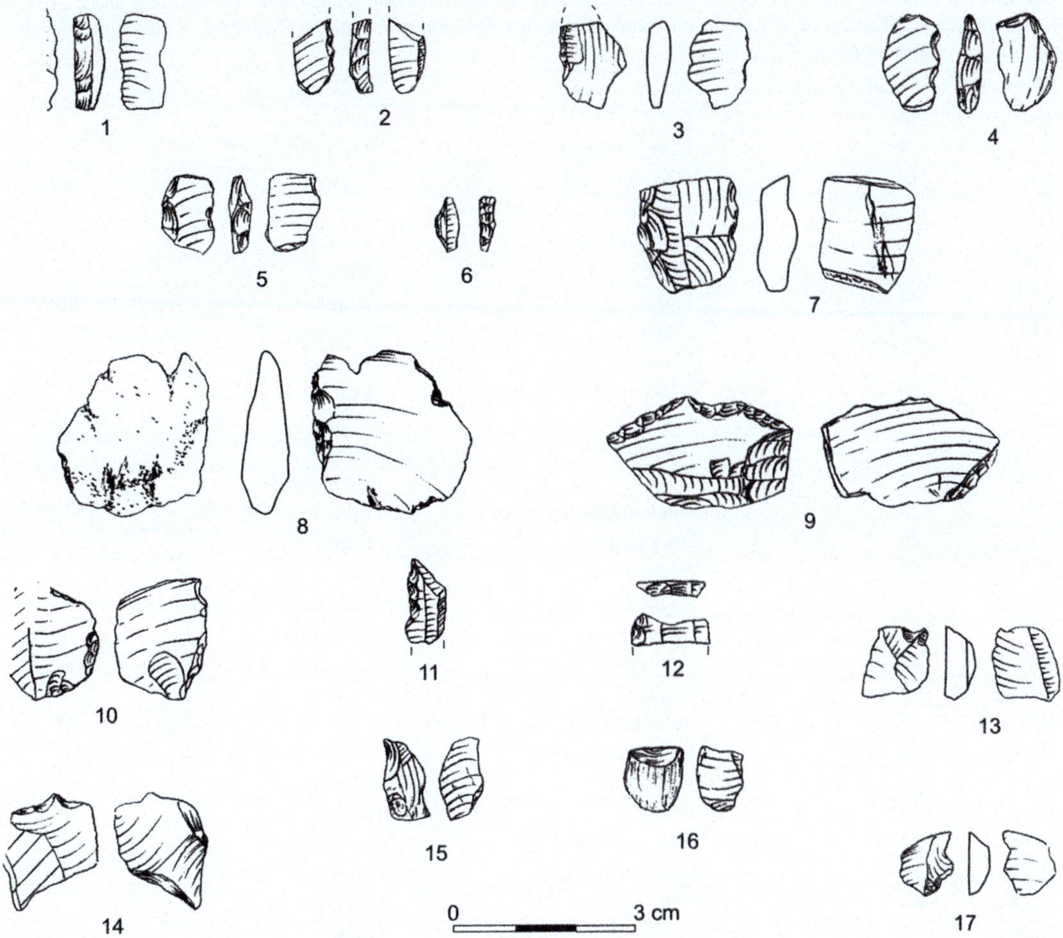

Fig. 22.10 Drawings of the chipped stone artefacts recovered from Dive Site C at Aspros. From Kaczanowska and Kozlowski (2014a: Plate II). See Table 22.1 for the description of the respective pieces. Drawings by Janusz Kozłowski

purpose to put the recovery of lithics on a more systematic basis (See Bailey et al. Chap. 3, this volume). Second, there is the need to produce a high definition bathymetry of Dive Site C and the areas adjacent to it. Third, it would be productive to conduct acoustic seafloor mapping along the foot of the cliff on the north bank of the submerged river valley by using side-scan sonar and multibeam echo sounders. Work of this kind might even lead to the discovery of a submerged rock shelter in the aeolianite cliff. Taking the long view, one would also like to extend the reconnaissance work out to selected places closer to where the shoreline once stood some 12,000 years ago, which will call for more technical dives going down to greater depths in the water.

Our work has demonstrated two important outcomes: first that it is possible to engage in purposeful and successful discovery of underwater sites using a predictive approach based on the location and geomorphology of known archaeological sites on dry land and secondly that it is possible for divers to identify and recover even the smallest types of lithic artefacts from the seabed. Both outcomes hold out the prospect of successful future discoveries of underwater archaeological material.

Table 22.1 This table lists some of the lithic artefacts recovered at Dive Site C and shown in Figs. 22.9 and 22.10 (after Kaczanowska and Kozlowski 2014a), with details of artefact type and raw material. Numbers in the raw material column refer to different types of flint

Number	Catalogue number	Artefact type	Raw material	Comments
I:1	24	Core	Flint 7	
I:2	9	Core fragment	Flint 9	
I:3	11	Core fragment	Flint 7	
I:4	8	Fragment of splintered piece	Flint 10	
I:5	21	Blade fragment	Flint 4	
I:6	31	Flake	Flint 5	
I:7	34	Flake	Burnt	Top of the cliff
I:8	35	Flake	Patinated	Top of the cliff
I:9	25	Flake	Flint 11	
I:10	10	Blade fragment	Flint 1	
I:11	2	Chip	Flint 9	
I:12	3	Chip	Flint 9	
I:13	1	Chip	Quartzite	
I:14	23	Blade fragment	Flint 8	
II:1	26	Backed piece	Flint 12	
II:2	50	Backed piece	Patinated	
II:3	51	Backed piece	Flint 8	
II:4	20	Segment	Flint 4	
II:5	29	Segment	Flint 8	
II:6	44	Segment	Flint 1	
II:7	33	Sub-crested retouched blade	Flint 2	Top of the cliff
II:8	32	Retouched flake	Burnt	Top of the cliff
II:9	6	Retouched flake	Flint 3	
II:10	36	Retouched flake	Radiolarite 1	
II:11	37	Retouched blade	Patinated	
II:12	38	Retouched blade	Flint 5	
II:13	19	Bec	Flint 7	
II:14	7	Bec on splintered piece	Flint 5	
II:15	30	Retouched chip	Flint 1	
II:16	22	Retouched chip	Flint 7	
II:17	28	Retouched chip	Flint 2	

The finds were recovered from the base of the underwater cliff except where indicated in the final comments column, where 'top of cliff' refers to the top of the underwater cliff. The numbers in column 1 correspond to the drawings in Figs. 22.9 and 22.10

Acknowledgements The investigation at Aspros would not have had a positive outcome without the major contributions of the following individuals: Alan Blum; Tim Turnbull, who acted as divemaster; Duncan Howitt-Marshall, who contributed his experience of underwater archaeology; Janusz Kozlowski and Malgorzata Kaczanowska, who undertook the detailed analysis of the lithic artefacts; Pavlos Flourentzos and Jay Noller. Our appreciation goes to INSTAP for the support of the fieldwork on Cyprus.

References

Ammerman AJ (2010) The first Argonauts: toward the study of the earliest seafaring in the Mediterranean. In: Anderson A, Barret JH, Boyle KV (eds) The global origins and development of seafaring. McDonald Institute for Archaeological Research, Cambridge, pp 81–92

Ammerman AJ (2013) Tracing the steps in the fieldwork at the sites of Aspros and Nissi Beach on Cyprus. In: Ammerman AJ, Davis TW (eds) Island archaeology and the origins of seafaring in the eastern Mediterranean. Proceedings of the Wenner-Gren workshop held at Reggio Calabria on October 19–21, 2012. Part one. Eurasian prehistory, vol 10, pp 117–138

Ammerman AJ (2014a) Aspros and Nissi Beach: tracing the steps in the study of early sites on Cyprus and the origins of seafaring in the Mediterranean world. In: Kaczanowska M, Kozłowski JK, Kaminska-Szymczak J (eds) Contributions to the archaeology of early Cyprus. Polish Academy of Arts and Sciences, Krakow, pp 7–26

Ammerman AJ (2014b) Setting our sights on the distant horizon. In: Ammerman AJ, Davis TW (eds) Island archaeology and the origins of seafaring in the eastern Mediterranean. Proceedings of the Wenner-Gren workshop held at Reggio Calabria on October 19–21, 2012. Part two. Eurasian prehistory, vol 11, pp 203–236

Ammerman AJ, Florentzos P, McCartney C, Noller J, Sorabji B (2006) Two new early sites on Cyprus. In: Report of the Department of Antiquities, Cyprus, pp 1–22

Ammerman AJ, Florentzos P, Gabrielli R, McCartney C, Noller J, Peloso D, Sorabji B (2007) More on the new early sites on Cyprus. In: Report of the Department of Antiquities, Cyprus, pp 1–26

Ammerman AJ, Florentzos P, Higham T, McCartney C, Turnbull T (2008) Third report on early sites on Cyprus. In: Report of the Department of Antiquities, Cyprus, pp 1–32

Ammerman AJ, Howitt Marshall D, Benjamin J, Turnbull T (2011) Underwater investigations at the early sites of Aspros and Nissi Beach on Cyprus. In: Benjamin J, Bonsall C, Pickard C, Fischer A (eds) Submerged prehistory. Oxbow, Oxford, pp 263–271

Ammermans AJ, Noller JS (2005) New light on Aetokremnos. World Archaeol 37:533–543

Bailey GN (2013) Early seafaring and the archaeology of submerged landscapes. In: Ammerman AJ, Davis TW (eds) Island archaeology and the origins of seafaring in the eastern Mediterranean. Proceedings of the Wenner-Gren workshop held at Reggio Calabria on October 19–21, 2012. Part one. Eurasian prehistory, vol 10, pp 99–114

Balme J, O'Connor S (2014) Early modern humans in island Southeast Asia and Sahul: adaptive and creative societies with simple lithic industries. In: Dennell R, Porr M (eds) Southern Asia, Australia and the search for human origins. Cambridge University Press, Cambridge, pp 164–174

Bar-Yosef O (1998) The Natufian culture in the Levant, threshold to the origins of agriculture. Evol Anthropol Issues News Rev 6(5):59–177

Bar-Yosef Mayer DE, Kahanov Y, Roskin J, Gildo H (2015) Neolithic voyages to Cyprus: wind patterns, routes and mechanisms. J Island Coast Archaeol 10:412–435

Boyd B (2006) On 'sedentism'in the later Epipalaeolithic (Natufian) Levant. World Archaeol 38(2):164–178

Broodbank C (2006) The origins and early development of Mediterranean maritime activity. J Mediterr Archaeol 19(2):199–230

Cherry J (1990) The first colonization of the Mediterranean islands: a review of recent research. J Mediterr Archaeol 3:145–221

Kaczanowska M, Kozlowski J (2014a) Underwater finds from dive site C at Aspros. In: Kaczanowska M, Kozłowski JK, Kaminska-Szymczak J (eds) Contributions to the archaeology of early Cyprus. Polish Academy of Arts and Sciences, Krakow, pp 59–66

Kaczanowska M, Kozlowski J (2014b) Reconstruction of lithic assemblages from Nissi Beach (Cyprus). In: Kaczanowska M, Kozłowski JK, Kaminska-Szymczak J (eds) Contributions to the archaeology of early Cyprus. Polish Academy of Arts and Sciences, Krakow, pp 67–106

Kaczanowska M, Kozlowski J, Kaminska-Szymczak J (2014) Analysis of the artifact collections from the littoral zones of Aspros. In: Kaczanowska M, Kozłowski JK, Kaminska-Szymczak J (eds) Contributions to the archaeology of early Cyprus. Polish Academy of Arts and Sciences, Krakow, pp 27–57

Knapp AB (2013) The archaeology of Cyprus: from earliest prehistory through the bronze age. Cambridge University Press, Cambridge

Lambeck K, Rouby H, Purcell A, Sun Y, Sambridge M (2014) Sea level and global ice volumes from the last glacial maximum to the Holocene. Proc Natl Acad Sci 111(43):15296–15303

Simmons AH (1999) Faunal extinction in an island society: pygmy hippopotamus hunters of Cyprus. Kluwer, New York

Yalçinkaya I, Otte M, Kozlowski JK, Bar-Yosef O (2002) La grotte d' Öküzini: Évolution du Paléolithique final du sud-ouest de l'Anatolie. Études et Recherches Archéologiques de l'Université de Liège 96. Liège, Belgium

Zazzo A, Lebon M, Quiles A, Reiche I, Vigne JD (2015) Direct dating and physico-chemical analyses cast doubts on the coexistence of humans and dwarf hippos in Cyprus. PLoS One 10(8):e0134429. https://doi.org/10.1371/journal.pone.0134429

Israel: Submerged Prehistoric Sites and Settlements on the Mediterranean Coastline—the Current State of the Art

23

Ehud Galili, Baruch Rosen, Mina Weinstein Evron, Israel Hershkovitz, Vered Eshed, and Liora Kolska Horwitz

Abstract

Inundated archaeological sites dating from the Middle Palaeolithic to the Pottery Neolithic periods have been exposed off the Mediterranean coast of Israel, mainly the northern Carmel coast. The bulk of the sites represents in situ Neolithic settlements dating from the tenth to the seventh millennia BP, including the Pre-Pottery Neolithic site of Atlit-Yam and the Pottery Neolithic sites of Kfar Samir, Kfar Galim, Tel Hreiz and Neve-Yam. These are some of the best-preserved underwater settlements in the world with excellent preservation of human and animal remains, plant materials used as food and animal fodder, basketry, wood used in building construction and for making bowls and a wide range of flint, bone and ground-stone artefacts. The sites include rectangular stone dwellings, the earliest known stone-built water wells in the world, megalithic structures of probable ritual significance, numerous human burials, many in stone-lined graves or cists, and reveal details of village layout including a separation between domestic and graveyard areas. The human remains provide pathological evidence for the earliest known case of tuberculosis and for malarial infection. These settlements are the earliest known examples of the typical Mediterranean fishing village, with a subsistence economy based on crop cultivation, domestic animals, some hunting of wild animals, a significant emphasis on marine fishing and the earliest known evidence for the production of olive oil. This combination of resources contributed to the establishment of year-round sedentary, Mediterranean fishing villages. These features owe their preservation and discovery to the location of the settlements on the shoreline. Rising sea level eventually forced their abandonment and sealed

E. Galili (✉)
Zinman Institute of Archaeology, University of Haifa, Haifa, Israel

Israel Antiquities Authority, Haifa, Israel
e-mail: galilish@netvision.net.il

B. Rosen
Independent researcher, Atlit, Israel
e-mail: resenbar@netvision.net.il

M. W. Evron
Zinman Institute of Archaeology, University of Haifa, Haifa, Israel
e-mail: evron@research.haifa.ac.il

I. Hershkovitz
Department of Anatomy and Anthropology, Faculty of Medicine, Tel-Aviv University, Tel Aviv, Israel
e-mail: anatom2@tauex.tau.ac.il

V. Eshed
Israel Antiquities Authority, Jerusalem, Israel
veshed@gmail.com

L. K. Horwitz
National Natural History Collections, Faculty of Life Science, Berman Building, E. Safra Campus, The Hebrew University, Jerusalem, Israel
lix1000@gmail.com

G. Bailey et al. (eds.), *The Archaeology of Europe's Drowned Landscapes*, Coastal Research Library 35, https://doi.org/10.1007/978-3-030-37367-2_23

the remains in anaerobic conditions under a protective cover of marine sand, until recent disturbance by storms and sand-mining exposed parts of the submerged ancient land surface to archaeological discovery.

Keywords

Atlit-Yam · Underwater settlements · Palaeocoastlines · Sea level · Domestication · Pre-Pottery Neolithic · Pottery Neolithic

23.1 Introduction

Israel has a long tradition of underwater archaeology and has produced some of the best preserved and best studied underwater Stone Age settlements in the world, with a total of 23 sites of which 19 are in situ settlements or sectors of settlements (Table 23.1). They are concentrated on the Mediterranean littoral in northern Israel and are mostly of Neolithic date lying at depths rang-

Table 23.1 Chronological chart showing main archaeological periods and cultures of the southern Levant, with particular reference to the underwater sites of the Carmel and Galilee coasts from 10,000 to 6000 cal BP

Cal years BP	Archaeological Period	Archaeological Culture	Submerged Sites/Settlements
6800–5600	Chalcolithic		
7700–6800	Late Pottery Neolithic/Early Chalcolithic	Wadi Rabah	Kfar-Samir (S, Centre & N)
			Hishuley Carmel (S & N)
			Kfar-Galim (S & N)
			Nahal Galim
			Hahoterim
			Tel Hreiz (S & N)
			Megadim
			Atlit north bay
			Neve-Yam south
			Neve-Yam west
			Neve-Yam Temanun Island
			Habonim
Pre-7700	Prehistoric		Carmel Head
7800–7700	Early Pottery Neolithic	Lodian	Neve-Yam north
8400–7800		Yarmukian	
9000–8400	Pre-Pottery Neolithic	Pre-Pottery Neolithic C	Atlit-Yam
10,500–9000		Pre-Pottery Neolithic B	
11,500–10,500		Pre-Pottery Neolithic A	
14,500–11,500	Epipalaeolithic	Natufian	Nahal Beit Haemeq
22,000–14,500		Kebaran	Caesarea south anchorage
47,000–22,000	Upper Palaeolithic		
160,000–47,000	Middle Palaeolithic	Middle-Late Mousterian	Kfar Samir centre
250,000–160,000	Middle Palaeolithic		
1,500,000–250,000	Lower Palaeolithic		

Modified after Gopher (2012) and Bar-Yosef and Garfinkel (2008)

Chronological boundaries in some cases are approximate or time transgressive across different sites and regions. The Wadi Rabah culture is considered as Late Pottery Neolithic by some scholars and Early Chalcolithic by others. The Palaeolithic sites are unstratified finds, Neve-Yam Temanun is a human burial, and Neve-Yam west is a stone tumulus. All the other sites are settlements. The date of the Carmel Head (also referred to as Carmel Nose) stone cairns is uncertain, although based on their depth and distance from the shoreline they may be Pottery Neolithic or earlier ing from the modern intertidal zone down to −12 m, the earliest in situ sites occurring at greatest depth and furthest offshore. Submerged sites from earlier periods are rare and comprise lithic scatters of Middle and Upper Palaeolithic material. In this chapter we summarise the palaeogeographical context of the material and the geological factors contributing to its survival and discovery, describe the underwater sites and the finds recovered from them and assess their significance in relation to sea-level rise and the archaeology of the wider region.

23.2 Palaeogeographical Context and Preservation Conditions

The Mediterranean coast of Israel is c. 195 km long, oriented north–south in northern Israel and north–east to south–west in southern Israel. It is slightly curved with mostly sandy beaches in the south, coastal erosional cliffs in the centre and rocky and sandy areas in the north on the Carmel and Galilee coasts. The offshore shelf that would have been exposed at a sea-level regression of −120 m is about 40 km wide in the south, tapering to 10 km in northern Israel and narrowing further to a few kilometres along the coast of Lebanon and Syria (Fig. 23.1; see Galili et al. 2017a for further details).

The area of new land exposed at low sea levels would have offered a significant and relatively productive increment of land, with fertile coastal plains and rivers. Geomorphological features identified on the submerged shelf that provide clues to the palaeoenvironmental and landscape context of human activity are 'kurkar' ridges (formed of calcareous-cemented, aeolian, quartz sandstone), palaeosols, river channels and variable sediment cover dominated by sand but with intermittent areas of clay or sandy clay. The submerged kurkar ridges represent consolidated sand dunes that formed along palaeocoastlines during lower sea-level still stands. The sand content is high in carbonate resulting in rapid cementation to form a hard, aeolianite rock (see Ammerman, this volume for other examples). These ridges run parallel to the coast and when on dry land can impede drainage, creating wetland environments with clay sediments in the lowlands behind and between them. In some locations, stream channels cut through the ridges, creating narrow valleys or canyons. Sandy-loam palaeosols often red in colour (known locally as 'hamra') developed on the kurkar ridges, and often between them (in central Israel) when the shelf was a subaerial landscape. Submerged prehistoric remains recovered off the Carmel coast are embedded in the upper layer of palaeosol clay sediments of terrestrial origin, which date to the Upper Pleistocene and Early Holocene, 14,563–9534 cal BP (Sneh and Klein 1984). The water retentive properties of the clay may have been a valuable environmental feature for the inhabitants of the ancient landscape. They also offer the best conditions for preserving organic artefacts after the settlements had been inundated by sea-level rise.

Tidal range on this coast is minimal, 0.5 m or less, but wind fetch is large, extending for up to 2000 km to the west, generating large wave-swells during westerlies and waves up to 10 m in height. Though much of the wave energy is dissipated by the relatively shallow slope of the continental shelf and the submerged kurkar ridges, storm disturbance is a significant factor. Despite their potentially destructive effects, storms can actually enhance the discovery of submerged features by periodically removing areas of sand cover and exposing the underlying palaeo-land-surface, as discussed below.

In this region, the main source of sediments accumulated on the sea floor is the large volume of quartz sand discharged by the River Nile and carried by longshore currents along the coastline of the Sinai and the Levant as far north as Haifa Bay in northern Israel (Zviely et al. 2007). Most of this is deposited as sand along the inner shelf, with sediments containing a higher proportion of fluvial clay originating from the local coastal streams deposited further offshore. The sand cover plays an important role in protecting the underlying archaeological deposits and features of the original palaeo-land-surface. The best conditions for discovering such inundated sites are near the shore along the Carmel coast, where the

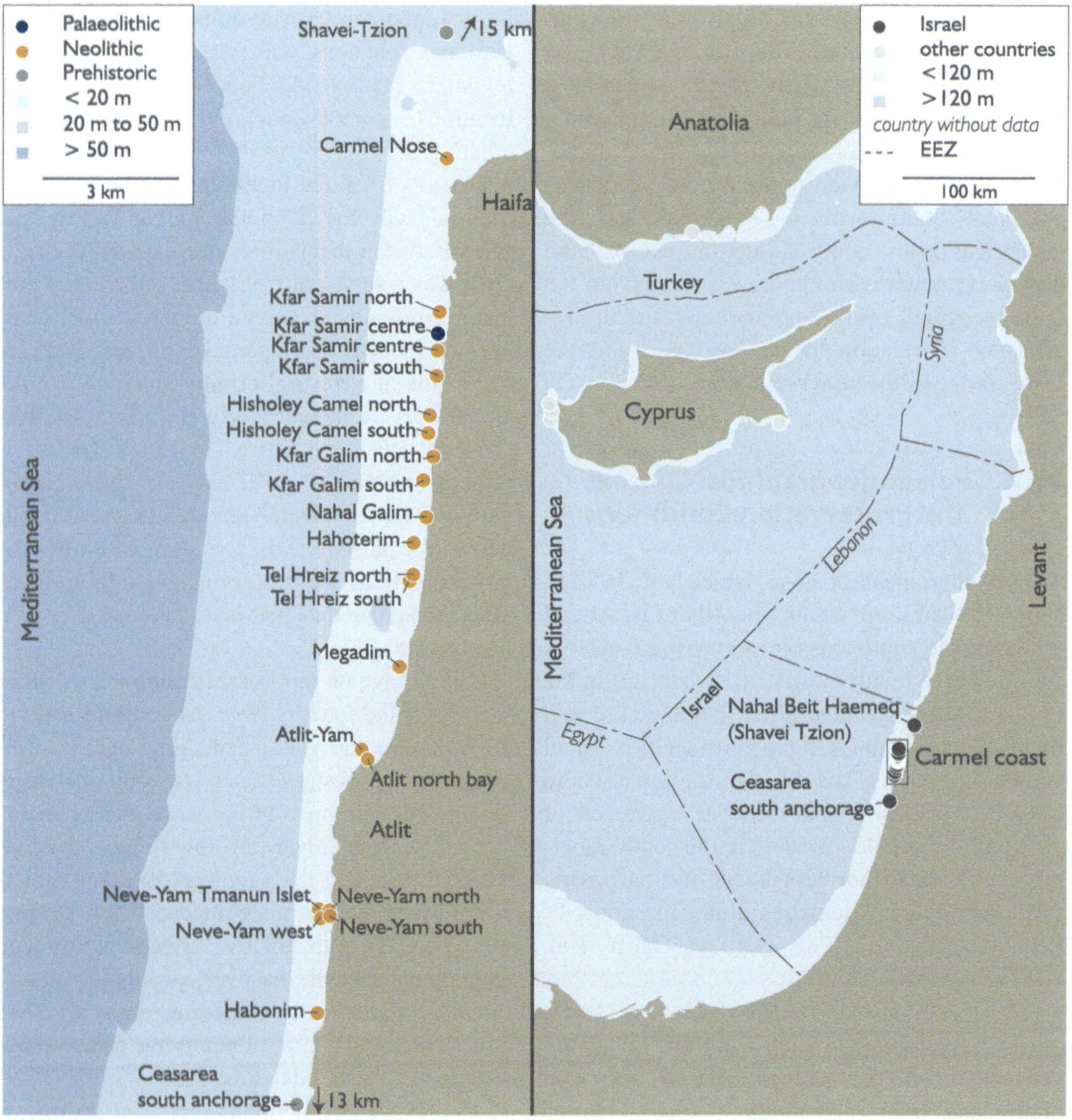

Fig. 23.1 Location map showing distribution of underwater sites. Site information from the SPLASHCOS Viewer http://splashcos-viewer.eu. Drawing by Moritz Mennenga

sand cover is thick enough (1–3 m) to provide protective cover, but thin enough that periodic storms or other disturbances can remove it to expose archaeological features on the underlying land surface. On the Galilee coast further to the north, the continental shelf is narrower and rocky, and the sediment cover thinner, with poorer conditions for site preservation and discovery, while to the south, the sand cover is thicker with fewer opportunities for the exposure of the underlying palaeo-land-surface.

It is this factor of protective sand cover subject to periodic disturbance, more than any other, that accounts for the preservation and exposure of sites on the sea bottom. Coupled with a long tradition of monitoring the shoreline and the underwater environment for signs of disturbance after storms by members of the public, sports divers and professional archaeologists, this has resulted in the discovery of a concentration of underwater settlements in northern Israel (Wreschner 1977a, b; Galili et al. 1988). In recent decades, human activities and especially sand mining have accelerated the removal of sand cover. This has enhanced the potential for discovery of archaeological sites but also increased the threats to their long-term preservation.

23.3 Archaeological Context

The southern Levant has a long history of human occupation, and the region is widely considered to have provided one of the primary corridors for human dispersal and expansion out of Africa both for archaic humans after c. 1.4 Ma and for anatomical moderns c. 200 ka (Enzel and Bar-Yosef 2017; Hershkovitz et al. 2018). The earliest sites on, or close to, the Carmel coast are the open-air site of Evron Quarry dated to c. 800 ka (Ron et al. 2003) and the later Mount Carmel caves with long sequences of Lower, Middle and Upper Palaeolithic as well as Epipalaeolithic material extending back to at least 500 ka (Garrod and Bate 1937; Shea 2003; Weinstein-Evron 2015). The role of coastlines and the coastal plain in these early periods is unclear, and the earliest reliable evidence is flint artefacts found directly in association with beach deposits from the MIS 5e high sea-level stand, which is dated at c. 134–116 ka by a combination of biostratigraphic indicators (principally the presence of the mollusc *Persististrombus latus* (previously *Lentigo latus* and *Strombus bubonius*) and radiometric dates (Ronen et al. 2008a, b; Galili et al. 2007, 2017e; Mauz et al. 2013; Benjamin et al. 2017; Porat et al. 2018). Scattered flint artefacts have also been recovered from the hamra palaeosols embedded between or within the kurkar ridges on the coast, and these too are attributed to the same period of high sea level (Ronen 1977; Ronen and Chernikov 2010). Despite the evidence of sites in MIS 5e shoreline settings, the nature of the activities practised by their occupants, in particular whether or not they exploited marine resources, is unclear. Occasional pieces of mammalian bone recovered in the MIS 5e beach deposits indicate terrestrial hunting, while marine shells in these sites may either be anthropogenic or natural in origin. Marine mollusc shells are also present in inland sites of about this period including Skhul, Qafzeh and Sefunim caves (Bar-Yosef Mayer et al. 2009), indicating visits to the shoreline or exchange relations with people living on the coast, but the shells were collected to make ornaments and are not definitive evidence that molluscs served as dietary items.

The region is also of importance as one of the earliest centres for the development of plant and animal domestication. This process is now thought to have begun over wide areas of the Near East perhaps as early as the Pre-Pottery Neolithic A period (PPNA) c. 11,000–10,500 cal BP, with progressive expansion of crop cultivation and animal domestication and a concomitant reduction in hunting and gathering throughout the following PPNB and PPNC periods (c. 10,500–8400 cal BP) (Simmons 2007; Zeder 2011). The subsequent Pottery Neolithic period (PN) lasted from about 8400–6800 cal BP and has been sub-divided into several major cultural phases: from oldest to youngest, the Yarmukian, Jericho IX, the Lodian and the Wadi Rabah cultures, though there may have been some overlap between them (Table 23.1; Gopher and Gophna 1993; Gopher 1995, 2012; Garfinkel 1999). People lived in small sedentary villages dependent on animal husbandry (of caprines, cattle and pigs) and agriculture (cereals and legumes). Hunting had largely ceased as can be seen by the marked decrease in wild fauna and arrowheads (Gopher and Gophna 1993; Simmons 2007). Sickle blades are abundant and are shorter and wider than those found in the PPN, perhaps indicating a wider range of cultivars (Barkai and Gopher 2012).

Few Neolithic sites are known on the present-day Israeli-Gazan coastal plain—exceptions are Ashkelon (PPNC) and later PN sites of Ziqim, Nizzanim and Qatif Y2 (Gaza, Palestinian Authority), and this situation has changed little since the 1990s (Gopher and Gophna 1993;

Gopher 1995). The paucity of Neolithic sites on the present-day coast may be in part due to sea-level rise during the early Holocene, which inundated areas of the coastal plain where Neolithic village settlements might have been located (e.g. Galili and Weinstein-Evron 1985; Galili 1985, 2004; Galili et al. 1988, 2005a; Sivan et al. 1999). The discovery of underwater settlements as described below confirms this interpretation and demonstrates the importance of underwater investigations in filling apparent gaps in the archaeological sequence as recorded on dry-land sites.

In total, there are records of 23 submerged prehistoric sites in Israel, including 1 Middle Palaeolithic, 2 Epipalaeolithic, 19 Neolithic/Early Chalcolithic and 1 of uncertain period (possibly dating to the Late Pleistocene-Early Holocene).[1] The Neolithic sites belong to three cultural phases: the Pre-Pottery Neolithic C (PPNC), represented by the site of Atlit-Yam, c. 9400–8000 cal BP; the early Pottery Neolithic site of Neve-Yam North belonging to the Lodian (Jericho IX) culture, dated by association to the first half of the seventh millennium BP; and 17 Pottery Neolithic/Early Chalcolithic sites belonging to the Wadi Rabah culture dated between c. 7800 and 7300 cal BP. The Pottery Neolithic sites are located close to the present shore (1–200 m offshore) at depths of 0–5 m, while the older Atlit-Yam site is located further offshore (200–400 m) and in deeper water (8–12 m below sea level).

23.4 Submerged Middle and Upper Palaeolithic Sites

23.4.1 Kfar Samir Centre

Located on the southern municipal beach of the town of Haifa (Fig. 23.1), this Middle Palaeolithic site, most likely belonging to the Middle-Late Mousterian, comprises hundreds of scattered flint implements, orange in colour, embedded in a clay palaeosol, some produced by the Levallois technique (Galili and Weinstein-Evron 1985). These artefacts may have been washed out by stream action from the nearby Mount Carmel. Some of the retouched artefacts show two stages of patination, indicating that they may have been re-used by the later Neolithic inhabitants of the site (see PN sites below).

23.4.2 Caesarea South Anchorage

This site is located at the south-east corner of the south anchorage at Caesarea, at 1–2 m depth, some 20–50 m offshore (Fig. 23.1). The remains are embedded in a hamra soil that was exposed on the sea bottom after a storm (Galili 1985, p 22, 27; SPLASHCOS viewer site 23). The finds consist of 12 flint implements, including 6 microlithic bladelets, a few flakes, core fragments and an exhausted core. These finds are probably associated with the terrestrial Epipalaeolithic sites of Heftzibah and Nahal Hadera containing similar flint industries found on the surface in exposures of red loam some 1000–1500 m to the south-east (Ronen et al. 1975; Ronen and Kaufman 1976).

23.4.3 Nahal Beit Haemeq (Shavei Tzion)

The site is located between the modern settlements of Shavei-Tzion and Nahariya, at a water depth of 5–8 m and 150–250 m offshore (Fig. 23.1). A palaeosol was exposed on the sea bottom. Skull remains of a large herbivore (probably wild cattle) were discovered in the clay (Galili 1985: p 26, 30; L. Horwitz pers. comm. 1991). Several non-diagnostic flint artefacts were recovered close to the skull. The similar geological settings of the clay palaeosol to that of the Carmel coast (see above) and the proposed age of the Carmel coast palaeosol (Upper Pleistocene to Early Holocene) suggests an Epipalaeolithic or Neolithic cultural association for the material.

Given the time that has elapsed and the likely impact of underwater currents and pre-inundation

[1] Some of these 23 sites are sectors of larger sites, e.g. Neve-Yam North and South, and these sectors have separate entries on the map and in the SPLASHCOS Viewer.

alluvial processes, these finds may not be in situ and lack features such as hearths or structures. They offer similar evidence as found in the open-air sites associated with the MIS 5e beach deposits (Ronen et al. 2008a; Galili et al. 2017e; Porat et al. 2018). The small number of finds in the MIS 5e sites and in the inundated sites described above suggests that they were most likely seasonal or short-lived camps, leaving behind scanty remains. This contrasts with inland caves such as those on Mount Carmel, which were never inundated, and were repeatedly visited over long periods of time, resulting in deep sequences of occupation and the accumulation of large assemblages of stone artefacts and faunal remains.

23.4.4 Submerged Stone Mounds on the Carmel 'Nose' Rocky Surface

This locality is of unknown date, but we include it here as it may be pre-Neolithic. It comprises several man-made stone piles that were discovered west of Cape Carmel (Galili 1981, pp 56–57; Galili et al. 2019a) on a flat, rocky sea floor c. 2 x 2 km known as the submerged Upper Carmel 'Nose' (Fig. 23.1, see also Fig. 23.24). These round stone features (2–4 m in diameter, up to 0.8 m in height) are composed of rounded, undressed limestone rocks of various sizes (15–30 cm). The piles are scattered over an area of 800 × 800 m at depths of 4–7 m. The lack of small finds in and around these stone arrangements makes dating impossible. These structures require further research to determine their period and function. Given their depth and the local sea-level changes, they may be dated as Pottery Neolithic/Chalcolithic or they may be ballast from later shipwrecks.

23.5 Submerged Neolithic Villages: Pre-Pottery Neolithic Atlit-Yam

Atlit-Yam is located in the North Bay of Atlit, 10 km south of Haifa, 200–400 m offshore, 8–12 m below sea level (Fig. 23.2), and is well dated by radiocarbon dates on charcoal and waterlogged plant remains to the PPNC period c. 9400–8000 cal BP (Table 23.2). The site covers approximately 40,000 m^2 and offers a unique opportunity to examine a large settlement and its surrounding terrain (Galili 2004).

The site is covered by 1–2 m of sand, parts of which are randomly removed by natural storms. Thus, exposure of different sectors of the site is accidental, unpredictable and cannot be preplanned. During exposure, the site erodes and finds may shift or be damaged. After significant storms, the site was surveyed by scuba diving to locate newly exposed areas and a rapid rescue and conservation operation, including excavation and documentation, was carried out. After several decades, separate documentation in isolation of many different areas of the site has resulted in enough 'jigsaw' pieces to provide quite a complete picture of the overall site pattern.

Excavation of the site was carried out both manually and using a dredging system within a laid-out grid (Fig. 23.3), and finds were taken ashore for sieving, conservation and more detailed analysis (see Galili et al. 2017b for further detail).

23.5.1 Architectural Remains

These comprise the foundations of several rectangular dwellings built of two rows of undressed kurkar stones, hearths, 2 megalithic ritual structures, stone-built water wells and about 20 circular stone-lined storage pits (Fig. 23.2). Other architectural remains identified are straight walls up to 60 m long (Fig. 23.2, Inset B, area F), a pair of thick parallel walls 1 m apart (Fig. 23.2, Inset A, 54; Fig. 23.4), a mudbrick wall (Fig. 23.2, area A), concentrations of burnt mudbricks and stone-paved floors (Galili and Weinstein-Evron 1985; Galili et al. 1993; Galili 2004, pp 42–55, 357–366; Galili and Rosen 2011a). The layout of the settlement indicates a cluster of family units with dwellings, courtyards, storage pits and open spaces between them which could have been used for crop cultivation or keeping of domestic animals. This is in marked contrast to the con-

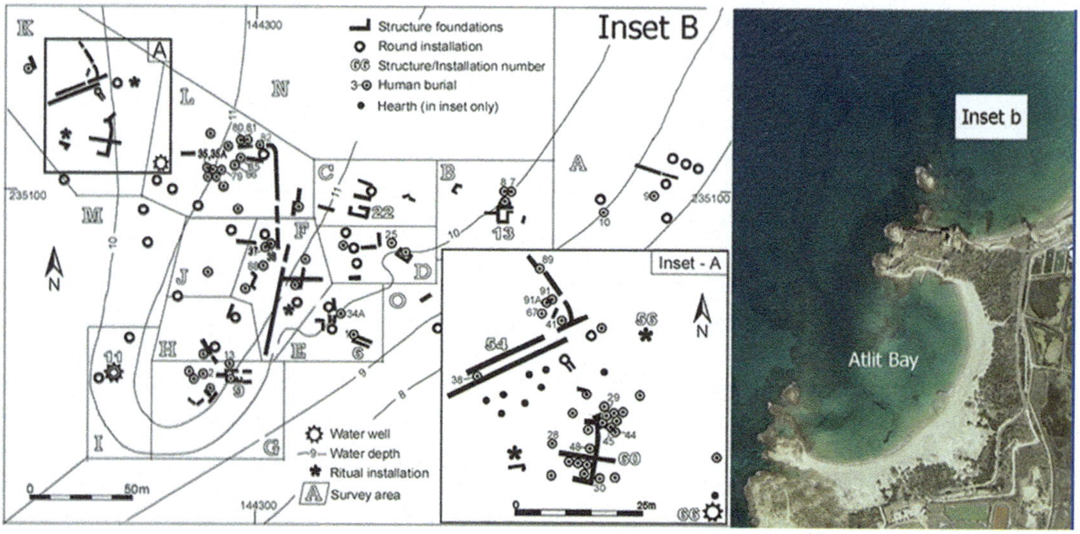

Fig. 23.2 Location and site plan of Atlit-Yam site. R: Aerial view of Atlit Bay showing the location of the site (Inset B); L: Plan of the site showing the location of dwelling structures, human burials, wells and other features. Drawing by Ehud Galili

temporaneous coastal site of Ashkelon, where dwelling pits, hearths and a wall were identified suggesting that this was a seasonal camp site (Garfinkel and Dag 2008).

23.5.1.1 Water Wells

Unique finds at Atlit-Yam are the circular stone-built features 0.8–1.5 m in diameter. Three of these were excavated and identified as water wells, representing the earliest constructed water wells in the world, while the others may represent water wells or storage pits (Galili and Nir 1993; Galili 2004, pp. 55–63, 367–368; Galili and Rosen 2011b).

The largest and most fully excavated well (well no. 11, area I) is a cylindrical shaft, 5.7 m deep and 1.5 m in diameter (Fig. 23.5). The uppermost three courses projected above the pre-existing land surface and were still in situ and visible on the sea floor when first discovered. In antiquity, the upper part of the well shaft was dug through clay sediments and lined with c. 28 courses of undressed stones. The base of the well was then dug for about 1.5 m into the underlying kurkar bedrock in order to tap into the freshwater aquifer (Fig. 23.6). The fill of the well contained soft sandy clay, and kurkar stones of various sizes, many of them heat-fractured, numerous animal remains, flint and stone artefacts and considerable amounts of waterlogged and charred plant remains. The bottom 2 m of fill contained layers of large kurkar stones and almost no animal bones. These large stones are interpreted as an attempt to raise the base of the well to tap into a higher level of the freshwater aquifer in response to sea-level rise and saltwater seepage. Three radiocarbon dates from this lowest fill give a date range of 8540–7930 cal BP, representing the latest possible age for the well construction (Table 23.2; Galili 2004).

The upper layers of the fill consist of similar components to the lower ones, in addition to numerous animal bones, some partly articulated, indicating that food waste was dumped there. The finds suggest that in this period the well had ceased to function as a freshwater source, most likely because of further sea-level rise and salinization, and served as a rubbish dump.

A second well (well 66, area L) was partially excavated to a depth of 1 m and shows similar features, with radiocarbon dates at the base of 8977–8187 cal BP (Table 23.2). In 2011, during the SPLASHCOS field school, a third well (well

Table 23.2 Radiocarbon dates: RT=Weizman Institute, Israel (Segal and Carmi 1996; Boaretto E. personal communication-unpublished report 2005); PITT=Pittsburgh, USA; PTA=Pretoria, South Africa, Beta=Beta Analytic Inc., Miami, Florida, HV=Hanover Radiation Laboratories USA

Site	Lab reference	^{14}C Age BP	Calibrated age 2σ range (95.4%) Cal BP	Material	Provenance
Atlit-Yam	PTA 3950	8000±90	9112–8598	Charcoal	Area B, 13
	RT 707	8140±90	9402–8775	Charcoal	Area B, 13
	RT 944A	7670±85	8628–8340	Charcoal	Area E, 10A
	RT 944C	7610±90	8590–8206	Charcoal	Area E, 10A
	PITT 0622	7550±80	8518–8186	Charcoal	Area E, 10A
	RT 1431	7300±120	8374–7933	Wood	Area I, Well 11
	RT 2479	7460±55	8379–8181	Wood	Area I, Well 11
	RT 2477,78	7605±55	8540–8336	Wood	Area I, Well 11
	RT 2475	7465±50	8376–8187	Wood	Area I, Well 11
	RT 2495,93	7755±55	8628–8420	Wood	Area L, Well 66
	RT 2489	7880±55	8977–8555	Wood	Area L, Well 66
	RT 2681	6580±35	7563–7427	Charcoal	Area E, 32
	RT 3038	8000±45	9010–8659	Charcoal	Area K, 54
	RT 3043	7250±45	8169–7980	Charcoal	Area K, 56
	RT 2497,96	8170±55	9281–9007	Charcoal	Area E, 65
Neve-Yam	HV 4256	6310±395	7951–6323	Charcoal	Northern area
	RT 1723	6390±70	7430–7174	Charcoal	Southern area, graveyard
	RT 1724	6565±70	7579–7326	Charcoal	Southern area, graveyard
Kfar Samir	Beta 82851	5860±140	7146–6323	Wood	Central, Well 13
	RT 682B	6470±130	7613–7030	Wood	Central, Well 5
	RT 682A	6670±160	7786–7320	Wood	Central, Well 3
	PTA 3820	6830±80	7844–7520	Wood	Central, Well 5
	PTA 3821	6830±160	7964–7431	Wood	Central, Well 3
	Beta 82850	6940±60	7930–7667	Wood	Central, Pit 10
	Beta 82845	6080±70	7162–6783	Olive pit	Central, 6
	Beta 82846	6210±150	7424–6751	Olive pit	Central, 6
	Beta 82847	6210±80	7288–6897	Olive pit	Central, 6
	Beta 82848	6230±80	7310–6930	Olive pit	Central, 6
	Beta 82715	6500±70	7560–7273	Olive pit	Central, 6
	RT 1898	5790±55	6729–6466	Olive pit	Central, 6
	RT 1930	5870±70	6877–6497	Olive pit	Central, 6
	Beta 82843	6100±60	7162–6793	Olive pit	Central, 7
	Beta 82844	6290±60	7413–7128	Olive pit	Central, 7
	RT 1929A	5630±55	6531–6299	Olive pit	Central, 7
	RT 1929	5870±70	6877–6497	Olive pit	Central, 7
	Beta 82849	6350±90	7434–7019	Wood	Central, 8
	RT 855	6420±120	7568–7027	Mat fragment	Central, 8
	RT 1360	7230±80	8275–7872	Wooden bowl	South
	Beta 433765	6890±30	7791–7667	Wood	Central, Well 13
Tel Hreiz	RT 799A	7330±120	8379–7952	Wood	Northern
	PTA 3460	6310±70	7418–7025	Wood	Northern
	RT 779B	6260±150	7440–6791	Wood	Northern
	RT 2480	6150±30	7159–6958	Wood	Northern
Megadim	PTA 3652	7960±70	8983–8484	Clay	Southern
	PTA 3648A	6310±70	7418–7025	Bone	Southern
	PTA 4339A	6270±50	7305–7018	Bone	Southern

(continued)

Table 23.2 (continued)

Site	Lab reference	^{14}C Age BP	Calibrated age 2σ range (95.4%) Cal BP	Material	Provenance
Kfar Galim	RT 1748	5985±55	6950–6676	Wood	Central, Well
	RT 1749	5985±70	7000–6662	Wood	Central, Well
	RT 1750	6890±50	7835–7623	Wood	Northern, branch

Calibration according to OxCal v 4.3.2 (Stuiver and Reimer 1993; Bronk Ramsey 2017; Reimer et al. 2013). Provenance refers to the area of the site and the feature number shown on the site plans

Fig. 23.3 Atlit-Yam structure 10a (concentration of fish bones and wheat grains) during the course of excavation. Photo by Ehud Galili

80, area M) was discovered and excavated to a depth of 30 cm (Galili et al. 2017b).

23.5.1.2 Megalithic Structures

Two were identified and consist of kurkar stones, some of which may have been partly shaped. The northern structure (Fig. 23.2, Inset A, 56) comprises large standing stones up to 1.8 m high, arranged in a circle with more stone slabs (up to 1 × 1 × 0.3 m) lying horizontally to the west (Fig. 23.7; Galili 2004, pp. 48–50, 363–364; Galili and Rosen 2011a). Tens of cup marks were recovered in and around the structure, as well as traces of freshwater vegetation. The nature of the structure and the finds suggest that it was used for cultic purposes, probably associated with fresh water. About 20 m west of this structure, there are two parallel walls, 1 m apart, forming a corridor leading to the general area of the megalithic structure (Fig. 23.2, Inset A, 54). This corridor may have been part of the megalithic complex. Palaeo-astronomical calculations suggest that the orientation of the corridor could have been associated with the sunset during the longest day

Fig. 23.4 Two parallel stone walls (structure no. 6) in area E, Atlit-Yam in 10 m of water. Photo by Ehud Galili

Fig. 23.5 Well no 11 at Atlit-Yam, showing a diver inspecting the upper mouth of the well. Photo by Itamar Grinberg

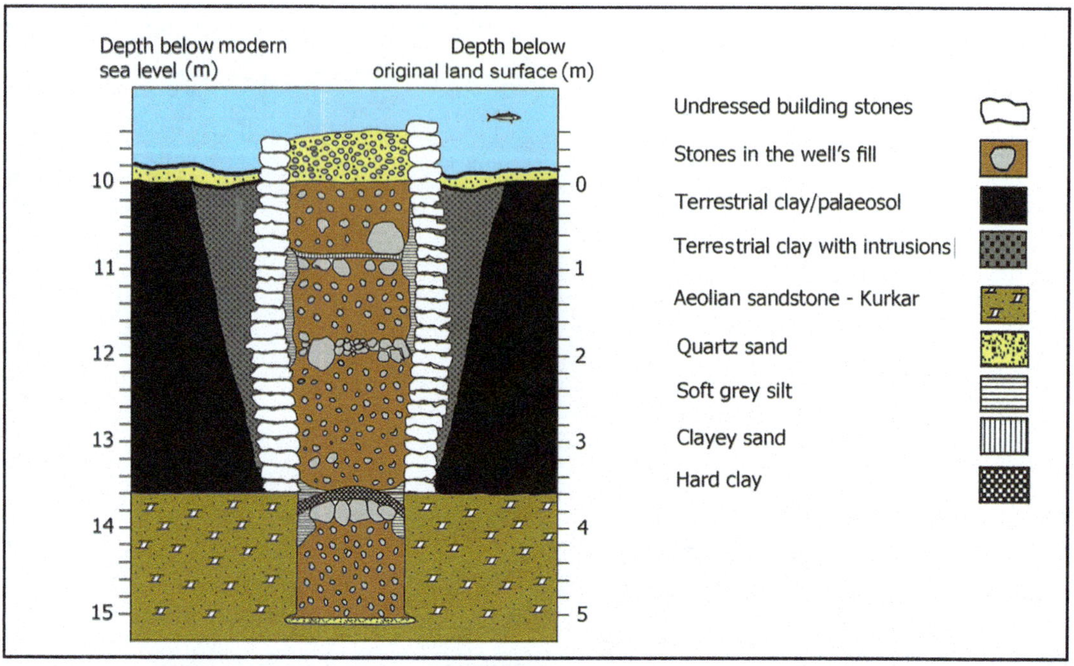

Fig. 23.6 Cross section of well no 11 at Atlit-Yam, showing the stone lining, the nature of the fill and the surrounding sediments. Note that the well has been dug 1.5 m into the kurkar bedrock. Note also the layers of stones near the base, which looks like an attempt to raise the base of the well to offset sea-level rise and salinization of the well. See text for further discussion. Drawing by Ehud Galili

of the year (Clive Ruggles pers. comm. 2010). The southern megalithic structure (Fig. 23.2, inset B, area E) is composed of three large stones, each c. 1.7 m long and weighing 450–550 kg. Two of them are anthropomorphic in nature, having a circumscribed groove at one end to delineate a schematic head.

23.5.2 Material Culture

23.5.2.1 Flint and Ground-Stone Artefacts

At least three spatially discrete areas with different flint assemblages, indicating different activities, are distributed across the site (Galili 2004: pp. 95–126, 382–423): (1) a chipping floor or workshop with evidence of the manufacture, hafting and repair of tools such as arrowheads and axes, and with a high percentage of naviform cores with two striking platforms indicating specialized production of blades (area D); (2) an assemblage with a relatively high percentage of bifacial tools (areas K, L), indicative of extensive wood working (Barkai and Galili 2004; Barkai 2011), perhaps associated with production of water craft (Yerkes et al. 2014); and (3) an assemblage of mainly flint waste and exhausted cores deliberately dumped in the upper levels of well 11 after it had ceased to be used as a water source (area I).

In general terms, the flint assemblage at Atlit-Yam is similar to other PPN sites of the region, indicating a reduction over time in naviform cores and in the use of long blades manufactured from them, and a reduction in the percentage of arrowheads (Fig. 23.8: B, C). Two other features are worth noting. The first is the presence of flint daggers (Fig. 23.8: D). These are also found at other coastal sites like PPNC Ashkelon and PN Zikim but are rare in inland Neolithic sites. Given their geographic distribution, it is possible that they were particularly associated with exploitation of marine resources in activities such as spear fishing or gutting and processing of fish (Galili et al. 2004).

Fig. 23.7 The megalithic structure at Atlit-Yam showing standing stones arranged in a circle. (**a**) The structure after excavation. Photo by Itamar Grinberg; (**b**) Artist's reconstruction by the Israel Antiquities Authority

The second feature is the presence of denticulate sickle blades produced from long delicate blades extracted from naviform cores (Fig. 23.8, A). These blades were previously dated to the early PN (Yarmukian Culture). Some items recovered from the PPNC levels of 'Ain Ghazal (Jordan), were considered as intrusive from the overlying Yarmukian layers (Rollefson 1990; Rollefson et al. 1992; Garfinkel 1993). Their presence at Atlit-Yam, which is a single period PPNC site, demonstrates for the first time their use in the PPNC period.

The ground-stone assemblage includes mainly containers such as large basins made of kurkar stone, polished bowls made of limestone and cup marks in slabs and boulders. Other finds are polished pebbles, hammer stones and grinding slabs. The reliance on stone containers in Atlit-Yam, compared to the submerged PN sites of Neve-Yam and Tell Hreiz, is probably due to the absence of

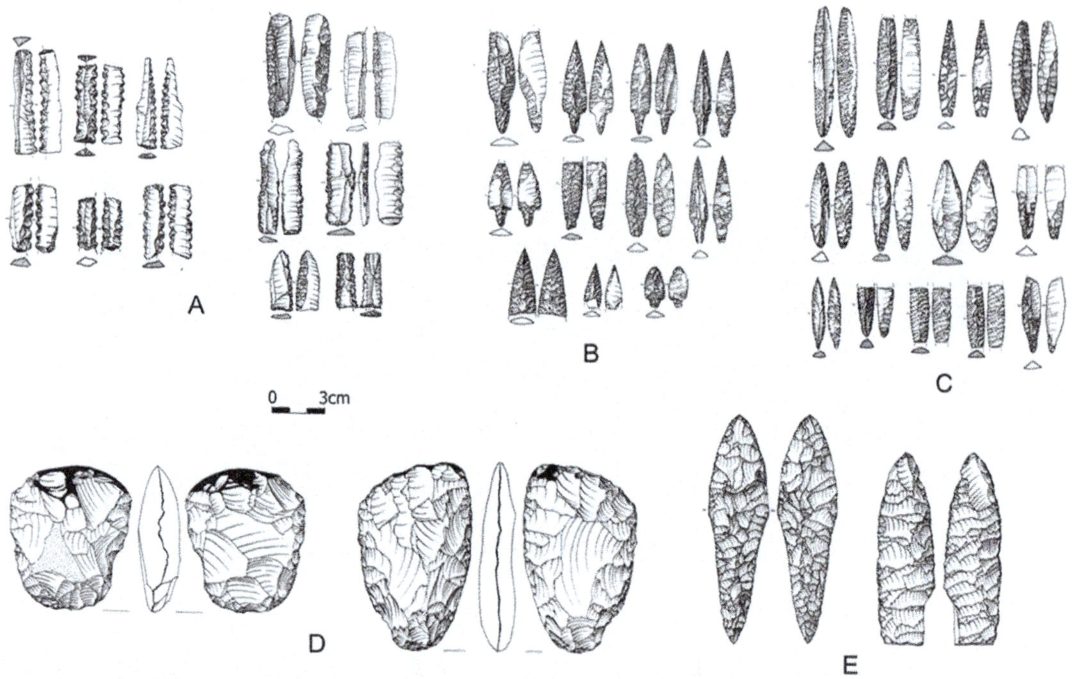

Fig. 23.8 Selection of artefacts from Atlit-Yam. (**a**) Sickle blades; (**b**) Byblos arrowheads; (**c**) Amuq arrowheads; (**d**) daggers; (**e**) bifacial axes. Courtesy of Israel Antiquities Authority

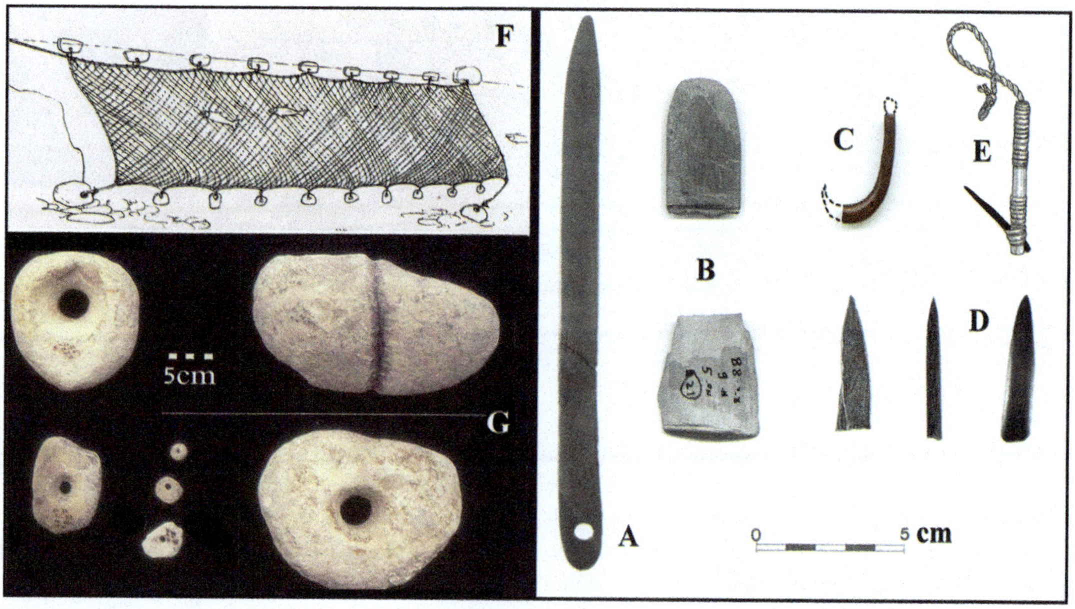

Fig. 23.9 Items of fishing gear at Atlit-Yam. (**a**) bone needle; (**b**) bone gauges for producing fishing nets; (**c**) bone fish hook; (**d**) bone barbs of composed fishing hooks; (**e**) artist's reconstruction of a barbed fishing hook; (**f**) artist's impression of a fish net with sinkers attached; (**g**) stone net sinkers (large perforated and circumscribed end sinkers and small perforated bottom-rope sinkers). Photo by Ehud Galili, reconstruction by the Israel Antiquities Authority

pottery at the earlier site. Perforated stones recovered in the site may have been used as fishing-net sinkers (Fig. 23.9: G; Galili 2004, pp. 73–95, pp 375–380; Galili et al. 2013).

23.5.2.2 Figurines and Ornaments

Ornamented stone includes an anthropomorphic figurine depicting a female buttock, a phallus, decorated pendants, decorated discs and rings and a large rounded stone with two lines of regularly spaced holes that may have been used as a gaming board (Fig. 23.10, A). In one of the megalithic structures (No. 67), two of the kurkar stones bear circumscribed grooves, possibly representing anthropomorphic figures, as noted above.

23.5.3 Plant and Animal Remains

Thanks to the conditions of preservation, with material sealed in the sediment fill of abandoned wells or storage pits and further protected in anaerobic conditions by marine inundation, Atlit-Yam has provided an unusually rich, diverse and well-preserved collection of plant and animal remains, providing many new insights into the subsistence practices of the period.

23.5.3.1 Plant Remains

The plant material mostly comes from the fill of well 11 and consists of carbonised and waterlogged seeds or other parts of 91 species (Fig. 23.11: A) as well as tree branches and pol-

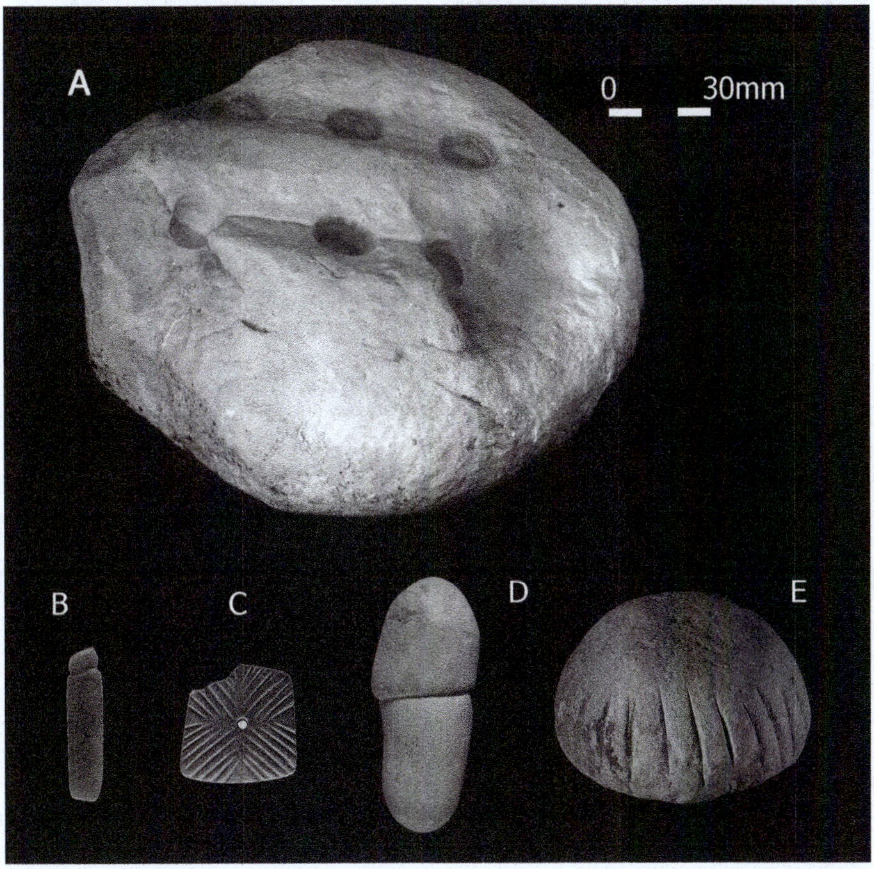

Fig. 23.10 Selection of stone items from Atlit-Yam. (**a**) Gaming board made of limestone; (**b**) decorated pendant made of calcite; (**c**) decorated item (button?) made of limestone; (**d**) limestone phallus; (**e**) decorated item made of limestone (miniature anvil?). Photos by Ehud Galili

Fig. 23.11 Selection of items showing the preservation of uncharred organic materials. (**a**) seeds of figs and wheat grains; (**b**) olive pits; (**c**) remains of flax; (**d**) remains of grain weevil. Photos by Joseph Galili

len grains (Galili and Weinstein-Evron 1985; Galili et al. 1993; Weinstein-Evron 1994; Kislev et al. 2004). Domesticated crops include wheat (*Triticum aestivum*), barley (*Hordeum vulgare*), lentils (*Lens lentis*) and flax (*Linum usitatissimum*). Flax fibres were also recovered, perhaps used for making fish nets or fishing lines (Fig. 23.11: C). Seeds of wild fruit species include almonds (*Amygdalus communis*), figs (*Ficus carica*), grapes (*Vitis sylvestris*) and a single date (*Phoenix theophrasti*). Additionally, various seeds of plants commonly used as fodder were present (Galili et al. 2002, p. 182). The plant assemblage includes 'obligatory' weeds, plant species adapted to growing alongside cultivated crops, indicating that the plant ecosystem had already been modified by crop cultivation, and there are also 27 remains of the grain weevil (Fig. 23.11: D), an obligatory grain pest (*Sitophilus granarius*), one of the earliest records in the Near East (Hartmann-Shenkman et al. 2015). One of the stone-built pits contained masses of waterlogged plant material identified as straw, probably used for animal fodder.

23.5.3.2 Animal Remains

Bones of terrestrial vertebrates (over 8000) are well preserved and include domestic goats (*Capra hircus*) and sheep (*Ovis aries*), cattle (*Bos taurus*), domestic pigs (*Sus scrofa domestica*) and dogs (*Canis familiaris*) (Horwitz and Tchernov 1987; Galili et al. 1993, 2002, 2004; Horwitz and Ducos 2005). Some of the cattle and goats resemble their ancestral wild forms suggesting an early stage of domestication. Wild species include wild boar (*Sus scrofa ferus*), mountain gazelle (*Gazella

gazella), Persian fallow deer (*Dama dama mesopotamica*), carnivores especially red fox (*Vulpes vulpes*), reptiles, rodents and amphibians. In the early phases of the settlement, wild animals and those in the initial stages of domestication were dominant, especially cattle. In contrast, in the fill of the well, representing a slightly later occupation stage, remains of juvenile animals were predominant, especially those of domestic goats, sheep, pigs and dogs. In both assemblages, butchery damage was evident with numerous cut and chop marks (Greenfield et al. 2013), but few of the bones were burnt.

More than 6000 fish remains (bones, scales, teeth) were recovered, representing seven families of Mediterranean marine fish but with the Grey triggerfish (*Balistes carolinensis*) accounting for the great majority. Some fish bones were blackened by fire, indicating direct association with human processing activities, and many fish bones were found concentrated in close association with archaeological remains, indicating that they were in situ food remains rather than more recent intrusions of natural death assemblages. A number of artefacts including the stone net sinkers referred to above reinforce the evidence for fishing as an important subsistence activity (Fig. 23.9). The size distribution of the fish remains suggests that fishing nets were used, and the pelagic fish bones recovered imply that offshore fishing with boats took place (Zohar et al. 1994, 2001; Galili et al. 2004).

The faunal assemblage from PPNC Ashkelon had a similar composition to that of Atlit–Yam, including a substantial fish component, suggesting a similar combination of hunting, herding and fishing (Hesse and Rookis 2008; Lernau 2008). However, the architectural remains suggest that Atlit-Yam was a permanent sedentary settlement, while Ashkelon was probably a seasonal site (see above).

23.5.4 Human Burials

Atlit-Yam yielded a very large number of human burials, with an estimated number of 63 individuals. Burials were found in all parts of the site. Most of the skeletons were found in organized graves adjacent to structures, but rarely within them; nevertheless, isolated bones were also found in different areas of the site. The latter probably represent disturbed graves due to build-

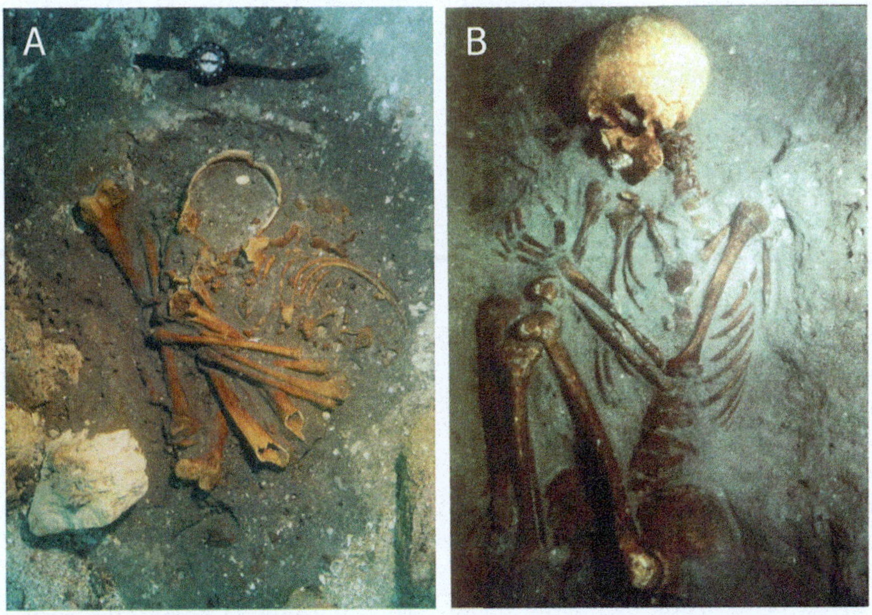

Fig. 23.12 Burials at Atlit-Yam: (**a**) primary burial of a woman in flexed position, (**b**) woman and infant, both with evidence of infection by tuberculosis. Photos by Ehud Galili

ing and digging activities or renovation of structures from time to time (Galili et al. 2005b; Eshed and Galili 2011). Burials were mostly primary (Fig. 23.12). In six cases there was evidence of secondary burials. Of the 37 primary burials, 24 contained a single individual, 11 contained 2 individuals and 2 contained 3 individuals. Most of the deceased were found in a flexed or semi-flexed position, with no clear pattern in orientation. Skulls were usually intact, but a few skeletons had no skull, and in one burial only the skull was found. Graves were simple pits dug into the clay and had no external markings. In some cases, a few small-sized stones were scattered at random on top of, or around, the pit. Most burials (N = 45) were in the north-west area of the site (Fig. 23.2: areas K and L). Twenty-three of these were adjacent to a single rectangular structure (no. 60) and were associated with hearths. It may well be that this group of burials is associated in some way with a nearby megalithic structure (Fig. 23.2, Inset A: 56) and a two-wall corridor located c. 30 m to the north (see Fig. 23.2, Inset A: 60, 56 and 54).

Fifteen burials had grave offerings, five had flint axes, five had ground-stone tools, two had bone artefacts, and two had cattle horn cores and a calcite crystal. It seems that the type of offering was gender-based, stone axes with males and grinding stones with females (Barkai and Galili 2004).

Numerous skeletal pathologies were identified (Hershkovitz and Galili 1990; Galili et al. 2005b; Eshed et al. 2010). These include auditory exostosis, an ear pathology evident in 25% of the male skeletons, caused by diving or frequent swimming in cold water, perhaps indicating an engagement in underwater spear fishing. Thalassemia is also indicated, suggesting that the inhabitants of Atli-Yam were less susceptible to malaria, a mosquito-borne disease commonly associated with coastal swamps. Tuberculosis is confirmed by aDNA analysis (Hershkovitz et al. 2008; Donoghue et al. 2009) and is the earliest confirmed report of this disease in humans. An aDNA analysis of the Atlit-Yam cattle bones proved negative for the presence of *Mycobacterium bovis*, the bovine form of tuberculosis, suggesting the spread of the disease by human contact rather than from domestic cattle.

23.6 Submerged Pottery Neolithic Sites

There are 18 submerged PN sites. Some are sectors of sub-divided large sites. From north to south, they are as follows: Kfar Samir (north, centre and south), Hishuley Carmel (north and south), Kfar-Galim (north and south), Nahal Galim, Hahoterim, Tel Hreiz (north and south), Megadim, Atlit north bay, Neve-Yam (north, south, Temanun island and west) and Habonim. Based on radiocarbon dates (Table 23.2) and material culture, 16 are attributed to the Wadi Rabah culture, variously considered as late PN (Gopher and Gophna 1993; Gopher 2012) or early Chalcolithic (Garfinkel 1999). Recent C14 analysis of the Hishuley Carmel north site (not published here) attributes this site to the Middle Chalcolithic Period. The Neve-Yam north site includes pottery typical of the Lodian culture, which predates the Wadi Rabah culture (Table 23.1; Gopher and Gophna 1993; Galili 2004; Galili et al. 2017c).

23.6.1 Methods

These submerged Pottery Neolithic sites are located in shallow-water, high-energy marine environments in the intertidal and surf zones (0–7 m deep), posing methodological and logistical problems that required the development of novel survey and excavation methods (Galili et al. 1993, 2017b; Galili 2004). The eastern portions of these sites are usually 0–1 m deep, too shallow to apply the underwater excavation methods developed and described for sites in deeper water like Atlit-Yam. Waves interfere with the excavation and the visibility is poor. We adopted a strategy of allowing the sea to do the job of removing the overlying sandy sediments, as at Atlit-Yam. Most of the archaeological material was collected after storms, either during underwater surveys often undertaken by snorkelling, or sometimes even as washed up material on the shoreline. A few limited excavations were carried out in the deeper portions of these sites, at Kfar Samir central sector (water wells and pits), Kfar Galim and Megadim (stone-lined pits) and Neve-Yam (stone-built graves). The protocols for excavation and post-excavation treatment of material followed that for Atlit-Yam.

23.6.2 Architectural Remains and Features

After five decades of site exposure and documentation, considerable portions of the PN villages of Neve Yam, Tel Hreiz, Kfar Galim and Kfar Samir are available. Thus, the general nature and layout of these sites can be reconstructed. At Neve-Yam and Tell Hreiz, rectangular dwellings, round pits and stone-built graves were recovered. At the other sites, notably at Kfar Samir, stone paving slabs were found and features such as pits and water wells (the former also at Kfar Galim). In addition, hearths, c. 0.5 m in diameter, some lined with stones and containing fragments of burnt bone and charcoal, were also recovered at Kfar Samir. Tel Hreiz revealed evidence of rectangular structures, stone paving and hearths containing charred remains of wood and animal bones. At Hishuley Carmel, two stone-paved, elliptical features constructed of upright stone slabs were exposed near the shore at 0.8 m depth. Their fill contained a 15-cm-thick layer with thousands of olive stones preserved by waterlogging. At Tel Hreiz, two concentrations of vertical wooden poles made from tree branches were recovered. They are probably foundations of fences, huts or cabins. At this site, a stone-built wall >100 m long was discovered, lying parallel to the present coast at 3–4 m depth. It was built of boulders up to 1 m in their maximum dimension and may have served as a seawall to protect the village from wave action and rising sea level (Galili and Rosen 2013; Galili et al. 2019b).

The Neve-Yam south site showed a layout sub-divided into two distinct areas, the northern one with remains of dwellings and domestic activities, while the southern one represents an organized burial ground (Fig. 23.13 and 23.14, B). In the north, there were foundations of rectangular structures and wall fragments (c. 0.5 m in width) built of two rows of undressed sandstone (Fig. 23.15). Also found were pits, paved surfaces made of small undressed stones, stone slabs and postholes, probably evidence of hut structures. In the south, cist graves built of standing stones arranged in an elliptical pattern and covered with stone slabs were found, all with an east–west orientation forming an organized pattern (Fig. 23.14, A, B). Between the graves, large concentrations of charred wheat, barley and lentils were recovered, possibly representing burial offerings or the remains of ceremonial meals. Concentrations of burnt mudbricks were found there as well (Galili et al. 2009).

23.6.2.1 Water Wells

At several PN sites, there were circular structures 1–1.5 m in diameter, made of undressed stones, representing storage pits and water wells (Galili and Weinstein-Evron 1985; Galili et al. 1997; Galili et al. 2016a). Excavation of the pit fills yielded potsherds, flints and sometimes faunal remains and plant fragments. Cylindrical structures (up to 1.2 m high) consisting of several courses of undressed stones were found above some of these stone structures at Kfar Galim and Nahal Galim (Fig. 23.16, C). These probably represent the super-structure of water wells. At Kfar Samir, three water wells were found at a depth of −5.5 m, some 200 m offshore. They were constructed of alternating courses of wooden beams and undressed stones, mostly limestone pebbles. One (well no. 3) had a rectangular opening 1 × 0.8 m and was excavated to a depth of 2 m without reaching the bottom (Fig. 23.16, A, B). With depth, this well widens and becomes more circular in plan. In its lower part, two courses of

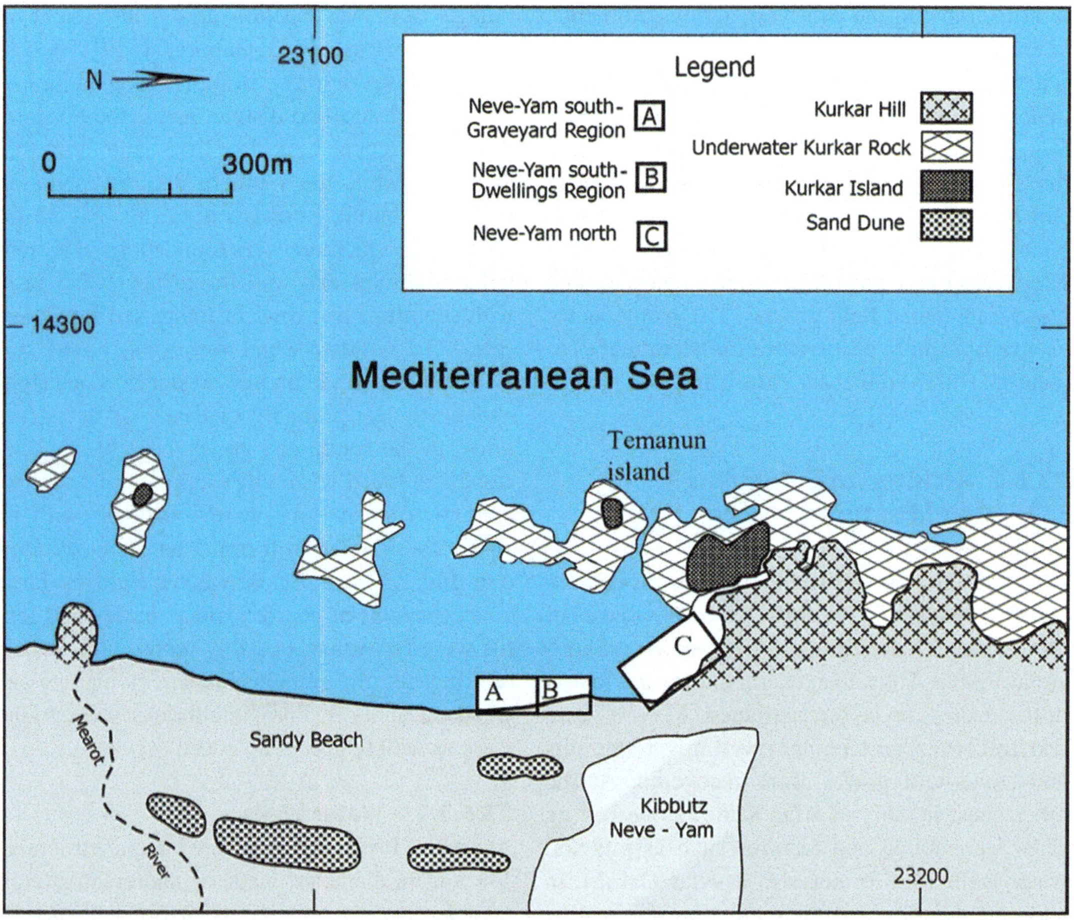

Fig. 23.13 Plan of the Neve-Yam site, showing the separation between the graveyard (**a**) and the contemporaneous dwelling area of the settlement (**b**). Also shown are environmental and geomorphological features in the surrounding area. Drawing by Ehud Galili

stones were laid between the wooden beams. The fill consisted of clay, small stones, a few bird bones, olive stones, potsherds, flint flakes and probable straw remains, with a date range of 7964–7320 cal BP (Table 23.2; Galili and Weinstein-Evron 1985; Galili et al. 2018). Similar wells were recovered at Kfar Galim at 2–3 m depth. These wells are some of the earliest known wooden structures in the world.

23.6.2.2 Olive Oil Production

At Kfar Samir central and south sectors, Hishuley Carmel, Tel Hreiz and Hahoterim, concentrations of thousands of olive stones, many of them crushed, were found in fill of pits dug into the clay and are interpreted as representing various stages in the extraction of olive oil (Galili et al. 1989, 1997; Galili and Sharvit 1994–5; Galili and Rosen 2007, 2011a; Galili et al. 2018). At Kfar Samir, a pit paved with stone pebbles and filled with mostly crushed olive stones, was excavated. Large stone basins and basket-work strainers (discussed below) were recovered and are thought to have been associated with the processing of the olives. These finds represent the earliest known evidence for the production of olive oil and an exemplary case of the secondary products revolution.

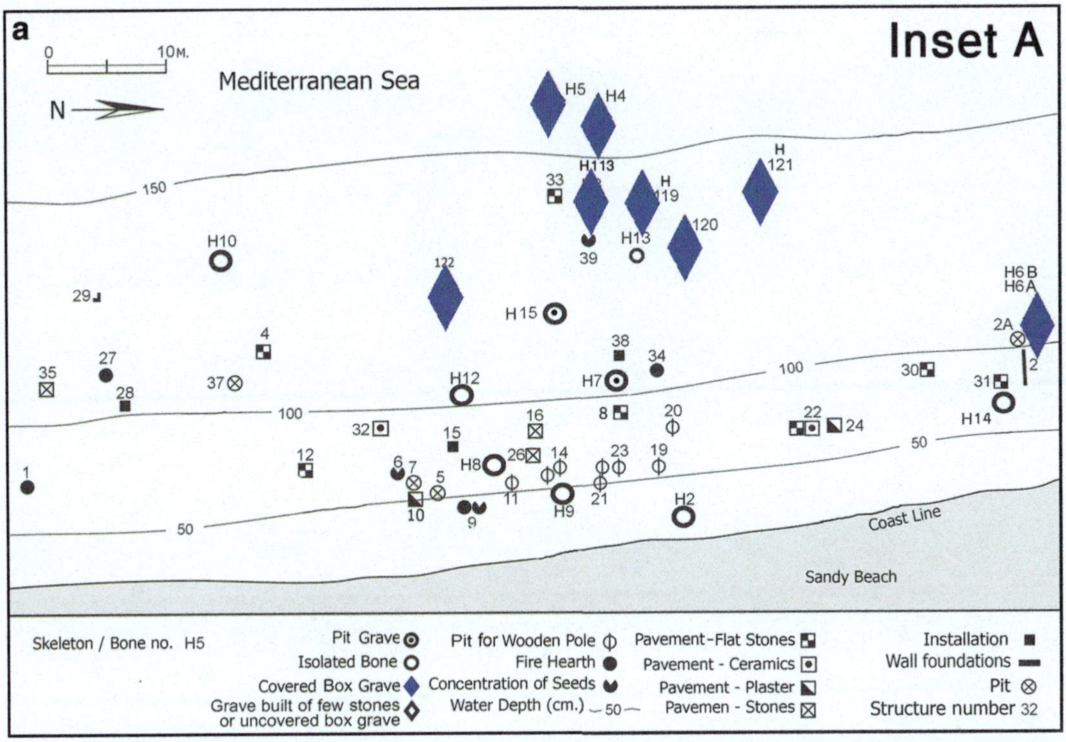

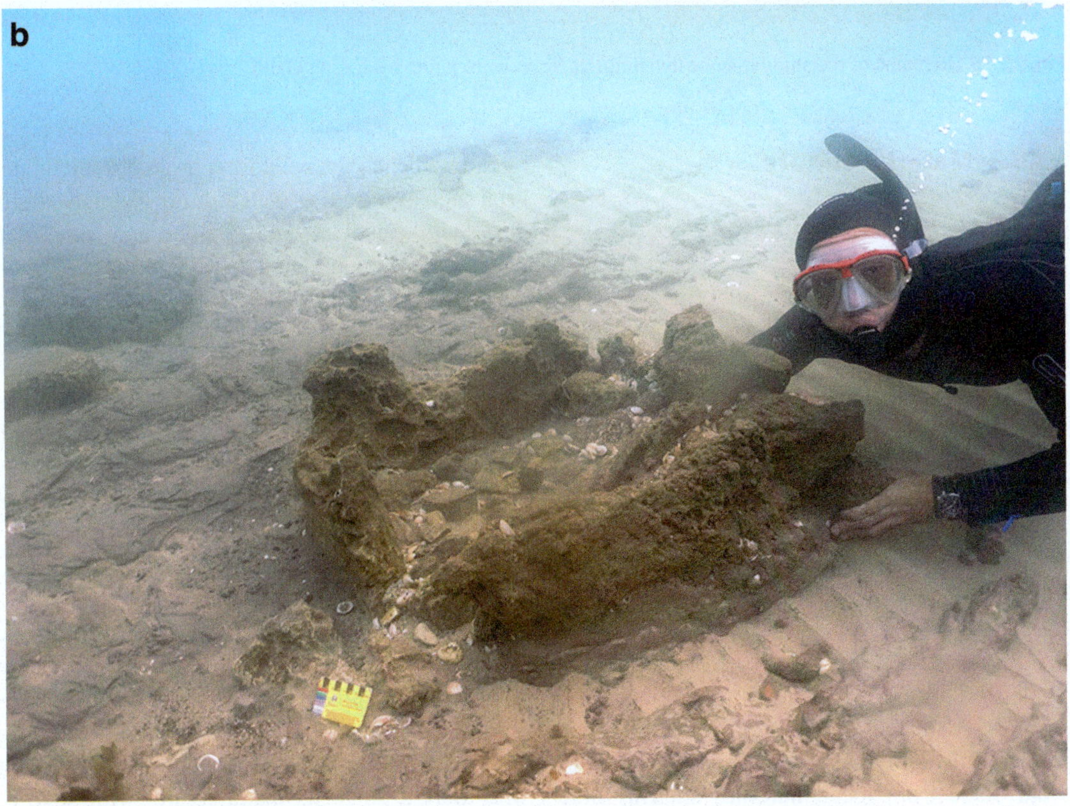

Fig. 23.14 Detailed plan of features in the Neve-Yam graveyard (**a**), showing graves, storage pits, dwelling and other features. Inset A refers to the areas marked in Figure 23.13. Drawing by Ehud Galili, (b) small stone-built box grave, probably of a child, Photo by Ehud Galili

Fig. 23.15 Remains of rectangular stone dwellings at Neve-Yam. Photo by Joseph Galili

23.6.3 Material Culture

23.6.3.1 Flint, Ground Stone and Pottery

Flint artefacts include flake tools and polished bifacials (adzes, chisels and axes), sickle blades and additional blade tools. Neve-Yam south had numerous narrow plano-convex chisels, adzes and rectangular backed and truncated sickle blades. Noteworthy in all PN sites is the absence of arrowheads (Galili 2004, pp. 245–255, 438–449). Ground-stone artefacts found in all PN sites were made of limestone, sandstone (kurkar) and especially basalt (mortars, grinding stones, chalices, troughs and cup marks). Stone containers are less abundant than at Atlit-Yam, probably associated with the introduction of pottery production. Also, the use of basalt contrasts with PPNC Atlit-Yam, where ground-stone artefacts were made from local materials such as limestone, for delicate small vessels, and kurkar, for large and robust bowls). This difference may reflect greater reliance on crop processing in the PN, and hence the need for a more durable material such as basalt, as well as more extensive trade ties with the hinterland (Galilee and the Golan Heights) where basalt is found (Galili et al. 1993; Galili 2004, pp. 189–193; 461–462).

The largest pottery assemblage is from Neve-Yam South, with painted and incised bowls, spouted vessels, hole-mouth jars, bow-rim jars and pithoi, including a sherd decorated with incised fish and a herringbone design (Fig. 23.17; Galili 2004, pp. 172–189 450–460).

23.6.3.2 Wooden Artefacts and Basketry

At Kfar Samir, a complete wooden bowl made from *Ceratonia siliqua*, the carob tree or St John's bread and dated to 8275–7872 cal BP was found in a pit (Fig. 23.18: A; Table 23.2; Galili and Schick 1990). The bowl shows evidence of the tools used in its manufacture, with adze or chisel marks on the outer surface, a rougher finish on the interior and the base and the rim smoothed and polished. Pieces of unidentified branches and

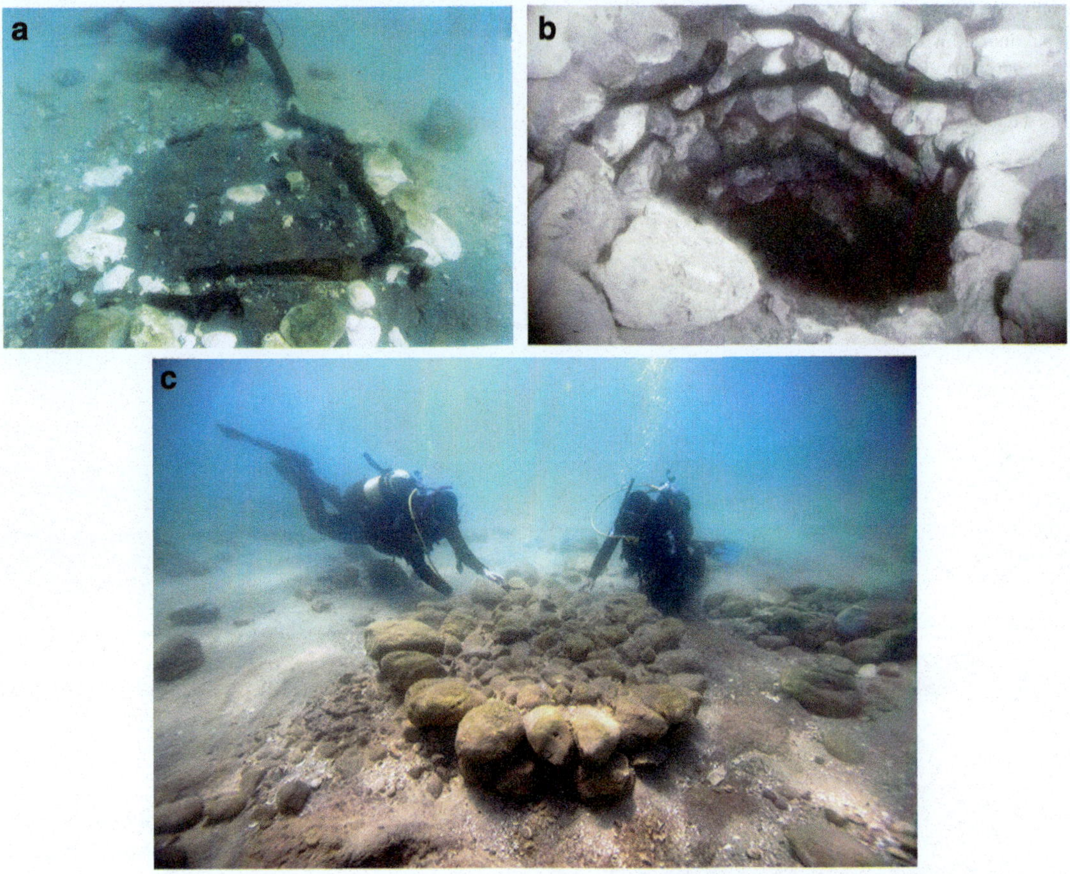

Fig. 23.16 Water-well 3 at Kfar Samir before excavation (**a**) and after excavation (**b**). The alternating layers of wooden branches and stones are clearly visible in (**b**). Photos by Ehud Galili, (c) superstructure of a water well from Nahal Galim, Photo by Itamar Grinberg

straw, perhaps remains of a mat or a basket, were found next to it. A second find from another part of the site is a fragment of a pot-shaped wooden bowl with a flat base, a straight wall, part of the rim and a knob-like handle with a narrow lateral perforation (Galili et al. 2007). In one of the Kfar Galim wells, three sections of wood (each c. 25 cm long and 15 cm in diameter) showed cut marks at both ends, suggesting pre-forms intended to produce wooden bowls similar to those described above (Fig. 23.18: B).

At Kfar Samir, a pit 0.9 m in diameter and 0.55 m deep was exposed on the sea bed at a depth of 2.3 m, and excavation revealed a fill of grey clay containing dozens of olive stones (Fig. 23.11: B) and pieces of a braided basket preserved by waterlogging (Fig. 23.19: A). The basket was made from stems 3–5 mm thick by the method of 'alternate pair twining' (Galili et al. 1997) and included a round piece, probably the base, and parts of the wall. Because of its association with the olive stones and its similarity to modern basketry items used to strain the oil from pressed olives (Fig. 23.19: B), this item is thought to have been used for a similar purpose. At Tel Hreiz, a woven circular ring of plant material may have been the rim of a basket. Also, at Kfar Samir a unique piece of basketry was found in an unlined pit dug in the upper clay level. After laboratory treatment, it turned out to be a mat, 7 × 16 cm, made of bundles of unidentified material, perhaps rushes or straw, and constructed using the technique of

Fig. 23.17 Potsherd with incised fish figures and herringbone designs from Neve-Yam. Photo by Ehud Galili

Fig. 23.18 Wooden artefacts from PN sites. (**a**) Complete bowl from Kfar Samir made from the wood of a carob tree; (**b**) section of wood from the fill of a well at Kfar Galim, showing cutmarks at both ends, possibly intended for the production of a wooden bowl. Photos by Ehud Galili

'coiled basketry with intricate stitch' (Galili and Schick 1990).

23.6.3.3 Figurines

At Neve-Yam south, several symbolic items were recovered near the burial ground (Milevski et al. 2016; Galili et al. 2016b), including an engraved oval kurkar slab with a schematic design of a human head and anthropomorphic figurines made of bone, clay and greenstone (Fig. 23.20).

23.6.4 Plant and Animal Remains

23.6.4.1 Plant Remains

At Kfar Galim and Tel Hreiz, numerous waterlogged plant remains were found, while at Neve-Yam south, three large concentrations of charred seeds, possibly associated with the burials, were recovered: (1) c. 300 cc of seeds mixed with clay, dominated by domesticated lentils (*Lens culinaris* var. *microsperma*) (Kislev et al. 2004), some infested with beetle pests, probably *Bruchus* sp; (2) c. 100 cc of grains mixed with clay in a hearth built of burnt mudbricks, and dominated by domesticated barley (*Hordeum sativum* = *H. vulgares* L.), along with several seeds of domesticated emmer (*Triticum dicoccum*), one seed of

Fig. 23.19 Items of basketry. (**a**) Archaeological specimen found next to a cache of olive pits; (**b**) modern item of similar design used as an oil strainer. Photos by Ehud Galili

Fig. 23.20 Anthropomorphic figurines from Neve-Yam south. (**a**) Bone figure with herringbone designs, photo by Ehud Galili; (**b**) figure made of green stone, courtesy of the Israel Museum

Viciae and a few seeds of Liliaceae; and (3) some 300 cc of seeds consisting of pulses: pea (cf. *Pisum vicia narbonensis*), vetch (*Lathyrus* sect. *cercula*), horse bean (*Vicia faba* var. *minor*), lentil (*Lens culinaris* var. *microsperma*), along with domesticated flax (*Linum usitatissimum*) and a few wild plants or weeds (*Galium*, *Lolium* and *Lilicaeae*) (Galili et al. 2018).

23.6.4.2 Animal Remains

At Neve-Yam and Tel Hreiz, and to a lesser extent at Kfar Galim, faunal remains are abundant, indicating the dominance of domestic animals—sheep, goat, cattle and pigs, with only a few remains of wild species—gazelle (*Gazella* sp.) Persian fallow deer (*Dama dama mesopotamica*), Eurasian badger (*Meles meles*), possibly also wild boar (*Sus scrofa ferus*) and wild birds, attesting to ongoing trapping/hunting as a minor subsistence activity. Isolated remains from the other sites include mallard duck (*Anas platyrynchos*) and the Palestine molerat (*Spalax ehrenbergi*) (Horwitz et al. 2002, 2006). Remains of dog (*Canis familiaris*) are notable at Tel Hreiz and Kfar Galim. Butchery marks indicate that animals were processed and consumed on-site using a variety of stone tools; kill-off patterns of goats are dominated by immature animals, indicating an emphasis on meat (Greenfield et al. 2006). Fish bones are present at Kfar Galim, Tel Hreiz and Neve-Yam south but are few—mostly Sparidae, Serranidae and the freshwater *Tilapia* sp. This scarcity is a notable contrast to Atlit-Yam but may indicate only that fish bones are underrepresented because bones at the PN sites were collected by hand without the use of sieves.

Fig. 23.21 Stone-built cist grave at Neve-Yam. (**a**) As it appeared on the sea bed; (**b**) plan. Drawing and photo by Ehud Galili

23.6.5 Human Burials

Human remains were found at Neve-Yam and Tel Hreiz. Most of the skeletal remains consist of primary in situ burials, while some were represented by scattered bones, most likely due to post-depositional disturbance of burials. At Tel Hreiz, two disturbed burials were recovered in pits dug in the clay. Additionally, two stone-built cists (0.6 × 1.3 m) built of stone slabs probably represent graves but have not been excavated. Eleven in situ human burials were excavated at Neve-Yam south, concentrated in a relatively small area (40 × 70 m; Galili et al. 2009). Eight are stone-built graves, some partially eroded by the sea, consisting of an oval burial chamber lined with undressed stones, covered by stone slabs and with an east–west orientation (Fig. 23.13: B; Fig. 23.21). Three others are in simple pits dug in the clay, while traces of four additional individuals are represented by scattered remains. Four skeletons were fully flexed, and four partially flexed. Six additional stone-built cist structures, most probably graves, were found at the site but not excavated. The estimated number of individuals at Neve-Yam is 15 (Eshed et al. 2004; Eshed and Galili 2011), including those represented by isolated or scattered remains, comprising eight children (0–10 years), one adolescent (10–18 years) and six adults (1 female, 3 males, 2 indeterminate). Oblique tooth wear in one individual suggests use of the teeth as tools (Eshed et al. 2010). No grave offerings were recovered, but nearby hearths, paved surfaces and concentrations of charred seeds may indicate activities such as ceremonial meals associated with the burial rite.

In addition, a primary burial of an infant (1–2 years old) was found on the Temanun Island (Octopus Island), an islet c. 140 m from the coast opposite the submerged site of Neve-Yam and was associated with possible grave goods—a broken pottery vessel with an elongated nozzle and a flint flake (Galili et al. 2016c). At the time of the burial, the islet was a kurkar hill on the shoreline (Fig. 23.22).

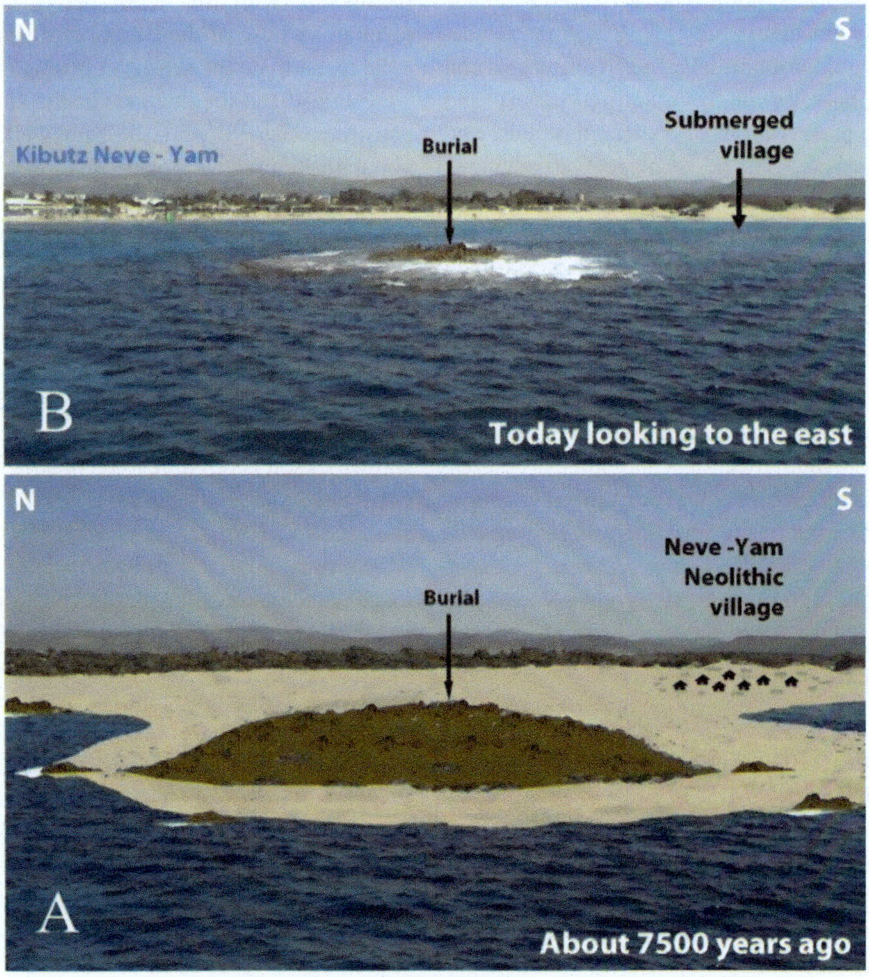

Fig. 23.22 View of the burial site on Temanun Island looking towards Neve-Yam. Upper (**b**): the present day; lower (**a**): reconstruction at the time of burial. Photos and reconstruction by Ehud Galili

23.7 Discussion

The submerged Neolithic villages off the Carmel coast have revealed a wealth of new data on the Pre-Pottery and Pottery Neolithic/Chalcolithic period of the region. In some respects, the finds are no different from what has been found in terrestrial sites of the same period—a similar range of flint artefacts, ground-stone tools, ornaments, figurines and ceramics, faunal remains and structures made from undressed stone or mudbrick. In other respects, these sites highlight features that are uniquely the result of the original site location close to the shoreline and to subsequent submergence by sea-level rise and the conditions of underwater preservation. The recovered evidence is not only unusual in preserving organic materials that would not normally survive in terrestrial deposits, but also in the quantity, variety and excellent preservation of other materials and features.

This evidence includes the use of wood in structures and for making artefacts; the construction of stone-lined and stone- and wood-lined water wells, which also provide good index points for measuring sea-level change; large numbers of human graves with well-preserved skeletal remains providing unusually rich details of burial practices, palaeodemography and health; stone-built dwellings and other structures, which together with the locations of graves show the wider layout of the settlement including evidence in the PN period of a separately demarcated burial ground; generally good preservation of bone; and the survival of a wide range of plant remains.

23.7.1 Water Wells and Sea-Level Change

The wells are of special interest not only for the details of their construction and their contents but also as evidence that already in these early periods people were able to improve the water supply by creating an artificial, permanent source of water. In a region where water is a limiting factor on the duration of human habitation, this technological development would have been an important contribution to sustaining permanent settlement in one place and to improving the reliability of the subsistence economy.

In addition, the wells provide valuable additional information on sea-level rise, palaeoenvironment and human adaptation to the changing coastal environment. The natural slope of the groundwater table on the Israeli coastal plain is about 1:1000 (Kafri and Arad 1978), and studies of numerous wells of all periods show that excavation about 0.5 m below the water table provides a sufficient column of fresh water (Nir and Eldar-Nir 1986, 1987, 1988). By combining these two parameters with the depth below modern sea level at the base of the Neolithic wells, it is possible to estimate the sea-level position when they were in use and the distance of the site from the coastline. The PPNC Atlit-Yam well bottom is about 15.5 m below modern sea level, indicating a sea-level position of c. −16 m. By the same method, the later PN well at Kfar Samir gives a sea-level position of c. −7 to −9 m (Galili et al. 1988, 2005a).

Combining these estimates with other natural and cultural indicators of sea level, such as wave-cut notches, abrasion platforms, maritime installations and clusters of anchors originating from shipwrecks, it is possible to reconstruct a general, estimated, sea-level curve (Fig. 23.23; Galili et al. 2017a;). This sea-level curve, in its turn, can be combined with bathymetric and geological maps to reconstruct changes in coastline configuration (Fig. 23.24; Galili et al. 1988, 2005a).

The curve shows how sea level has risen progressively and continuously from a depth of nearly −40 m 10,000 years ago. It continued to rise during the occupation of Atlit-Yam and the PN sites at a rate of 11–13 mm per year. About 6500 years ago, the rate of rise slowed to c. 2.5–3.5 mm per year, reaching the modern level during the Middle Bronze Age at about 4000 years BP. The maps show how the coastline at the beginning of occupation at Atlit-Yam was some 1.5–2.5 km west of its present position and consisted of an indented waterline with bays and

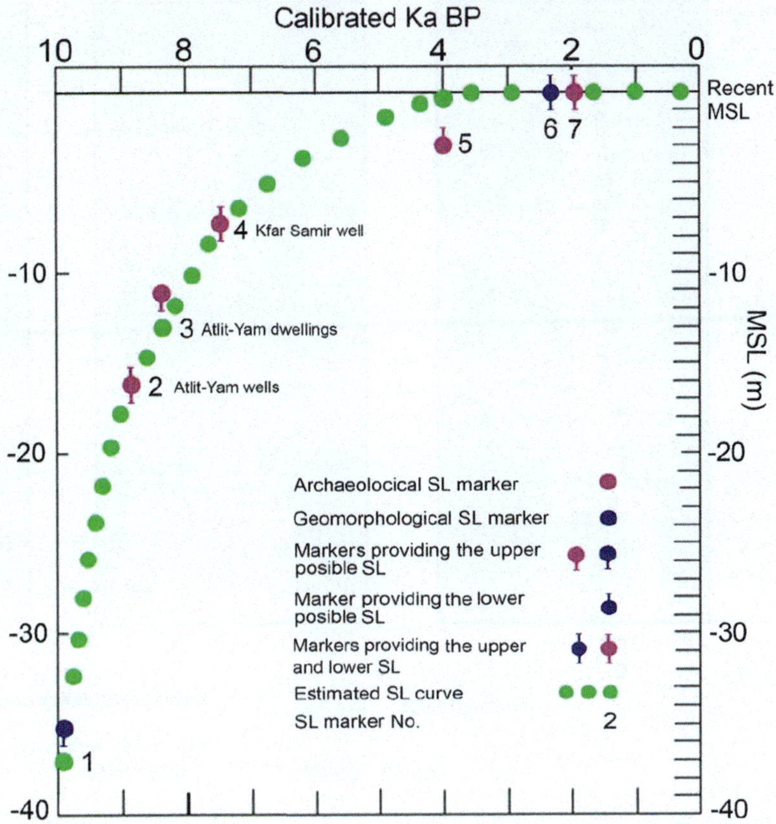

Fig. 23.23 Estimated SL curve for the past 10,000 years on the Carmel coast as indicated by archaeological and geomorphological SL markers: (1) terrestrial clay palaeosol; (2) Atlit-Yam water wells; (3) Kfar Samir water well; (4) Kfar Samir dwellings; (5) Middle-Bronze Age (Byblos Type) stone anchors; (6) erosion platforms and wave-cut notches; (7) Roman-Byzantine rock-cut features. Drawn by Ehud Galili

lagoons (Fig. 23.24). In the later period during the subsequent PN period, sea-level rise created a smoother sandy shoreline with a series of small offshore islands. These too disappeared shortly afterwards with continued sea-level rise creating a coastline configuration much as today.

The relatively rapid rise of sea level during the occupation of the Neolithic sites must have caused ongoing problems and required active human adaptation to the changing conditions. These would have begun with progressive salinization of the water wells. According to the evidence of well 11 at Atlit-Yam, initially the occupants tried to counter this tendency by putting rocks and other materials in the bottom of the well in an attempt to exploit the upper, less saline part of the water table. Eventually, the well fell out of use and was used as a rubbish pit. Finally, with further sea-level rise, the whole settlement was abandoned. Pareschi et al. (2006, 2007) have suggested that a tsunami associated with the eruption of Mount Etna caused the abandonment of Atlit-Yam. But there is no evidence of tsunami deposits, nor of sudden and catastrophic changes such as damaged structures, whole animal carcases or evidence of traumatic injury, such as might be expected to result from a tsunami event (Galili et al. 2008). The slow and inexorable rise of sea level provides a sufficient explanation for progressive change in site use and eventual abandonment, while such damage as is present can be accounted for by the post-depositional effects of marine erosion.

23.7.2 Burial Customs

The large number of burials makes it possible to chart changes through time and adds new information to what is already known from contemporaneous sites elsewhere in the Southern Levant. For example, the Atlit-Yam site shows similari-

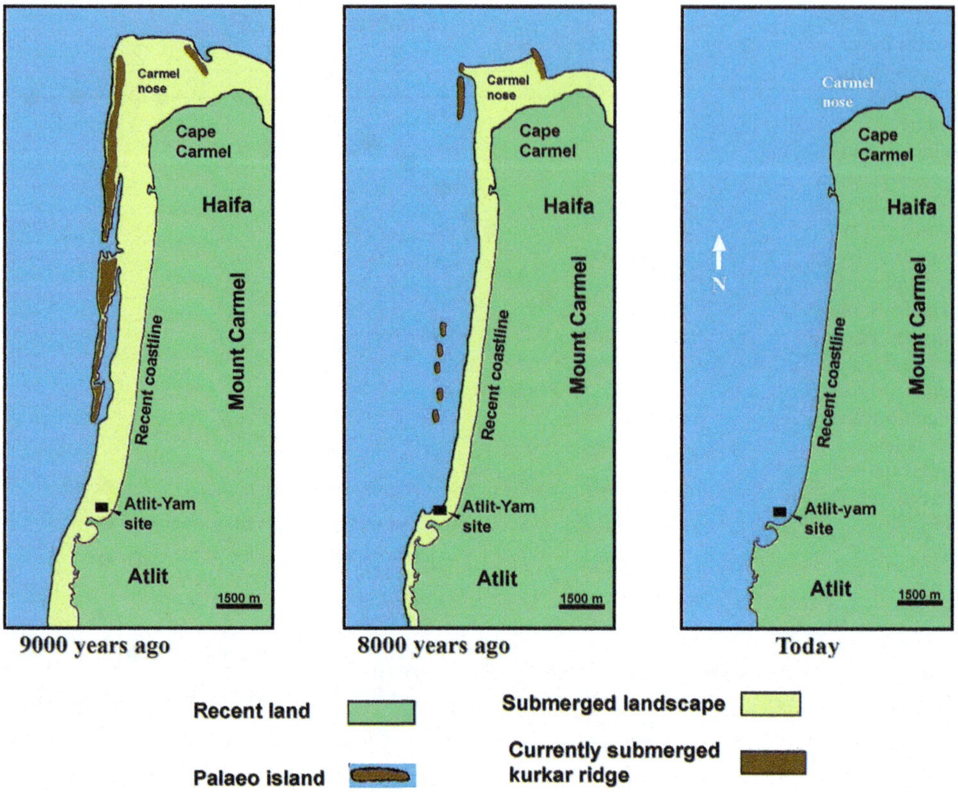

Fig. 23.24 Reconstruction of changes in the shoreline configuration of the Carmel coast over the past 9000 years. Drawn by Ehud Galili

ties in burial practices to the terrestrial PPNC site of 'Ain Ghazal in Jordan (Galili et al. 2005b; Eshed and Galili 2011). However, Atlit-Yam is unique in the large number of primary burials and the grave goods deposited with the dead, including flint axes and ground-stone items. Also, the extensive layout shows that the burials are widely distributed across the site—in open spaces between dwellings, close to walls and occasionally inside structures. By the time of the PN period, Neve-Yam and Tel Hreiz show evidence of burials in stone cist-like graves, a feature that is lacking at Atlit-Yam and other PPNC sites.

Similar graves are known from PN sites in the wider region such as Byblos and Tabaqat Al-Buma (Stekelis 1972; Dunand 1973; Banning 1995). However, Neve-Yam shows a new feature not previously recorded at any other PN site in the region, and that is the organisation of space within the settlement to achieve a clear separation between the areas of domestic activity and an area reserved for burials. This feature of a separate graveyard, containing stone-built cist graves in an organized pattern, is an innovation of the Wadi Rabah culture. It may have resulted from intensive soil-disturbing activities (planting, digging wells and foundations of structures) associated with long-lasting sedentism, agriculture and the burying of the dead, provoking competition for limited subterranean space and the need to resolve the conflict between the needs of the living and the needs of the dead (Galili et al. 2009, 2017c).

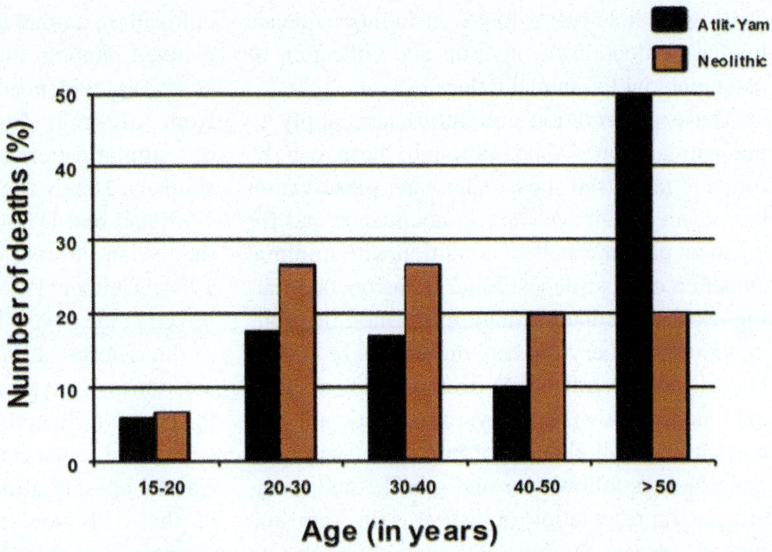

Fig. 23.25 Comparison of male mortality patterns at Atlit-Yam with Neolithic sites in the southern Levant. Drawing by Vered Eshed

23.7.3 Palaeodemography and Health

The large sample of human skeletal material made available from the many burials at these underwater sites and the generally excellent bone preservation have enabled a comparison of mortality profiles and produced evidence of pathologies indicating the presence of infectious diseases. The mortality pattern of the Atlit-Yam male population shows greater longevity than males in the PN and greater longevity than inland PPN populations (with higher mortality values in the over 50-year age cohort) (Fig. 23.25; Eshed and Galili 2011). This can be attributed to a balanced nutrition resulting from the more stable and varied food supply available to the inhabitants of Atlit-Yam and their access to marine resources, compared to their inland contemporaries. This is also consistent with indications that marine resources were less intensively exploited or less abundantly available along the Carmel coastline during the PN period.

Evidence of infectious diseases includes the presence at Atlit-Yam of the oldest known record of tuberculosis, and of thalassemia, a mutation that makes people less susceptible to malaria, and may reflect their adaptation to the malaria-infested coastal environment (Hershkovitz et al. 1991). Malaria is an infectious disease typically associated with mosquito-infested marshes and wetlands in coastal regions and continues to be a major health concern at the present day (Learmouth 1977; World Health Organization 2017). Malaria may also be tied to the increased population density associated with sedentary agriculture (Smith and Horwitz 2007; Eshed et al. 2010) and to the intensified construction of water wells and pools of standing water within, or close to, the settlements.

23.7.4 Organic Preservation

The wide range of organic remains preserved on these sites is perhaps their most distinctive feature, and the one feature, more than any other, that demonstrates the unique conditions of preservation in underwater settlements. Foremost are the botanical remains—wood, plant fibres, twigs, reeds and seeds. These demonstrate the use of wood as a building material and in the manufacture of artefacts such as bowls and basketry. The recovery of seeds and other plant materials extends the range of known edible plants foods, including evidence for the production of olive oil and collection of plant material for animal fodder.

These preservation conditions also apply to the animal bones, and although these can be found in terrestrial sites, underwater preservation has ensured their recovery in abundance and for the most part in excellent condition with minimal evidence of post-depositional abrasion facilitating extensive documentation of modifications including butchery and tool marks. These assemblages demonstrate the diversity of animals exploited, mostly the farmyard domestic animals but with a small element of hunting of wild animals such as fallow deer and gazelle, and a significant representation of fish at Atlit-Yam and perhaps also at the later PN sites although fish bones may be under-represented in the latter because of the recovery methods used. The condition of the bones and plants has also facilitated biometric and morphological studies, and these, together with the time span represented by the different sites and deposits, have demonstrated a transition from wild to domestic forms. The animals represented by bones from the surface collections at Atlit-Yam, which are the earliest deposits, resemble their wild progenitors, whereas the bones from the Atlit-Yam wells, which represents the last phase of occupation at this site, and from the PN sites, are all fully domestic forms.

23.7.5 The Mediterranean Fishing Village and Diet

One of the most interesting revelations to emerge from these investigations is the demonstration of a hitherto completely unsuspected and unrecorded maritime component to the earliest development of animal and plant domestication in one of the world's major centres of agricultural origins. The Levantine and south-eastern Turkish seashores are the closest coastal environments to the inland core areas of animal and plant domestication. At the end of the tenth millennium BP, a mixed mode of subsistence that included crop cultivation, animal husbandry and marine fishing evolved among inhabitants on the Levantine coast, as evidenced by the material recovered from Atlit-Yam (Galili et al. 2002, 2004). Hints of a similar pattern are present at Ashkelon on the southern Israeli coast (Perrot and Gopher 1996; Garfinkel and Dag 2008) and at Ras Shamra on the Syrian littoral (Van Zeist and Bakker-Heeres 1984; Helmer 1989). The presence of such a mixed mode of subsistence at Atlit-Yam fits many of the criteria outlined by Butzer (1996) as characteristics of a typical Mediterranean diet. During the eighth millennium cal BP, olive oil extraction was added to the economy of the PN sites on the Carmel coast (Galili et al. 1997). The subsistence of these PN settlements was characterized by increased reliance on farming and animal husbandry, reduction in the exploitation of marine resources and hunting and intensive use of secondary animal products (milk products, wool fibres). Later still, during the sixth millennium cal BP, more cultivated plants were introduced, such as the domesticated grape, which enabled the production of wine in the Levant (Zohary and Hopf 2000). By 5000 cal BP, what is today commonly termed 'the typical Mediterranean diet' as defined by Butzer (1996) was fully developed.

Examining the seasonality of the different subsistence activities in Atlit-Yam in different months over the year (sowing, harvesting, animal husbandry, gathering and fishing) demonstrates that it was possible to efficiently combine land-based food procurement activities with the exploitation of marine resources, with minimum overlap of the essential activities. Together with the invention of artificial permanent sources of water (water wells), these diverse year-round food procurement activities made seasonal mobility unnecessary and enabled a sedentary year-round occupation on the coast (Galili et al. 2004).

The development of the Mediterranean fishing village during the transition to a fully fledged agricultural economy is of great importance and raises a number of further questions: the significance of marine and coastal resources in helping to support year-round sedentism and the development of sedentary agriculture; the question of how widespread this phenomenon may have been on other coastlines adjacent to other centres of plant and animal domestication in the Levant and Anatolia; and the role of such coastal economies in promoting the dispersal of agricultural economies further westwards around the Mediterranean Basin (e.g. Galili et al. 2002).

23.7.6 Preservation Conditions, Exploration Methods and Future Challenges

Remaining questions concern the reasons for why so much of the evidence of these settlements has survived the potentially destructive impact of wave action and marine currents, why such a concentration of evidence has been preserved and discovered on this one short stretch of coastline in northern Israel and what the prospects are for the discovery of similar evidence elsewhere or at greater depth.

The factors that have contributed to the preservation and discovery of so much evidence in northern Israel can be summarised as follows:

(1) Many of the features were dug below the surface when the site was in use—human burials, well shafts, storage pits, dwelling foundations—ensuring protection from the worst destructive effects of waves and marine currents as the sea rose over the abandoned settlements.
(2) Accumulations of sediment fill in some of these sub-surface features, such as well shafts, during the lifetime of the settlement, ensured rapid burial and additional protection of discarded artefacts and organic remains.
(3) Rapid accumulation of protective sand covered many sites before they were inundated by sea-level rise and protected them before material could be washed away or destroyed.
(4) Use of stones in the building of structures and the lining of wells and graves provided added protection from, or resistance to, water damage.
(5) Further accumulation of thick deposits of marine sand, once the sites had been inundated, further sealed and protected them until exposed to discovery by partial removal of this sand cover in recent times.

This last factor probably accounts for the particular concentration of sites on the Carmel coastline, as emphasised at the beginning of this chapter, with a sand cover that was neither too thin to act as a protective cover nor too thick to be easily removed by storm action to expose the underlying land surface. Elsewhere, conditions are different. In Haifa Bay, for example, there are extensive palaeolandscapes that are covered with layers of sand up to 20 m thick. Preserved settlements that have never been exposed may exist below this sand cover. Conversely, on a rocky land surface, such as the 'Carmel nose' plate and off the Galilee coast, sites are likely to have been exposed to marine erosion and destructive post-depositional processes, and what remains is likely to be harder to identify because of marine fouling and biogenic rock growth. In central and southern Israel, the kurkar rock is poorly consolidated and fragile, resulting in considerable erosion, coastal retreat and the creation of coastal escarpments and cliffs tens of metres high (Galili and Zviely 2018). This means that any coastal Neolithic or earlier settlements that were present in these regions would have been vulnerable to destruction during sea-level rise.

Other locally variable factors are the relative attractiveness of the peri-coastal environment for permanent settlements. The presence of palaeosols between the kurkar ridges and the accessible aquifer of high groundwater close to the shoreline may have been a particular attraction on the Carmel coast, especially for societies dependent in part on agriculture.

Whether similar archaeological evidence has survived on earlier palaeocoastlines at deeper levels is unclear. It is not only the logistics of investigating the sea bed that escalate with increasing depth. As discussed in Galili et al.

(2017a, c) and Galili (2017), the length of time that palaeocoastlines were exposed as dry land, before being inundated again by subsequent sea-level rise, decreases with increasing depth, reducing the length of time available for material to accumulate in one place to create a visible archaeological site. Also, the cultural material deposited as one goes farther back in time is likely to consist of less easily detectable traces of human activity, such as scatters of stone tools associated with temporary hunting camps. In addition, conditions for preservation or visibility are likely to be poorer because of more prolonged exposure to marine erosion on the sea bed.

Experience on the Carmel coast suggests that the most productive approach to site discovery is to concentrate on water depths of 1–15 m, where the thickness of the sand cover provides the optimum compromise between protection and exposure of the underlying land surface, and to follow up chance exposures in the most promising areas with systematic underwater survey and excavation. Sometimes the fringes of such sites can be identified on the shore edge, and some of the underwater settlements, notably Neve-Yam and Tel Hreiz, were first discovered in this way (Wreschner 1977a, b, 1983; Ronen and Olami 1978; Olami 1984). Survey and excavation have therefore been concentrated in the first instance along the shoreline and in shallow water, with regular monitoring for erosion of sand cover, especially after storms. This Israeli model for discovering and studying submerged prehistoric settlements (Galili et al. 2019a) has proved very successful in the past, and it is likely that more remains will be discovered along the coastline of Israel. In terms of future management, priority should be given to searching for exposed and disturbed sites in these shallow areas, not only because of the likelihood of site preservation and discovery there but because these are the areas most vulnerable to natural and human-related destructive activities and most in need of protection or salvage work.

As for the discovery of settlements of similar age on other coastlines further afield, that will depend on whether similar conditions of site protection and exposure are present, and of course on regular monitoring of the offshore environment and the development of underwater methods of investigation similar to the practices that have been developed in Israel over the past 40 years.

23.8 Conclusion

The underwater Neolithic settlements of northern Israel provide a striking example of the sorts of evidence that can be preserved underwater and the ways in which that evidence can reveal quite different and unexpected information about coastal adaptations and early developments in social organisation, subsistence economy and technology. The Carmel coast sites have produced some of the earliest evidence in the world for the digging of water wells, the production of olive oil, technology based on the use of wood in building construction and artefact production, sedentary settlements based on a combination of fishing and early developments in plant and animal domestication and details of burial practices including early evidence for the organisation of space within settlements to create separate graveyards. Moreover, most of this evidence owes its survival to the conditions of preservation and discovery afforded by submergence underwater and protection by marine sediments and to the existence of settlements on, or close to, an ancient shoreline at a time when sea levels were lower than present. Little of this evidence would have any chance of being preserved in terrestrial deposits on the present-day land surface or in the hinterland. However, the evidence also demonstrates that the attractions and benefits of living in a shoreline location, notably a more diverse and stable food supply capable of supporting sedentary settlements, came at a price. That price was increased exposure to malaria, and the threat of inexorable sea-level rise, which altered the ecological conditions of the marine environment, caused progressive salinization of the water supply and ultimately forced the abandonment of the settlements.

23.9 Management of the Underwater Cultural Heritage

The Israel Antiquities Authority is in charge of the protection and management of Israeli cultural heritage, including underwater antiquities.

Acknowledgments We wish to thank the EU through COST Action TD0902 SPLASHCOS, the National Geographic Research Foundation, CARE Archaeological Foundation, MAFCAF Foundation, the Dan David Foundation, The Shelby White and Leon Levy Program for Archaeological Publications and Sandy and Joseph Lepelstat for their financial support of the research described here. The Israel Antiquities Authority, the Israel Prehistoric Society and Haifa University are acknowledged for their financial and administrative support of the underwater excavations and H. Rosenstein and Fantasea for providing photographic equipment. We also wish to express our appreciation to the divers and archaeologists who participated in the underwater excavations, to the many researchers who subsequently contributed to the study of the finds, to S. Ben-Yehuda and B. Galili for producing the drawings, to J. Galili and I. Grinberg for taking the underwater photographs, to M. Rasovsky (treatment of organic material) and to D. Segal, I. Carmi and E. Boaretto of the Weizmann Institute (radiocarbon dating). Special thanks are due to Geoff Bailey for discussion of the factors contributing to site preservation and discovery and for his useful comments and comprehensive editing of the manuscript.

References

Banning EB (1995) Herders or homesteaders? A Neolithic farm in Wadi Ziqlab, Jordan. Bibl Archaeol 58:2–13

Bar-Yosef O, Garfinkel Y (2008) The prehistory of Israel. Human cultures before writing. (In Hebrew)

Bar-Yosef Mayer DE, Vandermeersch B, Bar-Yosef O (2009) Shells and ochre in Middle Paleolithic Qafzeh cave, Israel: indications for modern behavior. J Human Evol 56:307–314

Barkai R (2011) The evolution of Neolithic and Chalcolithic woodworking tools and the intensification of human production: Axes, adzes, and chisels from the southern Levant. In: Davis V, Edmonds M (eds) Stone axe studies III. Oxbow Books, Oxford, pp 39–54

Barkai R, Galili E (2004) The PPNC bifacial tool industry from the submerged site of Atlit-Yam, Israel. Eurasian Prehist 1:139–162

Barkai R, Gopher A (2012) Flint assemblages from Nahal Zehora II: Techno-typological changes during the PN. In: Gopher A (ed) Village communities of the Pottery Neolithic Period in the Menashe Hills, Israel. Archaeological investigations at the sites of Nahal Zehora, vol II. Emery and Claire Yass Publications in Archaeology, Institute of Archaeology, Tel Aviv University, Tel Aviv, pp 757–869

Benjamin J, Rovere A, Fontana A, Furlani S, Vacchi M, Inglis RH, Galili E, Antonioli F, Sivan D, Miko S, Mourtzas N, Felia I, Meredith-Williams M, Goodman-Tchernov B, Kolaiti E, Anzidei M, Gehrels R (2017) Late quaternary sea-level changes and early human societies in the central and eastern Mediterranean Basin: an interdisciplinary review. Quater Int 449:29–57

Bronk Ramsey C (2017) Methods for summarizing radiocarbon datasets. Radiocarbon 59(2):1809–1833

Butzer KW (1996) Ecology in the long view: settlement histories, agrosystemic strategies, and ecological performance. J Field Archaeol 23:141–150

Donoghue HD, Hershkovitz I, Minnikin DE, Besra GS, Lee OYC, Galili E, Greenblatt CL, Lemma E, Spigelman M, Kahila Bar-Gal G (2009) Biomolecular archaeology of ancient tuberculosis: Response to "Deficiencies and challenges in the study of ancient tuberculosis DNA" by Wilbur et al. (2009). J Archaeol Sci 36:2792–2804

Dunand M (1973) Fouilles de Byblos V: L'architecture, les tombes, le matériel domestique: des origines néolithiques à l'avènement urbain. Paul Guethner, Paris

Enzel Y, Bar-Yosef O (2017) Quaternary of the Levant. Cambridge University Press, Cambridge

Eshed V, Galili E (2011) Palaeodemography of southern Levantine Pre-Pottery Neolithic populations: regional and temporal perspectives. In: Pinhasi R, Stock J (eds) Human bioarchaeology of the transition to agriculture. Wiley, Chichester, pp 403–428

Eshed V, Gopher A, Gage TB, Hershkovitz I (2004) Has the transition to agriculture reshaped the demographic structure of prehistoric populations? New evidence from the Levant. Am J Phys Anthropol 124:315–329

Eshed V, Gopher A, Pinhasi R, Hershkovitz I (2010) Paleopathology and the origin of agriculture in the Levant. Am J Phys Anthropol 143:121–133

Galili E (1981) Underwater erosion processes, unpublished field study report submitted to M Inbar, Department of Geography, The University of Haifa, 69 pp. (In Hebrew)

Galili E (1985) Clay exposures and archaeological finds on the sea bottom, between Haifa and Atlit. Unpublished MA Thesis, Dept. of Maritime Civilizations, University of Haifa. (In Hebrew)

Galili E (2004) Submerged settlements of the ninth to seventh millennia BP off the Carmel Coast. Unpublished PhD thesis, University of Tel Aviv. (In Hebrew, English abstract)

Galili E (2017) Book review: prehistoric archaeology on the continental shelf. A global review. J Island Coastal Archaeol 12(1):147–149. https://doi.org/10.1080/15564894.2016.1225323

Galili E, Nir Y (1993) The submerged Pre-Pottery Neolithic water well of Atlit-Yam, Northern Israel, and

its palaeoenvironmental implications. The Holocene 3:265–270
Galili E, Rosen B (2007) Artifacts made of plant material, Kfar-Samir, submerged Neolithic site off the south coast of Haifa. Hadashot Arkheologiyot 119
Galili E, Rosen B (2011a) Submerged Neolithic settlements off the Mediterranean coast of Israel. In: Benjamin J, Bonsall C, Pickard C, Fischer A (eds) Submerged prehistory. Oxbow, Oxford, pp 272–286
Galili E, Rosen B (2011b) Submerged Neolithic settlements off the Mediterranean Carmel coast of Israel and water mining in the Southern Levant. Neo-Lithics 2(10):47–52
Galili E, Rosen B (2013) Atlit Yam- preliminary report. Hadashot Arkheologiyot 125
Galili E, Schick T (1990) Basketry and a wooden bowl from the Pottery Neolithic submerged site of Kefar Samir. Mitekufat Haeven. J Israel Prehist Soc 23:65–68 (in Hebrew), 142–151 (in English)
Galili E, Sharvit J (1994–5) Evidence of olive oil production from the submerged site at Kfar Samir, Israel. Mitekufat Haeven. J Israel Prehist Soc 26:122–133
Galili E, Weinstein-Evron M (1985) Prehistory and paleoenvironments of submerged sites along the Carmel coast of Israel. Paléorient 11:37–52
Galili E, Zviely D (2018) Geo-archaeological markers reveal magnitude and rates of Israeli coastal cliff erosion and retreat. J Coast Conserv. https://doi.org/10.1007/s11852-018-0644-7
Galili E, Weinstein-Evron M, Ronen A (1988) Holocene sea-level changes based on submerged archaeological sites off the northern Carmel coast in Israel. Quater Res 29:36–42
Galili E, Weinstein-Evron M, Zohary D (1989) Appearance of olives in submerged Neolithic sites along the Carmel coast. Mitekufat Haeven. J Israel Prehist Soc 22:95–97
Galili E, Weinstein-Evron M, Hershkovitz I, Gopher A, Kislev M, Lernau O, Horwitz LK, Lernau H (1993) Atlit-Yam: a prehistoric site on the sea floor off the Israeli coast. J Field Archaeol 20:133–156
Galili E, Stanley DJ, Sharvit J, Weinstein-Evron M (1997) Evidence for earliest olive-oil production in submerged settlements off the Carmel coast, Israel. J Archaeol Sci 24:1141–1150
Galili E, Rosen B, Gopher A, Horwitz LK (2002) The emergence and dispersion of the Eastern Mediterranean fishing village: evidence from submerged Neolithic settlements off the Carmel coast, Israel. J Medit Archaeol 15:167–198
Galili E, Lernau O, Zohar I (2004) Fishing and marine adaptations at Atlit-Yam, a submerged Neolithic village off the Carmel coast, Israel. Atiqot 48:1–34
Galili E, Zviely D, Weinstein-Evron M (2005a) Holocene sea-level changes and landscape evolution on the Northern Carmel coast (Israel). Mediterranée 1(2):1–8
Galili E, Eshed V, Gopher A, Hershkovitz I (2005b) Burial practices at the submerged PPNC site of Atlit-Yam, northern Coast of Israel: what do they tell us about the final phase of the Pre-Pottery Neolithic Culture? Bull Am School Orient Res 339:1–19
Galili E Rosen B, Boaretto E (2007) Haifa, Kafr Samir. Hadashot Arkheologiyot 119. http://www.hadashot-esi.org.il/Report_Detail_Eng.aspx?id=552&mag_id=112
Galili E, Horwitz LK, Hershkovitz I, Eshed V, Salamon A, Zviely D, Weinstein-Evron M, Greenfield H (2008) Comment on "Holocene tsunamis from Mount Etna and the fate of Israeli Neolithic communities" by Maria Teresa Pareschi, Enzo Boschi, and Massimiliano Favalli. Geophy Res Lett 35:L08311. https://doi.org/10.1029/2008GL033445
Galili E, Eshed V, Rosen B, Kislev ME, Simchoni O, Hershkovitz I, Gopher A (2009) Evidence for a separate burial ground at the submerged Pottery Neolithic site of Neve-Yam, Israel. Paléorient 35(1):31–46
Galili E, Zemer A, Rosen B (2013) Ancient fishing gear and associated artifacts from underwater explorations in Israel – a comparative study. Archaeofauna 22:145–166
Galili E, Cvikel D, Benjamin J, McCarthy J (2016a) Haifa, Kefar Samir (Dado beach). HA-Excavations and Surveys in Israel 128
Galili E, Rosen B, Orrelle E, Yaroshevich A, Horwitz LK (2016b) Symbolic artefacts from the submerged Neolithic village of Neve-Yam and contemporaneous Israeli sites as *fossil directeur* for the Wadi Rabah culture. Israel Explor J 66(2):129–150
Galili E Eshed V Rosen B, Horwitz LK (2016c) Neve-Yam, Temanun Island. Hadashot Arkheologiyot 128. http://www.hadashot-esi.org.il/Report_Detail_Eng.aspx?id=25124
Galili E, Nir Y, Vachtman D, Mart Y (2017a) Physical characteristics of the continental shelves of the East Mediterranean Basin, submerged settlements and landscapes – actual finds and potential discoveries. In: Flemming NC, Harff J, Moura D, Burgess A, Bailey GN (eds) Submerged landscapes of the European continental shelf: quaternary paleoenvironments. Wiley, Chichester, pp 377–403
Galili E, Benjamin J, Hershkovitz I, Weinstein-Evron M, Zohar I, Eshed V, Cvikel D, Melamed J, Kahanov Y, Bergeron J, Ruggles C, Ronen A, Horwitz LK (2017b) Atlit-Yam: A unique 9000 year old village submerged off the Carmel coast, Israel – the SPLASHCOS field school (2011). In: Bailey GN, Harff J, Sakellariou D (eds) Under the sea: archaeology and palaeolandscapes of the continental shelf. Springer, Cham, pp 85–102
Galili E, Horwitz LK, Rosen B, Eshed V (2017c) Submerged pottery neolithic settlements off the Mediterranean coast of Israel: subsistence, material culture and the development of separate burial grounds. In: Bailey GN, Harff J, Sakellariou D (eds) Under the sea: archaeology and palaeolandscapes of the continental shelf. Springer, Cham, pp 105–130
Galili E, Benjamin J, Hershkovitz I, Weinstein-Evron M, Zohar I, Eshed V, Cvikel D, Melamed J, Kahanov Y, Bergeron J, Ruggles C, Ronen A, Horwitz LK (2017d) Atlit-Yam: a unique 9000-year-old prehistoric vil-

lage submerged off the Carmel coast, Israel –The SPLASHCOS Field School 2011. In: Bailey GN, Harff J, Sakellariou D (eds) Under the sea: archaeology and palaeolandscapes of the continental shelf. Springer, Cham, pp 85–102

Galili E, Ronen A, Mienis HK, Horwitz LK (2017e) Beach deposits containing Middle Paleolithic archaeological remains from northern Israel. Quater Int 464(Part A):43–57

Galili E, Cvikel D, Benjamin J, Langgut D, McCarthy J, Cavanagh M, Sapir Y, Weinstein-Evron M, Chaim S, Rosen B, Kolska Horwitz L (2018) The archaeology and paleoenvironment of the submerged Pottery Neolithic settlement of Kfar Samir. Paleorient 44(2):113–132

Galili E, Kolska Horwitz L, Rosen B (2019a) The Israeli model for the detection, excavation and research of submerged prehistory. TINA Maritime Archaeological Periodical, Periodical Publications of the Turkish Archaeological Foundations, pp 31–69

Galili E, Benjamin J, Eshed V, Rosen B, McCarthy J, Kolska Horwitz L, Biehl PF (2019b) A submerged 7000-year-old village and seawall demonstrate earliest known coastal defence against sea-level rise. PLOS ONE 14(12):e0222560

Garrod DAE, Bate DMA (1937) The stone age of Mount Carmel: excavations at the Wady El-Mughara, vol I. Clarendon Press, Oxford

Garfinkel Y (1993) The Yarmukian culture in Israel. Paléorient 19(1):115–134

Garfinkel Y (1999) Neolithic and Chalcolithic pottery of the southern Levant. Qedem 39. Institute of Archaeology, Hebrew University, Jerusalem

Garfinkel Y, Dag D (eds) (2008) Neolithic Ashkelon. Qedem 47. Institute of Archaeology, Hebrew University, Jerusalem

Gopher A (1995) Early pottery-bearing groups in Israel – the Pottery Neolithic period. In: Levy TE (ed) The archaeology of society in the Holy Land. Facts on File, New York, pp 205–225

Gopher A (2012) The pottery neolithic in the Southern Levant – a second Neolithic revolution. In: Gopher A (ed) Village communities of the Pottery Neolithic Period in the Menashe Hills, Israel. Archaeological investigations at the sites of Nahal Zehora, vol II. Emery and Claire Yass Publications in Archaeology, Institute of Archaeology, Tel Aviv University, Tel Aviv, pp 1525–1575

Gopher A, Gophna R (1993) Cultures of the eighth and seventh millennia BP in the southern Levant: a review for the 1990's. J World Prehist 7:297–353

Greenfield H, Galili E, Horwitz LK (2006) The butchered bones from Newe Yam. Mitekufat Haeven. J Israel Prehist Soc 36:173–200

Greenfield H, Cheney T, Galili E (2013) A taphonomic and technological analysis of the butchered animal bone remains from Atlit Yam, a submerged PPNC site off the coast of Israel. Poster presented at 11th Meeting of Archaeozoology of South West Asia (ASWA), June 23rd–28th, Haifa

Hartmann-Shenkman A, Kislev ME, Galili E, Weiss EA (2015) Invading a new niche: obligatory weeds at Neolithic Atlit-Yam, Israel. Veg Hist Archaeobot 24(1):9–18. https://doi.org/10.1007/s00334-014-0498-3

Helmer D (1989) Le dévelopment de la domestication au Proche-Orient de 9500 à 7500 BP: les nouvelles données d'El Kowm et de Ras Shamra. Paléorient 15:111–121

Hershkovitz I, Galili E (1990) 8000-year-old human remains on the sea floor near Atlit, Israel. Human Evol 5:319–358

Hershkovitz I, Bing B, Speirs M, Galili E, Kislev M, Edelson G, Hershkovitz A (1991) Possible congenital anaemia in prehistoric coastal inhabitants of Israel. Am J Phys Anthropol 85:7–13

Hershkovitz I, Donoghue H, Minnikin D, Besra G, Lee O, Gernaey AM, Galili E, Eshed V, Greenblatt C, Lemma E, Bar-Gal GK, Spiegelman M (2008) Detection and molecular characteristics of 9000-year-old *Mycobacterium tuberculosis* from a Neolithic settlement in the Eastern Mediterranean. PLoS ONE 3(10):e3426. https://doi.org/10.1371/journal.pone.0003426

Hershkovitz I, Weber GW, Quam R, Duval M, Grün R et al (2018) The earliest modern humans outside Africa. Science 359(6374):456–459. https://doi.org/10.1126/science.aap8369

Hesse B, Rookis D (2008) The mammal bones. In: Garfinkel Y, Dag D (eds) Neolithic Ashkelon. Qedem 47, pp 251–254

Horwitz LK, Ducos P (2005) Counting cattle: trends in Neolithic Bos frequencies from the southern Levant. Revue de Paleobiologie 10:209–224

Horwitz LK, Tchernov E (1987) Animal exploitation at the submerged PPNB site of Atlit. Mitekufat Haeven. J Israel Prehist Society 20:72–78

Horwitz LK, Galili E, Sharvit J, Lernau O (2002) Fauna from five submerged Pottery Neolithic sites off the Carmel coast. Mitekufat Haeven. J Israel Prehist Soc 32:147–174

Horwitz K, Lernau O, Galili E (2006) Fauna from the Pottery Neolithic site of Neve-Yam. Mitekufat Haeven. J Israel Prehist Soc 36:139–171

Kafri U, Arad A (1978) Paleohydrology and migration of the groundwater divide in regions of tectonic instability in Israel. Geol Soc Am Bull 89:1723–1732

Kislev ME, Hartmann A, Galili E (2004) Archaeobotanical and archaeoentomological evidence from a well at Atlit-Yam indicates colder, more humid climate on the Israeli coast during the PPNC, farming, and social complexity in the Pre-Pottery Neolithic of the southern Levant: a review and synthesis. J World Prehist 16:361–440

Learmouth ATA (1977) Malaria. In: Howe GM (ed) A world geography of human diseases. Academic Press, London, pp 61–108

Lernau O (2008) The fish bones. In: Garfinkel Y, Dag D (eds) Neolithic Ashkelon. Qedem 47, pp 263–268

Mauz B, Hijma MP, Amorosi A, Porat N, Galili E, Bloomindale J (2013) Aeolian beach ridges and their significance for climate and sea level: Concept and insight from the Levant coast (East Mediterranean). Earth Sci Rev 121:31–54
Milevski Y, Getzov N, Galili E, Horwitz LK (2016) Some iconographic motifs from the 6th-5th millennia BC in the Levant and Mesopotamia: clues for cultural connections. Paleorient 42:135–149
Nir Y, Eldar-Nir I (1986) Eustatic sea level changes and their possible influence on the Nile littoral-cell beaches. In: Raukas A (ed) Evolution and dynamics of seacoasts in conditions of relative sea level oscillation: excursion guides and abstracts, International Symposium, Aug. 20-26, 1986. International Symposium. Academy of Sciences of the Estonian SSR, Institute of Geology, Tallinn, p 63
Nir Y, Eldar-Nir I (1987) Ancient wells and their geoarchaeological significance in detecting tectonics of the Israel Mediterranean coastline region. Geology 15:3–6
Nir Y, Eldar-Nir I (1988) Construction techniques and building materials used in ancient water wells along the coastal plain of Israel. In: Marinos PG, Koukis GC (eds) Engineering geology of ancient works, monuments and historical sites. Balkema, Rotterdam, pp 1765–1771
Olami Y (1984) Prehistoric carmel. Israel Exploration Society, Jerusalem
Pareschi MT, Boschi E, Favalli M (2006) Lost tsunami. Geophys Res Lett 33:L22608. https://doi.org/10.1029/2006GL027790
Pareschi MT, Boschi E, Favalli M (2007) Holocene tsunamis from Mount Etna and the fate of Israeli Neolithic communities. Geophys Res Lett 34:L16317
Perrot I, Gopher A (1996) A Late Neolithic site near Ashkelon. Israel Explor J 46:145–166
Porat N, Jain M, Ronen A, Horwitz LK (2018) A contribution to late Middle Paleolithic chronology of the Levant: New luminescence ages for the Atlit Railway Bridge site, coastal plain, Israel. Quater Int 464(Part A):32–42
Reimer PJ, Bard E, Bayliss A, Beck JW et al (2013) IntCal13 and marine13 radiocarbon age calibration curves 0-50,000 years cal BP. Radiocarbon 55(4):1869–1887. https://doi.org/10.2458/azu_js_rc.55.16947
Rollefson GO (1990) Neolithic chipped stone technology at 'Ain Gazal, Jordan: the status of the PPNC phase. Paléorient 16:119–124
Rollefson G, Simmons A, Kafafi Z (1992) Neolithic cultures at Ain Ghazal, Jordan. J Field Archaeol 19:443–470
Ron H, Porat N, Ronen A, Tchernov E, Horwitz LK (2003) Magnetostratigraphy of the Evron member – implications for the age of the Middle Acheulian site of Evron quarry. J Human Evol 44:633–639
Ronen A (1977) Mousterian sites in red loam in the Carmel coastal plain. Eretz Israel (Moshe Stekelis) 13:183–190
Ronen A, Chernikov I (2010) The Mousterian of the red loam on the Carmel coast (Israel). Acta Universitatis Wratislaviensis No 3207. Studia Archeologiczne 41:1–22
Ronen A, Kaufman D (1976) Epi-Palaeolithic sites near Nahal Hadera, Central Coastal Plain of Israel. Tel Aviv 3(1):16–30
Ronen A, Olami Y (1978) Atlit map. Archaeological Survey of Israel, Jerusalem
Ronen A, Kaufman D, Gophna R, Bakler N, Smith P, Amiel A (1975) The Epipalaeolithic site of Hefziba, central coastal plain of Israel. Quärtar 26:54–72
Ronen A, Neber A, Mienis H, Horwitz LK, Frumkin A, Boenigk W, Galili E (2008a) A Mousterian occupation on an OIS 5e shore near the Mount Carmel Caves, Israel. In: Sulgostowska Z, Tomaszewski AJ (eds) Man, millennia, environment. Studies in honour of Romuald Schild. Institute of Archaeology and Ethnology, Polish Academy of Sciences, Warszawa, pp 197–205
Ronen A, Horwitz LK, Mienis H, Galili E, Zviely D (2008b) Mousterian sites on the Carmel Coast. Kadmoniot 134:75–81 (In Hebrew)
Segal D, Carmi I (1996) Reḥovot Radiocarbon Date List V.'Atiqot, pp 79–106
Shea JJ (2003) The middle Paleolithic of the East Mediterranean Levant. J World Prehist 17:313–394
Simmons AH (2007) The Neolithic revolution in the near east: transforming the human landscape. University of Arizona Press, Tucson, AZ
Sivan D, Widowinski S, Lambeck K, Galili E, Raban A (1999) Changes in sea-level along the Israeli coast during the Holocene – combination of eustatic, isostatic and tectonic factors. Archaeol Nat Sci, Bulletin no. 7, April 1999, pp 80–84. (In Hebrew)
Smith P, Horwitz LK (2007) Ancestors and inheritors: a bio-cultural perspective of the transition to agro-pastoralism in the Southern Levant. In: Cohen MN, Crane-Kramer GMM (eds) Ancient health. Skeletal indicators of agricultural and economic intensification. University Press of Florida, Gainesville, pp 207–222
Sneh Y, Klein M (1984) Holocene sea level changes of the coast of Dor, southeast Mediterranean. Science 226:831–832
Stekelis M (1972) The Yarmukian Culture of the Neolithic Period. Magnes Press, Jerusalem
Stuiver M, Reimer PJ (1993) Extended C-14 data-base and revised Calib 3.0 C-14 age calibration program. Radiocarbon 35(1):215–230
Van Zeist W, Bakker-Heeres JAH (1984) Archaeobotanical studies in the Levant 2: Neolithic and Halaf levels at Ras Shamra. Palaeohistoria 26:151–170
Weinstein-Evron M (1994) Biases in archaeological pollen assemblages: case studies from Israel. AASP Contrib Ser 29:193–205
Weinstein-Evron M (2015) The case of Mount Carmel: the Levant and human evolution, future research in the framework of World Heritage. In: Sanz N (ed) Human

origin sites and the World Heritage Convention in Eurasia, HEADS 4, vol 1. UNESCO, Paris, pp 72–92
World Health Organization (WHO) (2017) World Malaria Report 2017. Geneva, Switzerland. http://www.who.int/malaria/publications/world-malaria-report-2017/en/
Wreschner EE (1977a) Neve-Yam, a submerged Late Neolithic settlement near Mount Carmel. Eretz Israel 13:260–271
Wreschner EE (1977b) Sea level changes and settlement location in the coastal plain of Israel during the Holocene. Eretz Israel 13:277–282
Wreschner EE (1983) The submerged Neolithic village 'Neve-Yam' on the Israeli Mediterranean coast. In: Masters PM, Flemming NC (eds) Quaternary coastlines and marine archaeology. Academic Press, London, pp 325–333
Yerkes RW, Galili E, Barkai R (2014) Activities at final Pre-Pottery Neolithic (PPNC) fishing village revealed through microwear analysis of bifacial flint tools from the submerged Atlit-Yam site, Israel. J Archaeol Sci 4:120–128
Zeder MA (2011) The origins of agriculture in the Near East: New data, new ideas. Curr Anthropol 52(S4):S211–S235
Zohar I, Dayan T, Spanier E, Galili E, Lernau O (1994) Exploitation of gray triggerfish (*Balistes carolinensis*) by the prehistoric inhabitants of Atlit-Yam, Israel: a preliminary report. In: Van Neer W (ed) Fish exploitation in the past. Proceedings of the 7th Meeting of the ICAZ Fish Remains Working Group. Annalen Zoologische Wetenschapen; vol. 274. Koninklijk Museum voor Midden-Africa, Tervuren, België, pp 231–238
Zohar I, Dayan T, Galili E, Spanier E (2001) Fish processing during the early Holocene: a taphonomic study. J Archaeol Sci 28:1041–1053
Zohary D, Hopf M (2000) Domestication of plants in the Old World, 3rd edn. Oxford University Press, Oxford
Zviely D, Kit E, Klein M (2007) Longshore sand transport estimates along the Mediterranean coast of Israel in the Holocene. Mar Geol 238:61–73

Part IV

The Legal, Historical and Industrial Context of Underwater Heritage

The Legal, Historical and Industrial Context of Underwater Heritage: Introduction

24

Hans Peeters, Geoff Bailey, and Nena Galanidou

Abstract
Interactions between commercial and industrial exploitation of the seabed and archaeological and scientific investigation have been at the heart of developments in the understanding of Europe's submerged landscapes and prehistory since at least the early twentieth century. This introduction considers some of the ways in which that relationship has evolved since that time, including the adoption of international laws under the aegis of United Nations Conventions, the development of close relationships between Dutch fishermen operating beam-trawl fishing nets in the North Sea and a network of private collectors specialising in Pleistocene fossils and artefacts, the imposition of European Union regulations on offshore industrial projects to include monitoring of underwater archaeology and palaeoenvironments and most recently the incorporation of seabed mapping and underwater cultural heritage in the European Union's 2020 Blue Growth agenda. These developments have played an important role in the growth of knowledge about the underwater cultural heritage notwithstanding the potentially destructive effects of offshore industrial activity. The impact of economic growth and industrial exploitation in the coastal zone, coupled with sea-level rise, is likely to intensify the threats to the underwater cultural heritage in the coming decades, posing new challenges as well as opportunities in the further development of relationships between industrial operators, government agencies and scientific and archaeological researchers.

Keywords
Beam-trawl fishing · Blue Growth · Environmental impact assessment · International law · Underwater cultural heritage · UNESCO convention

24.1 Introduction

One of the major objectives of the SPLASHCOS (Submerged Prehistoric Archaeology and Landscapes of the Continental Shelf) network as set out in its original Memorandum of Understanding in 2009 was to promote the field

H. Peeters (✉)
Groningen Institute of Archaeology, University of Groningen, Groningen, the Netherlands
e-mail: j.h.m.peeters@rug.nl

G. Bailey
Department of Archaeology, University of York, York, UK

College of Humanities, Arts and Social Sciences, Flinders University, Adelaide, SA, Australia
e-mail: geoff.bailey@york.ac.uk

N. Galanidou
Department of History and Archaeology, University of Crete, Rethymno, Greece
e-mail: galanidou@uoc.gr

G. Bailey et al. (eds.), *The Archaeology of Europe's Drowned Landscapes*, Coastal Research Library 35, https://doi.org/10.1007/978-3-030-37367-2_24

of submerged prehistoric archaeology and landscapes to heritage professionals, government agencies, commercial organisations, policy makers and a wider public, as well as to the wider community of professional archaeologist and scientists, and to encourage cross-border integration and collaboration in this aspect of the underwater cultural heritage. To that end, a variety of activities were organised and documents produced including guidance on the relationship between marine industry and submerged prehistoric archaeology (Satchell 2013) a major workshop held at Esbjerg in 2013 with the North Sea offshore industry: 'Offshore Industry and Archaeology: A Creative Relationship', a variety of other dissemination activities, new projects and publications (see in particular Satchell 2017; Tidbury et al. 2017; Missiaen et al. 2017), the production of a major position paper for the European Marine Board, the advisory group for European marine policy in all its many dimensions (Flemming et al. 2014), and contributions to international workshops, notably a session on Maritime Heritage and Blue Growth at the European Maritime Day held at Athens in June 2015 Full details are available on the SPLASHCOS website (https://www.splashcos.org/).

Some indication of the ways in which the underwater cultural heritage (UCH) is managed and the national and regional institutions responsible for research and management is given in the preceding chapters, most of which have sections on the management of UCH. Many chapters refer to the threat posed by the expansion of offshore industrial and commercial activity, especially commercial fisheries, drilling for oil and gas, extraction of aggregates, development of infrastructure for renewable energy such as wind farms, construction of harbours, tunnels and bridges, and laying of pipelines and cables on the seabed. In some areas, changes in the marine environment, whether resulting from climate change or pollution, are an added threat to the survival of underwater remains. These threats are of course, double-edged, because engineering works can lead to the exposure and discovery of submerged landscapes and archaeology that would otherwise have remained deeply buried and out of sight, as well as to their destruction. Some chapters provide examples of important archaeological discoveries that have resulted from cooperation with industry, notably in Denmark, Germany, Norway, the Netherlands and Britain (Bailey et al., Chap. 3, this volume; Jöns et al., Chap. 5, this volume; Glørstad et al., Chap. 6, this volume; Peeters and Amkreutz, Chap. 8, this volume; Bailey et al., Chap. 10, this volume).

Most countries discussed in this volume are members of the European Union, and much of the regulatory framework for UCH has been developed at European level with a series of EU directives on the management of UCH in relation to offshore industries. The EU is also responsible for other initiatives which take a basin-wide and continent-wide approach intended to ensure sharing of good practice across State borders and supports the development of a unified cross-border approach, for example, in the mapping of offshore geology and bathymetry, and notably in the mapping of the underwater cultural heritage through the SPLASHCOS Viewer. The need for robust and well-informed regulations for the management and protection of the UCH is ever-present and likely to grow.

The purpose of this section is to bring together a group of three chapters that focus in more detail on the legal, regulatory and historical issues that inform the relationship between research, offshore industrial activities and government agencies concerned with UCH. All three chapters are the result of presentations originally given at the Esbjerg conference of 2013 referred to above. We begin with the universally applicable framework of international legal principles that informs all governmental regulations and offshore industrial activity (Dromgoole, Chap. 25, this volume). The remaining chapters offer two national case studies, which provide insights, respectively, into the regulatory framework in the UK and the licensing regime for offshore industrial activity (Pater, Chap. 26, this volume), and the long history of mutual dependency and co-evolution of interests in the Netherlands between trawler fishermen, private collectors and the acquisition of new knowledge about the submerged landscape (Maarleveld, Chap. 27, this volume). Here, we summarise the key themes and add some additional commentary on the Blue Growth agenda of the EU, which is providing an important driver for the integration of all these interests in policies for sustainable growth and management of maritime interests on Europe's coasts.

Three themes recur throughout these discussions. The first is the explicit recognition that submerged prehistoric landscapes and archaeological remains on and beneath the seabed are an important part of the underwater cultural heritage (along with shipwrecks, airplanes and other sort of cultural materials of later periods), and this is increasingly acknowledged in the formal wording of national and international treaties, plans, policies and legislation. The second is the productive relationship that can develop between commercial and scientific interests despite the potentially destructive impact of offshore activity. The third is the degree of inter-relationship between the management and protection of the underwater cultural heritage, especially the prehistoric component, and that of the natural environment. At the same time, there are clearly unresolved ambiguities and uncertainties at the boundaries between different jurisdictions, and ongoing political sensitivities and tensions between commercial and scientific objectives.

24.2 The Legal Framework

The starting point for this discussion is the 1982 United Nations Convention on the Law of the Sea (LOSC) and the 2001 United Nations Convention on the Protection of the Underwater Cultural Heritage (UCHC). The LOSC has achieved very nearly universal ratification (by 168 States, notable exceptions in the European arena being Turkey and Israel (see http://www.un.org/depts/los/reference_files/status2018.pdf). The UCHC came into force in 2009 and has so far been ratified by 61 States including many of the coastal states in southern Europe (see http://www.unesco.org/eri/la/convention.asp?KO=13520&language=E&order=alpha). Notable exceptions relevant to this volume are Israel, Turkey and Greece in the south-east, and Ireland, the UK, the Netherlands, Denmark, Germany, Sweden and Norway in the north-west, although important principles of UCHC are incorporated in European Union (EU) regulations and the national policies of most of these States. There are also general principles of international law that are applicable.

Dromgoole (Chap. 25, this volume) sets out the issues at stake, highlights the definitions in law of terms such as the 'continental shelf, 'territorial waters', and the 'exclusive economic zone' (EEZ), clarifies what they do or do not cover in terms of UCH and draws attention to the gaps in legal coverage and how these can be addressed.

One of the principal difficulties with the LOSC is that the Articles relating to UCH (namely, Articles 149 and 303) were originally intended to regulate the intentional removal of underwater cultural heritage such as the looting of shipwrecks, rather than the possibility of inadvertent damage to UCH by industrial activities, and it is the latter which is of greatest relevance to submerged prehistoric remains. Moreover, these LOSC Articles only apply to certain areas of the sea. The distinction between 'intentional removal' of UCH and 'inadvertent damage' remains an important one in law as well as a source of ambiguity.

According to the LOSC, every State has full legal rights over what takes place within its own territorial waters, defined as the area out to a 12-mile limit[1] from coastal baselines, and over its own flag vessels when they travel beyond that limit. Article 303 also allows States to declare a contiguous zone extending out for a further 12 miles (i.e. to the 24-mile limit from coastal baselines) and imposes a general duty on States to cooperate in this zone but only with respect to the removal of UCH, not to inadvertent damage to UCH by industrial activity.

[1]Throughout this discussion, the term 'mile' refers to the nautical mile, equivalent to 1.15 miles or 1.852 km (see also Dromgoole, Chap. 25, this volume).

Beyond the 24-mile limit, the LOSC allows certain rights over the continental shelf but only with respect to natural resources. In law the continental shelf is defined as the area extending out to 200 miles from coastal baselines, regardless of the physical width of the continental shelf (or beyond 200 miles for States that have a very wide physical continental shelf) or to the mid-point of the shelf between States that share the intervening waters of a shallow sea, such as the States around the borders of the North Sea.

Article 149 of the LOSC also defines a zone called 'the Area' which lies beyond the limit of the continental shelf and is beyond the jurisdiction of any one State. Article 149 specifies that objects of archaeological and historical interest in that zone should be preserved. This provision clearly refers to an area of seabed that is too deep to be of relevance to submerged prehistory and can only apply to sunken wrecks of ships and planes or cultural objects lost overboard in transit. States can also declare the area from the 12-mile limit out to the edge of the continental shelf as an EEZ, but this only gives rights over natural resources and specifically excludes cultural materials on or beneath the seabed.

The result of this legislation is a jurisdictional 'gap' between the 24-mile outer limit of the contiguous zone and the outer limit of the continental shelf. It was filling this gap that originally motivated the creation of the 2001 UCHC. The Articles of the UCHC set out important principles, including a definition of the UCH as 'all traces of human existence having a cultural, historical or archaeological character which have been partially or totally under water, periodically or continuously, for at least 100 years' (Article 1), a definition that covers shipwrecks as well as submerged prehistoric remains, and treasure hunting as well as inadvertent damage. Other principles are the obligation to report cases of UCH, to preserve the UCH, with in situ preservation as the first option in cases of potential damage, and promotion of training, public access and cooperation between States. Many States have not yet ratified the UCHC, mainly because of political sensitivities about imposing restrictions on economically important activities on the continental shelf, and there remain 'constructive ambiguities' in some of its provisions that are open to different interpretation and the creation of loopholes for avoiding regulation.

Nevertheless, as Dromgoole points out, States that have not yet ratified the UCHC do have various legal methods at their disposal under international law and the LOSC, both direct and indirect, over activities relating to UCH in the jurisdictional gap. These include control of the activities of the flag vessels of the State in question when operating beyond territorial waters, and regulation of foreign companies that are exploiting natural resources through national and EU regulations governing the provision of licences for offshore activity.

A succession of EU directives has imposed legal requirements on EU member States by specifying the need to include consideration of UCH in environmental impact assessments that have to be carried out in advance of offshore development projects (Salter et al. 2014; Jöns et al., Chap. 5, this volume). Also to be mentioned in this context is the 1992 Valletta Convention for the Protection of the Archaeological Heritage of Europe of the Council of Europe, which includes submerged prehistory. But this Convention does not carry the same legal force as international treaties or EU directives and applies only within the 12-mile limit of territorial waters.

What the UCHC does is to impose a legal duty to make full use of existing regulatory methods. In practice, many States that have not yet ratified the UCHC apply these methods, and an example of how this works in a national context is provided by Pater (Chap. 26, this volume). Nevertheless, the ratification of the UCHC would give force to the management and protection of the UCHC in those States that have not yet ratified it.

24.3 Licensing and Regulation of Offshore Developments

Environmental impact assessments (EIAs) in advance of major offshore work and regulation and monitoring of their implementation are now enshrined in EU (European Union) legislation.

Moreover, the legislation explicitly includes underwater archaeological heritage. Pater (Chap. 26, this volume) examines how this process works in the UK (English) context to regulate and promote the relationship between commercial activity and the management of the UCH.

In the UK the term 'environment' is now widely used to include the 'historic environment', which refers to all interactions between people and places over time and to all surviving physical remains of past human activity whether visible, submerged or buried. Physical remains are referred to as 'heritage assets', and their significance is judged in relation to their heritage interest, whether archaeological, architectural, artistic or historic. Historic England (HE) is the UK government department responsible for the marine and terrestrial historic environment and carries out various functions in that capacity. In the marine sphere, it provides strategic oversight over areas of the marine environment associated with offshore industrial development and commissions specialist reports to assess the current state of the art in the investigation of submerged cultural remains in their sedimentary and palaeoenvironmental context. It provides advice to offshore developers, planning authorities responsible for granting licences for industrial development and archaeological consultants employed by both types of organisation. It can impose conditions on licences to ensure that mitigation work is carried out appropriately, that any palaeoenvironmental and archaeological data are collected to professional standards, that reporting procedures are adhered to and that technical reports are archived in a publicly accessible repository. HE also delivers or commissions training programmes to ensure continuing professional development in relation to new developments in archaeology and related disciplines. The actual funding of the environmental impact assessment that precedes the granting of a licence, and any follow-up work once the development is approved, is the responsibility of the developer.

The first step in a licence application is for the industrial company to appoint archaeological consultants to provide a desk-based EIA, which will assess the likely potential impact of the development work on known or anticipated heritage assets in light of current knowledge. This document is usually produced in consultation with HE and is accompanied by a Written Scheme of Investigation (WSI), which sets out desk-based sources of information, methodologies for the interpretation of geophysical data and the specific methods of data collection that will inform archaeological interpretation, once licence consent has been granted. These documents are reviewed by professional archaeological consultants employed by the planning authority and HE, and they will also draft the final report once work is completed.

Pater's outline of course presents the UK (English) approach. However, comparable approaches exist elsewhere, notably in the Netherlands, Denmark and Germany (Bicket et al. 2014; Salter et al. 2014). Moreover, given the similar context of offshore industrial activity in the North Sea, joint initiatives involving cross-border collaboration in developing the research base are clearly desirable. One example of such an initiative is the North Sea Prehistory Research and Management Framework (Peeters et al. 2009), which has facilitated close collaboration between researchers and heritage professionals in the UK and the Netherlands. However, sustained collaboration has proved difficult to maintain, not so much because of differences of national perspective but rather because of a fragile dependency on dedicated individuals (Salter et al. 2014).

In principle, the regulatory processes described above provide a strong framework for ensuring proper consideration of the UCH in offshore developments and the monitoring and implementation of data collection to professional standards. However, a regulatory framework can only ever be as good as the manner in which it is implemented in practice in given circumstances and must inevitably be vulnerable to commercial and central government pressures to minimise costs and delays. Counterbalancing that tendency is the publicity and public interest that can come with proper attention to archaeological and palaeoenvironmental discoveries made during commercial operations and an enhanced reputation for the companies involved for community engagement and sustainable management of seabed resources. Moreover, the cost of mitigation work is often a very small fraction of the total budget for large offshore projects and easily accommodated provided that the archaeological and palaeoenvironmental research in question is built into the work plan of the offshore development at an early stage (see Satchell 2013 for a general overview, and Peeters and Amkreutz, Chap. 8, this volume, and Moree and Sier 2015 for the Yangtze Harbour project).

24.4 The History of Fishing Technology and Its Archaeological Impact

A different perspective on the relationship between commercial and scientific interests is illustrated by the history of beam-trawl fishing in the Netherlands. Here voluntary agreements between private collectors and industrial operators, particularly trawler fishermen but also in more recent decades companies that excavate sand and gravel aggregates, have evolved over a long period outside the legal and regulatory framework discussed above. The beam trawl has been widely used by the Dutch fishing fleet in the North Sea since early in the twentieth century. Originally devised for inshore shellfishing, it was later extended to offshore trawling from the 1950s onwards. The technique is ideally suited to flat, sandy surfaces and involves dragging a solid beam along the seabed that disturbs the surface and collects everything in its path, but it is also highly damaging. Beam-trawl fishing, despite its destructive tendencies, has resulted in many important discoveries in the North Sea, especially remains of extinct fauna (Glimmerveen et al. 2004, 2006; Mol et al. 2006; Van Kolfschoten and Van Essen 2004. Peeters and Amkreutz, Chap. 8, this volume), resulting in one of the largest collections of offshore finds known anywhere and on a scale unmatched in comparison with the fishing industries of Ireland, the UK, France or Belgium. What is less well-known is the history of the relationship between Dutch fishermen and the collection of fossils and antiquities. Maarleveld (Chap. 27, this volume) examines this history, highlighting a unique conjunction in the Netherlands of regional geomorphology, fishing traditions, palaeontological finds and museum collection-policies that goes back to the nineteenth century.

The history of these relationships begins in the estuary of the Scheldt River. This has long been confined between dykes to mitigate the threat of flooding, resulting in downcutting of deep pits into the underlying Pleistocene and pre-Pleistocene geological strata and the recovery of animal fossils originating in these geological deposits. This was already attracting the attention of palaeontologists in the nineteenth century and soon resulted in the offer of payments to the fishermen of the Scheldt estuary to cast their nets into these deep pits to bring up more fossils. A formal relationship was established in the 1950s with the National Museum for Geology and Mineralogy in Leiden. This resulted in widespread publicity in cinema newsreels, the extension of 'bone fishing' to the North Sea with the introduction of more powerful boats and the beam trawl, an explosion in the number of discoveries, a community of private collectors feeding a commercial market in palaeontological fossils, and the creation of a journal, *Cranium*, with a respectable scientific reputation (Maarleveld, Chap. 27, this volume). These developments were taking place long before the creation of the UNESCO Conventions of 1982 and 2001. Many of the finds come from beyond the boundary of national territorial waters, and in any case, as Maarleveld notes, because they are palaeontological specimens, they also lie outside the jurisdiction of the 1970 UNESCO Convention on the Means of Prohibiting and Preventing the Illicit Import, Export and Transport of Ownership of Cultural Property.

Although these developments were concerned mainly with the collection of palaeontological specimens, the close connection between the National Museum for Geology and Mineralogy and the National Museum of Antiquities, also in Leiden, eventually led to the search for worked bone and antler in fishing expeditions, and many of these finds are in private collections, although it appears that this has not given rise to a commercial market. More recently, human bone specimens have been recovered, and the interest of collectors has extended to the search for stone and flint artefacts in the spoil from shellfish dredging or aggregates unloaded at Dutch harbours. A famous example of the latter is the recovery of a hand-axe from an aggregate load that could be traced back to the collection area and eventually led to the location and investigation of the A240 site in English territorial waters (Bailey et al., Chap. 10, this volume; Pater, Chap. 26, this volume).

In recent years, the use of the beam trawl has been the subject of fierce discussions between marine ecologists and the industry because of the damage caused by widespread 'ploughing' of the seabed and the collection of everything in the path of the beam trawl, a process that also destroys the upper layer of any archaeological deposits, and attempts have been made to ban its use. Despite opposition to beam trawling and the availability of less destructive technologies such as the use of electrical pulses, the EU has recently ruled in favour of continued use of the beam trawl (as of February 2019, Maarleveld, Chap. 27, this volume).

The reality is that the large archive of palaeontological and archaeological material recovered by the Dutch fishing fleet would have been lost to science had it not been for the use of the beam trawl and the interest of private collectors. Even without information on the depositional context, these finds remain valuable as indicators of potentially undisturbed sites or zones of archaeological significance, as evidence for the available animals and terrestrial environments present on the now-submerged landscape and their exploitation by human communities, and a reminder of the incomplete nature of the on-shore record. Bone remains can also contribute valuable intrinsic information in themselves, such as evidence of bone pathologies in human remains and through direct application to individual specimens of radiocarbon dating, aDNA and stable isotope analysis (Peeters and Amkreutz, Chap. 8, this volume).

This long history of collection is also testament to a successful cooperation between industry and collectors, and a pointer to the ways in which future cooperation might evolve. Recovery of archaeological and palaeontological specimens in fishing nets and in the spoil from sand dredging and gravel extraction will likely continue as an incidental by-product of commercial activities if not an intentional activity, regardless of the regulations discussed above. And these finds will continue to provide valuable new information. Data on the location of finds has occasionally been recorded, and that is beginning to be applied more widely with benefits to the scientific interpretation of the finds and to the location of target areas worthy of future investigation or to be avoided or carefully managed in future offshore developments. New networks are also being formed to develop this collaborative agenda (Peeters and Amkreutz, Chap. 8, this volume).

24.5 Blue Growth

The concept of Blue Growth is the EU term for a phenomenon that is being recognised across the world, namely, the relative fragility of the marine and coastal environment and the huge demographic pressures that are being placed on it in a world where a major part of the population already lives along its coastlines and depends on its resources for employment and recreation, with all that this implies in terms of infrastructural development, tourism, employment opportunities and further population growth. It is also a zone that is uniquely vulnerable to climate change, being in the frontline of sea-level rise that is predicted for the coming centuries. The need for coherent planning informed by policies for the sustainable management of natural ecosystems and cultural resources in this heavily crowded strip of land and sea is paramount and has obvious relevance to the protection of the cultural heritage both on land and under water.

The EU's agenda for Blue Growth is set out in its 'Report on Blue Growth: Enhancing Sustainable Growth in the EU's Marine, Maritime Transport and Tourism Sector' date 7 June 2013

(Danellis 2013). The document articulates the principles that have informed the maritime dimension of the Europe 2020 strategy and guidelines for financial investment. These include a recognition of the fragility of marine ecosystems, the likely role of the seas and oceans in future global economic growth, the importance of integrated management of all the activities focussed on the coastal zone—fisheries, tourism, transport, energy, biotechnology and mineral mining—and the value of regional centres of expertise and cross-border initiatives. Importantly in the context of the themes discussed in this volume, it calls for 'the uniform mapping of the seabed … so that the information can be accessed by research centres, universities, and public institutions' and the development of 'plans to map and survey wrecked ships and submerged archaeological sites, which form an important part of the Union's historical and cultural heritage[and] the need to facilitate the understanding and study of such sites and help prevent the despoliation to which they are being subjected, thus enabling them to be properly preserved' (Danellis 2013, paragraphs 21 and 22, see also https://ec.europa.eu/maritimeaffairs/policy/blue_growth_en, accessed 16 June 2019 for updates).

It is too early to say what impact this agenda is having on the development of policies and investments in relation to the underwater cultural heritage, but the fact that UCH is explicitly on the EU agenda presents both an opportunity and a challenge to the research community and governmental heritage agencies working in collaboration to define how UCH should be defined, managed and monitored. The results reported in this volume and the companion volume of Flemming et al. (2017) form a step on the road to that next stage of development.

24.6 Conclusions

The overwhelming impression of this section is the multiple ways in which the growth of knowledge about submerged prehistory has evolved in concert with the expansion of industrial and commercial activity on the seabed and the development of international laws and regulations for the protection of the underwater cultural heritage. The present system of protection and regulation is by no means fool proof, and there remain gaps in coverage and the risk of illegal or unrecorded destruction of valuable information. Offshore industrial activity is inherently destructive or potentially so, but that is famously true of all excavation whether conducted by professional archaeologists to the highest current standards or by industrial dredging and bore holes, whether under water or on dry land. Moreover, industrial activities often expose deeply buried archaeological remains and land surfaces to discovery, and the spoil from offshore activities such as aggregate extraction can yield finds of intrinsic value even if the original context is no longer known. At the same time, the relationship with offshore industries can carry risks of complacency, of allowing the research agenda to be defined by the nature of the finds exposed during commercial activities, or of reliance on such activities for the discovery of remains from the deeper and less easily accessible areas of the continental shelf (Sturt et al. 2017). The complexities of the interplay between government regulation, commercial activity and scientific research will continue to provide both challenges as well as opportunities for new discoveries and the development of new research agendas. As Dromgoole (Chap. 25, this volume) emphasises, the key to further progress in these relationships lies in dialogue between developers, archaeologists and government agencies, voluntary agreements and codes of practice, and joint research initiatives, as much as in legislation.

References

Bicket A, Firth A, Tizzard L, Benjamin J (2014) Heritage management and submerged prehistory in the United Kingdom. In: Evans AM, Flatman JC, Flemming NC (eds) Prehistoric archaeology on the continental shelf: A global review. Springer, New York, pp 213–232. https://doi.org/10.1007/978-1-4614-9635-9_12

Danellis S (2013) Report on Blue Growth: Enhancing sustainable growth in the EU's marine, maritime transport and tourism sectors (2012/2297(INI)) Committee

on Transport and Tourism, European Parliament, A7-0209/2013
Flemming NC, Cagatay MN, Chiocci FL, Galanidou N, Jöns H, Lericolais G, Missiaen T, Moore F, Rosentau A, Sakellariou D, Skar B, Stevenson A, Weerts H (2014) Land beneath the waves: Submerged landscapes and sea level change. A joint geoscience-humanities strategy for European continental shelf prehistoric research. Chu NC, McDonough N (eds) Position Paper 21 of the European Marine Board, Ostend, Belgium.
Flemming NC, Harff J, Moura D, Burgess A, Bailey GN (eds) (2017) Submerged landscapes of the European continental shelf: quaternary paleoenvironments. John Wiley & Sons, Chichester
Glimmerveen J, Mol D, Post K, Reumer JWF, Van der Plicht H, De Vos J, Van Geel B, Van Reenen G, Pals JP (2004) The North Sea project: the first palaeontological, palynological, and archaeological results. In: Flemming NC (ed) Submarine prehistoric archaeology of the North Sea: research priorities and collaboration with industry. CBA Research Report 141. Council for British Archaeology, York, pp 43–52
Glimmerveen J, Mol D, Van der Plicht J (2006) The Pleistocene reindeer of the North Sea — initial palaeontological data and archaeological remarks. Quatern Int 142/143:242–246
Missiaen T, Pieters M, Maes F, Kruiver P, De Maeyer Ph, Seys J (2017) The SeArch project: towards an assessment methodology and sustainable management policy for the archaeological heritage of the North Sea in Belgium. In: Bailey GN, Harff J, Sakellariou D (eds) Under the sea: archaeology and palaeolandscapes of the continental shelf. Springer, Cham, pp. 415–424
Mol D, Post K, Reumer JWF, Van der Plicht J, De Vos J, Van Geel B, Van Reenen G, Pals J-P, Glimmerveen J (2006) The Eurogeul — first report of the palaeontological, palynological and archaeological investigations of this part of the North Sea. Quatern Int 142/143:178–185
Moree JM, Sier MM (eds) (2015) Interdisciplinary research programme Maasvlakte, vol 2. BOOR, Rotterdam
Peeters H, Murphy P, Flemming NC (2009) North Sea prehistory research and management framework (NSPRMF) 2009. RACM, Amersfoort
Salter E, Murphy P, Peeters H (2014) Researching, conserving and managing submerged prehistory: national approaches and international collaboration. In: Evans AM, Flatman JC, Flemming NC (eds) Prehistoric archaeology on the continental shelf: a global review. Springer, New York, pp 213–232. https://doi.org/10.1007/978-1-4614-9635-9_9
Satchell J (2013) Marine industry and submerged prehistoric archaeology. Published online at https://www.splashcos.org/sites/splashcos.org/files/downloads/SPLASHCOS_Marine_Industry_Guide.pdf
Satchell J (2017) Education and engagement: developing understanding and appreciation of submerged prehistoric landscapes. In: Bailey GN, Harff J, Sakellariou D (eds) Under the sea: archaeology and palaeolandscapes of the continental shelf. Springer, Cham, pp 391–402
Sturt F, Dix J, Grant MJ (2017) The history of industry-linked research in English waters: lessons for the future. In: Bailey GN, Harff J, Sakellariou D (eds) Under the sea: archaeology and palaeolandscapes of the continental shelf. Springer, Cham, pp 425–436
Tidbury L, Satchell J, Momber G (2017) Arch-Manche: using archaeological, palaeoenvironmental, historic and artistic resources in coastal management. In: Bailey GN, Harff J, Sakellariou D (eds) Under the sea: archaeology and palaeolandscapes of the continental shelf. Springer, Cham, pp 403–413
Van Kolfschoten T, van Essen H (2004) Palaeozoological heritage from the bottom of the North Sea. In: Flemming NC (ed) Submarine prehistoric archaeology of the North Sea. Research priorities and collaboration with industry. CBA Research Report 141. Council for British Archaeology, York, pp 70–80

Continental Shelf Archaeology and International Law

25

Sarah Dromgoole

Abstract

This chapter outlines the international legal framework concerning the protection of submerged prehistoric resources from human activities that may cause inadvertent harm. It focuses on what is the core question from a legal perspective: to what extent does a coastal State have the jurisdictional power under international law to regulate such activities in its offshore waters with a view to protecting material of archaeological significance? As will become clear, this question is a complex one, requiring reference to three sources of international law: (i) general principles of international law, (ii) the 1982 UN Convention on the Law of the Sea and (iii) the 2001 UNESCO Convention on the Protection of the Underwater Cultural Heritage. Questions relating to the way that the relevant treaties define underwater cultural heritage, and the potential implications this may have for submerged prehistoric archaeology, though interesting, fall outside the scope of this chapter.

S. Dromgoole (✉)
School of Law, University of Nottingham, University Park, Nottingham, UK

Keywords

12-mile territorial limit · Jurisdictional gap · UN Convention on the Law of the Sea · Underwater cultural heritage · UNESCO Convention on the Protection of the Underwater Cultural Heritage

25.1 Introduction

Over the years a great deal has been written about international law in the context of marine archaeology, but for the most part, it has been addressed to the question of how to protect shipwrecks from treasure hunters, that is, from those who deliberately interfere with wrecks in order to recover items of commercial value (see, e.g. Churchill and Lowe 1988; Dromgoole 2013). From a legal perspective as well as an archaeological one, prehistoric submerged landscapes give rise to quite different questions from shipwrecks. The main threat to such landscapes is not from treasure hunters but rather from industrial and other human activities that may inadvertently damage or destroy the archaeological potential of such landscapes. Consequently, the legal means of countering the threat require specific consideration and analysis.

Prehistoric landscapes and associated archaeological remains are located on (or under the surface of) the geological continental shelf. This

G. Bailey et al. (eds.), *The Archaeology of Europe's Drowned Landscapes*, Coastal Research Library 35, https://doi.org/10.1007/978-3-030-37367-2_25

area is being subjected to increasingly intense levels of human activity that impact upon the seabed, particularly trawling by fishermen, dredging for marine aggregates, wind farm development, pipeline and cable-laying and hydrocarbon exploration and exploitation. Such activities are by no means confined to inshore waters, in other words the territorial sea and internal waters over which the coastal State has sovereignty. The North Sea Basin, an area with significant prehistoric archaeological potential, offers ample examples of offshore commercial activity taking place all across the Basin that interfere with the seabed (Peeters et al. 2009, p. 5). Gaffney et al. describe in quite striking terms the impact that trawling activity, over time, has had on prehistoric deposits on the Dogger Bank, which lies in the southern North Sea, 60 or so miles from the UK coast (Fig. 25.1). The physical infrastructure supporting the North Sea offshore oil and gas industry—rigs, pipelines and associated installations—is already immense (Fig. 25.2), and a new generation of energy-related infrastructure of substantial scale is likely to be put in place over the next two decades. The United Kingdom, for example, has ambitious plans for a rapid extension of wind farm developments beyond the 12-mile limit, as well as for the development of CO_2 storage facilities deep within the seabed on the continental shelf (see the Crown Estate website at http://www.thecrownestate.co.uk).

This chapter outlines the international legal framework relating to the protection of submerged prehistoric resources from human activities that may cause inadvertent harm. It focuses on the core question from a legal perspective which is to what extent does a coastal State have the power to regulate such activities in its offshore waters, in other words those beyond the 12-mile limit, with a view to protecting material of archaeological significance, or potential archaeological significance? This question is technically complex, requiring reference to three sources of international law: general principles of international law, the 1982 UN Convention on the Law of the Sea and the 2001 UNESCO Convention on the Protection of the Underwater Cultural Heritage.

25.2 Relevant Principles of International Jurisdiction

To regulate activities in any geographical space, be it on land or at sea, a State must have the requisite jurisdictional power under international law. There are various general principles of international jurisdiction that provide bases for a State to take lawful regulatory action, and there are also some specific principles relating to jurisdiction in the maritime sphere. In the context of this paper, both have relevance. The two general principles of international jurisdiction that are relevant are the so-called 'territorial' and 'nationality' principles (see Churchill and Lowe 1988 and Brownlie 2008).

Under the 'territorial' principle, a State has jurisdiction to regulate activities that take place within its territory; this jurisdiction extends to the regulation of the activities of foreign parties within the territory (subject to certain exceptions). By virtue of this principle, the State can regulate all activities taking place within its territorial boundaries, and these boundaries are set by the outer limits of the territorial sea. The territorial sea—extending up to 12 nautical miles (nmi)[1] from coastal baselines, generally low-water mark—is the maritime territory of the State and the State has sovereignty and jurisdiction over all activities taking place within its territorial sea, subject to various rules of international law. The main such rule is that foreign ships have a right of 'innocent passage' through the territorial sea. Ships engaged in activities that may affect underwater cultural heritage (UCH) are unlikely to be engaged in innocent passage because innocent passage requires 'continuous and expeditious' passage. Therefore, generally speaking, the coastal State has the right to regulate activities affecting UCH taking place within the 12-mile limit, including activities conducted by foreigners.

[1]A nautical mile is defined as 1852 m or 1.15 miles (Briney 2018, https://oceanservice.noaa.gov/facts/nauticalmile_knot.html); therefore, 12 nmi is equivalent to 13.8 miles or 22.2 km. All references to miles in this chapter are to nautical miles unless stated otherwise.

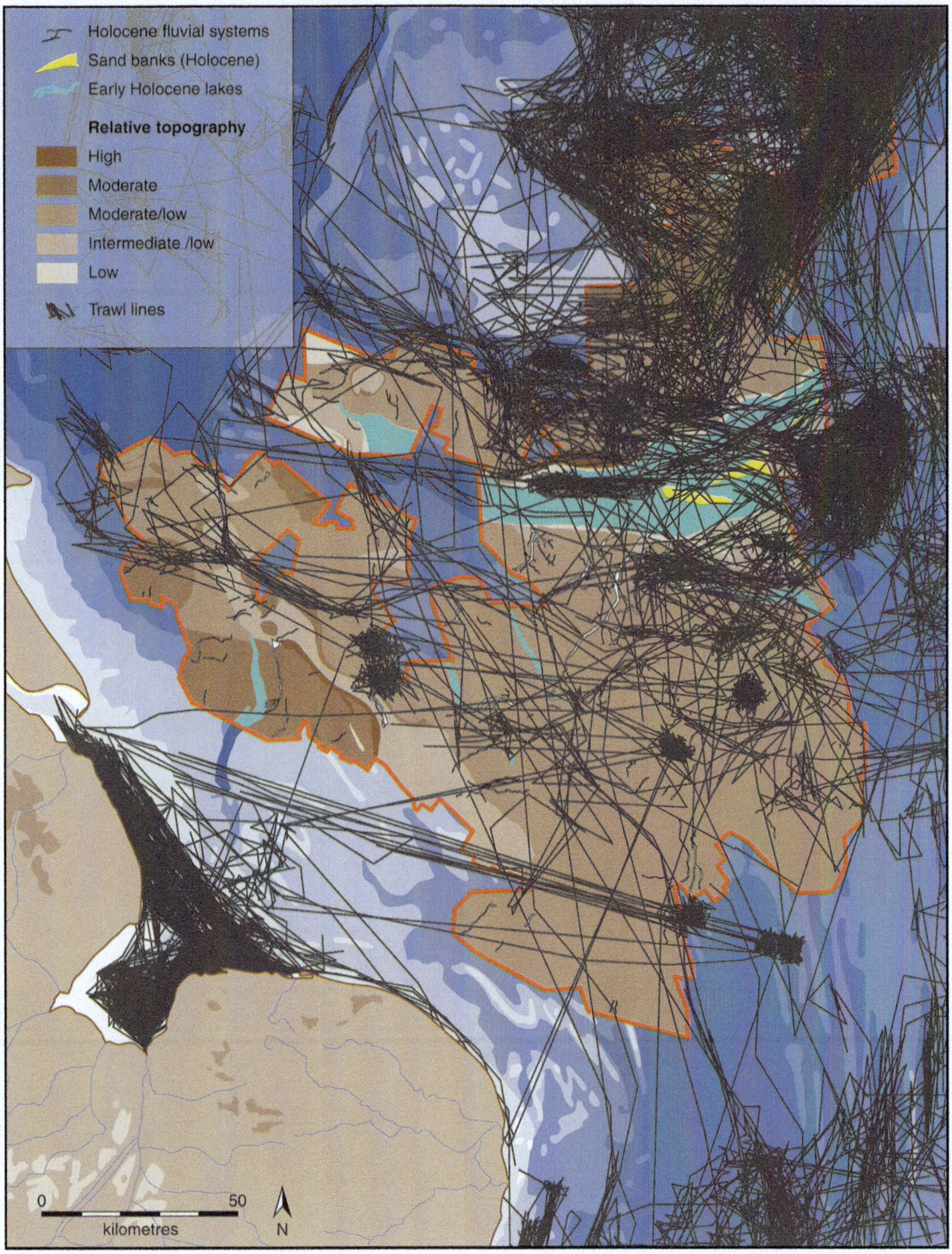

Fig. 25.1 The remarkable intensity of modern trawling activity in the southern North Sea as recorded in 2007 for UK fishing vessels >15 m. After Gaffney et al. (2009, p. 154, Fig. 5.17). Image created by the North Sea Palaeolandscapes Project and Henry Buglass. Courtesy of Vince Gaffney and the Council for British Archaeology

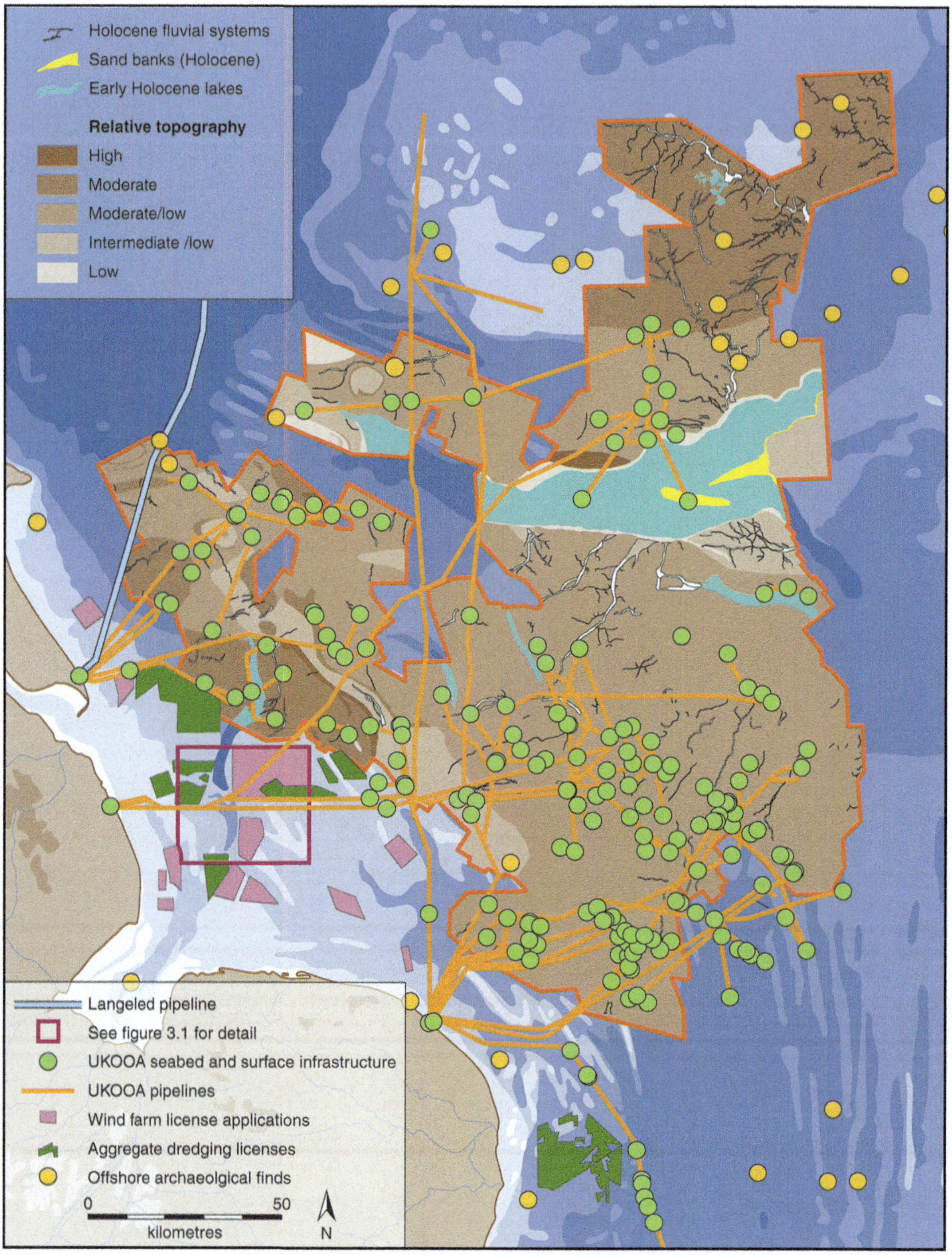

Fig. 25.2 The intense use of space within one sector of the North Sea, showing pipelines, seabed infrastructure, areas licensed for aggregate extraction and wind farms and topographic features. After Gaffney et al. (2009, p. 156, Fig. 5.19). Image created by the North Sea Palaeolandscapes Project and Henry Buglass. Courtesy of Vince Gaffney and the Council for British Archaeology

Under the 'nationality' principle of international jurisdiction, a State has the right to regulate the activities of its own flag vessels and nationals, wherever in the world they happen to be. This means that a State can extend its jurisdictional arm out beyond its territorial boundaries but only for the purpose of controlling the activities of its own flag vessels and nationals (including corporate entities). In a maritime context, this is a useful principle when it comes to finding ways of regulating activities in international waters beyond the 12-mile territorial limit.

Aside from these general principles of public international law, there are also principles relating specifically to maritime jurisdiction. These are set out in the United Nations Convention on the Law of the Sea, the so-called Constitution for the Oceans.

25.3 The Convention on the Law of the Sea 1982

25.3.1 Background

The Convention on the Law of the Sea 1982 (LOSC for short) is a comprehensive treaty establishing the international legal framework governing human activities in the oceans. It is the longest international treaty that has ever been created, and apart from the Charter establishing the United Nations, it is also the most successful. Unlike many other treaties, the Convention on the Law of the Sea is now close to universal ratification. With the exception of a small handful of States (notable non-parties are the United States, Turkey, Peru, Israel and Iran), it has strong international support.

The Convention divides the oceans up into a number of maritime zones and establishes a legal regime for each of these zones, which sets out the rights of States (Fig. 25.3). Under the Convention, coastal States (in the sense of States with a coastline) are afforded some limited rights and jurisdiction in their offshore waters, in other words waters beyond the 12-mile territorial limit. However, for the most part, the rights and jurisdiction in these offshore areas relate only to natural resources and to the exploration and exploitation of those resources. Despite the close link that exists in practice between the natural and historic environment, sunken vessels, their contents and other man-made objects found at sea are not regarded as natural resources for the purposes of the Convention on the Law of the Sea (Dromgoole 2013, pp. 29–30). In the context of submerged landscape archaeology, archaeologists are interested in much more than merely the

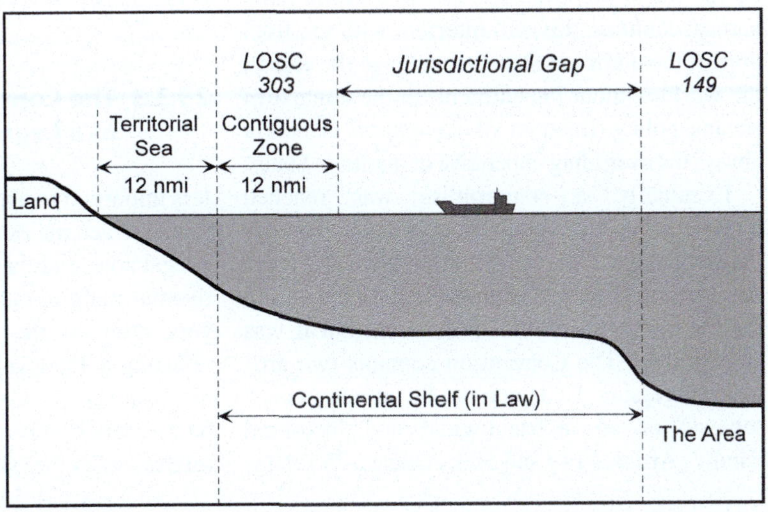

Fig. 25.3 Maritime zones, showing the jurisdictional 'gap' between the areas of seabed covered by Article 149 and Article 303 of the United Nations Convention on the Law of the Sea. Distances are in nautical miles. Not to scale. See text for further discussion and definitions

man-made objects that may be found within the landscape. Nonetheless, the whole provision for UCH in international law is predicated on the basis that there is a clear dichotomy between the natural and historic environments.

25.3.2 Specific Provision for UCH: Articles 149 and 303

When the Convention on the Law of the Sea was negotiated in the 1970s, appreciation of the nature and value of cultural resources located on the seabed was extremely limited and the main concern of the international community (or, to be more precise, a small section of that community) was to find means for protecting ancient shipwrecks and other antiquities from treasure hunters and souvenir seekers, primarily in the Mediterranean region. At the time, a few Mediterranean States argued that coastal States should be afforded the right to regulate activities affecting UCH over the whole of their continental shelf. On the other hand, a number of other States—maritime powers such as the United States, the Netherlands and the United Kingdom—argued that the rights of a coastal State over the continental shelf should be confined to natural resources only. What these and other maritime States were concerned about, and continue to be concerned about, is that if the rights of States to regulate activities in their offshore waters are extended beyond traditionally narrow confines, this will interfere with the freedom of the maritime States to navigate the seas of the world without impediment. In the context of oceans policy, freedom of navigation is nearly always the overriding imperative of maritime States.

Eventually, a compromise was reached between the two sides, although it was heavily tipped in favour of the maritime powers. Some specific provision was included in the Convention for the purposes of UCH protection, but it was very limited. The Convention contains two articles addressing UCH, or, to be more precise, 'objects of an archaeological and historical nature', Articles 149 and 303. Article 149 relates to the maritime zone referred to as 'the Area', which is the deep seabed beyond the limits of 'national jurisdiction' (in practice meaning beyond the limits of the continental shelf as legally defined: see further below). The other provision, Article 303, gives coastal States the right to exercise control over the removal of UCH in a relatively narrow strip of sea, immediately adjacent to the territorial sea, referred to as the contiguous zone (Fig. 25.3; LOSC, Art. 303(2)). Significantly, Article 303 also establishes a general duty under international law on all States to protect UCH in all zones and to cooperate for that purpose (LOSC, Art. 303(1)).

In light of Articles 149 and 303, the position under the Convention on the Law of the Sea can be summarised as follows. By virtue of its sovereignty over the territorial sea, the coastal State has the right to regulate activities affecting UCH out to the 12-mile territorial sea limit; by virtue of Article 303, it also has the right to regulate the removal of UCH in the contiguous zone. This zone, extending from 12 to 24 miles from baselines, is one which coastal States are free to declare if they wish; it does not exist unless they declare it. The Convention also makes special provision for UCH in the deep seabed area in Article 149. However, between 24 miles from baselines and the outer limit of the continental shelf as legally defined, there is a 'gap' in the provision made by the Convention—it makes no specific provision for UCH located in this gap (Fig. 25.3).

25.3.3 The Continental Shelf as a Legal Concept

It is important to appreciate that there is a distinction between the extent of the continental shelf as a legal zone and the extent of the geological continental margin. Generally speaking, when lawyers refer to the continental shelf, they are referring to the legal zone of that name; this does not necessarily equate to the geological continental margin or indeed to the section of that margin referred to as the 'shelf'.

The geological continental margin extends under the sea in widely varying ways. For some States, the continental margin falls away rapidly to the deep seabed; other States have a relatively broad continental margin; and some, a very extensive one (extending 300 miles or more from coastal baselines).

Under the Convention on the Law of the Sea, as a matter of principle, the maritime zone known as the continental shelf for all coastal States extends to at least 200 miles from coastal baselines (LOSC, Art. 76(1)), no matter how broad their physical continental margin (subject to delimitation between States with opposing coasts that are less than 200 miles apart). Therefore, even where a State has a very narrow physical continental margin or, indeed, no margin at all to speak of, it still has the rights accorded by the Convention under the legal regime for the continental shelf out to 200 miles.

Those States whose physical continental margin is broader than 200 miles (known as 'broad-margin' States) will have the rights afforded under the legal regime for the continental shelf to the physical extent of their continental margin. The 'continental margin' comprises a 'shelf', a 'slope' down to the deep seabed and a 'rise' (the area of sediment that accumulates at the bottom of the slope). In the case of broad-margin States, all three geological features are included within the maritime zone referred to as the continental shelf.

The inner limit of the continental shelf as a maritime zone is set at the outer limit of the territorial sea, in other words at 12 miles from baselines. So, the legal continental shelf starts at the 12-mile limit and extends to 200 miles or to the extent of the physical continental margin in the case of broad-margin States. This particular maritime zone is said to exist ipso facto, in other words as a matter of fact rather than as a matter of law, and therefore there is no need for the coastal State to formally claim it. The same is the case for the territorial sea.

The rights afforded to the coastal State under the continental shelf regime relate to the seabed and subsoil of the zone and to the natural resources of the seabed and subsoil; they do *not* relate to the water column above. The natural resources of the zone are primarily oil and gas, in other words non-living mineral resources, but they also include living resources in so far as they constitute sedentary species (as defined in LOSC, Art. 77(4)), such as oysters, clams and seaweed.

As far as the water column is concerned, the area from the 12-mile territorial sea limit to 200 miles from coastal baselines may be claimed by the coastal State as an exclusive economic zone (EEZ). The Convention on the Law of the Sea permits coastal States to claim an EEZ if they wish and, where they do so, affords the coastal State rights and jurisdiction over the natural resources of that zone. These resources include the fish in the water column and renewable energy resources generated from wind, waves and currents, but not items of cultural heritage lying on or beneath the seabed nor, indeed, items of cultural heritage found floating in the water column, such as a message in a bottle. To emphasise an important point made earlier, man-made objects are not regarded as natural resources for the purposes of the Convention on the Law of the Sea and the whole provision for UCH in international law is predicated on the basis that there is a clear dichotomy between the natural and historic environments (Dromgoole 2013, pp. 29–30).

One of the reasons that the legal maritime zones can be confusing is that a number of the zones overlap. For example, where coastal States claim an EEZ, their continental shelf and EEZ, as maritime zones, will overlap out to 200 miles. Also, if the coastal State claims a contiguous zone, that zone will fall within the EEZ if they claim an EEZ and will also be within the area of their legal continental shelf as well. The Convention on the Law of the Sea has different sections ('parts') setting out the regime for each of the zones, and it may be that two or even three parts of the Convention will apply to any particular spot in the ocean. In practice, generally this is not problematic as the parts have been carefully constructed to dovetail with one another. In the remaining sections of this chapter, references to the continental shelf are references to this zone as legally defined unless otherwise indicated.

25.3.4 The Jurisdictional 'Gap'

The 'gap' left by Articles 149 and 303 of the Convention on the Law of the Sea, identified above, relates to the continental shelf between 24 miles and the outer limit of the continental shelf, be that 200 miles from coastal baselines or further out to sea in the case of broad-margin States. In this area, which in the case of broad-margin States has the potential to be extensive, the coastal State is afforded no direct right by the Convention on the Law of the Sea to regulate activities affecting UCH. The rights it is afforded in this area relate to natural resources only. All the rights and jurisdiction afforded to the coastal State in the gap are closely tied to the natural resources of the area and to matters relating to the exploration and exploitation of those resources.

Even before the ink was dry on the text of the Convention on the Law of the Sea, it was clear that Articles 149 and 303 of the Convention were inadequate to protect UCH in the oceans and that something needed to be done to 'plug the gap' on the continental shelf, particularly to enable coastal States to regulate treasure hunting and other activities targeting UCH—shipwrecks specifically—located on the increasingly accessible 'shelf' section of their geological continental margin. Even more particularly, what was necessary was a means to enable them to regulate the activities of foreign flag vessels and nationals. As pointed out at the beginning of this chapter, States have the right to regulate the activities of their own flag vessels and nationals, whichever legal zone they are in. The difficulty has been in finding a lawful means of regulating the activities of foreign vessels and nationals beyond the outer limit of the territorial sea. It should be noted that while activities targeting UCH in the 12–24-mile contiguous zone can be regulated by the coastal State by virtue of the Convention on the Law of the Sea, Art. 303(2), this provision permits the coastal State only to regulate the removal of UCH and therefore does not address circumstances where activities may need to be regulated to avoid inadvertent harm to UCH.

In the early 1980s, the Council of Europe attempted to create a treaty to plug the gap, which is known as the Draft European Convention of 1985. However, this initiative was unsuccessful, and the Draft Convention was never formally adopted. In the late 1980s, and in the 1990s, the problem of unregulated treasure hunting became more acute with developments in deep-water technology, and eventually UNESCO focused its attention on the matter and created the UNESCO Convention of 2001. This Convention came into force in 2009 and, at the time of writing, has been ratified by 61 parties (see http://www.unesco.org/eri/la/convention.asp?KO=13520&language=E&order=alpha). In the European context, a good number of States littoral to the Mediterranean Sea and the Black Sea have become parties to the Convention (including Italy, Croatia, Bulgaria and the Ukraine, all represented in Part III of this volume). However, the picture is more mixed when it comes to the Atlantic margins and the North Sea, the focus of Part II of this volume. Here, only Portugal, Spain, France and Belgium are currently States Parties. In the Baltic Basin (the focus of Part I of the volume), the only State to have ratified the Convention to date is Lithuania.

25.4 The UNESCO Convention 2001

25.4.1 Introduction

Fundamentally, the original purpose of the UNESCO initiative was to create jurisdictional mechanisms to permit States that are parties to the Convention—operating individually and collaboratively—to regulate activities targeting UCH located beyond 24 miles, in other words, to plug the gap identified above, in order to regulate, in particular, commercial treasure hunting on the continental shelf. In fact, the Convention does much more than this, and among other things, it has the potential to impact on human activities that may only inadvertently cause harm to UCH, including fishing, dredging, pipeline and cable-laying and wind farm development.

During the UNESCO negotiations, there was some debate about the extent to which such activities should fall within the remit of the Convention. While there is clearly the potential for them to cause harm to UCH (a term specifically used by the Convention and defined in Art. 1(1)), the importance to national economies of activities such as these means that any potential interference with them is a matter that is politically highly sensitive. Moreover, the negotiators recognised that attempts to regulate the multitude of activities that take place in the marine zone would be a task well beyond the remit of a UNESCO-sponsored treaty. A distinction was therefore drawn between activities 'directed at' UCH, in other words activities that have UCH as their primary object (UNESCO Convention, Art. 1(6)), and activities 'incidentally affecting' UCH, in other words activities that do not have UCH as their primary object, or at least one of their objects, but may nonetheless disturb or cause damage to UCH (UNESCO Convention, Art. 1(7)). While a decision was made to focus on regulating the former, the Convention does not ignore the impact of activities that may inadvertently disturb, or otherwise damage, UCH. It does, however, tread carefully in relation to this issue.

As far as treasure hunting and other activities 'directed at' UCH are concerned (which include archaeological interventions), the Convention attempts to effect a compromise between those States that called for direct jurisdiction over UCH on their continental shelves and those States—namely, the maritime powers—who did not want coastal States to be afforded any further rights than those established under the Convention on the Law of the Sea. The relevant articles of the UNESCO Convention are Articles 9 and 10, which apply to the continental shelf and the EEZ (on the significance of their application to the EEZ, see Dromgoole 2013, pp. 288, fn. 45).

In light of the political difficulty of reaching agreement over this issue, these articles set out a complex web of reporting, notification and protection procedures. Not only are the articles complex, but they are deeply controversial politically. The reason for this is that they have been deliberately designed so that they can be interpreted in two different ways: one that gives coastal States no greater rights than they would have under the Convention on the Law of the Sea and another that would give them greater rights (Dromgoole 2013, pp. 299–300). It was hoped that the so-called constructive ambiguities in the text would form the basis for a compromise that would be acceptable to all States; instead, the fact that the provisions can be interpreted as giving coastal States greater rights on the continental shelf than they have under the Convention on the Law of the Sea is one of the features of the Convention that has led to a reluctance on the part of the maritime powers to ratify the Convention.

25.4.2 Principles and Objectives of the UNESCO Convention

Before considering what the UNESCO Convention has to say about activities incidentally affecting UCH, the core principles and objectives of the Convention (enshrined in Article 2 and elaborated upon in the Rules set out in the treaty's Annex) should be noted:

(i) The Convention is designed to ensure that activities 'directed at' UCH are conducted in accordance with internationally accepted archaeological standards. Those standards, enshrined in the Convention's Annex, have been widely praised, including by the maritime powers.
(ii) The overall objective of the Convention is to ensure that UCH is preserved for the benefit of humanity as a whole.
(iii) The Convention adopts the precautionary principle that preservation in situ should be the starting point when the competent national authorities are called upon to consider the future of any particular UCH site. This means that deliberate interference should be permitted only where justified for sound scientific reasons.

(iv) Where authorisation has been given for the recovery of UCH, the material recovered as far as possible must be kept together as a collection and made accessible to the general public and researchers.
(v) A cardinal principle of the Convention is that UCH must not be commercially exploited and the application of salvage law to the recovery of UCH is also severely restricted.
(vi) Responsible non-intrusive public access to UCH in situ is encouraged.
(vii) States Parties to the Convention *must* cooperate with one another in the interests of UCH protection.

25.4.3 Approach Taken to Activities 'Incidentally Affecting' UCH

How, then, does the UNESCO Convention deal with activities *incidentally affecting* UCH? The core provision on this is Article 5; indeed, this is the *only* article of the Convention devoted to such activities. The article places States Parties under a specific duty with respect to such activities:

> Each State Party shall use the best practicable means at its disposal to prevent or mitigate any adverse effects that might arise from activities under its jurisdiction incidentally affecting UCH.

Article 5 does not make reference to any specific maritime zones but instead applies to activities 'under [the] jurisdiction' of States Parties. What a State Party must do under the article is use the 'best practicable means at its disposal' to prevent or mitigate adverse effects.

The obligation in Article 5 is relatively 'soft'. The provision does not give States any new jurisdictional rights, but simply requires them to make the best use of the rights and jurisdiction already available to them to prevent or mitigate adverse effects. Although the principle of preservation in situ is one of the cornerstones of the Convention, it is recognised that in some circumstances preservation in situ simply may not be practicable and that mitigation of adverse effects may be all that is possible. It should be noted that when deliberate interference with UCH is found to be necessary to mitigate the negative impact of activities 'incidentally affecting' UCH, the interference will amount to an activity 'directed at' UCH and thus will become subject to the authorisation procedures in the Convention and will need to be conducted in accordance with the Rules in the Annex. There is no requirement that States Parties to the Convention establish new or additional regulatory frameworks; instead, Article 5 requires that they make use of, and develop, the regulatory frameworks that are already in place.

Although Article 5 is the only article of the UNESCO Convention devoted to activities incidentally affecting UCH, consideration of this article alone is insufficient to appreciate the extent of the potential impact of the Convention on activities incidentally affecting UCH taking place on the continental shelf. The Convention's specific regulatory scheme for this zone, set out in Articles 9 and 10, must also be taken into account.

The fact that Articles 9 and 10 have the potential to impact on activities incidentally affecting UCH is obscured by the general complexity of these articles. However, a close reading of the reporting provision set out in Article 9(1) will show that it extends to the reporting of discoveries by anyone undertaking activities on the continental shelf, not just by those who intend to engage in activities directed at UCH. Exactly who will be obliged to report in practice depends on the way that the coastal State Party interprets the provision. For example, it seems likely that any States in northern Europe signing up to the UNESCO Convention will interpret it strictly in accordance with the Convention on the Law of the Sea; as a result, this means that probably they would require only the reporting of discoveries by their own flag vessels and nationals apart from circumstances where flag vessels and nationals of other States fall within their regulatory powers by virtue of the fact that they are engaged in activities relating to the exploration and exploitation of the natural resources of the continental shelf or EEZ (see further below).

For commercial operators, a duty to report discoveries may not, of itself, be particularly worrisome. What is likely to be of greater concern is the potential consequences of making a report. Under the UNESCO Convention, any report of a discovery will trigger various procedures set out in Articles 9 and 10 relating to the notification of interested States and the protection of the UCH in question. However, any measures taken by States Parties under these articles—in so far as they relate to activities incidentally affecting UCH—should be regarded as an implementation of the general duty with respect to such activities set out in Article 5. That duty relates to activities 'under [the] jurisdiction of States Parties'. It therefore reinforces the fact that a State may only take action to regulate activities to the extent that it already has the jurisdiction to take such action.

This leads to the question: when does a State have jurisdiction to take regulatory action with respect to activities 'incidentally affecting' UCH under general international law, in other words outside the UNESCO Convention's framework, when those activities are being undertaken beyond the 12-mile territorial limit?

25.5 Jurisdictional Position Under General International Law: Summary

The jurisdictional position under general international law with respect to activities 'incidentally affecting' UCH is essentially threefold:

(i) If, by virtue of the 'nationality' principle of international jurisdiction, the coastal State has jurisdiction over a vessel, a company or an individual working on the continental shelf, or planning to work on the continental shelf, then it will have the right to regulate the activities in the interests of UCH protection.

(ii) Where the coastal State has jurisdiction over the activities of foreign companies on the continental shelf/in the EEZ by reason of the fact that they are engaged in the exploration and exploitation of the natural resources of these zones, then the coastal State probably has a right to require the reporting of UCH finds, and the appropriate treatment of any such finds, under the terms of the licence that it grants to that company. Justification for the imposition of reasonable conditions designed to protect UCH is probably provided by the fact that the Convention on the Law of the Sea imposes a general duty on States to protect UCH in all maritime zones under Article 303(1). Greece and Norway appear to have been the first States to have imposed such conditions on contractors working on the continental shelf, but as far as it is possible to tell, the practice appears to be quite prevalent today, certainly in northern Europe (Dromgoole 2013, pp. 266–267; see Pater, Chap. 26, this volume, for an English example, and Chapters in Parts I and II, this volume, for further detail).

(iii) In the case of activities incidentally affecting UCH which do not fall under (i) or (ii) above, and therefore are not subject to direct coastal State jurisdiction, in some circumstances, the coastal State may be able to make indirect use of the jurisdiction afforded to it in respect of the natural resources of the continental shelf and EEZ in order to protect UCH. For example, if the activities amount to drilling into or other disturbance of the seabed, or if they amount to the dumping of waste onto the seabed, then the coastal State may well have a right to regulate them and, by doing so, to indirectly afford protection to UCH (Dromgoole 2013, pp. 267–272).

The three methods of regulating activities set out above are available to States under general international law, in other words under the international legal regime existing outside the

UNESCO Convention. What the UNESCO Convention does by way of addition to this position is to impose a legal duty on States Parties to that Convention to make full use of these jurisdictional methods to prevent or mitigate adverse effects on UCH. It should be noted that the methods are available to all States in any event, but the problem is that they do not necessarily make use of them. It is worth noting too that, by applying to archaeological heritage 'located in any area within the jurisdiction of the Parties', the Council of Europe's regional treaty, the 1992 Valletta Convention, has had the effect of encouraging some States to make full use of the jurisdictional tools available to them in their offshore zones in the interests of UCH protection.

25.6 A Special Case: Pipelines and Cables

The pipeline and cable industries must be distinguished from other offshore industries because they are afforded a privileged position under the Convention on the Law of the Sea. Among other things, this means that the nationals and flag vessels of foreign States are entitled to lay submarine cables and pipelines on the continental shelf of a coastal State, generally speaking without impediment by the coastal State (LOSC, Art. 79). Nonetheless, there is a significant proviso to this general rule in the case of pipelines and that is that the delineation of their course is subject to the consent of the coastal State (LOSC, Art. 79(3)). Again, given the general duty on States under Article 303 of the Convention on the Law of the Sea to protect UCH in all maritime zones, it would appear to be justifiable for the coastal State to require that new pipelines avoid areas of particular archaeological sensitivity.

25.7 Concluding Remarks

Since the emergence of interest in marine archaeology in the 1950s, the debate concerning UCH and international law has focused almost exclusively on the question of how to regulate the activities of those that deliberately target UCH and, more specifically, shipwrecks. Until such time as the 2001 UNESCO Convention fulfils its fundamental objective of establishing a globally effective regime for the conduct of activities 'directed at' UCH, this question will continue to attract attention. However, in view of the relentlessly increasing pressures on the marine environment as a result of offshore human activity, as well as burgeoning interest in the archaeological potential of submerged landscapes, it is likely that greater attention will be paid in the future to activities 'incidentally affecting' UCH and to State practice relating to the regulation of such activities in the interests of UCH protection. This should help to bring greater clarity and certainty to the question of what States can, and cannot, do legitimately under international law to protect UCH from the adverse effects that such activities may have. Of course, much can be done without recourse to law by means of voluntary agreements and codes of practice. In this respect, the building of understanding through dialogue between offshore developers and archaeologists has already resulted in a number of important joint research initiatives in the North Sea in particular (see Peeters and Amkreutz, Chap. 8, this volume; Bailey et al., Chap. 10, this volume) and has been a major strand in the SPLASHCOS agenda (https://www.splashcos.org/outreach/collaboration). Such a dialogue has a vital role to play in future offshore research.

Acknowledgements This chapter arises from a paper delivered at the SPLASHCOS international conference and networking event, 'Offshore Industry and Archaeology: A Creative Relationship', held in Esbjerg, Denmark, in 2013. I thank Thijs Maarleveld for the invitation to participate in the original conference and Geoff Bailey for the invitation to turn the paper into a chapter for this volume.

References

Briney A (2018) How are nautical miles measured? https://wwwthoughtcocom/what-are-nautical-miles-1435097, updated 16 March 2018, Accessed 25 Jan 2019

Brownlie I (2008) Principles of public international law. Oxford University Press, Oxford

Churchill R, Lowe V (1988) The law of the sea. Manchester University Press, Manchester

Dromgoole S (2013) Underwater cultural heritage and international law. Cambridge University Press, Cambridge

Gaffney V, Fitch S, Smith D (2009) Europe's lost world: the rediscovery of Doggerland, CBA Research Report 160. Council for British Archaeology, York

Peeters JHM, Murphy P, Flemming NC (2009) North Sea prehistory research and management framework (NSPRMF) 2009. Cultural Heritage Agency, Amersfoort

Seabed Infrastructure Projects, Underwater Cultural Heritage and The Environmental Assessment Process: The UK Example

26

Christopher Pater

Abstract

This chapter focuses on the legal and regulatory requirements imposed on development projects in the marine offshore areas of England, in particular the requirement to collaborate with professional archaeologists in completing Environmental Impact Assessments that include underwater cultural heritage. This chapter explains the role of Historic England in providing independent advice to developers and regulatory bodies for all aspects of the historic environment. It explains how this advice informs the preparation and approval of planning applications for offshore developments, the implementation of effective survey campaigns that include archaeological and palaeoenvironmental objectives as part of the planning application, mitigation of potential damage, protection of new discoveries and the delivery of training programmes for government administrators, archaeological consultants and industrial partners. The benefits and results of this approach are presented in the context of offshore developments such as wind farms, the laying of electricity interconnector cables, the extraction of gravel deposits and the dredging of shipping channels into ports.

Keywords

Historic England · Marine infrastructure · Historic environment · Environmental Impact Assessment · Written Scheme of Investigation · Archaeological Exclusion Zone · Marine planning · Wind farms

26.1 Introduction

This chapter focuses on the government regulations in force in the UK, and specifically in England, that require commercial and industrial companies working on the seabed to complete an Environmental Impact Assessment (EIA) in advance of development work and to carry out any mitigation required to rescue or protect environmental features of significance. The scale and intensity of offshore development in response to modern needs, especially in the sectors of sea transport and renewable energy, and the areas of seabed that can be affected by such developments are large and growing in number, emphasising the need for adequate regulation in relation to the underwater heritage (Figs. 26.1, and 26.2).

'Environment' in this context is explicitly taken to include the cultural heritage, including not only easily recognisable artefacts such as shipwrecks but also features of the prehistoric environment and archaeology such as topographic features, sediment sequences and stone

C. Pater (✉)
Historic England, London, UK
e-mail: chris.pater@HistoricEngland.org.uk

G. Bailey et al. (eds.), *The Archaeology of Europe's Drowned Landscapes*, Coastal Research Library 35, https://doi.org/10.1007/978-3-030-37367-2_26

Fig. 26.1 Construction work in progress at the London Gateway Port development on the Essex coast in the Thames Estuary, giving an impression of the area affected and the potential impact on the historic environment. © Historic England Archive, Damian Grady

artefacts associated with the pre-inundation landscape—the 'historic environment' in English terminology. The historic environment is defined as follows:

> …all aspects of the environment resulting from the interaction between people and places through time, including all surviving physical remains of past human activity, whether visible, buried or submerged. Those elements of the historic environment – buildings, monuments, sites or landscapes – that have been positively identified as holding a degree of significance meriting consideration are called 'heritage assets'… Significance is the value of a heritage asset to this and future generations because of its heritage interest. That interest may be archaeological, architectural, artistic or historic. (HM Government and Devolved Administrations 2011, p. 21)

The offshore area that falls within UK jurisdiction, the UK Marine Area, includes the national territorial waters extending from the high water mark on the shoreline out to the limit of 12 nautical miles, the additional area within the Exclusive Economic Zone (EEZ), which extends out a further 188 nautical miles beyond the territorial limit—or to the median line with an adjacent maritime State, as in the North Sea—and for any areas of continental shelf that extend beyond the EEZ, which is the case for an area extending to the west of Scotland (see Dromgoole, Chap. 25, this volume, for further discussion of these terms, and HM Government and Devolved Administrations, 2011, p. 3, footnote 1 for the definition of the UK Marine Area).

Historic England (HE) is the government agency responsible for recording and researching the historic environment, operating under the UK Government Department for Digital, Culture, Media and Sport. It is the national curator for the marine (and terrestrial) historic environment and provides advice to developers, their archaeological consultants and regulators such as planning

Fig. 26.2 The Westermost Rough offshore wind farm under construction, located 8 km offshore of the Yorkshire coast near the Humber estuary, giving some idea of the infrastructure involved in wind farm installations. The total array covers an area of 32 km^2 with 35 turbines, three of which are visible in this photograph, connected by 53 km of submarine cables, and an offshore electricity substation, visible on the far left. The red ship is a specialist vessel for servicing the turbines. © Historic England Archive, Alun Bull

authorities for projects within English inshore and offshore areas. Government legislation on cultural heritage, as enshrined in the National Heritage Act of 1980 and subsequently amended, is enacted by the UK parliament in London and applies throughout the UK including the Devolved Administrations of Scotland, Wales and Northern Ireland (the latter currently suspended). However, the implementation of the law and any new legislation for cultural heritage rests with the Devolved Administrations and is carried out by different executive agencies in England, Scotland, Wales and Northern Ireland (see Bailey et al. Chap. 10, this volume; Westley and Woodman, Chap. 11, this volume). Here I deal only with the arrangements in England as a case study in the relationship between a government agency, planning authorities, offshore companies and specialists in archaeology and related disciplines and how these different partners work together to achieve the best possible outcome for the protection of the underwater heritage and the discovery of new information.

26.2 The Role of Historic England

Historic England is the public service established by law to provide specialist and independent advice to regulatory public bodies such as the Planning Inspectorate for nationally significant infrastructure projects or other industrial activities such as aggregate extraction, including the period before an application is submitted for examination by a regulatory authority. In doing so, Historic England provides advice directly to commercial developers and associated consultants, so that all issues to do with cultural heritage are fully considered in EIAs, which are required in support of applications for regulatory consent. However, the appreciation within UK public policy that our shared environment should be considered inclusive of cultural heritage is relatively recent:

> The United Kingdom's seas represent a valuable facet of our cultural heritage. There are the thousands of shipwrecks and remains of aircraft that one might immediately think of, but there are also drowned prehistoric landscapes containing artefacts that are more than 8,000 years old… (House of Commons Environment, Food and Rural Affairs Committee Marine Environment. Sixth Report of Session 2003–2004)

The above quote shows a broad interpretation of the marine environment and reflects the empowerment of Historic England to provide advice to government about underwater and offshore cultural heritage, following a jurisdictional amendment in 2002 to the National Heritage Act which included underwater archaeology in the territorial waters of the UK. A core responsibility is therefore to explain, protect and manage heritage assets that might be impacted by proposed development projects.

Historic England holds information and data to inform decision-making at all levels through a publicly available national archive that currently has 46,000 records including terrestrial as well as marine material (see https://www.pastscape.org.uk/, accessed 3 November 2018). These records include a diverse set of data on underwater heritage including shipwrecks, aircraft losses and other seabed anomalies of possible archaeological interest (see Bailey et al. Chap. 10, this volume, for discussion of these records in relation to prehistoric material). These records provide a key source of information to inform desk-based assessments conducted at the early stage by developers planning seabed development projects. For England, the Planning Act 2008 and the Marine and Coastal Access Act 2009 are key legislative measures that have expanded Historic England involvement in the decision-making process throughout the English sectors of the UK Marine Area (Pater and Oxley 2014).

Once a development project is submitted for regulatory permission, it is possible to secure specific conditions as part of the consent, which specify archaeological works to be completed to professional standards and thereby deliver agreed mitigation for any anticipated impacts, whether these are direct, as in the removal of identified features of possible archaeological interest, or indirect, as in the disturbance of sedimentary conditions that may cause exposure of cultural heritage materials. Crucially, this appreciation of what constitutes the underwater cultural heritage is acknowledged both strategically, that is, as government intentions regarding seabed development programmes in general, and for individual projects.

26.3 Marine Planning and the Historic Environment

While it is possible for individual development projects to identify mitigation measures to offset significant impact to identified cultural heritage assets, it is important that overall planning and management measures do not isolate individual components of the marine environment for attention while ignoring others. It is therefore important to emphasise that cultural heritage and other concepts such as seascape are included in the UK High Level Marine Objectives (HM Government and Devolved Administrations 2009) and within the UK Marine Policy Statement (HM Government and Devolved Administrations 2011), the latter produced as a specific requirement of the Marine and Coastal Access Act 2009. The clear explanation provided in these government publications about what comprises the marine environment, inclusive of cultural heritage, is an important matter. It is also important to note that the Marine Management Organisation (an executive non-departmental public body, sponsored by the UK Department for Environment, Food and Rural Affairs) acting as the marine planning authority, now applies such policy considerations in producing marine plans for the English sectors of the UK Offshore Marine Area (https://www.gov.uk/government/organisations/marine-management-organisation. Accessed 3 November 2018).

Other recent government policy initiatives to promote a marine science agenda have also recognised how studies directed at prehistoric environments can expand our knowledge and thereby support our understanding about future climate change (HM Government and Devolved Administrations 2010).

Other programmes that have strengthened the management of the marine historic environment include projects commissioned through the Aggregates Levy Sustainability Fund (2002–2011), such as the programme of Regional Environmental Characterisation (REC) surveys, which included archaeological objectives. For example, the Humber REC study was a multidisciplinary study of the geology, biology and archaeology of an area of 11,000 km^2 off the east coast of England and demonstrated multiple benefits for effective coordination of geophysical and geotechnical data acquisition (see http://nora.nerc.ac.uk/15037/, accessed 3 November 2018).

26.4 Working with Seabed Industries

The approach taken by the UK reflects differences between what the State is responsible for, strategic assessment to support leasing of areas for development, and the studies that a private developer should complete to assess the risk associated with a particular project and its impact—both positive and negative—on cultural heritage.

At the strategic level, determining how the historic environment including prehistoric landscapes might be affected by development work is informed by the European Commission's directives on Strategic Environmental Assessments (SEA Directive 2001: http://ec.europa.eu/environment/eia/sea-legalcontext.htm. Accessed 3 November 2018). At the individual project level, the EIA follows the criteria of the European Commission's directive on Environmental Impact Assessment (EIA Directive 1985 and subsequent amendments: http://ec.europa.eu/environment/eia/eia-legalcontext.htm. Accessed 3 November 2018). The SEA approach adopted within the UK is for assessment of specified development interests that the government wishes to promote, for example, licensing of oil and gas exploration and production, wind, tidal range and current devices and sub-seabed gas storage options (Department of Energy and Climate Change 2016). Such exercises are desk-based assessments of existing information conducted by consultants appointed by the responsible government department and do not necessarily include dedicated marine survey campaigns. However, mitigation principles will be defined, for example, how it might be possible to use archaeological analysis to generate new information about prehistoric environments in locations where modern seabed developments are planned.

At the individual project level, for example, construction of an offshore wind farm, the approach taken by the UK Government and Devolved Administrations is that private developers are responsible for commissioning marine survey to inform their proposed development. There is therefore a division between the strategic assessment exercises conducted by the State, which are relevant to a particular industrial sector, and the risk that rests with the developer to complete an EIA and apply for consent for a defined project. In situations without an automatic requirement for an SEA to be completed, industry-led initiatives that follow similar assessment criteria can be implemented to produce baseline environmental information, for example, Regional Environmental Assessments used by the marine sand and gravel dredging sector (see http://www.bmapa.org/regulation_and_management/regional_assessment.php, accessed 3 November 2018).

In support of its role, Historic England has produced or supported the production of guidance material for seabed developers on how they should include archaeology within EIAs. In most instances this guidance is thematic and tailored to the specific requirements of particular marine industries, such as marine sand and gravel (aggregate) dredging, the ports sector and offshore renewable energy infrastructure (Wessex Archaeology 2007; Oxford Archaeology 2008; Firth 2013; Cooper and Gane 2016). For example, when an EIA concludes that harmful or other damaging impact is possible, consent conditions can be imposed on the development to steer the delivery of mitigation work within a legal framework. Examples of such conditions are as follows:

- Archaeological interpretation of geophysical and geotechnical data acquired by developers will be subject to analysis and interpretation by professionally accredited and experienced archaeological contractors. Technical reports produced within defined time periods should be agreed and published results made publicly available through an online national archive system.
- Archaeological Exclusion Zones are to be declared by the developer (i.e. Consent Holder) so that identified seabed anomalies of known or possible archaeological interest are avoided by all project contractors conducting seabed-impacting operations.
- Reporting procedures should be adopted by all project contractors, so that if, during delivery of the project, any further materials of possible archaeological interest are encountered, appropriate action can be taken by all parties.

Historic England is also aware that seabed development projects generate a data legacy which can be used to support further archaeological research. For example, seismic data acquired by developer-led marine surveys in the North Sea between Dogger Bank and around the margins of the Outer and Inner Silver Pits was subsequently used to generate new interpretations of palaeoenvironmental landscape sequences (Gaffney et al. 2007). We therefore hope to encourage new partnerships with academic research interests that align with Historic England's published research priorities (see https://historicengland.org.uk/images-books/publications/he-corp-plan-2017-20/, accessed 3 November 2018), the maritime archaeological research agenda for England (Ransley et al. 2013), the North Sea Prehistory Research and Management Framework (Peeters et al. 2009) and the UK Marine Science Strategy (HM Government and Devolved Administrations 2010). Such collaborative action should maximise the potential of the data archives generated by developers and curated through the Marine Data Exchange (see http://www.marinedataexchange.co.uk/, accessed 3 November 2018). International partnerships are also to be promoted to examine submerged prehistoric environmental conditions and to develop effective approaches to assessment, as demonstrated to date by the results of effective collaborative exercises, for example, the former REC survey programme and development-led archaeology work programmes (Dix and Sturt 2011; Salter et al. 2014).

26.5 Marine Development and Assessment of Underwater Cultural Heritage

When a seabed development project requires an EIA to be completed, the assessment will examine the diverse nature of the historic environment as might be encountered using a variety of investigative techniques. The first phase of analysis, conducted by consultants appointed by the developer, is to review relevant published and unpublished documents and reports (i.e. grey literature) which may have been produced for other research purposes or as part of assessment exercises for other developments. It is also worth emphasising that for marine development projects, especially those concerned with offshore renewable energy infrastructure, considerable expenditure is invested in primary data acquisition of seabed features as a necessary part of the construction project, which will then inform mitigation programmes that may be required as a result of the EIA conclusions.

The data gathered through survey programmes should provide complete coverage of the proposed marine development area. For example, for an offshore wind farm, the area should include the turbine array and the electricity export cable route to shore. It is also important to determine how change, as brought about by the proposed development, might have an impact on the significance of any cultural heritage. In this regard, it is relevant to highlight that not all impacts are necessarily negative; a development project can make a positive impact through supporting the publication of professionally produced information that expands knowledge about the historic environment, and this is recognised in the published SEA Environmental Report for Offshore Energy (Department of Energy and Climate Change, 2016).

26.5.1 The Written Scheme of Investigation

A key part of Historic England's advice to developers during the early stages of project planning is to explain the importance of designing survey campaigns to include archaeological objectives, so that data generated are sufficient to support archaeological interpretation and analysis, for example, to reveal sub-seabed prehistoric sedimentary sequences of possible archaeological interest. During this early stage, Historic England discusses with the developer and their professional archaeological consultants the scope of a document known as an archaeological Written Scheme of Investigation (WSI). The use of WSIs is well established for terrestrial development projects and has now been adapted to serve marine development activities.

A standard aspect of EIA exercises, particularly for offshore wind farms, is to consider a wide range of engineering designs that could be used to deliver the total energy output limit for the development, should permission be obtained. For example, different electricity-generating wind turbines of different capacities could be employed using different foundation designs, each with a different impact on the cultural heritage. Consent for the development will be informed by the worst-case scenario—the design that has the greatest potential impact. It usually follows from this that the developer has to commission additional high-resolution marine survey to inform delivery of the final project design. The preparation of a draft or outline WSI as a supporting document within the application for consent is a crucial step to be taken by the developer to demonstrate how archaeological analysis and interpretation will be completed as part of the construction programme.

The draft WSI is prepared concurrently with the production of the EIA and should set out methods of archaeological analysis to be employed if the project is approved. This is a crucial condition for the granting of consent for the project. Thereafter, during the post-consent period, once the final design for the project is selected, the WSI will be used to support the archaeological interpretation of survey data commissioned at all stages of the project, including pre-construction, construction and post-construction.

The WSI should include several components. First, it will detail desk-based sources of information in local and national archives such as records of prehistoric material, as derived from the original EIA, about the character of the historic environment. Secondly, The WSI will include methodologies for the interpretation of geophysical data such as side-scan sonar, multibeam bathymetry, magnetometer and sub-bottom seismic profiling, all of which can produce different visualisations of seabed topography and sub-seabed sedimentary structures. These in their turn can indicate the presence of buried landscape features such as former river and terrace systems with potential to contain organic sediments of palaeoenvironmental significance and archaeological material.

Finally, it will produce survey-specific method statements on how data collected will inform archaeological interpretation, for example, the use of shallow seismic data collected along the route of electricity export cables. It is from these method statements that information is generated to produce the final technical reports.

The draft technical reports will be reviewed by local and national curators, that is, archaeologically trained personnel employed by local planning authorities and Historic England, to check that it complies with established professional archaeological practice (see https://www.archaeologists.net/, accessed 3 November 2018), and a final report will be produced by the consultant employed by the developer and formally agreed in accordance with a specified time frame with the regulatory authority. Failure by the Consent Holder (i.e. the company responsible for the development) to comply with the conditions of consent, such as archaeological mitigation, carries the risk of enforcement action against the developer.

A key instrument to demonstrate completion of these defined tasks is deposition of completed reports, within an agreed time frame, with a UK online archive system known as OASIS (Online AccesS to the Index of archaeological investigations (see http://oasis.ac.uk/pages/wiki/Main, accessed 3 November 2018) (for examples see Table 26.1). From here reports can be uploaded as publicly available documents to the website maintained by the Archaeological Data Service (ADS) (see archaeologydataservice.ac.uk/archives/view/greylit/, accessed 3 November 2018).

Table 26.1 Examples of completed geoarchaeological reports and submission of OASIS forms to enable access through local Historic Environment Records (HERs) and their respective national Heritage Bodies within the UK

Offshore wind farm	Geoarchaeological reports	OASIS ID
Gunfleet Sands (Thames Estuary)	Archaeological recording of geotechnical cores	Englishh1-87577 (2007); Englishh1-87573 (2008); Maritime2-169743 (2013)
Hornsea ONE (North Sea)	Geoarchaeological assessment report	Maritime2-324357 (2018)
Hornsea TWO (North Sea)	Assessment of borehole and vibrocore logs	Wessexar1-164773 (2013)
Humber Gateway (North Sea)	Assessment of borehole and cone penetrometer tests	Wessexar1-155988 (2013)
London Array (Thames Estuary)	Organic material recovered during benthic fish surveys	Gifford1-198856 (2015); gifford1-271153 (2016)
Lynn and Inner Dowsing (North Sea)	Geoarchaeological assessment. Final report	Wessexar1-59263 (2009)
Westermost Rough (North Sea)	Analysis of borehole and vibrocore material	hampshir2-157098 (2013)

Effective mitigation following the procedures outlined above can produce new knowledge and understanding and, in some cases, has resulted in publication in the specialist literature, for example, the data generated by the Dudgeon offshore wind farm (Brown et al. 2018).

26.6 Seabed Development and Other Mitigation Measures

The preparation of a project-specific archaeological WSI starts the process of mitigation after consent has been granted, so that identifiable seabed anomalies and sub-seabed features identified in the EIA exercise of potential archaeological or historic interest are dealt with in an agreed manner. It is a shared goal between any seabed developer and Historic England that a primary strategy to reduce risk of impact is by avoidance and in situ protection. This approach is delivered using Archaeological Exclusion Zones (AEZs), which are spatially delineated around the anomaly or other feature of interest. While this approach might be readily deliverable for a feature such as a shipwreck, the preferred approach to deliver appropriate mitigation for areas of seabed containing prehistoric landscape features is through the production of models of sedimentary sequences based on sub-bottom geophysical data and corroborated with geotechnical material (e.g. data from boreholes and vibrocores). In some instances, early involvement in survey planning, as part of WSI implementation, allows additional data to be acquired that directly support palaeoenvironmental analysis. This type of mitigation for identified elements of submerged prehistoric landscapes is appropriate given that access to such buried features could be compromised in the future if seabed infrastructure such as cables and foundations are likely to be left in situ over a wide area of seabed.

To support the production of sedimentary deposit models, Historic England commissioned guidance on offshore geotechnical investigations and archaeological analysis through the former Collaborative Offshore Wind Research into the Environment (COWRIE) Group. An important aspect of this guidance was to explain how the acquisition of geotechnical material should be planned to include archaeological objectives and to deliver mitigation through the geoarchaeological interpretation and modelling of palaeoenvironmental features (Gribble and Leather 2011). Established working practice for palaeoenvironmental analysis of geotechnical material frequently includes phased interpretation of materials, from visual inspection of logged sediment sequences in cores through to radiocarbon dating and identification of pollen, microcharcoal, molluscs and ostracods. For example, geotechnical vibrocore material obtained from the NEMO submarine electricity interconnector cable project at a location 12 km off Ramsgate (Kent) revealed the presence of successive environments including early Holocene freshwater conditions within a possible wooded river valley comprised predominantly of pine and hazel trees and considered typical of the Early Mesolithic period (Wessex Archaeology 2015).

Although a sedimentary deposit model is designed to illustrate submerged prehistoric landscape complexity, site-specific prehistoric finds do occur, such as the discovery in 2008 of 28 flint handaxes originating from in situ contexts and estimated to be around 250,000 years old; the handaxes derived from gravel dredged from a licensed marine aggregate area 13 km off Great Yarmouth in the southern North Sea. Following this discovery, and in partnership with the dredging operator, Historic England supported a number of studies to determine the extent of the archaeological materials within the active dredging area and also to devise appropriate monitoring procedures to determine whether handaxes in primary context existed adjacent to the dredging zone within Area 240 (Wessex Archaeology 2011; Russell & Tizzard 2011; Salter et al. 2014; Tizzard et al. 2014, 2015; see also Bailey et al., Chap. 10, this volume). The original find location in Area 240 remains subject to a voluntary exclusion zone to avoid any further direct impact by dredging operations. This approach of working with industry is essential, as archaeological protection through area-based statutory designation is not possible within Area 240, since the finds occur in the absence of any evidence of human-built structures, which is a key criterion within existing legislation for historic environment protection in England.

26.7 Reporting of Archaeological Discoveries

Once consent has been obtained for an offshore development project, there will be specified conditions attached to the consent, which must be delivered by the developer. To address a situation in which an archaeological site or other material of archaeological interest is unexpectedly encountered during construction works, a reporting protocol is used to ensure appropriate action, including the following:

- The site is avoided by all subsequent seabed-impacting activities associated with delivery of the consented development to reduce risk of further disturbance.
- If the particular activity resulted in the recovery, inadvertently, to the surface of material of possible historic or archaeological interest, the protocol will explain how to identify such materials and how to correctly store the items.
- Procedures for notifying the relevant parties internal and external to the development project to reduce delay in subsequent decision-making.

To assist effective implementation, guidance has been published specifically for the offshore renewable energy sector to inform the preparation of reporting procedures, so that archaeological reporting protocols can follow a standard format, which can be adapted to individual projects (The Crown Estate 2010, 2014). The reporting protocol can be used at any stage such as during preliminary offshore environmental assessment exercises. For example, the protocol was used to report archaeological interests associated with peat blocks obtained from a benthic trawl survey campaign for possible offshore wind farm development on the Dogger Bank (central North Sea). The subsequent programme of palaeoenvironmental assessment revealed evidence of a now drowned terrestrial environment from around 12,000–8000 years ago (Russell and Stevens 2014).

26.8 Professional Development and Training

To support ongoing professional development, Historic England has either delivered or commissioned training events to ensure that curators (both national and local), developers and archaeological consultants have the opportunity to maintain skills and knowledge about how the historic environment is included within development assessment exercises. Technical training is also highly relevant given the speed of technological change in survey data-gathering platforms and processing software. To support this objective, to date, the following courses have been commissioned:

- Marine renewable energy developments and the historic environment (Department for Continuing Education, University of Oxford) in March 2014.
- Marine geotechnical training course for archaeologists (National Oceanographic Centre, Southampton) in January 2015 (see https://www.southampton.ac.uk/coars/services/geoarchcpd.page, accessed 3 November 2018).
- Marine geophysical survey techniques for archaeologists (National Oceanographic Centre, Southampton) in January 2017.
- Coastal development and the historic environment (Historic Environment Local Management programme) in September 2018 (see https://historicengland.org.uk/services-skills/training-skills/helmtraining/, accessed 3 November 2018.

Overall, the publication of thematic advice for the marine renewable power generation sector now equips developers with the information necessary to select viable strategies for data capture to deliver the necessary mitigation. It also gives curators the confidence to review and question the technical reports produced. These courses have helped to build on published guidance, such as that available for marine geophysical survey data (Plets et al. 2013). However, it is likely that HE will have to re-commission these training events since techniques both in data capture and processing are constantly changing and also to ensure that new staff members within curatorial bodies, consultancies and the developers themselves are familiar with the effective inclusion of cultural heritage as part of project planning, examination and delivery.

26.9 Conclusions

Historic England is the national curator for the marine environment and it provides advice to developers, their archaeological consultants and regulators for projects within the English inshore and offshore areas. It has supported the publication of guidance targeted at different marine industry sectors and will continue to commission projects to address key issues as necessary to support effective management of the historic environment. It is therefore important that such effort is described in terms of a constructive approach based on an understanding of how change, such as new seabed infrastructure projects, may affect the historic environment (see https://historicengland.org.uk/advice/constructive-conservation/conservation-principles, accessed 3 November 2018). Such an approach is based on identifying risk to known and unknown features of the historic environment within the development area. Central to this approach is to ensure that survey campaigns are effectively planned to increase our understanding about the historic environment through the analysis of developer-generated survey data. This matter is reflected in national policy used to support regulatory decision-making, so that completed assessments deliver effective description and analysis of environmental data including cultural heritage.

This approach is also based on the premise that the positive outcomes from offshore development in terms of new knowledge and wider understanding are dependent on completion of defined programmes of work to professional archaeological standards. Such an approach not only has scientific value but should also encourage greater public understanding and appreciation of the underwater cultural heritage and the ways in which its study contributes to a better understanding of environmental change. A good example is the recent British media interest in the evidence of a prehistoric landscape found during survey work for a proposed offshore wind farm in the southern North Sea (BBC Online at https://group.vattenfall.com/uk/newsroom/news-press-releases/pressreleases/stories/will-the-secrets-of-doggerland-be-revealed-by-unique-find-at-north-sea-site).

Much of the work described above is coming about as the result of a huge investment in renewable sources of energy, particularly offshore wind farms, in response to climate change. It is therefore highly appropriate that examination of the cultural heritage associated with these developments fully embraces analysis of prehistoric environments and contributes to the modelling of past climates, conditions as experienced by past peoples and a better appreciation of how the environment changes over time. In Historic England, we see this as an important exercise that can be used to support public communication and explanation about how environments change and why action is necessary to address climate change as it affects the world we live in today.

Acknowledgements Thanks are due to Edward Salter (formerly of English Heritage and now of The Crown Estate, London) for help in preparing the presentation given at the SPLASHCOS Conference 'Offshore Industry and Archaeology: a creative relationship' in Esbjerg, Denmark, on 14 March 2013, to Thijs Maarleveld for his original invitation to the conference and to Geoff Bailey, Hans Peeters and an anonymous reviewer for helpful comments that have improved the text.

References

Brown A, Russell J, Scaife R, Tizzard L, Whittaker J, Wyles SF (2018) Lateglacial/early Holocene palaeoenvironments in the southern North Sea Basin: new data from the Dudgeon offshore wind farm. J Quat Sci 33(6):597–610

Cooper V, Gane T (2016) The assessment and Management of Marine Archaeology in port and harbour development. Wessex Archaeology, Salisbury

Department of Energy and Climate Change (2016) UK offshore energy strategic environmental assessment (UK OESEA3) environmental report. The Stationary Office, London. https://www.gov.uk/government/collections/offshore-energy-strategic-environmental-assessments

Dix JK, Sturt F (2011) The relic Palaeo-landscapes of the Thames estuary. MALSF, London

Firth A (2013) Historic environment guidance for wave and tidal energy. Fjordr, Wiltshire

Gaffney V, Thomson K, Fitch S (eds) (2007) Mapping Doggerland: The mesolithic landscapes of the Southern North Sea. Archaeopress, Oxford

Gribble J, Leather S (2011) Offshore geotechnical investigations and historic environment analysis: guidance for the renewable energy sector. COWRIE, London

HM Government and Devolved Administrations (2009) Our seas: a shared resource, high-level marine objectives. Department for Environment, Food and Rural Affairs, London. https://www.gov.uk/government/publications/our-seas-a-shared-resource-high-level-marine-objectives

HM Government and Devolved Administrations (2010) UK marine science strategy. Department for Environment, Food and Rural Affairs, London. https://www.gov.uk/government/publications/uk-marine-science-strategy-2010-to-2025

HM Government and Devolved Administrations (2011) UK marine policy statement. The Stationary Office, London. https://www.gov.uk/government/publications/uk-marine-policy-statement

Oxford Archaeology (2008) Cumulative impact assessment and the marine historic environment. Oxford Archaeology/George Lambrick Archaeology and Heritage, Oxford

Pater C, Oxley I (2014) Developing marine historic environment management policy: the English Heritage experience. Mar Policy 45:342–348

Peeters H, Murphy P, Flemming N (2009) North Sea prehistory research and management framework. Rijksdienst voor het Cultureel Erfgoed, Amersfoort

Plets R, Dix J, Bates R (2013) Marine geophysics data acquisition, processing and interpretation guidance notes. English Heritage, London

Ransley J, Sturt F, Dix J, Adams J, Blue L (eds) (2013) A maritime archaeological research agenda for England, CBA Research Report, vol 171. Council for British Archaeology, York

Russell JW, Stevens CJ (2014) Palaeoenvironmental assessment of peat samples. The Crown Estate, London
Russell JW, Tizzard L (2011) Seabed prehistory: site evaluation techniques (area 240) synthesis. Wessex Archaeology, Salisbury
Salter E, Murphy P, Peeters H (2014) Researching, conserving and managing submerged prehistory: national approaches and international collaboration. In: Evans AM, Flatman JC, Flemming NC (eds) Prehistoric archaeology on the continental shelf: a global review. Springer, New York, pp 151–172
The Crown Estate (2010) Model clauses for archaeological written schemes of investigation. The Crown Estate, London. https://www.wessexarch.co.uk/sites/default/files/field_file/4_WSI%20Renewables.pdf
The Crown Estate (2014) Protocol for archaeological discoveries: offshore renewables projects. Wessex Archaeology, Salisbury. https://www.thecrownestate.co.uk/media/1782/ei-protocol-for-archaeological-discoveries-offshore-renewables-projects.pdf
Tizzard L, Bicket AR, Benjamin J, De Loecker D (2014) A middle Palaeolithic site in the southern North Sea: investigating the archaeology and palaeogeography of area 240. J Quat Sci 29(7):698–710
Tizzard L, Bicket AR, de Loecker D (2015) Seabed prehistory: investigating the palaeogeography and early middle Palaeolithic archaeology in the southern North Sea, Wessex archaeology report 35. Wessex Archaeology, Salisbury
Wessex Archaeology (2007) Historical environment guidance for the offshore renewables energy sector. COWRIE, London
Wessex Archaeology (2011) Licence area 240: archaeological monitoring of dredging activity. Wessex Archaeology, Salisbury
Wessex Archaeology (2015) Nemo link: archaeological stage 4 analysis. Wessex Archaeology, Salisbury

27 Beam Trawls and Bones: A Reflection on Dutch Fisheries

Thijs J. Maarleveld

Abstract

This chapter deals with fishing and archaeology. Knowledge held by fishermen has contributed to underwater archaeology's great moments. It is comparable to 'local' knowledge on land, although the locales may be far offshore. To some extent, fishing interests and the management of underwater cultural heritage are at odds but hardly as much as sometimes claimed. Future cooperation with fishermen is of the essence, as the fishing industry has been an essential informer for the development of archaeology offshore, all over the world, and continues to be so. This chapter explores how the development of fishing techniques over the last 150 years has informed prehistoric archaeology of the European continental shelves, notably of the North Sea. It does so through a historical analysis of technological development in its social setting and by highlighting some developments in Dutch fishing communities. It puts collecting of bones and trade in antiquities in perspective. It is mostly concerned, however, with the contingent knowledge base of archaeology and therefore informs archaeological epistemology.

T. J. Maarleveld (✉)
Maritime Archaeology Programme, University of Southern Denmark, Esbjerg, Denmark
e-mail: t.maarleveld@sdu.dk

Keywords

Archaeological epistemology · Fishing techniques · Industrial collaboration · Palaeontology · Private collectors · Underwater cultural heritage

27.1 Introduction

This volume provides an overview of the prehistoric archaeology of the European continental shelves. It gives an overview of what we know. Obviously, it occasionally addresses what we do not know or only can infer, but the core is hard evidence. Hard evidence is tangible and measurable. But nevertheless, it is subject to all kinds of contingencies, trends and influences that are not directly measurable or in evidence. So is its interpretation. Doubtless, that is true for all scientific evidence, but here we focus on prehistoric evidence offshore.

What we know of the past is strongly qualified by what is preserved and what is not. That is relatively straightforward, and archaeologists have, therefore, always been preoccupied by physical and taphonomic processes that define preservation of archaeological phenomena, stratigraphy of archaeological sites and physical changes of landscapes (Flemming et al. 2017a, b). Also, it has long been recognised that abandonment, loss or purposeful deposi-

G. Bailey et al. (eds.), *The Archaeology of Europe's Drowned Landscapes*,
Coastal Research Library 35, https://doi.org/10.1007/978-3-030-37367-2_27

tion results in very different forms of preservation as well. Such considerations are quite central to a theoretical understanding of archaeological knowledge. In acknowledgement of Michael Schiffer's seminal book of 30 years ago (Schiffer 1987), all these processes that determine what we know or can know are commonly referred to as *formation processes of the archaeological record*. But there is more to it than past behaviour or physical preservation. Evidence only exists when it is recognised as such (Maarleveld 2010).

That is what this chapter is about. The logic of scientific discovery (cf. Popper 1935), and the way in which scientists operate, means that they focus on creating discoveries through purposeful action. Offshore, this leads to systematic data capture strategies (Bailey et al. 2017; Missiaen et al. 2017). But research-oriented survey is not by any means the only process through which we create the building blocks of what we know. Harnessing the evidence that industry produces as a by-product of their activities has been fruitful as well (Gaffney et al. 2007, 2009). And for the future, it is undoubtedly important to integrate data capture and research strategies in impact mitigation of offshore development (Maarleveld 2003, 2006; Vos et al. 2015; Pater, Chap. 26, this volume). After all, that is the means by which much archaeological activity is funded and how archaeological services and endeavours are preponderantly organised (Firth 2015). A precondition, however, is that evidence is recognised.

An open mind and awareness of what to look for—and what to be struck by!—is of prime importance for all contingent or systematically organised data gathering. With that in mind, this chapter explores the relationship between contemporary fishing and the knowledge of submerged prehistory that is central to this book. It focuses on the North Sea and on Dutch fisheries. Through the discussion of techniques and mentalities in their historical setting, it reflects on the contingent ways in which archaeological knowledge comes about.

27.2 Setting and Problem Definition

Amongst prehistorians, the trawler *Colinda* is undoubtedly the most famous modern fishing vessel by far. It was, after all, this vessel that brought up the lump of peat in which the barbed antler point of the Leman and Ower Banks was discovered in 1931 (Fig. 27.1; see also Bailey et al. Chap. 10, this volume, Fig. 10.13). The find was recognised as important and promptly published in the influential anthropological journal *Man*, where it was described as a harpoon of Mesolithic date (Burkitt 1932). The barbed point has been cited as archetypal evidence of submerged prehistory ever since (e.g. Clark 1952,

Fig. 27.1 The barbed antler point of the Leman and Ower Banks. It was discovered in a lump of peat that was fished up by the trawler Colinda, registered in Lowestoft as LT 382 in 1931. (Adapted from Clark 1975; see also Bailey et al., Chap. 10, this volume, Fig. 10.13)

p. 45; Louwe Kooijmans 1972; Clark 1975, p. 134; Momber 2011, p. 92). But in a way, *Colinda* represents whole fleets of fishing vessels that have impacted prehistoric archaeological remains and that have influenced scientific research.

A regular flow of fishing finds has marked the gradual extension of our knowledge of submerged early prehistory (e.g. Bjerck 1995, p. 133; Sanger 1995, p. 341) and particularly in the North Sea Basin (Peeters and Amkreutz, Chap. 8, this volume). As elsewhere, some of the artefacts are of flint or stone. Bone and antler objects, however, are far more prominent.

Why do bone artefacts pop up in fishing nets, whereas hand axes and stone maces are mostly found in dredged material? There are several explanations for this. Some are related to deep time, to the specific geological circumstances that embedded the finds after deposition and subsequent preservation. Equally pertinent, however, are the contemporaneous processes that lead to discovery. Some of these processes are general, some are specific to dredging, and some are specific to the fishing industry. No finds become known if no one heeds or registers them. No dredging finds occur in areas where no dredgers are active. Moreover, such finds depend on the way marketable aggregates are separated from overburden and waste and whether separation or transport allows for human observation. Here, we focus on fishing. Fishing techniques differ and have different impacts and results. But the background, knowledge and mind-set of the fishing crews differ as well and are equally important. Both are relevant to understanding the distribution of finds. They are relevant to what we know and what we miss. Both are also relevant to the relationship between archaeologists, fishermen, environmental planning and heritage management.

27.3 Fishing Techniques

Colinda was a drifter, a relatively large trawler that mostly used a long trawl of drifting nets with negligible impact on the seafloor. She was a modern, industrial ship with a steam engine, registered at the English port of Lowestoft, owned by a shipping company and manned by a hired captain and crew. It is likely that she occasionally engaged in trawling for flatfish and cod with a so-called otter trawl. Many drifters occasionally did, and it would explain why a lump of peat was brought up in the first place. In fishing with an otter trawl, two otter boards are fastened to a funnel-shaped net (Fig. 27.2). Under motion the boards pull sideways, opening the net. It is a technique that is still widely used today in a wide variety of fisheries. In the North Sea, it used to be the dominant technique for bottom trawling. During the late 1950s and 1960s, the otter trawl gave way to a new technique, the beam trawl, the use of which until then had been limited to shellfishing. In a beam trawl, the mouth of the funnel-shaped net is held open by a solid metal beam, attached to a shoe or skid on either side. The skids slide over the sea bottom. For several decades the beam trawl defined Dutch North Sea fisheries. It was effectively the producer of most North Sea palaeontological remains.

Under the beam of the beam trawl and between its two skids, tickler chains are mounted (Fig. 27.3). The chains disturb flatfish and make them swim into the net. Besides tickling fish, this gear obviously catches on any irregularity. Fishermen who are active in environments with rocky outcrops have therefore continued to prefer otter boards, which are less likely to get stuck. For the sandy bottom of the shallow North Sea, however, the beam trawl became paramount. The tickler chains have thoroughly impacted the sediment, stirring up any individual object at (or close to) the surface. Ever fresh material was removed from outcropping contexts. The top layer of the North Sea bottom was systematically ploughed through, removing all sessile life and collecting every conceivable kind of litter into the beam trawling nets. Now in itself, this is not enough to explain the substantial number of stray finds that Dutch fishermen in particular have come up with.

Everything that is caught in a net and comes on board a fishing vessel, but which is not fish, will immediately be dumped. That is to say, it

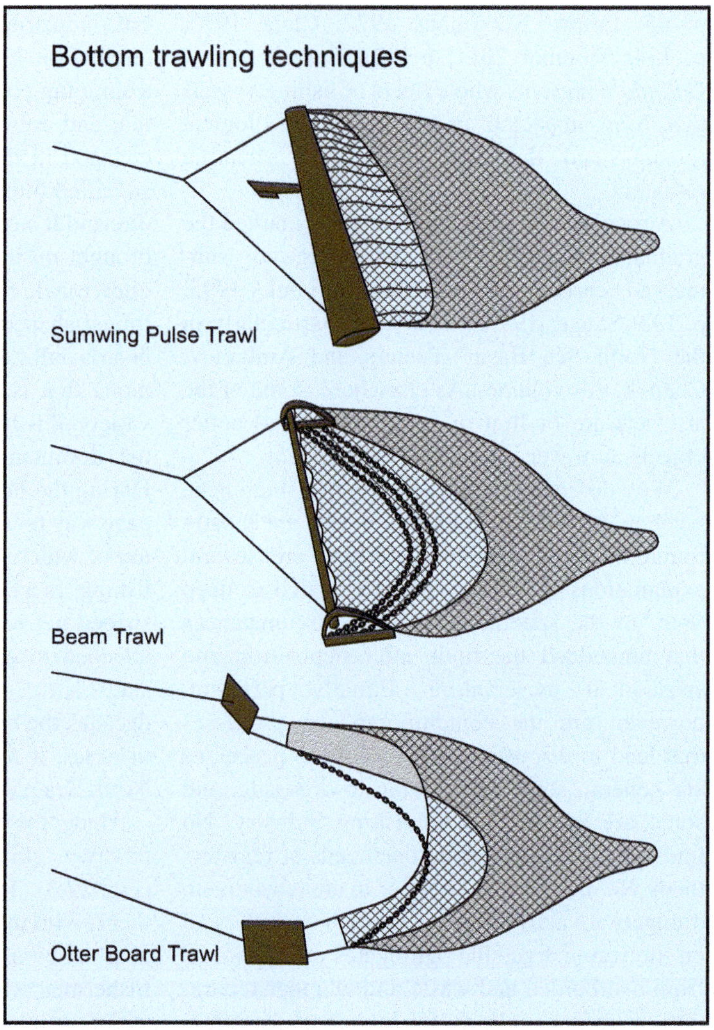

Fig. 27.2 Bottom trawling techniques used in North Sea cod and flatfish fisheries. The otter trawl was paramount until it was replaced by the beam trawl in the 1960s. The heavy tickler chains plough through the bottom surface and call for great traction power. During the last 15 years, the beam trawl is gradually being replaced by hovering wings, like the Sumwing and electric pulse ticklers instead of tickler chains. This means less resistance and a reduction in fuel consumption. A side effect is less bottom impact. Drawing by Thijs Maarleveld

will immediately be dumped unless it is remarkable and therefore remarked on. Peat pellets, wood chips, stones, bricks, bottles and cans are not heeded, and many a hand axe may have gone unnoticed as just another stone. But bones are different, at least amongst the fishermen from Stellendam, Urk, Texel and a few other places. These are three fishing communities in the Netherlands that are closely connected, despite their location in the south, the middle and the north of the country, respectively. The connections between the three communities are reinforced by the fact that they engage in similar types of fishery, organised through similar structures of ship ownership, supply and profit sharing (Van den Bossche et al. 1999). Moreover, these communities share the same evangelic reformed Protestant creed.

At some point, all three communities consciously engaged in deliberate fishing for bones as an attractive supplementary catch. The story of bone fishing does not, however, start in the North Sea fisheries. It started in the estuarine channels of the Scheldt, the hinterland of the Stellendam fishermen, not in the North Sea itself. From the mid-twentieth century onwards, bone fishing in the Scheldt estuary included some notable fishermen from the town of Zierikzee. It started several generations earlier, however, when fossilised bones were brought up by the traditional sailing

Fig. 27.3 In beam trawling, two funnel-shaped nets are deployed on either side of the trawler. Both are pulled over the sea bottom by great engine force and hauled simultaneously. The metal beam with skids as well as so-called tickler chains can be clearly discerned. Photo by Thijs Maarleveld

craft of the Scheldt estuary, wooden ships no longer than 15 m, with simple bottom tackle and a small crew (Van Beylen 1993; Van Ginkel 1991). From around 1910 onwards, small combustion engines had started to be mounted in these vessels, but certainly not in all, and they were hardly used for more than manoeuvring and running to and from a harbour. Unlike the drifting and lugging industry, the structure of these fisheries or those of Urk and Texel hardly changed until after the closure of the Zuiderzee, the economic crisis of the 1930s and the Second World War (Van Ginkel 1993; Dorleijn 1982–1996).

27.4 Geographical Setting and Quaternary Geology

To understand the meaning of the estuarine fisheries in relation to prehistoric archaeological research, it is useful to consider the geology of the area. The landscape of the Netherlands, of which the estuary is a part, is a very young landscape. It is flat. There is no rock. There are hardly any outcrops of early Quaternary or earlier deposits, apart from the area around Maastricht in the far southeast of the country (see also Peeters and Amkreutz, Chap. 8, this volume). In fact, almost all surface deposits in the country and its adjacent sea are much younger (Laban 1995; Berendsen 1996). This is clearly true for recent marine and riverine clays and sands as well as the peat areas and polders. The oldest recognisable landforms derive from the Saalian glaciation and are no more than 100–150,000 years old, but most cover sands date from after the Last Glacial Maximum (20,000 BP). With the area being a subsidence basin, sediments are, however, very deep. In principle, the deep stratigraphy gives a detailed geological record. But in practice it is hardly accessible, unless through coring, deep (=wet) excavation or deep natural erosion. The latter phenomenon is one that is carefully monitored in a country that is prone to

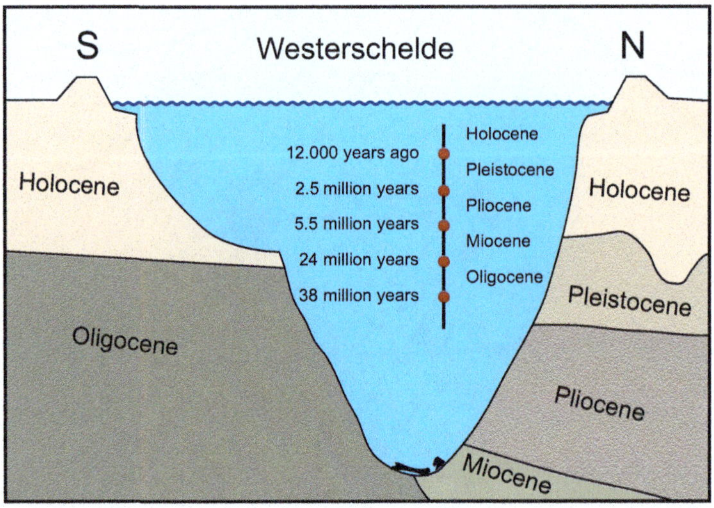

Fig. 27.4 The Scheldt estuary lies on the southern border of the subsidence basin that characterizes the geology of the Netherlands. Erosion pits in the estuary therefore cut through the entire Holocene and Pleistocene sequence but also expose older layers. On the southern side of the Westerschelde, the Holocene immediately overlies the Oligocene. Fossils from all intermediate periods accumulate in the deepest parts of the erosion pits. Drawing by Thijs Maarleveld, after Wetsteyn 2001

and conscious of the risks of dike subsidence and flooding. The estuarine channels of the Scheldt, close to the Belgian border and at the rim of the subsidence basin, are a case in point. Contained between dikes that protect the land from the tides, these channels have cut ever more deeply into ancient sediments. At some points, the erosion channels are more than 40 m deep (Fig. 27.4).

From 1874 onwards, these deep erosion pits have enjoyed the full attention of geologists and palaeontologists. In that year Dr. J.C. de Man acquired the first fossil bones from an Arnemuiden fisherman for the Scientific Society of Zeeland (Zeeuwsch Genootschap der Wetenschappen, later the Royal Society, Koninklijk Zeeuwsch Genootschap der Wetenschappen). He and many of his successors developed into keen palaeontological specialists of Pleistocene mammals and regularly paid local fisherman to purposefully cast their nets into the deep pits in search of fossils. A large collection was thus built up until around 1930 (Van Nieulande et al. 2017). It contributed enormously to the definition of the Pleistocene fauna, one to two million years ago. Black, strongly fossilised bones of land mammals such as sabre-tooth tiger, early horses, deer and elephants, as well as sea mammals such as seals, walruses and whales are the oldest examples. In everyday language, the Early Pleistocene fauna is referred to as the *black bone fauna*, despite the fact that several environments are represented (Hooijer 1957; Drees 1986). Incidental finds date even earlier, from the Pliocene or even the Miocene, up to around 10 million years ago. Apparently, reworking gathered together fossils from different ages in the deepest parts of the erosion pits. Fishing in these deep erosion pits thus contributed to science in a major way (Van der Vlerk 1938).

27.5 The Leiden Connection

Much to the chagrin of some of Dr. de Man's successors, the active involvement and curation of the Royal Society of Zeeland waned somewhat after the 1930s (Moraal 1964, p. 45). But more fossils were being found, both in deep excavation and in erosion pits where fishermen cast their nets. Not all Zeeland fishermen changed to North Sea fisheries and the beam trawl. While specialising and modernising in different ways, some

Fig. 27.5 The owners of the shellfish trawler ZZ8 *Wilhelmina*, the Schot family of Zierikzee, established a close relationship with the staff of the Natural History Museum in Leiden and started to organise yearly fossil hunts in the tidal inlets of the Scheldt estuary in the 1950s. Similar trips had occasionally been organised with the *YE7*, a fishing vessel from Yerseke as early as 1926. Photo courtesy of Kotterfoto

remained in the estuary, focusing especially on oyster and mussel culture. And some continued their interest in fossils and prehistoric bones, although this was certainly not a very general phenomenon. In pursuit of rare specimens, the Schot family, for instance, who run a mussel enterprise in Zierikzee, used the small beam or rake trawl of their ship in the deeper locations (Fig. 27.5). In pursuit of scientific expertise to identify their finds, they found ready enthusiasm with the then young keeper of Pleistocene palaeontology at the National Museum for Geology and Mineralogy in Leiden. As a consequence, they developed a close cooperation with the museum's staff in the 1950s. The cooperation was formalised into a 'society' with a yearly fieldtrip. The society was called 'Kor en Bot', which means 'Beam-trawl and Bone'.

In 1960 the two scientists directly involved, G. Kortenbout van der Sluijs (Fig. 27.6) and C.J. Overweel, secured a substantial grant. The society went fishing for bones for several months, adding substantially to the understanding of the fauna in Tiglian and Pretiglian times, around 2.5 million years ago. Scientifically, the expedition was a success. A recent quantitative analysis reconfirms this (Scager et al. 2017). But what is perhaps more important in the present context is that it got extensive public exposure. A short but beautifully made documentary in the 'Polygoon' newsreel was shown as a curtain raiser in all cinemas in the Netherlands in 1961. As this is before the media explosion, it is likely that very many would have seen it and been impressed by it. Many fishermen saw their trade depicted on the wide screen—a rare event—and will have realised that bones that they might find could be of interest. Their consciousness was raised, part of their mind-set defined.

Around the same time, collectors of 'black bones' from the Zeeland gullies became aware of toolmarks. Rare finds such as an early Mesolithic antler axe fished from the Westerschelde in 1957 were published by archaeologists (Elzinga 1960).

Fig. 27.6 G. Kortenbout van der Sluijs was the palaeontologist behind the promotion of deliberate fishing for fossil bones. There is a deliberate rhyme between his name Kortenbout and the society Kor en Bot. Notable was his expedition of 1961. Here he is photographed in August 1973. Photo by C. Kotvis, courtesy of Zeeuwse Bibliotheek

Such finds aroused substantial interest amongst prehistorians, as the early prehistory of the western part of the Netherlands was largely unknown at the time.

27.6 The Boom of the 1960s to the 1980s

It is interesting to consider that this developing awareness occurred at a time when many fishermen were experimenting, adopting the beam trawl and investing in larger and more powerful equipment. While 400 or 600 horsepower (HP) was still considered powerful for a fisherman's main engine in the 1960s, 4000 HP was more the norm in the early 1980s. Also, fishermen developed a better understanding of the extensive fishing grounds of the North Sea and got to know how the beam trawl not only efficiently harvests any fish close to the sea bottom but also disturbs other objects that may be at or close to the bottom surface. Given the background of experience in Zeeland and the consciousness raised by the newsreel in the cinemas, many a find was reported to the palaeontologists at the museum in Leiden as the beam trawlers ploughed their way through much of the North Sea bottom, including those parts where concentrations of large mammal bones occur (Fig. 27.7).

Reports and deliveries sky-rocketed; so much so in fact, that the curators were instructed by management to be very selective in what to accept. Mammoth bones, after all, are recognised more easily, but take up far more space, than the bones of small rodents. The interest, however, had become endemic, and besides the museum in Leiden, other collections were built up, in local museums, in topical displays and in private hands. A very active study group emerged: the study group for Pleistocene mammals, collecting, publishing and exchanging rare fossils. Established in 1983, it continues to be more active than ever. Jointly with the Geoscience Institute at the University of Utrecht, the study group publishes the journal *Cranium*, which in the 30 years of its existence has built up a solid scientific reputation, integrating and revitalising the early initiatives of the Royal Society of Zeeland (Moraal 1991). Fishing for bones in the North Sea has thus contributed at least as much to science as fishing for bones in the deep erosion gullies of Zeeland. The fossils span a similar range of time, although most North Sea finds relate to the later phases of the Pleistocene, whereas finds from the estuary go back further.

27.7 Going Commercial

Before coming back to the relevance for archaeology, a few words should be said on collecting policies. By far the most iconic specimens and the largest amount fished from the North Sea are

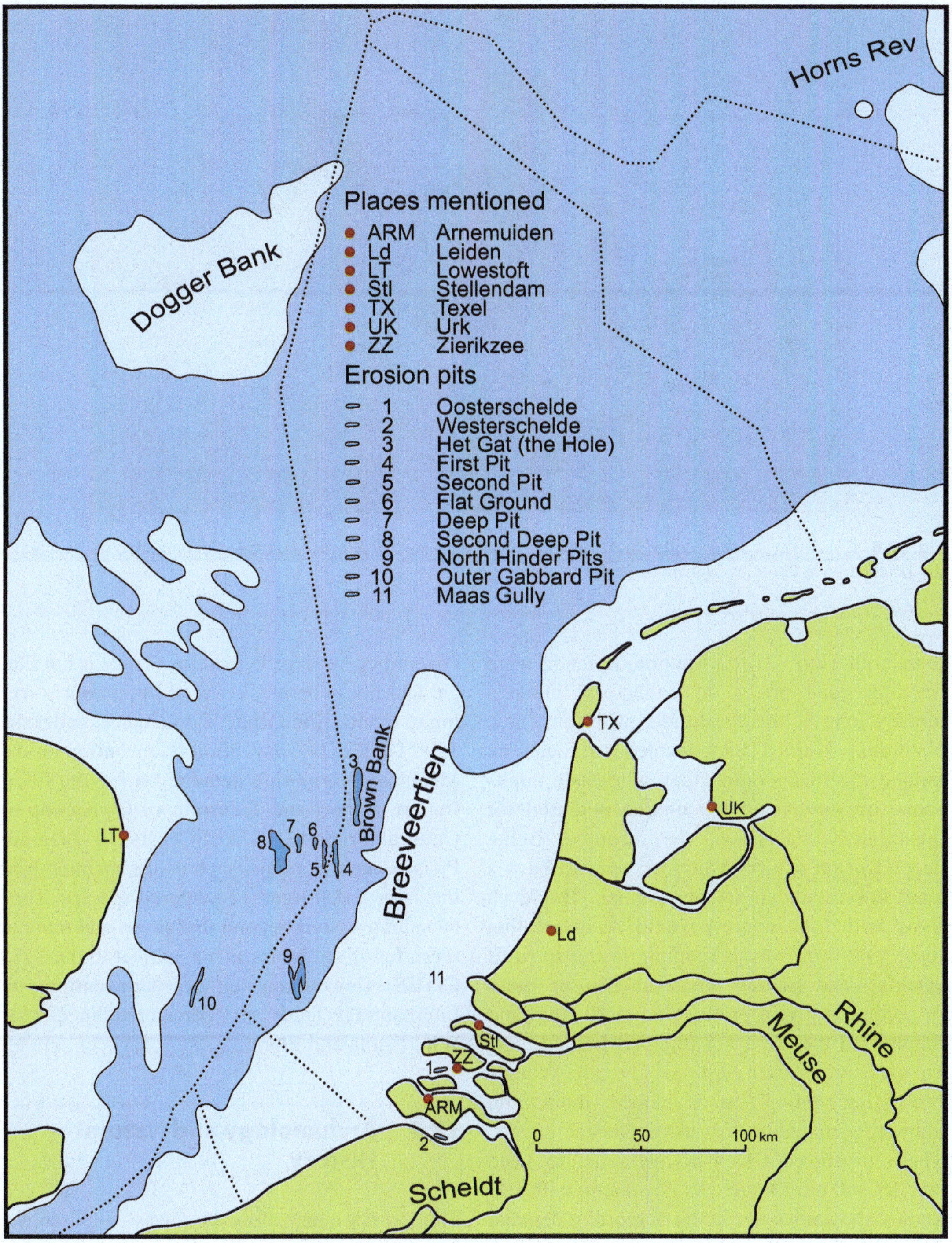

Fig. 27.7 The North Sea with depth contours at 15 and 35 m. The erosion pits of the North Sea that produce most bones are indicated with the colloquial names given by Dutch fishermen. The dredging areas of the Maas gully and the aggregate dredging areas from which other material is recovered, such as Palaeolithic hand axes, are not indicated. For the location of erosion pits in the Westerschelde, the reader is referred to Post and Reumer (2016). The most important cities and fishing ports mentioned in the text are indicated. Drawing by Thijs Maarleveld

Fig. 27.8 Fisherman and bone collector Adrie Vonk (fishing vessel *TX1 Klasina J*) brings a mammoth bone to shore in the Texel harbour. Photo by Martijn de Jonge

mammoth (Fig. 27.8). Mammoth teeth were fetching good prices on collectors' markets already long before the Leiden Museum (now Naturalis) decided they wanted to focus on unique specimens rather than wholesale duplicates, irrespective of the unique potential for quantitative analysis of the resource. Being dependent on nature, fishermen tend to have a keen interest in all its phenomena. But even those with little interest would be aware that these teeth were worth keeping, in the spirit of catching and trading whatever fish or other resource can give a profit. Mammoth teeth and femurs are therefore not dumped immediately but put aside as extra earnings. Over the years, a processing routine was developed, integrating knowledgeable collectors as purchasing agents. These instructed the fishermen also to heed smaller and rarer bones. As any stamp collector knows, the unique specimen is more in demand than the item that is already represented. Integrating the activities in the other ports, sorting, describing and trading gradually centred on Urk, from where the worldwide market in fossils and minerals is fed (Fig. 27.9). Needless to say, that fossils and minerals are not generally covered by heritage legislation. Trade is not illegal and not generally covered by measures trying to reduce illicit trade, for instance, under the 1970 UNESCO Convention (Convention on the Means of Prohibiting and Preventing the Illicit Import, Export and Transport of Ownership of Cultural Property) (UNESCO 1970). Nor are Pleistocene mammals, such as the mammoth or the sabre-tooth tiger, endangered species. They have long passed beyond that stage, and trade in these fossilised bones is not subject to the 1973 CITES Convention either (Convention on International Trade in Endangered Species of Wild Fauna and Flora).

27.8 Archaeology and Natural History

The Leiden connection, as it was called above, had an important side effect. The National Museum for Geology and Mineralogy and its researchers had a close cooperation with the archaeologists at the National Museum of Antiquities, based in the same city. They helped in petrological analysis of archaeological finds

Fig. 27.9 Albert Hoekman at Urk runs a trading business in North Sea fossils and ships them worldwide. Photo by Cees Baars

and generally exchanged information if this was deemed relevant, whether in relation to inland find spots or to fishing finds. In 1970, for instance, the spectacular discovery in the Oosterschelde of the remains of a Roman temple and a plethora of votive altars dedicated to the goddess Nehalennia was brought to archaeological attention through this connection. A major expedition followed, largely based on the fishing methods that had proven themselves in the yearly fieldtrips with the 'Kor and Bot' society. The findspot, after all, was very similar: the remains had gathered in a deep erosion pit in the Oosterschelde, together with much older fossils (Louwe Kooijmans 1971). It was the fishing vessel of skipper Bout from Tholen, *TH6 Johanna Cornelia*, which was used for the operation.

More important in the present context is that occasional worked bone and antler implements were identified amongst the many fossils that the palaeontologists collected. They confirmed the implication of the barbed antler point of the *Colinda*, that at least part of the North Sea floor had been inhabited by humans before sea level rose dramatically, and they were systematically studied (Louwe Kooijmans 1972). Many other finds followed. They include not only implements but also 'debitage': pieces of bone and antler that are the discarded waste from the production of artefacts (Fig. 27.10). Some pieces continued to be offered to the National Museum of Antiquities, but others ended up in the many other collections (e.g. Erdbrink 1982; Maarleveld 1984; Glimmerveen et al. 2004; Peeters et al. 2009; Peeters and Amkreutz, Chap. 8, this volume). In private hands, the artefacts became subject to the same practices of exchange that prevail with collections of minerals and fossils. There seems to be no indication, however, that a truly commercial market for these worked bone artefacts has arisen.

With greater awareness about the mysterious existence of 'Doggerland', however, and the possibility of occasional decorated pieces being found, it is in no way inconceivable that an anonymous market exists, even at present, or will arise. Hopefully, collectors will continue to realise that find-context means everything for an archaeological object's scientific value and that even if the find-context is not apparent in detail, the statistical value of every single item is paramount in an area like this (Van der Plicht et al. 2016).

Two things are clear. Beam trawling has thoroughly impacted the bottom of the North Sea. In

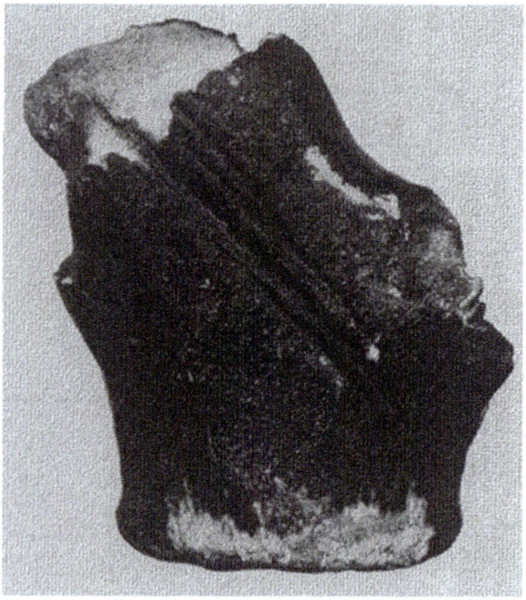

Fig. 27.10 An antler base with grooves stands out amongst the many bones fished from the North Sea bottom. This specimen is from the 'Second Deep Pit' in the Brown Bank area. It is refuse or debitage from the production of bone splinters that are transformed into awls, harpoons or other artefacts. This groove-and-splinter technique is known from early in human prehistory. It is hard to date typologically. Photo by Jan Pauptit, from Maarleveld (1984)

the process many prehistoric artefacts have surfaced, and these are of major scientific significance, not only because, without controlled excavation in the area, they are the best tangible evidence for human activity but also because they represent such a welcome addition to the repertoire of early material culture as we know it (Louwe Kooijmans 1972; Verhart 1995; Verhart 2001). Collections and stray finds from dry land may be more plentiful but are largely composed of fewer types and more durable materials. This is not only a result of preservation but also a result of the selective biases of collection. Bone and antler do not survive under all circumstances; in the North Sea, they do. But also, they are favoured by collection. On board, other materials are quickly thrown out. Due to the processes involved and the knowledgeable awareness of the fishermen described here, bone and antler get special treatment, at least on board Dutch vessels. Although there are exceptions, most flint and stone artefacts from the North Sea have been collected from the sieves and dumps of aggregate and shell dredging, not through fishing (Tizzard et al. 2015). Beam trawls favour bones.

27.9 Fishing and Future Finds

Since the 1980s, bottom-impacting fishing techniques have been much debated worldwide. The use of ever stronger gear and engines, and the fact that slow-growing bottom vegetation is as much affected as fish populations, has made it a major concern for environmentalists (Roberts 2007). Policies were devised to ban these fisheries from specific zones or to limit them to less powerful ships. In North Sea fisheries, the norm for larger trawlers is now 2000 HP, almost half of what it was in the 1980s. The fishing techniques themselves have also been under development. Ground tackle like the beam trawl that plods and ploughs with great force is gradually being phased out in favour of tackle of a more hovering kind, including re-designed otter boards, hovering wing-beams and electrical ticklers instead of mechanical ones (Fig. 27.2). Reduction of fuel consumption may be the strongest driving force of this technical development, but the reduction of impact on the sea bottom is a welcome side effect. In that light, the recent decision of the European Union (13 February 2019) to put a ban on electrical ticklers and 'pulse fishing' while approving the less energy effective and more bottom-impacting techniques of mechanical beam trawling seems to be counterintuitive and inspired by the traditional and sometimes irrational fear of technology so comprehensively described by J.H.J. van der Pot (Van der Pot 1985).

With the gradual disappearance of the beam trawl, the heyday of bone fishery may be coming to an end (Vonk 2011). Archaeological layers will be impacted less. This means that better strategies can be devised to investigate them and realise their potential (Peeters et al. 2009). In the Scheldt estuary, the erosion of the northern gullies has been checked by damming and coastal protection, but important fossils continue to be fished from the deeper pits all the same (Reumer et al. 2005). The southernmost gully, the Westerschelde, is another matter. Deep erosion continues and continues to be monitored as a function of the maintenance of the shipping channel that leads to Antwerp. Deliberate fishing for bones continues as well, with the Schot family and the Royal Zeeland Society focusing on a particularly deep pit near Terneuzen, and other initiatives addressing a range of other erosion pits (Wilde 2007; Post and Reumer 2016). As always, this is not part of normal fishing procedure. In the North Sea, there is another exception, where active fishing for bones and artefacts has been used as what is technically termed an 'Activity directed at Underwater Cultural Heritage' (Maarleveld et al. 2013; see also Dromgoole, Chap. 25, this volume). This expedition was integrated in the monitoring of sand extraction for the Rotterdam Harbour construction (Kuitems et al. 2015).

All in all, there is every reason for archaeology and heritage management to continue and improve their relationship with both the fishing industry and avocational collectors (Firth et al. 2013). The beam trawl may disappear, but occasional finds will still pop up with other fisheries, just as the *Colinda* point did. Moreover, fishermen are extremely knowledgeable informants on the sea environment, in the North Sea as elsewhere. They are therefore an important partner not only in any potential research initiatives but also in efforts to protect the archaeological and historical elements involved (Maarleveld et al. 2013, pp 267–271). The relationship with collectors is another matter. It will never be without tensions, due to the overlapping yet different interests of scientists, private collectors and public institutions with a brief to streamline heritage management. But that is nothing new in the history of archaeology (Fagan 1975; Schnapp 1996; Watson and Todeschini 2006). In view of the knowledge base discussed in this book, fishermen and collectors have so far played a significant role. For the future we can only hope that this role will not be compromised by the development of an anonymous market that will foster the loss or falsification of an object's pedigree in pursuit of confidentiality—in pursuit of exclusive possession instead of sharing.

References

Bailey GN, Harff J, Sakellariou D (eds) (2017) Under the sea: archaeology and palaeolandscapes of the continental shelf. Springer, Cham

Berendsen HJA (1996) De vorming van het land, Inleiding in de geologie en de geomorfologie. Van Gorcum, Assen

Bjerck HB (1995) The North Sea Continent and the pioneer settlement of Norway. In: Fischer A (ed) Man and sea in the Mesolithic. Oxbow, Oxford, pp 131–144

Burkitt MC (1932) A Maglemose harpoon dredged up recently from the North Sea. Man 32:118

Clark JGD (1952) Prehistoric Europe. The economic basis. Methuen, London

Clark G (1975) The earlier Stone Age settlement of Scandinavia. Cambridge University Press, Cambridge

de Wilde B (2007) Fieldtrip to Westerschelde. North Sea Prehistory Newsletter 2

Dorleijn P (1982–1996) Van gaand en staand want. De zeilvisserij voor en na de afsluiting van de Zuiderzee. Van Kampen, Bussum/Van Wijnen, Franeker

Drees M (1986) Kritische kanttekeningen bij de naam 'zwarte botten fauna'. Cranium 3(2):103–120

Elzinga G (1960) Een Lyngby-bijl uit Zeeland. Westerheem 9:102–106

Erdbrink DP (1982) Red deer keratic artefacts in Dutch collections. Berichten Rijksdienst voor het Oudheidkundig Bodemonderzoek 32:103–137

Fagan BM (1975) The rape of the Nile: tomb robbers, tourists, and archaeologists in Egypt. Scribner, New York

Firth A (2015) Risks, resources and significance: navigating a sustainable course for marine development-led archaeology. Bulletin of the Australasian Institute for Maritime. Archaeology 39:1–8

Firth A, McAleese L, Anderson R, Smith R, Woodcock T (2013) Fishing and the historic environment. (EH6204. Prepared for English Heritage). Wessex Archaeology, Salisbury

Flemming NC, Harff J, Moura D, Burgess A, Bailey GN (eds) (2017a) Submerged landscapes of the European continental shelf: Quaternary paleoenvironments. Wiley, Chichester

Flemming NC, Harff J, Moura D (2017b) Non-cultural processes of site formation, preservation and destruction. In: Flemming NC, Harff J, Moura D, Burgess A, Bailey GN (eds) Submerged landscapes of the European continental shelf: Quaternary paleoenvironments. Wiley, Chichester, pp 51–82

Gaffney V, Thomson K, Fitch S (2007) Mapping Doggerland: the Mesolithic landscapes of the southern North Sea. Archaeopress, Oxford

Gaffney V, Fitch S, Smith D (2009) Europe's lost world: the rediscovery of Doggerland, CBA Research eport 160. Council for British Archaeology, York

Glimmerveen J, Mol D, Post K, Reumer JWF, Van der Plicht H, De Vos J, Van Geel B, Van Reenen G, Pals JP (2004) The North Sea project: the first palaeontological, palynological and archaeological results. In: Flemming NC (ed) Submarine prehistoric archaeology of the North Sea. CBA Research Report 141. Council for British Archaeology, York, pp 43–52

Hooijer DA (1957) Mammals and correlation of the 'Black Bones' of the Schelde estuary with those of the Red Grag. Geologie en Mijnbouw, new series 19:255–256

Kuitems M, Van Kolfschoten T, Busschers F, De Loecker D (2015) The geoarchaeological and palaeontological research in the Maasvlakte 2 sand extraction zone and on the artificially created Maasvlakte 2 beach – a synthesis. In: Moree JM, Sier MM (eds) Interdisciplinary Archaeological Research Programme Maasvlakte 2, Rotterdam. Boorrapporten 566. Gemeente Rotterdam, Rijksdienst voor het Cultureel erfgoed. Port of Rotterdam, Rotterdam, pp 351–398

Laban C (1995) The Pleistocene glaciations in the Dutch sector of the North Sea. A synthesis of sedimentary and seismic data. Proefschrift (PhD) thesis, University of Amsterdam, Amsterdam. ISBN 90-9008587-4

Louwe Kooijmans LP (1971) Oudheidkundige Boomkorvisserij op de Oosterschelde. Westerheem 20:151–188

Louwe Kooijmans LP (1972) Mesolithic bone and antler implements from the North Sea and from the Netherlands. Berichten Rijksdienst voor het Oudheidkundig Bodemonderzoek 20(21):27–73

Maarleveld TJ (1984) 1983, Derde jaar archeologie in Nederlandse wateren. WVC, Rijswijk

Maarleveld TJ (2003) Mitigation as archaeological strategy. The Bulletin of the Australasian Institute for Maritime. Archaeology 27:135–139

Maarleveld TJ (2006) Underwater approaches. Towards mitigation of large-scale disturbances. In: Rensink E, Peeters H (eds) Preserving the early past: investigation, selection and preservation of Palaeolithic and Mesolithic sites and landscapes. Nederlandse Archeologische Rapporten 31, pp 127–137

Maarleveld TJ (2010) Fish and "Chips of Knowledge": Some thoughts on the biases of the archaeological record. In: Bekker-Nielsen T, Bernal Casasola D (eds) Ancient nets and fishing gear. Universidad de Cádiz, Cádiz, pp 256–274

Maarleveld TJ, Guérin U, Egger B (eds.) (2013) Manual for activities directed at underwater cultural heritage. A guide on the rules annexed to the UNESCO 2001 Convention on the Protection of the Underwater Cultural Heritage. UNESCO, Paris

Missiaen T, Sakellariou D, Flemming NC (2017) Survey strategies and techniques in underwater geoarchaeological research: An overview with emphasis on prehistoric sites. In: Bailey GN, Harff J, Sakellariou D (eds) Under the sea: archaeology and palaeolandscapes of the continental shelf. Springer, Cham, pp 21–37

Momber G (2011) Submerged landscape excavations in the Solent, southern Britain: climate change and cultural development. In: Benjamin J, Bonsall C, Pickard C, Fischer A (eds) Submerged prehistory. Oxbow, Oxford, pp 85–98

Moraal JM (1964) De Mens van Zeeland. Grondboor en Hamer 18(2):44–49

Moraal JM (1991) Het Koninklijk Zeeuws Genootschap der Wetenschappen. Cranium 8(/2):88

Peeters H, Murphy P, Flemming N (eds) (2009) North Sea Prehistory Research and Management Framework (NSPRMF) 2009. RCE, Amersfoort

Popper K (1935) Logik der Forschung. Springer, Wien

Post K, Reumer JWF (2016) History and future of paleontological surveys in the Westerschelde Estuary (Province of Zeeland, the Netherlands). DEINSIA 16:1–9

Reumer JWF, Mayhew DF, Van Veen JC (2005) Small mammals from the Late Pliocene Oosterschelde dredgings. DEINSIA 1:103–118

Roberts C (2007) The unnatural history of the sea. The past and future of humanity and fishing. Island Press, Washington

Sanger D (1995) Mesolithic maritime adaptations: the view from North Americas. In: Fischer A (ed) Man and sea in the Mesolithic. Oxbow, Oxford, pp 335–349

Scager DJ, Ahrens H-J, Dieleman FE, Van den Hoek Ostende LW, De Vos J, Reumer JWF (2017) The Kor & Bot collection revisited, with a biostratigraphic interpretation of the Early Pleistocene Oosterschelde Fauna (Oosterschelde Estuary, the Netherlands). DEINSEA 17:16–31

Schiffer MB (1987) Formation processes of the archaeological record. University of New Mexico Press, Albuquerque

Schnapp A (1996) The discovery of the past: the origins of archaeology. The British Museum Press, London

Tizzard L, Bicket AR, de Loecker D (2015) Seabed prehistory: investigating the palaeogeography and Early Middle Palaeolithic archaeology in the southern North Sea. Wessex Archaeology Report 35. Wessex Archaeology, Salisbury

UNESCO 1970 Convention on the Means of Prohibiting and Preventing the Illicit Import, Export and Transfer of Ownership of Cultural Property – 1970 http://portal.unesco.org/en/ev.php-URL_ID=13039&URL_DO=DO_TOPIC&URL_SECTION=201.html. Accessed 4 Oct 2017

Van Beylen J (1993) De hoogaars en de visserij van Arnemuiden. Hedeby, Leeuwarden

Van den Bossche M, Van Donselaar P, Van Haselen H, Piers R (1999) De Nederlandse Visserijsector: Economische Betekenis en Structuur. Delft University Press, Delft

Van der Plicht J, Amkreutz LWSW, Niekus MJLT, Peeters JHM, Smit BI (2016) Surf'n turf in Doggerland: Dating, stable isotopes and diet of Mesolithic human

remains from the southern North Sea. J Archaeol Sci Rep 10:110–118. https://doi.org/10.1016/j.jasrep.2016.09.008
Van der Pot JHJ (1985) Die Bewertung des technischen Fortschritts. Eine systematische Übersicht der Theorien. Van Gorcum, Assen/Maastricht
Van der Vlerk IM (1938) Nederland in het IJstijdvak. Oratie, Leiden
Van Ginkel R (1991) Elk vist op zijn tij: Een Zeeuwse maritieme gemeenschap, Yerseke. Walburg Pers, Zutphen, pp 1870–1914
Van Ginkel R (1993) Tussen Scylla en Charybdis. Een Ethnohistorie van Texels vissersvolk. Het Spinhuis, Amsterdam, pp 1813–1932
Van Nieulande F, Bosselaers M, Loeff H, Bekker JP (2017) Natuurhistorische voorwerpen http://www.kzgw.nl/index.php/collecties/natuurhistorische-voorwerpen. Accessed 4 Oct 2017
Verhart LBM (1995) Fishing for the Mesolithic. The North Sea: a submerged Mesolithic landscape. In: Fischer A (ed) Man and sea in the Mesolithic. Oxbow, Oxford, pp 291–302
Verhart L (2001) De zee neemt, de zee geeft. Spectaculaire vondsten uit de Noordzee. Westerheem 50(3):102–108
Vonk A (2011) Vissers vangen bot. Unpublished presentation Noordzeedagen, Leeuwarden
Vos PC, Bunnik FPM, Cohen KM, Cremer H (2015) A staged geogenetic approach to underwater archaeological prospection in the Port of Rotterdam (Yangtzehaven, Maasvlakte, The Netherlands): A geological and palaeoenvironmental case study for local mapping of Mesolithic lowland landscapes. Quatern Int 367(5):4–31. https://doi.org/10.1016/j.quaint.2014.11.056
Watson P, Todeschini C (2006) The Medici conspiracy: the illicit journey of looted antiquities, from Italy's tomb raiders to the world's greatest museums. Public Affairs, New York
Wetsteyn B (2001) Fossielen in de Westerschelde. Schelde Nieuwsbrief 27:4

Appendix: The SPLASHCOS Viewer

This Appendix sets out the programs and metadata used to create the SPLASHCOS Viewer and the maps derived from them.

Data Categories

For the sake of simplicity, the primary data are in spreadsheet format, organised by Nation State as a series of Excel files. The data have been compiled, for the most part, by the authors of the national overviews presented in this volume and have been merged at the Lower Saxony Institute for Historical Coastal Research (Niedersächsisches Institut für historische Küstenforschung—NIhK), where the central database is maintained. The data are hosted online by MARIS (The Marine Information Service of the Netherlands), which is responsible for the pan-European coordination and management of marine and oceanographic data (https://www.maris.nl/). The archaeological data can be uploaded to a local working environment using the web map service at http://geoservice.maris2.nl/wms/project/splashcos?service=WMS&request=GetCapabilities.

If there are major changes to the data, these will be transmitted to the online host at MARIS. The online version is available at http://splashcos.maris2.nl/ and also at http://splashcos-viewer.eu/.

The primary categories of data are given in Table 1. Like all such data compilations, this one has both limitations and benefits. Ideally, a compilation of data records of this sort should meet four conditions:

(1) A clearly set-out list of data categories or attributes, with their corresponding conventions and definitions—the metadata. The data attributes and definitions should be sufficiently comprehensive and flexible enough to cater for variations in the data. If the list of attributes is too short or the definitions are too narrow, this will impose a straitjacket that may distort or oversimplify the original data. If the categories are too numerous and their definitions too wide, this may limit the usefulness of the data for making generalisations.
(2) Consistent application of attributes and definitions, especially where many people are involved in inputting data. Experience shows that different inputters may have different interpretations of categories and definitions and that definitions may 'wander' over time. This is a major issue where many people are involved over extended periods of time.
(3) Central oversight and quality control of the source data. If the original data are poorly described, incomplete or inaccurate, this will compromise their usefulness.
(4) A mechanism for updating the data records to keep abreast of new discoveries or to facilitate later entry of missing data.

Regarding point (1), we believe that our categories and definitions are appropriate to the purpose. The structure of the worksheet also has

G. Bailey et al. (eds.), *The Archaeology of Europe's Drowned Landscapes*, Coastal Research Library 35, https://doi.org/10.1007/978-3-030-37367-2

Table 1 The layout of the worksheet used in the SPLASHCOS Viewer and the column labels

A	SPLASHCOS Site Number	K	Dating Method
B	Country	L	Minimum Depth
C	Site Name	M	Maximum Depth
D	Local Site Number	N	Reference Datum (Chart Datum, OD)
E	Type of Site	O	Presence of Organics (Y/N/Unknown)
F	Further information	P	Further Details
G	Archaeological Date	Q	Url Reference (if available, e.g. online records and archives)
H	Archaeological Date (General)	R	Bibliographic Reference (published literature)
I	Earliest Date (cal BC)	S	Longitude
J	Latest Date (cal BC)	T	Latitude

Letters refer to the Excel columns. Letters refer to relevant columns of the Excel worksheet. A–D are self-explanatory. E: type of site. Single unstratified find; single stratified find; collection of unstratified finds; collection of stratified finds; settlement (in situ cultural layer, features); grave(s); depot find/offering; rock art/carving; other; unknown. F: further information. This is a free-format field and may include additional descriptors about the nature of the site such as 'fish trap', 'footprints', 'submerged forest', 'megalith' and so on. G: archaeological date. Stone Age; prehistoric; Palaeolithic; Lower Palaeolithic; Middle Palaeolithic; Upper Palaeolithic; Late Palaeolithic; Late Palaeolithic/Early Mesolithic; Mesolithic; Early Mesolithic; Middle Mesolithic; Late Mesolithic; Late Mesolithic/Early Neolithic; Neolithic; Early Neolithic; Middle Neolithic; Late Neolithic; Late Neolithic/Early Bronze Age; Bronze Age. H: archaeological date (general). Prehistoric; Palaeolithic; Palaeolithic-Mesolithic; Mesolithic; Mesolithic-Neolithic; Neolithic; Neolithic-Bronze; Bronze Age; Metal Age. K: dating method. Stratigraphy; typochronology; dendrochronology; radiocarbon dating; OSL dating; unknown. P: further details. This is another free-format field and allows for the entry of additional summary information about the site and its finds and qualifying comments about the application of the pre-set categories in the drop-down menus where these do not fit well or where the archaeological material comprises more than one category. Q–R are self-explanatory. S and T present data on location in decimal degrees

built-in provision for grouping information into fewer categories, where desirable for summary analyses, or widening of definitions to cater for ambiguities and uncertainties, especially in the chronological fields. The free-format fields of the worksheet also allow for the incorporation of additional detail and qualifying comments appropriate to individual sites and artefacts.

Point (2) is obviously a potential problem for a multi-national data set such as this one, involving many different individuals from different backgrounds and intellectual traditions familiar with different sorts of archaeological records and working across a variety of cultural, intellectual and linguistic boundaries. The original meetings of WG1 involved discussion of these issues and an attempt to forge a common understanding, but some inconsistency is inevitable, especially as more individuals not involved in the original discussions have been drafted in to help with data collection. Some classifications may ultimately come down to a matter of individual judgement or opinion.

In keeping with the discussion about definitions and boundaries set out in Bailey et al. (Chap. 1, this volume), the Viewer data include artefacts and other archaeological features in the intertidal zone where these unambiguously meet the criteria for underwater status, namely, archaeological material originally deposited on a dry land surface or abandoned in the immediately adjacent shallow water of a palaeoshoreline, which has subsequently been inundated by relative sea-level rise. The Viewer records exclude, or are intended to exclude, archaeological material that is on the shoreline but has never been inundated or which has been displaced into the intertidal zone by erosion from adjacent land surfaces. However, different authors take different attitudes to how much of the intertidal material is included in the Viewer data, depending on regional shoreline conditions and the nature of the available archaeological record, and the reader is advised to consult the national overviews for further clarification.

Other uncertainties arise from the question of whether collections of artefacts are to be regarded as stratified or unstratified, in situ or in secondary position, and representative of settlements and cultural features, or not. Some authors have included isolated finds of submerged animal bones or tree stumps as evidence of underwater preservation, especially where archaeological finds are rare, in order to include every residual scrap of relevant information that may provide clues to future archaeological discoveries. Other authors with more underwater finds at their disposal have included such materials only where they are clearly associated with artefacts or other evidence of human involvement such as butchery marks on animal bones. As always in cases of uncertainty, the interested reader is advised to look at the additional information in the free-format fields and to consult the source literature.

As for point (3), the individual authors or teams of authors for each national data set are ultimately responsible for their national records, with relatively little opportunity for the exercise of a central audit, beyond the identification of obvious anomalies noted by the team at the NIhK. Missing data is a perennial problem of all databases and nothing can be done to repair such omissions where they are due to the absence of relevant observations in the original field observations and records. However, collection of new data plays a big role in the future of the viewer, and provision for this, which relates to point (4) above, is built into the arrangements for the curation of the database (see below).

Summary Statistics

Summary statistics are set out in Table 2 and show the distribution of sites by country and site category. Information is also present on the depth categories at which sites were found below sea level and the number of sites with recovery of organic remains (bones and/or wood). Two features should be noted. The first is the heavy weighting towards single artefacts, which dominate the counts for most countries. The second is the marked differential between the numbers of sites in different countries. Denmark has by far the largest number of sites, closely followed by other countries in NW Europe. This variation undoubtedly reflects a wide range of factors. Some of these have to do with conditions of preservation and varying interest in and availability of funding and resources for the investigation of the prehistoric finds of the submerged landscape. Other variables have to do with the arrangements in different countries for reporting such finds to local museums and heritage agencies and for recording that information in a central data repository. This is very likely a significant factor in Denmark, where single finds are numerically dominant. Further comment on the statistics for individual countries can be found in the chapters in this volume.

Digital Maps

One data field in the Viewer data that we have made considerable efforts to fill for every record is information on the location of finds using the same system of geographical coordinates—latitude and longitude expressed in decimal degrees[1]. This is to facilitate the display of distribution maps at different scales ranging from pan-European to national and regional scales. The majority of the site distribution maps presented in this volume are derived from the Viewer records in this way.

The creation of the distribution maps has been mainly realised with Quantum GIS (QGIS Development Team 2018). The map bases come from different sources. Land and marine areas are in most cases based on the data available at http://naturalearthdata.com/; in the case of France and Norway, the source is diva-gis.org, available at https://gadm.org/. The borders of each State and its exclusive economic zone are from http://marineregions.org and bathymetry from the EMODnet portal at http://www.emodnet.eu/. ASCII grids can be downloaded and were used to calculate the elevation contours with SAGA GIS

[1]In some individual cases, the coordinates were drawn inaccurately to ensure protection from unwanted attention. All coordinates are in WGS84 EPSG 4326. Some authors delivered the data in national specific systems. These coordinates were re-projected to WGS84 by the standard procedure in QGIS.

Table 2 Distribution of archaeological sites by country and region, showing site type, depth and presence of organic remains

Country	SF	CF Uns	CF Strat	In situ	Other	<5m	5–10m	10–40m	>40m	Organic	Total
Baltic and Norway											
Denmark	1306	12	0	299	89	93	21	5	0	156	1706
Sweden	1	36	0	7	0	27	15	0	0	12	44
Germany	44	69	0	25	3	59	9	1	0	21	141
Norway	2	80	0	0	1	83	0	0	0	2	83
Poland	4	0	0	0	0	0	0	0	0	3	4
Subtotal	**1357**	**197**	**0**	**331**	**93**	**262**	**45**	**6**	**0**	**194**	**1978**
%	***68.7***	***10***	***0***	***16.7***	***4.7***	–	–	–	–	–	***100.1***
Atlantic and North Sea											
Netherlands	8	22	0	1	0	0	0	3	0	18	31
Belgium	1	2	0	0	0	3	0	0	0	1	3
Britain (England/Scotland/Wales)	155	35	22	21	26	213	6	9	1	36	259
Ireland (Rep. Ireland/N. Ireland)	11	35	0	1	3	41	0	0	0	9	50
France	0	0	78	12	85	82	2	1	0	3	175
Portugal	0	0	0	5	1	4	0	0	0	3	6
Subtotal	**175**	**94**	**100**	**40**	**115**	**343**	**8**	**13**	**1**	**70**	**524**
%	***33.4***	***17.9***	***19.1***	***7.6***	***22***	–	–	–	–	–	***100***
Med. and Black Sea											
Gibraltar	0	0	0	0	2	1	0	1	0	0	2
Spain	3	1	0	4	5	12	1	0	0	7	13
Italy	0	1	0	1	6	4	1	0	0	3	8
Malta	0	0	0	0	0	0	0	0	0	0	0
Croatia	2	10	0	17	2	10	1	0	0	7	31

Greece	2	3	0	15	8	3	0	3	2	3	28
Bulgaria	0	0	0	19	3	10	9	0	0	9	22
Ukraine	0	1	0	0	0	0	0	0	0	0	1
Cyprus	0	3	0	1	0	0	0	1	0	0	4
Turkey	0	5	0	1	0	5	0	0	0	1	6
Israel	0	1	0	18	4	20	2	0	0	17	23
Subtotal	**7**	**25**	**0**	**76**	**30**	**65**	**14**	**5**	**2**	**47**	**138**
%	***5.1***	***18.1***	***0***	***55.1***	***21.7***	–	–	–	–	–	***100***
Grand total	**1539**	**316**	**102**	**447**	**238**	**670**	**67**	**24**	**3**	**311**	**2640**
%	***58.3***	***12***	***3.9***	***16.9***	***9***	–	–	–	–	–	***100.1***

SF single find, *CF Uns* collection of unstratified finds, *CF Strat* collection of stratified finds; in situ, undisturbed culture layer. Organic: finds of bone and/or wood. Data from http://splashcos-viewer.eu/. Note that these figures are only for sites that have been formally recorded in national databases and that the totals, especially for single finds and unstratified finds, are minimum figures (http://www.saga-gis.org). All the shape files, including the ones with the sites, were then combined in a GIS and further reworked in a graphics editing program (Adobe Illustrator). In this last step of the graphic processing, the colours were adapted to a style guide devised by Rolf Kiepe (NIhK).

The Future of the SPLASHCOS Viewer

To secure the future of the Viewer, the following points are important. There must be long-term maintenance of the database, permanent accessibility and the possibility of updates, and these are the responsibility of the NIhK. However, equally important is the gathering and testing of new data. Here, it is necessary that the members of the SPLASHCOS community and all other archaeologists who deal with this topic take on the responsibility of filling gaps in the existing records or submitting new records. NIhK can take on the responsibility for administration of the Viewer, but the quality of the information provided in the Viewer can only be as good or as bad as the information supplied by specialists familiar with the details of the material at the national and regional level. It should be relatively easy to adapt the data to the categories defined in the Viewer. Contact information for uploading new data is available on the Viewer web page. Longer-term objectives are to review in a comprehensive manner existing publications and data sources and to tap into regions of Europe and the wider Mediterranean where data is currently not available, objectives that are likely to require additional funding. The ultimate goal is to ensure the long-term curation of the data, updating in light of new information, and wide availability to a variety of users including other research disciplines.

Index of Sites

G. Bailey et al. (eds.), *The Archaeology of Europe's Drowned Landscapes*, Coastal Research Library 35, https://doi.org/10.1007/978-3-030-37367-2

Subject Index

G. Bailey et al. (eds.), *The Archaeology of Europe's Drowned Landscapes*,
Coastal Research Library 35, https://doi.org/10.1007/978-3-030-37367-2

G

T

The manufacturer's authorised representative in the EU is Springer Nature Customer Service Centre GmbH, Europaplatz 3, 69115 Heidelberg, Germany. If you have any concerns regarding our products, please contact ProductSafety@springernature.com

Printed and bound by CPI Group (UK) Ltd, Croydon, CR0 4YY

22/09/2026

02220980-0013